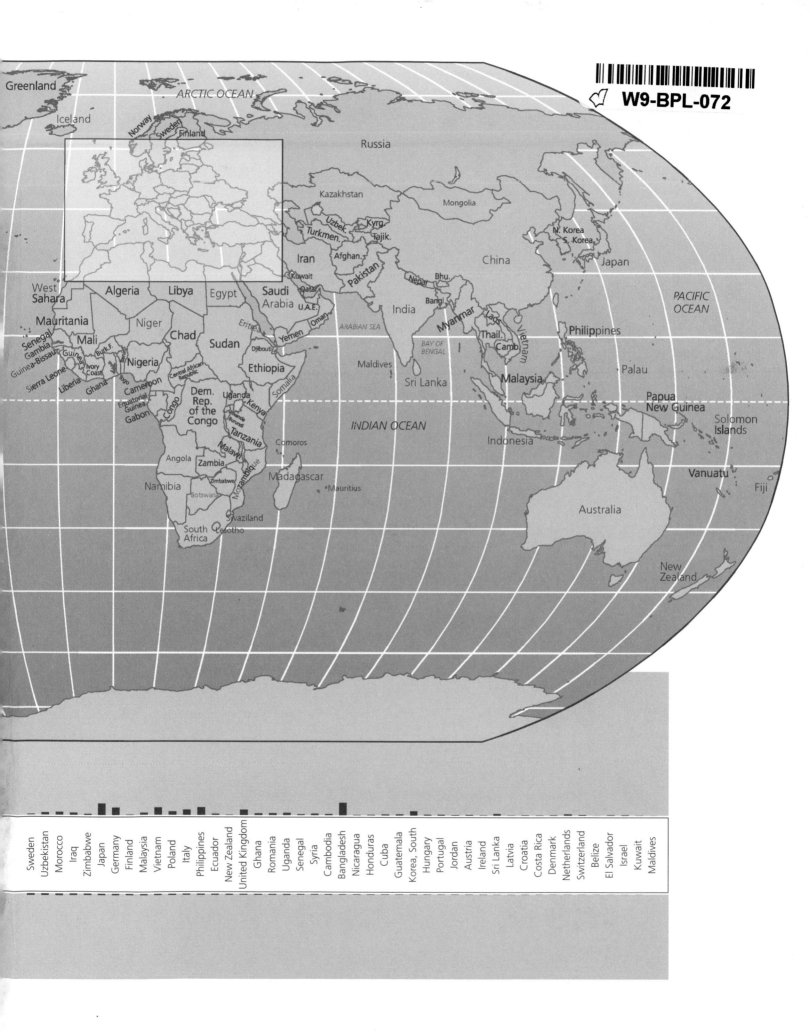

Environment

THE SCIENCE BEHIND THE STORIES

Scott Brennan

Huxley College of the Environment
Western Washington University

Jay Withgott

Portland, Oregon

PEARSON

Benjamin Cummings

San Francisco • Boston • New York
Cape Town • Hong Kong • London • Madrid • Mexico City
Montreal • Munich • Paris • Singapore • Sydney • Tokyo • Toronto

Acquisitions Editor: Chalon Bridges
Development Editor: Susan Teahan
Senior Production Editor: Corinne Benson
Senior Art Editor: Donna Kalal
Photo Program Manager: Travis Amos
Senior Art Development Editor: Russell Chun
Project Editor: Susan Minarcin
Associate Project Editor: Alexandra Fellowes
Editorial Assistant: Alissa Anderson
Senior Marketing Manager: Josh Frost
Managing Editor, Production: Erin Gregg
Manufacturing Manager: Pam Augspurger
Composition: The GTS Companies
Production Project Manager: Doña Uhrig
Text Design: Mark Ong
Cover Designer: Mark Ong
Illustrations: Fineline Illustrations, Inc.
Media Producer: Aaron Gass
Cover Printer: Phoenix Color Corporation
Text Printer: Quebecor World Versailles

Library of Congress Cataloging-in-Publication Data

Brennan, Scott R.
    Environment: the science behind the stories / Scott Brennan, Jay Withgott.
        p. cm.
    Includes index.
    ISBN 0805344276
        1. Environmental sciences. I. Withgott, Jay. II. Title.

GE105.B74 2004
363.7--dc22   2003068922

ISBN 0-8053-4427-6

www.aw-bc.com   1 2 3 4 5 6 7 8 9 10 -QWV-08 07 06 05 04

About the Authors

 Scott Brennan teaches environmental science, policy, and journalism at Western Washington University's Huxley College of the Environment. He has worked as an environmental journalist, photographer, lobbyist, and consultant and once taught ecology inside the maximum security unit of the Washington State Penitentiary. His work as a teacher, writer, and scholar has been recognized by the Society of Professional Journalists, Duke University, Washington State University, and Western Washington University. Western Washington University has identified him as embodying the university's "best practices in pedagogy, instructional technologies, and expanded learning environments." Scott has cultivated his expertise in the intersection of environmental science and public policy by serving as Executive Conservation Fellow of the National Parks Conservation Association in Washington, D.C., and by working as a consultant to the U.S. Department of Defense Environmental Security Office at the Pentagon. He has worked closely with a wide array of community organizations devoted to the advancement of natural resource stewardship, conservation, education, and sustainable community development. His research interests include science education and communication, national park and wilderness management, ecotourism and curriculum design and innovation. He is a frequent guest lecturer at universities throughout western North America and, when not at work, is likely to be found exploring his home hills, the Chuckanut Range, and the North Cascades of Washington state. He lives in the Whatcom Creek Watershed in Bellingham, Washington.

 Jay H. Withgott is a science and environmental writer with a background in scientific research and teaching. He holds a B.A. from Yale University in History and Studies in the Environment, an M.S. in Biological Sciences from the University of Arkansas, and an M.S. in Ecology and Evolutionary Biology from the University of Arizona. He has published nearly 20 scientific papers on topics in ecology, evolution, animal behavior, and conservation biology and on research involving insects, snakes, and birds. He has also taught university-level laboratory classes in ecology, ornithology, vertebrate diversity, anatomy, and general biology. As a science writer, Withgott has authored articles for a variety of journals and magazines including *Science, New Scientist, Current Biology, Conservation in Practice,* and *Natural History*. He combines his scientific expertise with his past experience as a reporter and editor for daily newspapers, to make science accessible and engaging for general audiences.

Environment

Using This Book to Your Advantage

To the Student
We created this text with a number of special features to help you learn environmental science.
Use this quick reference guide to help you make the most of the materials in this book.

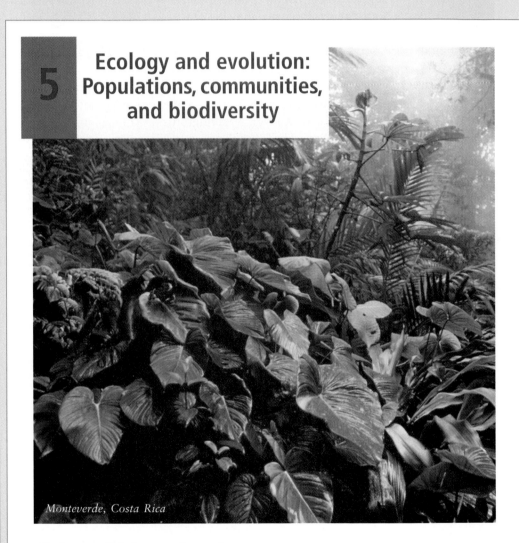

5 **Ecology and evolution: Populations, communities, and biodiversity**

Monteverde, Costa Rica

Prepare for what's ahead.
Chapter outlines offer concise bulleted lists of important topics to make studying easier.

This chapter will help you understand:

- How evolution generates biodiversity
- Concepts of speciation and extinction
- Outlines of the biodiversity "crisis"
- Fundamental ecological concepts and principles
- Units of ecological organization
- Fundamentals of population ecology, including carrying capacity and limiting factors
- Food webs, trophic levels, and ecological communities
- Species interactions, including predation, competition, parasitism, and mutualism
- Primary and secondary succession
- Challenges for biodiversity conservation

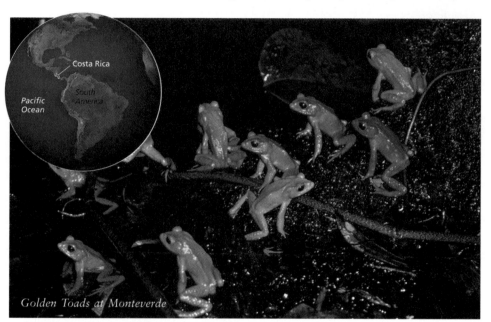

Golden Toads at Monteverde

Central Case: Striking Gold in a Costa Rican Cloud Forest

> **"I must confess that my initial response when I saw them was one of disbelief and suspicion that someone had dipped the examples in enamel paint."**
> —*Dr. Jay M. Savage, describing the golden toad in 1966*

> **"What a terrible feeling to realize that within my own lifetime, a species of such unusual beauty, one that I had discovered, should disappear from our planet."**
> —*Dr. Jay M. Savage, describing the golden toad in 1998*

During a 1963 visit to Central America, biologist Jay M. Savage heard rumors of a previously undocumented toad living in Costa Rica's mountainous Monteverde region. The elusive amphibian, according to local residents, was best known for its color: a brilliant yellow-orange that seemed almost golden. Savage was told the toad was hard to find because it appeared only in spring, during the early part of the region's rainy season.

Monteverde means "green mountain" in Spanish, and the name couldn't be more appropriate. The village

of Monteverde sits on a bench beneath the verdant slopes of mountains known as the Cordillera de Tilarán, which receive more than 400 cm (157 in) of annual rainfall. Some of the lush forests above Monteverde, which begin around 1,600 m (5,249 ft, just under a mile high), are known as **lower montane rainforests.** These forests are also known as **cloud forests** because much of the moisture they receive arrives in the form of low-moving clouds that blow inland from the Caribbean Sea. Monteverde's cloud forest was not fully explored at the time of Savage's first visit, and researchers who had been there described the area as pristine, with a rich bounty of ferns, liverworts, mosses, clinging vines, orchids, and other organisms that thrive in cool, misty environments. Savage knew that such conditions create ideal habitat for many amphibians, including toads.

In May of 1964, Savage organized an expedition into the muddy mountains above Monteverde to try to document the existence of this previously unknown species in its natural habitat. Late in the afternoon of

Make a connection.
Each chapter opens with a central case study that takes you into the heart of a complex environmental issue, and introduces you to environmental issues that impact the lives of real people. Throughout the chapter, the central case study, or similar issues raised by the central case study, reinforce key concepts and provide you with a new perspective and an understanding of our world today.

Discover how real science is done.

The Science behind the Stories boxes highlight how scientists make discoveries, analyze data, and generate and test models and hypotheses.

S The Science behind the Story

The Last Mass Extinction

Throughout the history of life on Earth, individual species such as the golden toad have occasionally died off, and new ones have appeared. On five occasions, however, huge numbers of species went extinct in a geologic instant. The last mass extinction occurred 65 million years ago, the date that marks the dividing line between the Cretaceous and the Tertiary periods, or the K-T boundary for short. About 70% of the species then living, including the dinosaurs, disappeared.

When he first started working at Bottaccione Gorge, Walter Alvarez had no idea he would help discover what killed off the dinosaurs. An American geologist, Alvarez was working for the summer in northern Italy, developing a new method to determine the age of sedimentary rocks. But he and his father, physicist Luis Alvarez, would soon come up with a dramatic theory about what caused the best-known mass extinction. Alvarez chose Bottaccione Gorge because it formed an ideal geological archive: Its 400 m (1,300 ft) walls are stacked like layer cake with beds of rose-colored limestone that

A colossal asteroid impact 65 million years ago is thought to have caused the Cretaceous-Tertiary mass extinction.

formed between 100 million and 50 million years ago from dust that had settled to the bottom of an ancient sea. While analyzing the many layers of the Bottaccione Gorge, Walter Alvarez noticed a 1 cm (.4 in) thick band of reddish clay sandwiched between two layers of limestone. The older layer just below it was packed with fossils of globotruncana, a sand-sized animal that lived in the late Cretaceous period. The newer layer just above it instead contained just a few scattered fossils of a cousin of globotruncana, typical of sedimentary rock formed in the early Tertiary period. What Alvarez found

Figure 5.26 Costa Rica has protected a wide array of its diverse communities. This protection has stimulated the nation's economy through ecotourism. Here, visitors experience a walkway through the forest canopy in one of the nation's parks.

Figure 5.27 Ecological restoration attempts to reverse the effects of human disruption and restore communities to their natural state. Here, a crew plants trees in the Puriscal region of Costa Rica, for a Costa Rican environmental group that brings farmers and nursery workers together to reforest the land.

Weighing the Issues:
How best to conserve biodiversity?

Most people view national parks and ecotourism as excellent ways to help keep ecological systems intact. Yet the golden toad went extinct despite having a reserve established to protect it, and invasive species do not pay attention to park boundaries. What lessons can we take from this about the conservation of biodiversity?

Altered communities can be restored to their former condition

Creating public parks and defining areas in which to preserve natural communities are well-tested ways to ~~~ ~iodivers~ ~~hancing~ ~~~ ~~~ ~an-

of prairie habitat by planting native prairie plants, weeding out invaders and competitors, and introducing controlled fire to mimic the natural prairie fires that historically maintained this community. Most such efforts at ecological restoration so far have been made in developed nations because more natural areas have been lost in these countries and because the effort to restore native communities can be resource-intensive. However, ecological restoration is taking place in Guanacaste Province in Costa Rica as well, where scientists are restoring dry tropical forest from grazed pasture. The more human population grows and development spreads, the more ecological restoration may become a prime conservation strategy.

Exercise your critical thinking skills.
Weighing the Issue discussion questions appear throughout the chapter and ask you to think critically about the issues that have been raised. Answering them will help you reinforce key concepts and help you to put them into your own words.

Environment

Consider both sides of the question.

Viewpoints feature the opinions of two guest writers and afford you an opportunity to consider multiple perspectives on controversial issues.

VIEWPOINTS

Conservation of Monteverde

What lessons, if any, can we learn from ecological changes that have occurred in the Monteverde since the discovery of the golden toad in 1964?

Four Lessons from Monteverde

There are four lessons to be learned from the Monteverde experience. First, a few committed people can have an impact. The Monteverde was originally occupied by Tico and Quaker families, pioneers in the usual sense of the word—people who farmed for a living, but struggled to create their farms in a beautiful, but primitive environment. George Powell, who is now actively involved in green macaw conservation efforts and studies of bellbird migrations, converted a local Monteverde pioneer, Wolf Guindon, into an ardent conservationist. It was Powell who convinced the international conservation organizations to fund land purchases of the Monteverde Cloud Forest Preserve. Guindon became the chief ranger of the Preserve.

Second, practical conservation efforts must take into account local social aspirations. Although hard for a conservationist to watch, pioneers are just trying to make a hard life better by converting natural landscapes to farmland. Conservationists must recognize this to work with their neighbors. For example, the Monteverde Cloud Forest Preserve allows its springs to be tapped for local water supplies, runs educational programs, and leases a peak for television and radio transmission towers. The Monteverde Conservation League runs a tree nursery and an agroforestry outreach program for local farmers.

Third, conservation can lead to local economic success. As adults, the children of the agricultural pioneers at Monteverde had no place to claim for their own and faced leaving the area for employment elsewhere, or finding alternative livelihoods. Many of my Costa Rican friends stayed in Monteverde to build an ecotourism industry, and are now hoteliers, restauranteurs, or natural history guides.

Finally, even if every bit helps, local conservation is not enough. Human use of fossil fuels is making the world on average a warmer place. Deforestation is changing climate at regional scales. Conservationists must plan reserves that can adapt to changes in the coming century or more species will follow the golden toad into extinction.

Robert Lawton is a forest ecologist at the University of Alabama-Huntsville, who has worked in Monteverde for over 25 years.

Conservation Successes in the Shadow of the Golden Toad

Given the discouraging news about conservation problems worldwide, it wouldn't be hard to become pessimistic about what has happened in Monteverde, Costa Rica. After all, in spite of the country's international reputation as a "green republic," deforestation rates are higher in Costa Rica than almost anywhere in the world, isolating Monteverde as an ecological island perched on the continental divide.

The last 15 years has also seen widespread changes within Monteverde: a proliferation of new houses, hotels, restaurants, and an explosion of ecotourism, with tens of thousands of people per year visiting what was once a rural community. Along with the Golden Toad, other amphibian and reptile species have disappeared, introduced species have invaded, and lowland species have moved upslope as Monteverde becomes warmer and drier.

Yet there have also been impressive achievements in conservation at Monteverde. In 1979, poaching of large mammals and birds was commonplace, and species such as tapirs and guans were rare. Now the very people who hunted with rifles use binoculars instead as they lead natural history tours. Tapirs and guans are more common today than they have been for more than half a century. As the Monteverde Cloud Forest Preserve has grown ten-fold in area, clearings on the Atlantic slope have reverted to lush forest. The Guacimal River, formerly rancid due to waste dumped by the community dairy plant, is much cleaner now.

On both global and local scales, the most enduring impact of Monteverde has been the education of the public about environmental values. I like to think that whatever negative local impact the steady onslaught of ecotourists may have on resplendent quetzals and howler monkeys, it is more than compensated by inspiring people to appreciate tropical forests and their own natural heritage. If so, the conservation gains at Monteverde may help save other tropical and temperate zone habitats worldwide.

Nathaniel Wheelwright is Professor of Biology at Bowdoin College in Brunswick, Maine, director of the Bowdoin Scientific Station on Kent Island, New Brunswick, and co-editor of *Monteverde: Ecology and Conservation of a Tropical Cloud Forest* (Oxford University Press, 2000).

Get the big picture.
Our art will walk you, step-by-step, through complicated processes.

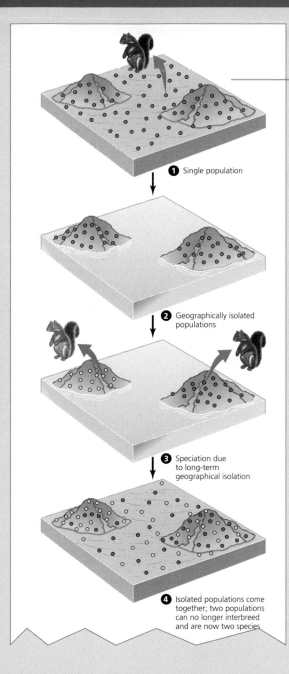

① Single population

② Geographically isolated populations

③ Speciation due to long-term geographical isolation

④ Isolated populations come together; two populations can no longer interbreed and are now two species

Appendix A
Some Basics on Graphs

The collection of data to understand trends in various phenomena and to test ideas is a core part of the scientific endeavor. Data are numbers presented within a context, and, as such, help scientists to gain insights and draw conclusions. The basic tool of scientists for expressing data is the graph, and the ability to read graphs and plot data is one that students must cultivate early in their study of the various sciences. This appendix provides basic information on three of the most common types of graphs—line, scatter, and bar—and the rationale for the use of each.

Line Plot

A line plot is drawn when the data set involves a sequence of some kind, for example, a sequence through time or across distance (see Figure A.1; see figure 5.12 a and b in the main text). Line plots are appropriate when both the x and the y axes represent continuous, numerical data.

Using a line plot allows you to see increasing or decreasing trends in the data. A curved line indicates that the y variable is increasing or decreasing more quickly for some values of x.

A useful graphing technique is to plot two data sets together on the same graph (Figure A.2; see figure 5.17 in the main text). This allows you to compare the trends in the two data sets to see whether and how they seem to be related.

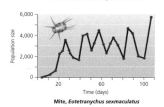

Yeast cells, *Saccharomyces cerevisiae*

Mite, *Eotetranychus sexmaculatus*

Figure A.1

Figure A.2

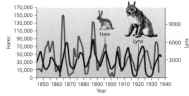

GRAPH IT

A graphing tutorial in the appendix will help you read and interpret the graphs throughout the book. Build your own graphing skills and broaden your understanding of how data can be used with the 14 interactive graphing activities you will find on the website and the CD-ROM.

Environment

Test your understanding.

At the end of the chapter you will find review and discussion questions to help you prepare for tests.

REVIEW QUESTIONS

1. What does the term "biodiversity" encompass?
2. How many species are there in the world?
3. What does a phylogenetic tree show?
4. Name three organisms that have gone extinct.
5. To what levels of biological organization does the science of ecology pertain?
6. What is the difference between a species and a population? A population and a community?
7. Contrast the concepts of habitat and niche.
8. What are the differences between population size, population density, and population distribution? Use examples from this chapter in your answer.
9. List and describe all the major population characteristics discussed in this chapter. Explain how each shapes population dynamics.
10. Could any species undergo exponential growth forever? Explain your answer.
11. Describe how limiting factors relate to carrying capacity.
12. Explain the difference between K–strategists and r–strategists. Can you think of examples of each that were not mentioned in the chapter?
13. Contrast Clements's and Gleason's views of ecological communities.
14. How does parasitism differ from predation?
15. What effects does competition have on the species involved?
16. Name one mutualistic relationship that affects your day-to-day life.
17. Explain and contrast primary and secondary terrestrial succession.
18. What has Costa Rica's experience been with parks and ecotourism?
19. Why are amphibians considered indicators of environmental quality?

DISCUSSION QUESTIONS

1. How has Earth come to have so many species? Contrast the two modes of speciation discussed.
2. How are the phylogenetic trees in Figure 5.4 similar to a family geneaology? How are they different? What kind of information can we learn from them?
3. What types of species are most vulnerable to extinction, and what kinds of factors threaten them? Can you think of any species that are threatened with imminent extinction today? What reasons lie behind their endangerment?
4. How did precipitation, runoff, population density, and population distribution affect the harlequin frog?
5. Can you think of one organism not mentioned as a keystone species that you believe may be a keystone species? For what reasons do you suspect this? Can you think of an organism that you would guess is not a keystone species? What reasoning lies behind your answer? How could you experimentally test whether an organism is a keystone species?
6. Why do scientists consider invasive species to be a problem? What makes a species "invasive," and what ecological effects can invasive species have?
7. Describe the evidence in this chapter that supports scientists' assertion that changing temperatures and precipitation led to the extinction of the golden toad and to population crashes for many other amphibians in Monteverde.
8. What are the advantages of ecotourism for a country like Costa Rica? Can you think of any disadvantages?

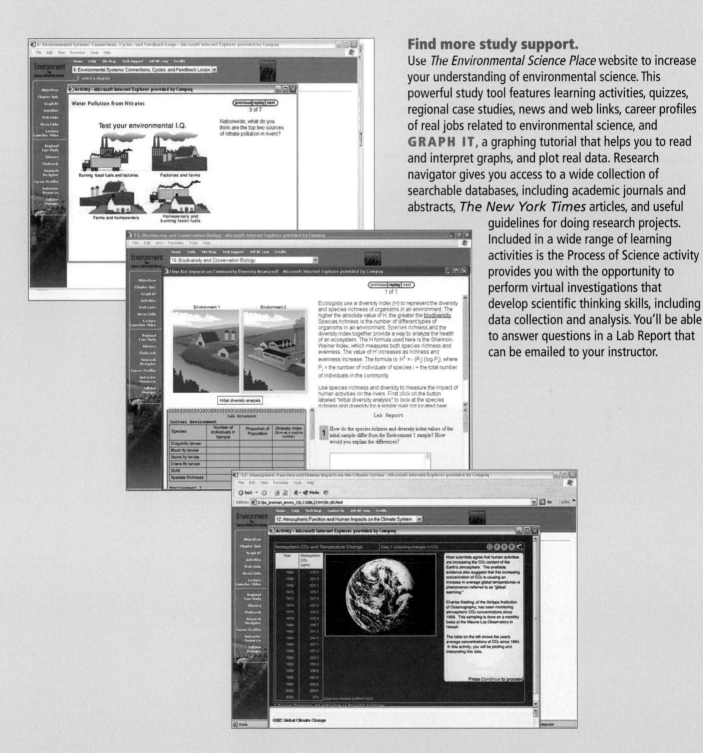

Find more study support.

Use *The Environmental Science Place* website to increase your understanding of environmental science. This powerful study tool features learning activities, quizzes, regional case studies, news and web links, career profiles of real jobs related to environmental science, and **GRAPH IT**, a graphing tutorial that helps you to read and interpret graphs, and plot real data. Research navigator gives you access to a wide collection of searchable databases, including academic journals and abstracts, *The New York Times* articles, and useful guidelines for doing research projects. Included in a wide range of learning activities is the Process of Science activity provides you with the opportunity to perform virtual investigations that develop scientific thinking skills, including data collection and analysis. You'll be able to answer questions in a Lab Report that can be emailed to your instructor.

Environment

Look ahead.
Career profiles highlight real people working in jobs related to environmental science. Additional career profiles are featured on the website.

CP-3

CAREER PROFILE — Ecotourism Specialist

Tourism with a conscience? Ecotourism is just that: responsible travel to natural areas—travel that conserves the environment and preserves local cultures. Ecotourism is travel with a social motive.

One of the big advantages of ecotourism is that it benefits local economies directly. The monies generated go, not to big tour companies and outfitters, but instead to locally owned lodges and homegrown tour guides and artisans. Tourist dollars also help fund the management of the natural areas that are visited. Equally important to supporting local economies is the way ecotourists see the areas they are visiting. Small groups stay at "eco-lodges" and "green hotels" (environmentally friendly lodging), travel with educational guides, and most important, stay on the trails, where, ecotourism manager Eileen Gutierrez says, "they take only pictures and leave no footprints."

Although ecotourism is still a relatively new industry, the term was coined in the early 1980s, when a handful of conservation biologists asserted that local economies could benefit from tourist dollars while maintaining local cultures and environments. Bringing tourists to visit an area responsibly has become a popular way to conserve natural areas and to allow travelers a chance to see destinations that they might otherwise never have a chance to see. In 2002, a United Nations summit in Quebec named that year "The International Year of Ecotourism," signaling the official worldwide acceptance of the trade and adding significance to ecotourism initi-

Eileen Gutierrez works for Conservation International, based in Washington, D.C. As an ecotourism manager whose regional specialty is mainland Asia, her job is to promote economic development and conservation in those areas by creating ecotourism initiatives. "We work in conservation. Our foundation deals with

> **66** [Ecotourists] take only pictures and leave no footprints. **99**

diverse areas worldwide that we call 'hot spots'—highly biodiverse but threatened," explains Gutierrez. "We work with communities in and around protected areas whose threats are socioeconomic in nature. We help provide alternatives to their livelihood that can benefit conservation."

As an ecotourism expert, Gutierrez works on a policymaking level, designing and developing fundraising for ecotourism efforts. These tourism and development guidelines are carefully initiated because although they are designed to benefit communities, they could also easily threaten them. "Ecotourism plans for parks and protected areas and provides guidelines on development and how to zone for ecotourism. We design facilities (eco-lodges and hotels) to be more harmonious with the environment. We also analyze how community involvement and participation might take place and be encouraged," says Gutierrez.

Training for ecotourism is diverse. More schools are offering specialized degrees in environmental s... and e...

grounds among people working to further ecotourism's cause is almost endless. "Ecotourism employs biologists, social anthropologists, and people with business administration degrees working within the field," says Gutierrez. "The four key pillars for success in ecotourism are knowledge of international development issues, a solid foundation in business and economics, an understanding of community development issues, and an understanding of biodiversity and conservation."

These broad requirements mean that anyone interested in ecotourism must be prepared to get a broad education. Working in ecotourism might mean managing endangered areas, as Gutierrez does, or it might involve such varied jobs as managing an eco-lodge or conducting tours and educational programs. For Gutierrez, this kind of variety in her day-to-day job is exactly what motivates her. "Although ecotourism is not an extremely well defined field right now, there's a lot of groundwork that's being done, and the creativity involved in that is great. I've had the chance to develop methodologies for ecotourism assessments, and it has been exciting to be a part of that."

For more information on a career in ecotourism, contact:
The International Ecotourism Society
733 15th Street NW, Suite 100
Washington, D.C. 20005
Telephone: (202) 347-9203
Fax: 202-387-7915
http://www.ecotourism.org

Also...
U... Nations ...mer...

Preface

We live in extraordinary times. Human impact on our environment has never been so intensive or so far-reaching. The future of Earth's systems and of our society depends more critically than ever on the way we interact with the natural systems around us. Fundamental conditions in nutrient cycling, biological diversity, atmospheric composition, and climate are changing at dizzying speeds. Yet thanks to environmental science, we now understand more than ever our environment, the way our planet's systems function, and the ways we influence these systems. Understanding environmental science illuminates not only human-induced problems but the tremendous opportunities we have before us for affecting positive change.

The field of environmental science captures the very essence of this unique moment in history. This interdisciplinary field integrates the natural sciences with the social sciences, studying both the workings of our planet and the workings of our own species. Environmental science draws upon the methods and findings of numerous established academic disciplines, from ecology to geology to chemistry to economics to political science to ethics. This interdisciplinary pursuit stands at the vanguard of the current need to synthesize our increasingly narrow academic disciplines and to incorporate their contributions into a big-picture understanding of the world and our place within it.

We wrote this book because we feel that the vital importance of environmental science in today's world makes it imperative to engage, educate, and inspire a broad audience of today's students—the citizens and leaders of tomorrow. We felt that this called for implementing the very best in modern teaching approaches and for clarifying the ways in which the scientific process can inform human efforts. We also aimed in this book to maintain a balanced approach and to encourage critical thinking, while fleshing out the social debate over many environmental issues. Finally, we aimed to avoid gloom and doom and instead provide hope and solutions. These several aims guided our crafting of the main text and are also reflected by several of the features that make our book unique:

- **Central Case Study integration.** Our teaching experiences and feedback from colleagues across the continent clearly reveal that students' interest is best captured by compelling story-telling about real people and real places. Providing narratives with concrete detail also greatly aids in teaching abstract concepts, because students have a tangible framework with which to incorporate new ideas. While many textbooks these days serve up case studies in isolated boxes, we have chosen to integrate each chapter's central case study into the main text, weaving information and elaboration throughout the chapter. In this way the concrete realities of the people and places of the central case study are used to help illustrate the topics we cover. We are gratified that students and instructors during this book's development and review have consistently applauded this approach, and we hope it can help bring about a new level of effectiveness in environmental science education.

- **The Science behind the Stories.** Our goal is not simply to present students with facts but to engage them in the scientific process of testing and discovery. To do this we have included an extended discussion of the scientific method and the social context of science in the opening chapter, and we have described hundreds of real-life studies throughout the main text. We also feature in each chapter The Science behind the Story, boxes that elaborate upon particular studies important to the topic of each chapter, guiding readers through the details of the research that led to the findings. In this way we show not merely what scientists discovered, but how they discovered it. We expect this feature to enhance students' comprehension of each chapter's material. We also expect it to deepen their understanding of the scientific process itself—a key component of effective citizenship in today's science-driven world.

- **Viewpoints.** Although environmental science is an objective pursuit that is separate from environmental advocacy, many instructors and students have expressed the feeling that some existing textbooks in this field have conveyed an unwarranted bias toward environmental protection and against development. We have strived to present a balanced picture of environmental issues, one always informed by the best science that bears upon those issues. Moreover, to assure that students are exposed to opinions held by advocates on multiple sides of key issues, we present in each chapter the Viewpoints feature, paired essays authored by invited experts that present different points of view on particular questions of importance. These help assure that, beyond our own synthesis of the issues, students receive a taste of informed arguments directly from individuals who are actively involved in debates on environmental issues.

- **Weighing the Issues.** Because the multifaceted issues in environmental science often lack black-and-white answers, critical thinking skills are necessary to help navigate through the gray areas at the juncture of science, policy, and ethics. We have aimed to encourage and help develop these skills with our end-of-chapter questions for review and discussion and with our Weighing the Issues feature. Several Weighing the Issues questions are dispersed throughout each chapter, serving as stopping points for students to absorb and reflect upon what they have read and wrestle with some of the dilemmas of the complex field of environmental science.

- **An emphasis on solutions.** The complaint we have most frequently heard from students of environmental science courses is that the deluge of environmental problems can seem overwhelming. In the face of so many problems, students often come to feel that there is no hope or that there is little they can personally do to make a difference. We have aimed to counter this impression by drawing out throughout the text innovative solutions that have been implemented, that are working, or that can be tried in the future. While not painting an unrealistically rosy picture of the challenges that lie ahead, we have tried to instill hope and encourage action. Problems are portrayed as challenges and dilemmas as opportunities. Indeed, for every problem that human carelessness has managed to create, human ingenuity can devise one—and likely multiple—solutions.

Environment: The Science behind the Stories has grown directly from our professional experiences in teaching, research, and writing. Scott Brennan has taught environmental science to thousands of undergraduates and has developed an intimate feeling for what works in the classroom. His knowledge and experience has profoundly shaped the pedagogical approach taken in this book. Jay Withgott has synthesized and presented science to a wide readership as a science writer, as well as conducted scientific research and teaching at the university level. His experience in distilling and making accessible the fruits of scientific inquiry has deeply influenced the execution and presentation of material.

We have also been guided in our efforts by input from our professional colleagues and from many dozens of instructors from around North America who have served as reviewers for our chapters and as advisors in focus-group meetings arranged by Benjamin Cummings. The participation of so many learned and thoughtful experts has improved this book in countless ways.

We sincerely hope that our efforts will come close to being worthy of the immense importance of our subject matter. We invite you, professors and students alike, to let us know how well we have achieved our goals and where you feel we have fallen short. We are committed to turning out a second edition that is even better than the first, and we would value your feedback. Please write the authors in care of Benjamin Cummings Publishing, 1301 Sansome Street, San Francisco, California, 94111.

At this most historic time to study environmental science, we are honored to serve as your guides in the quest to better understand our world and ourselves.

Scott Brennan and Jay Withgott

INSTRUCTOR SUPPLEMENTS

Instructor's Guide and Test Bank
0-8053-4431-4

Use this helpful instructor's resource to prepare your course and test your students. You will find chapter objectives, lecture outlines, key terms, and teaching tips, additional web and audio-visual resources, and assignable text-related questions for every chapter. Included in this manual is a comprehensive printed test bank that features multiple choice, matching, true/false, short answer, and essay questions. Over 1,000 questions, which are also available in Test-Gen-EQ software.

Instructor's Resource CD-ROM with Video
0-8053-4432-2

Enhance your lecture using this cross-platform CD-ROM, which includes all illustrations from the text, plus key photos and tables. Each image is available, with and without labels, in PowerPoint® or as a jpeg. Also included are 5–10 minute video clips from *National Geographic,* which show scientists pursuing solutions to environmental problems. These video clips are also available on VHS video cassette and may be ordered separately.

Lecture Launcher Videos
0-8053-4437-3

Launch your lectures for each chapter using these dynamic 5–10 minute video clips from *National Geographic,* which are available on VHS video cassette.

Transparency Acetates
0-8053-4430-6

Includes 300 full-color acetates of illustrations, photos, and tables from the text.

Computerized Test Bank
0-8053-4433-0

Prepare your tests using this Test-Gen-EQ software. With more than 1,000 test questions, the Test Bank features multiple choice, matching, true/false, short answer, and essay questions on a cross-platform CD-ROM that allows you to generate tests with a user-friendly interface ion which you can view, edit, sort, and add your own questions.

Course Management Options: CourseCompass and Blackboard

These flexible, easy-to-use course management tools allow professors to combine their own material with the material in the Environmental Science Place to create dynamic, online learning environments. Professors can post their syllabi, assign tutorials, customize quizzes, automatically grade them, and track the results instantly in a gradebook.

STUDENT SUPPLEMENTS

Study Guide
0-8053-8148-1

Get a jump on studying and preparing for tests with this comprehensive study guide that features chapter overviews and outlines, key terms, practice exercises and answers, and solutions to selected end-of-chapter questions from the text.

Environmental Science Place Website

Use this powerful study tool to help you succeed in your course. Features include a wide range of learning activities, practice quizzes, a glossary and flashcards to build and test your comprehension of environmental science concepts. With **GRAPH IT**, a series of 14 interactive graphing tutorials, you can learn to read and interpret graphs and plot data. Research navigator provides useful guidelines for doing research projects, and provides links to a wide collection of searchable databases, including academic journals and abstracts, *The New York Times* articles, and Link Library. Broaden your knowledge of environmental issues using the news and web links and regional case studies, and acquaint yourself with jobs related to environmental science and the educational requirements to pursue them.

Acknowledgments

A textbook is the product of *many* more minds and hearts than one might guess from the names on the cover. The two of us have been unimaginably fortunate to have been carried along by the tremendous staff of Benjamin Cummings and by the guidance of a small army of experts in the field of environmental science who generously gave of their time and expertise. While we alone, as authors, bear responsibility for any inaccuracies, the strengths of this book result from the collective labor and dedication of innumerable people.

We would first like to thank our acquisitions editor, Chalon Bridges. Chalon's commitment to making this book the best it could be has been inspiring to us, and her unremitting enthusiasm enabled all of us on the team to overcome many obstacles and, indeed, to relish the challenges. In addition, her extensive contact with instructors across North America helped us refine our pedagogical direction. The approach, design, and essence of this book owe a great deal to Chalon's astute guidance and vision.

Developmental editors rarely get the acclaim they deserve, which is why we are particularly eager to single out Susan Teahan's extensive contributions to this book, which went above and beyond the call of duty. Every single aspect of this volume benefited from her input and oversight. Susan's vigilant attention to detail, her sound judgment, her discerning sense of the needs of students, and her dedication and hard work have improved this book more than any of us can truly appreciate.

One great strength of this volume we feel is its art program, and for this we are largely indebted to Senior Art producer Russell Chun. Russell's ability to transform our nebulous suggestions into clear and insightful illustrations that gave life to key points from the text continuously amazed us. Photo manager Travis Amos helped provide the other half of our visual impact, with his consummate skill in coming up with arresting images that repeatedly said a thousand words. The talented sleuthing of photo researcher Kristin Piljay and Ira Kleinberg further strengthened the photo program. Senior art editor Donna Kalal oversaw the art program in full and guided the art through production with alacrity.

Editorial assistance was provided by project editor Susan Minarcin, who headed up the supplements and appendices, and by editorial assistant Alissa Anderson, who ably guided the Viewpoints through completion. Our thanks to Alissa and to associate editor Alexandra Fellowes for managing the reviewing process. We are indebted to the unstoppable Kay Ueno, director of development, for her clear-headed management, criticism, and encouragement. Copy editor Sally Peyrefitte provided thorough coverage and improvement of our text. We also thank Elizabeth Fogarty, Moira Lerner-Nelson, Aaron Gass, and Nora Lally-Graves.

We would like acknowledge the authors of our excellent supplements: Dawn Ford from the University of Tennessee at Chattanooga for the Instructors Manual, Sherri Morris from Bradley University for the Test Bank, and Heidi Marcum from Baylor University for the Student Study Guide. Special thanks to Andrew Corbett for his outstanding development and production of **GRAPH IT**.

A number of significant contributions were made by April Lynch, Etienne Benson, Susan Borowitz, Christy Brigham, and Eric Simon. Other contributions were offered by Malina Mamagonian, Dan Ferber, Scott Norris, Virginia Gewin, Arlyn Christopherson, Diana Siemens, Laura Brancella, and Christine Freeman. Instructors that provided contributions include Patricia Smith from Valencia Community College, Brian Shmaefsky from Kingwood College, and Jennifer Warner from University of North Carolina at Charlotte.

Once the text manuscript was ready, production editor Corinne Benson was there. In fact, she was generally there before we were ready, waiting on us; we are grateful to her for running a tight ship and offering just the right balance of warm encouragement and stern whip-cracking. Erin Gregg oversaw the design process. Finally, manufacturing manager Pam Augspurger saw our book through the final phases of production.

Of course, none of this has any impact on education without the marketing staff to get the book into your hands. Senior marketing manager Josh Frost, marketing development manager Susan Winslow, and field marketing specialist Jeff Howard brought their considerable sales and marketing talent to our book throughout the development process. And last but surely not least, the many field representatives who help communicate our vision and deliver our product to instructors are a vital link in the chain.

This vast supportive structure functions effectively only with sound direction from the top. We thank vice president Frank Ruggirello for his support and his hard-nosed business sense. And we thank Benjamin Cummings president Linda Davis for her steadfast commitment to this project; we are grateful and proud to publish with her company.

In the list that follows, we acknowledge the many professors, instructors, and outside experts who helped us maximize the quality and accuracy of our presentation through their chapter reviews, class testing, or related services. If the thoughtfulness and thoroughness of these reviewers are any indication, we feel confident that the teaching of environmental science is in excellent hands! We also would like to thank the individuals who wrote our Viewpoints essays, each of whom is credited along with his or her essay.

Scott thanks the students, faculty, and staff at Western Washington University's Huxley College of the Environment, in particular Brad Smith, Gigi Berardi, Jack Hardy, John Miles, Paul Olmstead, Andrew Bodman, Kris Bulcroft, Karen Casto, Floyd McKay, Carmen Werder, and Justina Brown. In addition, Brooke Simler of COMPASS made critical contributions to the early development of the book's key features, including the Central Case and The Science behind the Story concepts, and Kate Koch and Nausheen Mohamedali assisted with research. Scott would also like to thank Korby Lenker, Sean Cosgrove, and Jonathan Duncan for their countless, critical contributions over the years.

Scott dedicates this book to his parents, Jack and Melinda Brennan, and to his students—all of who have taught him many precious lessons. Jay dedicates this book to his wife Susan Masta, who endured its writing with tremendous patience and sacrifice and provided loving support and sustenance throughout.

Scott Brennan and Jay Withgott

Reviewers

Mary E. Allen, *Hartwick College;* Dulasiri Amarasiriwardena, *Hampshire College;* Gary I. Anderson, *Santa Rosa Junior College;* Joeseph A. Arruda, *Pittsburg State University;* Timothy J. Bailey, *Pittsburg State University;* Stokes Baker, *University of Detroit;* David Bass, *University of Central Oklahoma;* Timothy J. Bell, *Chicago State University;* William Berry, *University of California—Berkeley;* Kristina Beuning, *University of Wisconsin—Eau Claire;* Richard Drew Bowden, *Allegheny College;* Nancy Broshot, *Linfield College;* David Brown, *California State University—Chico;* Lee Burras, *Iowa State University;* Charles E. Button, *University of Cincinnati Clermont College;* Jon Cawley, *Roanoke College;* Linda Chalker-Scott, *University of Washington;* Sudip Chattopadhyay, *San Francisco State University;* Luke W. Cole, *Center on Race Poverty & the Environment;* Darren Divine, *Community College of Southern Nevada;* Iver W. Duedall, *Florida Institute of Technology;* Margaret L. Edwards-Wilson, *Ferris State University;* Anne H. Ehrlich, *Stanford University;* Thomas R. Embich, *Harrisburg Area Community College;* W. F. J. Evans, *Trent University;* Jiasong Fang, *Iowa State University;* M. Siobhan Fennessy, *Kenyon College;* Linda Mueller Fitzhugh, *Gulf Coast Community College;* Doug Flournoy, *Indian Hills Community College—Ottumwa;* Johanna Foster, *Johnson County Community College;* Chris Fox, *Catonsville Community College;* Nancy Frank, *University of Wisconsin—Milwaukee;* Robert Frye, *University of Arizona;* Sandi Gardner, *Triton College;* Marcia L. Gillette, *Indiana University Kokomo;* Jeffrey J. Gordon, *Bowling Green State University;* John G. Graveel, *Purdue University;* Cheryl Greengrove, *University of Washington;* Amy R. Gregory, *University of Cincinnati Clermont College;* Sherri Gross, *Ithaca College;* David E. Grunklee, *Hawkeye Community College;* Mark Gustafson, *Texas Lutheran University;* Daniel Guthrie, *Claremont College;* David Hacker, *New Mexico Highlands University;* Greg Haenel, *Elon University;* David Hassenzahl, *University of Nevada Las Vegas;* Joseph Hobbs, *University of Missouri at Columbia;* Catherine Hooey, *Pittsburg State University;* Jonathan E. Hutchins, *Buena Vista University;* Juáña Ibanez, *University of New Orleans;* Walter Illman, *University of Iowa;* Gina Johnston, *California State University—Chico;* Stanley Kabala, *Duquesne University;* Carol Kearns, *University of Colorado—Boulder;* Dawn Keller, *Hawkeye Community College;* Tom Kozel, *Anderson College;* Frank T. Kuserk, *Moravian College;* William R. Lammela, *Nazareth College;* Michael T. Lares, *University of Mary;* John Latto, *University of California—Berkeley;* Joseph Luczkovich, *East Carolina University;* Jennifer Lyman, *Rocky Mountain College;* Ian R. MacDonald, *Texas A & M University;* Kenneth Mantai, *State University of New York—Fredonia;* Patrick S. Market, *University of Missouri—Columbia;* Steven R. Martin, *Humboldt State University;* John Mathwig, *College of Lake County;* Allan Matthias, *University of Arizona;* Robert Mauck, *Kenyon College;* Debbie McClinton, *Brevard Community College;* Paul McDaniel, *University of Idaho;* Gregory McIsaac, *Cornell University;* Dan McNally, *Bryant College;* Richard McNeil, *Cornell University;* Mike L. Meyer, *New Mexico Highlands University;* Patrick Michaels, *Cato Institute;* Chris Migliaccio, *Miami Dade Community College;* Kiran Misra, *Edinboro University of Pennsylvania;* Mark Mitch, *New England College;* Brian W. Moores, *Randolph-Macon College;* James T. Morris, *University of South Carolina;* Sherri Morris, *Bradley University;* William M. Murphy, *California State University—Chico;* Rao Mylavarapu, *University of Florida;* Jane Nadel-Klein, *Trinity College;* Muthena Naseri, *Moorpark College;* Michael J. Neilson, *University of Alabama—Birmingham;* Moti Nissani, *Wayne State University;* Richard B. Norgaard, *University of California—Berkeley;* Niamh O'Leary, *Wells College;* Brian O'Neill, *Brown University;* Eric Pallant, *Allegheny College;* Phillip J. Parker, *University of Wisconsin—Platteville;* Daryl Prigmore, *University of Colorado;* Loren A. Raymond, *Appalachian State University;* Barbara C. Reynolds, *University of North Carolina—Asheville;* Thomas J. Rice, *California Polytechnic State University;* Gary Ritchison, *Eastern Kentucky University;* Mark G. Robson, *University of Medicine and Dentistry of New Jersey;* Carlton Lee Rockett, *Bowling Green State University;* Armin Rosencranz, *Stanford University;* Robert E. Roth, *The Ohio State University;* Christopher T. Ruhland, *Minnesota State University;* Ronald L. Sass, *Rice University;* Richard A. Seigel, *Towson University;* Wendy E. Sera, *NDAA's National Ocean Service;* Maureen Sevigny, *Oregon Institute of Technology;* Linda Sigismondi, *University of Rio Grande;* Jeffrey Simmons, *West Virginia Wesleyan College;* Jan Simpkin, *College of Southern Idaho;* Patricia L. Smith, *Valencia Community College;* Douglas J. Spieles, *Denison University;* Bruce Stallsmith, *University of Alabama at Huntsville;*

Richard J. Strange, *University of Tennessee;* Robert A. Strikwerda, *Indiana University at Kokomo;* Richard Stringer, *Harrisburg Area Community College;* Ronald Sundell, *Northern Michigan University;* Bruce Sundrud, *Harrisburg Area Community College;* Max R. Terman, *Tabor College;* Adrian Treves, *Wildlife Conservation Society;* Charles Umbanhowar, *St. Olaf College;* G. Peter van Walsum, *Baylor University;* Callie A. Vanderbilt, *San Juan College;* Elichia A. Venso, *Salisbury University;* Rob Viens, *Bellevue Community College;* Maud M. Walsh, *Louisiana State University;* Phillip L. Watson, *Ferris State University;* Richard D. Wilk, *Union College;* James J. Winebrake, *Rochester Institute of Technology;* Jeffrey S. Wooters, *Pensacola Junior College;* Zhihong Zhang, *Chatham College.*

Class Testers

David Aborne, *University of Tennessee at Chattanooga;* Dr. Barret, *Prairie State College;* Morgan Barrows, *Saddleback College;* Henry Bart, *LaSalle University;* James Bartalome, *University of California at Berkeley;* Christy Bazan, *Illinois State University;* Dr Richard Beckwitt, *Framingham State College;* Dr. Elizabeth Bell, *Santa Clara University;* Peter Biesmeyer, *North Country Community College;* Donna Bivans, *Pitt Community College;* Evert Brown, *Casper College;* Christina Buttington, *University of Wisconsin, Milwaukee;* Tait Chirenje, *Richard Stockton College;* Reggie Cobb, *Nash Community College;* Ann Cutter, *Randolph Community College;* Lola Deets, *Pennsylvania State Erie;* Ed DeGrauw, *Portland Community College;* Mrs. Dockstader, *Monroe Community College;* Dr. Dee Eggers, *University of North Carolina, Asheville;* Jane Ellis, *Presbyterian College;* Paul Fader, *Freed Hardeman University;* Joseph Fail, *Johnson C Smith University;* Brad Fiero, *Pima Community College, West Campus;* Dane Fisher, *Pfeiffer University;* Chad Freed, *Widener University;* Sue Glenn, *Gloucester County College;* Sue Habeck, *Tacoma Community College;* Mark Hammer, *Wayne State University;* Michael Hanson, *Bellevue Community College;* David Hassenzahl, *Oakland Community College;* Kathleen Hornberger, *Widener University;* Paul Jurena, *Univ. of Texas at San Antonio;* Dawn Keller, *Hawkeye Community College;* Dr. David Knowles, *East Carolina University;* Erica Kosal, *Wesleyan College;* John Logue, *University of Southern Carolina Sumter;* Keith Malmes, *Valencia Community College;* Nancy Markee, *University of Nevada, Reno;* Bill Mautz, *Univ. of New Hampshire;* Julie Meents, *Columbia College;* Mr. Getchell, *Mohawk Valley Community College;* Lori Moore, *Northwest Iowa Community College;* Elizabeth

Pixley, *Monroe Community College;* John Novak, *Colgate University;* Brian Peck, *Simpson College;* Sarah Quast, *Middlesex Community College;* Dr. Roger Robbins, *East Carolina University;* Mark Schwartz, *University of California at Davis;* Julie Seiter, *University of Nevada, Las Vegas;* Brian Shmaefsky, *Kingwood College;* Diane Sklensky, *Le Moyne College;* Mark Smith, *Fullerton College;* Patricia Smith, *Valencia Community College East;* Sherilyn Smith, *Le Moyne College;* Jim Swan, *Albuquerque Technical Vocational Institute;* Amy Treonis, *Creighton University;* Dr. Darrell Watson, *The University of Mary Hardin Baylor;* Barry Welch, *San Antonio College;* Susan Whitehead, *Becker College;* Roberta Williams, *University of Nevada, Las Vegas;* Justin Williams, *Sam Houston University;* Tom Wilson, *University of Arizona.*

Viewpoint Essayists

Jock Anderson, *World Bank;* Jay Brandes, *University of Texas at Austin;* Ignacio Chapela, *University of California—Berkeley;* Timothy L. Cline, *Population Connection;* Paul Craig, *University of California—Davis;* Richard Denison, *Environmental Defense;* Thomas Flint, *AgFARMation;* Christopher H. Foreman, Jr., *University of Maryland;* Pete Geddes, *Foundation for Research on Economics and the Environment;* Eban Goodstein, *Lewis and Clark College;* Susan Hock, *National Renewable Energy Laboratory;* Peter Kareiva, *The Nature Conservancy;* Debra Knopman, *RAND Corporation;* Robert Lawton, *University of Alabama—Huntsville;* Antonio Lazcano, *Universidad Nacional Autónoma de México;* Michael Leech, *International Game Fish Association;* Jane Lubchenco, *Oregon State University;* David McIntosh, *Natural Resources Defense Council;* Norman Myers, *Oxford University;* Sara Nicolas, *American Rivers;* Al Norman; Warren Porter, *University of Wisconsin—Madison;* Daryl Prigmore, *University of Colorado at Colorado Springs;* William Rathje, *University of Arizona;* Terry Roberts, *Foundation for Agronomic Research;* Florence Robinson, *North Baton Rouge Environmental Association;* Lori Saldaña, *San Diego Mesa College;* Debra Saunders, *San Francisco Chronicle;* Mark Schaffer, *Defenders of Wildlife;* Gary Sirota; Samuel Staley, *The Buckeye Institute for Public Policy;* Marian K. Stanley, *American Chemistry Council;* Douglas Sylva, *Catholic Family and Human Rights Institute;* Paul H. Templet, *Louisiana State University;* Frederick Troeh, *Iowa State University;* Indra K. Vasil, *University of Florida;* Nathaniel T. Wheelwright, *Bowdoin College;* Robert Williams, *Oil and Gas Journal.*

Brief Contents

Detailed Contents

PART ONE

Foundations of Environmental Science

1 An introduction to environmental science

Our Island, Earth

This chapter will help you understand:

- What environmental scientists mean by the term *environment*

- The interdisciplinary nature of environmental science

- The scientific method and how science operates

- Some of the pressures on the global environment

- Natural resources and their importance to human life

- How population pressures and natural resources interact

- The concepts of sustainability and sustainable development

Our Island, Earth

People who have viewed Earth from space say our home planet resembles a small blue marble suspended against a vast inky black backdrop. Although few of us will ever get to witness that sight directly, photographs taken by astronauts do indeed convey a sense that Earth is small, isolated, and fragile. It may seem vast to us as we go about our lives on its surface, but from the astronaut's perspective it is apparent that Earth and its natural systems are not unlimited. From this perspective, it becomes clear that as our population, our technological powers, and our consumption of resources increase, so does our ability to alter our planet and damage the very systems that keep us alive.

Our environment is the sum total of our surroundings

A photograph of Earth reveals a great deal, but it does not convey the complexity of the human environment. This **environment** (from the French *environner,* "to surround") is more than water, land, and air; it is the sum total of our surroundings. It includes all of the **biotic factors,** or living things, with which we interact. It also includes the **abiotic factors,** or nonliving things, with which we interact. Our environment includes the continents, clouds, and ice caps you can see in the photo of Earth from space, and the animals, plants, forests, and farms that comprise the landscape around us. In a more inclusive sense, it also encompasses our built environment, the structures, urban centers, and living spaces humans have created. In its most inclusive sense, our environment additionally consists of the complex web of scientific, ethical, political, economic, and social relationships and institutions that shape our daily lives.

People most commonly use the term *environment* in the first, narrow sense—of a natural or non-human world that is outside of and apart from the human world. This is unfortunate, because it masks the very important fact that humans exist within the environment and are a part of nature. As one of many of species of animals on Earth, we share the same dependence upon a healthy functioning planet as do all others. The limitations of language make it all too easy to speak of humans and nature, or the human world and the environment, as though they are separate and do not interact. The fundamental insight of environmental science, however, is that we are part of the natural world and that our interactions with it matter a great deal. We will want to keep this in mind as we progress through this text, whose purpose, like environmental science itself, is to further our understanding of human interactions with the environment.

Understanding our interactions with the environment is important for several reasons. First, we depend upon our environment for air, water, food, shelter, and everything else essential for living. Second, our actions modify our environment, whether we intend them to do so or not. Many of our actions have brought us beneficial changes, such as longer lifespans, material wealth, mobility, and leisure time. However, numerous others have spawned undesirable changes, from air and water pollution to soil erosion to species extinction, that compromise human well-being and can threaten human life. In addition, understanding our relationship with our surroundings is crucial for a well-informed view of our place in the world and a mature awareness that we are one species among many on a planet full of life. The elements of the environment were functioning long before humans appeared, and we would be wise to realize that we need to keep these elements in place. Finally, we need to understand our interactions with the environment because such knowledge is the essential first step toward devising solutions to our most pressing environmental problems.

It can be daunting to reflect on the sheer magnitude of environmental dilemmas that confront us today, but with these problems also come countless opportunities for devising creative solutions. The topics studied by environmental scientists are today's most centrally important issues to our world and its future. Right now, global conditions are changing more quickly than ever, scientists are gaining knowledge more rapidly than ever, and the window of opportunity for acting to solve problems is still open. With such bountiful challenges and opportunities, this particular moment in history is indeed an exciting time to be studying environmental science.

Natural resources are vital to our survival

An island by definition is finite and bounded, and its inhabitants must cope with limitations in the materials they need. On our island, Earth, human beings, like all living things, ultimately face environmental constraints. Specifically, there are limits to our **natural resources,** which are the various substances and energy sources we need in order to survive. Some resources, such as the plants we harvest for food and shelter, can regenerate over time, but only if we are careful not to use them up too quickly or destructively. Other resources, such as the

**Renewable
natural resources**

- Sunlight
- Wind energy
- Wave energy
- Geothermal energy

- Agricultural crops
- Fresh water
- Forest products
- Soils

**Nonrenewable
natural resources**

- Crude oil
- Natural gas
- Coal
- Gold, silver, and
  other metals

Figure 1.1 Natural resources lie along a continuum from perpetually renewable to nonrenewable. Perpetually renewable resources, such as sunlight, will always be there for us. Nonrenewable resources, such as oil and coal, exist in limited amounts that could one day be gone. Other resources, such as timber, soils, and food crops, can be renewed on intermediate time scales, if we are careful not to deplete them.

crude oil we extract to produce gasoline, are finite. Once we use them up, they are no longer available.

Natural resources that are virtually unlimited or that are replenished by the environment over relatively short periods of hours to weeks to years are known as **renewable natural resources.** Some renewable resources, such as sunlight, wind, and wave energy, are perpetually available, whereas others, such as timber and soil, may take decades to renew themselves. Other resources, such as oil, are in limited supply and are formed much more slowly than we use them. These are known as **nonrenewable natural resources.** We can view the renewability of natural resources as a continuum (Figure 1.1). Some renewable resources may turn nonrenewable if we deplete them too drastically. For example, overpumping groundwater can deplete water reserves in underground aquifers and turn a lush landscape into a desert. Populations of animals and plants we harvest from the wild may be renewable if we do not overharvest them, but may become extinct if we do. In recent years, our consumption of the natural resources we depend on has increased greatly, driven by increased affluence and the growth of the largest human population in history.

Human population growth has shaped our relationship with natural resources

For nearly all of human history, only a few million people populated Earth at any one time. Although past popula-

tions cannot be calculated precisely, Figure 1.2 gives some idea of just how recently and suddenly our population has grown past 6 billion. Two events caused especially remarkable increases in population size. The first event was the transition from a hunter-gatherer lifestyle to an agricultural way of life, which occurred around 10,000 years ago and is known as the **Neolithic**, or **agricultural**, **revolution.** As people began to grow their own crops, raise domestic animals, and live sedentary lives in villages, they found it easier to meet their nutritional needs. As a result, they began to live longer and to produce more children who survived to adulthood. The second major event, known as the **industrial revolution** (Chapter 2), began during the mid-1700s. It entailed a shift from rural life, animal-powered agriculture, and manufacturing by craftsmen to an urban society powered by fossil fuels such as coal and oil. The industrial revolution also introduced improvements in sanitation and medical technology and increased agricultural production through fossil fuel-powered equipment and synthetic fertilizer (Chapter 9). Such developments enabled our dramatic rise in population.

Thomas Malthus and population growth At the outset of the industrial revolution in England, population growth was regarded, overall, as a good thing. For parents, high birth rates meant more children to support them in old age; for society, a greater pool of labor for factory work. However, Thomas Malthus (1766–1834), a British economist, had quite a different opinion.

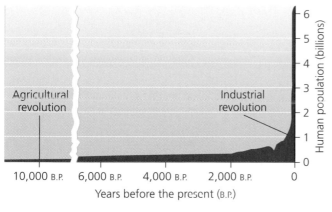

Agricultural revolution

Industrial revolution

10,000 B.P.   6,000 B.P.   4,000 B.P.   2,000 B.P.   0

Years before the present (B.P.)

(a) World population growth

(b) Urban society

Figure 1.2 (**a**) For almost all of human history our population has been low and relatively stable. It increased significantly as a result first of the agricultural revolution and later of the industrial revolution. (**b**) Our skyrocketing population has given rise to congested cities, such as this one in Java, Indonesia.

Malthus claimed that unless the rapid growth of the population was controlled by laws, rules, or other social strictures, the number of people would outgrow the available food supply until starvation, war, and disease arose and reduced the population (Figure 1.3).

Malthus's most influential work, *An Essay on the Principle of Population,* published in 1798, argued that a growing population would eventually be checked either by limits on births or increases in deaths. Limits on

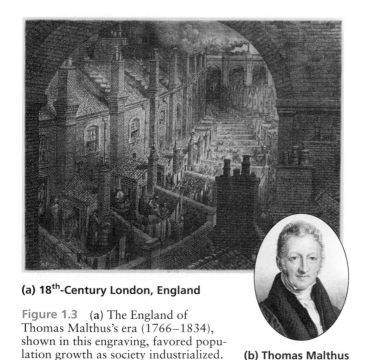

(a) 18ᵗʰ-Century London, England

Figure 1.3 (**a**) The England of Thomas Malthus's era (1766–1834), shown in this engraving, favored population growth as society industrialized. (**b**) Malthus argued that population growth could lead to disaster.

(b) Thomas Malthus

births might involve "moral restraint" (or abstinence) and contraception. If such restrictions were not implemented soon enough, Malthus wrote, an increase in the number of deaths through famine, plague, and war would be the catastrophic consequence.

Malthus's thinking was shaped by the rapid urbanization and industrialization he witnessed during the early years of the industrial revolution, but debates over his views continue today. As you will see in Chapter 7 and throughout this book, global population growth has indeed helped spawn famine, disease, and social and political conflict. However, increasing material prosperity has also helped bring down birth rates—something Malthus did not foresee.

Paul Ehrlich and "The Population Bomb" Biologist Paul Ehrlich of Stanford University has been called a *neo-Malthusian* because he too has warned that population growth will have disastrous effects on human well-being. In his 1968 book, *The Population Bomb,* Ehrlich predicted that a rapidly increasing human population would cause widespread famine and conflict that would consume civilization by the end of the 20ᵗʰ century. Ehrlich, like Malthus, argued that the human population was growing much faster than was our ability to produce and distribute food. He too claimed that population control was the only way to prevent massive starvation and civil strife.

Although human population has nearly quadrupled in the last 100 years—the fastest it has ever grown over just one century (see Figure 1.2)—Ehrlich's predictions have not come true on the scale he initially predicted.

This is due, in part, to the agricultural advances made in recent decades (Chapter 9). As a result, Ehrlich and other neo-Malthusians have revised their predictions accordingly and now warn of a delayed, but still impending, global food crisis. Some critics have scoffed at this revision, arguing that Ehrlich's dire predictions are not likely to be realized in the foreseeable future. Other observers maintain that the consequences he predicts are inevitable unless we act to prevent them.

Garrett Hardin and "The Tragedy of the Commons"
Whereas Malthus and Ehrlich focused on how population growth would affect the availability of resources, the late Garrett Hardin, of the University of California, Santa Barbara, has analyzed how people use resources. Hardin's best-known essay, "The Tragedy of the Commons," first published in the journal *Science* in 1968, disputed the economic theory that unfettered exercise of individual self-interest will, in the long term, serve the public interest. Hardin bases his argument on a scenario described in a pamphlet published in 1833.

According to Hardin's essay, a public pasture, or common, that is open to unregulated grazing will eventually be destroyed. Destruction of the pasture will occur because each person whose animals graze there will be motivated to increase the number of his or her animals in the pasture until overgrazing causes the pasture's food production to collapse (Figure 1.4). Because no one person owns the pasture, no one has the incentive to expend effort taking care of it, and everyone takes what he or she can until the resource is depleted. Some have suggested that private ownership can address this problem. Others argue that those who use the commons can cooperate in enforcing the responsible use of its resources. For the most part, however, it has been argued that the dilemma justifies government regulation of the use of pastures and other resources held in common by the public, from forests to clean air to clean water.

Figure 1.4 Unregulated environments that offer limited resources freely to the public are prone to be depleted by the process Garrett Hardin dubbed "the tragedy of the commons."

Weighing the Issues:
The Tragedy of the Commons

Imagine you make your living fishing for lobster. You are free to boat anywhere and to set out as many traps as you like. Your harvests have been good, and there is nothing stopping you from increasing your number of traps. However, all the other lobster fishers are thinking the same thing, and the fishing grounds are getting crowded. Catches decline year by year, until one year the fishery crashes, leaving you and all the others with catches too meager to support your families. Some people call for dividing the waters and selling access to individuals plot-by-plot, while others urge the fishers to team up, set quotas among themselves, and prevent newcomers from entering the market. Still others are urging the government to get involved and pass laws regulating how much fishers can catch. What do you think is the best way to combat this tragedy of the commons and restore the fishery?

Environmental science can help us avoid mistakes made by past civilizations

While it remains to be seen whether the direst predictions of Malthus and Ehrlich will come to pass for today's global society, we already have historical evidence that civilizations can crumble when population pressure overwhelms resource availability. Easter Island is the classic case of this (see The Science behind the Story), but it is not the only example. Numerous great civilizations have fallen after depleting resources from their environments, and each has left devastated landscapes in its wake. The Greek and Roman empires and the Mayan and Inca civilizations of the New World all show evidence of such a trajectory. Plato wrote of the environmental degradation accompanying ancient Greek cities, and today further evidence is accumulating from research of the archaeologists, historians, and paleoecologists who study past societies and landscapes. The arid deserts of today's Middle Eastern countries, for example, were far more vegetated when the great ancient civilizations thrived there; at that time these regions were lush enough to support the very origin of agriculture.

Today we are confronted with news and predictions of environmental catastrophes on a regular basis, but it is often difficult to evaluate the reliability of such reports. It is even harder to evaluate the causes and effects of environmental change. Perhaps most difficult of all is to devise solutions to environmental problems. Studying environmental science will outfit you with the tools that can help you evaluate information on environmental change and think critically and creatively about possible actions to take in response.

Because addressing environmental problems is a complex endeavor, environmental science is a multifaceted field that brings together many disciplines in the **natural sciences** (disciplines that study the natural world) and the **social sciences** (disciplines that study human interactions and institutions). To make decisions and solve problems, we need to have accurate information about our environment and interpret it reasonably, and this is what study within the natural sciences enables us to do. Decision-making and problem-solving also involve weighing values and understanding human behavior, thus requiring the social sciences. Let us examine this broad field we call environmental science and then explore the basics of the process and methods of science.

The Nature of Environmental Science

Environmental scientists aim to understand how the natural world works, how the environment affects us, and how we affect the environment. They also try to determine how we might develop solutions to the many environmental problems we increasingly face. Indeed, for many environmental scientists, developing solutions

is a large part of what motivates their work. The solutions themselves (such as new technologies, public-policy decisions, or resource management strategies) are best thought of as applications of environmental science. However, the study of such applications and their consequences is, in turn, also part of environmental science.

Environmental problems are perceived differently by different people

Environmental science arose in the latter half of the 20th century as an attempt to understand environmental problems and their origins. An **environmental problem**, stated simply, is any undesirable change in the environment. However, the perception of what constitutes an undesirable change may vary from one person or group of people to another, or from one context or situation to another. A person's age, gender, class, race, nationality, employment, and educational background can all affect whether he or she considers a given environmental change to be a "problem."

For instance, modern-day industrial societies are more likely to view the spraying of the pesticide DDT as a problem than those same societies viewed it in the 1950s, because today more is known about the health dangers of pesticides (Figure 1.5, page 10). At the same time, a person living today in a malaria-infested village in Africa or India may welcome DDT usage if it kills mosquitoes that transmit malaria, which is viewed as a more immediate health threat. Thus an African and an American who have each knowledgeably assessed the pros and cons may still, because of the differences in their circumstances, differ in their judgement of DDT's severity as an environmental problem.

In other cases, different types of people may vary in their awareness of problems. For example, in many cultures women are responsible for collecting water and fuel wood. Therefore, they may be the first to perceive environmental degradation affecting these resources, while men in the same area simply might not "see" the problem. Furthermore, in most societies, information about environmental health risks tends to reach wealthy people more readily than poor people.

Thus, who you are, where you live, and what you do can have a huge effect on how you perceive your environment, how you perceive and react to change, and what impact those changes may have on how you live your life. In Chapter 2, we will examine the diversity of human values and philosophies and consider their effects on how we define environmental problems.

The Lesson of Easter Island

The haunting statues of Easter Island were erected by a sophisticated civilization that collapsed after depleting its resource base and devastating its island environment.

Easter Island is one of the most remote spots on the globe, located in the Pacific Ocean 3,750 km (2,325 mi) from South America and 2,250 km (1,395 mi) from the nearest inhabited island. When the first European explorers reached the island (today called Rapa Nui) in 1722, they found a barren landscape populated by fewer than 2,000 people, who lived in caves and eked out a marginal existence from a few meager crops. Explorers also observed that the desolate island featured hundreds of gigantic statues of carved stone, evidence that a sophisticated civilization had once inhabited the island.

We know today that at an earlier time the island had indeed been lushly forested, with all the appeal of a South Pacific paradise, and had supported a prosperous society with a population of 6,000 to 20,000. Tragically, however, this once-flourishing society overused its resources, cutting down all its trees, which in turn brought starvation and conflict to its people and destroyed its own civilization. Today Easter Island stands as a parable and a warning for what can happen when a population grows too large and consumes too much of the limited resources that support it.

How do we know what we do about Easter Island's history? Since the time of its discovery, historians and anthropologists have wondered how people without wheels

or ropes, on an island without trees, managed to move statues 10 m (33 ft) high weighing 90 metric tons (99 tons) as far as 10 km (6.2 mi) from the quarries where they were chiseled to the coastal sites where they were erected. The explanation, scientists have discovered, lay in the fact that the island did not always lack trees, and its people were not always without rope.

To solve the mystery of Easter Island's past, scientists excavated samples of sediments from the bottom of three of the island's volcanic crater lakes. These researchers, who included John Flenley of the University of Hull in Great Britain, drilled cores deep into the mud of the lakes and examined preserved ancient grains of pollen in the sediments. Because pollen

grains vary from one plant species to another, scientists, by identifying specific pollen grains, can reconstruct, layer by layer, the history of vegetation within a region through time. By analyzing pollen grains, Flenley and other researchers found that when Polynesian people arrived (possibly by A.D. 300, or A.D. 690 at the latest), the island was covered with palm trees. By examining the palm pollen under scanning electron microscopes, the researchers determined that the now-vanished palms were related to the Chilean wine palm, a tall and thick-trunked tree. Archaeologists located ancient palm nut casings in caves and rock niches, and a geologist found carbon-lined channels in the soil that matched root channels typical of the Chilean wine palm. Furthermore, scientists deciphering the island people's script on stone tablets discerned characters etched in the form of palm trees.

Pollen analysis showed that trees other than palms had been common. The island had supported at least a scrubby forest and perhaps a dense rainforest, scientists inferred. However, pollen analysis also revealed that starting around A.D. 750 tree populations had declined and ferns and grasses had become more common. By A.D. 950, the trees were largely gone, and around A.D. 1400 overall pollen values plummeted to extremely low levels, indicating severe lack of vegetation. The same sequence of events occurred about two centuries later at the other two lake sites, which were higher and more remote from village areas. Researchers first hypothesized that the pattern of deforestation was due to climate change, but evidence instead supported the hypothesis that the people had gradually denuded their own island. Furthermore, researchers came across old nut casings scarred with teeth marks of rats, suggesting that the rodents, which were most likely transported from Polynesia, had prevented new palms from sprouting.

The palms and other trees would have been used for fuel wood, as building material for houses and canoes, and presumably to move the stone statues. Several anthropologists in recent

years have experimentally tested hypotheses about ways the islanders might have moved monoliths down from the quarries, by hiring groups of men to recreate the feat. Some methods have actually worked. All involve using numerous tree trunks as rollers or sleds, as well as great quantities of rope, the only likely source of which would have been the fibrous inner bark of the hauhau tree, a species that today is near extinction.

With the trees gone, soil would have eroded away more easily—a phenomenon confirmed in the bottom of Easter Island lakes, where large quantities of sediment had accumulated. Faster runoff of rainwater would have meant less fresh water available for drinking. Runoff and erosion would have degraded the islanders' agricultural land, lowering yields of crops, such as bananas, sugar cane, and sweet potatoes. Reduced agricultural production would have led to starvation and subsequent population decline.

Archaeological evidence supports the scenario of environmental degradation and civilization decline. Food remains, which can be aged by radiocarbon dating (Chapter 4), shifted over the years. Besides their crops, early islanders feasted on the bounty of the sea, including fish, sharks, turtles, octopus, shellfish, and seabird eggs. Analysis of islanders' diets in the later years, however, indicated that little seafood was consumed. Evi-

dently, with the trees gone the islanders could no longer build the great double-canoes their proud Polynesian ancestors had used for centuries to fish and travel among islands. Indeed, the Europeans who arrived at Easter Island in the 1700s observed only a few old small canoes and flimsy rafts made of reeds. As resources declined, the islanders' main domesticated food animal, the chicken, became more valuable; archaeologists found that later islanders had kept their chickens in stone fortresses with entrances designed to prevent theft. The once prosperous and peaceful civilization fell into clan warfare, as revealed by unearthed skeletons and skulls with head wounds and by artifacts of weapons made of obsidian, a hard volcanic rock.

Is the story of Easter Island as unique and isolated as the island itself, or does it hold lessons for our world today? Like the Easter Islanders, we are all stranded together on an island with limited resources. The Earth may be vastly richer in resources than was Easter Island, but Earth's human population is much greater. It is clear that although the Easter Islanders could see that they were depleting their resources, they could not seem to stop. Whether we can learn from the history of Easter Island and act more wisely to conserve the resources on our island, Earth, is entirely up to us.

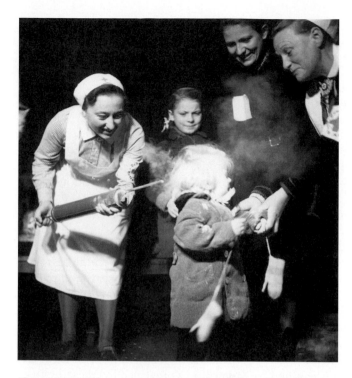

Figure 1.5 How a person or a society defines an environmental problem can vary with time and circumstance. In Germany in 1945, the health hazards of the pesticide DDT were not yet known so children were doused with the chemical to treat head lice. Today, knowing of its toxicity to people and wildlife, many developed nations have banned DDT outright. However, in some developing countries where malaria is a threat, DDT is still welcomed to combat mosquitoes that carry the disease.

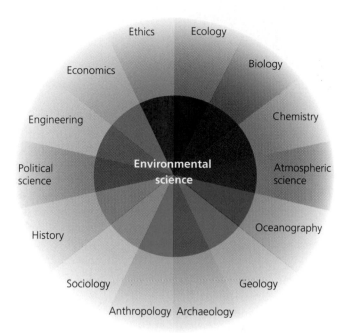

Figure 1.6 Environmental science is a highly interdisciplinary pursuit, involving input from many different established fields of study across the natural sciences and social sciences.

Environmental science provides interdisciplinary solutions to environmental problems

Studying and proposing solutions to environmental problems requires expertise from many disciplines, including ecology, earth sciences, economics, political science, demography, ethics, and others. Environmental science is thus an **interdisciplinary** field: one that borrows techniques from numerous, more traditional fields of study and brings together into a broad synthesis research results from these fields (Figure 1.6). Traditional established disciplines are valuable because scholars of these disciplines delve deeply into topics, uncovering new knowledge and developing expertise in a particular area. Interdisciplinary fields, in contrast, are valuable because specialized knowledge from different disciplines is consolidated, synthesized, and made sense of in a broad context to better serve the multifaceted interests of society.

Environmental science is especially broad because it encompasses not only the natural sciences but also many social sciences. For this reason, many people prefer using the term *environmental studies* to describe this academic umbrella. Whichever term one chooses to use, the field reflects many diverse perspectives and sources of knowledge.

Just as an interdisciplinary approach to studying environmental issues can help us better understand them, an interdisciplinary approach to addressing environmental problems can produce effective and lasting solutions. One example is the dramatic improvement in one aspect of air quality in the United States that has taken place over the past few decades. Ever since automobiles were invented, lead had been added to gasoline to make cars run more smoothly, even though medical professionals knew that lead emissions from tail pipes could cause serious health problems, including brain damage and premature death.

In 1970 air pollution was severe, and motor vehicles accounted for 78% of lead emissions in the United States. In the following years, engineers, physicians, atmospheric scientists, and politicians all merged their knowledge and skills into a process that eventually resulted in the banning of leaded gasoline. Whereas in 1975 only 13% of gas sold in the United States was unleaded, by 1996 this number had reached 100%, and the nation's largest source of atmospheric lead emissions had been completely eliminated.

Environmental science is not the same as environmentalism

Although many environmental scientists are interested in solving problems, it would be incorrect to confuse environmental science with environmentalism, or environmental activism. They are not the same. Environmental science is the pursuit of knowledge about the workings of the environment and our interactions with it, whereas **environmentalism** is a social movement dedicated to protecting the natural world—and, by extension, humans—from undesirable changes brought about by certain human choices (Figure 1.7).

Although environmental scientists may study many of the same issues environmentalists care about, as scientists they attempt to take an objective approach as they integrate the work of ecologists, ethicists, economists, political scientists, and practitioners of many other disciplines. In this textbook we will explore the perspectives and findings of all these disciplines at one

Figure 1.7 Environmental scientists and environmental activists play very different roles. Some scientists have become activists to promote solutions to environmental problems, but most of them try hard to keep their advocacy separate from their pursuit of objective scientific work.

time or another, and often simultaneously. First, we will begin with a brief overview of how science works and how scientists go about this enterprise that brings us so much valuable knowledge.

The Nature of Science

Modern scientists describe **science** (from the Latin *scire*, "to know") as a systematic process for learning about the world and testing our understanding of it. The term *science* is also commonly used to refer to the accumulated body of knowledge that arises from this dynamic process of observation, testing, and discovery. Richard Feynman, recipient of the 1965 Nobel Prize in physics, described new discoveries as the "gold" that scientists seek.

In addition, people sometimes use the word *science* in a still broader sense, including within its bounds the practical application of scientific knowledge to problem-solving. Foremost among the applications of science are its use in developing technology and its use in informing policy and management decisions (Figure 1.8). It is important to understand that technology is not the same as science; rather, it is a product of science. Often technology is also a tool that enables further science to be conducted. While many scientists are motivated simply by a desire to know how the world works, others are motivated by the potential for developing useful applications.

Environmental science encompasses the two meanings of science outlined above; it is a dynamic yet systematic way of studying the world, and it is also the body of knowledge accumulated from this process. Like science in general, environmental science does not strictly include its applications, but it is often motivated by them.

Why does science matter? The late astronomer and author Carl Sagan wrote the following in his 1995 treatise, *The Demon Haunted World: Science as a Candle in the Dark*:

> We've arranged a global civilization in which the most crucial elements—transportation, communications, and all other industries; agriculture, medicine, education, entertainment, protecting the environment; and even the key democratic institution of voting—profoundly depend on science and technology. We have also arranged things so that almost no one understands science and technology. This is a prescription for disaster. We might get away with it for a while, but sooner or

(a) Prescribed burning

(b) Methanol-powered fuel cell car

Figure 1.8 Scientific knowledge can be applied in the form of policy and management decisions, and in the form of technology. (a) Prescribed burning, as shown here in Ouachita National Forest, Arkansas, requires that fires are kept under careful control. This management practice to restore healthy forests is informed by research into forest ecology. (b) Energy-efficient automobiles like this methanol-powered fuel cell car from Daimler–Chrylser, is a technological advance made possible by materials and energy research.

later this combustible mixture of ignorance and power is going to blow up in our faces. . . . Science is an attempt, largely successful, to understand the world, to get a grip on things, to get hold of ourselves, to steer a safe course.

Sagan and many other scientists before and since have argued that science is essential if we hope to sort fact from fiction and develop solutions to the problems—environmental and otherwise—that we face today.

Scientists test ideas by weighing evidence

How can we tell whether the warnings of impending environmental catastrophes—or any other claims, for that matter—are based on scientific thinking? Scientists examine statements or claims about the workings of the world by designing tests to determine whether they are supported by evidence. Ideas can be refuted by evidence but can never be absolutely proven, so, strictly speaking, scientific testing amounts to attempting to disprove ideas. If a particular statement or explanation is testable and resists repeated attempts to disprove it, scientists are more likely to accept it as a useful and true explanation. Scientific inquiry thus consists of an incremental approach to the truth.

The scientific method is the key element of science

Scientists generally follow a process called the **scientific method.** A technique for testing ideas with observations, it involves several assumptions and a more-or-less consistent series of interrelated steps. There is nothing mysterious about the scientific method; it is merely a formalized version of the procedure any of us might naturally take, using common sense, to figure out a problem.

However, the scientific method is a theme with variations, and scientists pursue their work in many different ways. Because science is an active, creative, imaginative process, an innovative scientist may find good reason to stray from the traditional scientific method when a particular situation demands it. In addition, scientists from different fields may approach their work somewhat differently because they deal with different types of information. A natural scientist, such as a chemist, will conduct research quite differently from a social scientist, such as a sociologist. Because environmental science includes both the natural and social sciences, in our discussion here we use the term *science* in its broad sense, to include both. Despite their many differences, scientists of all persuasions broadly agree on the fundamental elements of the process of scientific inquiry.

The scientific method relies on the following assumptions:

- The universe functions in accordance with fixed natural laws that do not change from time to time or from place to place.
- All events arise from some cause or causes and, in turn, cause other events.
- We can use our senses and reasoning abilities to detect and describe the natural laws that underlie the cause-and-effect relationships we observe in nature.

Individual researchers or teams of researchers follow the scientific method as they investigate questions that interest them. However, all scientific work takes place within the context of a community of peers, and to have any impact, a researcher's work must be published and made accessible to this community. Thus below we first summarize the scientific method as practiced by individual scientists or research teams and then discuss some elements of the larger process as it takes place on the level of the scientific community as a whole.

As practiced by individual researchers or research teams, the scientific method generally consists of the following steps (Figure 1.9), although not always in the exact order presented here.

Make observations Advances in science typically begin with the observation of some phenomenon the scientist wishes to explain. Observations can be made either directly through the five senses, or indirectly with the assistance of instruments or tools, such as microscopes, that amplify the senses. Observations set the scientific method in motion and also function throughout the process.

Ask questions Curiosity is a fundamental human characteristic. This is evident to anyone who has observed the explorations of a young child in a new environment. Babies want to touch, taste, feel, watch, and listen to anything that catches their attention, and as soon as they can speak, they begin asking questions. Scientists, at least in this respect, are kids at heart. Why are certain plants or animals less common today than they once were? Why are storms becoming more severe and flooding more frequent? What is causing excessive algae growth in local ponds? Do pesticide impacts on fish or frogs indicate that humans may be affected in the same ways? All of these are questions environmental scientists

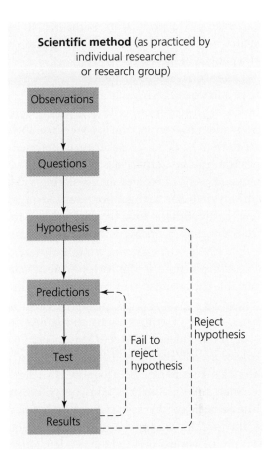

Figure 1.9 The scientific method is the observation-based hypothesis-testing approach that scientists use to learn how the world works. This diagram is a simplified generalization that, although useful for instructive purposes, cannot convey the true dynamic and creative nature of science. Moreover, researchers from different disciplines may pursue their work in ways that legitimately vary from this model.

have asked and attempted to answer. We will ask similar questions and search for answers to those questions throughout this book.

Develop a hypothesis Scientists attempt to answer their questions by devising explanations that they can test. A **hypothesis** is an educated guess that explains a phenomenon or answers a scientific question. For example, a scientist investigating the question of why algae are growing excessively in local ponds might notice chemicals being applied on farm fields nearby. The scientist might then state a hypothesis as follows: "Agricultural fertilizers running into ponds cause the amount of algae in the ponds to increase."

Make predictions The scientist next uses the hypothesis to generate **predictions**, which are specific statements that can be directly and unequivocally tested. In our algae example, a prediction might be: "If agricultural fertilizers are added to a pond, the quantity of algae in the pond will increase."

Test the predictions Predictions are tested one at a time by gathering evidence that could potentially refute the prediction and thus refute the hypothesis. Often the strongest form of evidence comes from experimentation. An **experiment** is an activity designed to test the validity of a hypothesis; it involves manipulating **variables,** or conditions that can change. For example, a scientist could test the hypothesis linking algal growth to fertilizer by selecting two identical ponds and adding fertilizer to one while leaving the other in its natural state. In this example, fertilizer input is the **independent variable,** or the variable the scientist manipulates, while the quantity of algae that results is the **dependent variable,** one that depends on the fertilizer input and is measured. If the two ponds are identical except for a single variable (the fertilizer input), then any differences that arise between the two ponds can be attributed to that single variable. Such an experiment is known as a **controlled experiment** because the scientist controls for the effects of all variables except the one whose effect he or she is testing. In our example, the pond left unfertilized serves as a control, an unmanipulated point of comparison for the manipulated treatment pond. Whenever possible, it is best to replicate one's experiment, that is, to stage multiple tests of the same comparison of control and treatment. Our scientist could perform a replicated experiment on, say, 10 pairs of ponds, adding fertilizer to one of each pair.

Experiments can establish causal relationships, showing that changes in an independent variable cause changes in a dependent one. However, experiments are not the only way of testing a hypothesis. Sometimes a hypothesis can be convincingly addressed through **correlation,** or searching for relationships among variables. For instance, let's say our scientist surveys 50 ponds throughout a county, 20 of which happen to be fed by fertilizer runoff from nearby farm fields and 30 of which are not. Let's also say he or she finds seven times as much algal growth in the fertilized ponds as in the nonfertilized ponds. The scientist can say that algal growth is correlated with fertilizer input; that is, that one tends to increase along with the other. Although this type of evidence is weaker than that the causal demonstration experimentation can provide, sometimes it is the best approach, or the only feasible one. For example, in studying the effects of global climate change, we could hardly add carbon dioxide to ten Earth-like planets and ten control planets and compare the results.

Analyze and interpret results Scientists record data, or information, from their studies. They particularly value **quantitative data,** information that can be expressed using numbers, because numbers provide precision and are easy to compare. The scientist running the fertilization experiment, for instance, might quantify the area of water surface covered by algae in each of the ponds or might measure the dry weight of algae contained in a certain volume of water taken from each. Even with the precision that numbers provide, however, the scientist's results may not be clear-cut. The measurements between ponds may vary only slightly, and if our scientist used replicates and tested 10 pairs of ponds, different replicates could have yielded different results. The scientist must therefore analyze the data using statistical tests. With these mathematical methods, scientists can objectively and precisely determine the strength and reliability of patterns they find, by determining probabilities. **Probability** is a numerical expression of the likelihood that a particular conclusion is true. Because some results may differ from others simply by chance, determining probability is crucial to data analysis. In quantitative scientific work, probabilities accompany each and every conclusion. (Chapter 10 describes probability in more detail.)

A fair amount of research, however, especially in the social sciences, involves data that is by its nature **qualitative,** or not expressible in terms of numbers. Research involving historical texts, personal interviews, surveys, detailed examination of case studies, or descriptive observation of behavior can all include qualitative data on which statistical analyses may not be possible. Such studies can still be scientific in the broad sense, because their data can be interpreted systematically using other accepted methods of analysis.

Weighing the Issues:
Replicates and Data Analysis

Let's say our scientist testing for the effects of agricultural fertilizer on algal growth used experimental replicates, testing 10 pairs of lakes. If the scientist finds that in all 10 pairs, the fertilized treatment pond grows more algae than the unfertilized control pond, he or she would likely conclude that the results support the hypothesis that fertilizer encourages algal growth. If, however, treatment ponds grew more algae in only 5 of the replicates, while control ponds grew more algae in the other 5, there would clearly be no reason to conclude that fertilizer had any effect. However, what if 7 of the 10 treatments grew more algae than controls—what do you think the scientist should conclude? What if 8 did? What if 9 did? What if 8 treatment ponds showed 10% more growth than the control ponds they were paired with, but the remaining 2 control ponds showed 300% more growth than their paired treatments? Such cases require statistical analysis and the calculation of probabilities, so that we can judge levels of confidence to assign to our conclusions. Given possibilities like this, can you explain why scientists believe replicates are important?

If a hypothesis is refuted by experimentation, the scientist will reject it and may develop a new educated guess to replace it. If experiments result in failure to reject the hypothesis, this lends support to the hypothesis but does not prove it is correct, in the sense of a formal logical or mathematical proof. The scientist may choose to generate new predictions in order to test the hypothesis in a different way and better assess its likelihood of being true. Thus, the scientific method loops back on itself, often giving rise to repeated rounds of hypothesis-revision and new experimentation (see Figure 1.9).

If repeated tests result in failure to reject a particular hypothesis, evidence in favor of the hypothesis thereby accumulates, and the researcher may eventually conclude that the idea is well supported. To garner strong support for a particular hypothesis, though, one would ideally want to test **alternative hypotheses,** or different potential answers for the question of interest, as well. For instance, our scientist might propose and test an alternative hypothesis that algae increase in ponds because of a decrease in the numbers of fish or invertebrate animals that eat algae. It is possible, of course, that both hypotheses may be correct and that each may explain some portion of the initial observation that local ponds were experiencing algal blooms.

There are different ways to test hypotheses

An experiment in which the researcher actively chooses and manipulates the independent variable is known as a **manipulative experiment** (Figure 1.10a). A manipulative experiment often provides the strongest type of evidence a scientist can obtain. In practice, however, some modes of scientific inquiry are more amenable to

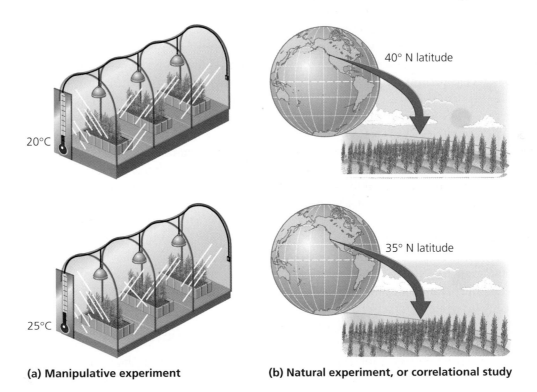

(a) Manipulative experiment

(b) Natural experiment, or correlational study

Figure 1.10 (a) A researcher wishing to test how temperature affects the growth of wheat might run a manipulative experiment in which wheat is grown in two identical greenhouses, one kept at 20°C (68°F) and the other kept at 25°C (77°F). (b) Alternatively, the researcher might run a natural experiment in which he or she compares the growth of wheat in two fields at different latitudes, a cool northerly location and a warm southerly one. Because it would be difficult to hold all other variables besides temperature constant, the researcher might want to collect data on a number of northern and southern fields and correlate temperature and wheat growth using statistical methods.

manipulative experimentation than others. Physics and chemistry tend to involve manipulative experiments, but many other fields deal with entities less easily manipulated than physical forces and chemical reagents. This is true of **historical sciences** such as cosmology, which deals with the history of the universe, or paleontology, which deals with the history of past life. It is difficult to manipulate experimentally a star thousands of light years away or the fossil tooth of a mastodon. Moreover, many of the most interesting questions in these fields revolve around the causes and consequences of particular historical events, rather than the behavior of general constants. Yet these disciplines are still science, every bit as rigorous and relevant to understanding our world as physics and chemistry.

Disciplines that do not quite fit the so-called "physics model" of science sometimes rely on **natural experiments** rather than manipulative ones (Figure 1.10b). For instance, an evolutionary biologist wondering whether animal species isolated on oceanic islands tend to evolve large body size over time might test the idea by comparing pairs of closely related species, in which one of each pair lives on an island and the other on the mainland of a continent. The biologist cannot run a manipulative experiment by placing animals on islands and continents and waiting long enough for evolution to do its work, but this is exactly what nature has already done for the biologist. Thus the experiment has in essence been conducted naturally, and it is up to the scientist to interpret the results.

In ecology, manipulative experimentation is sometimes, but not always, possible. The science of **ecology** deals with the distribution and abundance of organisms (living things), the interactions among them, and the interactions between organisms and their abiotic environments. It is one of the most important disciplines involved in environmental science. When possible, an ecologist tries to run a manipulative experiment. An ecologist who wants to measure the importance of a certain insect in pollinating the flowers of a given crop plant might, for example, fit some flowers with a device to keep the insects away while leaving other flowers accessible and later measure the fruit output of each group. Other questions that involve large spatial scales or long time scales may require natural experiments instead.

The social sciences often involve less experimentation than the natural sciences and depend more on careful observation and interpretation of patterns in data. A so-ciologist interested in how people from different cultures conceive the notion of wilderness might conduct a survey and analyze responses to the survey's questions, looking for similarities and differences across groups of people. An environmental historian might study old accounting records of a fur trade company to see how numbers of animals trapped and numbers of pelts sold varied through the years and whether these related to estimated population levels of the animals. An economist might take such data and relate it to broader economic trends or to trade in other commodities. (For example, does declining fur availability correlate with rising production of wool for clothing?) Such analyses may be either quantitative or qualitative, depending on the nature of the data and the researchers' particular questions and approaches.

However, even in the natural sciences some fields, such as taxonomy (the description and classification of organisms), are largely descriptive, relying on careful observation and measurement but not on experimentation.

Observational studies and natural experiments can show correlation between variables, but cannot demonstrate that one variable causes change in another, as manipulative experiments can. Not all variables are controlled for in a natural experiment, and thus a single result could give rise to several interpretations. However, correlative studies when done well can make for very convincing science, and they preserve the real-world complexity that often is not reflected in manipulative experiments. Furthermore, sometimes correlation is all we have. Because manipulations are often difficult on large scales, many of the most important questions in environmental science tend to be addressed, at least at first, with correlative data.

In addition, because of the large scale and the complexity of many questions in environmental science, few studies, whether manipulative or correlative, come up with neat and clean results. As such, scientists are not always able to give policymakers and society black-and-white answers to questions.

The scientific process does not stop with the scientific method

As we noted above, the scientific method pursued by an individual researcher or research team takes place within the broader context of the scientific community. Let's now look at how science works on the community level (Figure 1.11).

Figure 1.11 The scientific method (inner box) followed by individual researchers or research teams exists within the context of the overall process of science at the level of the scientific community (outer box). If a particular hypothesis garners extensive support from the work of many independent researchers, it may lead to the development of models and theories.

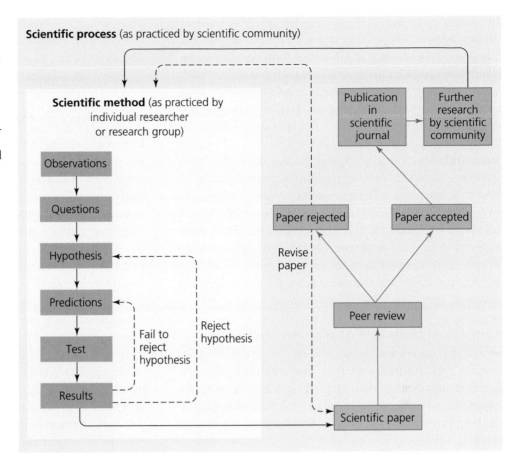

Peer review When a researcher's work is done and the results thoroughly analyzed, he or she writes up the findings and submits them to a journal for publication. Several other scientists specializing in the topic of the paper examine the manuscript and provide comments and criticism (generally anonymously) and judge whether the work merits publication in the journal. This procedure, known as peer review, is an essential part of the scientific process. Peer review is a valuable guard against faulty science contaminating the literature on which all scientists rely. However, because scientists are human and may have their own personal biases and agendas, politics can sometimes creep into the review process. Fortunately, just as individual scientists strive their best to remain objective in conducting their own research, the scientific community does its best to ensure fair review of all work. Winston Churchill once called democracy the worst form of government, except for all the others that had been tried. The same might be said about peer review; it is an imperfect system, yet it seems to be the best that has been proposed.

Conference presentations Scientists will frequently present their work at professional conferences, where scientists interact with colleagues and often receive informal comments on their research. When research has not yet been published, feedback from colleagues can help improve the quality of a scientist's work before it is submitted for publication.

Grants and funding A large portion of a research scientist's time is spent writing grant applications requesting money for research from private foundations or from government agencies, such as the National Science Foundation. Researchers rely on this type of funding, so grant-writing is a vital skill. Grant applications undergo peer review just as scientific papers do, and competition for funding can often be intense. Scientists' dependence on funding sources can also lead to potential conflicts of interest. A scientist who obtains data showing his or her funding source in an unfavorable light may be reluctant to publish such results for fear of losing funding—or worse yet, may be tempted to doctor the results. This situation can arise, for instance, when research is funded by industries to test their products for safety or environmental impact. While most scientists do not succumb to these temptations, and while most funding sources are careful not to pressure their scientists for certain results, when critically assessing a scientific study, one should always try to find out where the researchers obtained their funding.

Repeatability Sound science is based on doubt rather than on certainty, and on repeatability rather than on one-time occurrence. Even when a hypothesis appears to provide an accurate explanation of observed phenomena, scientists are inherently wary of accepting it. The careful scientist will test a hypothesis repeatedly in different ways before submitting the findings for publication. Following publication, other scientists may attempt to reproduce the results in their own experiments and analyses.

Models and theories If a hypothesis survives repeated testing by numerous research teams and continues to predict experimental outcomes and observations accurately, it may be modeled, and may potentially be incorporated into a theory. A **model** is a deliberately simplified representation of a complex natural (or social) system. Models synthesize known information and attempt to use it to represent how complex processes operate. In Chapter 12 we discuss climate models and the need for simplification. If a model is reliable and helps explain a broad array of phenomena, it may eventually be integrated into a theory. A **theory** is a widely accepted, well-tested explanation of one or more cause-and-effect relationships that has been extensively validated by a great amount of research. It differs greatly from a hypothesis in that a hypothesis is a single explanatory statement that may be refuted, while a theory consolidates many related hypotheses that have been tested and have not been refuted.

Note that the scientific use of the word *theory* is vastly different from the popular usage of the word. In everyday language, when we say something is "just a theory," we are suggesting it is a speculative idea without much substance. Scientists, however, mean just the opposite when they use the term; to them, a theory is a conceptual framework that effectively explains a phenomenon and has undergone extensive and rigorous testing, such that confidence in it is extremely strong. For example, Darwin's theory of evolution by natural selection has been supported and elaborated by many thousands of studies over 150 years of intensive research. Such research has shown repeatedly and in great detail how plants and animals change over generations, or evolve, to express characteristics that best promote survival and reproduction. Because of its strong support and explanatory power, evolutionary theory is the central unifying principle of modern biology.

Science may go through "paradigm shifts"

It is worth noting that science as we have described it is not the only means for accumulating knowledge. In fact, some cultures do not use the scientific method at all, but this does not mean their understanding of the world is uninformed or incorrect. In addition, it is crucial to realize that science is not infallible and that interpretations of results obtained via the scientific method can later be proven incorrect. Thomas Kuhn's 1962 book *The Structure of Scientific Revolutions* argued that science goes through periodic revolutions, dramatic upheavals in thought, in which one scientific **paradigm**, or dominant view, is abandoned for another. An example is the shift from the ancient Earth-centered view of the solar system to the modern sun-centered view. Before the 16th century, scientists believed that Earth was at the center of the known universe, and they made elaborate and accurate measurements explaining the movements of planets from that viewpoint. Their data fit the theory quite well, yet the theory eventually was disproved by Nicolaus Copernicus, who showed that the sun-centered view explained the planetary data even better. A similar paradigm shift occurred in the 1960s, when geologists accepted the theory of plate tectonics, because evidence for the movement of continents and the action of tectonic plates had accumulated and become overwhelmingly convincing.

Understanding how science works, ideally and in reality, is paramount to comprehending how scientific ideas and interpretations change through time as new information accrues. This is especially important in environmental science, a young field that is changing exceedingly rapidly as we learn huge amounts of new information, as the impacts of human activity on the planet multiply, and as lessons from the consequences of our actions, good and bad, become apparent. Because so much remains unstudied and undone, and because so many issues we cannot foresee now are likely to appear in the near future, environmental science will remain an exciting frontier for you to explore as a student, as an informed citizen, and perhaps as a future environmental scientist. Next we will take a quick tour of ways in which environmental scientists are studying and addressing some of the major environmental issues in our world today.

Table 1.1 Selected Environmental Trends 2000–2002

Environmental indicator	Trend/Current status (2000–2002 data)
Global fossil fuel consumption	Increased by 1% in 2001 to highest level ever
Global air and ocean temperatures	In 2001 reached second-highest level since late 1800s
Oil spills from civilian sources	50,000 tons in 2001, the least spilled since 1968
Global automobile numbers	555 million cars, increasing by 40 million each year
Global bicycle production	100 million produced in 2000, the first increase since 1995
Global cropland degradation	~1.5 billion ha (20% of world cropland) are degraded as of 2002
Farmland paved each year	2.5 million acres in the United States, 500,000 acres in China
Irrigated land suffering salinization (excessive amounts of salt due to method of irrigation)	20% of world's total irrigated land
Annual pesticide use	Increased 15 times since 1950
Pesticide poisoning	3 million people poisoned, 200,000 people killed in 2001
Aquaculture/fish farming production	Increased by more than 400% from 1984–2000
Global per capita grain production	1.843 billion tons, down from high of 1.880 billion tons in 1997
Global meat production	Record high of 237 million tons in 2001
Total human population	6.157 billion in 2001, increase of 77 million over 2000 total
Population growth	95% of growth occurs in poor, developing countries

Data adapted from Worldwatch Institute, *Vital Signs 2002: The trends that are shaping our future.*

Environmental Science and the State of the World

Throughout this book you will see examples of environmental scientists asking questions, developing hypotheses, conducting experiments, gathering and analyzing data, and drawing conclusions about both stable environmental processes and environmental changes and their causes and effects. This pursuit of knowledge requires drawing on a sound understanding of the natural environment and the human activities that affect it, and keeping abreast of environment-related discoveries, issues, and trends. Table 1.1 outlines some key environmental indicators and their current status, using data from 2000–2002.

Human population growth lies at the root of many environmental changes

The ways we modify our environment are diverse, but one factor has, in some manner, influenced nearly all of our modifications: the steep and sudden rise in human population (Chapter 7). Our numbers have nearly quadrupled in the past 100 years, passing 6 billion in 1999 and 6.3 billion in 2003, representing the fastest population growth in our species' history. Furthermore,

of the 79 million people added to the planet each year (that's 217,000 per day), most are born in the poorest and least technologically developed nations. Today, the rate of population growth is slowing, but our absolute numbers continue to increase and to shape our interactions with one another and with our environment. Understanding human population growth is central to environmental science, because human population growth can potentially intensify nearly every environmental problem.

Increased agricultural production has been a triumph but also entails serious costs

Human population has grown so large in part because of our mostly successful efforts to expand and intensify the production of food (Chapter 8 and Chapter 9). Since the agricultural revolution, advancing technology has enabled us to grow more and more food per unit of land. These advances in agriculture must be counted as one of humanity's great achievements. However, they have come at some cost, including massive use of chemical fertilizers and pesticides and widespread conversion of natural habitats. We have converted nearly half the planet's land surface for agriculture of some kind—11% for farming, 11% for forestry, and 26% for grazing. Farmland today is itself undergoing rapid alteration, as erosion, climate change, and poorly managed irrigation destroy 5–7 million ha (12.5–17.5 million acres) of productive cropland each year (Chapter 8). Millions more are lost to the spread of cities and suburbs.

In response to agricultural problems, people have developed and promoted soil conservation, high-efficiency irrigation, organic agriculture, and genetically engineered "super-crops." Proponents of genetically modified crops argue that super-crops could feed the world's hungry while lessening environmental degradation (by, for instance, reducing the need for pesticides and fertilizers). Critics, however, maintain that any such gains could be short-lived and that engineered genes could pose threats to human health, wild relatives of crop plants, and natural systems.

Weighing the Issues:
Genetically Engineered Crops to
Feed the World

If the proponents of genetically engineered crops are right, these crops could boost food production while decreasing the need to apply chemicals and irrigation, thus making our resource use more efficient and our agriculture safer. "Super-crops" could, they say, enable us to feed more of the world's people. Do you think that being able to feed more people is necessarily a good thing? In what ways might it be good, and in what ways might it not be good?

One of the major drawbacks of large-scale agriculture is pesticide use, which is rooted in the need to protect crop plants from the insects and other organisms that feed on them and the weeds that compete with them. Modern industrial agriculture's practice of dedicating huge swaths of land to single crops encourages outbreaks of such "pest" organisms. Furthermore, fast-reproducing organisms, such as insects and microbes, can quickly evolve resistance to pesticides, rendering the chemicals useless and forcing us to develop new ones that are more powerful. Ever-more powerful chemicals, however, may be ever more toxic to humans and to the many other organisms in our environment. Chapter 9 and Chapter 10 explore the tradeoffs involved in chemical use and the development of more environmentally friendly approaches to pest management.

Pollution comes in many forms, with diverse consequences

Pesticides from farms are just one type of the many synthetic chemicals widespread in our environment today.

Figure 1.12 Indoor and outdoor air pollution contribute to millions of premature deaths each year, and environmental scientists are working to reduce this problem in a variety of ways.

Scientists in the field of **toxicology** (Chapter 10) study the health effects of such chemicals on humans and other organisms. Environmental toxicologists who study nonhuman organisms often view the objects of their studies like the proverbial canary in the coal mine whose death would indicate that air in the mine was getting too dangerous to breathe. Toxic effects on nonhuman animals, the logic goes, may warn of similar effects on humans.

Artificial chemicals produced by our farms, our industries, and our individual actions pollute land, water, and air. Scientists have recently calculated that indoor air pollution kills more than a million people each year in poor countries and that outdoor air pollution kills 40,000 annually in Austria, France, and Switzerland alone (Chapter 11). At the same time, technological advances and new laws have greatly reduced the pollution emitted by industry and automobiles in wealthier countries (Figure 1.12). Such progress provides hope that many forms of environmental degradation can be partially reversed when there is the political will to implement change.

Political will is crucial if we are to address the looming specter of climate change (Chapter 12). Scientists have firmly concluded that human activity is altering the composition of the atmosphere and that these changes are affecting Earth's climate. Since 1750, coinciding with the start of the industrial revolution, atmospheric carbon dioxide concentrations have risen by 31%, to a level not present in at least 420,000 years. Methane concentrations have increased by 151%, nitrous oxide concentrations by 17%, and ozone in the lower atmosphere by 36%. All of these gases absorb heat in the lower atmosphere and warm Earth's surface, which is likely responsible for glacial melting, sea-level rise, impacts on wildlife and crops, and increased episodes of destructive weather. The U.S. government has resisted international efforts to rein in these pollutants and halt climate change, even though American scientists have been at the forefront of climate change science. Other nations, however, are beginning to address the problem, as are some state governments in the United States.

Aquatic resources have been neglected but are now receiving more attention

Understanding climate and climate change involves study of the oceans as well as the atmosphere. The socioeconomic impact of the fluctuations in ocean temperature we call El Niño and La Niña has heightened the attention to ocean science in recent years (Chapter 13). In addition, both oceans and freshwater systems (Chapter 14) face serious pollution problems. For example, scientists have learned that sulfur compounds emitted into the air by coal and oil combustion can cause water pollution hundreds of miles away. The large quantities of trash, oil, and other pollutants released into rivers, lakes, and oceans pose additional dangers to aquatic life. During a recent National Coastal Cleanup, an annual volunteer event sponsored by the Center for Marine Conservation, volunteers collected 1.3 million kg (2.9 million lbs) of garbage from U.S. beaches.

Even worse than pollution for ocean life, marine ecologists are finding, is overfishing (Figure 1.13). Most ocean fisheries around the world today are overexploited and in decline, and some have already become depleted, harming not only marine ecosystems but also the people who fish and depend on the sea's bounty for their livelihoods. Fortunately, scientists are finding that establishing protected reserves for marine life can help restore depleted fish populations. Freshwater fish could use similar protection; as of 2002, approximately 20% of the 10,000 species of freshwater fish estimated to exist in North America at the time of European settlement had become extinct or were threatened with extinction because of environmental change.

Earth's biodiversity is gravely threatened

The combined impact of human actions such as overharvesting, habitat alteration, climate change, pollution, and the introduction of alien species has driven many aquatic and terrestrial species out of large parts of their

Figure 1.13 Although oceans cover most of the planet, the problems they face are frequently less visible and less understood than terrestrial problems. Depletion of fish populations through overharvesting threatens species with extinction, causes ripple effects through marine ecosystems, and hurts fishers who depend on catches for their livelihoods.

former ranges and toward the brink of extinction (Chapter 15). Today Earth's **biological diversity**, or **biodiversity**, the cumulative number and diversity of living things, is declining dramatically; many biologists say we are already in the midst of one of the greatest mass extinction events in Earth's history.

Biologist Edward O. Wilson has warned that the loss of biodiversity is our most serious and threatening environmental dilemma, because it is not the kind of problem that responsible human action can remedy, as is the case with pollution. Rather, the extinction of species is irreversible; once a species has become extinct, it is lost forever. Although there are ample reasons for concern about the state of global biodiversity, advances in conservation biology are enabling scientists and policymakers in many cases to work together to protect habitat, slow extinction, and safeguard endangered species (Figure 1.14).

We need to use resources, but we can use them in better ways

Pressure on Earth's plants and animals often stems from our exploitation of natural resources, but these resources are necessary for the continued maintenance of our civilization. For instance, without minerals and mining there would be no cars, no computers, no paved roads, and no metals of any kind. Along with their many benefits, however, mining and mineral processing also cause severe environmental problems. Mines can harm wildlife and force people from their homes, degrade

Figure 1.14 Human activities are pushing many organisms, including the panda, toward extinction. Efforts to save endangered species include captive breeding in zoos. However, such efforts will come to naught unless adequate areas of appropriate habitat are preserved in the wild.

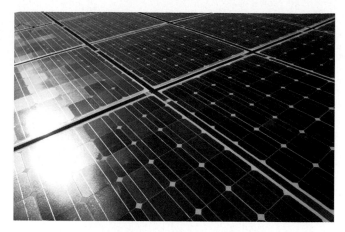

Figure 1.16 Our dependence on fossil fuels has caused a wide array of environmental problems. Although fossil fuels have powered our civilization since the industrial revolution, many renewable energy sources, such as solar energy that can be collected by panels like these, exist. Such alternative energy sources could be further developed for use now and in the future.

farmland and hunting grounds, and pollute air and water (Figure 1.15).

We face such tradeoffs with many environmental issues, and the challenge is to develop solutions that further our quality of life while minimizing harm to the natural environment that supports us. Recycling is one such solution, and it has increased dramatically in recent years; today 29% of the aluminum and 13% of the copper produced globally come from recycled sources (Chapter 17).

Energy resources are also central to our society's continued functioning. As human populations and affluence have grown in recent decades, so too has our energy

Figure 1.15 Our civilization depends on products manufactured from raw materials mined from the earth, but mining degrades landscapes, harms wildlife, and is a major source of pollutants around the world. Thankfully, there exist potential solutions for these problems.

consumption. Between 1950 and 2001, our numbers doubled, but global consumption of **fossil fuels** (nonrenewable energy sources, such as oil, gas, and coal produced by the decomposition and fossilization of ancient life) increased by more than four times (Chapter 17). Fossil fuel consumption has caused various kinds of air and water pollution and also has contributed to political instability in the world. Alternative energy sources are rapidly becoming more popular, however, and may bring new kinds of environmental changes in the future (Figure 1.16) (Chapter 20).

Another resource that has become limited because of recent population growth is land (Chapter 16). Densely populated countries such as South Korea, Taiwan, the Netherlands, and Bangladesh possess only 0.11–0.25 ha (0.27–0.62 acres) of land per person. Even residents in many areas of less densely populated countries such as Canada (32.4 ha [80 acres] per person) and the United States (3.4 ha [8.3 acres] per person) are feeling widespread pressure from suburban sprawl and the loss of natural areas. Such limitations make efficient management of land and its associated resources (such as forest products) more important than ever; this includes city planning as well as management of parks, reserves, and recreation areas.

In addition to improving the efficiency of our resource use, we also need to find better ways to dispose of the vast quantities of waste that our consumption of resources produces (Chapter 19). This waste comes in many forms, from the trash we as individuals throw out each week to the toxic chemical and radioactive waste products that many of our industries emit. Scientists, engineers, and policymakers are working on developing new ways to reduce the volume and harmful effects of waste.

Solutions to environmental problems must be global and sustainable

The nature of virtually all of these environmental issues is being changed by the set of ongoing phenomena commonly dubbed *globalization*. Our increased global interconnectedness in trade, politics, and the movement of people and of other species poses many challenging problems, but it also sets the stage for novel and effective solutions. Key to the success of any solution is that it be sustainable, or able to function for the long term. Our final chapter takes a wide-ranging look at emerging sustainable solutions, and because sustainability is a guiding principle of modern environmental science, it is a concept that you will encounter throughout this book (Chapter 20).

Are things getting better or worse?

Despite the myriad challenges we review in this book, and despite the kinds of trends shown in Table 1.1, many people maintain that the general conditions of human life are in fact getting better, not worse. Furthermore, some say that human ingenuity and Earth's vast natural resources will meet all of our needs indefinitely and see us through any difficulty. Such views are sometimes characterized as Cornucopian. In Greek mythology, cornucopia—literally "horn of plenty"—is the name for a magical goat's horn that overflowed with grain, fruit, and flowers. In contrast, people who predict doom and disaster for the world because of our impact on it have been called Cassandras, after the mythical Greek woman with the gift of prophecy whose dire predictions were not believed. As you proceed through this book, think about how one's perception of issues is affected by the timescale at which one considers them. Might conditions sometimes get both better *and* worse, depending on whether one is thinking in the short term or the long term?

Sustainability

The primary challenge in our increasingly populated world is how to live within the planet's means, such that Earth and its resources can sustain us and the rest of Earth's biota for the foreseeable future. *Sustainability*

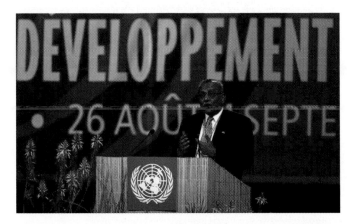

Figure 1.17 Sustainable development was the focus of the United Nations World Summit held in Johannesburg, South Africa, in 2002 (Chapter 20).

and *sustainable development* have become the catchwords for this concept. *Development* is a term economists use to describe the use of natural resources for economic purposes (rather than for simple subsistence, or survival) alone. Logging, farming, mining, and building homes and factories are all types of development, and all of them affect the environment and give rise to changes that environmental scientists study.

Sustainable development is the use of renewable and nonrenewable resources in a manner that satisfies our current needs without compromising future availability of resources (Figure 1.17). The United Nations defines sustainable development as development that ". . . meets the needs of the present without sacrificing the ability of future generations to meet theirs." Answering a simple question—"Can this activity continue forever?"—indicates whether a particular activity is sustainable. Sustainability depends, in large part, on the ability of the current human population to limit its environmental impact.

Conclusion

The prudent, sustainable use of resources and the relationship between resource use and population are themes that you will encounter in every chapter of this book. Indeed, they are themes we begin to encounter nearly everywhere, every day of our lives, once we start to examine and think attentively about the quality of our surroundings and the origins of the products we use. Finding effective ways of living peacefully, healthily, and sustainably on our diverse and complex planet will require a thorough scientific understanding of both natural and social systems. Environmental science helps us understand our intricate relationship with the environment and can inform our attempts to prevent and solve environmental problems.

It is important to keep in mind that identifying a problem is the first step in devising a solution to it. Many of the trends detailed in this book may cause us worry, but others give us reason to hope. One often-heard criticism of environmental science courses and textbooks is that too often they emphasize the negative. Recognizing the validity of this criticism, in this book we attempt to balance the discussion of environmental problems with a corresponding focus on potential solutions. Solving environmental problems can move us toward health, longevity, peace, and prosperity. Science in general, and environmental science in particular, can aid us in our efforts to develop balanced and workable solutions to the many environmental dilemmas we face today.

REVIEW QUESTIONS

1. What do renewable resources and nonrenewable resources have in common? How are they different?
2. How did the agricultural revolution affect human population size? Explain your answer.
3. How did the industrial revolution affect human population size? Explain your answer.
4. What did Thomas Malthus say about human populations and food supply?
5. What did Paul Ehrlich predict in his book *The Population Bomb*?
6. What is "the tragedy of the commons"? Explain how the concept might apply to an unregulated industry that is a source of water pollution.
7. How did the Easter Islanders cause their own civilization to collapse?
8. What is environmental science?
9. What are some of the disciplines involved in environmental science?
10. What is the difference between environmentalism and environmental science?
11. What are the two meanings of *science*, and what related concept do some people sometimes include under the term *science*?
12. Describe the scientific method. What is the typical sequence of steps?
13. Define and give an example of a dependent variable and an independent variable.
14. Explain the difference between a manipulative experiment and a natural experiment.
15. What needs to occur before a researcher's results are published? Why is this important?
16. What is meant by a scientific "paradigm shift"?
17. Give examples of five major environmental problems in the world today.
18. Explain the trade-offs involved in agriculture, mining, and energy production.
19. Why does Edward O. Wilson consider the loss of biodiversity our most severe environmental problem?
20. What is a Cornucopian? A Cassandra?
21. What is sustainable development?

DISCUSSION QUESTIONS

1. Is the photo of Earth from space shown at the opening of the chapter an adequate representation of the environment? Explain your answer. What impressions does the photo give you about Earth?
2. Many resources are renewable if we use them in moderation but can become nonrenewable if overexploited. Order the following resources on a continuum of renewability (see Figure 1.1), from most to least renewable: soils, timber, fresh water, food crops, and biodiversity. What factors influenced your decision? For each of these resources, what might comprise overexploitation, and what might comprise sustainable use?
3. Garrett Hardin described the overcrowding, overgrazing, and starvation that would result from unregulated pursuit of self-interest as a "tragedy." This

idea is the basis for many arguments in favor of regulating human interactions with the environment. Can you think of other examples in which unrestricted individual freedom to use public resources might cause problems? Can you think of examples where such freedom would be beneficial?

4. Why do you think the Easter Islanders did not or could not stop themselves from denuding their island of all its trees? Do you see similarities between the history of the Easter Islanders and the modern history of our society? Why or why not?

5. Can you think of an example of an application of science, and contrast it with the scientific research that lead to it?

6. Is it possible to answer all questions about the environment via strict application of the scientific method? Can you think of any specific questions that the scientific method would not help answer?

7. If the human population were to stabilize tomorrow and never surpass 7 billion people, would that solve any of our environmental problems? Which types of problems might be alleviated, and which might continue to get worse?

8. What environmental problem do *you* feel most acutely yourself? Do you think there are people in the world who do not view your issue as an environmental problem? Who might they be, and why might they take a different view?

Media Resources *For further review, go to the website* **www.envscienceplace.com** *or student CD-ROM, where you will find quizzes, flashcards, a glossary, additional interactive exercises, and links to relevant news and research sources. Also, on the website and CD-ROM is* **GRAPH IT**, *a series of interactive graphing tutorials to help you interpret graphs and plot data.*

2 Environmental ethics and economics: Values and choices

Kakadu National Park Wetlands, Australia

This chapter will help you understand:

- The influences of culture and worldview on the choices people make

- The history and evolution of major schools of thought in environmental ethics

- Precepts of classical and neoclassical economic theory and their implications for the environment

- Concepts of environmental economics and ecological economics

- Concepts of economic growth, economic health, and sustainability

- The relationships among ethics, economics, and environmental science

Protestors against Jabiluka Uranium Mine in Sydney, Australia

Central Case: The Mirrar Clan Confronts the Jabiluka Uranium Mine

"For some people, what they are is not finished at the skin, but continues with the reach of the senses out into the land. If the land is summarily disfigured or reorganized, it causes them psychological pain."
—*Barry Lopez, American nature writer*

"The Jabiluka uranium mine will improve the quality of the environment. The uranium resource there has been polluting the river system naturally, probably for thousands of years. . . . With the uranium resource removed and put to good use, the level of radioactivity will fall."
—*Michael Darby, Australian political commentator*

The remote Kakadu region of Australia's Northern Territory is home to several groups of Australian Aborigines, native people who lived there long before the British colonization of Australia. The region is also home to Kakadu National Park, a World Heritage Site recognized by the United Nations for its irreplaceable natural and cultural resources. In addition, the region's land contains large amounts of uranium, the naturally occurring radioactive metal valued economically for its

use as fuel in nuclear power plants, as the principal raw material in nuclear weapons, and as a component of medical and industrial tools. Uranium mining is a key part of the Australian national economy, accounting in 2002 for 7% of Australia's economic output.

Many of Australia's uranium deposits occur on Aboriginal lands, giving rise to conflict between corporations seeking to develop mining operations and the Aboriginal people living on the lands and trying to maintain their traditional culture. One such group is the Mirrar Clan, an extended family of Kakadu-area Aborigines currently numbering 27 members. The Mirrar have been living with the area's first uranium mine, known as the Ranger mine, on their land since the Australian government approved its development in 1978.

In recent years, the Mirrar have been fighting the proposed development of a second mine, Jabiluka, nearby on their land. The Mirrar see Jabiluka as a threat to their culture and religion, which are deeply tied to the landscape. Like other Australian Aborigines, they hold the landscape to be sacred, and they also depend on its resources for their daily needs. They also see the mine as a threat to their health and to the integrity of their environment, particularly given the numerous instances of radioactive spills at the Ranger mine. The proposed mine site is near their traditional hunting and gathering sites and is in the floodplain of a river that provides the clan food and water. Many of the Mirrar feared the mine would cause pollution problems on their land. They worried that contaminated water could be released into area creeks and that radioactive radon gas would be released from stored waste materials. Moreover, mindful of the geological faults that exist in the area, the Mirrar were concerned that the dam holding mine waste might one day catastrophically fail. Many environmental activists have joined the Mirrar's struggle; in mid-1998 nearly 3,000 people traveled to the Kakadu region to protest Jabiluka. The Mirrar's opposition to mining comes despite the economic benefits, in the form of jobs, income, development, and a higher standard of living, the mining company has promised.

The Mirrar Clan's campaign now appears to have succeeded. On September 5, 2002, Sir Robert Wilson, chief executive officer of the corporation that holds rights to the Jabiluka ore body, announced the cancellation of mining plans at Jabiluka, citing economic factors (declining world uranium prices) and ethical factors (concerns about developing the mine without Mirrar consent). Wilson added that the company plans to rehabilitate the Jabiluka site and restore damage done there during preliminary exploration and assessment of the uranium ore body. The Mirrar are still waiting to see these plans put in writing in a legally binding agreement, but it seems possible the lengthy debate may at last be close to resolution.

In formulating their approach to the mining proposal, the Mirrar and other Australians had to weigh economic, social, cultural, and philosophical questions as well as scientific ones. The story of mining and the Mirrar exemplifies some of the ways in which values, beliefs, and traditions interact with economic interests to influence the choices all of us must make about how to live within our environment.

Culture, Worldview, and Our Perception of the Environment

The Mirrar faced difficult choices. They were offered potentially substantial economic benefits, but they also felt mine development ran counter to their ethical re-

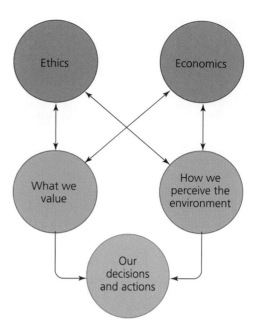

Figure 2.1 The disciplines of ethics and economics each deal (in very different ways) with what we value and with how we perceive our environment. These values and perceptions, together with assessments of costs and benefits, influence the decisions we make and the actions we take.

spect for their land. Such tradeoffs between economic benefits and ethical concerns crop up frequently in environmental issues.

Ethics and economics are sources of values

As we saw in Chapter 1, much of environmental science revolves around the study of Earth's natural systems, the ways they affect humans, and the ways humans affect them. As such, a firm understanding of the natural sciences is crucial. To address environmental problems, however, it is also necessary to understand how people perceive their environment, how people relate to their environment philosophically and pragmatically, and how they value certain elements of their environment. The fields of ethics and economics are two quite different academic disciplines, but each deals with questions of what we value and how we perceive the environment, which in turn influence our decisions and actions (Figure 2.1). Anyone trying to address an environmental problem must try to understand not only the natural science of how systems work, but also the values that lie behind human behavior.

Values, culture, and worldview influence a person's perception of the environment

Almost every action we take affects the environment. Growing food requires land, cultivation, and often irrigation. Building homes requires land, lumber, and metal. Manufacturing and fueling vehicles requires metal, plastic, glass, petroleum, and computer components. From nutrition to housing to transportation, we meet our needs by altering our surroundings. Our decisions about how we meet our needs, and how we manipulate and exploit our environment to do so, depend in part on rational assessments of costs and benefits. However, our decisions are also heavily influenced by the particular culture of which we are a part, and by the particular worldview we share. **Culture** can be defined as the overall ensemble of knowledge, beliefs, values, and learned ways of life shared by a group of people. Culture influences each person's perception of the world and his or her place within it, something described by philosophers as the person's **worldview.** A worldview reflects a person's (or a group's) beliefs about the meaning, purpose, operation, and essence of the world.

People with different worldviews can study the same situation and review identical data, yet draw dramatically different conclusions. For example, many well-meaning people supported the Jabiluka mine while many other well-meaning people opposed it. The officials, employees, and shareholders of the mining company, and those government officials who supported the mine, saw uranium mining primarily as a source of jobs, income, economic growth, and much-needed energy. They believed that all of these would benefit Australia in general and the Mirrar Clan in particular. Mine opponents, meanwhile, cited environmental problems, negative social consequences, and questions of civil rights and justice. Despite its economic benefits, uranium mining disturbs Earth's surface, can expose miners to elevated radiation levels, and can create air and water pollution. Besides these environmental problems, mining booms frequently are associated with social problems, including community disruption, alcoholism, and crime.

Many factors can shape our worldviews and perception of the environment

The traditional culture and worldviews of the Mirrar Clan have played a large role in its view of the proposed Jabiluka mine. For the Mirrar, their Northern Territory homelands are an immediate source of food, shelter, and economically valuable resources, but they also believe the landscape itself is sacred. Australian Aborigines see the landscape as the physical embodiment of stories that express the beliefs and values central to their culture. In other words, the Australian landscape to them is a sacred text, analogous to that of the Bible in Christianity, the Qu'ran in Islam, or the Torah in Judaism.

Aborigines, including the Mirrar, believe that they are descended from "sky heroes," spirit ancestors possessing both human and animal features. These spirit ancestors traveled throughout Australia, leaving trails of important signs and lessons in the landscape; the routes these spirit ancestors traveled are called "dreaming tracks." Modern Aborigines still engage in "walkabouts," long walks that retrace the dreaming tracks and allow Aborigines to retell the sacred stories. By explaining the origin of specific features in the landscape, dreaming track stories assign meaning to notable landmarks and help Aborigines construct detailed mental maps of their surroundings. The stories also serve to teach cultural lessons concerning important aspects of Aboriginal life (such as family relations, hunting, food gathering, and conflict resolution) and to define Aborigines' relationships with each other and with the environment. Dreaming-track stories passed from one generation to the next help maintain Aboriginal culture. The Mirrar who opposed the Jabiluka mine did so because they believed it would desecrate sacred sites and compromise their culture (Figure 2.2).

Weighing the Issues:
Uranium Mining in Bethlehem

Suppose a mining company discovered uranium beneath the site in Bethlehem believed to be the birthplace of Jesus—or near the Wailing Wall in Jerusalem, or the mosque in Mecca. What do you think would happen if the company then announced plans to open a mine there, assuring the public that environmental impacts would be minimal and that the mine would create jobs and stimulate economic growth? What aspects of this situation resemble that of the Mirrar case, and what factors are different? Explain your answers.

Religion is one of many factors that can shape people's worldviews and perception of the environment. A given social group may also share a particular view of its environment because its members have lived through similar experiences. For example, early European settlers in both Australia and North America viewed their environment as a hostile force because inclement weather, wild animals, and other natural forces frequently destroyed crops, killed livestock, and took settlers' lives. Such experiences were shared in stories and in songs and, coupled with the values inspired by religious teaching and secular education, shaped prevailing

(a) Ranger mine

(b) Jabiluka mine

Figure 2.2 The Ranger mine (**a**), located on Aboriginal lands amid sacred sites, has caused significant environmental impacts—enough to spark the Mirrar Clan's opposition to the proposed Jabiluka mine (**b**) whose development began nearby.

social attitudes in many frontier communities. The view of nature as a hostile force and an adversary to be overcome has passed from one generation to the next and still influences the way many North Americans and Australians view their surroundings.

A person's political ideology can also shape his or her attitudes toward the environment. One's view of the proper role of government will influence whether one wants government to intervene in the market economy and protect environmental quality, for instance. Votes on environmental matters before the U.S. Congress are frequently split along party lines. A person's worldview affects his or her political ideology, and a person's political ideology affects his or her worldview.

Economic factors also sway how people perceive their environment and how they make decisions regarding it. An individual with a strong interest in the outcome

of a decision that may result in that individual's private gain or loss is said to have a *vested interest*. Vested interests in and of themselves are neither good nor bad, but they can influence the way people see the world. For example, mining company executives have a vested interest in a decision to open an area to mining because a new mine could increase profits, to which executive income is frequently linked. Likewise, a company's shareholders would also have a vested interest in such a decision because the value of the shares they hold increases with profits. In each case, vested interests may lead people to view a proposed mine like Jabiluka primarily as a source of economically lucrative resources.

Throughout this book you will encounter a great deal of scientific data regarding the environmental impacts of our choices (where to make our homes, how to make a living, what to wear, what to eat, how to travel, how to spend our leisure time, and so on). You will see that culture, worldviews, ethics, and economics play critical roles in such decisions and can even influence the interpretation of scientific data. As we emphasized earlier, acquiring scientific understanding is only one part of the search for solutions to environmental problems. Attention to ethics and economics helps us understand why and how we value those things we value.

Environmental Ethics

The field of **ethics** involves the study of good and bad, of right and wrong. The term *ethics* can also refer to a person's or group's set of moral principles or values. Ethicists help clarify how people judge right from wrong, by elucidating the criteria, standards, or rules that people use in making these judgments. Such criteria are grounded in values—for instance, promoting human welfare, maximizing individual freedom, or minimizing pain and suffering. People of different cultures or with different worldviews may differ in their values, and thus they may differ in what specific actions they consider to be right and wrong. Nonetheless, ethicists generally maintain that there are objective notions of right and wrong that should hold across cultures.

Ethics overlaps in its standards with both religion and law, but it is distinct from both. For example, by almost any standard, slavery and the subsequent suppression of civil rights of African Americans in the United States were and are unethical. Yet for most of the history of European settlement in North America such treatment was legal, and Americans of some religions supported this discrimination while those of other religions opposed it. Unlike ethics, both law and religion rely on the force of authority, but both are also subject to outside scrutiny and evaluation on the basis of independent ethical standards of right and wrong.

Ethical standards may be viewed as tools for decision-making, as criteria that help differentiate right from wrong. One classic ethical standard is that of *virtue*, which, as the ancient Greek philosopher Aristotle held, centered on the ideas of justice and equal treatment of individuals. Another ethical standard is the *categorical imperative* proposed by Immanuel Kant, which roughly approximates Christianity's Golden Rule: to treat others as you would prefer to be treated yourself. A third standard is the principle of *utility*, as elaborated by British philosophers Jeremy Bentham and John Stuart Mill. The utilitarian principle holds that something is right when it produces the greatest benefits for the greatest number of people—that is, when the action maximizes overall benefits in comparison to other options. We employ such standards as tools when we apply ethics, consciously or unconsciously, to situations in everyday life.

Environmental ethics is the application of ethical standards to environmental questions

The application of ethical standards to environmental questions is known as **environmental ethics.** This relatively new branch of philosophy arose once people began to identify the environmental changes brought about by modern industrialization. Environmental ethics can clarify our thinking and help us decide among possible courses of action when we are faced with such issues as whether to proceed with development of a uranium mine near Aboriginal lands. Human interactions with the environment frequently give rise to ethical questions that may be difficult to resolve. Consider the following examples:

- Does the present generation have an obligation to conserve resources for the use of future generations? If so, how should we determine the effect such obligation should have on our decision-making today?
- Are humans justified in driving other species to extinction? Are we justified in causing other changes in ecological systems? If destroying a forest would drive extinct an insect species few people have heard of, but would create jobs for 10,000 people, would that action be ethically admissible? What if it were

an owl species, not an insect species? What if only 100 jobs would be created?
- Are there situations that justify exposing certain communities to a disproportionate share of pollution or other form of environmental degradation? If not, what actions are justified in remedying this problem?

We have extended ethical consideration to more entities through time

Answers to the questions above depend partially on what ethical standard(s) a person chooses to use. They also depend on how broad and inclusive the person's domain of ethical concern is. A person who feels responsibility for the welfare of insects may answer the second question very differently from a person whose domain of ethical concern ends with humans. Most of us feel ethical or moral obligations to some entities in the world, but by no means to all of them. Throughout Western history, people have gradually enlarged the array of entities they have felt deserve ethical consideration. The enslavement of human beings by other human beings was common in many societies until recently, for instance. Women in the United States were not given the right to vote until 1920 and still face lower pay for equal work. Consider, too, how little ethical consideration citizens of one nation sometimes extend to those of another on which their government has declared war. Many societies are only now beginning to embrace the principle that all human individuals be given equal ethical consideration.

As ethical consideration has expanded to include more people, ethical consideration of nonhuman entities has developed as well. Concern for the welfare of domestic animals is evident in humane societies and related organizations. Animal-rights activists voice concern for animals that are hunted, eaten, or used in laboratory testing. A great many people now accept that wild animals (at least those obviously sentient animals, such as large vertebrates, with which we share similarities) merit ethical consideration. Many people who are concerned with the welfare of wild animals identify especially with species threatened with extinction; it was ethical concern, as much as scientific information, that led to the passage of the U.S. Endangered Species Act in 1973. Today many environmentalists are concerned not just with endangered species but also with the well-being of whole natural communities. Some have gone still further, suggesting that all of nature—living things and nonliving things, even rocks—should be ethically

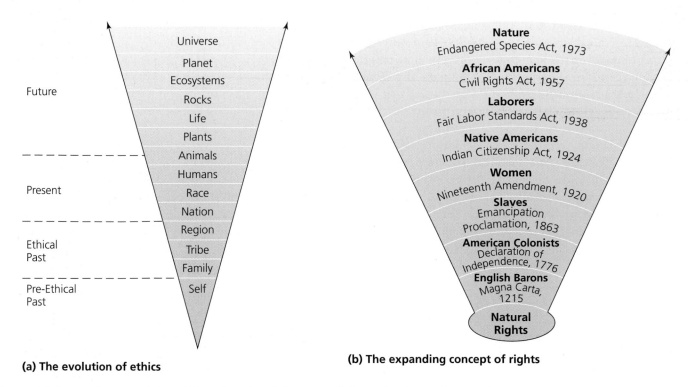

(a) The evolution of ethics

(b) The expanding concept of rights

Figure 2.3 As time has passed, people have broadened the scope of their ethical consideration for others. (a) We can view ethics progressing in a generalized and idealized way outward from the self. (b) This ethical expansion is reflected more concretely by certain key legal milestones in the expansion of rights granted by Britain and then the United States. *Source:* Roderick F. Nash, *The Rights of Nature,* Univ. of Wisconsin Press, 1989.

represented. The historian Roderick Nash illustrated this historical expansion of ethics in his 1989 book, *The Rights of Nature* (Figure 2.3).

What is behind this still-ongoing expansion? Most likely the increase in economic prosperity in Western cultures over the centuries, as more people found themselves with more leisure time and less anxiety about their own day-to-day survival, has helped enlarge our ethical domain. Science has no doubt also played a role. Ecology, as it has developed over the past half century, has made clear that all organisms are interconnected and that what affects plants, animals, and ecosystems can in turn easily affect humans. In addition, evolutionary biology over the past 150 years has made clear that humans are merely one species out of millions and have evolved subject to the same pressures as all other organisms. Ecology and evolution, then, have shown scientifically that humans do not stand apart from nature, but are part of it. This rejection of the common notion of the "dualism" of humans and nature is a crucial aspect of environmental science. Some environmental scientists maintain that recognizing that humans are a part of nature is a central prerequisite to solving our environmental problems.

Some philosophers see the expansion of ethical consideration as a key element of environmental ethics. They tend to view environmental ethics as a radically new way of thinking. Others consider environmental ethics to be merely the application of time-tested ethical rules to new environmental questions. As we consider the expansion of ethical domains, we should keep in mind that it may pertain mainly to modern European-based culture; for many non-Western cultures, particularly traditional hunter-gatherer cultures, expanded ethical domains are nothing new. The Mirrar are a case in point. However, it is worthwhile to examine the Western ethical expansion because it is tied to so many of our society's beliefs and actions regarding the environment. For convenience, people often simplify the continuum that Nash portrayed by subdividing it into three loosely conceived categories. These three ethical perspectives, or worldviews, are anthropocentrism, biocentrism, and ecocentrism.

Anthropocentrism **Anthropocentrism** is a human-centered view of our relationship with the environment. An anthropocentrist denies or ignores the notion that non-human entities can have rights. An anthropocen-

Figure 2.4 We can categorize people's ethical worldviews as anthropocentric, biocentric, or ecocentric. An anthropocentrist extends ethical standing only to humans, and judges actions in terms of their effects on humans. A biocentrist considers all living things, human and otherwise. An ecocentrist extends ethical consideration to living and nonliving components of the environment. The ecocentrist also takes a holistic view of the connections among these components, valuing the larger functional systems of which they are a part.

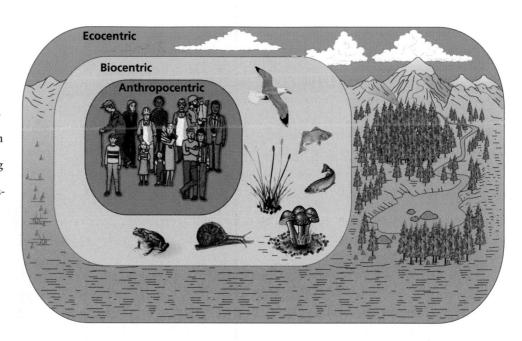

trist also measures the costs and benefits of actions solely according to their impact on people (Figure 2.4). To evaluate a human action that affects the environment, an anthropocentrist might use such criteria as impacts on human health, economic costs and benefits, and aesthetic concerns. For example, if the Jabiluka mine provided a net economic benefit while doing no harm to human health and having little aesthetic impact, the anthropocentrist might decide it was a worthwhile venture, even if it would likely drive some native species extinct. If protecting the area would provide spiritual, economic, or other benefits to humans now or in the future, an anthropocentrist might argue in favor of its protection. In the anthropocentric perspective, anything that does not provide some benefit to people, in terms of health, finances, nutrition, aesthetics, or otherwise, is considered to be of negligible value.

Biocentrism In contrast, **biocentrism** ascribes relative values to actions, entities, or properties on the basis of their effects on all living things, or on the integrity of the biotic realm in general (see Figure 2.4). In this perspective, all life has ethical standing. Thus, the biocentrist would evaluate an action in terms of its overall impact on living things, including—but not exclusively focusing on—human beings. In the case of the Jabiluka mine proposal, the biocentrist might oppose the mine if it posed a serious threat to the abundance and variety of living things in the area, even if it would create jobs, generate economic growth, and pose no threat to human health. While some biocentrists advocate equal consid-

eration of all living things, others advocate that some types of organisms should receive more than others.

Ecocentrism **Ecocentrism** considers actions in terms of their damage or benefit to the integrity of whole ecological systems, which by definition can consist of both biotic and abiotic elements and the relationships among them (see Figure 2.4). For an ecocentrist, the well-being of an individual organism—human or otherwise—is less important than the long-term well-being of a larger integrated ecological system. For example, an ecocentrist might approve of an action that harmed human health, caused economic losses, and took a number of lives, if such impacts were necessary to protect an entire species, community, or ecosystem. (We will study these concepts in Chapter 5 and Chapter 6.) In addition, ecocentrism is a more holistic perspective than biocentrism or anthropocentrism. Not only does it include a wider variety of entities, but it stresses the need to preserve the connections that tie the entities together into functional systems. Ecocentrism and biocentrism may sometimes conflict. An ecocentrist might favor reducing or eliminating a particular species from an ecosystem if it is foreign to the system and judged to be doing more harm than good, while a biocentrist might object to this.

Environmental ethics has ancient roots

Environmental ethics was first recognized as a distinct academic discipline in the early 1970s, but people have contemplated our relationship with, and possible responsibilities toward, Earth for thousands of years. Some of the Australian Aborigines' dreaming-track stories—many of them thousands of years old—touch on these questions. Aborigines ascribe ethical lessons to elements of their environment, and view their environment as a source of sacred teachings. As such, boulders, caves, patches of lichen, and other abiotic entities are portrayed as having moral significance worthy of contemplation and protection.

In the Western tradition, the ancient Greek philosopher Plato, writing in the fourth century B.C., also addressed what he considered to be humans' moral obligation to the environment:

> The land is our ancestral home and we must cherish it even more than children cherish their mother; furthermore, the Earth is goddess and mistress of mortal men, and the gods and spirits already established in the locality must be treated with the same respect.

Some ethicists and theologians have pointed to the Judeo-Christian religious tradition as a source of anthropocentric hostility toward the environment. They point out biblical passages such as, "Be fruitful and multiply, and fill the earth and subdue it; and have dominion over the fish of the sea and over the birds of the air and over every living thing that moves upon the earth." Such wording justified and encouraged an animosity toward nature that has characterized Western culture over the centuries, some scholars say. Judeo-Christian religious texts, however, have also been interpreted to encourage benevolent human stewardship over nature. Consider the directive, "You shall not defile the land in which you live. . . ." In fact, a 2003 poll showed that 56% of Americans supported environmental protection because they considered the environment to be "God's creation."

Although humans have held differing views of their ethical relationship with the environment for millennia, the developments leading to the rise of modern environmental ethics occurred much more recently. Environmental changes that became apparent during the industrial revolution intensified thought and debate about our species' relationship with the environment.

The industrial revolution inspired environmental philosophers

As the industrial revolution spread in the 19th century from Great Britain throughout Europe, North America, and much of the rest of the world, it amplified the intentional and unintentional impacts of human activities on the environment. In this period of social and economic transformation, agricultural economies became industrial economies, machines replaced or enhanced human and animal labor, and much of the rural human population moved into cities. Consumption of natural resources accelerated rapidly, and pollution increased dramatically. Thus, in addition to transforming society, the industrial revolution transformed the environment. In Great Britain and in other parts of the industrializing world, both transformations gave rise to a shift in human perception of the environment.

Many prominent British writers and philosophers of the time, including the art critic, poet, and writer **John Ruskin** (1819–1900), criticized industrialized cities as "little more than laboratories for the distillation into heaven of venomous smokes and smells" caused by the combustion of coal to fuel railroads, steamships, ironworks, and factories. Ruskin also complained that people had "desacralized" nature, valuing the material benefits the environment could provide but no longer appreciating its spiritual or aesthetic benefits. Motivated by similar concerns, a number of citizens' groups sprang up around England during the 19th century; these might be considered some of the first environmental organizations (Table 2.1).

Early environmentalism was not limited to Great Britain. In the United States during the first half of the 19th century, a philosophical movement called **transcendentalism** flourished, especially in New England,

Table 2.1 Early Environmental Organizations Formed in 19th-Century Great Britain		
Organization	Year established	Purpose
Scottish Rights of Way Society	1843	Protect walking paths in and near cities
Commons Preservation Society	1865	Preserve forests and other landscapes
Society for the Protection of Ancient Buildings	1877	Protect the built environment, especially historic buildings
Selborne League	1885	Protect rare birds, plants, and landscapes
Coal Smoke Abatement Society	1898	Improve urban air quality

where it was most ardently espoused by the American philosophers **Ralph Waldo Emerson** and **Henry David Thoreau.** The transcendentalists viewed nature as a direct manifestation of the divine, emphasizing the soul's oneness with nature and with God. While Ruskin and his contemporaries worked to mitigate the effects of industrialization, the transcendentalists worked to promote their holistic view of nature among the public. For example, they objected to what they saw as their fellow citizens' obsession with material things. The transcendentalist attitude toward the environment resembled that of the Mirrar in some respects; both traditions view everything in nature as a symbol or a messenger of some deeper truth. Although Thoreau viewed all of nature as divine, he also observed the natural world closely and came to understand it in the manner of a scientist; he was in many ways one of the first ecologists. His book *Walden*, recording his observations and thoughts while he lived at Walden Pond away from the bustle of urban Massachusetts, remains a classic of American literature.

The conservation and preservation movements arose around the turn of the 20th century

One admirer of Emerson and Thoreau was **John Muir** (1838–1914), a Scottish immigrant to the United States who eventually settled in California and made the Yosemite Valley his wilderness home. Although he chose to live in isolation in his beloved Sierra Nevada for long stretches of time, he nonetheless became politically active and won fame as a tireless advocate for the preservation of wilderness (Figure 2.5). Muir was motivated by the rapid deforestation and environmental degradation he witnessed throughout North America and by his belief that the natural world should be treated with the same respect that cathedrals and other places of worship receive. Today he is most strongly associated with the **preservation** ethic, which holds that we should protect the natural environment in a pristine, unaltered state.

Muir not only argued that nature deserved protection for its own inherent values (an ecocentrist argument), but also claimed that nature played a large role in human happiness and fulfillment (an anthropocentrist argument). "Everybody needs beauty as well as bread," he wrote in 1912, "Places to play in and pray in, where nature may heal and give strength to body and soul alike."

Some of the same factors that motivated Muir also inspired the first professionally trained American forester, **Gifford Pinchot** (1865–1946) (Figure 2.6). Both these men opposed the rapid deforestation and unregulated economic development of North American lands that occurred during their lifetimes. However, Pinchot took a more anthropocentric view of how and why nature should be valued than did Muir. He is today the person most closely associated with the **conservation** ethic, which holds that humans should put natural resources to

Figure 2.5 A pioneering advocate of the preservation ethic, John Muir is also remembered for his efforts to protect the Sierra Nevada from development and for his role in founding the Sierra Club, a leading environmental organization. Here Muir (right) is shown with President Theodore Roosevelt (left) in Yosemite National Park. After his 1903 wilderness camping trip with Muir, the president instructed his interior secretary to vastly increase protected areas in the Sierra Nevada.

Figure 2.6 Gifford Pinchot, the first chief of what would become the U.S. Forest Service, was a leading proponent of the conservation ethic. The conservation ethic holds that humans should utilize natural resources, but that in doing so, they should strive to ensure the greatest good for the greatest number for the longest time.

use, but that we have a responsibility to manage them wisely. Whereas preservation aims to preserve nature for its own sake and for the aesthetic, spiritual, symbolic, and recreational benefit of people, conservation focuses on the prudent and efficient extraction and use of natural resources for the benefit of present and future generations. The conservation ethic states that in using resources, humans should attempt to provide the greatest good to the greatest number of people for the longest time.

Pinchot and Muir came to represent different branches of the American environmental movement, and their contrasting ethical approaches—utilitarian versus spiritual—often pitted them against one another on policy issues of the day. Pinchot eventually founded what would become the U.S. Forest Service and served as its chief in Theodore Roosevelt's administration. Both Pinchot and Muir have left legacies that reverberate today in the different ethical approaches to environmentalism.

Weighing the Issues:
Preservation, Conservation, and the Jabiluka Uranium Mine

How might preservationists and conservationists respond to the question of whether uranium mining should occur in the homelands of the Mirrar people? What types of questions would each group ask? Which group would be more likely to support the mine? With which ethic do you identify? Explain your answers.

Aldo Leopold's land ethic arose from the conservation and preservation ethics

As a young forester and wildlife manager, **Aldo Leopold** (1887–1949) (Figure 2.7) began his career fully in the conservationist camp, having graduated from the Yale Forestry School that Pinchot helped found just as Roosevelt and Pinchot were advancing conservation on the national stage. As a forest manager in Arizona and New Mexico, he embraced the government policy of shooting predators, such as wolves, in order to keep up populations of deer and other game animals. At the same time, Leopold followed the development of ecological science. He eventually ceased to view certain species as "good" or "bad," and instead came to see that healthy ecological systems depend on the protection of all their interacting parts, including predators as well as prey. Drawing an analogy to mechanical maintenance, he wrote, "to keep every cog and wheel is the first precaution of intelligent tinkering."

Figure 2.7 Aldo Leopold, a wildlife manager and pioneering environmental philosopher, articulated a new relationship between people and the environment. In his essay *The Land Ethic,* he called on people to include the environment in their ethical framework.

It was more than science that drove Leopold away from an anthropocentric perspective toward a more holistic one. One day he shot a wolf, and when he reached the animal, Leopold was transfixed by "a fierce green fire dying in her eyes." The experience remained with him for the rest of his life and helped lead him to a more ecocentric ethical outlook. As an environmental philosopher and professor at the University of Wisconsin, Leopold argued that humans should view themselves and "the land" as members of the same community and that humans are obliged to treat the land in an ethical manner. In his 1949 essay *The Land Ethic,* he wrote,

> All ethics so far evolved rest upon a single premise: that the individual is a member of a community of interdependent parts. . . . The land ethic simply enlarges the boundaries of the community to include soils, waters, plants, and animals, or collectively: the land. . . . A land ethic changes the role of *Homo sapiens* from conqueror of the land-community to plain member and citizen of it. . . . It implies respect for his fellow-members, and also respect for the community as such.

Leopold intended that the land ethic would help guide decision-making. "A thing is right," he wrote, "when it tends to preserve the integrity, stability, and beauty of the biotic community. It is wrong when it tends otherwise." Leopold died before seeing *The Land Ethic* and his best-known book, *A Sand County Almanac,* in print, but today many view him as the single most eloquent and important philosopher of environmental ethics.

Deep ecology is a recent philosophical extension of environmental ethics

One philosophical perspective that goes beyond even Leopold's ecocentrism is **deep ecology**, established in the 1970s. Proponents of deep ecology describe the movement as resting on principles of "self-realization" and biocentric equality. They define self-realization as the awareness that humans are inseparable from nature and that the air we breathe, the water we drink, and the foods we consume are both products of the environment and integral parts of us. Biocentric equality is the precept that all living beings have equal value and that because we are truly inseparable from our environment, we must protect all other living things as we would protect ourselves.

Ecofeminism draws parallels between male attitudes toward nature and toward women

Deep ecology and mainstream environmentalism helped drive people to extend their domain of ethical concern outward during the 1960s and 1970s. This was also the time when social movements, such as the civil rights movement and the feminist movement, were gaining prominence. A number of feminist scholars saw parallels in human behavior toward nature and men's behavior toward women. The degradation of nature and the social oppression of women shared common roots, these scholars asserted. Ecological feminism, or **ecofeminism,** argues that the patriarchal (male-dominated) structure of society—which traditionally grants more power and prestige to men than to women—is a root cause of both social and environmental problems. Ecofeminists hold that a worldview traditionally associated with women, which interprets the world in terms of interrelationships and cooperation, is more in tune with nature than a worldview traditionally associated with men, which interprets the world in terms of hierarchies and competition. Ecofeminists maintain that a male tendency to try to dominate and conquer what men hate, fear, or do not understand has historically been exercised against both women and the natural environment.

Environmental justice seeks equal legal treatment for all races and classes in environmental matters

Our society's domain of ethical concern has been expanding, not just from men to women and from human to nonhuman elements of the environment, but from rich to poor and from majority races and ethnic groups to minority ones. Such expansion of ethical concern involves applying a moral sense of fairness and equality. If society agrees on standards of fairness and equal treatment, these can be codified in law. Such notions have given rise to the **environmental justice** movement. The U.S. Environmental Protection Agency (EPA) defines environmental justice as "the fair treatment and meaningful involvement of all people regardless of race, color, national origin, or income with respect to the development, implementation, and enforcement of environmental laws, regulations, and policies."

A protest in the early 1980s by African Americans in Warren County, North Carolina, against a toxic waste dump in their community is widely seen as the beginning of the environmental justice movement (Figure 2.8). The state had chosen the county with the highest percentage of African Americans as the site of the dump, prompting Warren County residents to suspect "environmental racism." Environmental justice grew to prominence in the early 1990s as more people across the United States began fighting environmental hazards in their communities. This movement—in contrast to earlier environmental movements—was made up primarily of low-income people and minorities.

Figure 2.8 Communities of poor people and people of color have suffered more than their share of environmental problems in the United States and around the world, a situation that has given rise to the environmental justice movement. Shown in this photo are demonstrators protesting a toxic waste dump in Warren County, North Carolina.

The movement was fueled by the perception that poor people and minorities are exposed to a greater share of pollution, hazards, and environmental degradation than richer people and whites. Consider, for example, where the polluting facilities are located in the town in which you grew up.

In 1983, the U.S. General Accounting Office (GAO) examined the distribution of hazardous waste disposal facilities in eight southeastern states. The GAO found that three of four commercial toxic waste landfills were located in communities where the population of minorities was higher than that of whites, while the fourth landfill was located in a community that was 38% African American. In contrast, minorities made up only 20% of the region's population. In 1987 the United Church of Christ Commission for Racial Justice found that the percentage of minorities in areas with toxic waste sites was twice that of areas without toxic waste sites. Researchers looking at environmental dangers such as air pollution, lead poisoning, pesticide exposure, and workplace hazards have reached similar conclusions.

Weighing the Issues:
Environmental Justice

Racial variables aside, very poor communities suffer from pollution more so than very wealthy communities. Nationally, communities with the greatest number of commercial hazardous-waste facilities have some of the highest proportions of minority residents.

Do you think the status quo is acceptable, or should something be done to ensure that poor communities are no more polluted than wealthy ones? Should government officials or industry attempt to eliminate this inequality? If so, what should be done, and how could your recommendation be implemented?

In 1994, President Bill Clinton issued an executive order requiring all federal agencies to ensure that their environmental actions do not discriminate—intentionally or unintentionally—on the basis of race or income. Today the environmental justice movement has broadened to encompass transportation equity (who gets the clean electric trains and who gets the polluting diesel buses), redevelopment of abandoned urban sites, worker health and safety, and urban parks.

The attempts of the predominantly white Australian government and uranium mining companies to open mines on the traditional lands of the Mirrar (Figure 2.9) have also been characterized as violations of environmental justice principles. In March, 2002, the Aus-

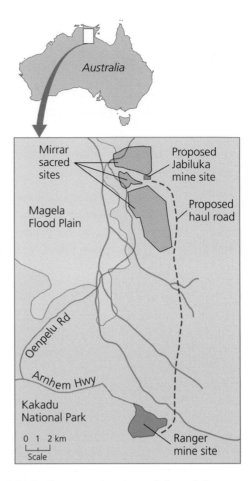

Figure 2.9 The Ranger mine site and that of the proposed Jabiluka mine lie in the midst of Aboriginal lands and sacred sites.

tralian Broadcasting Corporation reported that radioactive material from the Ranger mine had contaminated a stream on the Mirrar homeland and that uranium concentrations in the creek were 4,000 times higher than allowed by law. According to an Aboriginal representative, this was the fourth such violation in less than 3 months.

Uranium mining has been a source of environmental justice concerns in North America as well. Native Americans of the Dene people in Canada's Northwest Territories and Saskatchewan have long suffered health effects from employment as uranium miners with minimal safeguards. In the southwestern United States, from 1948 through the late 1960s, uranium mining employed a large number of Native Americans, including members of the Navajo nation. Uranium mining had already been linked to health problems and premature death, yet awareness of radiation and its risks was practically nonexistent among the miners. At the time, the Navajo language did not even have a word for *radiation,* and neither the U.S. government nor

Figure 2.10 Native Americans in Canada and the United States, such as the Navajo people, have also suffered from the adverse effects of uranium mining and mine waste.

the mining industry provided information or safeguards to the miners until nearly two decades after mining began. Many Navajo families even built homes and bread-baking ovens out of the abundant waste rock produced as a byproduct of the mining process, not knowing it was radioactive (Figure 2.10).

Cases of lung cancer began to appear among Navajo miners in the early 1960s, but scientific studies of the effects of radiation on miners at the time excluded Native American workers. The decision to include only white miners in the studies was attributed to the researchers' desire to study a "homogeneous population." A later generation of Americans perceived this as negligence and discrimination, and their desire for justice gave rise to the Radiation Exposure Compensation Act of 1990, a federal law that provided compensation for Navajo miners who suffered health effects as a result of their unprotected work in the mines. Today an estimated 1,000 abandoned uranium mine shafts remain on Navajo lands. These developments illustrate the ongoing interplay between changing ethical values and resultant policymaking, which we will examine further in Chapter 3. First, however, we will explore the domain of economics, which, like ethics, addresses people's values and is widely used as a tool for policy.

Weighing the Issues:
Environmental Ethics and Political Decisions

How might a politician who was concerned with the economy of the Kakadu region and the wishes of the Mirrar Clan balance a desire to provide jobs in the area with a desire to protect the region's cultural integrity? Ex-

plain several scenarios that might enable an elected official to achieve both goals.

Economics: Old Approaches, New Approaches, and Environmental Implications

People who opposed the Jabiluka mine did so largely on the basis of ethical concerns stemming from Aboriginal culture and on the basis of worries about environmental impacts. Few people challenged the mining plan on economic grounds; mine opponents generally recognized uranium as a valuable resource that generates jobs, income, and electricity, and did not dispute the importance of uranium exports to the Australian economy. Support for the mine was based primarily on economic factors. Such conflict between ethical and economic motivations is a common and recurrent theme in environmental issues worldwide.

Is there a tradeoff between economics and the environment?

Although measures to safeguard the environment may frequently mesh well with ethical considerations, we often hear it said that environmental protection works in opposition to economic health. Is this necessarily the case? Growing numbers of economists assert that there is no such tradeoff—that in fact, environmental protection is *good* for the economy. Which view one takes can depend on whether one thinks in the short term or the long term, and whether one holds to traditional economic schools of thought or to newer ones that view the human economy as coupled to the nonhuman environment.

Economics is the study of our use of scarce resources for competing purposes

According to British economist Lionel Robbins (1898–1984), economics is "the science which studies human behavior as a relationship between ends and scarce means which have alternative uses." In other words, **economics** is the study of how we decide to use scarce resources to provide goods and services in the face of demand for them. By this definition, many, if not all, environmental problems are also economic problems that can intensify as population and per-capita resource consumption increase. Pollution may be viewed as a problem relating to human use of limited supplies of clean air, water, or soil, for example. Indeed, the word *economics* and the word *ecology* come from the same Greek root, *oikos*, meaning "household." In its broadest context, the human "household" is Earth itself. Economists traditionally have studied the household of human society, and ecologists the broader household of all life.

Several types of economies exist today and have borrowed from one another

An **economy** is a social system that converts resources into **goods**, material commodities manufactured for and bought by individuals and businesses; and **services**, work done for others as a form of business. Several kinds of economies exist in today's world. The oldest type is the survival economy, or **subsistence economy.** People in subsistence economies—who still comprise much of the human population—meet most or all of their daily needs directly from nature and do not purchase or trade for most of life's necessities.

A second type of economy is the **capitalist market economy.** In this system, buyers and sellers interact to determine which goods and services to produce, how much to produce, and how these should be produced and distributed. Capitalist economies are often contrasted with state socialist economies, or **centrally planned economies,** in which a nation's government determines how to allocate resources in a top-down manner. In today's world, capitalism has become predominant over socialism. However, a pure capitalist market economy would operate without government intervention. In reality, all so-called capitalist market economies today, including that of the United States, have borrowed much from state socialism and are in fact hybrid systems. In modern capitalist market economies, governments typically intervene for several reasons:

- To eliminate unfair advantages held by single buyers or sellers
- To manage the commons (Chapter 1)
- To mitigate pollution
- To provide safety nets (for people in their old age, in response to natural disasters, and so on)
- To provide other social services, such as national defense, medical services, and education

Many societies function somewhere along a continuum between a pure subsistence economy and a capitalist market economy. For example, the Mirrar still acquire essential food and water directly from their environment, but they purchase many other necessities.

The environment and the economy are intricately linked

All types of human economies exist within the larger environment and depend upon it in several important ways. Human economies are systems that receive inputs from the nonhuman environment, process these in complex ways so as to enable human society to function, and then use the environment as a receptacle for outputs of waste from this process. Human economies are thus *open systems* (Chapter 6), connected with the larger environmental system of which they are a part. Earth, in turn, is a materially *closed system,* so the inputs Earth can provide to human economies are ultimately limited. Although these interactions between human economies and the nonhuman environment are readily apparent, traditional schools of thought in economics have long overlooked the importance of these connections. Indeed, most conventional economists today still adhere to a worldview that interprets the environment as a subset of the human economy (Figure 2.11a), and this worldview continues to drive most policy decisions. However, modern economists belonging to the fast-growing fields of environmental economics and ecological economics explicitly accept that human economies are subsets of the environment and that the two are crucially interdependent (Figure 2.11b).

Economic activity utilizes resources from the environment. Natural resources (Chapter 1) are the various substances and forces we need in order to survive: the sun's energy, the fresh water we drink, the trees that provide our lumber, the rocks that provide our metals, and the fossil fuels that power our machines and produce our plastics. Without Earth's natural resources, there would clearly be no human economies and, in fact, no human beings.

We obtain natural resources (which we can think of as "goods" produced by nature) from the environment, but the environment also naturally functions in a manner that supports economies. It naturally purifies air and water, cycles nutrients, provides for plants to be pollinated by animals, and serves as a receptacle and recycling system for the waste generated by our economic activity. Such essential services, often called **ecosystem services** (Table 2.2), support the life that makes our economic activity possible. Many ecosystem services represent the nuts and bolts of our survival, and others enhance our quality of life. For instance, nonmaterial amenities, such as aesthetic beauty and opportunities for

Figure 2.11 Modern environmental and ecological economists view economic activity very differently from economists of the more conventional neoclassical school. (**a**) Standard neoclassical economics focuses on processes of production and consumption between households and businesses, viewing the environment only as a "factor of production" that helps enable the production of goods. (**b**) Environmental and ecological economists view the human economy as existing within the natural environment, receiving resources from it, discharging waste into it, and interacting with it through various ecosystem services.

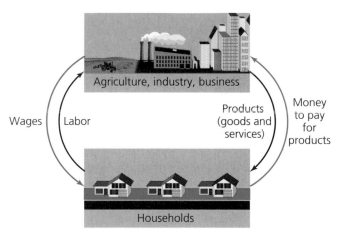

(a) Conventional view of economic activity

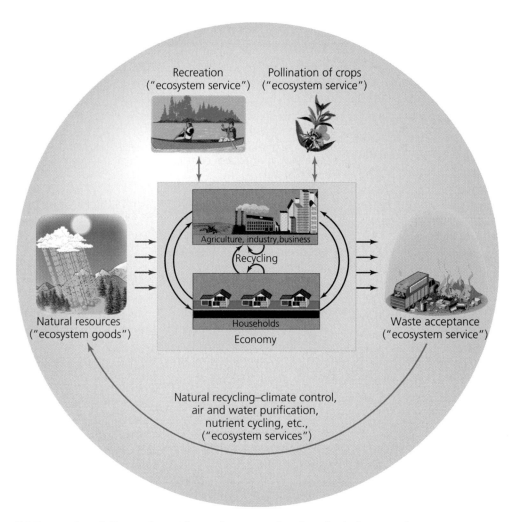

(b) Economic activity as viewed by environmental and ecological economists

Table 2.2 Ecosystem Services and Functions

Ecosystem service*	Ecosystem functions	Examples
Gas regulation	Regulation of atmospheric chemical composition	CO_2/O_2 balance, O_3 for UVB protection, and SO_x levels
Climate regulation	Regulation of global temperature, precipitation, and other biologically mediated climatic processes at global or local levels	Greenhouse gas regulation, DMS production affecting cloud formation
Disturbance regulation	Capacitance, damping and integrity of ecosystem response to environmental fluctuations	Storm protection, flood control, drought recovery and other aspects of habitat response to environmental variability mainly controlled by vegetation structure
Water regulation	Regulation of hydrological flows	Provisioning of water for agricultural (such as irrigation) or industrial (such as milling) processes or transportation
Water supply	Storage and retention of water	Provisioning of water by watersheds, reservoirs and aquifers
Erosion control and sediment retention	Retention of soil within an ecosystem	Prevention of loss of soil by wind, runoff, or other removal processes, storage of silt in lakes and wetlands
Soil formation	Soil formation processes	Weathering of rock and the accumulation of organic material
Nutrient cycling	Storage, internal cycling, processing and acquisition of nutrients	Nitrogen fixation, N, P and other elemental or nutrient cycles
Waste treatment	Recovery of mobile nutrients and removal or breakdown of excess or xenic nutrients and compounds	Waste treatment, pollution control, detoxification
Pollination	Movement of floral gametes	Provisioning of pollinators for the reproduction of plant populations
Biological control	Trophic-dynamic regulations of populations	Keystone predator control of prey species, reduction of herbivory by top predators
Refugia	Habitat for resident and transient populations	Nurseries, habitat for migratory species, regional habitats for locally harvested species, or overwintering grounds.
Food production	That portion of gross primary production extractable as food	Production of fish, game, crops, nuts, fruits by hunting, gathering, subsistence farming or fishing
Raw materials	That portion of gross primary production extractable as raw materials	The production of lumber, fuel or fodder
Genetic resources	Sources of unique biological materials and products	Medicine, products for materials science, genes for resistance to plant pathogens and crop pests, ornamental species (pets and horticultural varieties of plants)
Recreation	Providing opportunities for recreational activities	Eco-tourism, sport fishing, and other outdoor recreational activities
Cultural	Providing opportunities for non-commercial uses	Aesthetic, artistic, educational, spiritual, and/or scientific values of ecosystems

*Ecosystem 'goods' included in ecosystem services.
Reprinted with permission: Robert Costanza et al., "The value of the world's ecosystem services and natural capital," *Nature*, May 1997.

recreation, contribute to human well-being and economic activity. Such amenities have made tourism an important economic force in many countries, including Australia, Costa Rica (Chapter 5), and the Maldives (Chapter 12).

While the environment enables economic activity by providing ecosystem goods and services, economic activity can affect the environment in return. For example, if we consume natural resources too quickly, those resources may become depleted, and environmental systems may be harmed as a result. Environmental systems may also suffer if economic activity produces too much pollution. These two processes, resource depletion and pollution, can lead to breakdowns in the way the environment functions, which can in turn negatively affect human economies. However, these facts have only recently become widely recognized. Let's briefly examine how economic thought has changed over the years, tracing the path that is now beginning to lead human economies to become more compatible with natural systems.

Adam Smith and other philosophers founded classical economics

Just as economies and the environment are closely connected, and just as economics and ecology share roots, economics and ethics also share a common intellectual heritage. When economics began to develop as a distinct discipline during the mid-18th century, many philosophers of morals were interested in the relationship between individual decisions and societal well-being. Some argued that individuals acting in their own self-interest would harm society as a whole. Others believed that such behavior would benefit society as long as the behavior was constrained by the rule of law and private property rights and operated within fairly competitive markets. The latter view was first fully articulated by the Scottish philosopher **Adam Smith** (1723–1790). Known today as the father of classical economics, Smith believed that when people are free to pursue their own economic self-interest in a competitive marketplace, the marketplace will behave as if guided by "an invisible hand" that ensures their actions will benefit society as a whole. In his 1776 book *Inquiry into the Nature and Causes of the Wealth of Nations,* Smith wrote:

> It is not from the benevolence of the butcher, the brewer, or the baker that we expect our dinner, but from their regard to their own self-interest. [Each individual] intends only his own security, only his own gain. And he is led in this by an invisible hand to promote an end which was no part of his intention. By pursuing his own interests he frequently promotes that of society more effectually than when he really intends to.

Smith's philosophy remains a pillar of free-market thought today. The laissez-faire policies it has spawned, however, have been widely criticized by those who feel that market capitalism exacerbates inequalities between rich and poor, contributes to environmental degradation, and should be constrained and regulated by democratic government.

Weighing the Issues:
Adam Smith and the Invisible Hand

In what ways might individuals help or harm society by pursuing their own economic interests? How might Smith's reasoning apply to the case of the Jabiluka mine? Explain your answers.

Neoclassical economics incorporates human psychology and behavior

Economists inspired by Smith's ideas subsequently took more quantitative approaches and incorporated human psychology into their work. Modern **neoclassical economics** focuses on consumer choices and the psychological factors underlying those choices, explaining market price in terms of consumer preferences for units of particular commodities. In neoclassical economic theory, buyers desire the lowest possible price whereas sellers desire the highest possible price. This conflict between buyers and sellers results in a compromise price being reached and the "right" quantity of commodities being bought and sold. This is often phrased in terms of **supply,** the amount of a product offered for sale at a given price, and **demand,** the amount of a product that people will buy at a given price if they are free to do so. Theoretically, when prices go up, demand drops and supply increases; and when prices fall, demand rises and supply decreases. Thus the market automatically moves toward an equilibrium point, a price at which supply equals demand, such that their curves on a graph intersect (Figure 2.12).

Several aspects of neoclassical economics contribute to environmental problems

Today's capitalist market systems are organized largely in accordance with the precepts of neoclassical economics, and these precepts have profound implications for the environment. Although modern economic systems have generated unprecedented material wealth, employment, and other desirable outcomes, they have also contributed to a variety of contemporary environmental problems. We will explore four fundamental assumptions of neoclassical economics and their implications for the environment:

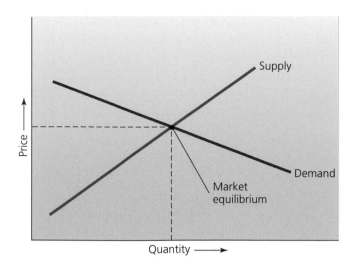

Figure 2.12 This basic supply-and-demand curve illustrates the relationship between supply, demand, and market equilibrium, the "balance point" at which demand is equal to supply.

- Resources are infinite or substitutable.
- Long-term effects are discounted.
- Costs and benefits are internal.
- Growth is good.

Resources are infinite or substitutable Neoclassical economic models generally treat workers and other resources either as being infinite or as being largely "substitutable and interchangeable." In other words, once we have depleted a resource—natural, human, or otherwise—we should be able to find a replacement for it. Human resources can substitute for financial resources, for instance, or manufactured resources can substitute for natural resources. Theory allows that the substituted resource may be less efficient and become more costly, but some degree of substitutability is generally assumed.

Certainly it is true that many resources can be replaced; our societies have transitioned from manual labor to animal labor to steam-driven power to fossil fuel power, and may yet transition to renewable power sources, such as solar energy. As we noted above, however, Earth's material resources are ultimately limited. As we saw in Chapter 1, nonrenewable resources, such as fossil fuels, can be depleted, and many renewable resources can be used up as well if we exploit them faster than they can be replenished. This is what happened to the Easter Islanders (Chapter 1) who harvested their forests faster than the forests could regrow.

Weighing the Issues:
Substitutability and the Environment

Neoclassical economic theory represents resources as being substitutable and interchangeable. Can you think of a natural resource that might be difficult to replace with a substitute? What problems might arise from the assumption that all resources, including clean air and water, are substitutable? Can you think of examples that violate any of the other three assumptions? Explain your answers.

Long-term effects are discounted Although few people would dispute that resources are *ultimately* limited, many have assumed that their depletion will take place so far in the future that there is no need for current generations to worry. For economists in the neoclassical tradition, an event far in the future counts much less than one in the present; in economic terminology, future effects are "discounted." In discounting, short-term costs and benefits are granted more importance than long-term costs and benefits, causing policy to play down long-term consequences of decisions we make today. Some governments and businesses use a 10% annual discount rate to make cost-benefit decisions regarding resource use. This means that the long-term value of a stand of ancient trees worth $500,000 would drop by 10% each year and, after 10 years of discounting, would be worth only $174,339.22. In this view, the more quickly the trees are cut, the more they are worth.

Costs and benefits are internal A third assumption of neoclassical economics is that all costs and benefits associated with a particular exchange of goods or services are borne by the individuals engaging in the transaction. In other words, it is assumed that the costs and benefits of a transaction are "internal" to the transaction, experienced by the buyer and seller alone, and do not affect other members of society.

However, in many situations this is simply not the case. For example, pollution from a manufacturing process can harm people living nearby. In such cases, someone—often taxpayers not involved in the production of the pollution—will pay the costs of alleviating it. The market does not take the costs of this pollution into account. Costs or benefits of a transaction that involve people other than the buyer or seller are known as **externalities.** A positive externality is a benefit enjoyed by someone not involved in a transaction, and a negative externality, or **external cost,** is a cost borne by someone

Figure 2.13 An Indonesian boy wading in a polluted river suffers external costs. External costs are costs not borne by the buyer or seller; they may include water pollution, aesthetic harm, human health problems, property damage, harm to aquatic life, aesthetic degradation, declining real estate values, and other problems.

not involved in a transaction (Figure 2.13). Negative externalities often harm groups of people or society as a whole, while allowing certain individuals private gain. External costs may include the following:

- Human health problems
- Property damage
- Declines in desirable elements of the environment, such as a decrease in the number of fish in a stream
- Aesthetic harm, such as that resulting from dirty air or water
- Worry and anxiety experienced by people downstream from or downwind of a pollution source
- Declining real estate values resulting from the above problems

By ignoring external costs, economies can create a false idea of the true costs of particular choices and can unjustly subject innocent parties to the consequences of transactions in which the parties do not participate. The existence of external costs is one reason governments have become active in developing environmental regulations such as those we will study in Chapter 3. Unfortunately, external costs are difficult to account for and to eliminate. It is tough to assign a monetary value to illness, premature death, or degradation of an aesthetically or spiritually significant site.

Weighing the Issues:
Quantifying External Costs

If pollution caused by the manufacture of a particular product were known to cause one death each year, how much should we spend to reduce the pollution causing the problem? This is equivalent to asking how much one human life is worth. Would it matter if the life at stake were that of an infant, a college student, or a retiree? These are the kinds of questions we face when attempting to quantify external costs.

Growth is good A fourth assumption made by the neoclassical economic approach is that economic growth is required to keep employment high and to maintain social order. The argument goes something like this: If the poor within a society view the wealthy as the source of their suffering, and if the poor are unemployed and desperate to meet their basic needs, revolution may result. One way of defusing this situation is to promote economic growth, which should create opportunities for the poor to meet their needs and perhaps become wealthy themselves.

The idea that economic growth is good has also been encouraged over the centuries by the concept of material "progress," which has been espoused by Western cultures since the Enlightenment. The modern-day United States, many people point out, represents the clearest example of this worldview that "more and bigger" is always better. We hear constantly in business news of increases in an industry's output or percentage growth in a country's economy, with increases touted as good news and decreases, stability, or even a small drop in the *rate* of growth presented as bad news. Economic growth has become the quantitative yardstick by which progress is measured.

Is the growth paradigm good for us?

The rate of economic growth in recent decades has been unprecedented in human history, and the world economy is seven times the size it was a half century ago. All measures of economic activity—trade, rates of production, amount and value of goods manufactured—are higher than they have ever been and are still increasing. This growth has brought many people much greater material wealth (although it has not done so evenly, and gaps between rich and poor remain great and have increased in many countries).

To the extent that economic growth is a means to an end—a tool with which we can achieve greater human happiness—it can be a good thing. However, many observers today worry that growth has become an end in itself and is no longer necessarily the best tool with which to pursue human happiness. Critics of the growth paradigm have often noted that runaway growth resembles the multiplication of cancer cells, which of course eventually overwhelm and destroy the system (the organism) within which they grow. These critics fear that runaway economic growth will likewise destroy the economic system on which we all have become dependent. Resources for growth are ultimately limited, they argue, so nonstop growth is not sustainable and will fail as a long-term strategy.

Defenders of traditional economic approaches reply that Cassandras (Chapter 1) have been saying for decades that limited resources would doom growth-oriented economies, and yet most of these economies are still expanding like gangbusters. Thomas Malthus (Chapter 1 and Chapter 7) warned two centuries ago that increasing populations would run up against absolute limits on resources. British classical economist David Ricardo (1772–1823) reasoned that as more and more high-quality farmland became occupied, profits from the lower-quality farmland that was left would progressively decrease, eventually halting growth. German philosopher Karl Marx (1818–1883) held that limits on economic growth would occur in the form of working-class revolutions (such as Russia's Bolshevik Revolution and China's Communist Revolution) against the capitalist leisure class.

So why then, 150 to 200 years after such predictions were made, are we seeing the world's most rapid growth of material wealth in human history? A prime reason is technological innovation. In case after case, improved technology has enabled us to push back the limits on growth. Better technology for extracting minerals, fossil fuels, and groundwater has effectively expanded the amount of these natural resources available to us. Technological developments such as automated farm machinery, fertilizers, and chemical pesticides have allowed for more food to be grown per unit area of land, boosting agricultural output. Faster, more powerful machines in our factories have enabled us to turn the higher rates of resource extraction and agricultural production into greater rates of manufacturing. Such increases in efficiency explain why many of the dire predictions of those concerned about resource limitations have only partially come true (Figure 2.14).

Weighing the Issues:
A Natural Gas Pipeline Across Peru

In 2003, the Inter-American Development Bank approved a $135 million loan to support construction of a natural gas pipeline across the nation of Peru, from its rainforest interior to its desert coastline. The pipeline would be one of the most significant capital investments in Peru's history.

Environmentalists are convinced the pipeline, which will run deep into the Peruvian Amazon, will degrade the rainforest environment, harm its plants and animals, and threaten the traditional way of life of the people indigenous to the rainforest. In addition, the pipeline terminal would be built in the buffer zone of Peru's Paracas Marine Reserve. Economic studies, however, estimate that the project could increase Peru's gross domestic output by 0.8% annually over the next 30 years. Peru is one of the poorer nations of the Western Hemisphere, and its people could use the jobs and income the project could bring. What questions would you ask to determine for yourself whether projected economic gains justify projected environmental and social disruption in this case? Might your answer differ according to the ethical standard you use? How so?

Figure 2.14 Advances in technology have enabled people to push back the natural limits on growth and continue expanding their economies. For example, innovations in machinery for petroleum extraction, shown here, have allowed us to extract more fossil fuels than ever before, enabling us to push our economic productivity far beyond what was possible a century or two ago.

Economists disagree on whether economic growth is sustainable

Can we conclude, then, that infinite improvements in technology are possible and that we will never run into shortages of resources? At one end of the spectrum are those Cornucopians (Chapter 1), economists, business-people, and policymakers who believe technology can solve everything—a philosophy that has greatly influenced economic policy in market economies over the past century.

At the other end of the spectrum, ecological economists argue that a couple of centuries is not a very long period of time and that history suggests that civilizations do not in the long run overcome their environmental limitations. Consider the Maya, the Sumerians, and the Easter Islanders, for example. **Ecological economics**, which has emerged as a discipline only in the past decade or so, applies the principles of ecology and systems thinking (Chapter 5 and Chapter 6) to the description and analysis of economic systems. Earth's natural systems generally operate in self-renewing cycles, not in a linear or progressive manner. Ecological economists advocate sustainability in human economies and therefore see natural systems as good models for that purpose. To evaluate an economic system's sustainability, ecological economists ask, "Could we continue this activity forever and be happy with the outcome?" The ecological economist's approach thus requires us to take a long-term perspective on the impacts of our choices. Most ecological economists argue that the growth paradigm will eventually fail and that if nothing is done to rein in population growth and increased consumption of resources, depleted natural systems will plunge human economies into ruin. Many of these economists advocate economies that do not grow and do not shrink, but are stable. Such "steady-state" economies are intended to mirror natural ecological systems.

In the middle of the spectrum are **environmental economists.** They tend to agree with ecological economists that human economies are unsustainable if population growth is not reduced and if we do not find ways to use resources more efficiently. Environmental economists, however, maintain that we can accomplish these changes and attain sustainability within our current economic systems. By retaining the principles of neoclassical economics but modifying them to address environmental challenges, environmental economists argue that we can keep our economies growing and that technological improvements can continue to improve efficiency. Thus, whereas ecological economists call for rev-olution, environmental economists call for reform. Environmental economists were the first to recognize and develop ways to tackle the problems of external costs and discounting. In the 1950s and 1960s a nonprofit, nonpartisan organization called Resources for the Future began laying the groundwork for solutions by pioneering ways to weigh the true costs and benefits associated with resource use. These environmental economists blazed a trail, and the ecological economists then went further, proposing that sustainability requires revolutionary changes leading to a steady-state economy.

A steady-state economy is a revolutionary alternative to growth

The idea of a **steady-state economy,** one that does not grow and does not shrink but remains stable, did not originate with the rise of ecological economics in the past decade or two. Back in the 19th century, British philosopher **John Stuart Mill** (1806–1873) addressed the issue in some detail. Mill believed that as resources became harder to find and extract, economic growth would slow and eventually stabilize. Economies would carry on in a state in which individuals and society subsist on steady flows of natural resources and on savings accrued during occasional productive but finite periods of growth. Such a model, in fact, matches the economies of Australian Aborigines and many other traditional societies throughout the world before the global expansion of European culture.

Modern proponents of a steady-state global economy, such as the American economist Herman Daly, are not so sanguine that a steady-state economy will evolve on its own from a capitalist market system. Instead, most believe we will need to radically rethink our assumptions and fundamentally change the way we conduct economic transactions in order to achieve a steady state.

Those resisting such notions often assume that an end to economic growth will mean an end to increases in the quality of life. Ecological economists, however, argue that quality of life can continue to rise under a steady-state economy and, in fact, may be more likely to do so. Technological advances will not cease just because economic growth stabilizes, they argue; improvements in efficiency from technology and from behavior (for instance, by greater use of recycling) together can decouple growth from wealth. Thus, material wealth and human happiness can and should continue to rise after economic growth has leveled off.

The Science behind the Story

Gross Domestic Product vs. Genuine Progress Indicator

For years, economists have assessed the economic health of a nation by calculating its **Gross Domestic Product (GDP)**, the total monetary value of final goods and services produced in the country each year.

However, there are problems with using this measure of economic activity to represent a nation's economic well-being. For one, GDP does not account for the nonmarket values we are discussing in this chapter. For another, contrary to general assumption, GDP is not necessarily an expression of *desirable* economic activity. In fact, GDP can increase whether the economic activities driving it help or hurt the environment or society. For example, a large oil spill in a national park might have the effect of increasing a country's GDP, because oil spills require cleanups, which cost money and, as a result, increase the production of goods and services. Uranium mining on the Mirrar homelands, including the costs of cleaning up radiation leaks and repairing damage done to sacred sites, would add to the Australian GDP because of the many monetary transactions to which cleanup and medical problems would give rise.

Some economists have attempted to develop more sensitive economic indicators to differentiate between desirable and undesirable economic activity. One alternative to the GDP is the **Genuine Progress Indicator (GPI)** introduced in 1995 by Redefining Progress, a nonprofit organization that develops economic and policy tools to promote accurate market prices and sustainability. The GPI has not yet gained widespread acceptance, but it has

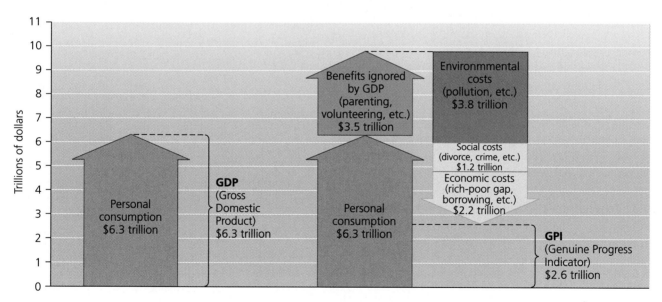

Gross domestic product (GDP) sums together all economic activity, whether good or bad. It does not account for benefits such as volunteerism or for external costs such as environmental degradation and social upheaval. The genuine progress indicator (GPI) does account for these factors and, as a result, can often be quite different in value from the GDP. Shown here are values for GDP and GPI for the United States in the year 2000. Data from Clifford Cobb, Mark Glickman, and Craig Cheslog, *The Genuine Progress Indicator 2000 Update*, Redefining Progress Issue Brief, 2001.

Attaining long-term sustainability will certainly require the reforms that environmental economists advocate, and may well require the fundamental shifts in thinking, values, and behavior that ecological economists say is necessary (see The Science behind the Story). How can these goals be attained in a world whose economic policies are still largely swayed by a Cornucopian worldview that barely takes the environment into account at all? While keeping in mind that ecological and environmental economic approaches are still actively being developed, we will now take a look at some strategies for sustainability that have been offered so far.

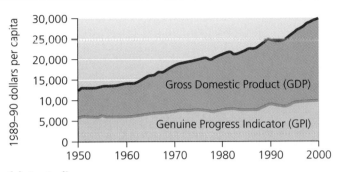

(a) Australia

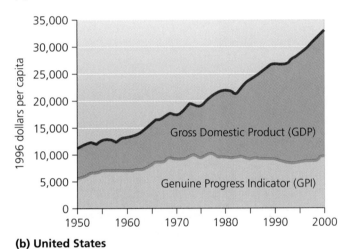

(b) United States

Although the GDPs of both Australia (**a**) and the United States (**b**) have increased dramatically since 1950, the GPIs of these two countries have not. GPI advocates suggest that this discrepancy means that we are spending more money than ever, but that our lives are not that much better. Data from Genuine Progress Indicator, The Australia Institute, 2000. *Genuine Progress Indicator,* Redefining Progress, 2001.

ates between economic activity that increases societal well-being and economic activity that decreases it. Thus, whereas the GDP increases when fossil fuel consumption increases, the GPI declines because of the adverse environmental and social impacts of such consumption, including air and water pollution, increased road congestion and traffic accidents, and global climate change. As an example, see the figure in this box comparing calculated per capita (per person) GPI and GDP changes in the United States and Australia from 1950 to 2000. The two countries' GDPs have increased as a result of increased economic activity. Their GPIs have also increased, but not nearly as rapidly.

The GPI is not the only alternative to the GDP. The United Nations uses a tool of its own called the Human Development Index, which is calculated by using assessments of a nation's standard of living, life expectancy, and education. Another proposed alternative to the GDP is the Index of Sustainable Economic Welfare (ISEW), developed by ecological economists Herman Daly and John Cobb in their 1989 book, *For the Common Good.* The ISEW is based on income, distribution of wealth, natural resource depletion, benefits associated with volunteerism, and environmental degradation.

generated a great deal of discussion that has drawn attention to the weaknesses of the GDP.

The GPI takes conventional economic activity into account and also reflects positive additions to the economy that do not have

to be paid for, such as volunteer work. It also subtracts the negative effects of crime, pollution, and other detrimental factors. In short, the GPI summarizes many more forms of economic activity than does the GDP and differenti-

Ecosystem goods and services can be given monetary values

As we have noted, economies receive from the environment vital resources and ecosystem services. However, any survey of environmental problems in the world

today—deforestation, biodiversity loss, pollution, collapsed fisheries, climate change, and so on—makes it immediately apparent that our society often mistreats the very systems that keep it alive and healthy. Why is this? From the economist's perspective, humans exploit natural resources and systems in large part because the market assigns these entities no quantitative monetary value or, at best, assigns values that underestimate their true worth.

Think for a minute about the nature of some of these services. The aesthetic and recreational pleasure we get from natural landscapes, whether wildernesses or city parks, is something of real value. Yet this value is hard to quantify and appears in no traditional measures of economic worth. Alternately, take Earth's water cycle (Chapter 6), by which rain fills the reservoirs from which we get drinking water, rivers give us hydropower and take away our waste, and water evaporates to the air, purifying itself of contaminants and readying itself to fall as rain again. This cycle is of immense value, and yet because that value is not generally quantified, the market imposes no financial penalties when humans interfere with this cycle. Because the market does not assign

(a) Existence values

(b) Use values

Figure 2.15 Accounting for nonmarket values such as those shown here may help us to make better environmental and economic decisions.

(c) Option values

(d) Aesthetic values

(e) Scientific values

(f) Educational values

(g) Cultural values

Table 2.3 Values Modern Market Economies Generally Do Not Address

Nonmarket value	Is the worth we ascribe to things . . .
Use value	that we use directly
Option value	that we do not use now but might use later
Aesthetic value	for their beauty or emotional appeal
Cultural value	that sustain or help define our culture
Scientific value	that may be the subject of scientific research
Educational value	that may teach us about ourselves and the world
Existence value	simply because they exist, even though we may never experience them directly (e.g., an endangered species in a far-off place).

value to ecosystem services, debates such as that over the Jabiluka mine often involve comparing apples and oranges—in this case, the intangible cultural, ecological, and spiritual arguments of the Mirrar versus the hard numbers of mine proponents.

To resolve this dilemma—and to assess the true costs and benefits of natural systems and the human actions taken to protect them—environmental and ecological economists have sought ways to quantify the value of ecosystem services. They have tried to assign them **nonmarket values,** values that are not usually included in the price of a good or service (Table 2.3 and Figure 2.15). Assessing these nonmarket values and weighing them against some of the more tangible costs of preventing environmental problems (such as installing pollution control devices) constitute difficult and relatively new pursuits.

One technique, called **contingent valuation,** involves using surveys to determine how much people would be willing to pay to protect a resource or to restore it after damage has been done. Such an exercise was conducted with a mining proposal in the Kakadu region in the early 1990s that preceded the Jabiluka proposal. The Kakadu Conservation Zone, a government-owned 50-km^2 (19-mi^2) plot of land entirely surrounded by Kakadu National Park, was either to be developed for mining or to be preserved and added to the park. To determine the degree of public support for environmental protection versus mining, a government commission sponsored a contingent valuation study to determine how much Australian citizens valued keeping the Kakadu Conservation Zone preserved and undeveloped. Researchers interviewed 2,034 citizens, asking them how much money they would be willing to pay to stop mine development.

The interviewers first presented the respondents two scenarios. The "major impact" scenario was based on predictions of environmentalists, who held that mine development would cause great harm to the area. The "minor impact" scenario was based on predictions of the mining company, which held that development would have few downsides. After presenting both scenarios in detail, complete with photographs, the interviewers asked the respondents how much their households would pay, if each scenario, in turn, was correct. The respondents on average said their households would pay $80 per year to prevent the minor impact scenario and $143 per year to prevent the major impact scenario. Multiplying these figures by the number of households in Australia (5.4 million at the time), the researchers found that preservation was "worth" $435 million annually to the Australian population un-

der the minor impact scenario, and $777 million under the major impact scenario. Because both of these numbers exceeded the $102 million in annual economic benefits expected to be generated by development of the mine, the researchers held that preserving the land undeveloped and adding it to the park was worth more than mining it.

Because contingent valuation relies on survey questions and not on actual expenditures, critics complain that it produces inflated values people would not actually pay. Critics also point out that the method requires people to price things they have never bought or considered buying and that in such cases people will express idealistic rather than realistic values, knowing that they will not actually have to pay the price they name. In part because of such concerns about the method, the Australian government commission in the end decided not to use the Kakadu contingent valuation study results. (The mine was stopped instead as a result of Aboriginal opposition.)

Environmental economists have developed a number of other approaches to quantify the value of intangible entities. For instance, while contingent valuation measures people's *expressed* preferences, some methods aim to measure people's *revealed* preferences; that is, their preferences as revealed by data resulting from their actual behavior. The amount of money, time, or effort people expend to travel to parks for recreation, for example, has been used to measure how much people value parks. In another approach, economists have analyzed housing prices, comparing homes with similar characteristics but different environmental settings, to try to isolate the dollar value of landscapes, views, and peace and quiet. Economists have also assigned environmental amenities value by measuring the cost required to restore natural systems that have been damaged or to mitigate harm from pollution. Debate continues on the pros and cons of various approaches, as existing methods are improved and new methods are developed.

In 1997 researchers led by environmental economist Robert Costanza of the University of Maryland made an effort to calculate the overall economic value of all of the services that ecosystems provide, by reviewing studies using various valuation methods (see The Science behind the Story). The researchers came up with a figure of $33 trillion per year ($39 trillion in 2004 dollars)—nearly twice the combined gross national products of all nations in the world. Follow-up efforts have included a 2002 study, which concluded that the economic benefits of conserving the world's remaining natural areas outweighed the costs by a factor of 100 to 1.

Calculating the Economic Value of Earth's Ecosystems

To Robert Costanza, the problem was like an elephant in the living room that economists had ignored for decades: Earth's ecosystems provide essential life-support services, including arable soil, waste treatment, clean water, and clean air. However, economists had failed to account for how much those services contribute economically to human welfare, as is routinely done for conventional goods and services.

So Costanza, a professor at the University of Maryland, joined with 12 colleagues in June, 1996, at the National Center for Ecological Analysis and Synthesis in Santa Barbara, California, and combed the scientific literature. The team identified more than 100 studies that estimated the worth of such ecosystem services as purifying water, controlling greenhouse gases, pollinating plants, and cleaning up pollution. Authors of the studies that the team reviewed had estimated the values of particular ecosystem services in several ways, often by measuring people's hypothetical willingness to pay for them. Methods such as contingent valuation had frequently been used because people clearly value such aspects of natural systems as biodiversity and aesthetics, even though no one pays actual money directly and specifically for them. After

poring over studies that examined the value of 17 services provided by oceans, forests, wetlands, and other ecosystems, Costanza and his colleagues synthesized the results to provide the first comprehensive quantitative estimate of the global value of ecosystem services.

To better estimate the worth of the services, Costanza and his team reevaluated the data from the earlier studies using alternative valuation techniques. One method was to calculate the cost of replacing ecosystem services with technology. For example, marshes protect people from floods and filter out water pollutants. If a marsh were destroyed, the researchers would calculate the value of the services it had provided by measuring the cost of the levees and water-purification technology that would be needed to assume those tasks. The researchers then calculated the global monetary value of such wetlands by multiplying those totals by the global area occupied by the ecosystem. By calculating similar totals from coral reefs, deserts, tundra, and other ecosystems, they arrived at a global value for ecosystem services. By their calculation, the biosphere provides at least $33 trillion worth of ecosystem services each year—nearly twice the global gross domestic product (GDP) of $18 trillion. (In

2003 dollars, these sums are equivalent to $38.6 and $21 trillion, respectively.) Published in the journal *Nature* in 1997, their research paper, "The value of the world's ecosystem services and natural capital," ignited a firestorm of controversy.

Some environmental advocates and ethicists criticized the study, maintaining that it was a bad idea to put a dollar figure on priceless services such as clean air and clean water. The value of these services cannot be calculated, they held, because we would all perish if they disappeared. Environmental ethicist Timothy Weiskel of Harvard Divinity School contended that to make a commodity out of biodiversity confused "sacred space with the market place."

Some economists, meanwhile, disparaged the way the authors calculated value. Their methods, these economists claimed, were flawed on several counts. Replacement costs were not a legitimate way of determining value, they held, because nature's services, like other economic goods and services, are worth only what people will demonstrably pay for them. Critics also argued that totaling up the value of ecosystem services on small tracts of land is meaningless, because people make decisions about whether to preserve or ex-

Markets can fail

When they do not reflect the full costs and benefits of actions, markets can be said to fail. **Market failure** can occur when markets do not take into account the environment's good effects on economies (for example, ecosystem services), or when they do not reflect the bad effects of economic activity on the environ-

ment and thereby on people (external costs). Traditionally, market failure has been countered by government intervention. Governments can intervene in several ways. They can directly dictate limits on corporate behavior through laws and regulations. They can institute so-called *green taxes,* which penalize environmentally harmful activities. Or, govern-

ploit resources based on local considerations, not global ones.

To address the economists' criticisms, Costanza joined Andrew Balmford of Cambridge University in Great Britain and 17 other colleagues to conduct another analysis. They compared the benefits and costs of preserving wild nature with those of converting wild lands for agriculture, logging, or fish farming.

Again, they synthesized previous studies on ecosystem services, this time focusing on just five ecosystems: west African and Malaysian tropical forests, Thai mangrove swamps, Canadian wetlands, and Philippine coral reefs. In their paper, "Economic reasons for conserving wild nature," published in the journal *Science* in 2002, the team reported that a global network of

nature reserves that covered 15% of Earth's land surface and 30% of the ocean would be worth between $4.4 and $5.2 trillion—about 100 times the value of the same land were it to be converted for direct exploitative human use. That 100:1 benefit-cost ratio, they wrote, demonstrates clearly that "conservation in reserves represents a strikingly good bargain."

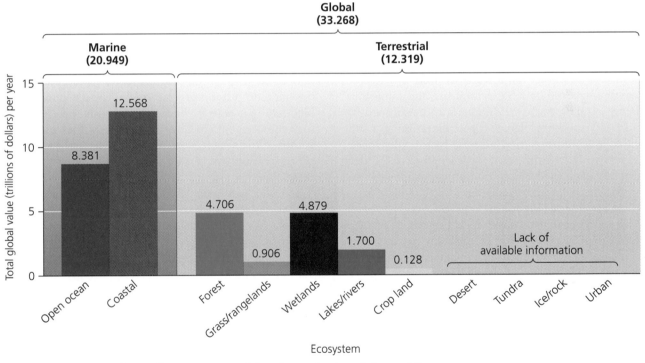

In 1996, Costanza and his colleagues estimated that the total value of the world's ecosystem services, including water supply, erosion control, nutrient cycling, and so forth, was approximately $33 trillion. Data from Costanza et al., "The value of the world's ecosystem services and natural capital," *Nature*, 1997.

ments can use alternative, innovative methods that put market mechanisms to work to promote fairness, resource conservation, and economic sustainability. We will examine legislation, regulation, green taxation, and market incentives in Chapter 3 as part of our policy discussion, but we will note a few examples of alternative methods here.

Ecolabeling and permit-trading are two of several ways to address market failure

Government intervention does not always effectively address market failure. Therefore, some economists have helped contrive ways to use aspects of the market itself to counteract market failure. In one technique, consumers are the key. Governments may allow manufacturers of

VIEWPOINTS

Environmental Justice

In 2001, environmental policy professor Mark Atlas published a study that questioned previous assertions that minority communities do not receive equal protection from pollution under environmental laws. Are such critiques of the environmental justice movement warranted, and have they affected the way environmental justice is pursued?

Studies of Social Injustice Do Not Rid Communities of Pollutants

Some academic studies conclude that environmental injustice is a problem, while others find no evidence that it occurs. Academicians and government researchers write treatises, crunch numbers in incomprehensible ways, and reach conclusions that are quite meaningless in the real world. So what?

No study or critique has had an impact on my community of Alsen, Louisiana, located just north of Baton Rouge. Since 1991 six permits have been issued for the construction of major environmental hazards in our community, even though we were already heavily impacted by a Superfund site and numerous chemical facilities. In 1999, napalm was shipped from San Diego to Baton Rouge for incineration at a site one mile from the nation's largest historically black university. By the time citizens learned of it, the decision had been made, with no public participation and no evidence of Executive Order 12898, which was intended to ensure such participation.

Despite numerous studies indicating that chemicals, such as benzene and chlorine, are toxic to exposed communities, government health agencies refuse to look at the real-world effects of cancer-causing pollutants, neurotoxins, mucous membrane irritants, and teratogens. They rely instead on data that is oftentimes outdated or irrelevant. For example, studies may focus on whether there are more waste sites or chemical plants in minority neighborhoods than in white neighborhoods or the exact height of a smoke stack, the width of its plume, and the concentrations and traveling speed of the pollutants it emits. Instead, we need to realize that exposure to chemical pollutants can harm babies *in utero*, increase rates of illnesses, and shorten the life span of the community.

We must look at the total picture: some communities are being exposed to higher levels of toxins than others. Until we focus on this simple fact, executive orders and academic treatises are meaningless.

Florence Robinson, founder of the North Baton Rouge Environmental Association, and until recently a professor of biology at Southern University, pioneered the correlation of zip code demographics with Toxic Release Inventory Data.

Environmental Justice Research Must Be Intellectually Honest

Mark Atlas's article has had a crucial, albeit indirect, impact on the environmental justice movement. The main result has been greater discretion among political elites regarding claims of inequity and racism.

Beginning with a landmark 1982 protest against a PCB landfill in Warren County, North Carolina, and extending into the Clinton presidency, the environmental justice movement enjoyed a long period of relatively uncontested dominance in the emerging national conversation. The movement contends that low-income and minority communities are unfairly blighted by pollution, betrayed by inadequate regulatory enforcement, and burdened by insufficient power within environmental policy processes. For several years, quantitative research seemed only to bolster these claims.

In 2001, however, Mark Atlas released his controversial article, "Rush to judgment: An empirical analysis of environmental equity in U.S. Environmental Protection Agency enforcement actions," in combination with other research and commentary, administered an important corrective. Only the most naive or devoted environmental justice enthusiast could fail to grasp serious flaws in the work Atlas criticized. Atlas's critique showed some environmental justice researchers to be not only incompetent methodologically but also ignorant of the EPA's enforcement regime. The researchers in question embraced fatally erroneous assumptions regarding just and unjust EPA practices.

The environmental justice movement retains political potency, but key institutions (including the mainstream media) are no longer so accepting of the movement's empirical claims. Greater intellectual honesty and sophistication might sharpen the movement's agenda and support its credibility.

Christopher Foreman, a professor in the School of Public Affairs at the University of Maryland at College Park, writes on the politics of regulation, public health, and governmental reform.

certain products to designate on their labels how the products were grown, harvested, or manufactured, so that consumers buying them are aware of the processes involved. This method, called **ecolabeling,** serves to tell consumers which brands use processes believed to be environmentally beneficial (or less harmful than others), and which do not. By favoring products that are ecolabeled, consumers can provide a powerful incentive for companies to switch to more environmentally friendly processes. The best-known example of this has been labeling cans of tuna as "dolphin-safe," a term that indicates that the methods used to catch the tuna prevent the accidental capture of dolphins in the process. Other examples include labeling organically grown foods, labeling furniture made of wood harvested through sustainable forestry, and labeling genetically modified foods (Chapter 9).

Ecolabeling is not the only technique environmental economists have devised to protect the environment and make prices reflect the true costs of production using market mechanisms. Another technique is to create markets in permits for environmentally harmful activities, thereby allowing companies to buy and sell the rights to conduct such activities. For instance, utilities and industries might trade permits to emit air pollution. In such a scheme, the government first determines an acceptable level of pollution, and then issues permits to existing industries. Each party may emit its permitted amount, but if it is able to reduce its amount of pollution, it receives credit for the amount it did not emit and can sell this credit to other industries that want to expand operations or are less able to control their emissions. Creating markets in permits provides companies with an economic incentive for finding ways to reduce emissions, because the sale of permits can bring in money.

We will discuss permit-trading further in Chapter 3. The method is not perfect. Large firms can hoard permits, deterring smaller new firms from entering the market and thereby suppressing competition. Nevertheless, permit-trading markets have shown great promise for preventing further environmental degradation while granting industries the flexibility to reduce their impacts in ways that are economically palatable to them.

Conclusion

Permit-trading, ecolabeling, and the valuation of ecosystem services are some of many recent developments that have brought economic approaches to bear on environmental protection and resource conservation. As economics becomes more environmentally friendly, it renews some of its historic ties to ethics. Environmental ethics, as we saw, has expanded people's sphere of ethical consideration outward to encompass other societies and cultures, nonhuman animals, and even nonliving entities that were formerly outside the realm of ethical concern. This ethical expansion involves the concept of distributional equity, or equal treatment—the aim of environmental justice. One type of distributional equity is equity among generations, concern by current generations for the welfare of future generations. Equity among generations is the basis for the notion of sustainable economic development, that which can be continued in perpetuity without degradation.

Unquestionably, sustainability (Chapter 20) is the key to an ethical treatment of future generations of humans, as well as the nonhuman environment. However, is sustainability a pragmatic pursuit for us, the generations alive right now? The answer largely depends on whether we believe that economic well-being and environmental well-being are opposed to one another, or whether we accept that they work in tandem. This choice largely depends on how we define economic well-being. Equating it with economic growth, as do the majority of today's economists, suggests that the well-being of the economy entails a tradeoff with the well-being of the environment. However, recognizing that economic well-being can exist and increase without growth enables us to envision human economies and environmental quality benefiting mutually from one another.

Although we have not yet arrived at the point where science, ethics, and economics are fully integrated, the decision to cancel mining plans at Jabiluka demonstrates how economics, ethics, and science can converge and play roles in solving important human problems. The convergence of science, ethics, and economics becomes evident in many other environmental issues. We have seen in this chapter some ways in which ethical and economic philosophies influence the values we attribute to our environment and the decisions we make about how to interact with our environment. Let's keep these perspectives in mind as we go on in future chapters to explore the environmental policy process, to examine several key scientific disciplines, and to survey the major topics and issues in environmental science.

REVIEW QUESTIONS

1. How do particular aspects of culture determine worldview and, ultimately, perceptions of the environment? Describe the effect of cultural considerations in the Mirrar response to the Jabiluka mine project.

2. What does the study of ethics encompass? What is the one principle to which ethicists generally adhere? How is ethics, as a discipline, different from religion and law?

3. Describe the three classic ethical standards discussed in this chapter. What is environmental ethics?

4. Why in Western cultures have ethical considerations expanded to include non-human entities?

5. Define the perspectives of anthropocentric, biocentric, and ecocentric worldviews. How would you characterize the worldview of people of the Mirrar Clan?

6. What did John Ruskin mean when he complained, during the industrial revolution in England, that people had "desacralized" nature?

7. Describe the "holistic" view of nature expounded by the American transcendentalists.

8. What is the difference between the preservation ethic and the conservation ethic?

9. Explain the contributions of John Muir and Gifford Pinchot in the history of environmental ethics.

10. In what respect was Aldo Leopold's view in *The Land Ethic* (1949) new and different from those of Muir and Pinchot? How did Leopold define the "community" to which ethical standards should be applied?

11. What is an economy? What is the difference between an *open system* and a *closed system,* and how does this bear on human economies?

12. Name four key contributions the environment makes to the economy.

13. Describe Adam Smith's metaphor of the "invisible hand." How have neoclassical economists refined classical economics?

14. Describe three ways in which critics hold that neoclassical economic views fail to accurately account for the value of the environment.

15. Compare and contrast the views of environmental economists with those of ecological economists on the "growth is good" paradigm of neoclassical economics.

16. What is a steady-state economy? Do you think this model is a practical alternative to the growth paradigm? Why or why not?

17. See Table 2.3. List examples of resources associated with seven kinds of non-market values that modern economics does not adequately account for.

18. What is contingent valuation and what are two of its weaknesses?

19. What is the GPI and how does it compare to the GDP?

DISCUSSION QUESTIONS

1. Do you believe that an introduction to environmental ethics and worldviews is an important part of a course in environmental science? Should an introduction to ethics and worldviews be a component of other science courses? Explain your answers.

2. Write a short essay describing your worldview as it pertains to your relationship with the environment. Feel free to borrow from any of the worldviews presented in this chapter. Do you feel that you fit into any single category that we discussed in this chapter? Why or why not?

3. How would you analyze the case of the Mirrar People and the proposed Jabiluka uranium mine from each of the following perspectives? In your description, list at least three questions that would likely be asked by a person of each perspective when attempting to decide whether or not the mine should be developed. Be as specific as possible and be sure to identify similarities and differences in approaches:

- Preservationist
- Conservationist
- Deep ecologist
- Environmental justice advocate
- Neoclassical economist
- Ecological economist

4. The Reverend Benjamin E. Chavis, Jr., former chairman of the National Association for the Advancement of Colored People (NAACP), has defined environmental racism as "racial discrimination in environmental policy making, enforcement of regulations and laws, and targeting of communities of color for toxic waste disposal and siting of polluting industries." Some critics of this view argue that the most polluted communities are simply the poorest; hazardous waste facilities will simply be placed in the areas with the lowest property values. These critics argue that the problems environmental justice advocates describe are really a result of economics and not racial discrimination. Are poor minority-based communities the targets of racist policies or simply the unintended victims of disinterested market forces? Should environmental hazards be equally distributed? How do you think these kinds of disputes might be prevented?

5. How would you balance the costs and benefits of pollution control in the following situation? Assume that there are no environmental laws in place.

A manufacturing facility located on a major U.S. river near your home, produces a popular electronic device that many U.S. residents use every day and that is produced nowhere else in the world. The manufacture and sale of this device provides jobs for 200 people in your community. The average salary for these workers is $42,000, and the plant pays $2 million in taxes to the local government each year. The city and county use this money to pay a substantial portion of the costs of operating public schools and the police department. Sales taxes from purchases made by plant employees and their families contribute an additional $1.5 million to local and state government coffers. In addition, the owner of the plant contributes $500,000 per year to local charities.

A recent study has revealed that the plant has been discharging large amounts of waste directly into the river running through the community. According to the authors of the study (which was published in a peer-reviewed and well-respected scientific journal) this pollution is causing the following adverse effects:

- A 15% increase in cancer rates among elderly residents of downstream communities, shortening their lives by an average of 5 years.
- A reduction in value of riverfront property by an average of 30%, which has reduced property tax revenues to local government by $200,000 per year.
- A 75% decrease in native fish populations, putting 25 commercial fishermen out of business and causing hundreds of fly-fishing tourists to seek other vacation destinations.
- An increase in community unease and unrest concerning the safety and aesthetic appeal of the river.

According to the plant owner and operator, the reason the facility is profitable and can stay in business is that there are no regulations mandating expensive treatment of the waste from the plant. If any such regulations were imposed on the plant, he says he would be likely to close the plant, lay off its employees, and relocate to a more sympathetic and business-friendly environment.

How would you recommend resolving this situation? What further information would you want to know before making a recommendation? In arriving at your recommendation, how did you weigh the costs and benefits associated with each of the plant's impacts?

Media Resources *For further review, go to the website* **www.envscienceplace.com** *or student CD-ROM, where you will find quizzes, flashcards, a glossary, additional interactive exercises, and links to relevant news and research sources. Also on the website and CD-ROM is* **GRAPH IT**, *a series of interactive graphing tutorials to help you interpret graphs and plot data.*

Environmental policy: Decision-making and problem-solving

Tijuana, Mexico and San Diego, California Border

This chapter will help you understand:

- The origins, history, and societal role of environmental policy

- The relationship between science, ethics, economics, and policy

- The institutions important to U.S. environmental policy

- The environmental policy process, including the role of citizens and interest groups

- Selected U.S. environmental laws

- The nature and institutions of international environmental policy

- How nations handle transboundary issues

Closed San Diego Beach

Central Case: San Diego and Tijuana's Sewage Pollution Problems and Policy Solutions

"Ignorance is an evil weed, which dictators may cultivate among their dupes, but which no democracy can afford among its citizens."
—*William Beveridge, British economist*

"It is the continuing policy of the Federal Government . . . to create and maintain conditions under which man and nature can exist in productive harmony and fulfill the social, economic, and other requirements of present and future generations of Americans."
—*Text of the National Environment Policy Act, 1969*

On November 23, 1996, officials closed all the public beaches in San Diego, California. The closures were necessary because stormwater runoff following heavy rains washed pollutants and contaminants into local rivers and coastal waters. This also occurred across the border in the Mexican city of Tijuana, whose aging sewage system became clogged with dirt, rocks, and trash, causing raw sewage to overflow into streets, canyons, and beaches. Such incidents, known as *rogue flows,* occur when heavy rain overwhelms the ability of

sewage treatment plants to process wastewater. Rogue flows had become so common in San Diego and Tijuana that local surfers and swimmers casually referred to the initial one of each rainy season as the "first flush."

The Tijuana River winds northwestward through the arid landscape of northern Baja California, Mexico, crossing the U.S. border south of San Diego. A river's **watershed** consists of all the land from which water drains into the river, and the Tijuana River watershed is home to two million people in the cities and towns of two nations. This particular **transboundary,** or international, watershed covers about 4,500 km² (1,750 mi²), with approximately 70% of its area in Mexico. On the Mexican side of the border, the river and the arroyos, or creeks, that flow into it are lined with farms, apartments, shanties, factories, and all kinds of urban development, including leaky sewage treatment plants and dump sites for toxic materials. Many of these sources release pollutants that heavy rains wash through the arroyos, into the Tijuana River, and eventually onto area beaches in both the United States and Mexico.

Although raw sewage has periodically flowed from Mexico into the United States via the Tijuana River for at least 70 years, this pollution problem has become much worse in recent years as the region's population has boomed, outstripping the capacity of sewage treatment facilities. Thousands of beach closures and advisories have resulted from these spills in recent years.

The problem is worse on the Mexican side, because most Mexican residents of the Tijuana River watershed live in poverty relative to their U.S. neighbors, and the pollution of their river directly affects their day-to-day lives. In poor neighborhoods like Loma Taurina, raw sewage has even spilled onto the streets during rogue flows. The rise of U.S.-owned factories, or *maquiladoras*, on the Mexican side of the border has contributed to the river's increased pollution, both through direct disposal of industrial waste and by attracting thousands of new workers to the already-crowded region.

As impacts increased, many people in the San Diego and Tijuana areas, from coastal residents to grassroots activists to businesspeople, pressed policymakers to do something. As we explore environmental policy in this chapter, we will periodically return to the Tijuana River watershed and touch on how, in recent years, citizens and policymakers together have addressed these problems.

Environmental Policy: An Introduction

We noted in Chapter 1 how people in different circumstances may perceive environmental changes differently. Then in Chapter 2 we evaluated some of the ethical and economic criteria people apply to judge whether a given change is desirable or undesirable. Whereas ethics and economics supply the values that help us determine which environmental changes are problems, science provides the information needed to delineate and understand the problems and to devise workable solutions to them. Science, ethics, and economics all influence the problem-solving process of environmental policymaking that we explore in this chapter (Figure 3.1).

Policymaking can play an important role in resolving problems like those of the Tijuana River watershed (Figure 3.2). Sewage-tainted rivers and beaches pose several threats. First, raw sewage carries pathogens, organisms that can cause illness in humans and other animals (Table 3.1). Second, untreated sewage can radically alter conditions for aquatic and marine life, lowering

concentrations of dissolved oxygen in the water and increasing mortality for many species. Third, beach pollution and closures cause economic damage by reducing tourism, recreation, and other economic activity associated with clean coastal areas. This is a major consideration both in Mexico and in southern California, whose beaches each year host 175 million visitors who spend over $1.5 billion. Many phenomena share this combination of effects—harming human health, altering ecological systems, and inflicting economic damage—and such phenomena generally come to be viewed as environmental problems.

When a society reaches broad agreement that an environmental problem exists, its leaders may often be persuaded to address the problem through the making of policy. A **policy** is a rule or guideline that directs individual, organizational, or societal behavior. Environmental policies are developed primarily by governments and are intended to regulate the behavior of individuals, corporations, and government agencies. We will examine the laws, political activities, and institutions involved in today's environmental policy process, both in the United States and abroad.

Environmental policy addresses issues of equity and resource use

Citizens have identified environmental problems and asked government officials to develop solutions since at least the first century, when Roman citizens complained about air quality problems caused by wood smoke. Today, our unprecedented population growth and resource consumption have generated unparalleled environmental problems. Furthermore, the market capitalist systems of modern constitutional democracies are driven by incentives for short-term economic gain over long-term social and environmental stability. As such, they provide little incentive for businesses or individuals to behave in ways that minimize environmental impact or that equalize costs and benefits among different parties. Such market failure (Chapter 2) has traditionally been viewed as justification for government intervention. Thus, environmental policy aims to protect the natural resources people use and promote equity in people's use of resources.

The tragedy of the commons One goal of environmental policy is to protect resources, held and used in common by the public, from depletion or degradation. As Garrett Hardin explained in his essay, "The Tragedy of the Commons" (Chapter 1), an unregulated commons will eventually become overused and degraded.

Figure 3.1 Policy plays a central role in the overall process of applying scientific knowledge to solve environmental problems. This process begins when research from the various disciplines involved in environmental science is synthesized and leads to some understanding of a given environmental problem and its potential solutions. Government uses this information and guidance in formulating policy, with input from citizens and from the private sector. Government policy includes laws, regulations, and market-based incentives that aim, for example, to restrict resource exploitation or to ensure fairness in resource use. Along with technological developments from the private sector and consumer choices exercised by citizens in the marketplace, policy can produce lasting solutions to environmental problems.

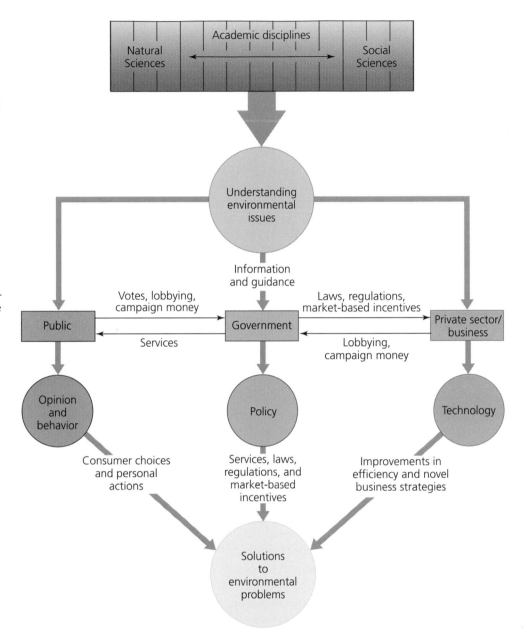

Therefore, he claimed, it is in our best interest to develop guidelines or policies for the use of the commons. Such guidelines might limit the number of animals each individual can place on the commons or might require individuals using the common pasture to pay for restoring and managing the shared resource. These two concepts, restriction of use and active management, are central to environmental policy today. In fact, however, the tragedy of the commons does not always play itself out as Hardin predicted. Some traditional societies had safeguards against exploitation, and in modern Western society users have occasionally joined together or teamed with government to prevent overexploitation. In addition, many cases do not meet Hardin's starting assumptions, since even on public lands, resources may not be equally accessible to all comers, but more accessible to established resource extraction industries. Nonetheless, the always-real threat of overexploitation of publicly held commons has been a driving force behind much environmental policy.

External costs Environmental policy also is developed to ensure that some people do not benefit from common resources in ways that harm others. One way to promote fairness is by dealing with external costs (Chapter 2), harmful impacts that result from market transactions but that are borne by people who are not involved in those transactions.

Figure 3.2 Pollution flowing into the Tijuana River as it snakes through its watershed affects Mexican residents of the watershed and U.S. citizens on San Diego County beaches. During high-water episodes that overwhelm treatment facilities, millions of gallons of raw sewage have flushed into the river and the Pacific Ocean.

For example, if a factory is allowed to discharge waste freely into a river, the owners of the factory may reap greater profits by avoiding the costs of waste disposal or recycling. Their benefit, however, becomes an external cost (in the form of water pollution, declining fish populations, aesthetic damage, or other problems)

that downstream users of the river will suffer. U.S.–owned *Maquiladoras* in the Tijuana River watershed dump waste that affects downstream users, such as the Mexican families who use the river water for washing (Figure 3.3). Likewise, the detergents these families use for washing further pollute the river, creating additional external costs for families further downstream and for beachgoers in Mexico and California.

Table 3.1 Pathogens and Health Effects Associated with Untreated Sewage

Pathogen	Human health effect
Cryptosporidium	Diarrhea, vomiting, cramps, dehydration
Giardia lamblia	Diarrhea, vomiting, cramps, dehydration
Salmonella typhi	Typhoid fever
Vibrio cholerae	Cholera
Poliovirus	Poliomyelitis
Shigella spp.	Dysentery
Entamoeba histolytica	Dysentery
Hepatitis A virus	Infectious hepatitis

Weighing the Issues:
Do We Really Need Environmental Policy?

Many free-market advocates who invoke the spirit of Adam Smith and the "invisible hand" maintain that environmental laws and regulations are an unnecessary government intrusion into private affairs. As you recall from Chapter 2, Smith argued that fully informed individuals who weigh the costs and benefits of a particular action and are free to exchange materials or services will benefit themselves and society by pursuing their self-interest. Do you agree? Can you think of exceptions to Smith's argument? Describe a situation in which an individual acting in his or her self-interest could harm society by

Figure 3.3 Pollution of a river raises many issues that have been viewed as justification for environmental policy. This woman washing clothes in the river may suffer from upstream pollution from factories, and her actions may cause further pollution for people living downstream.

causing an environmental problem. Could environmental policy rectify the situation? Are there other ways to solve the problem? What are the advantages and disadvantages of instituting environmental laws and regulations versus allowing free exchange of materials and services?

These goals of environmental policy—to protect resources against the tragedy of the commons and to promote equity by dealing with external costs—will become apparent in case after case explored in this book.

Many factors can hinder implementation of environmental policy

If the goals of environmental policy are so seemingly noble, why is it that environmental laws are so often challenged, environmental regulations frequently derided, and the ideas of environmental activists repeatedly ignored or rejected by policy-makers?

In the United States, much environmental policy has come in the form of regulations handed down from the federal government, which businesses and individuals may view as restrictive, bureaucratic, and unresponsive to human needs. Many landowners fear restrictions on their use of land due to protections of endangered species, for instance. Developers complain of time and money lost to the bureaucracy in obtaining permits; reviews by government agencies; the sometimes arbitrary application of regulations; surveys for endangered species; and required environmental controls, monitoring,

and mitigation. In the eyes of such citizens and business-people, environmental regulation all too often has meant inconvenience and/or economic loss.

Another reason for resistance to environmental policy involves the nature of environmental problems, which often develop slowly and gradually. The degradation of ecosystems and public health due to human impact on the environment are long-term processes. Human behavior, however, is often geared toward addressing short-term needs, even if they conflict with long-term needs—and this tendency is reflected in social institutions. Businesses usually opt for short-term economic gain over long-term considerations. The media has a short attention span traditionally based on the daily news cycle, whereby new and sudden one-time events are given more coverage than slowly developing long-term trends. Finally, politicians most often act out of short-term interest, since they depend on re-election every few years. For all these reasons, many environmental policy goals that seem admirable and that attract wide public support in theory may be hindered in their practical implementation.

U.S. Environmental Policy: An Overview

The United States can serve as a good starting point in understanding environmental policy strengths and limitations in constitutional democracies worldwide. First,

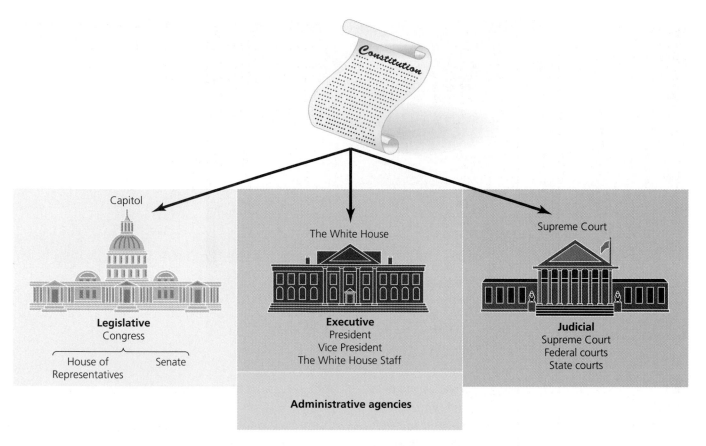

Figure 3.4 In the United States, federal powers are shared among three branches of government: the legislative, judicial, and executive. All three branches, plus the administrative agencies (sometimes known as the "fourth branch"), are responsible for aspects of developing and implementing environmental policy.

the United States historically has been a pioneer in creating and enforcing environmental laws. Second, U.S. policy has served as a model—both of success and of failure—for many other nations and international government bodies. Third, the United States, through its environmental policies and many other institutions, exerts a great deal of influence on the affairs of other nations. Finally, understanding U.S. environmental policy on the federal level can enable you more easily to understand environmental policy on the local, state, and international levels.

U.S. policy results from the three branches of government

Environmental policy, like all U.S. policy, results from processes involving the three branches of government established under the U.S. Constitution (Figure 3.4). Statutory laws are passed by the **legislative branch,** or Congress, consisting of the Senate and the House of Representatives.

The Tijuana River Valley Estuary and Beach Sewage Cleanup Act passed in 2000, for example, addressed the treatment of sewage emanating from the Tijuana River watershed and flowing into the United States.

Legislation is enacted (approved) or vetoed (rejected) by the president, who heads the **executive branch.** The president may also issue **executive orders,** which are specific legal instructions for government agencies.

The **judicial branch,** or judiciary, consisting of the Supreme Court and various lower courts, is charged with interpreting the law, which is necessary due to changing social factors, new technologies, and because Congress must write laws broadly to ensure that they apply to varied circumstances throughout the nation. Decisions rendered by the courts make up a body of law known as **case law.** Previous rulings serve as **precedents,** or legal guides, for later cases, steering and stabilizing judicial decisions through time. The judiciary has been an important arena for environmental policy. Grassroots environmental advocates and nongovernmental organizations have used **lawsuits,** or requests for judicial action,

Federal Agencies and Departments of the Executive Branch that Affect Environmental Policy	
Environmental Protection Agency (EPA)	Department of Housing and Urban Development (HUD)

Office of Management and Budget	Community Planning and Development (CPD)
	  Healthy Homes and Lead Hazard Control
	  Policy Development and Research (PDR)

U.S. Trade Representative	
Tennessee Valley Authority (TVA)	Department of Interior (DOI)
Nuclear Regulatory Commission	Bureau of Indian Affairs (BIA)
Council on Environmental Quality	Division of Energy and Mineral Resources Bureau of Land Management
Department of Treasury	Minerals Management Service National Park Service

Community Development Financial Institutions Fund	Office of Ethics
	  Fish and Wildlife Service
	  U.S. Geological Survey
	  Water Resources Division

Department of Commerce (DOC)	Department of Justice (DOJ)

National Institute of Standards and Technology
  National Oceanic and Atmospheric Administration
	Environmental and Natural Resources Division
	  Office of Legal Policy
	  Office of Public Affairs

Department of Defense (DOD)	Department of Labor (DOL)
Army Corps of Engineers	Mine Safety and Health Organization
Department of Energy	Department of State

Office of Environment, Safety, and Health
  Energy Efficiency and Renewable Energy Network
  Office of Civilian Radioactive Waste Management (OCRWM)
  Office of Fossil Energy
  Office of Health and Environmental Research
  Office of Environmental Management
	U.S. Agency for International Development

	Department of Transportation (DOT)
	Office of Pipeline Safety Office of Environment and Energy
Department of Health and Human Services (HHS)	Department of Agriculture (USDA)

Public Health Service
  Agency for Toxic Substances and Disease Registry (ATSDR)
  Centers for Disease Control and Prevention (CDC)
  Food and Drug Administration (FDA)
	Farm Service Agency
	  Forest Service
	  Natural Resources Conservation Service
	  National Rural Development Council
	  Food and Consumer Service

Figure 3.5 The many administrative agencies of the executive branch are the source of most U.S. environmental policy. *Source:* U.S. General Services Administration, Washington, D.C.

as tools to level the playing field somewhat with large corporations and agencies. Likewise, the courts have allowed recourse for businesses and individuals to challenge the constitutional validity of environmental laws they feel to be infringing on their rights.

Administrative agencies are the "fourth branch" of government

Once statutory laws are enacted, their enforcement and elaboration is assigned to an agency within the executive branch known as an **administrative agency** (Figure 3.5). Administrative agencies, which may be established by Congress or by presidential order, are sometimes nicknamed the "fourth branch" of government because they are the source of a great deal of environmental

policy, in the form of regulations. **Regulations** are specific rules that are based on the more broadly written statutory law passed by Congress and enacted by the president. In addition to issuing regulations, administrative agencies are responsible for monitoring compliance with statutory and administrative law and for enforcing the law when individuals or corporations violate it. In fact, administrative agencies have come to be the source of most U.S. environmental policy.

State and local policy also affects environmental questions

The structure of the federal government is mimicked at the level of the states, with their governors, legislatures, judiciaries, and agencies. Many states with dense urban populations such as California, New York, and Massachusetts have strong environmental laws and well-funded environmental agencies, while many less-populous states such as those of the interior West put relatively little priority on environmental protection. To safeguard public health in locations such as San Diego's beaches, California legislators in 1997 required state environmental health officials to set standards for testing waters for bacterial contamination. Officials issue an advisory, or warning, when bacterial concentrations in near-shore waters exceed health limits established in California law. In 1999, further legislation mandated another state agency to develop methods for locating the source of contamination episodes. As we proceed through our discussion of the federal government, keep in mind that important environmental policy also gets made at the state and local levels.

When state and federal jurisdiction overlap, state laws may be more strict, but not less strict, than federal laws. If state and federal laws are found to be in direct conflict, federal laws take precedence.

Some constitutional amendments bear on environmental law

Just as the Constitution clarified the relationship between states and the federal government, it also laid out several principles that have come to be especially relevant to environmental policy. One of these is the clause from the Fourteenth Amendment prohibiting states from denying "equal protection of its laws" to any person. This amendment provides the constitutional basis for the environmental justice movement (Chapter 2), which is based on the notion of equity toward people of all races, cultures, genders, and economic backgrounds.

The Fifth Amendment ensures, in part, that private property shall not "be taken for public use without just compensation." Courts have interpreted this prohibition, known as the **takings clause,** to ban not only the literal taking of private property but also what is known as regulatory taking. A **regulatory taking** occurs when the government, by means of a law or regulation, does not actually take possession of private property but instead deprives its owner of most or all economic uses of that property.

Many people cite the takings clause in arguing against environmental regulations that restrict land use and development on privately owned land. For example, many would contend that a law prohibiting a landowner from opening a hazardous waste dump in a residential neighborhood deprives the landowner of an economically

valuable use of the land and therefore violates the Fifth Amendment.

In a landmark court case in 1992, the Supreme Court ruled that a state land use law intended to "prevent serious public harm" violated the takings clause. The case, known as *Lucas v. South Carolina Coastal Council,* involved a developer named Lucas who in 1986 purchased beachfront property in South Carolina for $975,000. In 1988, before Lucas began to build on his land, the South Carolina legislature passed a law, the Beachfront Management Act, banning construction on eroding beaches. The state classified the Lucas property as an eroding beach and prohibited residential construction there. Believing that the takings clause had been violated, Lucas asked a state court to overrule the Beachfront Management Act and allow him to build homes on his property. Agreeing with Lucas, the state court ruled that he was entitled to $1.2 million in compensation. South Carolina appealed the case to the state supreme court, which overturned the lower court's decision. Lucas then appealed to the U.S. Supreme Court, which overturned the state supreme court and sided with the lower court, declaring that the state law deprived Lucas of all economically beneficial uses of his land. A decade later, Lucas's land holds houses, but the general issue of regulatory takings remains a contentious area of law (Figure 3.6).

Weighing the Issues:
The Takings Clause and Land Use Laws

What do you think of the way the Lucas case turned out? Do you agree with the Court that land use laws can in some circumstances violate the takings clause and that victims of such takings require compensation? Can you think of any better approaches to balance private property rights with protection of the public good in cases like this?

The first laws shaping U.S. environmental policy addressed public land management

The laws that comprise U.S. environmental policy were produced largely in three periods. Those laws enacted during the first period, from the 1780s to the late 1800s, dealt primarily with the management of public lands and accompanied the westward expansion of the nation. Many of the environmental laws of this period were intended to promote settlement and the extraction and use of the West's abundant natural resources.

Central among the early environmental laws were the **General Land Ordinances** of 1785 and 1787, which

Figure 3.6 Rapidly eroding coastal areas like this one in South Carolina have given rise to environmental policy debates in recent years. Should the government be able to prevent development in such areas where erosion, storms, and flooding pose a risk to property and human life? If so, does such a prohibition constitute a taking of private property rights? This was the question addressed in *Lucas v. South Carolina Coastal Council.*

gave the federal government the right to manage western lands and created a grid system for surveying them and readying them for private ownership. Between 1785 and the 1870s, the federal government worked hard to encourage settlement in the West by doling out as many of these lands as possible. Western settlement provided citizens with means to achieve prosperity, relieved crowding in Eastern cities, and expanded the nation's geographical reach at a time when the young United States was still jostling with European powers for control of the continent. Environmental policy of this era reflected the public perception that western lands were practically infinite and virtually uninhabited and their natural resources inexhaustible. The following are just a few of the laws typical of this era:

- The Homestead Act of 1862 allowed any citizen to claim 160 acres of public land by living there for 5 years, cultivating the land or building a home, and paying a $16 fee (Figure 3.7a). A waiting period of only 14 months was available to those who could pay $176 for the land.
- The Mineral Lands Act of 1866 provided free land to promote mining and settlement. It stated that mining could occur subject to local customs, with no governmental oversight. This system has not changed much in the 140 years since (Figure 3.7b).
- The Timber Culture Act of 1873 granted 160 acres to any citizen promising to cultivate trees on 40 of those acres (Figure 3.7c).

Such laws encouraged settlers, entrepreneurs, and land speculators to move west, hastening the closing of the frontier.

The second wave of U.S. environmental policy sought to address impacts of the first

In the late 1800s, as the West became more populated and its resources were increasingly exploited, public perception and government policy toward natural resources began to change. Laws of this period were intended to reduce and correct some of the environmental problems associated with westward expansion. In 1872 Congress designated Yellowstone as the first national park in the world. In 1891 Congress passed a law authorizing the president to create "forest reserves" out of existing public lands. These reserves were to be off-limits to logging in order to prevent overharvesting and to protect forested watersheds. Legal actions like these reflected the new understanding that the West's resources were not inexhaustible but instead required and deserved legal protection. These laws enabled the creation, over the next few decades, of a national park system and a national forest system that still stand as global models. Land management policies continued through the 20th century, targeting soil conservation in the Dust Bowl years (Chapter 8) and extending through the Wilderness Act of 1964, which sought to preserve still-pristine lands "untrammeled by man, where man himself is a visitor who does not remain."

Figure 3.7 (a) Settlers such as these took advantage of the federal government's early environmental policies, including the Homestead Act of 1862. (b) Early mining activities on public lands were loosely regulated. The Mineral Lands Act of 1866 required that mining could occur "subject . . . to the local customs or rules of miners," rather than in a way that would protect the environment.

(a) Settlers in Custer County, Nebraska, circa 1860

The third wave of U.S. environmental policy responded largely to pollution problems

Further social changes in the mid to late 20th century gave rise to the third major period of U.S. environmental policy. In a more densely populated country driven by technology, heavy industry, and intensive resource consumption, Americans found themselves in a better economic position, but living amid dirtier air, dirtier water, and increasing amounts of waste and toxic chemicals. Public priorities shifted, and during the 1960s and 1970s several events triggered increased awareness of environmental problems and brought about important changes in public policy.

One of the landmark environmental events of the 1960s was the publication of *Silent Spring,* a book by the American scientist and writer Rachel Carson (Figure 3.8), in 1962. *Silent Spring* awakened the American public to the negative ecological and health effects of pesticides and industrial chemicals. (The title refers to Carson's warning that pesticides might kill so many birds that few would be left to sing in future springs.) The publication of this book marked the beginning of the third wave of U.S. environmental policy; however, it was not alone in drawing public attention to the hazards of pollution.

The Cuyahoga River (Figure 3.9) also did its part to arouse the public. The Cuyahoga was so polluted with oil and industrial waste that the river itself caught fire near Cleveland, Ohio, more than half a dozen times during the 1950s and 1960s. This spectacle, coupled with an enormous oil spill off the Pacific coast near

(b) Nineteenth-century mining operation, Lynx Creek, Alaska

(c) Loggers falling a redwood tree, Washington

Figure 3.7 cont. (c) Early forest policy promoted rapid clearing of land and offered no incentives for replanting or habitat conservation.

Santa Barbara, California in 1969, moved the public to prompt Congress and the president to do more to protect the environment.

Today, in large part because of environmental policies enacted since the 1960s, pesticides such as DDT are more strictly regulated, and the nation's waters are considerably cleaner. Even the Cuyahoga River is in much better shape. The public enthusiasm for environmental protection that spurred such advances remains strong today; polls repeatedly show an overwhelming majority of Americans favoring environmental protection. Such support is evident each year in April, when millions of people worldwide celebrate **Earth Day** in thousands of locally based events featuring everything from speeches and lectures to demonstrations to hikes and birdwalks. In the more than 30 years since the first Earth Day, celebrated on April 22, 1970, participation in this event has grown markedly and has spread to nearly every country in the world (Figure 3.10).

NEPA guarantees citizens input into environmental policy decisions

Besides Earth Day, two federal actions marked 1970 as the dawn of the modern era of environmental policy in the United States. On January 1, 1970, President Richard Nixon signed the **National Environmental Policy Act (NEPA)** into law (Figure 3.11). NEPA created an agency called the Council on Environmental Quality and required that an **environmental impact statement**

Figure 3.8 Scientist, writer, and citizen activist Rachel Carson identified the problem of pesticide pollution in her 1962 book, *Silent Spring*.

(EIS) be prepared for any major federal action. An EIS is a report of results from detailed studies that assess the potential effects on the environment that would likely result from development projects or other actions undertaken by the government. In mandating these reports,

Figure 3.9 In a spectacular display of the need for better water pollution control, Ohio's Cuyahoga River caught fire several times during the 1950s and 1960s. The Cuyahoga was so polluted with oil and industrial waste that the river would burn for days at a time.

(a) Environmental Teach–In, Inc., Washington, D.C., 1970

(b) School children celebrating Earth Day, Katmandu, Nepal, 2002

Figure 3.10 (a) April 22, 1970, was the first Earth Day celebration. This public outpouring of support for environmental protection catalyzed the third wave of environmental policy in the United States. (b) Three decades later, Earth Day is celebrated by millions of people all across the globe, as is shown here in Nepal.

Figure 3.11 Richard Nixon served as president during the period when modern environmental policy took shape. Here he signs the National Environmental Policy Act (NEPA) into law on January 1, 1970.

NEPA made understanding human impacts on the environment a national priority.

NEPA's effects have been far-reaching. The EIS process forces government agencies and any businesses that contract with them to slow down and evaluate impacts on the environment before proceeding with a new dam or highway or building project. Although the process generally does not halt such projects, it does serve as a powerful disincentive for environmentally damaging development or activity. NEPA also gives ordinary citizens a say in the policy process by requiring that environmental impact statements be made publicly available and that public comment on them be solicited.

The creation of the EPA marked a shift in U.S. federal environmental policy

Six months after signing NEPA into law, Nixon issued an executive order that called for a new integrated approach to environmental policy based on the understanding that environmental problems are interrelated. "The Government's environmentally-related activities have grown up piecemeal over the years," the order stated. "The time has come to organize them rationally and systematically."

Nixon's order moved elements of the Federal Water Quality Administration, National Air Pollution Control Administration, Bureau of Solid Waste Management, Bureau of Water Hygiene, Bureau of Radiological Health, and others into a newly created agency, the **Environmental Protection Agency (EPA).** The order charged the EPA with conducting and evaluating research, monitoring environmental quality, setting standards (for

Figure 3.12 Most major laws in modern U.S. environmental policy were enacted in the 1960s and 1970s.

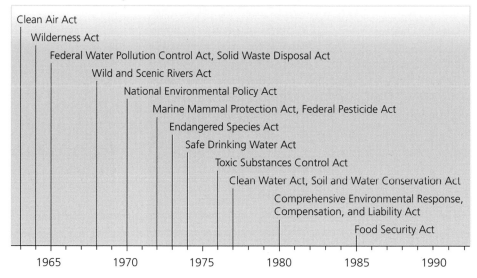

Key Environmental Protection Laws, 1963-1985

Clean Air Act

Wilderness Act

Federal Water Pollution Control Act, Solid Waste Disposal Act

Wild and Scenic Rivers Act

National Environmental Policy Act

Marine Mammal Protection Act, Federal Pesticide Act

Endangered Species Act

Safe Drinking Water Act

Toxic Substances Control Act

Clean Water Act, Soil and Water Conservation Act

Comprehensive Environmental Response, Compensation, and Liability Act

Food Security Act

1965   1970   1975   1980   1985   1990

minimal pollution levels, for instance), enforcing those standards, assisting the states in meeting standards and goals, and educating the public.

Other key laws followed, including the Clean Water Act

Ongoing public demand for a cleaner environment resulted in a number of key laws during this era that remain the linchpins of U.S. environmental policy (Figure 3.12). Today there are thousands of federal, state, and local environmental laws in the United States and thousands more abroad.

The legislation from this era applies to many of the biggest environmental issues in the United States today. For dilemmas like the Tijuana River's pollution problem, one of the key U.S. laws is the Clean Water Act of 1977. For much of U.S. history, water law was left to local and state governments. Some early exceptions include the Rivers and Harbors Act of 1899, which restricted dumping and discharges into navigable waters (Figure 3.13); the Public Health Service Act of 1924, which allowed the federal government to investigate cases of water pollution harming public health; and the Oil Pollution Act of 1924, which outlawed the dumping of oil into coastal waters.

Prior to the passage of legislation like the Clean Water Act, pollution problems were primarily subject to *tort law* (law addressing harm caused one entity by another), and individuals suffering external costs from pollution were limited to seeking redress through lawsuits. The ruling in the case *Boomer v. Atlantic Cement Company* in 1970 effectively ended tort law's viability as a tool for preventing pollution. The court decided that even though Albany, New York, residents were

suffering pollution from a neighborhood cement plant, the pollution could continue because the economic costs of controlling the pollution were greater than the economic costs of the damage the pollution caused.

Figure 3.13 In the 19th and early 20th century, harbors around the country, including that of New York City (shown here), were badly polluted. The Rivers and Harbors Act of 1899 was the first attempt to address such environmental problems.

The flaming waters of the Cuyahoga, however, indicated to many people that tough legislation was needed. With the Federal Water Pollution Control Acts of 1965 and 1972, and then the Clean Water Act, waterways finally began to recover. These acts regulated the discharge of wastes, especially from industry, into rivers and streams. The Clean Water Act also aimed to protect wildlife and to establish a system for granting permits for the discharge of pollutants.

Weighing the Issues:
Boomer vs. Atlantic Cement and
the End of the Tort Era

Do you agree with the court's decision in Boomer v. Atlantic Cement Company *to weigh costs of pollution versus costs of shutting down the plant? Why or why not? How easy do you think it is to calculate such costs? Can you think of any situations in which accurately calculating the costs of pollution would be difficult or perhaps impossible?*

The social context for environmental policy changes over time

Historians of environmental policy have suggested that these accomplishments, such as the Federal Water Pollution Acts and the Clean Water Act, occurred in the 1960s and 1970s because three factors converged. First, evidence of environmental problems became widely and readily apparent; second, people could visualize policies to deal with the problems; and third, the political climate was ripe, with a supportive public and leaders willing to act.

By the 1980s, the political climate in the United States had changed. While public support for the goals of environmental protection remained high, many citizens and policy experts felt that the legislative and regulatory means used to achieve environmental policy goals too often imposed economic burdens on businesses and personal burdens on individuals. Efforts were made to roll back or reform environmental laws between 1980 and 1992.

As the United States retreated from its leadership in environmental policy during the years surrounding the turn into the 21st century, other nations were drastically increasing their political attention to environmental issues. The 1992 Earth Summit at Río de Janeiro, Brazil, was the largest international diplomatic conference ever held, drawing representatives from 179 nations and unifying the leaders around the idea of sustainable development (Chapter 1 and Chapter 20). As the world's nations continue to feel the social, economic, and ecological effects of environmental degradation, environmental policy will without doubt become a more central part of governance and of everyday life in all nations in the years ahead.

Approaches to Environmental Policy

Criticism by people in the United States and elsewhere of environmental policy and the means by which it often has been implemented has spurred political scientists, economists, and others to come up with options for attaining environmental policy goals. The most widely developed alternatives contrast with the traditional regulatory approach as they involve the creative use of economic incentives.

Many environmental laws and regulations aiming to reduce pollution have simply set strict legal limits and threatened punishment for violations of these limits, in what is sometimes called a **command-and-control** or *end-of-pipe* approach.

The command-and-control approach has resulted in some definite successes, as evidenced by the cleaner air and water U.S. residents enjoy today. Without doubt the global environment would be in far worse shape were it not for government regulatory intervention. However, many people have grown disenchanted with the top-down, sometimes heavy-handed nature of the command-and-control approach, and it is clear that government intervention sometimes fails every bit as badly as markets can fail.

Sometimes government actions are well-intentioned but not well-enough informed, and lead to unforeseen consequences. The predator-control policies that Aldo Leopold first supported, and later opposed, are a classic example (Chapter 2); killing wolves and mountain lions may have saved a few livestock, but we now have enough ecological knowledge to know that it has created other problems, like a population explosion of deer that has changed the vegetative structure of forests across North America.

Policy can also fail if a government does not live up to its stated responsibilities to protect its citizens or treat them equitably. This may occur when interest groups (small groups of people seeking private gain) work against the larger public interest by seeking to influence government officials. Finally, the command-and-control approach can fail in the sense that it may generate opposition among citizens to government policy. If citizens view laws and regulations primarily as restrictions on the freedom of individuals, those policies will not last long in a constitutional democracy.

Green taxes offer financial rewards and punishments

The frequent failure of the command-and-control approach has led many economists to advocate the use of market-based incentives to correct for market failure. Such approaches attempt to "internalize" external costs by making those costs part of the original transaction, part of the overall cost of doing business. The most common types of incentive are so-called **green taxes**, charges on environmentally harmful activities and products. By taxing activities and products that cause undesirable environmental changes, a tax becomes a policy tool as well as simply a way to fund government. Green taxes have yet to gain widespread support in the United States, although similar "sin taxes" on cigarettes and alcohol have become tools of U.S. social policy. Taxes on pollution have been widely instituted, however, in Europe, where many nations have adopted the "polluter pays principle." This principle specifies that the price of a good or service should include its entire cost, including costs of environmental degradation that would otherwise be passed on as external costs.

In levying a green tax , a government must first assess the monetary value of a given reduction of pollution, and then set a tax rate per unit emission. A factory that pollutes a waterway would pay taxes based on the amount of pollution it emits. The idea is to give companies a financial incentive to reduce pollution by internalizing the cost of pollution, while allowing the polluter the freedom to decide how best to minimize its expenses. One polluter might choose to reduce its pollution if doing so is less costly than paying the taxes. Another polluter might find abating its pollution more costly, and could choose to pay the taxes instead—funds the government might then apply toward mitigating pollution in some other way.

Using market-based incentives in this way can thus protect the environment while minimizing overall costs to industry and easing concerns about the intrusiveness of government regulation. Such techniques also give industry incentives to develop new pollution-control technologies. Furthermore, they provide incentives for industry to lower emissions not merely to a level spelled out in a regulation, but to still lower levels. However, green taxes do have disadvantages. Among these are that much of the tax often gets passed on to consumers, that

this portion of the tax may affect low-income consumers relatively more than high-income ones, and that it can be difficult to measure an industry's pollution output.

Creating markets in permits can save money and produce results

A still more creative market-based approach is for government to sell to companies the right to pollute, by establishing markets in tradeable pollution permits. After determining the overall amount of pollution it will allow an entire industry to produce, the government can then issue permits to polluters that allow them to emit a certain fraction of that amount. Polluters would then be allowed to buy, sell, and trade these permits with other polluters. With such **marketable emissions permits**, governments create incentives for firms to reduce their pollution in a way that is compatible with market capitalism.

Suppose, for example, you were a plant owner with permits to release 10 units of pollution, but you found that you could become more efficient and release only 5 units of pollution instead. You would then have a surplus of permits, which might be very valuable to another plant that was having trouble reducing its pollution or that wanted to expand its production. In this case, you could sell your "extra" permits, which would meet the needs of the other factory and generate additional income for you, while preventing an increase in the total amount of pollution released. In addition, environmental organizations could buy up surplus permits and "retire" them, thus reducing the overall amount of pollution produced.

Such a system of marketable emissions permits has been in place in the United States, established by 1990 amendments to the Clean Air Act that mandated reduced emissions of sulfur dioxide that cause acid rain. Starting in 1995, permits were issued to power plants, allowing fewer emissions gradually year by year. Los Angeles had success in the 1990s with a similar program to reduce its smog. Marketable emissions permits generally end up costing both industry and government much less than a conventional system of regulation. Savings from the Clean Air Act permits have been estimated to add up to several billion dollars per year.

A tradeable emissions permitting system has also been widely discussed for managing carbon dioxide emissions internationally. Under the 1992 U.N. Framework Convention on Climate Change, and later the 1997 Kyoto Protocol (Chapter 12), nations have proposed setting up a system whereby nations could trade carbon permits, while the numbers of permits drop yearly by designated amounts. A number of countries, such as Canada, Denmark, Norway, Australia, New Zealand, and the European Union, have already begun preparing for such a system.

Market incentives are being tried widely on the local level

You may well have already taken part in transactions involving financial incentives as policy tools. Many municipalities charge residents for waste disposal according to the amount of waste they generate. Others place taxes or disposal fees on items like tires and motor oil that require costly safe disposal. Other cities give rebates to residents who get water-efficient toilets, because this may cost the city less than upgrading its sewage treatment systems. Similarly, electric utilities have offered discounts to customers buying electric-efficient lightbulbs and appliances, because doing so is cheaper for the utilities than expanding the generating capacity of their plants.

Weighing The Issues:
Environmental Policy Approaches

Imagine a factory is polluting a river near where you live, you decide to complain to your Congressperson, and she promises to try to help fix the problem. What would you urge your Congressperson do first? If her initial action does not address the problem, which approach to environmental policy would you recommend that she pursue next, and why?

Many subsidies are harmful

By far the most widespread economic policy tool is the subsidy. A **subsidy** is a government giveaway of cash or publicly owned resources, or a tax break, that is intended to encourage a particular activity. For well over a century, the U.S. government has used subsidies to encourage certain activities while discouraging others. Subsidies can be used to promote sustainable activities, but all too often they have been used to assist unsustainable ones. Subsidies judged to be harmful to the environment and to the economy total roughly $1.45 trillion yearly across the globe—an amount larger than the economies of all but five nations, according to British environmental scientist Norman Myers. The average U.S. taxpayer pays $2,000 per year in environmentally harmful subsidies, plus $2,000 more through increased prices for goods and through degradation of ecosystem services. U.S. subsi-

dies for industries and activities promoting automobile transportation alone amount to $1,700 per year, and the country's heavily subsidized gasoline is now cheaper than bottled water. The annual Green Scissors Report estimates that in 2002, $54 billion of U.S. taxpayers' money was budgeted for environmentally harmful subsidies. The Green Scissors Report is a project of 22 nongovernmental organizations such as Friends of the Earth, Taxpayers for Common Sense, and the U.S. Public Interest Research Group. Among the 78 subsidies highlighted in the 2002 Report are the following:

- *Hardrock Mining Act of 1872.* During 2000, mining companies extracted close to $1 billion in minerals from U.S. public lands without paying a penny in royalties to the taxpayers who own these lands. Since this law was enacted, the U.S. government has given away over $245 billion of mineral resources, and mining activities have polluted more than 40% of the watersheds in the West. The 130-year-old act still allows mining companies to buy public lands for $5 or less per acre.
- *Coal subsidies.* From 1984 to 2002 Congress has made $1.8 billion available to the coal industry through the "Clean Coal" Technology Program. In 2001 the House and Senate approved $2 billion more to promote coal combustion as a source of electricity, while paying for industry research that may lead to technologies to reduce certain types of air pollution associated with coal combustion.
- *U.S. Forest Service road-building subsidies.* The U.S. Forest Service manages taxpayer-owned forests for many uses. Logging is one of these uses, and it is heavily subsidized. Although many people assume that the corporations that harvest trees from public forests cover the costs of roads required to get the logs out, this is not the case. In fact, from 1992 to 1997 the Forest Service spent more than $387 million in tax dollars to build roads for logging companies (Figure 3.14).

Advocates of sustainable resource use have long urged that governments subsidize environmentally sustainable activities instead. To some extent this is being done; subsidies for developing renewable energy sources, such as solar and wind power, totaled $346 million in the United States in 1999, according to the U.S. Department of Energy. But this amount pales in comparison to the $54 billion figure quoted in the Green Scissors Report. How this balance may change in the future depends critically on the public's involvement in the political process.

Figure 3.14 When companies cut timber in U.S. national forests, roads must be built and maintained to enable access. The costs of these roads are paid not by the companies but by taxpayers' dollars.

The Environmental Policy Process

Anyone can become involved in helping ideas become public policy; in the U.S. system, it is true that each and every person has a political voice and can make a difference. Unfortunately, it is also true that money wields influence and that some people and organizations are more politically connected, and thus more influential, than others. We will explore some of the ways people make themselves influential as we examine the main steps of the policymaking process. Our discussion will pertain both to citizens at the grassroots level and also to large, well-connected organizations and corporations.

The environmental policy process begins when a problem is identified

The first step in the environmental policy process is to identify an environmental problem (Figure 3.15). Identifying a problem requires curiosity, observation, record keeping, and an awareness of our relationship with the

1 Identify problem

2 Identify specific causes of the problem

3 Envision solution and set goals

4 Get organized

5 Cultivate access and influence

6 Manage development of policy

Figure 3.15 Understanding the policy process is an essential element of solving environmental problems.

Spotting Sewage via Satellite

For decades, San Diego's sewage-contaminated coast remained a stubborn mystery—until scientists turned to the sky.

In the late 1990s, water quality along San Diego's southern beaches, which had started a steady decline in the 1960s, was supposed to be improving for surfers and swimmers. Some key sewage sources had been pinpointed along the U.S.-Mexico border, and the U.S. government had spent $260 million to build a new sewage treatment plant in the Tijuana River Valley and clean up coastal water.

Yet bacteria levels in coastal water sometimes remained surprisingly high. Beaches in San Diego County were either closed or posted with pollution warnings 877 times in 1998 and 685 times in 1999. The city of Imperial Beach, near the border, saw its beach closed 161 days in 1998 because of raw sewage contamination. Researchers and policymakers wondered what the earlier cleanup efforts had missed.

Were existing sewage treatment plants not operating as well as planned? Had past tracking efforts overlooked some pollution sources? And could new types of imaging technology help find the answer?

In 1999 local water quality officials got together with environmental scientists at Ocean Imaging, Inc., a company based near San Diego, and came up with a hypothesis—and a plan. There were probably undetected pollution sources along the U.S.-Mexico border, the researchers surmised, and those sources were large enough to cause widespread contamination. They also thought such pollution flows might be traceable by using an innovative type of airborne imaging technology, known as synthetic aperture radar, or SAR.

SAR was originally developed as a military tracking tool, but through a joint effort between Ocean Imaging and NASA, the technology is now used to study environmental problems. Once installed aboard an airplane or satellite, SAR instruments bounce

pulses (or radar signals) off Earth's surface and record the resulting echo. That echo changes in frequency, based on the substance being scanned—for example, clean ocean water will return one type of signal, whereas large bodies of spilled sewage or oil suspended in that water will usually return a different type. If a large flow of sewage is spilling directly into the ocean, the mixing of water and sewage will create an area of turbidity, where the water is stirred up, and that disturbance also returns a unique signal when tracked by SAR.

These different signals, when put together, allow researchers to develop a detailed image of the surface they are studying. Sewage flows or oil spills will often show up on SAR images as dark plumes or patches in ocean water. And SAR carries one other bonus—unlike other types of radar, SAR can operate through clouds or at night, making it invaluable in tracking something as unpredictable as a sewage spill.

Using SAR to scan the coast along the U.S.-Mexico border,

environment. For example, identifying and defining the contamination of beaches in southern California and Mexico requires understanding the ecological and health impacts of untreated sewage, being able to detect contamination on beaches, and understanding water flow dynamics and the relationship between the beaches, the Pacific Ocean, and the Tijuana River watershed (see The Science behind the Story).

Identifying specific causes of the problem is the second step in the policy process

Once an individual or group has defined a particular environmental change as a problem, discovering specific

causes of the problem is next on the agenda. A person seeking causes for pollution in the Tijuana River watershed might notice, for one thing, that transboundary sewage spills took on a more toxic and industrial dimension during the mid-1960s, when U.S.-based companies began opening *maquiladoras* on the Mexican side of the border. Advocates of the *maquiladora* system argue that these factories provide much-needed jobs south of the border while keeping companies' costs low by paying Mexican workers far less than U.S. workers. Critics argue that the factories are waste-generating, water-guzzling sources of environmental problems that are particularly difficult to deal with because of their transboundary nature.

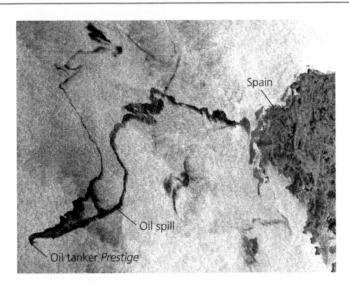

The same radar technology that has mapped San Diego's and Tijuana's sewage flows has also been used to track oil spills at sea. Here, a synthetic aperature radar image shows the sinuous trail of oil spilling from the tanker *Prestige* that broke apart off the Spanish coast in November, 2002.

scientists quickly found what they had predicted—a little-known source of almost entirely raw sewage dumping into the surf in northern Mexico. The sewage, most likely from residential areas, was flowing through a contaminated creek at a rate of 25–35 million gallons a day. Once it entered the ocean, the sewage flow showed up in the satellite image as a dark plume of disturbed water that, carried by currents, stretched north along the San Diego coast. Water samples taken at the same time confirmed that high levels of bacteria followed the plume's path.

Ocean Imaging's sewage findings, made public in 2000, gave San Diego badly needed answers to its pollution puzzle. Local environmental policymakers used the report to lobby for more sewage tracking and cleanup efforts along the border. The SAR research gave a major push to the Tijuana River Valley Estuary and Beach Sewage Cleanup Act, as well as the new cross-border master plan to improve water quality in Tijuana. Ocean Imaging is now involved in regular sewage detection efforts along the San Diego coast.

By unveiling what had been the invisible source of a persistent problem, radar-based pollution tracking has allowed San Diego to get a full grip on cleaning up its contaminated beaches. Now everyone from environmental scientists to surfers working for cleaner water can "actually visualize what is going on," as a Congressman from San Diego said when the first SAR sewage images were shown. "This takes water monitoring into this millennium."

The third step in the policy process is envisioning a solution

The better one can identify specific causes of a problem, the more effectively one will be able to envision solutions to it and later argue for implementing those solutions. A key question to ask is what particular changes might be required to eliminate the problem. In San Diego, citizen activists wanted Tijuana to better enforce its own pollution laws—something that, once clearly visualized, began to happen when San Diego city employees started training and working with their Mexican counterparts to keep hazardous wastes out of the sewage treatment system.

Getting organized is the fourth step in the policy process

When it comes to gaining the ear of elected officials and influencing policy, organizations are generally more effective than are lone individuals. The sole critic or crusader is easily dismissed as a crackpot or troublemaker, but a group of hundreds or thousands of individuals is not as easily dismissed. Furthermore, organizations are more effective at raising funds, which by U.S. law they are permitted to contribute to political campaigns.

As effective as organizations can be, it is important to remember that even individual citizens who are committed, informed, and organized can solve environmental

The Science behind the Story

Assessing the Environmental Impact of Treating Transboundary Sewage

In 1990 the United States and Mexico formally agreed to construct a water treatment plant in the U.S. portion of the Tijuana River valley. Before the South Bay International Water Treatment Plant (IWTP) could be built, however, the U.S. government was legally required by the National Environmental Policy Act (NEPA) to assess its environmental impact.

NEPA's environmental impact statement (EIS) process provides the framework for environmental impact assessment but it is not the only law that the IWTP's planners had to take into account. The federal Clean Water Act, for example, requires that harmful toxins and bacteria be removed from wastewater discharged into U.S. rivers, lakes, and oceans, and the Endangered Species Act protects species such as the Pacific pocket mouse, an inhabitant of the Tijuana River valley. International treaties also constrained the IWTP. For exam-

ple, one 1989 agreement between the United States and Mexico required that the plant be funded by the EPA, be built on U.S. territory, and treat at least 1,095 l/s (25 million gal per day). Finally, state and local laws regulated the impact of the plant on nearby communities and on the quality of California's coastal waters. Before construction could begin, all these constraints had to be addressed by the EPA and the U.S. section of the International Boundary and Water Commission, the organization that would own and operate the plant.

In 1991 an EIS draft for the IWTP was released for public comment. The draft provided a preliminary assessment of the plant's impact on biological and cultural resources, public health and safety, scenic and recreational values, water quality, and other environmental factors. Three years later, after extensive research and public discussion, a final EIS was released. In it, the EPA and International

Boundary and Water Commission endorsed the plant as the "preferred alternative"—the most environmentally friendly option—and the EPA confirmed the endorsement with an official Record of Decision.

In its Record of Decision, the EPA made several controversial choices about the IWTP's design and operation. One choice concerned opening the plant in phases. In the first phase, which was expected to last for 2 years, large solids and some suspended particles would be filtered out of the wastewater using an advanced primary treatment process. However, other pollutants—including toxic metals and bacteria—would remain untreated until facilities for a secondary treatment process known as activated sludge, where microorganisms convert highly toxic waste into a less toxic substance, were constructed. As long as those facilities remained uncompleted, the plant would be

problems. San Diego resident Lori Saldaña is one such citizen. Concerned about the Tijuana River's pollution, Saldaña reviewed plans for the international wastewater treatment plant (see The Science behind the Story) the U.S. government proposed to build, and concluded that it would merely shift the pollution from the river to the ocean, where sewage would be released 5.6 km (3.5 mi) offshore. Working with her local Sierra Club chapter, Saldaña protested the plant's design and participated in a lawsuit that forced the government to conduct further studies. The EPA finally agreed to a design change, but funding for it stalled in Congress; now all parties are exploring further options. For her efforts Saldaña has received awards and was appointed to a commission on border environmental issues by President Bill Clinton. After a decade of activism, Saldaña chose to run for the California State Assembly. As Margaret Meade, the renowned anthropologist, once re-

marked, "Never doubt that a small group of thoughtful, committed citizens can change the world—indeed it is the only thing that ever has."

Gaining access to political powerbrokers is the fifth step in the policy process

The fifth step in the policy process entails gaining access to and influence over the policymakers who have the power to help enact the changes you desire. People gain such access and influence through lobbying, campaign contributions, and the revolving door.

Lobbying Anyone who spends time or money trying to change an elected official's mind is engaged in **lobbying.** The term was originally used to describe the activities of corporate representatives who waited in the lobbies of Washington, D.C. establishments for opportunities to

releasing polluted water into the ocean, a clear violation of the Clean Water Act.

In response to the EIS and Record of Decision, two environmental groups brought a lawsuit against the EPA and International Boundary and Water Commission. In their suit, the groups argued that the EIS had failed to consider a type of secondary treatment facility known as a completely mixed aerated pond system. The lawsuit, which was eventually settled out of court, helped spur the EPA and International Boundary and Water Commission to conduct a supplemental EIS in which they considered seven different secondary treatment alternatives, including activated sludge and the pond system. Compared to activated sludge, the pond system offered several advantages. First, it produced a smaller volume of toxic byproducts. Second, monitoring of Tijuana's wastewater suggested that spikes of extremely high

toxicity would occasionally flood the treatment system, possibly upsetting its chemical and biological balance. Because the pond system contained a large quantity of stored water, it would be better able to absorb the effects of such spikes. Based on these considerations, the EPA eventually decided to endorse the pond system.

In the supplemental EIS, the EPA also reaffirmed its decision to release treated wastewater through the South Bay Ocean Outfall, a 5.8-km (3.6-mi) underwater tunnel with outlets 29 m (95 ft) below the ocean surface. The decision was based on a computer model of ocean currents and pollution levels off the California coast, which indicated that wastewater released through the outfall—once it had been treated to the secondary level—would be sufficiently diluted to meet federal and state pollution standards.

Even though the formal EIS process has now largely been

completed, scientific findings continue to shape the future of the IWTP. Tests conducted in 1997 and 1998 indicated that wastewater treated to the advanced primary level was still acutely toxic to fish and other marine life. The findings suggested that Tijuana's sewage would require more aggressive treatment than most wastewater generated in the U.S. In 1999, local activist Lori Saldaña and oceanographer Tim Baumgartner studied water quality above the 29-m (95-ft) deep outlet. They found a noticeable decrease in water quality compared to nearby areas. By posting their results on the Internet, Saldaña and Baumgartner have tried to urge policymakers to speed construction of the IWTP's secondary treatment facilities, which are now scheduled to be completed by 2007. The federal government is also funding a more formal monitoring process to ensure that the plant's outflow does not damage California's coastal environment.

talk with members of Congress. Although anyone can lobby, not everyone has the access, influence, and financial backing to lobby effectively. It can be much more difficult for an ordinary citizen to lobby than for the thousands of full-time professional lobbyists employed by the many businesses and organizations seeking a say in Washington politics.

Opponents of environmental advocacy often paint large environmental organizations, such as the National Audubon Society, as "special interest groups" with inordinate lobbying power in Washington. But according to a 2002 report published by *Fortune* magazine, these organizations are surprisingly weak. The highest-ranking environmental organization on *Fortune's* list, the Sierra Club, ranked only at #52, in the vicinity of such nonenvironmental groups as the Distilled Spirits Council of the United States and the National Association of Letter Carriers, and far back

from heavyweights such as the National Rifle Association, the American Association of Retired Persons, and the National Federation of Independent Businesses. In a further illustration, the American Petroleum Institute spends on lobbying nearly as much as the entire budgets of the top five U.S. environmental advocacy groups combined.

Weighing the Issues:
The Lobbying Power of Environmental Organizations

Does it surprise you that the highest-ranking environmental organization is ranked at #52 in the Fortune *survey of lobbying effectiveness? Would you have expected a higher or lower ranking? What kinds of factors do you think go into making an organization an influential lobbying force?*

Figure 3.16 People frequently move between working for industry and the government agencies that regulate industry, in a process that has been termed "the revolving door."

Government   The revolving door   Industry

Campaign contributions For those of us who can't spend our time hanging around Washington lobbies, supporting a candidate's reelection efforts with our money is another way to get our voice heard. Any individual can donate money to political campaigns. Because environmental laws often regulate the activities of corporations, corporations have a strong interest in shaping them. Corporations may not legally make direct campaign contributions; however, they are allowed to establish political action committees (PACs) for that purpose. Generally affiliated with corporations, environmental groups, and other organizations with an interest in election outcomes, PACs raise money and distribute it to political campaigns, helping like-minded candidates win elections, in hopes of gaining access to those individuals once they have become elected officials.

The revolving door Another way in which political access and influence is gained is when former employees of businesses in government-regulated industries take jobs with the very government agencies responsible for regulating their industry. Businesses also often hire former government bureaucrats who once regulated those businesses' industries. The movement of powerful officials between the private sector and government agencies is known as the **revolving door** (Figure 3.16). For example, the cabinet of the George W. Bush administration included a commerce secretary who was CEO of an oil and gas company, a transportation secretary who worked for a leading corporation in the transportation industry, an agriculture secretary who served on the board of the first company to market genetically engineered foods in the United States, a chief of staff who was a head lobbyist and CEO in the auto industry, and a national security adviser who served 10 years on the board of an oil company.

Supporters of the revolving door system maintain that private-sector leaders who take government jobs bring with them an intimate knowledge of their industry and can therefore benefit the nation by making especially well-informed policy. Critics of the system contend that industry leaders who become regulators will be biased toward assisting their industry and may water down or refuse to enforce existing regulations, thus hurting the interests of the taxpayers who pay their salaries.

Weighing the Issues:
Pros and Cons of the Revolving Door

Proponents of the revolving door assert that corporate executives who take government jobs regulating the industry they previously worked for are, because of their experience and inside knowledge, often the best-qualified people for the job. Critics of the system say taking a job regulating your former employer is a clear conflict of interest that undermines the effectiveness of the regulatory process. What do you think of each of these arguments? Would the citizens of the United States be better off with or without the revolving door?

Shepherding a solution from concept into law is the sixth step in the policy process

Whether you're a corporate lobbyist or a grassroots activist, once your organization has the access and clout required to shape decisions of policymakers, the better-known parts of the policy process come into play. Having gained access to elected officials and convinced them to hear your requests, you may be asked to prepare a **bill**, or draft law, that embodies the solutions you seek. Technically, anyone can draft a bill; the hard part is finding members of the House and Senate who are willing to introduce the bill and shepherd it from subcommittee through full committee and on to passage by the full Congress. Lobbying and the application of pressure in the media continue and intensify as the bill progresses through this process (Figure 3.17). If it passes through all of these steps, the bill may become law, but it can die in countless fashions along the way.

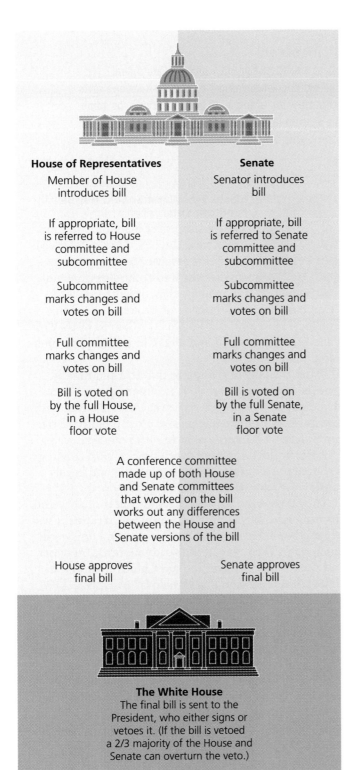

House of Representatives

Member of House
introduces bill

If appropriate, bill
is referred to House
committee and
subcommittee

Subcommittee
marks changes and
votes on bill

Full committee
marks changes and
votes on bill

Bill is voted on
by the full House,
in a House
floor vote

Senate

Senator introduces
bill

If appropriate, bill
is referred to Senate
committee and
subcommittee

Subcommittee
marks changes and
votes on bill

Full committee
marks changes and
votes on bill

Bill is voted on
by the full Senate,
in a Senate
floor vote

A conference committee
made up of both House
and Senate committees
that worked on the bill
works out any differences
between the House and
Senate versions of the bill

House approves
final bill

Senate approves
final bill

The White House
The final bill is sent to the
President, who either signs or
vetoes it. (If the bill is vetoed
a 2/3 majority of the House and
Senate can overturn the veto.)

Figure 3.17 Before a bill becomes a law, it must pass through a number of hurdles in both legislative bodies. If the bill passes the House and Senate, a conference committee must work out any differences between the House version and the Senate version before the bill is sent to the president. The president may then sign or veto the bill.

However, for a policymaker, the enactment of a law or the issuing of a regulation is not the end of the policy process. To describe the full process as it typically occurs, both before and after enactment, students of policy have identified several stages making up what is known as the **policy cycle:**

1. *Agenda setting:* perceiving and defining problems
2. *Policy formulation:* developing goals; examining strategies for action to meet goals
3. *Policy legitimation:* justifying policy action by mobilizing support; formally enacting policy
4. *Policy implementation:* administrating policy; putting programs into effect
5. *Policy evaluation:* assessing successes or failures of policy and programs
6. *Policy change:* revising goals and/or means as necessary

The policy cycle is a long and sometimes cumbersome process, but it has resulted in a great deal of effective governance in constitutional democracies in the United States and many other nations.

International Environmental Law

Although the details of the policy process vary from one nation to another, the fundamental steps and principles of the process are similar in all constitutional democracies. Environmental problems, by their nature, however, often are not limited to the confines of particular countries. As you recall, the problems along the Tijuana River are international in nature. In fact, the majority of the world's major river systems cross international borders. As the Tijuana River watershed illustrates, environmental problems pay no heed to political boundaries. Because U.S. law has no authority in Mexico or any other nation outside the United States, international law is vital to solving such transboundary problems.

Mexico and the United States are working together to improve water management

Often progress is made on international issues not through legislation per se but through creative agreements hammered out after a lot of hard work and diplomacy. Such was the case with the successful effort to develop a long-term plan to manage drinking water and wastewater

VIEWPOINTS

Public vs Private: In Whose Best Interest?

Developing satisfactory and long-term solutions to transboundary pollution presents challenges to both citizens and policymakers in affected countries. Frequently these situations give rise to debates over the best management approach—for example, public vs. private, or some combination thereof. In your view, what are the appropriate roles of the public and the private sectors in resolving the sewage pollution problems of the San Diego–Tijuana region?

Public/Private Partnerships Mean Faster and More Efficient Environmental Change

A policy that excludes private participation in public service projects is bad policy if the government is unwilling or unable to fund project implementation.

It is a philosophical debate, not a technical question: "Should the private sector, motivated by profit, be able to participate in providing public services that are traditionally the exclusive realm of the public sector?" The private sector is more innovative, efficient, and flexible. The public/private partnership model (PPV) also provides for full disclosure, transparency, public comment, oversight, and regulatory control. The private sector can more often implement projects faster, and at a lower cost, than a lumbering government. PPV models incorporate a benefit unavailable in public sector projects: the ability to regulate by contract (as well as by statute). Desired environmental benefits are attained, while profit invigorates the market.

The Bajagua Project, a proposed PPV development model wastewater treatment plant to be built in Tijuana, Mexico, is such a project. It can be built faster and for less than any public facility. Bajagua will utilize technology selected by the U.S. EPA. It must comply with NEPA and all U.S., Mexican, and California environmental laws. Bajagua must also satisfy all contractual obligations BEFORE payment is due from the government, more closely emulating the "pay for services rendered" transaction model that we utilize every day. A policy where the public gets the environmental benefit it desires before it pays is a good policy.

In 1993, Congress capped the amount to be spent on the International Wastewater Treatment Plant. The federal agency which built the primary module in 1999 exhausted its funding and now cannot build the secondary module. Yet, the San Diego beaches remain polluted, local economies depressed, and the federal agency remains in violation of the Clean Water Act. Clearly, PPV offers the only real possibility of success for positive environmental change.

Gary L. Sirota, Esq. is a complex business and environmental litigator, international PPV infrastructure development consultant, specializing in legislative finance and policy analysis.

Clean, Safe Water Should Not Be a For-Profit Enterprise

The persistent problem of water pollution near the U.S.-Mexico border is partly the result of breakdowns in engineering and technology, including inadequately designed sewage treatment plants and missing or broken collection pipelines. However, it also reflects a shared history of water mismanagement and a lack of cooperative binational planning efforts.

Water is vitally important to this arid and rapidly growing region, yet over 90% is imported, used once, then dumped into the sea via sewage outfalls. Renegade sewage flows from these pipelines often pollute the river and beaches. Ultimately, to have sufficient clean water for basic health and future development, both countries will need to reconsider this pattern of use and begin conserving, reusing, and reclaiming water.

The best solutions for protecting water quality and ensuring adequate supply and sufficient treatment will be based on international cooperation and long-term planning projects. Japan is already investing in water and wastewater projects in Baja California and the United States is beginning to help by using U.S. EPA grants to develop long-term master plans for improving water quality in the region.

However, changes in attitude and water use require support from governmental, research, and community organizations. Border residents need to understand not only their rights to the water they use, but also their role in keeping it clean and using it responsibly.

Relying on private for-profit companies to solve water problems—especially if they do not involve the community in the planning and decision-making process—can result in higher costs and fees that many families cannot afford, and will do little to change the public mindset. Public access to clean, safe water is a basic human right and is needed to protect public health. If companies fail to make a profit, who will step in to ensure that safe drinking water is available and sewage is treated?

Lori Saldaña is an environmental policy researcher, writer, and community activist, specializing in water issues along the U.S.-Mexico border. She is a candidate for State Assembly in California (March 2004).

in the Tijuana metropolitan area. This master planning process, funded by the U.S. Environmental Protection Agency (EPA) under Congressional direction, began in January 2002 and involved the Comision Estatal de Servicios Publicos de Tijuana (CESPT), the city agency that manages water and wastewater in Tijuana; the Mexican National Water Commission; the State Water Commission for Baja California; and the North American Development Bank. The resulting Tijuana Master Plan for Water and Wastewater Infrastructure intends to address the root of the region's water shortage and pollution problems, including the shortage of drinking water and possibilities for its reuse; water infrastructure; wastewater collection and transport; and wastewater treatment.

In 2002 the U.S. Congress also provided funding and authority for the EPA to work with CESPT to upgrade Tijuana's sewer system. The Tijuana Sewer Rehabilitation Project, known as Tijuana Sana ("Healthy Tijuana") was approved in late 2001. This cooperative transboundary pollution prevention program will attempt to repair leaky sewer pipes in Tijuana. The $43-million 4-year project will replace 131 km (81 mi), or 7.5%, of Tijuana's sewer pipes. If it succeeds, the project should reduce or eliminate the most severe sewage spills into the Tijuana River.

International environmental law includes conventional law and customary law

As the case of the Tijuana River illustrates, solving environmental dilemmas often requires international cooperation. Several principles of international law and a number of international organizations have arisen in response to this realization. Whereas U.S. law arises from the Constitution and the Bill of Rights, international environmental law is a bit more nebulous in its origins and authorities.

International law known as **conventional law** arises from conventions, or treaties, that nations agree to enter into. Examples (discussed in Chapter 11 and Chapter 12) include the Montreal Protocol, a 1987 accord among more than 160 nations to reduce the emission of airborne chemicals that deplete the ozone layer; and the 1997 Kyoto Protocol to reduce fossil fuel emissions that contribute to global climate change, which in late 2003 was nearing international ratification without the United States. To deal with the Tijuana River situation, in 1990 Mexico and the United States entered into a treaty and agreed to build an international wastewater treatment plant to handle excess sewage from Tijuana. The plant began operating just north of the border in 1997 and collects and treats up to 95 million l (25 million gal) of sewage each day (Figure 3.18). In this case the treaty process worked, but the results fell short of

Figure 3.18 In July of 1990, the United States and Mexico entered into a treaty and agreed to build the International Wastewater Treatment Plant (IWTP) to handle excess sewage from Tijuana. The IWTP began operating just north of the border in 1997 and treats up to 95 million l (25 million gal) of Mexican sewage each day. The IWTP has provided great benefits, but it reached its capacity within 3 years of opening.

expectations. Despite the benefits it provides, the plant reached its capacity within 3 years of opening and then began discharging material to the ocean that did not meet safety standards established by U.S. law.

Other international law arises from long-standing practices, or customs, held in common by most cultures; this is known as **customary law.** Four principles underlie customary law:

- *Good neighborliness* is based on an understanding that no nation should use its natural resources in a way that adversely affects any other nation.
- *Due diligence* is based on a belief that every nation should respect and protect the rights of other states through the prevention and reduction of pollution.
- *Equitable resource use* requires that no nation use more than its share of a natural resource.
- *The principle of information and cooperation* requires that a nation provide notice and other information to nations when its actions might affect the interests and affairs of those nations.

Several organizations shape international environmental policy

Despite the fact that there is no real mechanism for enforcing international environmental law, there are several international organizations whose actions can influence—either through funding or through applying peer pressure and encouraging media attention—the behavior of other nations. These organizations include the United Nations, the World Bank, and a wide variety of nongovernmental organizations (NGOs).

The United Nations sponsors large and active environmental agencies

On October 24, 1945, representatives of 50 countries founded the United Nations (U.N.). The purpose of this organization is "to maintain international peace and security; to develop friendly relations among nations; to cooperate in solving international economic, social, cultural and humanitarian problems and in promoting respect for human rights and fundamental freedoms; and to be a centre for harmonizing the actions of nations in attaining these ends."

Several agencies within the United Nations shape international environmental policy, most notably the United Nations Environment Programme (UNEP), created in 1972, which helps nations understand and solve environmental problems. Today it is based in Nairobi, Kenya, and its mission is sustainability, enabling countries and their citizens "to improve their quality of life without compromising that of future generations."

UNEP's extensive research and outreach activities provide a wealth of information useful to policymakers and scientists throughout the world, and have provided a great deal of the data cited throughout this book.

The World Bank holds the purse strings for development projects

The United Nations can encourage cooperation and promote awareness of environmental problems, but the World Bank has hold of the purse strings. Established in 1944 and based in Washington, D.C., the World Bank is one of the globe's largest sources of funding for economic development. This institution can shape environmental policy through its funding of major development projects, including dams, irrigation infrastructure, and other undertakings. In 2002 the Bank provided over $19.5 billion in loans for projects aimed to benefit the poorest people in the poorest countries around the world. Despite its admirable mission, the World Bank has been widely criticized for funding unsustainable projects that cause more environmental problems than they solve.

In 1987 the World Bank was reorganized, and an environmental office was established. In the years since, the Bank has made assessing environmental impacts of its projects a higher priority. In 2001 the World Bank issued its Environmental Strategy, a document intended as a guide to improve quality of life, promote sustainability, and protect regional and global commons. Providing for the needs of rapidly growing human populations in poorer countries while minimizing damage to the environmental systems people depend on can be a tough balancing act. The concept of sustainable development (Chapters 1 and 20) must be the guiding principle for such efforts, environmental scientists today agree.

The European Union is active in environmental affairs

Like the World Bank, the European Union (EU) was not established primarily with environmental problem-solving in mind. However, the treaty that established it held as one of its goals the promotion of solutions to environmental problems. Formed after World War II, the EU as of late 2003 contained 15 nations as members and was preparing for the addition of 13 more. It seeks to promote Europe's unity and its economic and social progress (including environmental protection) and to "assert Europe's role in the world." It is also devoted to removing trade barriers between member nations. Some national environmental regulations have been classified as barriers to trade because certain northern European nations have traditionally had more stringent environmental laws that have prevented the import and sale of environmentally harmful products from other member nations. Despite this potential problem, the EU's European Environment Agency works to solve problems arising from waste management, noise pollution, water pollution, air pollution, habitat degradation, and natural hazards.

Unlike the United Nations, which has no direct governing authority over its members, the EU can sign binding treaties on behalf of its members. Furthermore, the EU can enact regulations that have the same authority as the national laws in each member nation, as well as issue *directives*, which are more advisory in nature.

The World Trade Organization has recently attained surprising power

Whereas the United Nations and the EU have limited influence over nations' internal affairs, especially in the United States, the World Trade Organization (WTO) has real authority to impose financial penalties on nations that do not comply with its directives. These penalties can on occasion play a major role in shaping national and international environmental policy. The WTO, based in Geneva, Switzerland, was established in 1995, having grown out of a half-century-old international trade agreement. The WTO represents multinational corporations and is intended to promote free trade by reducing obstacles to international commerce and enforcing fairness among nations in trading practices.

Like the EU, the WTO has interpreted some national environmental laws as unfair barriers to trade. For instance, in 1995 the U.S. EPA issued new regulations requiring cleaner-burning gasoline in U.S. cities, following Congress's earlier amendment of the Clean Air Act. Brazil and Venezuela filed a complaint with the WTO, saying the new rules unfairly discriminated against the petroleum they exported to the United States, which did not burn as cleanly. The WTO agreed, ruling that despite the threat to human health in the United States, the EPA rules represented an illegal trade barrier. The ruling forced the United States to change its approach to regulating clean-burning gasoline. Not surprisingly, critics have frequently charged that the WTO aggravates environmental problems.

Nongovernmental organizations also exert influence

While national governments and a few key international organizations exert a great deal of influence over international environmental policy, so too do a number of NGOs that have grown to become international in scope. The structure, character, and operation of these advocacy groups are diverse. Some, such as The Nature Conservancy, focus on accomplishing conservation objectives on the ground (in its case, purchasing and managing land and habitat for rare species) without becoming politically involved. Other groups, including Conservation International, the World Wide Fund for Nature, Greenpeace, Population Connection, and many others, attempt to shape policy directly or indirectly through their work. As we will discover in later chapters, NGOs apply more funding and expertise to environmental problems, and conduct more research intended to solve them, than do many national governments.

Conclusion

Environmental policy is a problem-solving tool that requires an understanding of science, ethics, economics, and the political process. As we have seen in the case of the Tijuana River, science enabled us to identify the sources and impacts of the pollution, and ethics and economics enabled us to define the pollution as a problem worth rectifying. Through the hard work of concerned citizens interacting with their government representatives, the political process eventually produced promising solutions in the Tijuana River Valley Estuary and Beach Sewage Cleanup Act, and in binational agreements and management plans. We will draw on the fundamentals of environmental policy introduced in this chapter throughout the remainder of this book. By understanding these fundamentals, you will be well equipped to develop your own creative solutions to many of the challenging problems we will encounter.

REVIEW QUESTIONS

1. What is a transboundary environmental system?
2. What are three problems that pollution of beaches can cause?
3. How can science help us identify and define environmental problems?
4. Do we need environmental policy? Why or why not? List, describe, and critique two common justifications for environmental policy.
5. Explain the concept of external costs, and state why it is relevant to environmental policy.
6. List and describe the primary responsibilities of the legislative, executive, and judicial branches of the U.S. government.
7. What is the "fourth branch" of the U.S. government?
8. What is a regulatory taking?
9. What were the differences between the first, second, and third waves of environmental policy in U.S. history?
10. What did the National Environmental Policy Act accomplish? Do you agree that such an approach to environmental policy is appropriate? Why or why not?
11. Briefly describe the origins and the mission of the U.S. Environmental Protection Agency.
12. What is tort law? How might it help us solve environmental problems? How did the *Boomer v. Atlantic Cement* case affect the value of tort law as an environmental problem-solving tool?

13. What is the difference between a green tax, a subsidy, and a marketable emissions permit?
14. Describe the environmental policy process from identification of a problem through the enactment of a federal law.
15. What is political access and why is it an essential part of the environmental policy process?
16. What kinds of things can an individual citizen do to become influential in the policymaking process?
17. What is the difference between lobbying and making campaign contributions?
18. Explain the revolving door concept. What benefit might a business gain from hiring a former government regulator?
19. List and describe the six steps of the policy cycle.
20. What difficulties are present in the case of transboundary environmental problems but not present when a problem lies entirely within a single political jurisdiction?
21. What is the difference between conventional law and customary law?
22. List and describe the four principles of customary international law. Which of these principles applies to the Tijuana River case? Why?
23. What role does the United Nations play in shaping international environmental policy?
24. Why are environmental regulations sometimes considered to be unfair barriers to trade?

DISCUSSION QUESTIONS

1. Imagine that you live in the San Diego area and could not safely use beaches in your neighborhood because of water pollution that originated in Mexico. Who do you think should pay to prevent the pollution of your beaches? Should you? Should the Mexican government? Should the state of California or the U.S. government pay? Should *maquiladoras*? What are the pros and cons of each of these potential funding sources? Right now, a combination of state and federal funds from the United States and Mexico is being devoted to reducing rogue flows, but some people argue that it isn't right for taxpayers in the Midwest or New England to see their tax dollars go toward cleaning up southern California beaches. What do you think?

2. Now imagine that you live in Mexico in the Tijuana River watershed and depend on its water for your drinking, washing, and the irrigation of your garden. Who do you think should pay to prevent the pollution of your water supply?

3. Compare the main approaches to environmental policy—command-and-control, tort law, and market-based approaches. Can you name an advantage and disadvantage of each? Do you think any one approach is most effective? Could we do with just one approach, or does it help to have more than one?

4. Reflect on the causes for the transition in U.S. history from one type of environmental legislation to another. Now peer into the future and think about how life might be different in 25, 50, or 100 years. What would you speculate about the environmental laws of the future? What issues might they address? Might we have more or fewer environmental laws?

5. Think of one environmental issue that you would like to see solved through legislation. From what you've learned about the policymaking process, how do you think you could best approach pushing your ideas through the process?

6. Compare the roles of the United Nations, the European Union, the World Bank, the World Trade Organization, and nongovernmental organizations. If you could gain the support of just one of these institutions for a policy you favored, which would you choose? Why?

Media Resources *For further review go to the website* **www.envscienceplace.com** *or student CD-ROM, where you will find quizzes, flashcards, a glossary, additional interactive exercises, and links to relevant news and research sources. Also, on the website and CD-ROM is* **GRAPH IT**, *a series of interactive graphing tutorials to help you interpret graphs and plot data.*

4

From chemistry and energy to life

Exxon Valdez oil tanker in Prince William Sound, Alaska

This chapter will help you understand:

- Fundamentals of environmental chemistry
- Fundamentals of energy and matter
- Molecular building blocks of living organisms
- Photosynthesis, respiration, and chemosynthesis
- Hypotheses on the origin of life on Earth
- The theory of evolution by natural selection

Prince William
Sound, Alaska

North
America

Pacific
Ocean

Workers spraying fertilizer on oil coated beach along the Alaskan coastline

Central Case: Bioremediation of the *Exxon Valdez* Oil Spill

"There is a dramatic difference. . . . [I]t really cleaned the oil off the rock. It looked like someone brought in new rock."
—*EPA program manager Chuck Costa, describing experimental bioremediation results in 1989, shortly after the spill*

"The rush to bioremediation in Alaska was a function of the size of the problem and limited availability of options. . . . It did not turn out to be the silver bullet that many hoped it would be."
—*Alaska Department of Environmental Conservation report, 1993*

On March 24, 1989, the tanker *Exxon Valdez* struck a reef in Alaska's Prince William Sound and spilled 42 million liters (11 million gallons) of crude oil, which eventually coated 2,100 km (1,300 mi) of Alaskan coastline. It was the largest oil spill in U.S. history, and killed an estimated 100,000–400,000 seabirds, 2,600–5,500 sea otters, 200–300 harbor seals, and countless fish. The oil smothered intertidal plants and animals and defiled the area's relatively pristine

ecosystem. The local economy took a nosedive as hundreds of fishermen were thrown out of work and tourism plummeted.

The massive spill was met with a massive response. Thousands of workers employed by Exxon (now Exxon-Mobil) and by government agencies, together with local volunteers, launched a cleanup effort of unprecedented scope. The cleanup crews corralled the oil with booms, skimmed it from the water, soaked it up with absorbent materials, and dispersed it with chemicals. They pressure-washed the beaches, removed contaminated sand with backhoes and tractors, and even tried burning the oil.

Meanwhile, scientists used the opportunity to study the spill's impacts and to test a new method of mitigating them. They tried enlisting nature to help take care of the mess by encouraging naturally occurring bacteria to break down the oil. About 5% of the single-celled microbes present on Alaskan beaches feed on chemical compounds called hydrocarbons that are produced by the region's conifer trees. The hydrocarbons from the conifers are similar to the hydrocarbons that make up crude oil, so scientists expected that the microbes would be able to degrade the oil. Scientists from the EPA and Exxon decided to put the bacteria to work in a process called **bioremediation**, the attempt to clean up pollution by speeding up natural processes of biodegradation by living organisms.

Although the bacteria were presented with an abundant new food source in the form of oil washing up on the beaches, they were not immediately able to consume it. The oil contained too much of a good thing—too much carbon—but not enough nitrogen and phosphorus. To remedy the imbalance of nutrients, scientists applied a fertilizing mixture containing nitrogen and phosphorus to several beaches, leaving other areas of the shore as untreated controls. The fertilizing treatment seemed to work; bacterial numbers increased, and oil residues decreased visibly. Encouraged, scientists put the program into full swing, and by the end of the year workers had treated more than 113 km (70 mi) of contaminated beach with the fertilizer. They expanded the applications over the next two years.

Because there were many complicating factors, experts have debated how much the treatments sped up degradation of the oil (some say up to five times, some think not at all). However, the well-publicized *Valdez* operation served as a model effort, and today bioremediation is actively researched and increasingly applied. Bioremediation schemes have many practical limitations, but when they are feasible they can accomplish much good with a minimum of ecological disruption and a minimum of expense.

Chemistry and the Environment

The *Exxon Valdez* oil spill ignited a wide array of scientific, economic, political, and ethical concerns. Today many wildlife populations have recovered, but some have not; moreover, small amounts of oil remain throughout the region. The lawsuits against the company (some of which are still ongoing) testify to the economic and social concerns of fishermen of the region whose livelihoods were wrecked, and of cleanup workers who say their health was affected. In the political arena, the U.S. Congress in the year following the spill passed the Oil Pollution Control Act, which required the Coast Guard and the EPA to strengthen regulations on tankers and their operators.

At the root of all these diverse impacts, however, is the chemistry of the oil. It is the chemistry of crude oil that causes it to gum up the feathers of birds and the fur of mammals, ruining their insulating abilities and bringing on hypothermia. It is the chemistry of oil that allows it to float on water and accumulate on beaches. It is certain hydrocarbons from oil that, when dissolved in the water column or volatile in the air, are known to be harmful to wildlife and carcinogenic to humans. Yet, it is also the chemistry of crude oil that provides the energy that powers our remarkable civilization and modern way of life—a way of life that allows us the luxury to study, reflect on, and act on these very issues.

Indeed, examine any environmental issue and you will likely discover chemistry playing a central role. Chemistry is crucial to understanding how gases such as carbon dioxide and methane contribute to global climate change, how pollutants such as sulfur dioxide and nitric oxide cause acid rain, and how pesticides and other artificial compounds we release into the environment affect the health of wildlife and of people. Chemistry is central, too, to understanding water pollution and sewage treatment, atmospheric ozone depletion, hazardous waste and its disposal, and just about any energy issue.

Chemistry is also often central to developing solutions to environmental problems. Organisms have been used to clean up pollution via bioremediation, for instance, in a variety of situations in recent years. Hydrocarbon-consuming bacteria and fungi are often used to clean up soil beneath the many thousands of leaky gasoline tanks that threaten drinking water supplies. Other kinds of microbes are being used to degrade pesticide residues in soil. In addition, plants such as wheat, tobacco, water hyacinth, and cattails have been employed to clean up toxic waste sites by letting them draw heavy metal toxins, such as lead and cadmium, into their roots (see "The Science behind the Story"). Of course, the most familiar and widespread example of bioremediation is sewage treatment, which involves bacteria and, increasingly, filtration by wetland plants.

Weighing the Issues:
Pros and Cons of Bioremediation

Bioremediation techniques can be far less expensive, less environmentally intrusive, and more effective than conventional methods for cleaning up pollution. However, such techniques do not always work—and even when they do, they can sometimes be very slow, leave a job uncompleted, or introduce new problems. Imagine that

Figure 4.1 Each chemical element has a different number of protons, neutrons, and electrons. Carbon possesses 6 of each, nitrogen 7, and phosphorus 15. Protons and neutrons are held in each atom's nucleus, while electrons orbit in clouds around the nucleus.

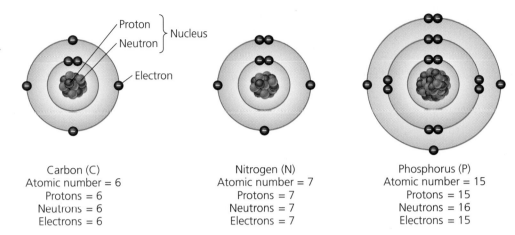

Carbon (C)
Atomic number = 6
Protons = 6
Neutrons = 6
Electrons = 6

Nitrogen (N)
Atomic number = 7
Protons = 7
Neutrons = 7
Electrons = 7

Phosphorus (P)
Atomic number = 15
Protons = 15
Neutrons = 16
Electrons = 15

an oil spill occurs tomorrow in a body of water near your college campus and that you are put in charge of the cleanup. What questions would you want to have answered before recommending bioremediation with bacteria?

Sometimes suitable plants or bacterial cultures are introduced to a site; sometimes naturally existing ones are fertilized, as was done at Prince William Sound; and sometimes the best course in mitigating pollution may be simply to monitor naturally occurring organisms as they do their work, without disrupting the system. Increasingly, scientists are seeking ways to take bioremediation to the next level: genetically engineering microbes and plants to become more efficient at the specific metabolic tasks we want to ask of them. Environmental chemists today are excited about the countless future applications of chemistry yet to be discovered that can help us solve environmental problems.

Atoms and elements are the chemical building blocks

To appreciate the complex chemistry of living things, or of such processes as bioremediation or commercial energy production, we must begin with a grasp of the fundamentals. The carbon, nitrogen, and phosphorus that played such key roles in the bioremediation of the oil spill in Prince William Sound are each **elements** (Figure 4.1). An element is a fundamental type of matter, a chemical substance with a given set of properties, which cannot be broken down into substances with other properties. Chemists currently recognize 92 elements that occur in nature, as well as more than 20 others that have been artificially created. Besides carbon and nitrogen, elements especially abundant in living organisms include hydrogen and oxygen (Table 4.1). Each element is assigned an abbreviation, or chemical symbol; refer to the periodic table of the elements (see Appendix) for more information.

Table 4.1 Earth's Most Abundant Chemical Elements, by Mass

Earth's crust	Oceans	Air	Organisms
Oxygen (O), 49.5%	Oxygen (O), 88.3%	Nitrogen (N), 78.1%	Oxygen (O), 65.0%
Silicon (Si), 25.7%	Hydrogen (H), 11.0%	Oxygen (O), 21.0%	Carbon (C), 18.5%
Aluminum (Al), 7.4%	Chlorine (Cl), 1.9%	Argon (Ar), 0.9%	Hydrogen (H), 9.5%
Iron (Fe), 4.7%	Sodium (Na), 1.1%	Other, <0.1%	Nitrogen (N), 3.3%
Calcium (Ca), 3.6%	Magnesium (Mg), 0.1%		Calcium (Ca), 1.5%
Sodium (Na), 2.8%	Sulfur (S), 0.1%		Phosphorus (P), 1.0%
Potassium (K), 2.6%	Calcium (Ca), <0.1%		Potassium (K), 0.4%
Magnesium (Mg), 2.1%	Potassium (K), <0.1%		Sulfur (S), 0.3%
Other, 1.6%	Bromine (Br), <0.1%		Other, 0.5%

Student Chemist Lets Plants Do the Dirty Work

Bacteria that break down oil are just one example of organisms that scientists are putting to work to clean up environmental pollutants. Green plants can help, too.

When soil is contaminated with heavy metals from mining, manufacturing, oil extraction, or military facilities, the standard solution is to dig up tons of soil and pile it into a hazardous waste dump. Bulldozing so much dirt, however, can release toxic chemicals into the air, waste soil, leave gaping holes, and cost as much as $2.5–7.5 million per hectare ($1–3 million per acre). As an alternative, scientists are developing methods of phytoremediation, using plants (*phyto* means plant) to remediate, or detoxify, contaminated soils. "You end up with a less expensive solution, and a greener solution, to the problem," says phytoremediation expert Peter Goldsbrough of Purdue University in Indiana.

One researcher making advances in phytoremediation is Marc Burrell, a sophomore at Rice University in Houston, Texas.

Rice University student Marc Burrell has done novel scientific research to find new ways to get plants to clean up toxic heavy metals from contaminated soil.

While still in high school in Wisconsin, Burrell, with Goldsbrough and other mentors, began researching how to get plants to remove toxic lead from soil.

Working under Greg and Maria Begonia at Jackson State University in Mississippi, Burrell ran lab experiments to test how wheat could be made to draw up lead from soil. Normally, lead is not ac-

cessible to plants, because it is tied up in compounds such as lead carbonate and lead oxide, which are not water soluble. But chemicals called chelating agents can bind to lead and make it water soluble so that plant roots can draw it up into the plant. Burrell's greenhouse experiments showed that adding the chelating agent ethylenediaminetetraacetic acid (EDTA) to the soil increased wheat's uptake of lead by about 300,000 times.

Burrell predicted that adding an acid would enhance the effect, because an acid's hydrogen ions would help dissociate lead compounds, freeing more lead ions to bind to EDTA. His experiments supported this hypothesis; when Burrell added acetic acid to the EDTA treatment, the plants took up three times as much EDTA than was the case when only the chelating agent alone was added—a total of 1% of the plants' dry biomass.

Next, Burrell wanted to better understand how genes and proteins affect lead uptake, with an eye toward bioengineering plants to accumulate lead. To deal with toxic metals they take up by acci-

Elements are composed of **atoms,** the smallest components of an element that maintain the chemical properties of that element. Atoms consist of a nucleus made of **protons** (positively charged particles) and **neutrons** (particles lacking electric charge). The atoms of each element have a defined number of protons, called the atomic number (elemental carbon has six protons in its nucleus; thus, its atomic number is 6). An atom's nucleus is surrounded by negatively charged particles known as **electrons**, which balance the positive charge of the protons.

Although all atoms of a given element consist of the same number of protons, they do not necessarily contain the same number of neutrons. Those with differing numbers of neutrons are called **isotopes** (Figure 4.2). Many elements found in living organisms, including carbon, hydrogen, oxygen, and nitrogen, occur as different isotopes. Chemically, isotopes of an element behave almost identically, but they have different physical properties because they differ from one another in mass. This fact has turned out to be very useful. Scientists have been able to use isotopes to study a number of phenomena, including the age of matter, the flow of nutrients within and among organisms, and the movement of organisms from one geographic location to another (see "The Science behind the Story"). Isotopes are denoted

dent, many plants produce and send chelating agents called *phytochelatins* to drag the metals to vacuoles, or empty cellular sacs, where they can be stashed away without harm. Phytochelatins were thought to work with many metals, but no one had tested their interaction with lead.

Working under Heather Owen at the University of Wisconsin-Milwaukee, Burrell experimented with mustard plants engineered by geneticists who had knocked out the gene responsible for producing phytochelatins. He grew engineered plants in lead-contaminated soil next to normal plants in similar soil. He found that the plants engineered not to produce phytochelatins were more susceptible to lead poisoning and died sooner. This strongly suggested that phytochelatins do, in fact, squirrel away lead into vacuoles.

Burrell then worked with tobacco engineered to overexpress the protein for producing phytochelatins. With Hillel Fromm of the University of Leeds, United Kingdom, he found that presenting the tobacco with lead caused the plants to produce still more phytochelatins, increasing the amount they could accumulate. Finally, Burrell assisted Owen in her research on phytoferritins, proteins that help regulate iron. Plants need iron for photosynthesis, but any excess is harmful and must be sequestered in vacuoles. Burrell and Owen experimented and found that phytoferritins could help sequester lead as well as iron.

Such research results are beginning to be applied in the field at contaminated sites. Once plants have accumulated metals, they can be harvested and put through a smelting procedure to recover the metals. Alternatively, the plants can be dried and disposed of at a hazardous waste site.

Phytoremediation is still a new pursuit, and it faces some hurdles. One hurdle is time; individual plants can take in only so much of a substance, and it may require 5 to 20 years of repeated plantings before the soil's metal content is reduced to an acceptable level. Also, cleanup is limited to the depth of soil the plants' roots will reach, and the metals need to be in a water-soluble form. Finally, plants that accumulate toxins can potentially harm insects that eat the plants and, in turn, animals that eat the insects.

Despite such obstacles, phytoremediation is catching on. At military bases in Iowa, Tennessee, and Nebraska, the U.S. Army Corps of Engineers is using vegetation in artificial wetlands to minimize contamination of groundwater by ammunition. A Virginia company, Edenspace, has used plants to extract lead from residential sites, arsenic from military and energy facilities, zinc and cadmium at EPA Superfund sites, and tungsten from abandoned mines. It has even extracted radioactive uranium, strontium, and cesium from U.S. military sites and the Chernobyl nuclear reactor site in Ukraine. An Iowa company, Ecolotree, has used poplar trees, legumes, and grasses at 55 landfills, wastewater treatment sites, agrochemical spill areas, and other locations in the United States and Europe. Such efforts are at the forefront of "green chemistry" today, thanks to the endeavors of bright and hard-working young researchers like Marc Burrell.

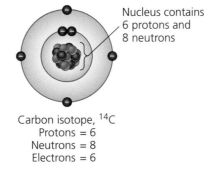

Carbon isotope, ^{14}C
Protons = 6
Neutrons = 8
Electrons = 6

Figure 4.2 Atoms of many elements exist as different isotopes, which differ in mass. Shown here is one isotope of carbon, carbon-14 (^{14}C).

by the elemental symbol preceded by the combined number of protons and neutrons in the atom. For example, ^{14}C (carbon-14) is an isotope of carbon with 8 neutrons (plus 6 protons) in the nucleus rather than the normal 6 neutrons of ^{12}C (carbon-12).

Besides differing in the number of neutrons, atoms may gain or lose electrons to become **ions,** electrically charged atoms or combinations of atoms. Ions are denoted by the elemental symbol followed by the electronic charge. For instance, a common ion used by mussels and clams to form shells is Ca^{2+}, a calcium atom that, due to the loss of two electrons, has a charge of positive 2 (Figure 4.3, page 96).

How Isotopes Reveal Secrets of Earth and Life

Isotopes, those alternate versions of chemical elements, have become one of the most powerful instruments in the environmental scientist's toolkit. They have enabled scientists interested in the past to date ancient materials, reconstruct the climate of past ages, and study the lifestyles of prehistoric humans. They have also allowed scientists focused on the here-and-now to work out photosynthetic pathways, measure animals' diets and condition, and trace nutrient flows through organisms and ecosystems.

Radiocarbon dating is the isotope use that is most familiar to most people. Because organic molecules are based on carbon, organisms are loaded with this element. The most abundant carbon isotope is ^{12}C, but ^{13}C and ^{14}C also occur. ^{14}C is radioactive and occurs in organisms in the small proportion that it occurs in the atmosphere. Once an organism dies, no new ^{14}C is incorporated into its structure, and the radioactive decay process gradually reduces its store of ^{14}C, converting the atoms to ^{14}N (nitrogen-14).

The decay is slow and steady enough to act as a kind of clock, so that scientists can date a fossil (or any old material containing carbon) by measuring the percent-

age of carbon that is ^{14}C and matching this value against the clocklike progression of decay. In this way, archaeologists and paleontologists have dated prehistoric human remains; charcoal, grain, and shells found at ancient campfires; and bones and frozen tissues of recently extinct animals, such as mammoths. Scientists may also age a fossil by radiocarbon dating the rock, peat, or sediment that surrounds it. Glacial ice sheets of the most recent ice age have been dated from ^{14}C analysis of trees overrun by the advance of glaciers, and ice can be aged directly by measuring the ^{14}C in air bubbles trapped during its formation. The half-life (time it takes for one-half of a sample to decay) of ^{14}C is 5,730 years. Radiocarbon dating thus is not useful for very old fossils, because virtually all ^{14}C is gone. However, ^{14}C is very useful for items under 50,000 years old.

For dating older items, other isotopes can be used; uranium-238 (with its half-life of 4.5 billion years) has been used to date some very early fossils. For aging geological formations, potassium-argon dating is useful (potassium-40 decays to argon-40). Oxygen-18 has been widely used to infer changes in climate and sea level.

Some isotopes are radioactive and decay, but others are stable

and occur in nature in constant ratios. For instance, 99.63% of environmental nitrogen is nitrogen-14, whereas nitrogen-15 occurs at a frequency of 0.37%. Ratios of isotopes are called isotopic signatures, and it is by analyzing these signatures that scientists gain valuable information.

For example, organisms tend to retain ^{15}N in their tissues, while ^{14}N is readily excreted. Thus animals higher in the food chain, and animals that are starving, show isotopic signatures biased toward ^{15}N. One pioneer of isotope science is Keith Hobson, an ecologist with Environment Canada and the University of Saskatchewan. Hobson has used nitrogen signatures to analyze the diets of seabirds and marine mammals, to show that geese fast while nesting in the Arctic, and to trace artificial contaminants in food chains.

Hobson and many other scientists have also used carbon isotopes for ecological studies. Plants produce food through one of three photosynthetic pathways, and the isotopic signature of carbon in plants varies predictably with the plant's pathway. Grasses have higher proportions of ^{13}C (carbon-13) than oak trees, for instance, whereas cacti have intermediate values. Plants' isotopic

Atoms bond to form molecules and compounds

Atoms can bond together in chemical reactions and form **molecules,** combinations of two or more atoms. Molecules may contain one element or several. Common molecules containing only a single element include oxygen gas (O_2) and nitrogen gas (N_2), both of which are abundant in air and dissolved in seawater. If the atoms in a molecule are composed of two or more ele-

ments, the molecule is called a **compound.** Water is a compound; composed of two hydrogen atoms bonded to one oxygen atom, it is denoted by the chemical formula H_2O. Another key compound is carbon dioxide, consisting of one carbon atom bonded to two oxygen atoms; its chemical formula is CO_2.

Atoms may be held together in different ways. If atoms in a molecule are uncharged, they may share electrons, generating a **covalent bond.** The uncharged atoms of carbon and oxygen in carbon dioxide form a covalent bond.

signatures are incorporated into the tissues of animals when animals eat them, and this signal passes up through the food chain; thus, carbon isotopes can tell ecologists what an animal has been eating. Archaeologists have used signatures in human bone to determine when ancient people switched from a hunter-gatherer diet to an agricultural one.

Carbon signature data can sometimes even tell a scientist where an animal has been. Nectar-feeding bats have been shown to move seasonally between communities dominated by cacti to communities dominated by trees, for instance. Similar movements have been inferred for migrating warblers, for elephants hunted for ivory, and for the oceanic movements of seals and salmon.

Recently, researchers have begun to use isotopes to track movements of birds and other animals that migrate thousands of miles. This is possible because the isotopic signature of hydrogen from rainfall varies systematically across large geographic regions. This signature gets passed from rainwater to plants, and from plants to animals, leaving a fingerprint of geographic origin in an animal's tissues. Hobson and his colleagues in 1998 used hydrogen in combination with carbon to

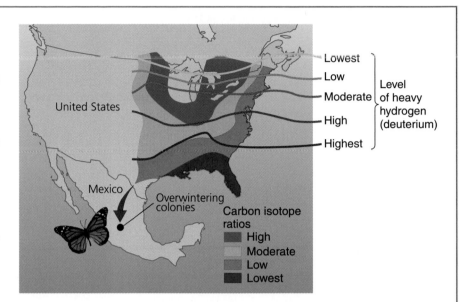

Plants in different geographic areas show different isotopic ratios of elements such as carbon and hydrogen. When caterpillars of monarch butterflies eat plants, the caterpillars incorporate into their tissues carbon and hydrogen in the isotopic ratios present in the plants. When these caterpillars metamorphose into butterflies and migrate away, they carry these isotopic signals with them, providing scientists clues to their origin. Shown is a map of isotopic ratios across eastern North America produced from measurements of monarchs in the summer. The five color bands show decreasing ratios of 13C to 12C from north to south. The five silver lines show increasing ratios of 2H (heavy hydrogen or deuterium) to hydrogen from north to south. By measuring carbon and hydrogen isotope ratios in monarchs wintering in Mexico, and matching the combination of these numbers against this map, researchers were able to pinpoint the geographic origin of many of the butterflies. *Source*: L.I. Wassenaar and K.A. Hobson, Proceedings of the National Academy of Sciences of the United States of America, 1998.

pinpoint the geographic origins of monarch butterflies that had migrated from throughout the United States and Canada to communal roosts in Mexico—important information for their conservation.

Other elements show similar standing patterns of variation in nature that have not yet been used or even discovered, researchers say, so there remains much more we can learn in the future from the use of these subtle chemical clues.

Oppositely charged ions often form **ionic bonds** in which an electrical attraction holds the ions together. These associations are not considered molecules, but instead are called **ionic compounds** or **salts.** Table salt (NaCl) contains ionic bonds between positively charged sodium ions (Na^+) and negatively charged chloride ions (Cl^-).

Elements, molecules, and compounds can also come together without chemically bonding, in mixtures. Homogenous mixtures of substances are called solutions, a term most often applied to liquids but also applicable to

some gases and solids. Air in the atmosphere is a solution formed of constituents like nitrogen, oxygen, water, carbon dioxide, methane (CH_4), and ozone (O_3). Metal alloys like brass are solutions. Human blood, ocean water, and plant sap are all solutions. Crude oil at high pressure may carry natural gas in solution, but often contains other substances distributed unevenly; it is a heavy liquid mixture of many kinds of molecules consisting primarily of carbon and hydrogen atoms. Its physical properties vary with its temperature, pressure, and composition.

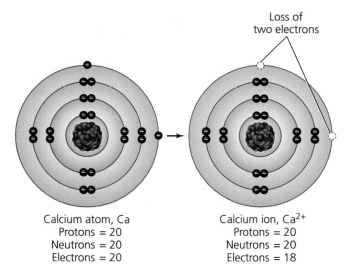

Calcium atom, Ca
Protons = 20
Neutrons = 20
Electrons = 20

Calcium ion, Ca^{2+}
Protons = 20
Neutrons = 20
Electrons = 18

Figure 4.3 Ions are atoms that are electrically charged due to the gain or loss of electrons. Here a calcium atom loses two electrons to become an ion with a charge of positive 2.

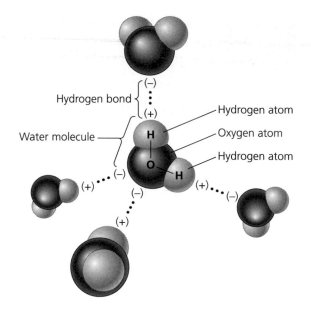

Figure 4.4 Water is a unique compound, one that has several properties crucial for life. Hydrogen bonds give water molecules cohesion by enabling them to adhere loosely to one another.

The chemical structure of the water molecule facilitates life

Water dominates Earth's surface, covering over 70% of the globe, and its abundance is a primary reason Earth is hospitable to life. Scientists think life originated in water and stayed there for 3 billion years before moving onto land. Today every land-dwelling creature remains critically tied to water for its existence.

The water molecule's amazing capacity to support life results from its unique chemical properties. A water molecule's single oxygen atom is bonded to its two hydrogen atoms at an angle, with more electrons remaining near the oxygen atom. Thus the oxygen end of the molecule has a partial negative charge and the hydrogen ends have partial positive charges. Because of this configuration, water molecules can adhere to one another in a special type of interaction called a hydrogen bond, in which the oxygen atom of one water molecule is attracted to one or two hydrogen atoms of another (Figure 4.4). Hydrogen bonds are most stable in ice, somewhat stable in liquid water, and broken in water vapor.

These loose connections between molecules give water several properties important in supporting life and stabilizing Earth's climate. First, water molecules in liquid form exhibit strong cohesion. (Think of how water droplets stick together.) This cohesion facilitates the transport of chemicals, such as nutrients and waste in plants and animals. Second, hydrogen bonding provides water molecules a great capacity to resist temperature change. Initial heating weakens hydrogen bonds

between molecules but does not speed molecular motion; as a result, water can absorb a large amount of heat with only small changes in temperature. This quality helps stabilize systems against change, whether those systems are organisms, ponds and lakes, or climate systems. In addition, water molecules in ice are farther apart than molecules in liquid water, resulting in a solid that is less dense than the liquid (Figure 4.5a)—the reverse pattern of most other compounds, which become more dense as they freeze. This is why ice floats on liquid water—a phenomenon that can have an insulating effect that prevents entire water bodies from freezing solid in winter. Moreover, water can readily dissolve many other molecules, including many of the chemicals that are necessary for life (Figure 4.5b). It follows that most biologically important solutions involve water.

Weighing the Issues:
Water's Properties for Life

Water has several special properties that make it accommodating to life. Can you generate examples of specific ways in which each of these properties might help a particular organism, such as a fish in a pond? Can you think of any ways in which any of water's properties might also bring some harm to the fish?

(a) Why ice floats on water

Figure 4.5 (**a**) Ice floats in liquid water because it is less dense. This is because in ice, molecules are connected to neighboring molecules by stable hydrogen bonds, forming a spacious crystal lattice. In liquid water, hydrogen bonds frequently break and re-form, and the molecules are closer together. (**b**) Water is often called the "universal solvent" because it can dissolve so many chemicals. Seawater holds sodium and chloride ions, among others, in solution.

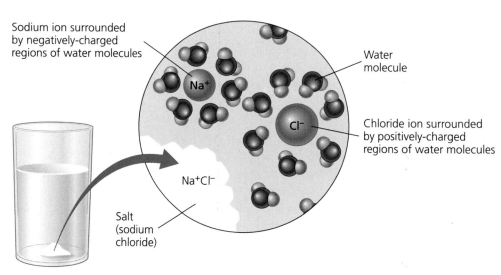

(b) Water as a solvent; how water dissolves salt

Hydrogen ions determine acidity

In any aqueous solution, a small number of water molecules dissociate, each forming two ions, a hydrogen ion (H^+) and a hydroxide ion (OH^-). Pure water contains equal numbers of these ions, and we say that this water is neutral. In most aqueous solutions, however, there is an imbalance in the concentrations of these two ions. Solutions in which the H^+ concentration is greater than the OH^- concentration are **acidic**, while solutions in which the OH^- concentration is greater than the H^+ concentration are **basic**.

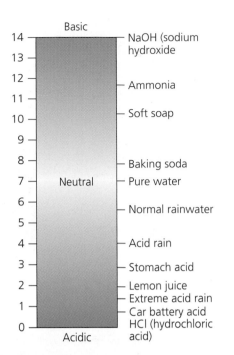

Figure 4.6 The pH scale measures how acidic or basic a solution is. Pure water calibrates the scale at its central value of 7. Naturally dissolved minerals make normal rainwater slightly acidic, but air pollution leading to acid rain has brought rain's pH even lower.

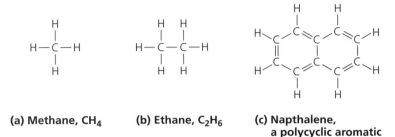

(a) Methane, CH₄ **(b) Ethane, C₂H₆** **(c) Napthalene, a polycyclic aromatic hydrocarbon**

Figure 4.7 Hydrocarbons are a major class of organic compound, and mixtures of them make up fossil fuels, such as crude oil. The simplest hydrocarbon is methane (**a**). Many hydrocarbons consist of linear chains of carbon atoms with hydrogen atoms attached; one of the shortest ones is ethane (**b**). Volatile hydrocarbons with multiple rings, such as naphthalene (**c**), are called polycyclic aromatic hydrocarbons (PAHs).

The pH scale (Figure 4.6) was devised to quantify the acidity or basicity of solutions, and runs from 0 to 14. Solutions with pH less than 7.0 are acidic, those with pH greater than 7.0 are basic, and those with pH of 7.0 are neutral. Each step on the scale represents a tenfold difference in hydrogen ion concentration. Thus a substance with pH of 6 contains 10 times as many hydrogen ions as a substance with pH of 7, and a substance with pH of 5 contains 100 times as many hydrogen ions as one with pH of 7. Figure 4.6 shows pH values for a number of common substances. The pH of rainwater, streams, ponds, and other bodies of water has become an important environmental concern since the increase in the acidity of precipitation due to industrial air pollution (Chapter 11). Rain in some parts of the northeast and midwest United States now regularly dips to pH values of 4 or lower.

Matter is composed of organic and inorganic compounds

In addition to their need for water, living things also depend on organic compounds, which they create and of which they are created. **Organic compounds** consist of carbon atoms (and, generally, hydrogen atoms) joined by covalent bonds, and may include other elements, such as nitrogen, oxygen, sulfur, and phosphorus. The unusual ability of carbon to build elaborate molecules has resulted in millions of different organic compounds showing various degrees of complexity. Because of the diversity among organic compounds and their importance in living organisms, chemists differentiate organic compounds from inorganic compounds, which lack carbon-to-carbon bonds.

Crude oil and the petroleum products are made up of organic compounds called *hydrocarbons*. **Hydrocarbons** consist solely of atoms of carbon and hydrogen (Figure 4.7). The simplest hydrocarbon is methane (CH₄), the key component of natural gas; it has one carbon atom bonded to four hydrogen atoms. Adding another carbon atom and two more hydrogen atoms gives us ethane (C₂H₆), the next simplest hydrocarbon. The smallest hydrocarbons (those consisting of four or fewer carbon atoms) are in a gaseous state at normal temperatures. Larger hydrocarbons are liquids, and those over 20 carbon atoms long are normally solids.

The bacteria used in the bioremediation of petroleum spills do not actually consume the entire hydrocarbon molecules they attack. Rather, the bacteria, facilitated by oxygen, degrade hydrocarbons by pulling carbon atoms from them. The complex hydrocarbon structures degrade into simpler ones, or into their simplest components, hydrogen and carbon. Some of the hydrocarbons that bacteria degrade are among the ones biologists worry about most. For example, polycyclic aromatic hydrocarbons (PAHs) (see Figure 4.7c), volatile molecules with a structure of multiple carbon rings, can evaporate from spilled oil and gasoline and can mix with water. In either context they pose a health

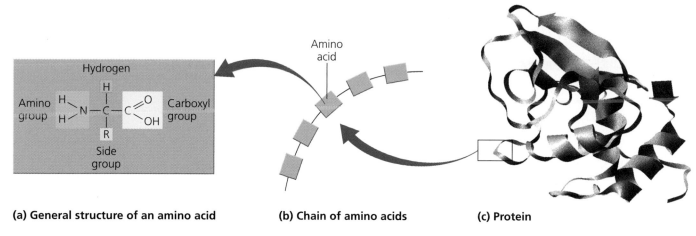

(a) General structure of an amino acid   **(b) Chain of amino acids**   **(c) Protein**

Figure 4.8 Proteins are huge polymers vital for life. They are made up of chains of amino acids (a, b), and fold up into complex convoluted shapes (c) that help determine their functions.

hazard to wildlife and people. The eggs and young of fish and other aquatic creatures are often most at risk. PAHs also occur in particulate form in various combustion products, including cigarette smoke, wood smoke, and charred meat.

Macromolecules are building blocks of life

Just as the carbon atoms in hydrocarbons may string themselves together in chains, other organic compounds can sometimes combine with one another to form long chains of repeated molecules. Some of these chains, called **polymers**, play key roles as building blocks of life. Three types of polymers are essential to life: proteins, nucleic acids, and carbohydrates. Lipids, a fourth important class of molecule, are also fundamental to life but are not considered polymers. All four of these molecules are called **macromolecules** because of their especially large size.

Proteins Proteins are made up of long chains of amino acids, which are organic molecules consisting of a central carbon linked to an acidic carboxyl group (—COOH), a basic amine group (—NH₂), and an organic side chain unique to each type of amino acid (Figure 4.8a). Organisms use 20 different amino acids to build proteins. A protein's identity is determined by the particular combination and sequence of amino acids that comprise it, as well as the shape the protein molecule assumes as it folds up. Protein molecules typically have highly convoluted shapes, with certain parts of the chain exposed and others hidden inside the folds. A protein's folding pattern affects its function, because the position of each chemical

group helps determine how it interacts with cell surfaces and with other molecules.

Comprising the majority of an organism's matter, the various types of proteins serve many diverse functions. Some help produce tissues and provide structural support for the organism; for example, animals use proteins to generate skin, hair, muscles, and tendons. Some proteins help store energy, and others transport substances. Still others function as components of the immune system, defending the organism against foreign attackers. Additionally, some proteins act as hormones, molecules that serve as chemical messengers for communication within an organism. Finally, proteins can also serve as **enzymes**, molecules that catalyze, or promote, certain chemical reactions. Bacteria used for bioremediation use specialized enzymes to break down hydrocarbons, for instance, just as we use enzymes for digesting our food.

Nucleic acids Nucleic acids direct the production of proteins. The two nucleic acids—**deoxyribonucleic acid (DNA)** and **ribonucleic acid (RNA)**—carry the hereditary information for living organisms and are responsible for passing traits from parents to offspring. Nucleic acids are comprised of nucleotides, molecules that each contain a sugar, phosphate group, and a nitrogenous base (Figure 4.9a). DNA is composed of four nucleotides, each with a different nitrogenous base: adenine (A), guanine (G), cytosine (C), and thymine (T). RNA is similar to DNA in structure, except that its sugar group is ribose (instead of deoxyribose), thymine is replaced by uracil (U), and RNA is generally single-stranded while DNA is double-stranded. Within DNA or

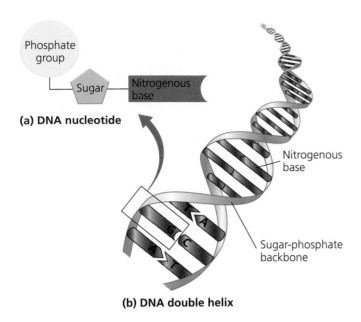

(a) DNA nucleotide

Phosphate group

Sugar

Nitrogenous base

Nitrogenous base

Sugar-phosphate backbone

(b) DNA double helix

Figure 4.9 DNA is the molecule that carries genetic information from parent to offspring across the generations. The information is coded in the sequence of nucleotides (**a**), small molecules that pair together like rungs of a ladder that twists into the shape of a double helix (**b**). Long stretches of DNA that code for particular proteins and perform particular functions are called *genes*. An entire DNA molecule is a chromosome.

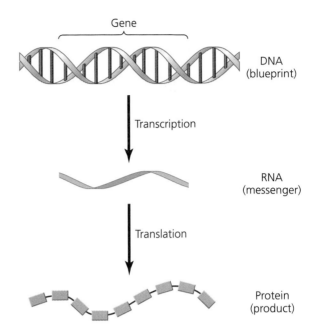

Gene

DNA (blueprint)

Transcription

RNA (messenger)

Translation

Protein (product)

Figure 4.10 DNA serves as the blueprint for synthesizing the proteins that help build organisms. The genetic information in DNA (functional stretches of which are called *genes*) instructs the creation of RNA through a process called *transcription*. Messenger RNA then directs the synthesis of proteins through a process called *translation*.

RNA, individual nucleotides are linked together to form extremely long chains, with a sugar and phosphate backbone and nitrogenous base pairs. Adenine (A) always pairs with guanine (G), and cytosine (C) always pairs with thymine (T). In DNA, the paired base chains can be pictured as rungs of a ladder, with the ladder twisted into a spiral, giving the entire molecule a shape called a double helix (Figure 4.9b).

In the process of transcription, the hereditary information in the nucleotide sequence of DNA is rewritten in a molecule of RNA. Then, during the process of translation, RNA directs the order in which amino acids are assembled to build proteins. The proteins influence the structure and maintenance of the organism (Figure 4.10). The genetic information from DNA is passed down from one generation to another, as the strands replicate during cell division and gamete (egg or sperm) formation.

Many **genes,** units of hereditary information, are arrayed along still longer strands of DNA that make up chromosomes. (A chromosome is, in fact, an entire DNA molecule.) In most organisms, the genome—the set of all of an organism's genes—is divided into separate chromosomes. Different types of organisms have different numbers of genes and chromosomes. Bacteria have a single circular chromosome, for instance, whereas humans have 46 linear ones.

Carbohydrates Carbohydrates constitute a third class of biologically vital polymer. These organic compounds consist of atoms of carbon, hydrogen, and oxygen (Figure 4.11). Simple carbohydrates, called sugars or monosaccharides, have structures, or skeletons, that are three to seven carbons long, and formulas that are some multiple of CH_2O. Glucose ($C_6H_{12}O_6$), for instance, is one of the most common and important sugars, fueling plant and animal cells. Glucose also serves as a building block for complex carbohydrates, or polysaccharides. Plants use starch, a glucose-based polysaccharide, to store energy. Animals eat plants to tap into the stored starch.

In addition, both plants and animals use complex carbohydrates to build structure. Insects and crustaceans form hard shells from the carbohydrate chitin. Cellulose, the most abundant organic compound on Earth, is a complex carbohydrate found in the cell walls of leaves, bark, stems, and roots. Cellulose, like starch, is composed of glucose molecules.

Lipids Lipids constitute a fourth type of macromolecule. Lipids are a chemically diverse group of compounds that

Figure 4.11 Like proteins, nucleic acids, and lipids, carbohydrates are macromolecules. The monosaccharide glucose (**a**) is the simplest and most abundant carbohydrate, and is a vital energy source for organisms. Linked glucose molecules form starch (**b**), an important polysaccharide.

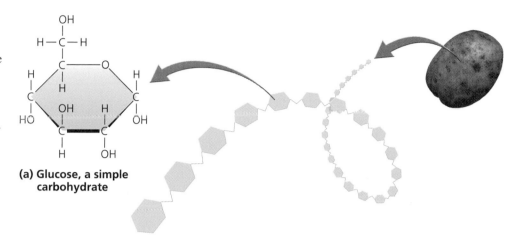

(a) Glucose, a simple carbohydrate

(b) Starch, a polysaccharide

are classified together because they do not dissolve in water. Lipids include fats, phospholipids, waxes, and steroids:

- *Fats and oils* are a convenient form of energy storage, especially for mobile animals. Their hydrocarbon structure somewhat resembles gasoline, a similarity echoed in their function: to effectively store energy and release it when burned.
- *Phospholipids* are similar to fats but contain one water-repellant side and one water-attracting side. This characteristic allows them, when arranged in a double layer, to make up the primary component of animal cell membranes.
- *Waxes* are lipids digestible by some but not all organisms, and they can have a structural function (for instance, in beeswax).
- *Steroids* are used in animal cell membranes and in production of hormones, including the sex hormones estrogen and androgen, which are vital in several aspects of sexual maturation.

Organisms use cells to compartmentalize macromolecules

All living things are composed of **cells**, the most basic unit of organismal organization. Organisms range in cellular complexity from single-celled bacteria to plants and animals that contain millions of cells. Cells vary greatly in size, shape, and function.

Biologists classify organisms into two groups based on the structure of their cells. All multicellular animals are **eukaryotes** (Figure 4.12). The cells of eukaryotic organisms consist of an outer membrane of lipids and an inner fluid-filled chamber containing **organelles**, internal structures that perform specific functions. These internal structures include (among others) **ribosomes**, organelles that synthesize proteins, and **mitochondria**, where energy is extracted from sugars and fats. Eukaryotes also have within each of their cells a membrane-enclosed **nucleus** that houses DNA.

In contrast to eukaryotes, **prokaryotic** organisms are much simpler. Prokaryotes are generally single-celled, and their cells lack organelles and a nucleus. All bacteria are prokaryotes. Bacteria are diverse and are ubiquitous in the environment, and of course they do far more than attack oil spills. Many types of bacteria perform functions vital to human life, for instance, aiding in digestion and preventing the buildup of harmful wastes.

Energy Fundamentals

Creating and maintaining organized complexity, whether of an ecosystem or an organism or a cell, requires energy. Energy is needed to organize matter into complex forms such as biological polymers, to build and maintain cellular structure, and to power the interactions that take place among species in communities. Indeed, energy is somehow involved in nearly every biological, chemical, and physical event.

But what, exactly, is energy? An intangible phenomenon, **energy** is that which can change the position, physical composition, or temperature of matter. A sparrow in flight expends energy to propel its body through the air. When the sparrow lays an egg, its body uses energy to create the calcium-based eggshell and color it with pigment.

Animal cell | Plant cell

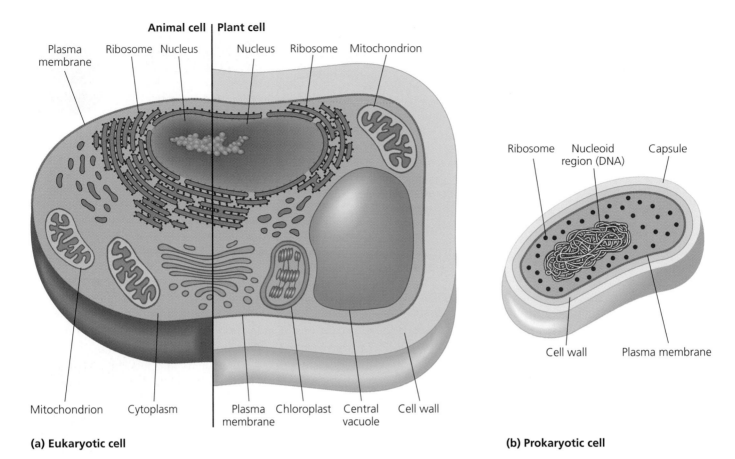

(a) **Eukaryotic cell**

(b) **Prokaryotic cell**

Figure 4.12 Cells are the smallest unit of life that can function independently. (**a**) Eukaryotic cells contain organelles such as mitochondria and chloroplasts, as well as a membrane-enclosed nucleus that contains DNA. Plant cells (right) have rigid cell walls of cellulose while animal cells (left) have more flexible cell membranes. (**b**) Prokaryotic cells are simpler, lacking organelles and an enclosed nucleus.

The sparrow sitting on its nest transfers energy from its body in heating the developing chicks inside its eggs. Some of the most dramatic releases of energy in nature do not involve living things; think of volcanoes erupting or tornadoes sweeping across the plains.

On the molecular level, it becomes clear that two attributes of energy—temperature and motion—are actually quite similar. Temperature is the measure of the motion of molecules within a solid, liquid, or gas (molecules move faster at higher temperatures); therefore, change in temperature is really just a shift in the degree of motion on an extremely small scale.

Scientists differentiate between two types of energy: **potential energy**, energy of position, and **kinetic energy**, energy of motion (Figure 4.13). Consider river water held behind a dam. In preventing water from moving downstream, the dam causes the water to accumulate potential energy. When the dam gates are opened, the potential energy is converted to kinetic energy, in the form of the motion of water rushing downstream.

Such energy transfers take place on the atomic level every time a chemical bond is broken or formed. **Chemical energy** is potential energy held in the bonds between atoms. Bonds differ in their amount of chemical energy, depending on their constituent atoms. Converting a molecule with high-energy bonds (such as the carbon-to-carbon bonds of petroleum products) into molecules with lower-energy bonds (such as those in water or carbon dioxide) releases energy by changing potential energy into kinetic energy, enabling motion, action, or heat. Just as our automobile engines split the hydrocarbons of gasoline to release chemical energy, our bodies split glucose molecules to form carbon dioxide and water for the same reason.

Energy is always conserved . . .

Although energy can change from one form to another, it cannot be created or lost; this means that the total energy in the universe remains constant and is said to

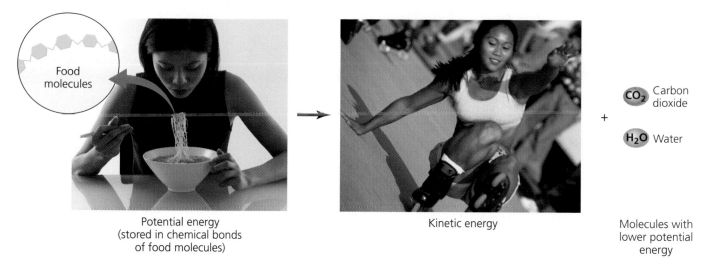

Figure 4.13 Energy is released when potential energy is converted to kinetic energy. Potential energy in the food we eat (**a**), becomes kinetic energy when we exercise (**b**).

be conserved. Scientists have dubbed this principle **the first law of thermodynamics.** The potential energy of the water behind a dam will equal the kinetic energy of its eventual movement and impact with the riverbed. Similarly, the potential energy in a log of firewood is converted by burning to an equal amount of energy produced in heat and light. We obtain energy from the food we eat, which we expend in exercise, put toward the body's maintenance, or store as fat. We do not somehow create additional energy or end up with less than the food gives us. Earth receives energy from the sun and from deep within its core. The sun's energy flows through plants, food webs, and ecosystems, some energy being given off as waste heat, and some eventually being deposited as fossil fuels. All these individual systems can temporarily increase or decrease in energy, but the total amount in the universe remains constant.

. . . But energy changes in quality

Although the first law of thermodynamics requires that the overall amount of energy must be conserved in any process of energy transfer, **the second law of thermodynamics** states that the nature of energy will tend to change from a more-ordered state to a less-ordered state. The degree of disorder in a substance, system, or process is called **entropy,** and the second law of thermodynamics holds that systems tend to move toward increasing disorder. We have all observed the tendency of objects to deteriorate over time. Every organism eventually goes through the process of death and decay, losing its structure as it does so. A log of firewood—the highly organized and structurally complex product of many years of slow tree growth—transforms in the campfire to a residue of carbon ash, carbon smoke, and gases, plus the light and the heat of the flame (Figure 4.14).

Figure 4.14 The burning of firewood demonstrates energy transfer leading from a more-ordered to a less-ordered state. This increase in entropy reflects the second law of thermodynamics.

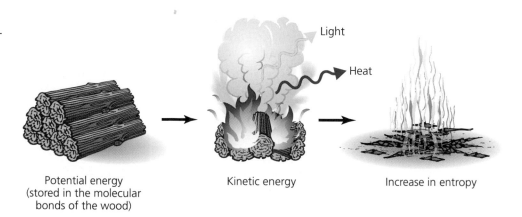

Figure 4.15 The sun sends our planet the entire range of radiation in the electromagnetic spectrum (although our atmosphere filters some radiation out).

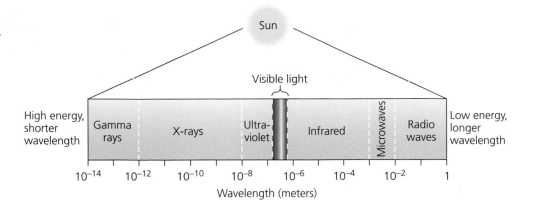

With the help of oxygen, the complex biological polymers making up the wood are converted into rudimentary molecules and heat and light energy, a motley assortment of disorganized simple entities.

Weighing the Issues:
Fossil Fuel Use and the Second Law of Thermodynamics

A perennial debate in our society is whether it is important to conserve fossil fuels and use them wisely. Does the second law of thermodynamics shed any light on this question? In what way?

The nature of energy helps determine how easily humans can harness an energy source for our own purposes. A source containing a concentrated amount of energy that is easily released is considered high-quality. Petroleum products and high-voltage electricity are high-quality energy sources. It is relatively easy for us to gain large amounts of energy from them efficiently. Sunlight and the heat stored in ocean water, in contrast, are considered low-quality energy sources. Every day the oceans absorb heat energy from the sun equivalent to that of 250 billion barrels of oil, but because this energy is spread out across such vast spaces, it is diffuse and difficult to harness.

Weighing the Issues:
Energy Quality and Energy Policy

Contrast the ease of harnessing high-quality energy, such as that of petroleum, with the ease of harnessing low-quality energy, such as that of heat from the oceans. How do you think this difference has affected our society's energy policy and energy sources through the years?

In every transfer of energy, some portion useable to us is lost. The inefficiency of some of the most common energy conversions that power our society can be surprising. An automobile's burning of gasoline is only about 16% efficient; the rest of the energy is converted to heat without powering the automobile. Incandescent lightbulbs are worse; only 5% of their energy is converted to the light that we use them for; the rest escapes as heat. Viewed in this context, the 15% efficiency of much current solar energy technology does not look bad at all.

It is important to note that although the second law of thermodynamics specifies that systems tend to move towards disorder, the order of an object or system can be increased through the input of additional energy from outside the system. This is precisely what living organisms do. Organisms maintain structure and function by consuming energy; they represent a constant struggle to maintain order and combat the natural tendency towards disorder.

Light energy from the sun powers most living systems

The energy that allows organisms to function and that flows through ecosystems comes primarily from the sun. The sun releases radiation across the entire electromagnetic spectrum, although our atmosphere filters much of this out, and we can see only some of this radiation as visible light (Figure 4.15). It is striking, in fact, just how small a portion of this spectrum humans can sense directly. Other organisms use slightly different portions of the spectrum. For instance, birds and many other animals can detect ultraviolet light.

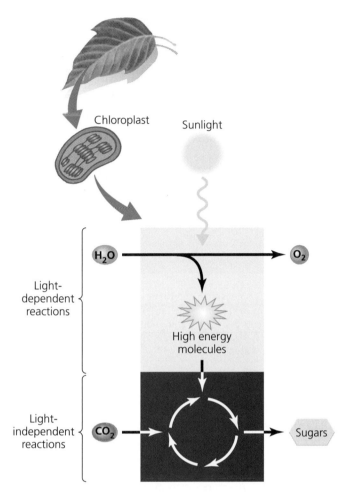

Figure 4.16 Plants use sunlight to convert carbon dioxide and water into sugars and oxygen in photosynthesis. As autotrophs, plants and cyanobacteria provide themselves and the many heterotrophs that eat them with energy for life.

Photosynthesis produces food for plants and animals

The sun's light energy powers most ecosystems on Earth by supplying energy to those organisms that can use it to produce their own food. Such organisms, called **autotrophs** or **primary producers**, are for the most part green plants and cyanobacteria. (Cyanobacteria are a class of bacteria named for their characteristic blue-green, or cyan, color.) Autotrophs turn light energy from the sun into chemical energy in a process called **photosynthesis** (Figure 4.16). In photosynthesis, sunlight powers a series of chemical reactions that convert carbon dioxide and water into sugars, thus transforming low-quality energy from the sun into high-quality energy the organism can utilize. It is an example of moving toward a state of lower entropy, and as such it requires a major input of outside energy.

Photosynthesis occurs within cell organelles called chloroplasts, where the light-absorbing pigment chlorophyll (which is what makes plants green) uses solar energy to initiate a series of chemical reactions called light-dependent reactions. During these reactions, water splits into hydrogen ions (H^+) and molecular oxygen (O_2), thus creating the oxygen that we breathe. The light-dependent reactions also produce small, high-energy molecules that are used to fuel a set of light-independent reactions, or dark reactions, which, as their names imply, can proceed without light. In the dark reactions, carbon atoms are stripped from carbon dioxide and link together to form sugars. Photosynthesis is a complex process, but the overall reaction can be summarized with the following equation:

$$6CO_2 + 12H_2O + \text{energy from the sun} \rightarrow C_6H_{12}O_6$$
$$\text{(sugar)} + 6O_2 + 6H_2O$$

The numbers preceding each chemical formula indicate the relative number of each molecule involved in the reaction. Note that the sum of the numbers on each side of the equation for each element are equal; that is, there are 6 C, 24 H, and 24 O on each side. This illustrates how chemical equations are balanced, with each atom recycled and matter conserved. No atoms are lost; they are simply rearranged among molecules. Note also that water appears on both sides of the equation. This is because for every 12 water molecules that are input and dissociated in the process, six water molecules are newly created. We can streamline the photosynthesis equation by showing only the net loss of six water molecules:

$$6CO_2 + 6H_2O + \text{energy from the sun} \rightarrow C_6H_{12}O_6$$
$$\text{(sugar)} + 6O_2$$

Thus in photosynthesis, water, carbon dioxide, and light energy from the sun are transformed to produce glucose (sugar) and oxygen. This is why green plants draw up water from the ground through their roots, suck in carbon dioxide from the air through their leaves, and need light to grow. With these ingredients, they create sugars for their own growth and maintenance and release oxygen as a by-product. This oxygen is a vital chemical for all of Earth's animals. In fact, animals appeared on Earth only after the planet's atmosphere had been supplied with oxygen by the earliest autotrophs, the cyanobacteria.

Cellular respiration releases chemical energy

The potential energy created by photosynthesis can later be used by the organism in a process called **cellular respiration.** To release the chemical energy of glucose, cells use the chemical reactivity of oxygen to split glucose into its constituent parts, water and carbon dioxide. The energy released during this process is used to form chemical bonds or to perform other tasks within the cell. The net chemical equation for cellular respiration is the exact opposite of photosynthesis:

$$C_6H_{12}O_6 \text{ (sugar)} + 6O_2 \rightarrow 6CO_2 + 6H_2O$$
$$+ \text{ energy from the sun}$$

This extraction of energy from glucose occurs in the autotrophs that created the glucose and also in the animals that gain glucose by consuming autotrophs. Organisms that consume autotrophs are called consumers, or **heterotrophs** (*hetero* means "different"), and include most animals, as well as fungi and microbes that decompose organic matter. In most ecosystems, plants or cyanobacteria form the base of a food chain through which energy passes to heterotrophs.

Tidal and geothermal forces also provide energy to Earth's systems

Although the sun is Earth's primary power source, it is not the only one. Another source is the gravitational pull of the moon, which causes ocean tides. This low-quality energy is weak and diffuse and is minor in comparison to radiative solar energy.

Another energy source is radiation emanating from inside Earth, powered by radioactivity. When we think of radioactivity, nuclear power plants and atomic weapons come to mind, but radioactivity is a natural phenomenon. It consists of the gradual release of high-energy rays or particles by certain unstable elements or isotopes as their nuclei decay. Such elements and isotopes gradually lose mass, changing into different elements or isotopes, until they become stable, nonradioactive elements.

Radiation from radioactive elements deep inside Earth heats the inside of the planet, and this heat gradually makes its way to the surface. There it drives plate tectonics (Chapter 6), heats magma that explodes from volcanoes, and warms groundwater, which in some regions of the planet shoots up out of the ground in the form of geysers (Figure 4.17). Called **geothermal energy,** this heating from deep within the planet is now being harnessed for power by some nations (Chapter 18). Long before we humans came along, however, geothermal energy was powering other biological communities.

Figure 4.17 This geyser in the Black Rock Desert of Nevada propels scalding water into the air, powered by geothermal energy from deep below ground. The bright colors on the rocks are from colonies of bacteria that thrive in the hot mineral-laden water.

Hydrothermal vent communities utilize chemical energy instead of light energy

Geysers occur here and there throughout the world, including on the floor of the ocean. Jets of geothermally heated water that emerge into the cold depths of the ocean are called **hydrothermal vents.** One of the amazing scientific discoveries of recent decades was the realization that entire communities of living organisms—wholly unknown and unsuspected just 30 years ago—thrive in the incredibly high-temperature, high-pressure conditions at these hydrothermal vents. Gigantic clams, immense tubeworms lacking digestive systems, and odd mussels, shrimps, crabs, and fish flourish in the seemingly hostile environment near scalding water that shoots out of tall chimneys of encrusted minerals (Figure 4.18).

These locations are so deep underwater as to completely lack sunlight, so their communities cannot fuel themselves through photosynthesis. Instead bacteria in deep-sea vents use the chemical-bond energy of hydrogen sulfide (H_2S) to transform inorganic carbon into organic carbon compounds in a process called **chemosynthesis.** The net reaction for chemosynthesis is

$$6CO_2 + 6\,H_2O + \text{chemical energy from } H_2S \rightarrow$$
$$C_6H_{12}O_6 \text{ (sugar)} + 6O_2 + \text{sulfates}$$

Notice that this reaction for chemosynthesis closely resembles the photosynthesis reaction. Although chemosynthesis requires energy from hydrogen sulfide whereas

Figure 4.18 (a) Hydrothermal vents on the ocean floor send spouts of hot mineral-laden water into the cold blackness of the deep sea. Amazingly, specialized biological communities thrive in these unusual conditions, scientists have discovered. (b) Odd creatures such as these giant tubeworms survive thanks to bacteria that produce food from hydrogen sulfide.

(a) Hydrothermal vent

(b) Giant tubeworms

photosynthesis uses light energy from the sun, both processes use water and carbon dioxide to produce sugar and oxygen, and both processes create potential energy that can be released during cellular respiration.

As in photosynthetic systems, energy from chemosynthesis enters the deep-sea-vent animal community when heterotrophs such as clams, mussels, and shrimp graze on chemoautotrophic bacteria. Energy is also transferred through a different relationship, in which bacteria and invertebrates exchange goods and services. Some tubeworms and clams house chemoautotrophic bacteria in specialized sac-like organs within their bodies; the wastes the sulfur-oxidizing bacteria discharge from the sacs provide their hosts with carbohydrates. (Now do you see why the tubeworms mentioned earlier do not need a conventional digestive system?) In this relationship the bacteria gain a stable and protective environment, while the invertebrates gain nutrients from the bacterial chemosynthesis. This ecological relationship of benefit to both parties is an example of mutualism, a phenomenon we will explore in the next chapter.

Hydrothermal vent communities caused excitement in the scientific community not only because they were newly discovered and unexpected, but also because some scientists believe they may help us understand how life itself originated.

The Origin of Life on Earth

How and where life originated is one of the most centrally important—and intensely debated—questions in modern biology. Scientists have developed various ideas about how life first appeared, but the question is still unresolved. In searching for the answer, scientists have learned a great deal about the history of life on Earth and about the likely conditions on early Earth—two areas of knowledge that help us approach the question of life's origin in an informed way.

Early Earth was a hostile place

Earth formed about 4.5 billion years ago in the same way as the other planets of our solar system; dispersed bits of material whirling through space around our sun were drawn by their own gravity into one another, coalescing into a series of spheres. For several hundred million years after planetary formation, there remained enough stray material in the solar system that Earth and the other young planets were regularly bombarded by large chunks of debris in the form of asteroids, meteorites, and comets. The largest impacts on Earth were probably so explosive as to vaporize the planet's newly formed oceans, some scientists calculate. Add to this the severe volcanic and tectonic activity, and it is clear that early Earth was a pretty hostile place (Figure 4.19). Any life that got underway on Earth during this bombardment stage might easily have been killed off. It was not until after most debris was cleared from the solar system that life was able to gain a foothold and keep it.

The fossil record has revealed to us much about the history of life

The earliest evidence of life on Earth comes from rocks about 3.5 billion years old. Although these earliest traces are still controversial, there is ample evidence that simple forms of life, such as single-celled bacteria, were present on Earth well over 3 billion years ago. Remains of these microscopic life forms (or their chemical by-products) have been preserved in some of the same ways as have

Figure 4.19 The young Earth on which life originated was a very different place from our planet today. Microbial life first evolved in an environment of sulfur-spewing volcanoes, intense ultraviolet radiation, frequent extraterrestrial impacts, and an atmosphere containing ammonia.

later, much larger, creatures, such as dinosaurs—by fossilization. Some of the early prokaryotic fossils resemble today's chemosynthetic bacteria that are anaerobic (not requiring oxygen). Thus life might have established itself without either oxygen or sunlight.

As organisms die, some are buried by sediment. Under the right conditions, the hard parts of their bodies—such as bones, shells, and teeth—may be preserved as the sediments are compressed into sedimentary rock. Minerals replace the organic material, leaving behind a **fossil**, an imprint in stone of the dead organism (Figure 4.20). In many thousands of locations around the world, geological processes over millions of years have buried sedimentary rock layers and later brought them to the surface, revealing assemblages of fossils representing plants and animals from different time periods. The cumulative body of fossils worldwide is known as the fossil record. The paleontologists who study the fossil record have learned to understand its biases; for example, organisms with hard parts are most likely to become fossilized, as are those living in, and therefore likely to be buried in, areas where sediment is deposited regularly, such as lakes and the edges of the ocean. With this understanding, paleontologists have used the fossil record to infer the history of past life on Earth.

The fossil record clearly shows that the species living today are but a tiny fraction of all the species that ever lived; the vast majority of Earth's species is long extinct. It also clearly shows how earlier types of organisms changed, or evolved, into later ones. Furthermore, it demonstrates that the overall number of species has increased through time; that there have been several

Figure 4.20 The fossil record has helped reveal the history of life on Earth. The numerous fossils of trilobites suggest that these animals, now extinct, were abundant in the oceans from roughly 540 million to 250 million years ago.

episodes of mass extinction, or simultaneous loss of great numbers of species; and that many organisms present early in history were smaller and simpler than modern plants and animals. In fact, the fossil record tells us that for most of life's history, microbes like the bacteria that consume hydrocarbons or the cyanobacteria that produce oxygen were the only life on Earth. It was not until the past 600 or so million years that large and complex organisms such as animals, land plants, and fungi appeared.

Paleontologists are not the only ones who use fossils; the crude oil with which we began this chapter is itself a kind of fossil. Plant and animal matter preserved upon sinking to the seafloor and buried in an area absent of oxygen may eventually become compressed and turn into the amorphous mixture of hydrocarbons we call fossil fuels. Coal, oil, and natural gas are the three forms (solid, liquid, and gas, respectively) of fossil fuel that we have used to power our civilization since the industrial revolution (Chapter 1). When we drive a car, ignite a stove, or flick on a light switch, we may be using energy from life that was buried millions of years ago.

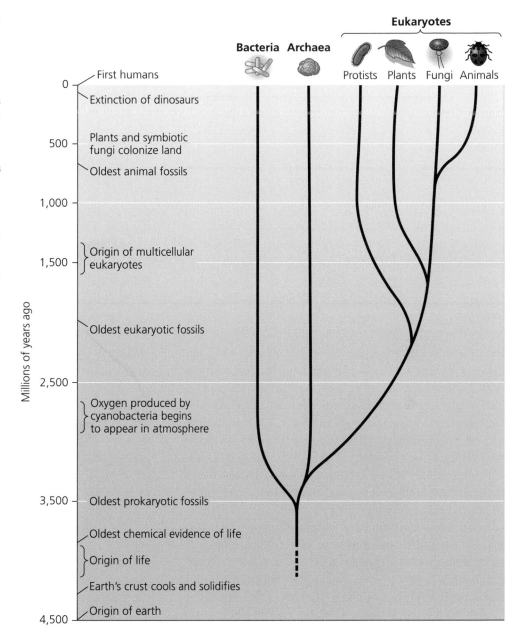

Figure 4.21 Microbes have lived on the planet for most of its 4.5-billion-year history. Large complex eukaryotes such as vertebrates, however, are relative newcomers. The fossil record and the analysis of present-day organisms and their genes allow scientists to reconstruct evolutionary relationships among organisms and to build a tree of life.

As you progress upward from the trunk of the tree to the tips of its branches, you are moving forward in time. Each fork denotes the divergence of major groups of organisms, each group of which in this greatly simplified diagram includes many thousands of species. The tree of life as understood by scientists today consists of three main groups— the bacteria, the recently discovered archaea, and the diverse eukaryotes. Protists and ancestral eukaryotes are poorly known, and their future study may well produce discoveries that further revise our understanding of life's history.

Present-day organisms and their genes can also help us decipher the history of life

Besides fossils, biologists also use present-day organisms to infer how evolution proceeded in the past. By comparing the genes and/or the external characteristics of organisms, scientists can work out patterns of relatedness. The systematists who specialize in this create branching trees, similar to family genealogies, that show the relationships among organisms and thus their history of divergence through time. As you follow such a tree from its trunk to the tips of its branches, you are proceeding forward through time, tracing the history of

life. A major advance was made in recent years as scientists discovered an entire new domain of life, the archaea, single-celled prokaryotes that are genetically very different from bacteria. Today most biologists view the tree of life as a three-pronged edifice consisting of the bacteria, the archaea, and the eukaryotes (Figure 4.21).

The air was very different on early Earth

Whereas some scientists try to infer life's history by building trees and analyzing the fossil record, others have tried to work out what environmental conditions were like on

Table 4.2 Components of Earth's Atmosphere

Today's atmosphere	Early atmosphere (older theory)	Early atmosphere (newer theory)
Nitrogen (N_2), 78.1%	Hydrogen (H_2)	Carbon dioxide (CO_2)
Oxygen (O_2), 21.0%	Ammonia (NH_3)	Nitrogen (N_2)
Argon (Ar), 0.9%	Methane (CH_4)	Carbon monoxide (CO)
Carbon dioxide (CO_2), 0.04%	Water (H_2O)	Hydrogen (H_2)
Other, <0.1%		

Earth around the time life first appeared. By deciphering geological evidence from ancient rocks and using our modern knowledge of chemistry, they have determined, for example, that active volcanism brought sulfur to the surface, and helped create salty oceans that appeared on the planet early on and covered much of it. They also have inferred that ultraviolet radiation was likely very intense, that asteroids and comets from space might have brought in water and possibly organic compounds, and that oxygen was largely lacking from the atmosphere until photosynthesizing microbes started producing it.

The atmosphere, in fact, was probably very different from the atmosphere today (Table 4.2). Whereas today's atmosphere is dominated by nitrogen and oxygen, many scientists during the 20th century thought that Earth's early atmosphere likely contained large amounts of hydrogen, ammonia (NH_3), methane, and water vapor. More recently, opinion has shifted; most scientists now infer that the atmosphere contained less hydrogen and more carbon dioxide, nitrogen, and carbon monoxide (CO) than previously thought. As we shall see, such debates are important to the question of how life originated on Earth.

Any explanation for the origin of life must make sense in the context of what we know about the conditions on early Earth and about the history of early life. More immediately, it must also clarify how simple inorganic chemicals first came together to form organic compounds, how the first cells were formed, and how replication originated with either DNA or RNA.

Several hypotheses have been proposed to explain life's origin

Most biochemists interested in life's origin think that life must have begun when inorganic chemicals linked themselves into small molecules and formed organic compounds. Some of these compounds gained the ability to replicate, or reproduce, themselves, while others found a way to group together to form early proto-cells. There is much debate and ongoing research, however, on the de-

tails of this process, especially concerning the location of the first chemical reactions and the energy source(s) that powered them.

Primordial soup: The heterotrophic hypothesis In the 1930s two scientists, J.B.S. Haldane and Aleksandr Oparin, working independently, advanced the idea that life evolved from a primordial soup of simple inorganic chemicals—carbon dioxide, oxygen, and nitrogen—dissolved in the surface waters of the oceans or the tidal shallows around oceanic margins. They suggested how simple amino acids might have formed under these conditions and how more complex organic compounds may have eventually followed, including simple ribonucleic acids that could replicate, giving rise to life in a very basic form. This hypothesis is termed *heterotrophic* because it proposes that the first life forms used organic compounds from their environment as an energy source.

This hypothesis has been traditionally favored, in part because early Earth was largely covered in water, so that an oceanic origin seemed likely. In addition, lab experiments in 1953 provided evidence (on a small scale) that the proposed process could work. Biochemists Stanley Miller and Harold Urey passed electricity through a mixture of water vapor, hydrogen, ammonia, and methane—mimicking conditions thought then to represent the early atmosphere—and were able readily to produce impressive amounts of organic compounds, including amino acids. If organic material could be assembled so easily in the lab, scientists asked, why not in nature? Subsequent experiments by other scientists confirmed Miller's and Urey's findings, but now that most scientists believe early atmospheric conditions were different, this hypothesis seems somewhat less likely to represent what actually happened.

Seeds from space: The extraterrestrial hypothesis A modification of the heterotrophic hypothesis posits that early chemical reactions on Earth may have received help from outer space. In the early 1900s Swedish chemist Svante Arrhenius proposed that microbes from space

VIEWPOINTS

What Is the Origin of Life on Earth?

There are several scientific hypotheses for the origin of life on Earth. Where do you think life on Earth began and what was the source of energy for the first organisms? What evidence supports your view?

On the Heterotrophic Origin of Life

The hypothesis that the first organisms were anaerobic heterotrophs is based on the assumption that life on Earth was preceded by a period during which the chemical precursors of cells were formed abiotically in the primitive Earth. In 1953, Stanley L. Miller, then at the University of Chicago, showed that electric discharges acting for a week over a mixture of methane, ammonia, hydrogen and water led to a complex mixture that included several amino acids. The ease of formation of many other molecules, including purines, pyrimidines, sugars, and membrane-forming compounds under reducing conditions in one pot, suggest these molecules were present in the prebiotic broth. Although there is no geological evidence of the environmental conditions of our planet at the time of the emergence of living systems, the possibility of a heterotrophic origin is also supported by the detection of molecules of considerable biochemical importance that are present in meteorites. The remarkable coincidence between the molecular composition of living organisms, the molecules synthesized in prebiotic experiments, and those present in meteorites is too striking to be fortuitous. Since many meteorites, comets, and other minor bodies must have collided with our planet during its early history, it is likely that the primitive soup was enriched by the contributions of extraterrestrial organic compounds.

Of course, the leap from amino acids that may have been present in the primitive soup to living cells is enormous. Scientists have proposed that the RNA world itself could have been preceded by simpler genetic polymers, also endowed with catalytic activity, and which depended on external sources of carbon and energy for their maintenance and multiplication. How they came into being is not yet understood, but the mere fact that we can pose these questions is, in itself, a major scientific achievement.

Antonio Lazcano is a Mexican biologist who is Professor at Universidad Nacional Autónoma de México, where he studies the earliest stages of cell evolution. He is currently President of the International Society for the Study of the Origin of Life.

Hydrothermal Vents Provide Clues to Life's Origin

Attempting to understand life's earliest common ancestor is a bit like piecing together what the Wright Brothers airplane was like from examining the Space Shuttle. The location for life's genesis is even more shrouded in mystery.

To peer back through 4 billion years of evolution, we have to rely on clues left in the fundamental biochemical pathways found in all forms of life. These pathways consist of a series of interlinked enzymes that produce or degrade organic compounds with cells and thus provide the "fuel" or raw materials necessary for life. Many of the enzymes involved in life's most fundamental metabolic processes have a metal sulfide at their core, which form the active sites where chemical reactions take place. Stripped of the protein matrix surrounding them, these active sites have a strikingly similar chemical structure to inorganic sulfide minerals such as pyrite (FeS_2). Scientists have hypothesized that these minerals could have formed the early precursors to enzymes.

On Earth, the greatest variety of such metal sulfides is found in hydrothermal vents and associated environments. Experiments have shown that hydrothermal conditions together with metal sulfide minerals can sustain a wealth of chemical reactions that are extremely similar to those found in our most essential biochemical pathways. Thus, if present-day enzymes began as reactions on sulfide mineral surfaces, they would have found a favorable home in these hot, dark environments. A general rule for the existence and support of life is that there must be a ready energy source. Hydrothermal systems abound in chemical and thermal energy, and present-day vent environments contain an abundance of organisms using this energy. Finally, hydrothermal environments are ubiquitous, protected from high-energy radiation from the early Sun, and, to some extent, asteroid bombardment. Although we may never be sure about who our earliest ancestor was, hydrothermal systems appear to be a good candidate for their home.

Dr. Jay A. Brandes is an Assistant Professor at the University of Texas Austin. His work involves investigating the processes that form and transform biologically important compounds in marine and terrestrial environments, and he is a founding member of the National Astrobiology Institute.

might have traveled on meteorites that crashed to Earth, seeding our planet with life. This idea was largely rejected in favor of Oparin's and Haldane's simpler explanation of amino acid synthesis on Earth. Even if amino acids or bacteria were to exist in space, scientists reasoned, their transport to Earth seemed implausible because the high temperatures that comets and meteors attain as they enter our atmosphere should destroy all biological compounds.

However, recent analyses of meteorites have caused many scientists to take a second look at the panspermia (meaning "seeds everywhere") hypothesis. The Murchison meteorite, a carbon-based meteorite that spewed rock fragments over the Australian countryside in 1969, was analyzed by NASA scientists and found to contain high concentrations of amino acids. This discovery seemed to demonstrate that amino acids within rock could survive an Earth impact intact. Recent experiments by astrobiologists have demonstrated that organic compounds and some bacteria can withstand a surprising amount of abuse. As a result, the long-distance travel of microbes through space and into our atmosphere seems more possible than previously thought (Figure 4.22).

Planetary scientists now believe that large asteroid impacts on one planet (such as Mars) can throw up so much material that some of it eventually may make its way to other planets (such as Earth), raising the possibility that planets might trade life. In 1996 a NASA team announced that a meteorite determined to be from Mars, ALH84001, showed evidence of fossilized bacteria—an indication of past Martian life. Although most scientists

Figure 4.22 The idea that life may have reached Earth from elsewhere is an old concept just beginning to be scientifically addressed. To test whether microbes can survive travel through space, the European Space Agency attached a container (bottom center, with *e*) of microbes to the outside of a module that orbited Earth. The container was opened in space, exposing the microbes to the harsh conditions. Once the module returned to Earth (see here), many microbes were found to have survived.

doubt the claim, the team has its supporters and is continuing to pursue the meteorite hypothesis. Astrobiological research in recent years also has made a strong case that comets have brought large amounts of water, and possibly organic compounds, to Earth throughout its history.

Weighing the Issues:
Microbes from Mars?

Let's say a meteorite falls to Earth tomorrow that scientists determine with absolute certainty is from Mars and contains living or fossilized microbial life. What, if anything, would that discovery suggest about the origin of life? In the wake of such an event, can you suggest how we might test whether life first began on Earth, or Mars, or both independently?

Life from the depths: The chemoautotrophic hypothesis While some scientists were lending support to the idea of life originating from the heavens, others were developing the concept of life originating from the deep sea. In the 1970s and 1980s several scientists, among them Jack Corliss, a discoverer of deep-sea hydrothermal vents, proposed such an alternative. The chemoautotrophic theory proposes that early life was formed at the scalding-hot deep-sea vent systems, where sulfur was abundant. In this hypothesis, the first organisms were chemoautotrophs, creating their own food from the hydrogen sulfide present at the vents.

Genetic analysis of the relationships of present-day organisms indicates that some of the most ancient ancestors of today's life forms likely lived in extremely hot, wet environments, suggesting that life may have originated in just such places. Additionally, the extreme heat of hydrothermal vents could act to speed up chemical reactions that link atoms together into long molecules, a necessary early step in life's formation. Such reactions might involve minerals from Earth's interior and carbon- and hydrogen-rich gases emerging from the vents. Indeed, scientists have shown experimentally that it is possible to form amino acids and begin the chain of steps leading to the formation of life under high-temperature, high-pressure conditions similar to those of deep-sea hydrothermal vents.

Organic polymers had to develop the ability to self-replicate

Whether they formed in deep-sea hydrothermal vents, on rocky shores, or in comet craters, early organic polymers had to develop the ability to self-replicate before life as we know it could commence. Contemplating the origins of self-replicating molecules involves a chicken-or-egg paradox. In order to replicate biological polymers, nucleic acids, typically DNA, are needed. However, in order to create DNA, enzymes, a type of biological polymer, are required. In order to get one, you would seem to need the other.

Scientists currently think that the first self-replicating molecule may have been RNA. RNA is unique in that it is able to act as both an enzyme and an information carrier. Early RNA molecules may have acted as their own enzymes, replicating themselves without the assistance of other polymers. Laboratory experiments have shown that when chains of RNA are added to a flask containing amino acids, the amino acids are capable of lining up along the RNA chain and forming a copy without enzymes. These results indicate that early self-replication of RNA is possible. Over time, scientists hypothesize, the RNA replicating process was modified, and DNA, which is more stable, came to dominate as the information carrier in animal and plant cells.

Cell precursors were another crucial step

The formation of cells must have been another key step in the origination of life, and ideas abound as to how this may have occurred. Cells have a partly permeable membrane, or wall, which allows them to maintain an internal environment different from their surroundings. The path to early production of cells may have included the chance formation of aggregates of abiotically produced polymers. Cell precursors that happened to contain RNA may have replicated. Such cell-like structures have formed spontaneously in laboratory experiments when certain abiotic molecules, including lipids, are mixed in a flask. These have been able to divide, and if enzymes are included, have been able to absorb substances from their surroundings, as cells do. Over time, simple unicellular organisms may have come into being, beginning the process of life as we know it.

The Theory of Evolution by Natural Selection

Although considerable uncertainty surrounds the way life first began on Earth, scientists do understand how from simple beginnings the world became populated with the remarkable diversity of organisms we see today.

We know that the process of biological evolution has brought us from a stark planet inhabited solely by microbes to a lush world of 1.5 million (and likely many more) species, from oak trees to fireflies to swordfish to mushrooms to flamingos to humans.

Understanding how organisms adapt to their environments and change over time is crucial not only for learning about the history of life, but also for appreciating ecology, which is such a central component of environmental science. Evolutionary processes are relevant to other aspects of environmental science as well, including pesticide resistance, agriculture, and environmental health and medicine.

The term *evolution* in the broad sense means change over time; however, scientists most commonly use the term to refer specifically to biological **evolution**—genetically based change in the appearance, functioning, and/or behavior of organisms across generations, often by the process of **natural selection.** The theory of evolution by natural selection is one of the best-supported theories in all of science, yet since its inception it has been socially controversial. As it was with the theory of planetary orbits around the Sun proposed by Copernicus and Galileo, many people outside the scientific community remain unconvinced of the validity of the theory of evolution or choose not to learn about it because they fear it may threaten their belief system. Practicing scientists sometimes disagree about the specific mechanisms thought to drive evolution in particular cases, or about the time scales on which it takes place, but this conventional scientific disagreement should not be equated with the socially driven debate among nonscientists. We will leave these controversies for other forums and instead will move on and explore the theory itself, because it is the foundational basis of all of modern biology.

Darwin and Wallace proposed the concept of natural selection

The idea that organisms change over time is an old one, but a satisfactory explanation for how biological evolution might work was a long time in coming. Both Charles Darwin and Alfred Russell Wallace in 1858 proposed the concept of natural selection as a mechanism for evolution, and as a way to explain the great variety of living organisms. Darwin and Wallace were each exceptionally keen naturalists from England who had studied plants and animals in such exotic locales as the Galapagos Islands and the Malay Archipelago.

In their concept of natural selection, Darwin and Wallace recognized that organisms face a constant

Table 4.3   Key Steps in the Logic of Natural Selection
• Organisms produce more offspring than can survive.
• Individuals vary in their characteristics.
• Many characteristics are inherited by offspring from parents.
Therefore,
• Some individuals will be better-suited to their environment than other individuals, and thus will survive and reproduce more successfully.
• By producing more offspring and/or offspring of higher quality, better-suited individuals transmit more genes to future generations than poorly suited individuals.
• Future generations will contain more genes, and thus more characteristics, of the better-suited individuals. Thus characteristics will evolve across generations through time.

struggle to gain sufficient food and shelter to survive and reproduce. Both men were influenced by the writings of Thomas Malthus (Chapter 1 and Chapter 7), who, you will recall, feared that population growth in industrialized England would outstrip resource availability and lead to widespread death and social upheaval. Extrapolating from Malthus's reasoning, Darwin and Wallace realized that if organisms produce more offspring than can possibly survive, then those individual offspring that have some edge over their competitors are more likely to survive and reproduce. Furthermore, whatever characteristics gave individuals a competitive edge might be inherited by their offspring and would therefore tend to become more prevalent in the population in future generations. Scientists at that time knew that traits were inherited, but did not know of Gregor Mendel's (considered the father of genetics) research showing a genetic basis for inheritance. Mendel's work, when rediscovered at the turn of the twentieth century, would provide further support for evolutionary theory.

Natural selection shapes organisms and organismal diversity

Natural selection is a simple concept that offers an astonishingly powerful explanation for the patterns apparent in nature. The idea of natural selection follows logically from a few straightforward premises (Table 4.3), among which is that individuals of the same species vary in their characteristics. Although not known in Wallace and Darwin's time, we now know that variation occurs because of differences in genes, the environments within which genes are expressed, and the interactions between genes and environment. As a result of this variation, certain individuals within a species will happen to be better suited to their environment than others and thus will be able to survive longer and/or reproduce more. Many characteristics are passed from parent to offspring through the genes, and a parent that is long-lived, robust, and produces many offspring will pass on genes to more offspring than a weaker, shorter-lived individual that produces only a few offspring. In the next generation, therefore, the genes of better-adapted individuals will be more prevalent than those of less well-adapted individuals. From one generation to another through time, species will evolve to possess characteristics that lead to better and better success in a given environment.

The tendency of some individuals to be more successful is often called, in nonscientific jargon, *survival of the fittest*. Note, however, that *fittest* here refers to individuals that are best suited to their environment, not necessarily those that are the strongest, fastest, or most aggressive. In many cases the fittest individual is one that can avoid competition or conflict by utilizing a new or previously unused resource. Biologists use the term **fitness** to refer both to the likelihood that an individual will reproduce and to the number of offspring an individual produces over its lifetime. A trait that confers greater fitness is called an **adaptive trait,** or an **adaptation.** A trait that reduces fitness is termed *maladaptive*. Recall that one reason for the success of bioremediation after the *Exxon Valdez* oil spill was that the region's bacteria were already accustomed to dealing with hydrocarbons produced by conifer trees. Bacteria of the region had adapted to this availability of hydrocarbons over many eons. Thus, evolution lent a helping hand to bioremediation of the spill.

Adaptation is a moving target

The environment to which an organism responds can change in both space and time, and animals and plants may move to new places and encounter new conditions. In either case, a trait that is adaptive in one location or season may be maladaptive in another. For instance, bacteria that feast on hydrocarbons will do well in an area where hydrocarbons occur naturally or where oil has been spilled, but not in an area devoid of hydrocarbons. Monkeyflower plants that line a desert stream in a wet year may disappear in a dry one, leaving plants adapted to more arid conditions to take their place. Differences in environmental conditions in time and space make adaptation a moving target for organisms, such that a species will likely never adapt perfectly to its environment.

Evidence of natural selection is all around us

The results of natural selection are all around us, visible with every adaptation of every organism (Figure 4.23). In addition, countless lab experiments, mostly with fast-reproducing organisms such as bacteria and fruit flies, have demonstrated rapid evolution of traits in controlled environments. The evidence for selection that may be most familiar to us is that which Darwin himself cited prominently in his work 150 years ago: that of human breeding of domesticated animals. In our dogs, our cattle and horses, and our breeds of pigeons, we have conducted natural selection artificially, by choosing animals with traits we like and breeding them in preference to those with variants we do not like. Through such selective breeding, we are often able to exaggerate particular traits we prefer. Consider the great number of dog breeds (Figure 4.23b), all of which comprise a single species; they can interbreed freely and produce viable offspring, yet we maintain striking differences between them by allowing only like individuals to breed with like.

The same principle applies to the many crop plants we depend on for food, all of which were domesticated from wild ancestors and specially bred over years, centuries, or millennia to create, for instance, sweet kernels of corn, numerous grains of wheat, and large heads of lettuce. Our entire agricultural system is in fact based on the process of **artificial selection,** or natural selection conducted under human direction.

Weighing the Issues:
Artificial Selection and "Survival of the Fittest"

Natural selection favors individuals that are best suited to the particular pressures of their environment—not necessarily the strongest, fastest, or most aggressive individuals. Consider how humans have applied selection to breed and domesticate animals, including cows, horses, dogs, and cats. Now compare these animals to their wild relatives. How do they differ? Do the evolved traits of the domesticated animals make sense in terms of the kinds of characteristics humans prefer?

Selection can operate in different ways

Selection, natural or artificial, can take three major forms (Figure 4.24). Selection that acts to drive a feature in one direction rather than another—for example, toward larger or smaller, faster or slower—is called **directional selection.** In deep-sea hydrothermal vent environments, organisms must withstand extremely high water pressure and temperature. Such hostile conditions may favor clams, mussels, shrimp, and crabs with thick shells over those with thin shells and result in thick-shelled animals' dominating the community.

In contrast, **stabilizing selection** acts to produce intermediate traits, in essence preserving the status quo. For example, whereas a thin shell of a shrimp or crab might break easily, a shell that is extremely thick could waste resources better used for feeding or reproduction. Under stabilizing selection, the crustacean species would evolve a shell that is neither too thick nor too thin.

Under **disruptive selection,** extreme traits are favored. For example, a tiny shrimp in a hydrothermal vent community might be able to hide from predators because of its small size. A very large shrimp would be conspicuous to predators but might avoid being eaten if its size allowed it to defend itself or if it simply were too large to consume. A medium-sized shrimp might be both visible to predators and also the right size for eating. In such a situation, natural selection would increase the relative numbers of small and large shrimp and reduce the number of medium-sized ones.

DNA is the source of heritability and variation

The effectiveness of natural selection in shaping organisms relies on adaptations passed on from generation to generation. For an organism to be able to pass a trait along to future generations—that is, for the trait to be

Figure 4.23 Natural selection has produced tremendous diversity among organisms in the wild. In the group of birds known as Hawaiian honeycreepers (**a**), different species have adapted to different food resources, habitats, or ways of life. The diversity in the shapes of their bills reflects differences in feeding behavior. Selection imposed by human beings has resulted in the numerous breeds of dogs (**b**). By breeding like with like and selecting for the traits we prefer, we have, within a single species, evolved breeds as different as great danes and chihuahuas.

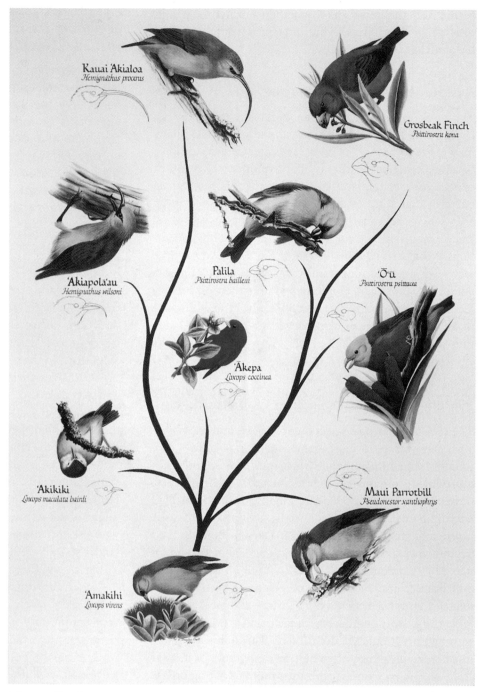

(a) Hawaiian honeycreepers

(b) Springer spaniel, golden retriever, mixed breed, Jack Russell terrier

Figure 4.24 Selection can act in different ways. From a starting population of animals with shells of a range of thicknesses, directional selection (**a**) could favor either thicker shells or thinner shells, stabilizing selection (**b**) could favor intermediate shells, or disruptive selection (**c**) could favor both thick and thin shells.

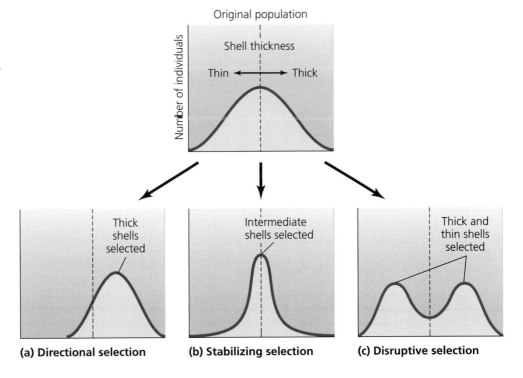

Original population

Number of individuals

Shell thickness

Thin ←→ Thick

Thick shells selected

Intermediate shells selected

Thick and thin shells selected

(a) Directional selection   **(b) Stabilizing selection**   **(c) Disruptive selection**

heritable—the genes in the organism's DNA must determine the trait. Accidental alterations that arise during DNA replication give rise to variation in living organisms. In the lifetime of a given organism, DNA will be copied millions of times by millions of cells. In all this copying and recopying, sometimes a mistake is made. Accidental changes in DNA are called **mutations** and range in magnitude from the deletion, substitution, or addition of single nucleotides to mistakes affecting entire sets of chromosomes (Figure 4.25). If a mutation occurs in a sperm or egg cell, it may be passed on to the next generation.

Mutations occur naturally but can be increased by exposure to certain chemicals and radiation. Polycyclic aromatic hydrocarbons in crude oil, for instance, are thought by health experts to induce mutations in DNA. Most mutations are either slightly harmful or of negligible effect, but the effects can range from deadly to beneficial. Those that are not lethal provide the genetic variation that drives natural selection.

Selection is a positive, creative force

Natural selection does not simply weed out unfit individuals. It helps to elaborate and diversify traits that in the long term may help lead to the formation of new species and eventually whole new types of organisms. In this way natural selection is a positive force of creation that has guided the wondrous flowering of biodiversity

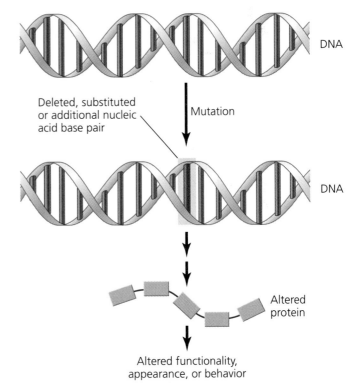

DNA

Deleted, substituted or additional nucleic acid base pair

Mutation

DNA

Altered protein

Altered functionality, appearance, or behavior

Figure 4.25 Mutations in DNA provide the raw material upon which natural selection acts. A mutation in a single nucleotide can potentially alter a protein by changing the identity of an amino acid. If alteration of the protein affects its functioning, then consequences could result for the organism's functioning, appearance, or behavior.

on our planet. We will explore how the process of evolution leads to the creation of new species as we continue our discussion of evolution in the next chapter.

Conclusion

The process of evolution has generated the tremendously diverse proliferation of life on our planet, a planet whose chemical and physical makeup seems particularly well-suited for living things. The prevalence of water has allowed carbon-based life to flourish here for well over 3 billion years, stemming from an origin that scientists are eagerly attempting to discover. Deciphering how life originated depends in part on understanding energy, energy flow, and chemistry. Knowledge in these areas also enhances our understanding of how present-day organisms interact with each other, how they relate to their nonliving environment, and how the environment itself functions. Energy and chemistry are in some way tied to nearly every significant process involved in environmental science.

Chemistry can also be a tool for finding solutions to environmental problems. Cleaning up chemical pollution through bioremediation with microbes or plants is only one example. Knowledge of chemistry can be a powerful ally, whether one is interested in analyzing agricultural practices, managing water resources, reforming energy policy, conducting toxicological studies, or finding ways to mitigate global climate change.

REVIEW QUESTIONS

1. What are the basic building blocks of matter? Give examples using chemicals common in living organisms.
2. Name four ways in which the chemical nature of the water molecule facilitates life.
3. What are the three classes of biological polymers, and what are their functions?
4. Name the two major forms of energy and give examples of each form.
5. State the first law of thermodynamics, and describe some of its implications.
6. What is the second law of thermodynamics, and how does it affect our interactions with the environment?
7. What are two sources of energy at the base of food chains?
8. What substances are produced by photosynthesis? By cellular respiration? By chemosynthesis?
9. What is a hydrothermal vent?
10. Name three things scientists have learned from the fossil record.
11. What are the three competing hypotheses for the origin of life on Earth?
12. Explain the premises and logical argument behind natural selection.
13. Name two examples of evidence for natural selection.
14. What are the three different types of selection, and which one is most likely to cause a new species to be formed?
15. What is the raw material for variation in living organisms?

DISCUSSION QUESTIONS

1. Under what conditions might bioremediation be a successful strategy, and when might it not be?

2. List three examples of energy cycling in your environment and, within those cycles, examples of energy transformation from one form to another.

3. Can you think of some examples of the interplay of increasing order and increasing entropy in biological systems?

4. The debate on the origin of life on Earth is heated and ongoing. What does each hypothesis entail? What evidence would convince you of the validity of one hypothesis over the other?

5. The theory of evolution by natural selection is heavily debated outside the scientific community. Why do you think this topic remains controversial? Do the ideas outlined by Darwin and Wallace make sense to you?

Media Resources *For further review go to the website* **www.envscienceplace.com** *or student CD-ROM, where you will find quizzes, flashcards, a glossary, additional interactive exercises, and links to relevant news and research sources. Also, on the website and CD-ROM is* **GRAPH IT**, *a series of interactive graphing tutorials to help you interpret graphs and plot data.*

5 Ecology and evolution: Populations, communities, and biodiversity

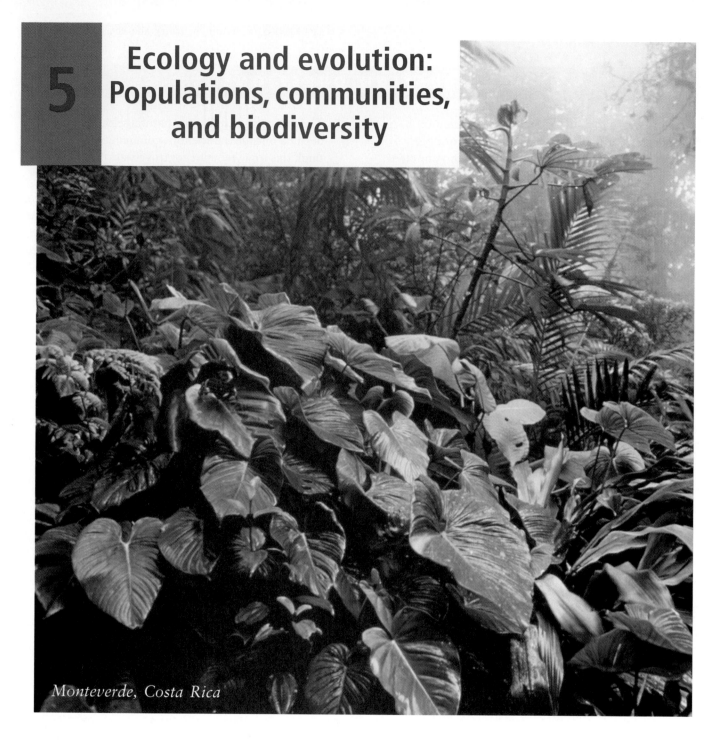

Monteverde, Costa Rica

This chapter will help you understand:

- How evolution generates biodiversity

- Concepts of speciation and extinction

- Outlines of the biodiversity "crisis"

- Fundamental ecological concepts and principles

- Units of ecological organization

- Fundamentals of population ecology, including carrying capacity and limiting factors

- Food webs, trophic levels, and ecological communities

- Species interactions, including predation, competition, parasitism, and mutualism

- Primary and secondary succession

- Challenges for biodiversity conservation

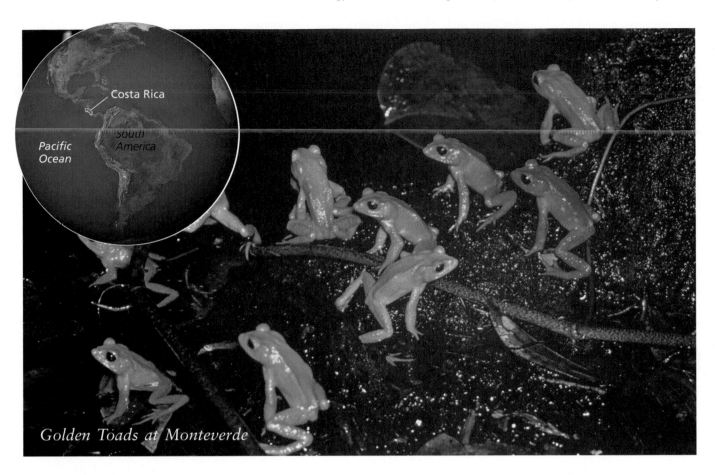

Golden Toads at Monteverde

Central Case: Striking Gold in a Costa Rican Cloud Forest

"I must confess that my initial response when I saw them was one of disbelief and suspicion that someone had dipped the examples in enamel paint."
—*Dr. Jay M. Savage, describing the golden toad in 1966*

"What a terrible feeling to realize that within my own lifetime, a species of such unusual beauty, one that I had discovered, should disappear from our planet."
—*Dr. Jay M. Savage, describing the golden toad in 1998*

During a 1963 visit to Central America, biologist Jay M. Savage heard rumors of a previously un-documented toad living in Costa Rica's mountainous Monteverde region. The elusive amphibian, according to local residents, was best known for its color: a brilliant yellow-orange that seemed almost golden. Savage was told the toad was hard to find because it appeared only in spring, during the early part of the region's rainy season.

Monteverde means "green mountain" in Spanish, and the name couldn't be more appropriate. The village of Monteverde sits on a bench beneath the verdant slopes of mountains known as the Cordillera de Tilarán, which receive more than 400 cm (157 in) of annual rainfall. Some of the lush forests above Monteverde, which begin around 1,600 m (5,249 ft, just under a mile high), are known as **lower montane rainforests.** These forests are also known as **cloud forests** because much of the moisture they receive arrives in the form of low-moving clouds that blow inland from the Caribbean Sea. Monteverde's cloud forest was not fully explored at the time of Savage's first visit, and researchers who had been there described the area as pristine, with a rich bounty of ferns, liverworts, mosses, clinging vines, orchids, and other organisms that thrive in cool, misty environments. Savage knew that such conditions create ideal habitat for many amphibians, including toads.

In May of 1964, Savage organized an expedition into the muddy mountains above Monteverde to try to document the existence of this previously unknown species in its natural habitat. Late in the afternoon of

May 14, he and his colleagues found what they were looking for. Approaching the mountain's crest, they spotted bright orange patches on the forest's black floor. In one area that was only 5 m (16.4 ft) in diameter, they counted 200 golden toads.

The discovery received international attention, making a celebrity of the tiny toad—which Savage named *Bufo periglenes* (literally, "the brilliant toad")—and making a travel destination of its mountain home. At the time, no one knew that the Monteverde ecosystem was about to be transformed. No one foresaw that the oceans and atmosphere would begin warming with global climate change (Chapter 12) and cause Monteverde's moisture-bearing clouds to rise, drying the forest. No one could guess that this newly discovered species of toad would become extinct in less than 25 years.

Evolution as the Wellspring of Earth's Biodiversity

The apparent extinction of the golden toad made headlines worldwide, but unfortunately it was not such an unusual occurrence. The extinction of species is a major dilemma in today's world—indeed, some say it is the single biggest environmental problem we face, because the loss of a species is irreversible. The lush biological diversity that makes Earth such a unique planet is being lost at an astounding rate. To understand why so many biologists are so deeply concerned with what is variously called a "biodiversity crisis" and a "mass extinction event," and to put current trends in context, we must take a further look at the evolutionary process and the history of life to this point.

In particular, we will look at the processes of speciation and extinction, which together determine how many life forms and what kinds of life exist on our planet. These are natural processes, but human action can profoundly affect the rates at which they occur. They are also processes that have ramifications for the ecological systems that ultimately support our society. Sound environmental science requires the incorporation of ecology, and ecology requires a solid understanding of evolution.

Biological diversity is the product of evolution

When Charles Darwin wrote about the wonders of a world full of diverse animals and plants, he conjured up the vision of a "tangled bank" of vegetation harboring all kinds of creatures. Such a vision fits well with the arching vines, dripping leaves, and mossy slopes of the tropical cloud forest of Monteverde. Indeed, tropical forests worldwide are teeming with life and harbor immense biological diversity (Figure 5.1).

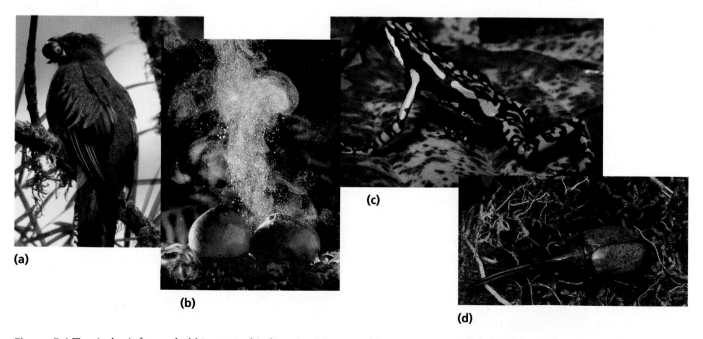

Figure 5.1 Tropical rainforests hold immense biodiversity. Monteverde's community includes creatures such as this **(a)** resplendent quetzal, **(b)** puffball mushroom, **(c)** harlequin frog, and **(d)** weevil.

Biological diversity, or **biodiversity** for short, is a catchall term meaning the sum total of all organisms in an area, taking into account the diversity of species, their genes, their populations, and their communities. A **species** is a particular type of organism, or more precisely, a population or group of populations whose members share certain characteristics and can freely breed with one another and produce fertile offspring. Scientists have described between 1.5 million and 1.8 million species, but there are many more still unnamed and undiscovered. Estimates for the total number of species in the world range up to 100 million, many of them in tropical forests. In this light, the discovery of a new toad species in Costa Rica in 1964 seems far less surprising. Although Costa Rica covers a tiny fraction (0.01%) of Earth's surface area, it is home to 5–6% of all species known to scientists. And of the 500,000 species scientists estimate exist in the country, only 87,000 (17.4%) have been inventoried and described.

Tropical rainforests such as Costa Rica's, however, are by no means the only places rich in biodiversity. Your own backyard is rich in species, too. Walk outside anywhere on Earth, even in a major city, and you will find numerous species within easy reach. They may not always be large and conspicuous like Yellowstone's bears or Africa's elephants, but they will be there. Plants poke up from cracks in asphalt in every city in the world, and even Antarctic ice harbors microbes. In a handful of backyard soil there may live several insect species, several types of mites, a millipede or two, many nematode worms, a few plant seeds, countless fungal spores, and millions upon millions of bacteria. A handful of soil can hold an entire miniature world of biodiversity.

Speciation is the process that produces new types of organisms

How did Earth come to have so many species? Whether there are 1.5 million or 100 million, such large numbers require scientific explanation. The process by which new species are generated is termed **speciation.** Individuals of a population that are in contact and able to reproduce with one another are able to mix their genes. A gene for a particular trait can spread through a population as individuals with the gene mate with other individuals and produce offspring with the gene. As a result, individuals within a population retain many similarities that unify them as a species. However, if a population is broken up into two or more populations that are isolated from one another, individuals from one population cannot reproduce with individuals from the others. When popula-

tions are kept separate for long periods of time, preventing the mixing of genes, this can lead to speciation.

Think about what will happen when a genetic mutation arises in the DNA of an organism in one of these isolated populations. When populations are separated, the mutations that occur in one cannot spread to the others. Over time, each population will independently accumulate its own set of mutations. Eventually, the two populations may diverge, or grow different enough, that their members can no longer mate with one another. Perhaps they no longer recognize one another as being the same species because they have diverged so much in appearance or behavior. Once this has happened there is no going back; the two populations cannot interbreed, and they have embarked on their own independent evolutionary trajectories as separate species. The populations will continue diverging in their characteristics, as chance mutations accumulate that confer traits causing the populations to become different in random ways. If environmental conditions happen to be different for the two populations, then in many cases the divergence can be accelerated by natural selection (Chapter 4). Through the speciation process, single species can generate multiple species, each of which can in turn generate more species by the same process.

Populations can be separated in many ways

When populations are physically separated from one another over a geographic distance, the process described above is termed **allopatric speciation** — the formation of species in separate locations (Figure 5.2). The long-term physical isolation of populations that can cause speciation can occur in various ways (Table 5.1). Glacial ice sheets may move across continents during ice ages and split populations in two. Major rivers may change course and do the same. Mountain ranges may rise and divide two regions and their organisms. Drying climate may partially evaporate lakes and, along with the aquatic life they contain, split the lakes into multiple smaller bodies of water. Moreover, warming or cooling temperatures may cause whole communities of vegetation to move slowly northward or southward, or upslope or downslope, creating new patterns of plant and animal distribution.

The island chain of Hawaii illustrates such mechanisms. Each Hawaiian island was formed by underwater volcanism and rose out of the sea, to be colonized by plants and animals from surrounding islands. Once on each new island, some of these organisms differentiated into new species. On a smaller scale, volcanism has led

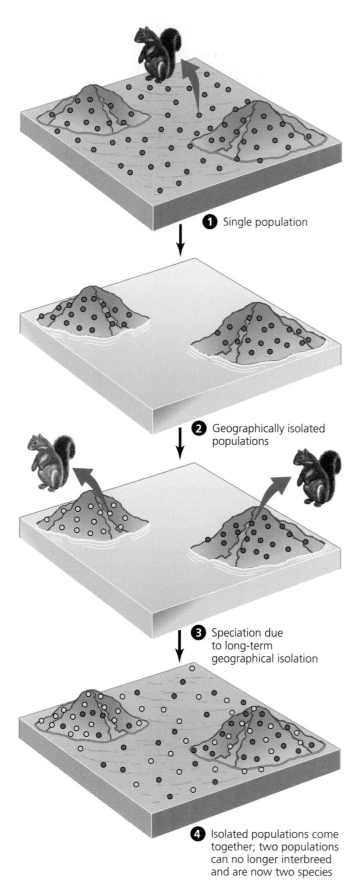

① Single population

② Geographically isolated populations

③ Speciation due to long-term geographical isolation

④ Isolated populations come together; two populations can no longer interbreed and are now two species

Table 5.1 Mechanisms of Population Isolation That Can Result in Speciation

Glacial ice sheets advance

Mountain chains are uplifted

Major rivers change course

Sea level rises, creating islands

Climate warms, pushing vegetation up mountain slopes and fragmenting it

Climate dries, dividing large single lakes into multiple smaller lakes

Ocean current patterns shift

Islands are formed in sea by volcanism

to divergence within each island, some biologists maintain. Recent lava flows that have wound down vegetated slopes have left patches of plants separated from one another by expanses of lava. Some fruit flies appear to have remained loyal to certain patches and to be diverging there. Regardless of the location and the mechanism of separation, in order for speciation to occur, populations must remain isolated for a very long time, generally over many thousands of generations.

The moment of truth for speciation is when isolated populations meet again

If the geological or climatic process that has isolated two populations reverses itself—if the glacier recedes, or the river returns to its old course, or warm temperatures turn cool again—then the isolated populations can come back together. This is the moment of truth for speciation. If the populations have not diverged enough, their members will begin interbreeding and reestablish gene flow, mixing those mutations that each population accrued while it was isolated. If the populations have diverged sufficiently, however, they will not interbreed, and two species will have been formed, fated to continue on their own evolutionary paths.

Figure 5.2 Much of Earth's diversity has been generated by allopatric speciation. In this process, some geographical barrier splits a population. Each isolated subpopulation accumulates its own independent set of genetic changes over time, until individuals become genetically distinct and unable to breed with individuals from the other population. The two populations now represent separate species, and will remain so even if the geographical barrier is removed and the new species intermix.

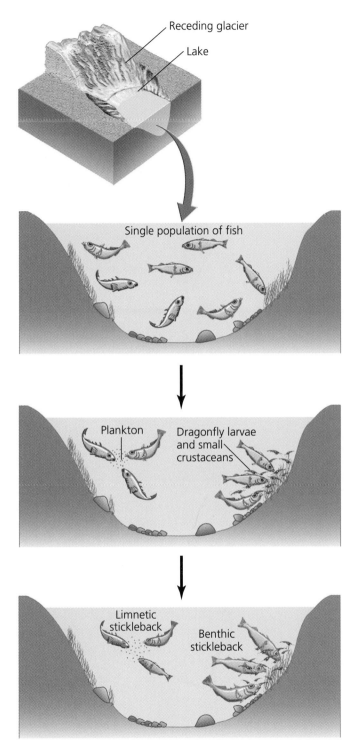

Figure 5.3 In sympatric speciation, populations split into two or more species while living together in the same area. Research on stickleback fish shows that fish colonizing lakes after glaciers recede have subsequently diverged into limnetic (open-water) forms that feed on plankton and benthic (bottom-dwelling) forms that feed on invetebrates on the substrate.

Not all speciation is allopatric

Allopatric speciation has long been considered the main mode of speciation, but some biologists today think that many species have arisen through **sympatric speciation**, or the formation of species together in a single location without geographical separation (Figure 5.3). Insects are the most likely candidates for this type of speciation. Many insects are specialized to feed only on certain types of plants. Monarch butterfly caterpillars feed only on milkweed plants, for instance, and thousands of kinds of beetles, aphids, and other insects likewise depend on single plant species. In one scenario for sympatric speciation, insects evolve to mate on the plant they feed on (called the host plant), while some insects of the same species gain a mutation for feeding and mating on another host plant. Two populations of insects may thus arise in the same area, one feeding and mating on one host plant and the other on a different species of host plant. If their mating is segregated enough, the populations may come to form two species. Evidence exists for such events in a number of cases, but garnering such evidence is difficult. Biologists, therefore, still actively debate how common sympatric speciation is relative to allopatric speciation.

Life's diversification results from numerous speciation events

Speciation tells us how one species gives rise to two or more. Repeated bouts of speciation have generated complex patterns of diversity at levels above the species level. Such patterns are studied by evolutionary biologists who investigate how the major groups of organisms arose and how they evolved the characteristics they show. How did we end up with plants as different as mosses, palm trees, daisies, and redwoods, or how did fish come to swim, snakes come to slither, sparrows come to sing, and mosquitoes come to suck? These evolutionary biologists may be interested, for example, in how the ability to fly evolved independently in birds, bats, and insects. To address such questions it is necessary to know how the major groups diverged from one another, and this pattern ultimately results from the history of individual speciation events.

Phylogenetic trees represent the history of divergence

We saw in Chapter 4 how the history of divergence can be represented in a treelike diagram (Figure 4.21). Such branching diagrams, called cladograms or **phylogenetic**

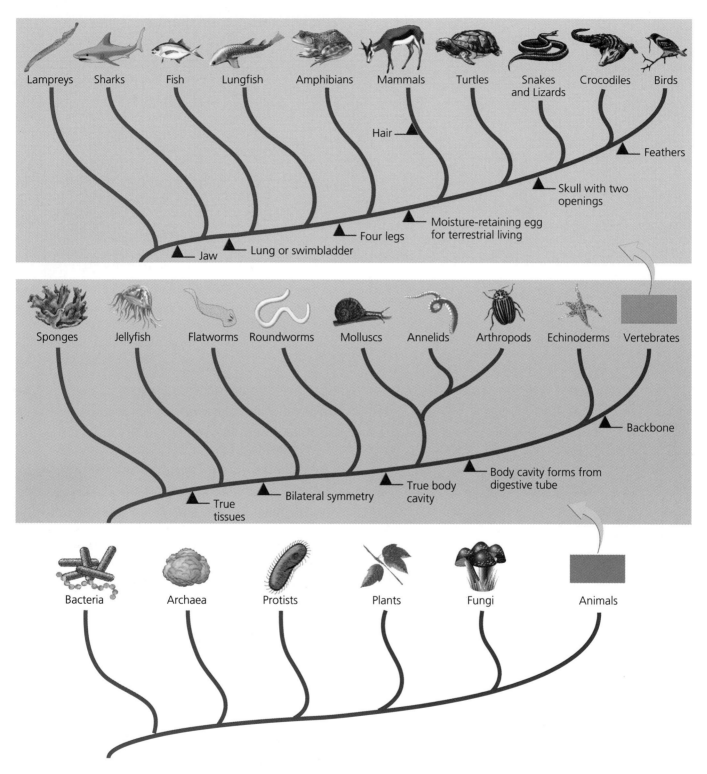

Figure 5.4 Phylogenetic trees show the history of life's divergence. Similar to family geneaologies, these trees indicate relationships among groups of organisms, as inferred from the study of similarities and differences among present-day creatures. They also allow traits to be mapped on, so that biologists can study how traits have evolved over time. The diagram here is a greatly simplified representation of relationships among a few major groups, one tiny portion of the huge and complex tree of life. Each branch results from a speciation event, and time proceeds upward from bottom to top.

Figure 5.5 Life has not always progressed from simple to complex during evolution. Many complex organisms, as those depicted in this Burgess Shale painting by Marella J. Sibbick, have gone extinct, taking their designs and innovations with them. The strange creatures found fossilized in the Burgess Shale of the Canadian Rockies in British Columbia are a case in point. They lived in marine environments 530 million years ago and vanished without a trace. Source: The National History Museum, London.

trees, illustrate scientists' hypotheses as to how divergence took place (Figure 5.4). Such trees can show relationships among species, among major groups of species, among populations within a species, or even among individuals.

Phylogenetic trees allow biologists to examine how traits of interest likely evolved. By mapping traits onto a tree with the organisms that possess them, one can trace how the traits themselves may have evolved. For instance, because the tree of life shows that birds, bats, and insects are distantly related, with many other flightless groups between them, it is far simpler to conclude that the three groups evolved flight independently than that the many flightless groups all lost an ancestral ability to fly. Because phylogenetic trees help biologists make such inferences about so many traits, they have become one of the modern biologist's most powerful tools.

Life has diversified in amazing ways

Life's history, as revealed by phylogenetic trees and by the fossil record, is complex indeed, but a few big-picture trends are apparent. As we mentioned in Chapter 4, life in its 3.5 billion years has evolved complex structures from simple ones, and large sizes from small

ones. However, these are only generalizations. Many organisms have in fact evolved to become simpler or smaller, when natural selection favored it. Many very complex life forms have disappeared (Figure 5.5). Despite the pretensions to grandeur that our own species often displays, it is easy to argue that Earth still belongs to the bacteria and other microbes, some of them little changed over eons.

Even fans of microbes, however, must marvel at some of the exquisite adaptations animals, plants, and fungi have evolved—the heart that beats so reliably for an animal's entire lifetime that we take it for granted; the complex organ system of which the heart is a part; the stunning plumage of a peacock in full display; the ability of each and every plant on the planet to lift water and nutrients from the soil, gather light from the sun, and turn it into food; the eyes with which we see (and that have evolved many times in diverse organisms); the staggering diversity of beetles and other insects; the human brain and its ability to reason. All these things and more have resulted from the process of evolution as it has generated new species and whole new branches on the tree of life.

Extinction and speciation together determine Earth's biodiversity

Although speciation generates Earth's biodiversity, it is only one part of the equation—for, as you recall, the vast majority of species that once lived are now gone. The disappearance of an entire species from the face of the Earth is called **extinction.** From studying the fossil record, paleontologists calculate that the average time a species spends on Earth is 1–10 million years. The number of species alive on Earth at any one time is equal to the number formed through speciation minus the number removed by extinction.

Some species are more vulnerable to extinction than others

Generally extinction occurs when environmental conditions change rapidly or severely enough that a species cannot genetically adapt to the change; natural selection simply does not have enough time to work. All manner of environmental events can cause extinction, from climate change to the rise and fall of sea level, to the arrival of new harmful species, to severe weather events such as extended droughts (Figure 5.6). In general, small populations and species narrowly specialized on some particular resource or way of life are most vulnerable to environmental change and potential extinction.

Figure 5.6 Until 10,000 years ago, the North American continent teemed with a variety of large mammals, including mammoths, camels, giant ground sloths, lions, sabre-toothed cats, and various types of horses, antelope, bears, and others. Nearly all of this megafauna went extinct suddenly about the time that humans first arrived on the continent. Scientists still debate whether overhunting by humans was the cause, but similar extinctions have occurred in other areas simultaneous with human arrival. Source: National Museum of Natural History, Smithsonian Institution.

The golden toad was a prime example of a vulnerable species. It was **endemic** to the Monteverde cloud forest, meaning that it occurred nowhere else on the planet. Endemic species face relatively high risks of extinction because all of their members belong to a single, sometimes small, population. At the time of its discovery, the golden toad was known to live only in a 4-km² (988-acre) area of Monteverde. It not only was restricted to a tiny patch of forest, but also required very specific conditions to breed successfully. During the spring at Monteverde, water collects in shallow pools formed within the network of roots that span the cloud forest's floor. It was in these pools that the golden toad gathered to breed, and it was there among the root-bound reservoirs that Jay Savage and his companions collected their specimens in 1964. Monteverde's climate provided the ideal habitat for the golden toad, but the tiny size of that habitat meant that any environmental stresses that deprived the toad of the resources it needed to survive might doom the entire world population of the species.

In the United States, a number of amphibians are similarly limited to very small ranges, and thus vulnerable to extinction. The Yosemite toad is restricted to a small area of the Sierra Nevada in California, the Houston toad occupies only a few areas of Texas woodland, and the Florida bog frog lives in a tiny region of Florida wetland. Many salamanders are even more geographically limited; fully 40 salamander species in the United States are restricted to areas the size of a typical county, and some of these live atop single mountains.

Earth has seen several episodes of mass extinction

Most extinction occurs gradually, one species at a time; this type of extinction is referred to as *background extinction*. However, Earth has seen five events of staggering proportions that killed off massive numbers of species at once. These episodes, called **mass extinction** events, have occurred at widely spaced intervals in Earth history and have wiped out anywhere from half to 95% of Earth's species each time.

The best-known mass extinction occurred 65 million years ago and brought an end to the dinosaurs (although birds are likely modern representatives of dinosaurs). Much evidence suggests that the impact of a gigantic asteroid caused this event, called the K–T event (see "The Science behind the Story," page 128).

The K–T event, as massive as it was, was moderate compared to the mass extinction at the end of the Permian period 250 million years ago (See Appendix A). Paleontologists estimate that 75–95% of all species on Earth may have perished during this event. Precisely what caused the event scientists don't know. The evidence for extraterrestrial impact is much weaker than it is for the K–T event, and other ideas abound. The hypothesis with the most support may be massive volcanism that threw into the atmosphere a worldwide blanket of soot and sulfur, smothering the planet, reducing sunlight and inducing severe climate gyrations.

The sixth mass extinction is upon us

Many biologists have concluded that Earth is currently in the throes of its sixth mass extinction event—and that we humans are the cause. The changes to Earth's environment set in motion by human population growth, resource use, and development have greatly altered conditions for many species, have driven many extinct already, and are threatening countless more. The alteration and outright destruction of natural habitats, the hunting and harvesting of species, and the introduction of invasive species from one place to another where they can harm native species—these processes and many more have combined to threaten Earth's biodiversity.

When we look around us, it may not appear as though a human version of an asteroid impact were taking place, but we cannot judge such things on a human timescale. On the geological timescale, extinction over 100 years or over 10,000 years appears every bit as instantaneous as extinction over a few days.

Weighing the Issues:
Are We Causing a Mass Extinction?

Some critics doubt that we are about to cause a mass extinction event, and criticize as alarmist the scientists who warn of this. What do you think can account for such disparity of viewpoints?

Other critics acknowledge human impacts on biodiversity but we should not be concerned about biodiversity loss or even about a new mass extinction. Thinking back to our discussion of ethics and economics in Chapter 2, can you think of some reasons we should be concerned about potential loss of biodiversity or extinction of species? If there are reasons for concern, do you think those reasons are substantial enough to justify measures to protect biodiversity and prevent extinction?

Units of Ecological Organization

The extinction of species, their generation through speciation, and other evolutionary mechanisms and patterns have substantial impact on ecology. In addition, it's often said that ecology provides the stage on which the play of evolution takes place. The two, it's clear, are tightly intertwined in many ways.

As we discussed in Chapter 1, ecology is the study of interactions among organisms and between organisms and their environments. Life occurs in a hierarchy of levels and ecologists study relationships on several of these levels (Figure 5.7). The atoms, molecules, and cells we reviewed in the previous chapter represent the lowest levels in this hierarchy. Aggregations of cells of a particular type form tissues, and tissues form organs, all housed within an individual living creature, or organism. A group of organisms of the same species that live in the same area is a population, and species are often composed of multiple populations. Communities are made up of multiple interacting species that live in the same area, and ecosystems encompass communities and the abiotic (nonliving) material with which their members interact.

The science of ecology deals with the organismal, population, community, and ecosystem levels. At the organismal level, ecology describes the relationships between organisms and their physical environments. Population ecology investigates the quantitative dynamics of how individuals within a species interact with one another. Community ecology focuses on interactions among species, from one-on-one interactions to complex interrelationships involving entire communities. Finally, ecology at the ecosystem level studies patterns such as energy and nutrient flow that are revealed when living and nonliving systems are studied in conjunction. As improving technologies allow scientists to learn more and more about the complex operations of natural systems on a global scale, ecologists are increasingly expanding their horizons beyond ecosystems to the entire biosphere and to Earth as a whole. In this chapter we explore ecology up through the community level; in the next chapter we continue with the ecosystem level.

Hierarchy of Life		
	Biosphere	The sum total of living things on Earth and the areas they inhabit
	Ecosystem	A functional system consisting of a community, its non-living environment, and the interactions between them
	Community	A set of populations of different species living together in a particular area
	Population	A group of individuals of a species that live in a particular area
	Organism	An individual living thing
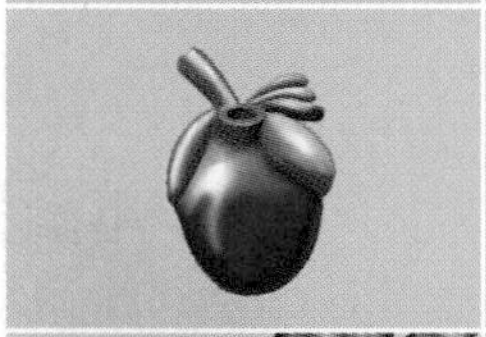	Organ	A structure in an organism composed of several different types of tissues and specialized for some particular function
	Tissue	A group of cells with common structure and function
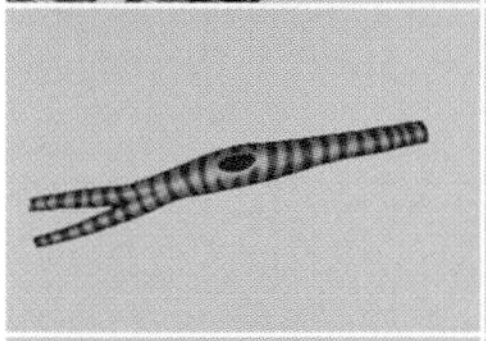	Cell	The smallest unit of living matter able to function independently, enclosed in a semi-permeable membrane
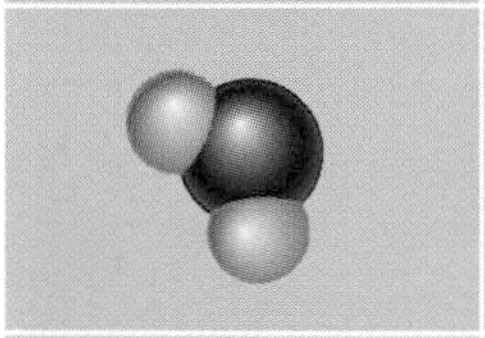	Molecule	A combination of two or more atoms chemically bonded together
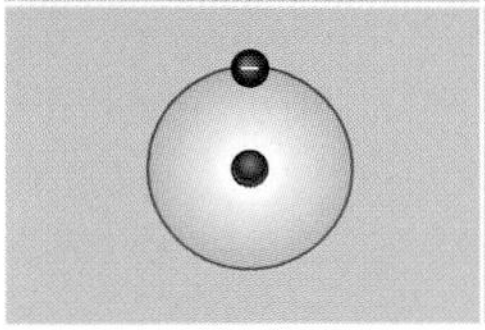	Atom	The smallest component of an element that maintains the element's chemical properties

Figure 5.7 Life occurs in a hierarchy of levels. Of these levels of biological organization, ecology includes the study of the organismal, population, community, and ecosystem levels.

Ecology on the Organismal Level

Each organism relates to its abiotic environment in ways that in the long run tend to maximize its survival and reproduction. One key relationship involves the specific environment in which an organism lives—its habitat. **Habitats** consist of both living and nonliving elements—of rock, soil, leaf litter, and humidity, as well as grasses, shrubs, and trees. Habitats are scale-dependent; a tiny soil mite may perceive its habitat as a mere square meter of soil. A vulture, in contrast, may view its habitat in terms of miles upon miles of hills and valleys that it easily traverses by air.

Organisms select habitats in which to live from among the range of options they encounter, a process usually called **habitat selection.** The criteria organisms use to select habitats can vary greatly. The soil mite might judge available habitats in terms of the chemistry, moisture, and compactness of the soil, and the percentage and type of organic matter. The vulture may ignore not only soil but also topography and vegetation, focusing solely on the abundance of dead animals in the area that it scavenges for food. Every species judges habitats differently because every species has different needs.

Habitat selection is important in environmental science because the availability and quality of habitat is crucial to an organism's well-being. Indeed, because habitats provide everything an organism needs, including food, shelter, breeding sites, and mates, the organism's very survival depends on the availability of suitable habitats. Sometimes this engenders conflict with humans who want to alter or develop a habitat for other purposes.

Another way in which an organism relates to its environment is through its niche. A species' **niche** is its functional role in a community. This includes not only its use of certain habitats, but also its consumption of certain foods, its role in the flow of energy and matter, and its interactions with other organisms. The niche is a multidimensional concept, a summary of everything that an organism does. Some mites decompose dead organic material and serve as food for larger spiders, while others prey on smaller organisms and serve as food for insects. An organism's habitat (where it lives) and the organism's niche (what it does) each reflect the adaptations of the species and thus are products of natural selection.

Population Ecology

Individuals of the same species inhabiting a particular area make up a population. Species may consist of multiple populations that are geographically isolated from one another. This is the case, for example, with a species characteristic of Monteverde—the resplendent quetzal, considered one of the world's most spectacular birds (see Figure 5.1). Although it ranges from southernmost Mexico to Panama, the resplendent quetzal lives only in high-elevation tropical forest, and it is absent from many nonmountainous areas. Furthermore, much of its forest habitat has been destroyed by human development. Thus the species today exists in many separate populations scattered across Central America. Humans, in contrast, have become more mobile than any other species and have spread into nearly every corner of the planet. As a result, it is difficult to define a distinct human population on anything less than the global scale; some would argue that in the ecological sense of the word, all six billion of us comprise one population.

Populations exhibit characteristics that can help predict their future dynamics

Whether one is considering humans or quetzals or golden toads, all populations show characteristics that help population ecologists make predictions about the future dynamics of that population. Attributes such as density, distribution, sex ratio, age structure, and birth and death rates all assist the ecologist trying to understand how a population may grow or decline. The ability to predict growth or decline is especially useful in monitoring and managing threatened and endangered species. It is also vital in applying to human populations (Chapter 7); understanding human population dynamics, their causes, and their consequences is one of the central elements of environmental science and one of the prime challenges for our society today.

Population size Expressed as the number of individual organisms present at a given time, **population size** may increase, decrease, undergo cyclical change, or remain the same over time. Extinctions are generally preceded by population declines. As late as 1987, scientists documented a golden toad population at Monteverde in excess of 1,500 individuals, but in 1988 and 1989 scientists sighted only a single toad. By 1990, the species had disappeared without a trace.

The passenger pigeon, now extinct, illustrates the extremes of population size. Once the most abundant bird in North America, flocks of passenger pigeons literally darkened the skies. An ornithologist of the early 19th century wrote of watching a flock of two billion birds 390 km (240 mi) long that took 5 hours to fly over. Passenger pigeons nested in gigantic colonies in the forests of the upper Midwest and southern Canada. Their great concentrations made them easy targets for market hunters, however, who gunned down thousands at a time and shipped them to market by the wagonload. By the end of the 19th century the passenger pigeon population had declined to such a low number that they could not form the large colonies they apparently needed to breed effectively. In 1914, the last passenger pigeon on Earth died in the Cincinnati Zoo, bringing the continent's most numerous bird to complete extinction within just a few decades.

Population density The flocks and breeding colonies of passenger pigeons showed high population density, another attribute that scientists assess to better understand populations. **Population density** describes the number of individuals within a population per unit area. For instance, the 1,500 golden toads counted in 1987 within 4 km^2 (988 acres) indicated a density of 375 toads/km^2. In general, larger organisms have lower population densities because they require more resources to survive.

Population density can affect a population and its members in many ways. High population density can make it easier for organisms to find mates and participate in other important social interactions, but it can also lead to conflict in the form of competition if space, food, or mates are in limited supply. Furthermore, overcrowded organisms may become more vulnerable to the predators that feed on them, and close contact among individuals can increase the transmission of infectious disease. For these reasons, organisms sometimes may leave an area when densities become too high. Low population density can provide organisms with more space and resources but can make it harder for them to find mates and companions.

The effects of overcrowding at high population densities is thought to have doomed the harlequin frog, an amphibian that disappeared from the Monteverde cloud forest at the same time as the golden toad. The harlequin frog had a narrow habitat preference, favoring *splash zones,* areas alongside rivers and streams that receive spray from waterfalls and rapids. As Monteverde's climate grew warmer and drier in the 1980s

The Last Mass Extinction

Throughout the history of life on Earth, individual species such as the golden toad have occasionally died off, and new ones have appeared. On five occasions, however, huge numbers of species went extinct in a geologic instant. The last mass extinction occurred 65 million years ago, the date that marks the dividing line between the Cretaceous and the Tertiary periods, or the K-T boundary for short. About 70% of the species then living, including the dinosaurs, disappeared.

When he first started working at Bottaccione Gorge, Walter Alvarez had no idea he would help discover what killed off the dinosaurs. An American geologist, Alvarez was working for the summer in northern Italy, developing a new method to determine the age of sedimentary rocks. But he and his father, physicist Luis Alvarez, would soon come up with a dramatic theory about what caused the best-known mass extinction. Alvarez chose Bottacione Gorge because it formed an ideal geological archive: Its 400 m (1,300 ft) walls are stacked like layer cake with beds of rose-colored limestone that

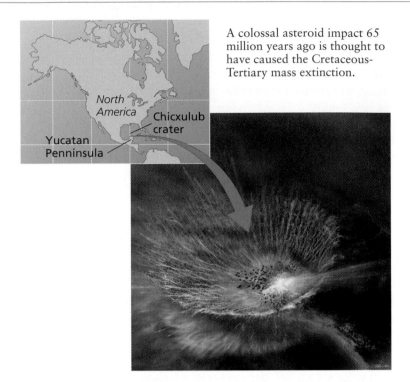

A colossal asteroid impact 65 million years ago is thought to have caused the Cretaceous-Tertiary mass extinction.

formed between 100 million and 50 million years ago from dust that had settled to the bottom of an ancient sea. While analyzing the many layers of the Bottaccione Gorge, Walter Alvarez noticed a 1 cm (.4 in) thick band of reddish clay sandwiched between two layers of limestone. The older layer just below it was packed with fossils of globotruncana, a sand-sized animal that lived in the late Cretaceous period. The newer layer just above it instead contained just a few scattered fossils of a cousin of globotruncana, typical of sedimentary rock formed in the early Tertiary period. What Alvarez found

and 1990s, water flow in rivers was reduced, and many streams dried up. The splash zones grew smaller and fewer, and harlequin frogs were forced to cluster together in what remained of the splash-zone habitat. Researchers J. Alan Pounds and Martha Crump recorded frog population densities up to 4.4 times higher than normal, with more than two frogs per every 1 m (3.3 ft) of stream. Such overcrowding likely made the frogs more vulnerable to disease transmission, predator attack, and especially assault from parasitic flies. From their field research, during which Pounds and Crump witnessed 40 frogs dead or dying, the researchers con-

cluded that these factors led to the harlequin frog's apparent extinction.

Population distribution It was not simply the harlequin frog's density, but also its distribution in space that led to its demise. **Population distribution**, or **population dispersion**, describes the spatial arrangement of organisms within a particular area. Ecologists define three distribution types: random, uniform, and clumped (Figure 5.8). In a **random distribution**, individuals are located haphazardly in space in no particular pattern. This type of distribution can occur when the resources an organ-

interesting was that the intermediate clay layer, formed just as dinosaurs were going extinct, had no fossils at all.

To see how quickly the K-T mass extinction event occurred, Walter and Luis Alvarez analyzed the intermediate clay layer to see how long it had taken to form, using the rare metal iridium as a molecular clock. Almost all of the iridium on Earth's surface comes from dust from tiny meteorites (shooting stars) that burn up in the atmosphere. Because the same amount of meteorite dust rains down each year, the Alvarezes measured iridium levels in the clay layer to determine how many years' worth of iridium had accumulated.

Iridium levels in the limestone up and down the gorge were typical for sedimentary rocks, about 0.3 parts per billion. In the clay layer, however, the Alvarezes were surprised to find levels 30 times higher. To make sure the finding was not unique to Bottaccione Gorge, they checked the K-T clay layer at a Danish sea cliff; it had 160 times more iridium than the surrounding rock. The Alvarezes boldly predicted that the excess iridium had come from a massive asteroid that smashed the earth, causing a global environmental catastrophe that had wiped out the dinosaurs.

To convince themselves and the science community that an asteroid impact did in fact cause the K-T event, the Alvarezes had to rule out other possible explanations. For example, the extra iridium could have come from seawater. Calculations, however, proved that seawater did not contain enough iridium to account for the high levels in the clay layers.

An asteroid 10 km (6.2 mi) wide strikes Earth, on average, every 100 million years. With each hit is an explosion with a force 1,000 times greater than the force of the 1883 explosion of the Indonesian volcano Krakatau, which scattered so much dust around the world that sunsets were multicolored for 2 years afterward. An asteroid impact 65 million years ago, they suggested, kicked up enough soot to blot out the sun for several years, which inhibited photosynthesis, causing plants to die off, food webs to collapse, and most animals, including dinosaurs, to die of starvation. Only a few smaller animals survived, feeding on rotting vegetation. When sunlight returned, plants sprouted from dormant seeds, and evolution began anew.

Published in the journal *Science* in 1980, the Alvarezes' theory was immediately attacked by other geologists, who claimed that spectacular volcanic eruptions more likely explained the high levels of iridium in the Bottoccione Gorge. But throughout the 1980s, scientists kept finding evidence in favor of the asteroid-impact theory. Iridium-enriched clay turned up at K-T layers around the world, as did bits of minerals called shocked quartz and stishovite, which, geologists have determined, form only under the extreme pressure of thermonuclear explosions and asteroid impact. The Alvarezes' theory was finally accepted in 1991, when scientists found a 65-million-year-old crater the expected size in the ocean off the coast of Mexico. Today, scientists generally agree that an asteroid impact 65 million years ago caused the extinction of the dinosaurs.

ism depends on are found throughout an area and other organisms do not exert a strong influence on where members of a population settle.

A **uniform distribution** is one in which individuals are evenly spaced out. This can occur when individuals hold territories or otherwise compete for space. For instance, in a desert where there is little water, plants may need a certain amount of space for their roots to gather adequate moisture; as a result, every plant of a species may be equidistant from other plants.

In a **clumped distribution**, the pattern most common in nature, organisms arrange themselves according to the availability of the resources they need to survive. Many desert plants grow in clumps or patches around isolated springs, or along arroyos that flow with water after rainstorms. During their mating season, golden toads were found clumped at seasonal breeding pools. Humans, too, exhibit clumped distribution; people frequently aggregate together in urban centers. Clumped distributions often indicate habitat selection.

Distributions can depend on the scale at which one measures them. At very large scales, for instance, all organisms show clumped or patchy distributions, because some parts of the total area they inhabit are bound to be

(a) Random

(b) Uniform

(c) Clumped

Figure 5.8 Individuals in a population can be spatially distributed over a landscape in three fundamental ways. (a) In a random distribution, organisms are dispersed at random through the environment. (b) In a uniform distribution, individuals are spaced evenly, at equal distances from one another. Territoriality can result in such a pattern. (c) In a clumped distribution, individuals occur in patches, concentrated more heavily in some areas than in others. Habitat selection or flocking to avoid predators can cause such a pattern.

more hospitable than others. Population distribution patterns can make species more or less vulnerable to population declines, as we saw with the harlequin frog.

Age structure Populations may consist of individuals of different ages. **Age distribution**, or **age structure**, describes the relative numbers of organisms of each age within a population. Age distribution can have a strong effect on rates of population growth or decline. Age structures are often expressed as a ratio of age classes, consisting of organisms (1) not yet mature enough to reproduce; (2) capable of reproduction, which represents potential population increase through births; and (3) beyond their reproductive years and likely nearing the ends of their lives. A population made up mostly of individuals past their reproductive stage will tend to decline over time. In contrast, a population that includes a large percentage of young organisms that will soon begin reproducing or that are in the midst of their reproductive years is likely to increase. A population with a more even age distribution will likely remain stable over time, as the reproduction of younger organisms keeps pace with the death of older ones.

Age structure diagrams, often called *age pyramids*, are visual tools that scientists use to show the age structure of a population (Figure 5.9). The width of each horizontal bar represents the relative size of each age class. A pyramid with a wide base has a relatively large age class that has not yet reached its reproductive stage, indicating a population much more capable of rapid growth. In this respect, the wide base of an age pyramid is like an oversized engine on a rocket—the bigger the booster, the faster the increase.

Sex ratios Besides age distribution, the sex ratio of a population can also help determine whether it will increase or decrease in size over time. A population's **sex ratio** is its proportion of males to females. In monogamous species in which each sex takes a single mate, a 50/50 sex ratio maximizes population growth, whereas an unbalanced ratio leaves many individuals of one sex without mates.

Populations may grow, shrink, or remain relatively stable

Now that we have outlined attributes of populations, we are ready to take a quantitative view of population change, by utilizing some simple mathematical concepts used by population ecologists and demographers (those who study human populations). Population growth, or decline, is determined by four factors:

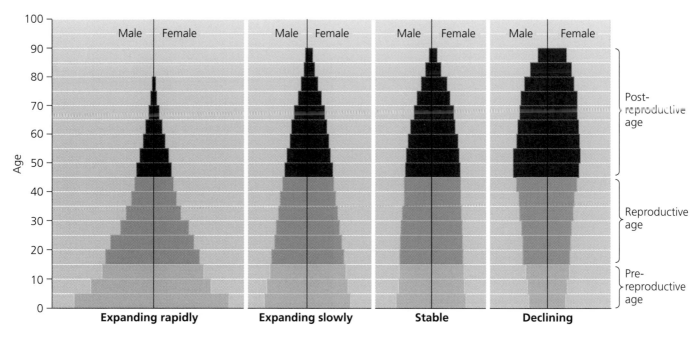

Figure 5.9 Age structure diagrams show the relative frequencies of individuals of different age classes in a population. Populations heavily weighted toward young age classes (at left) grow most quickly.

1. Births within the population
2. Deaths within the population
3. **Immigration** (arrival of individuals from outside the population)
4. **Emigration** (departure of individuals from the population)

To understand how a population changes, we measure its **growth rate**, which is the crude birth rate plus the immigration rate minus the crude death rate plus the emigration rate, each expressed as the number per 1,000 individuals per year:

(Crude birth rate + immigration rate) − (Crude death rate + emigration rate) = Growth rate

The resulting number tells us the net change in a population's size, per 1,000 individuals. For example, a population with a crude birth rate of 18 per 1,000, a crude death rate of 10 per 1,000, an immigration rate of 5 per 1,000, and an emigration rate of 7 per 1,000 would have a growth rate of 6 per 1,000:

(18/1,000 + 5/1,000) − (10/1,000 + 7/1,000) = 6/1,000

Thus, a population of 1,000 in one year will reach 1,006 in the next. If the population is 1,000,000, it will reach 1,006,000 the next year. These population increases are often expressed as percentages, which we can calculate using the formula:

Growth rate × 100%

Thus, a growth rate of 6/1,000 would be expressed as

6/1,000 × 100% = 0.6%

By measuring population growth in terms of percentages, scientists can compare increases and decreases in species that have far different population sizes. They can also thereby project changes that will occur in the population over longer periods, much like you might calculate the amount of interest your savings account will earn over time.

Unregulated populations tend to increase by exponential growth

When a population, or anything else, increases by a fixed percentage each year, it is said to undergo **exponential growth.** A savings account is a familiar frame of reference for describing exponential growth. If at the time of your birth your parents had invested $1,000 in a savings account earning 5% interest compounded each year, you would have only $1,629 by age 10, and $2,653 by age 20, but you would have over $30,000 when you turn 70. If you could wait just 10 years more, that figure would rise to nearly $50,000 (Table 5.2). Only $629 was added over the first decade, but approximately $19,000 was added between ages 70 and 80. The reason is that a fixed percentage of a small number is a small increase, but that

Table 5.2 Exponential Growth in a Savings Account with 5% Annual Compound Interest

Age (in years)	Principal
0 (birth)	$1,000
10	$1,629
20	$2,653
30	$4,322
40	$7,040
50	$11,467
60	$18,679
70	$30,426
80	$49,561

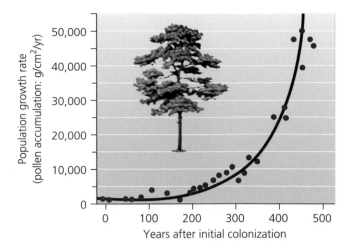

Figure 5.10 Although few species maintain exponential growth for very long in nature, some grow exponentially when colonizing an unoccupied environment or exploiting an unused resource. Scientists have used pollen records to determine that the Scots pine increased exponentially after the retreat of glaciers following the last ice age around 9,500 years ago. Go to **GRAPH IT** on the website or CD-ROM. Data from K. D. Bennett, Postglacial population expansion of forest trees in Norfolk, U. K., *Nature*, 1983.

same percentage of a large number produces a large increase. Thus as savings accounts (or populations) get larger, each incremental increase likewise becomes larger. Such rapid acceleration is a characteristic of exponential growth.

We can visualize changes in population size using **population growth curves.** The J-shaped curve in Figure 5.10 shows exponential increase. As Thomas Malthus realized, populations of all organisms will tend to increase exponentially unless they meet constraints, because each organism reproduces by a certain number, and as populations get larger there are more and more individuals reproducing by that number. If there are no external limits on growth, ecologists theoretically expect exponential growth.

Exponential growth usually occurs in nature when a population is small and environmental conditions are ideal for the organism in question. This most often occurs when organisms are introduced to a new environment. Mold growing on a piece of bread or fruit that's been left out or bacteria colonizing a recently dead animal are cases in point. But species of any size may show exponential growth under the right conditions. A population of the Scots pine, *Pinus sylvestris*, grew exponentially when it began colonizing the British Isles after the end of the last ice age (see Figure 5.10). Receding glaciers had left favorable habitat and conditions ideal for its exponential expansion.

Limiting factors restrain population growth

Exponential growth, however, rarely lasts long. If even a single species in the history of Earth had increased exponentially for very many generations, the planet's surface would be covered with it, and nothing else could

survive. Instead, every population eventually comes to be contained by **limiting factors,** which are physical, chemical, and biological characteristics of the environment that restrain population growth. The interaction of all of these factors will determine the **carrying capacity,** or maximum population size that a given environment can sustain.

Ecologists use the curve shown in Figure 5.11 to show how an initial exponential increase is slowed and finally brought to a standstill by limiting factors. Called the **logistic growth** curve, it rises sharply at first and then begins to level off as the effects of limiting factors become stronger. Eventually the force of these factors—which taken together are termed **environmental resistance**—stabilizes the population size at its carrying capacity.

In reality, the logistic curve is a simplified model, and populations can behave differently according to environmental conditions. Some may cycle indefinitely above and below the carrying capacity, some may show cycles that become less extreme and approach the carrying capacity, and others may overshoot the carrying capacity and then crash, fated either for extinction or recovery (Figure 5.12).

Many factors contribute to environmental resistance and influence a population's growth rate and carrying capacity. Space is one obvious such factor. If there is no physical room for additional organisms, it is unlikely that they will survive. Food is another factor that will limit the number of individuals in a population that a given environment can support. Other limiting factors for animals in a terrestrial environment might include the availability of mates, shelter, water, and suitable breeding sites; temperature extremes; prevalence of disease; and abundance of predators. Plants are often limited by amounts of sunlight and moisture and the type of soil chemistry, in addition to disease and attack from plant-eating animals. In aquatic systems limiting

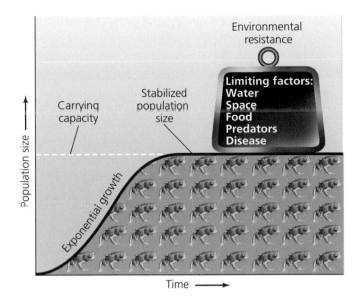

Figure 5.11 The logistic growth curve shows how populations may increase rapidly at first, then slow down, and finally stabilize at their carrying capacity. Carrying capacity is determined both by the biotic potential of the organism and by various external limiting factors, collectively termed *environmental resistance*.

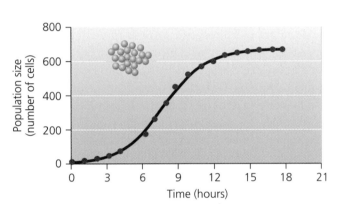

(a) Yeast cells, *Saccharomyces cerevisiae*

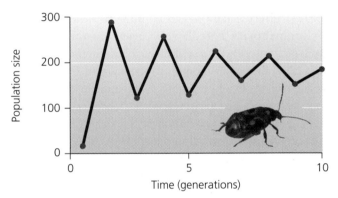

(c) Stored-product beetle, *Callosobruchus maculatus*

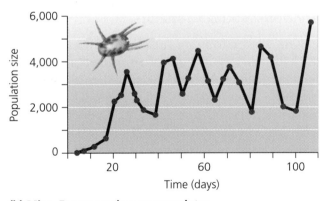

(b) Mite, *Eotetranychus sexmaculatus*

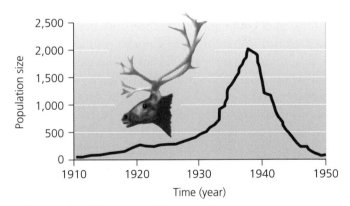

(d) St. Paul reindeer, *Rangifer tarandus*

Figure 5.12 Population growth in nature often departs from the stereotypical logistic growth curve, and can do so in several fundamental ways. **(a)** Yeast cells from an early lab experiment (Pearl 1927) show logistic growth that, like the Scots pine in Figure 5.10, closely matches the theoretical model. **(b)** Some organisms, like the mite shown here, show cycles in which population fluctuates indefinitely above and below the carrying capacity. **(c)** Population oscillations can also dampen, lessening in intensity and eventually stabilizing at carrying capacity, as in a lab experiment with the stored-product beetle (Utida 1967). **(d)** Populations that rise too fast and deplete resources may crash just as suddenly, like the population of reindeer introduced on the Bering Sea island of St. Paul. Data from R. Pearl, The growth of populations. *Quarterly Review of Biology,* 1927 (yeast). S. Utida, Damped oscillation of population density at equilibrium, *Researches on Population Ecology,* 1967 (stored product beetle). C.B. Huffaker, Experimental studies on predation: dispersion factors and predator-prey oscillations, *Hilgardia,* 1958 (mites). V.C. Scheffer, Rise and fall of a reindeer herd, *Scientific Monthly,* 1951 (reindeer).

factors might include salinity, sunlight, temperature, dissolved oxygen, fertilizers, and pollutants.

Sometimes one limiting factor may outweigh all others and restrict population growth. For example, scientists hypothesize that Monteverde's population of golden toads had plenty of space, food, and shelter but lacked adequate moisture. If moisture were the primary limiting factor, then increasing available moisture would have increased the carrying capacity of the habitat for the toads. Indeed, to determine limiting factors, ecologists often conduct experiments, and increase or decrease a hypothesized limiting factor to observe its effects on population size. Unfortunately in the case of the golden toad, such experiments could not be done before its disappearance.

Carrying capacities can change

Because limiting factors can be numerous, and because environments are complex and ever-changing, carrying capacity can change constantly. Nonetheless, carrying capacity is a crucially important concept in environmental science, and one that has great importance for questions of human population growth (Chapter 7). The human species illustrates another reason that carrying capacity is not necessarily a fixed entity. Although all organisms are subject to environmental resistance, they may be capable to some extent of altering their environment to reduce this resistance—and our species has proved particularly effective at this. For example, when our ancestors began to build shelters and use fire for heating and cooking, they reduced the environmental resistance of areas with cold climates and were able to expand into new territory. As environmental resistance decreases, either through the development of new technologies or through change driven by natural processes, the environment's carrying capacity for a particular species may increase. We humans have managed so far to increase the planet's carrying capacity for ourselves, but unfortunately for the golden toad, the environmental resistance to its population growth seemed to exert ever-increasing pressure during the late 1980s.

Weighing the Issues:
Carrying Capacity and Human Population Growth

As we discussed in Chapter 1, the global human population has shot up from fewer than 1 billion 200 years ago to more than 6 billion today and we have far exceeded our old carrying capacity. What factors enabled

an increase in Earth's carrying capacity for humans? Do you think there are limiting factors for the human population? What might they be? Do you think we can keep raising our carrying capacity in the future? Might Earth's carrying capacity for us decrease?

The influence of some factors on population depends on population density

Just as carrying capacity is not a fixed entity, the influence of limiting factors can vary with changing conditions. In particular, the density of a population can increase or decrease the impact of certain factors upon that population. Recall that high population density can help organisms find mates but can also increase competition and the risk of predation and disease. Such factors are said to be **density-dependent** factors; that is, their influence waxes and wanes according to the density of the population. The logistic growth curve in Figure 5.11 represents the effects of density dependence. The more population size rises, the more environmental resistance kicks in. **Density-independent** factors are those whose effects are constant regardless of population density. Factors such as temperature extremes and catastrophic events such as floods, fires, and landslides are examples of density-independent factors because they can eliminate large numbers of individuals without regard to their density.

Biotic potential and reproductive strategies vary from species to species

Limiting factors from an organism's environment are only half the story of population regulation, however. The other half comes from the attributes of the organism itself—for example, its ability to produce offspring. Organisms differ in their ability to produce offspring, and the term **biotic potential** refers to the innate reproductive capacity of a species. For example, a fish that has a short gestation period and lays thousands of eggs at a time has high biotic potential, whereas a whale that has a long gestation period and produces a single calf at a time has low biotic potential. The interaction between an organism's biotic potential and the environmental resistance to its population growth helps determine the fate of its population.

Giraffes, elephants, humans, and other large animals with low biotic potential produce a relatively small number of offspring and take a long time to gestate and raise each of their young. Species that take this approach to reproduction are known as **K-strategists**, devoting their energy and resources to caring for and protecting the relatively few offspring they produce during their lifetimes.

Organisms with high biotic potential include many fish, plants, frogs, insects, and others. Such organisms can produce a large number of offspring in a relatively short time, and these offspring do not require parental care after birth. Known as **r–strategists**, these species devote their energy and resources to producing as many offspring as possible. Rather than caring for their young, they leave their offspring's survival to chance. The golden toad was an r–strategist. Each adult female produced 200–400 eggs, and its tadpoles spent 5 weeks unsupervised in the breeding pools metamorphosing into adults.

K–strategists are so named because their populations tend to stabilize over time at or near their carrying capacity, and "K" is an abbreviation used to denote carrying capacity. The term "r–strategist" denotes the intrinsic growth rate. This is the rate at which a population would increase in the absence of limiting factors, and r strategists are intrinsically capable of rapid population growth.

K–strategists generally find their populations regulated by density-dependent factors such as disease, predation, and food limitation. In contrast, density-independent factors tend to regulate the population sizes of r–strategists, whose success or failure is often determined by large-scale environmental change. Many r–strategists frequently experience large swings in population size: rapid increases during the breeding season and rapid declines soon after, as unfit and unlucky young are removed from the population. For this reason scientists often have difficulty determining whether rapid population declines are part of natural cycles or a sign of serious trouble. For years scientists debated the golden toad's apparent extinction. Now that the disappearance has persisted for well over a decade, most agree that the toad's population crash was not part of a normal, repeating cycle.

Beyond Populations to Communities

In the late 1980s the golden toad and the harlequin frog were the most diligently studied species affected by changing environmental conditions in the Costa Rican cloud forest, but they were not alone. Once scientists began looking beyond the frog and toad populations to populations of other species at Monteverde, they began to notice more troubling changes. By the early 1990s not only had golden toads, harlequin frogs, and other organisms been pushed from their cloud-forest habitat into apparent extinction, but many species from lower, drier habitats had begun to appear in the higher, wetter cloud forest of Monteverde. These immigrant species included dry-tolerant birds such as blue-crowned motmots and brown jays. By the year 2000, 15 dry-forest species had moved into the cloud forest and begun to breed there. Several cloud-forest bird species that had been common in the past also had become less common. After 1987, 20 of 50 frog species vanished from one part of Monteverde, and ecologists later reported more such disappearances, including those of two lizards found in the cloud forest. Apparently, scientists hypothesized, the warming, drying trends that researchers were documenting (see "The Science behind the Story") were causing a change in the composition of the community.

Once we move from the level of the single population to consider groups of populations of different species, we have entered the realm of the community. A **community** is a group of populations of organisms that live in the same place at the same time. A population of golden toads, a population of resplendent quetzals, populations of ferns and mosses, together with all of the other interacting plant and animal populations in the Monteverde cloud forest, could be considered a community.

How cohesive are communities?

Ecologists have conceptualized communities in different ways. Early in the 20th century, botanist Frederick Clements promoted the view that communities are cohesive entities whose members remain associated over time and as environmental conditions change. Communities, he argued, are units with integrated parts, much like organisms are. Clements's view implied that the many varied members of a community share similar limiting factors and evolutionary histories.

Henry Gleason disagreed. Gleason, also a botanist, maintained that each species responds independently to its own limiting factors and that species can join or leave communities without greatly altering their composition. Communities, Gleason argued, are not cohesive units but temporary associations of individual species that could potentially reassemble themselves into different combinations.

Today ecologists side largely with Gleason, although most see validity in aspects of both men's ideas (Figure 5.13). One large-scale historical example that supports Gleason's interpretation of communities occurred after

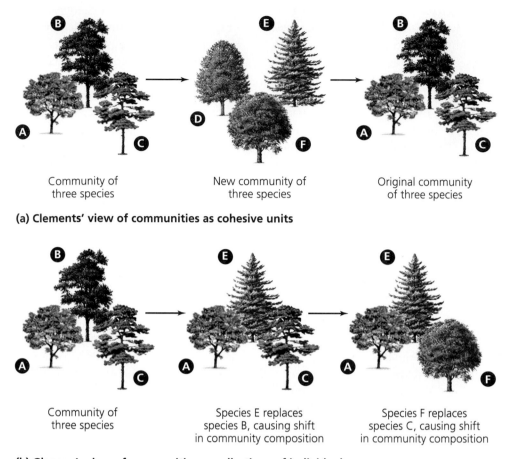

Community of three species

New community of three species

Original community of three species

(a) Clements' view of communities as cohesive units

Community of three species

Species E replaces species B, causing shift in community composition

Species F replaces species C, causing shift in community composition

(b) Gleason's view of communities as collections of individuals

Figure 5.13 (a) Frederick Clements viewed communities as cohesive units, tightly knit groups of species that remain associated over time. In this view, community change involves one distinct group replacing another as environmental conditions change. (b) Henry Gleason viewed communities as temporary collections of species that can independently reassemble themselves into different combinations. In this view, community change involves individual species joining or leaving communities one at a time, resulting in the formation of different species combinations over time.

the American chestnut, the dominant tree in many forests in eastern North America, was wiped out by an imported fungus. Chestnut blight killed nearly all mature chestnuts in the quarter-century preceding 1930. Rather than transforming the entire forest community, however, existing tree species, such as oaks and red maples, which had grown beneath the chestnuts, simply grew higher and became the dominant species, without causing much change to other plants in the community. Although most ecologists have adopted Gleason's perspective, they still find it useful to refer to communities by names that highlight certain key plants (such as *oak-hickory forest, tallgrass prairie,* and *pine-bluestem community*). Ecologists find labeling communities as though they were cohesive units to be a pragmatic tool, even though they know the associations could change radically decades or centuries hence.

A food web is a conceptual record of feeding relationships and energy flow

The interactions among members of a community are many and varied, but some of the most important involve who eats whom. As you learned in Chapter 4, the energy that drives such interactions comes ultimately from the sun via photosynthesis. This energy moves through the community when organisms feed on one another. Conceptual, visual representations of community feeding interactions are known as **food chains,** or **food webs,** (Figure 5.14a). Food chains are simplified linear representations, whereas food webs are generally more realistic—and complicated—because they show a greater array of relationships. Such diagrams show relationships between organisms at different **trophic levels,** or ranks in the feeding hierarchy (Figure 5.14b).

(a) Food web of an Eastern deciduous forest

Figure 5.14 Food webs are conceptual representations of feeding relationships in a community. This one **(a)** pertains to North America's eastern deciduous forest, and includes organisms on several trophic levels. In a food web diagram, arrows are drawn from one organism to another to indicate the direction of energy flow as a result of predation or herbivory. For example, an arrow leads from the grass to the cottontail rabbit to indicate that cottontails eat grass plants and derive nourishment from them. The arrow from the cottontail to the tick indicates that parasitic ticks derive nourishment from the skin of cottontails. **(b)**. Communities include so many species and are complex enough, however, that most food web diagrams are bound to be gross simplifications.

Producers Green plants are producers, or autotrophs ("self-feeders"), as we saw in Chapter 4, and function at the first trophic level. Terrestrial green plants, cyanobacteria, and phytoplankton (tiny aquatic algae) capture solar energy and use photosynthesis to produce sugars, whereas the chemosynthetic bacteria of hot springs and deep-sea hydrothermal vents use geothermal energy to produce food. Producers are the foundation of all energy and matter exchange that occurs in a community.

Consumers Organisms that consume producers are known as **primary consumers.** Grazing animals such

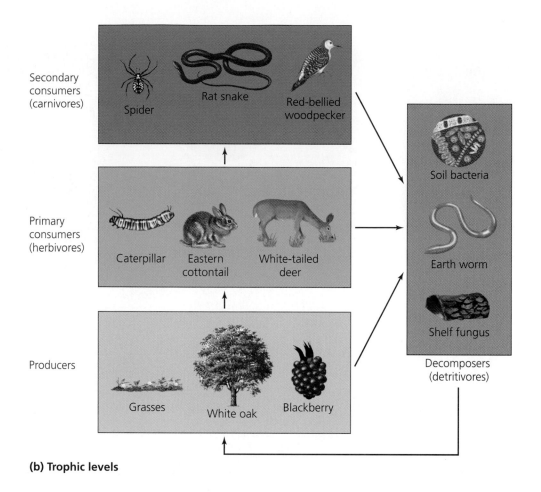

(b) Trophic levels

as deer and grasshoppers, for instance, are primary consumers. Such plant-eating animals are also known as **herbivores** and comprise the second trophic level in a food chain or food web. The third trophic level consists of **secondary consumers**, which prey on herbivorous animals. Wolves that prey on deer, for example, are considered secondary consumers, as are rodents and birds that prey on grasshoppers. Animals that eat other animals are termed **carnivores**. Animals that eat both plant and animal food are referred to as **omnivores**. Predators that eat at even higher trophic levels are known as **tertiary consumers**. Examples of tertiary consumers include hawks and owls that eat rodents that have eaten herbivorous grasshoppers.

Detritivores and decomposers **Detritivores** and **decomposers** consume non-living organic matter. Detritivores such as millipedes and soil insects eat the waste products or the dead bodies of other community members. Decomposers such as fungi and bacteria further break down non-living matter into simpler constituents that can then be taken up and used by plants. The community's recyclers, these organisms play an essential role

in food webs, making nutrients from organic matter available for reuse by living members of the community.

Some organisms play bigger roles in communities than others

"Some animals are more equal than others," George Orwell wrote in his 1945 book *Animal Farm*. Although Orwell was making wry sociopolitical commentary, his remark hints at a truth in ecology. In communities, ecologists have found, some species have greater impact on their fellow community members than do others. A species that has particularly far-reaching impact is often called a **keystone species** (Figure 5.15). A keystone is the wedge-shaped stone at the top of an arch that is vital for holding the structure together; remove the keystone, and the arch will collapse. In an ecological community, removal of a keystone species will have major ripple effects and will alter a large portion of the food web.

Often a large-bodied secondary or tertiary consumer near the top of a food chain is considered a keystone species. For instance, top predators in tropical rainforests, such as jaguars and harpy eagles control popu-

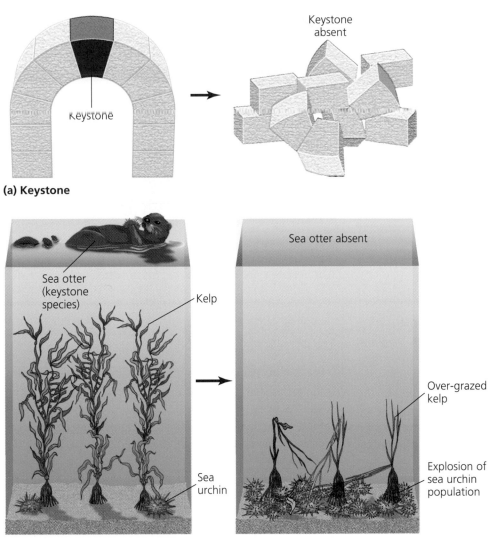

(a) Keystone

(b) A keystone species

Figure 5.15 (a) A keystone is the wedge-shaped stone at the top of an arch that holds its structure together. **(b)** A keystone species, such as the sea otter, is one that exerts great influence on a community's composition and structure. Sea otters consume sea urchins that eat kelp in marine nearshore environments of the Pacific. When otters are present, they keep urchin numbers down, which allows for lush underwater forests of kelp to grow and provide habitat for many other species. When otters are absent, urchin populations explode and the kelp is devoured, destroying habitat and depressing species diversity.

lations of herbivores, which would otherwise increase and greatly change the nature of the plant community. In the United States, wolves and mountain lions were largely exterminated by the middle of the 20th century, leading to population explosions of deer. Dense deer populations have subsequently overgrazed forest-floor vegetation and prevented tree seedlings from surviving, thereby causing major changes in forest structure.

The removal of top predators in the United States served as an unintentional and uncontrolled large-scale experiment. But ecologists have verified the keystone species concept in careful controlled experiments, too. The classic work is that of marine biologist Robert Paine, who established that the starfish *Pisaster ochraceus* has great impact on the community composition of tidepool organisms on the Pacific coast of North America. When *Pisaster* is present in this community, species diversity is high, with several types of barnacles, mussels, and algae. When *Pisaster* is removed, the mussels it

preys upon become dominant and displace other species, suppressing species diversity.

Some species attain keystone species status in ecologists' eyes not through what they eat, but by physically modifying the environment shared by community members. Beavers build dams and turn streams into ponds, flooding acres of dry land and turning them to swamp. Prairie dogs' burrows aerate the soil and serve as homes for other animals. Less-conspicuous organisms and organisms toward the bottom of a food chain could potentially be viewed as keystone species, too. Remove the fungi that decompose dead matter, or the insects that control plant growth, or the phytoplankton that are the base of the marine food chain, and a community will change very rapidly indeed. Because there are usually more species at lower trophic levels, however, it is less likely that any one of them alone might be a keystone species, because if one species is removed, other species may be able to perform its functions.

Table 5.3 Positive and Negative Effects of Species Interactions on Their Participants		
Type of interaction	Effect on species 1	Effect on species 2
Mutualism	+	+
Commensalism	+	0
Predation	+	−
Parasitism	+	−
Neutralism	0	0
Amensalism	−	0
Competition	−	−

Figure 5.16 Predator-prey interactions have numerous ecological and evolutionary consequences for both prey and predator. Here a fire-bellied snake feeds on a frog in Monteverde.

Weighing the Issues:
Keystone Species and Conservation

The federal government is funding a development project in your town and is gathering citizen input on three different options. Its environmental impact statement (Chapter 3) states that option 1 would likely result in the extermination of bobcats, a tertiary consumer in the community. Option 2, the EIS says, would probably kill off a species of pocket mouse, a primary consumer that is common in the community. Option 3 would likely eliminate a species of lupine, a plant that covers a large percentage of the ground in the present community.

You are a citizen desiring minimal change in the natural community, so your children can grow up in an area like the one you grew up in. What kind of information would you ask of an ecologist about the bobcat, pocket mouse, and lupine, so that you could decide which might most likely be a keystone species? If you instead had to provide citizen input without any further information, what would you advise the government?

Species interact in several fundamental ways

Species interactions lie at the heart of community ecology. Several major types of interactions occur among species; most prominent are predation, competition, parasitism, and mutualism. Predation describes one organism consuming another as food, and competition deals with the conflict between organisms vying for the same limited resources. In parasitism, one organism harms another, usually without immediately killing it, and in mutualism two or more species benefit from an association. Table 5.3 summarizes the positive and negative effects of each type of interaction for each participant. Parasitic and mutualistic relationships often occur between organisms that live in close physical contact with one another; these types of associations are called **symbioses**.

Predation Predation is a process in which one species, the **predator**, hunts, tracks, captures, and ultimately kills its **prey** (Figure 5.16). It is the interaction on which food webs are based. However, predation has consequences beyond simply who eats whom. By helping determine the relative abundance of predators and prey, predation rates affect community composition.

Predation also can sometimes drive population dynamics by causing cycles in population sizes. An increase in the population size of prey creates more food for predators, which may respond by surviving and reproducing more effectively. The predator population rises, and the additional predation that results drives down the population of prey. Fewer prey lead to the starvation of some predators, so that the predator population declines. This allows the prey population to begin rising again, starting the cycle anew. Most natural systems involve so many factors that such cycles don't last long, but in some cases we see extended cycles (Figure 5.17).

Predation also has evolutionary ramifications. Individual predators that are more adept at capturing prey will likely live longer, healthier lives and be capable of providing for their offspring better than less adept individuals. Thus predator species tend to evolve adaptations that make them better hunters. Prey, for their part, face an even stronger pressure from natural selection—the risk of immediate death. Thus predation pressure has caused organisms to evolve an elaborate array of defenses against being eaten (Figure 5.18).

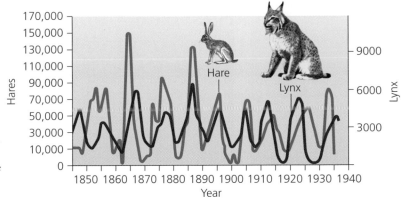

Figure 5.17 Predator-prey systems sometimes show paired cycles, with increases and decreases in one organism apparently driving increases and decreases in the other. Data from D. A. MacLulich, "Fluctuation in the numbers of varying hare *(Lepus americanus),*" University of Toronto, 1937.

Plants experience predation, although the term **herbivory** is generally used to describe animals eating plants. Like animal prey, plants have evolved a wide array of defenses against the animals that eat them. In most cases, the plants produce toxic or distasteful chemicals or arm themselves with thorns, spines, or irritating hairs. But some plants encourage particular animals to take up residence on them, such as ants that will protect the plant against the insects that eat it. Many plants respond to herbivory by releasing chemicals that attract enemies of the herbivore.

Competition Along with predation, competition has traditionally been viewed as one of the primary organizing forces in ecology. **Competition** describes a relationship in which multiple organisms seek the same limited resource. Competitive interactions can take place between members of two or more different species (**interspecific competition**) or between members of the same species (**intraspecific competition**). Competitors do not usually fight with one another directly; rather, most competition is more subtle and indirect, involving the consequences of one organism's ability to match or outdo the other in procuring resources. Those resources can include just about anything an organism might need to survive, including food, water, space, shelter, mates, and sunlight.

In a sense, we have already discussed intraspecific competition earlier in this chapter, without naming it as such. The density dependence that limits the growth of a population occurs because individuals of the same species compete with one another for limited resources,

(a) Cryptic coloration

(b) Warning coloration

(c) Mimicry

Figure 5.18 Natural selection on prey to avoid predation has resulted in many fabulous adaptations. Some prey hide from predators by crypsis, or camouflage, such as this gecko on tree bark **(a)**. Others are brightly colored to warn predators that they are toxic or distasteful, such as this blue poison frog **(b)**. Still others fool predators by mimicry. Some, like walking sticks imitating twigs, mimic for crypsis; others mimic toxic, distasteful or dangerous organisms, like this caterpillar **(c)**, which when disturbed swells and curves its tail-end and shows eyespots, to look like a snake's head.

Large-Scale Climate Change and Its Effects on Monteverde

Soon after the golden toad's disappearance, scientists began to investigate the potential role of climate change in driving cloud-forest species toward extinction. They had noted that the period from July 1986 to June 1987 was the driest on record in Monteverde, with unusually high temperatures and record-low stream flows. These conditions had caused the golden toad's breeding pools to dry up shortly after they filled in the spring of 1987, likely killing nearly all of the eggs and tadpoles present in the pools.

Scientists began reviewing reams of weather data and eventually found that the number of dry days and dry periods each winter in the Monteverde region had increased between 1973 and 1998. Biologists knew that local climate trends like this were bad news for amphibians like the golden toad and harlequin frog. Because amphibians breathe and absorb moisture through their skin, they are susceptible to dry conditions, high temperatures, acid rain, and pollutants concentrated by reduced water levels. Based on these facts, herpetologists J. Alan Pounds and Martha Crump in 1994 hypothesized that hot, dry conditions were to blame for increased adult mor-

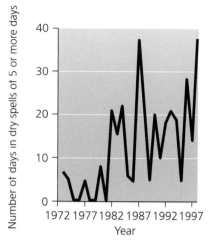

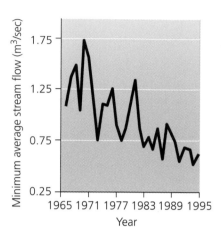

Warming and drying trends in Monteverde's climate may have contributed to the region's amphibian decline. Evidence gathered over 25 years shows (a) an increase in the annual number of dry days and (b) a decrease in the amount of annual stream flow. Data from J. A. Pounds, M. P. L. Fogden, and J. H. Campbell, "Biological response to climate change on a tropical mountain," *Nature*, 1999.

tality and breeding problems among golden toads and other amphibians.

Throughout this period scientists worldwide were realizing that the oceans and atmosphere were warming because of human release of carbon dioxide and other gases into the atmosphere. Global climate change, experts were learning, could produce varying effects on climate at the regional and local levels. With this in mind, Pounds and others concerned about Monteverde's changing conditions used the scientific literature on oceanog-

raphy and atmospheric science to analyze the effects of patterns of warming in the ocean regions around Costa Rica on Monteverde's local climate.

By 1997 these researchers had determined that Monteverde's cloud forest was becoming drier because the clouds that had given the forest its name and much of its moisture now passed by at higher elevations, where they were no longer in contact with the trees. The primary factor determining the clouds' altitude, the researchers determined, is nearby ocean tem-

so that this competition is more acute when there are more individuals per unit area (denser populations). Thus, intraspecific competition is really a population-level phenomenon.

Interspecific competition, however, gives rise to other phenomena of ecological importance. **Competitive exclusion** occurs when one species excludes another from resource use and eventually fully excludes it from the system. **Species coexistence** occurs when no outside influence disrupts the relationship between two species and the species live side by side in a certain ratio of population sizes. With species coexistence there

comes a stable point of **equilibrium**, where the population size of each remains fairly constant through time.

Species that coexist and use the same resources tend to adjust to their competitors in order to minimize competition with them. Individuals can do this by changing their behavior so as to use only some of the resources, while their competitor does likewise. In such cases the individuals are not fulfilling their entire niche, or ecological role. The full niche of a species is called its **fundamental niche** (Figure 5.19a). An individual that plays only part of its role because of competition is said to be displaying a **realized niche** (Figure 5.19b), or

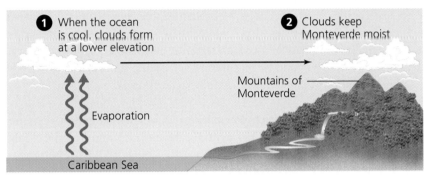

(a) Cool ocean conditions

When the ocean is cool, clouds form at a lower elevation
Clouds keep Monteverde moist
Mountains of Monteverde
Evaporation
Caribbean Sea

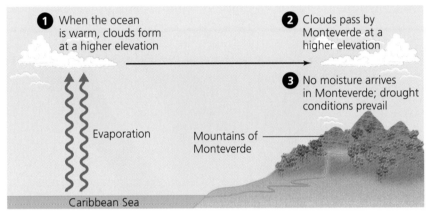

(b) Warm ocean conditions

When the ocean is warm, clouds form at a higher elevation
Clouds pass by Monteverde at a higher elevation
No moisture arrives in Monteverde; drought conditions prevail
Evaporation
Mountains of Monteverde
Caribbean Sea

Monteverde's cloud forest gets its name and life-giving moisture from clouds that sweep inland from the oceans. (**a**) When ocean temperatures are cool, the clouds keep Monteverde moist. (**b**) Warmer ocean conditions resulting from global climate change cause clouds to form at higher elevations and pass over the mountains, drying the cloud forest.

peratures; as ocean temperatures increase, clouds pass over Monteverde at higher elevations. Once the cloud forest's namesake water supply was pushed upward, out of reach of the mountaintops, the cloud forests began to dry out.

In a 1999 paper in the journal *Nature*, Pounds and two colleagues reported these findings.

Their conclusion—that broad-scale climate modification was causing local changes at the species, population, and community levels—explained a great number of events occurring at Monteverde. Rising cloud levels and decreasing moisture could explain not only the disappearance of the golden toad and harlequin frog, but also the concurrent population crashes in 1987, and subsequent disappearance of 20 species of frogs and toads from the Monteverde region. Amphibians that survived showed population crashes in each of the region's three driest years.

Pounds and his co-workers further described "a constellation of demographic changes that have altered communities of birds, reptiles and amphibians" in the area as likely additional consequences of this shift in moisture availability. As these mountaintop forests dried out, dry-tolerant species crept in, and moisture-dependent species were stranded at the mountaintops by a rising tide of dryness. Although species may in general be driven from one area to another by changing environmental conditions, if a species has nowhere to go, then extinction of populations and entire species may result.

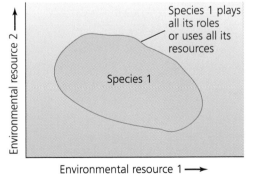

(a) Fundamental niche

Environmental resource 2
Species 1 plays all its roles or uses all its resources
Species 1
Environmental resource 1 →

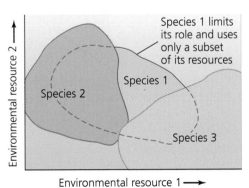

(b) Realized niche

Environmental resource 2
Species 1 limits its role and uses only a subset of its resources
Species 2
Species 1
Species 3
Environmental resource 1 →

Figure 5.19 An organism facing competition may be forced to play less of the ecological role or use less of the resources than it would in the absence of its competitor. With no competitors, an organism can exploit its full fundamental niche (**a**). When competitors restrict the things it can do or the resources it can use, however, then the organism is limited to a realized niche (**b**) that covers only a subset of its fundamental niche.

the portion of the fundamental niche that is actually realized.

Species can make similar adjustments over evolutionary time; that is, they can adapt to competition by evolving to use slightly different resources, or to use their shared resources in different ways. If two bird species eat the same type of seeds, one might come to specialize on larger seeds and the other to specialize on smaller seeds. Or one bird might become more active in the morning and the other more active in the evening, so as to avoid direct interference with one another. This process is called **resource partitioning**, because the species partition the resource they use in common by specializing in different ways (Figure 5.20). Resource partitioning may lead to **character displacement**, which occurs when competing species evolve physical characteristics that reflect their reliance on the portion of the resource they use. In becoming more different from one another, two species lessen their competition. Birds that specialize on larger seeds may eventually evolve larger bills that enable them to make best use of the resource, while birds specializing on smaller seeds may evolve smaller bills. This is precisely what has been found by recent research on the finches first described by Darwin on the Galapagos Islands.

Parasitism **Parasitism** is a relationship in which one organism, the **parasite**, depends on another, the **host**, for nourishment or some other benefit while simultaneously doing the host harm (Figure 5.21). Like predation, parasitism is generally good for one organism and bad for the other. Unlike predation, parasitism usually does not

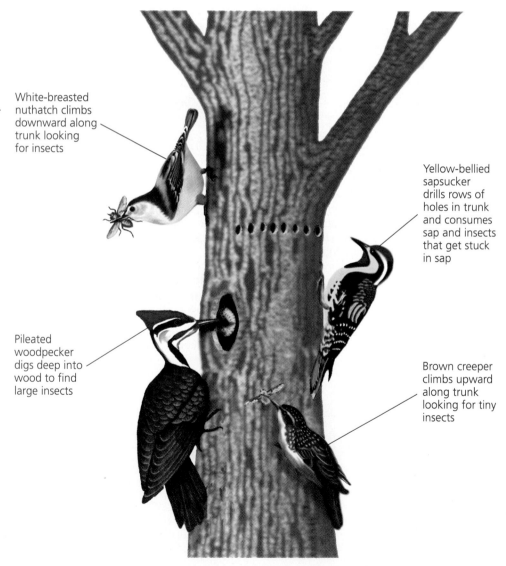

Figure 5.20 When multiple species compete for the same resource, they tend to partition the resource, each specializing on a different aspect. A number of birds including woodpeckers, creepers, and nuthatches feed on tree trunks, but they utilize different portions of the trunk, seeking different foods in different ways.

White-breasted nuthatch climbs downward along trunk looking for insects

Yellow-bellied sapsucker drills rows of holes in trunk and consumes sap and insects that get stuck in sap

Pileated woodpecker digs deep into wood to find large insects

Brown creeper climbs upward along trunk looking for tiny insects

result in one organism's immediate death, although it does often lead to the host's eventual death.

Many parasites live in close contact with their hosts. These parasites include ones we are familiar with as disease pathogens, such as the protist that causes malaria or the amoeba that causes dysentery. Parasites that live inside their hosts are called **endoparasites.** A tapeworm living in the digestive tract of its host is an endoparasite. **Ectoparasites,** in contrast, live on the exterior of their hosts. A tick, which attaches itself to the skin of its host, is an ectoparasite.

Other types of parasites are free-living and come into contact with their hosts only infrequently. Some birds, such as the cuckoos of Eurasia and the cowbirds of the Americas, parasitize the nests of other birds by laying eggs in them and letting the host bird raise the parasite's young. And there are insects, called parasitoids, that parasitize other insects. For instance, various parasitoid wasps lay eggs on caterpillars. When the eggs hatch, the wasp larvae burrow into the caterpillar's tissues and slowly consume them. The wasp larvae metamorphose into adults and fly from the body of the dying caterpillar. Insect parasitoids are inconspicuous, and there are likely many more parasitoid species than are currently known. In fact, some biologists hold that parasites of one kind or another may make up fully half the world's biodiversity.

Host-parasite interactions have interesting evolutionary consequences as well. Just as predators and prey evolve in response to one another, hosts and parasites can become locked in a duel of escalating adaptations. Such a situation has been termed an **evolutionary arms race.** Like rival nations racing to stay ahead of one another in military technology, host and parasite may repeatedly evolve new responses to the other's latest advance. In the long run, though, it may not be in a parasite's best interest to be too harmful to its host. A parasite might leave more offspring in the next generation—and thus be favored by natural selection—if it allows its host to live a longer time, or even to thrive. Some biologists hypothesize that in many cases, parasitic interactions have over time evolved into mutualistic ones.

Mutualism Mutualism is a relationship in which all participating organisms benefit from their interaction (Figure 5.22). Many mutualisms are symbiotic in nature, involving partners living closely together. Thousands of terrestrial plant species depend on mutualisms with fungi; plant roots and certain fungi together form associations called mycorrhizae, in which the plant provides energy to the fungus and the fungus assists the plant in absorbing nutrients from the soil. In the ocean, coral polyps, the small animals that build coral reefs, share beneficial arrangements with algae known as zooxanthellae. The coral provide housing and nutrients for the algae in exchange for a steady supply of food—90% of their nutritional requirements. Recall from Chapter 4 that chemosynthetic bacteria form mutualistic associations with some hydrothermal vent invertebrates.

You, too, are part of a symbiotic association. Your digestive tract is filled with microbes that help you digest food—microbes for which you in turn provide a place to live. Indeed, we all may owe our very existence to symbiotic mutualisms. It is now widely accepted that the eukaryotic cell itself originated from certain prokaryotic

Figure 5.21 Parasites harm their host organism in some way. Ectoparasites like ticks, for instance, suck fluids from mammals and may sometimes transmit disease.

Figure 5.22 In mutualism, organisms of different species benefit one another. An important mutualism for environmental science is pollination. This hummingbird is gathering nectar from flowers, but in the process transfers pollen between flowers, helping the plant reproduce. Pollination is of key importance to agriculture via the reproduction of crop plants.

VIEWPOINTS

Conservation of Monteverde

What lessons, if any, can we learn from ecological changes that have occurred in the Monteverde since the discovery of the golden toad in 1964?

Four Lessons from Monteverde

There are four lessons to be learned from the Monteverde experience. First, a few committed people can have an impact. The Monteverde was originally occupied by Tico and Quaker families, pioneers in the usual sense of the word—people who farmed for a living, but struggled to create their farms in a beautiful, but primitive environment. George Powell, who is now actively involved in green macaw conservation efforts and studies of bellbird migrations, converted a local Monteverde pioneer, Wolf Guindon, into an ardent conservationist. It was Powell who convinced the international conservation organizations to fund land purchases of the Monteverde Cloud Forest Preserve. Guindon became the chief ranger of the Preserve.

Second, practical conservation efforts must take into account local social aspirations. Although hard for a conservationist to watch, pioneers are just trying to make a hard life better by converting natural landscapes to farmland. Conservationists must recognize this to work with their neighbors. For example, the Monteverde Cloud Forest Preserve allows its springs to be tapped for local water supplies, runs educational programs, and leases a peak for television and radio transmission towers. The Monteverde Conservation League runs a tree nursery and an agroforestry outreach program for local farmers.

Third, conservation can lead to local economic success. As adults, the children of the agricultural pioneers at Monteverde had no place to claim for their own and faced leaving the area for employment elsewhere, or finding alternative livelihoods. Many of my Costa Rican friends stayed in Monteverde to build an ecotourism industry, and are now hoteliers, restauranteurs, or natural history guides.

Finally, even if every bit helps, local conservation is not enough. Human use of fossil fuels is making the world on average a warmer place. Deforestation is changing climate at regional scales. Conservationists must plan reserves that can adapt to changes in the coming century or more species will follow the golden toad into extinction.

Robert Lawton is a forest ecologist at the University of Alabama-Huntsville, who has worked in Monteverde for over 25 years.

Conservation Successes in the Shadow of the Golden Toad

Given the discouraging news about conservation problems worldwide, it wouldn't be hard to become pessimistic about what has happened in Monteverde, Costa Rica. After all, in spite of the country's international reputation as a "green republic," deforestation rates are higher in Costa Rica than almost anywhere in the world, isolating Monteverde as an ecological island perched on the continental divide.

The last 15 years has also seen widespread changes within Monteverde: a proliferation of new houses, hotels, restaurants, and an explosion of ecotourism, with tens of thousands of people per year visiting what was once a rural community. Along with the Golden Toad, other amphibian and reptile species have disappeared, introduced species have invaded, and lowland species have moved upslope as Monteverde becomes warmer and drier.

Yet there have also been impressive achievements in conservation at Monteverde. In 1979, poaching of large mammals and birds was commonplace, and species such as tapirs and guans were rare. Now the very people who hunted with rifles use binoculars instead as they lead natural history tours. Tapirs and guans are more common today than they have been for more than half a century. As the Monteverde Cloud Forest Preserve has grown ten-fold in area, clearings on the Atlantic slope have reverted to lush forest. The Guacimal River, formerly rancid due to waste dumped by the community dairy plant, is much cleaner now.

On both global and local scales, the most enduring impact of Monteverde has been the education of the public about environmental values. I like to think that whatever negative local impact the steady onslaught of ecotourists may have on resplendent quetzals and howler monkeys, it is more than compensated by inspiring people to appreciate tropical forests and their own natural heritage. If so, the conservation gains at Monteverde may help save other tropical and temperate zone habitats worldwide.

Nathaniel Wheelwright is Professor of Biology at Bowdoin College in Brunswick, Maine, director of the Bowdoin Scientific Station on Kent Island, New Brunswick, and co-editor of *Monteverde: Ecology and Conservation of a Tropical Cloud Forest* (Oxford University Press, 2000).

cells engulfing other prokaryotic cells and establishing mutualistic symbioses. The inner cells eventually evolved into what we now know as the cell organelles of eukaryotic cells, scientists have inferred. Biologists have found that the harder they look in nature, the more they see mutualisms. While these interactions have long been thought of as secondary in importance to competition and predation, biologists increasingly are recognizing them for the widespread and vitally important relationships that they are.

Not all mutualisms involve organisms living in close proximity. One of the most important mutualisms for environmental science involves free-living organisms that may only encounter each other once in their lifetimes. This is **pollination**, an interaction of key significance to agriculture and our food supply (Chapter 9). Bees, birds, bats, and other creatures help turn flowers into fruit when they transfer pollen (male sex cells) from one flower to ova (female cells) of another, fertilizing the female flower, which subsequently grows into a fruit. The pollinating animals may be visiting flowers for their nectar, which serves as a reward the plant uses to entice them. The pollinators get food, and the plants are pollinated and reproduce. Animals, especially bees and other insects, pollinate many of our crop plants. Various types of bees alone pollinate 73% of our crops, one expert has estimated—from soybeans to potatoes to tomatoes to beans to cabbage to oranges.

Commensalism and amensalism Two other types of species interaction get far less attention. **Amensalism** is a relationship in which one organism is harmed and the other is unaffected. In **commensalism**, one species benefits, and the other is unaffected.

Possible examples of amensalism include large trees shading smaller trees, penicillin and other antibiotics harming bacteria, and plants releasing poisonous chemicals that harm nearby plants (a phenomenon called allelopathy). However, amensalism has been difficult to pin down, since it is hard to prove that the organism doing the harm is not in fact besting a competitor for a resource. For instance, many cases of allelopathy can easily be viewed as one plant investing in costly chemicals in order to outcompete others for space.

One example of commensalism is the relationship between an epiphytic plant (a plant that lives on the surface of another plant) and its host. For instance, some orchids rely on tree limbs for support but do no harm and provide no benefit in exchange for the favor. Another occurs when the conditions created by one plant happen to make it easier for another plant to establish

and grow. For instance, palo verde trees in the Sonoran Desert create shade and leaf litter that allows the soil underneath them to hold moisture longer, creating an area that is cooler and moister than the surrounding sun-baked ground. Young plants find it easier to germinate and grow in these conditions, so seedling cacti and other desert plants generally grow up directly beneath "nurse" trees like palo verde. This phenomenon, called *facilitation*, is widely thought to be a crucial component of the process of succession.

Succession occurs after a disturbance or after a new substrate emerges

Scientists think the community-level changes at Monteverde are related to the drying climate of the past decade or two—a local pattern for Central America likely driven by **anthropogenic** (human-caused) global climate change. But communities change in their composition and structure naturally, too, because of the effects of species already present. Over a period of years to decades, most communities undergo a series of regular, predictable, and quantifiable changes ecologists call **succession**. Two types of succession occur in terrestrial (Figure 5.23) and aquatic (Figure 5.24) systems: primary succession and secondary succession.

In terrestrial communities, **primary succession** begins when a bare expanse of rock, sand, or sediment becomes newly exposed to the atmosphere. This may occur when glaciers retreat, volcanoes produce lava flows, lakes dry up, or sea level drops. Species that arrive first and begin the process of succession are referred to as **pioneer species.** Pioneer species generally have spores or seeds that can travel long distances. The pioneers best suited to colonizing bare and hostile rock are the mutualistic aggregates of fungi and algae known as **lichens.** Lichens succeed because their algal component provides food and energy via photosynthesis, while the fungus they contain can take a firm hold on bare rock and capture the moisture that both organisms need to survive.

As lichens become established and grow, they secrete acids that break down the rock surface to which they attach. The resulting waste material forms the beginnings of soil, and once soil begins to form, many organisms, including insects, small plants, and worms, may find the rocky outcrops slightly more hospitable. As these new organisms begin to arrive, they provide still more nutrients and habitat for future arrivals. As time passes, lichens will be displaced by a community of small plants, which will in turn be displaced by larger

Figure 5.23 Secondary succession occurs after a disturbance such as fire, farming, or landslides removes vegetation from an area. Here is shown a typical series of changes in a plant community of eastern North America following the abandonment of a farmed field.

Grass

Herbs, shrubs

Shrubs, poplar trees

Pines

Oaks, hardwoods

Time ⟶

plants and associated species. At each step, just as with the palo verde trees and young cacti in the Sonoran Desert, one plant or set of plants facilitates the establishment of the next.

These transitions between stages of succession will eventually lead to a community known as a **climax community**. Climax communities remain in place, with minimal modification, until some environmental change alters or displaces them. These communities vary depending on a location's temperature, precipitation, latitude, and many other abiotic factors.

Primary succession is not unique to terrestrial communities; it also occurs in aquatic systems. Lakes and ponds contain plenty of plants, some large and others microscopic. As these plants live, grow, reproduce, and die, they help fill in the water bodies in which they grow. Lakes and ponds that receive input from rivers, streams, and surface runoff receive decaying plant matter and sediments as well. As this occurs, water bodies undergo a gradual transition to terrestrial communities.

Secondary succession begins when some event disrupts or dramatically alters an existing community. Thus, terrestrial secondary succession starts not with bare rock or sand, but when a disturbance such as a fire, a hurricane, logging, or farming removes some or all of the biotic community from an area. Regardless of the severity of the disturbance, some vestiges of the previous biotic community will likely remain. The availability of these building blocks, and of other resources such as soils, will generally accelerate the secondary succession process. In an aquatic system, the construction of a beaver dam can act as a catalyst for secondary succession. A dam can convert a stream or river into a series of ponds, which may then undergo a typical aquatic successional sequence.

Once major disturbances occur and set secondary succession in motion, there is no guarantee that the community will ever return to its prior climax state. Many communities disturbed by humans have not returned to their former states. This is the case with vast areas of the Middle East that once were fertile enough to support farming but now are deserts.

Invasive species pose a new threat to community stability

Traditional concepts of succession entail cycles of community change involving sets of organisms understood to be native to an area. But what if a new organism arrives to an area? And what if this new organism has a strong impact, like that of a keystone species? These questions bring us to one of the central ecological forces in today's world: **invasive species**. This term refers to any species that spreads widely and rapidly becomes dominant in a community, interfering with the community's normal functioning. Most often, invasive species are exotic non-native species introduced from elsewhere in the world. Species become invasive when limiting factors that regulate their population growth are removed. Often plants and animals brought to one area from another may leave their predators, parasites, and competitors behind and be freed from these constraints on their population growth. If there happen to be few organisms in the new environment that can act as predators, parasites, or competitors, the introduced species can do very well.

The chestnut blight mentioned earlier that decimated eastern North America's mature chestnut trees was an invasive species—a fungus introduced accidentally

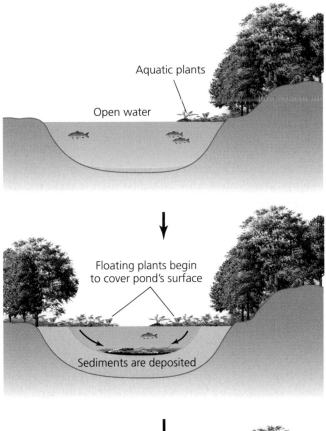

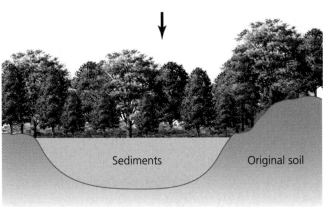

Figure 5.24 Primary aquatic succession occurs when plant growth gradually fills in a pond or lake and converts an aquatic system to a wet meadow and ultimately to a terrestrial system. Increased nutrient input can accelerate this process.

from Asia, where it attacked trees native to that area. The native trees had evolved defenses against the fungus, however, whereas the American chestnut had not. A different fungus caused similar destruction to elm trees in eastern North America in the early and mid-20th century. Dutch elm disease spread rapidly and virulently, killing off most of the American elms that once gracefully lined the streets of many U.S. cities. Other examples are almost too numerous to mention—European starlings and house sparrows have spread across

Figure 5.25 Species introduced to a new area that spread rapidly and come to dominate a community are called invasive species. The globalization of today's world has encouraged biological invasions that cause extensive change in ecological communities and cost our society billions of dollars. Here, purple loosestrife invades a wetland in central Washington.

the North American continent, while some North American birds are spreading in Europe. Grasses introduced in the American West by ranchers have overrun entire regions, pushing out native vegetation. Fish introduced into streams for sport compete with native fish, driving many to local extinction. Hundreds of island-dwelling animals and plants worldwide have been driven extinct by animals introduced by sailors—mammals such as goats, pigs, and rats, against which the isolated island creatures had never evolved defenses. The impact of introduced and invasive species on native species and ecological communities is growing with the ever-increasing mobility of humans and the globalization of our society (Figure 5.25).

Humans and the Conservation of Biodiversity

Natural change in communities (such as succession) has been going on as long as communities have existed, but today human development, resource extraction, and population pressure are speeding the rate of change and altering the types of change. The indirect effects of global climate change on the Monteverde community took a bit of scientific detective work to figure out, but it has been much easier to witness and document the pervasive phenomena of habitat destruction and alteration that result when people extract resources or carve out a place to live. The changes we induce in our environment cannot be fully understood in a scientific vacuum, however. The actions that threaten biodiversity have complex social, economic, and political roots, and environmental scientists appreciate that these aspects must be understood in order to develop solutions to problems that threaten the integrity of our world's ecological systems.

Fortunately, there are things people can do to forestall population declines of threatened species, to prevent habitat destruction and the alteration of communities, and to minimize the impact of the current mass extinction. Millions of people around the world are already taking actions to safeguard the biodiversity and ecological and evolutionary processes that make Earth such a unique place. Costa Ricans, for example, have been confronting the challenges to their nation's great biodiversity; the actions they have taken so far show what even a small country of modest means can do.

Social and economic factors affect species and communities

Many of the threats to Costa Rica's species and ecological communities result from past economic and social forces whose influences are still evident. European immigrants and their descendants viewed Costa Rica's lush forests as an obstacle to agricultural development, and timber companies saw them simply as a source of wood products. Costa Rica's leading agricultural products have long included beef and bananas, the production and cultivation of which require extensive environmental modification. Between 1945 and 1995, the country's population grew from 860,000 to 3.34 million, and the percentage of land devoted to pasture increased from 12% to 33%. With much of the formerly forested land converted to agriculture, the proportion of the country covered by forest decreased from 80% to 25%. In fact, in 1991 Costa Rica was losing its forests faster than any other country in the world—nearly 140 ha (350 acres) per day. As had occurred in the history of the United States, few people foresaw the need to conserve biological resources until it became clear that they were being rapidly lost.

Costa Rica has taken steps to protect its environment

During the 1950s a group of Quakers, Christian pacifists who opposed the U.S. military draft, emigrated from Alabama to Costa Rica and founded the village of Monteverde. The Quakers relied on milk and cheese for much of their economic activity, but they also set aside one-third of their land for conservation purposes. The Quakers' efforts, along with contributions from international conservation organizations, provided the beginnings of what is today the Monteverde Cloud Forest Biological Reserve. This privately managed 10,500-ha (26,000-acre) reserve was established in 1972 to protect the forest and its 2,500 plant species, 400 bird species, 500 butterfly species, 100 mammal species, and 120 reptile and amphibian species, including, for a brief time, the golden toad.

In 1970, the Costa Rican government and international representatives came together to create the country's first national parks and protected areas. The first parks centered on areas of spectacular scenery, such as the Poas Volcano National Park. Santa Rosa National Park encompassed valuable tropical dry forest, Tortuguero National Park contained essential nesting beaches for the green turtle, and Cahuita National Park was meant to protect a prominent coral reef system.

Initially the government gave the parks little real support. According to Costa Rican conservationist Mario Boza, park supporters in the early years faced a lack of power and resources to protect the areas. Five park guards, one vehicle, and no financial support constituted the sum total of resources originally provided to the parks.

Today government support is much greater. Fully 12% of the nation's area is contained in national parks, and a further 16% is devoted to other types of wildlife and conservation reserves. Costa Rica and its citizens are now reaping the benefits of the country's park system, and these benefits are economic as well as ecological. Because of its parks and its reputation for conservation, tourists from around the world now visit Costa Rica, a phenomenon called **ecotourism** (Figure 5.26). The ecotourism industry drew more than 1 million visitors to Costa Rica in 1999 (up from 780,000 in 1996), provides thousands of jobs to Costa Ricans, and is a major contributor to the country's economy. In 1999 alone, ecotourism increased by 9% in Costa Rica, while the global rate of increase was 3–5%. Today's Costa Rican economy provides $3,700 in per capita income. This economy is fueled in large part by commerce and tourism, whose contributions (40%) outweigh those of industry (22%) and agriculture (13%) combined.

Figure 5.26 Costa Rica has protected a wide array of its diverse communities. This protection has stimulated the nation's economy through ecotourism. Here, visitors experience a walkway through the forest canopy in one of the nation's parks.

Figure 5.27 Ecological restoration attempts to reverse the effects of human disruption and restore communities to their natural state. Here, a crew plants trees in the Puriscal region of Costa Rica, for a Costa Rican environmental group that brings farmers and nursery workers together to reforest the land.

Weighing the Issues:
How Best To Conserve Biodiversity?

Most people view national parks and ecotourism as excellent ways to help keep ecological systems intact. Yet the golden toad went extinct despite having a reserve established to protect it, and invasive species do not pay attention to park boundaries. What lessons can we take from this about the conservation of biodiversity?

Altered communities can be restored to their former condition

Creating public parks and defining areas in which to preserve natural communities are well-tested ways to preserve biodiversity while enhancing a region's economy, and nearly every nation in the world has now at least begun this process. However, with so much of Earth's landscape altered by human impact, it is hard to find areas that are truly pristine. This realization has given rise to another type of conservation effort, **ecological restoration** (Figure 5.27). The practice of ecological restoration is backed by the science of **restoration ecology.** Restoration ecologists research the historical conditions of ecological communities as they existed before humans altered them. They then try to devise ways to restore some of these areas to their natural "presettlement" conditions. For instance, in the United States nearly every last scrap of tallgrass prairie that once covered the eastern Great Plains and parts of the Midwest was converted to agriculture in the 19th century. Now a number of efforts are underway to restore small patches

of prairie habitat by planting native prairie plants, weeding out invaders and competitors, and introducing controlled fire to mimic the natural prairie fires that historically maintained this community. Most such efforts at ecological restoration so far have been made in developed nations because more natural areas have been lost in these countries and because the effort to restore native communities can be resource-intensive. However, ecological restoration is taking place in Guanacaste Province in Costa Rica as well, where scientists are restoring dry tropical forest from grazed pasture. The more human population grows and development spreads, the more ecological restoration may become a prime conservation strategy.

Conclusion

The golden toad and the cloud-forest community of Monteverde have helped illuminate many of the fundamentals of ecology and evolution that are integral to environmental science. The evolutionary processes of speciation and extinction help determine Earth's biodiversity. An understanding of how ecological processes work at the population level and the community level is crucial to our efforts to protect biodiversity threatened by the mass extinction event many biologists maintain is already underway.

Scientists have described amphibians such as the golden toad and harlequin frog as indicators of environmental quality because they are particularly sensitive to water pollution, climate change, and other stresses. Numerous factors today are interacting to drive amphibians and many other types of organisms out of their historic habitats and toward extinction. Global climate change; habitat alteration and destruction; the expansion of cities, suburbs, and farmland; the spread of invasive species; pollution; poaching; pesticides; and prejudices—all these and more play roles in biodiversity loss, and we will examine each of them in more detail in upcoming chapters.

Alleviating the problems that threaten biodiversity—as millions of citizens around the world are working to accomplish—requires science that untangles the complexities of ecological systems. In Chapter 6, the final chapter in this book's introductory section, we will explore some of the complex cycles and systems operating in ecosystems around the world.

REVIEW QUESTIONS

1. What does the term "biodiversity" encompass?
2. How many species are there in the world?
3. What does a phylogenetic tree show?
4. Name three organisms that have gone extinct.
5. To what levels of biological organization does the science of ecology pertain?
6. What is the difference between a species and a population? A population and a community?
7. Contrast the concepts of habitat and niche.
8. What are the differences between population size, population density, and population distribution? Use examples from this chapter in your answer.
9. List and describe all the major population characteristics discussed in this chapter. Explain how each shapes population dynamics.
10. Could any species undergo exponential growth forever? Explain your answer.
11. Describe how limiting factors relate to carrying capacity.
12. Explain the difference between K–strategists and r–strategists. Can you think of examples of each that were not mentioned in the chapter?
13. Contrast Clements's and Gleason's views of ecological communities.
14. How does parasitism differ from predation?
15. What effects does competition have on the species involved?
16. Name one mutualistic relationship that affects your day-to-day life.
17. Explain and contrast primary and secondary terrestrial succession.
18. What has Costa Rica's experience been with parks and ecotourism?
19. Why are amphibians considered indicators of environmental quality?

DISCUSSION QUESTIONS

1. How has Earth come to have so many species? Contrast the two modes of speciation discussed.
2. How are the phylogenetic trees in Figure 5.4 similar to a family geneaology? How are they different? What kind of information can we learn from them?
3. What types of species are most vulnerable to extinction, and what kinds of factors threaten them? Can you think of any species that are threatened with imminent extinction today? What reasons lie behind their endangerment?
4. How did precipitation, runoff, population density, and population distribution affect the harlequin frog?
5. Can you think of one organism not mentioned as a keystone species that you believe may be a keystone species? For what reasons do you suspect this? Can

you think of an organism that you would guess is not a keystone species? What reasoning lies behind your answer? How could you experimentally test whether an organism is a keystone species?
6. Why do scientists consider invasive species to be a problem? What makes a species "invasive," and what ecological effects can invasive species have?
7. Describe the evidence in this chapter that supports scientists' assertion that changing temperatures and precipitation led to the extinction of the golden toad and to population crashes for many other amphibians in Monteverde.
8. What are the advantages of ecotourism for a country like Costa Rica? Can you think of any disadvantages?

9. As Monteverde changed and some species disappeared, scientists reported that others moved in from lower, drier areas. If this is true, should we be concerned about the extinction of the golden toad and disappearance of other species from Monteverde? Explain your answer.

Media Resources *For further review go to the website* **www.envscienceplace.com** *or student CD-ROM, where you will find quizzes, flashcards, a glossary, additional interactive exercises, and links to relevant news and research sources. Also, on the website and CD-ROM is* **GRAPH IT**, *a series of interactive graphing tutorials to help you interpret graphs and plot data.*

Environmental systems: Connections, cycles, and feedback loops

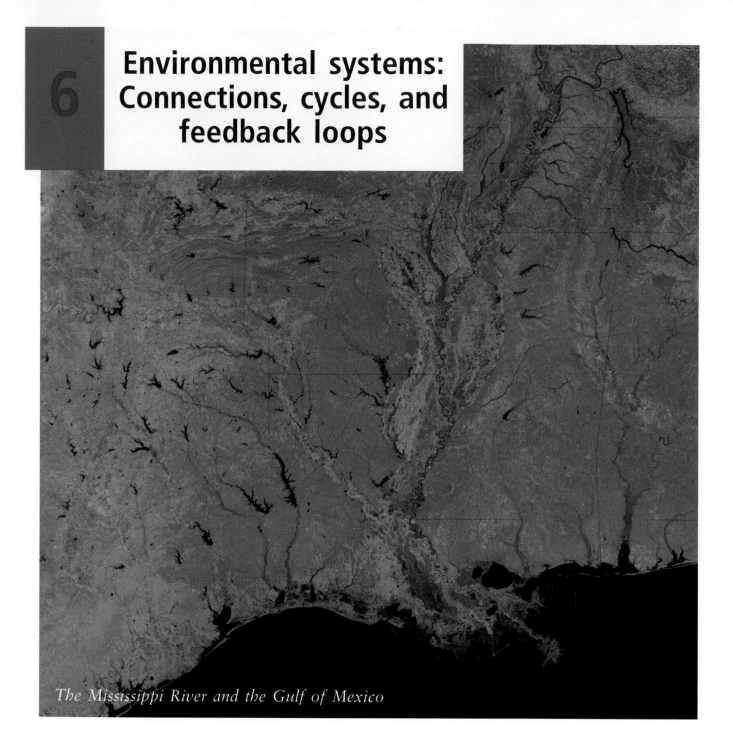

The Mississippi River and the Gulf of Mexico

This chapter will help you understand:

- The nature of systems and the fundamentals of systems thinking
- Ecosystem-level ecology
- Earth's biomes
- Plate tectonic and the rock cycles

- The hydrologic cycle
- The nitrogen, carbon, and phosphorus cycles
- The Gaia hypothesis

Fisherman in the Gulf of Mexico's "dead zone"

Central Case: The Gulf of Mexico's "Dead Zone"

"In nature there is no 'above' or 'below,' and there are no hierarchies. There are only networks nesting within networks."
—Fritjof Capra, theoretical physicist

"Let's say you put Saran Wrap over south Louisiana and suck the oxygen out. Where would all the people go?"
—Nancy Rabalais, biologist for the Louisiana Universities Marine Consortium

In mid-2002 newspapers and magazines reported a 22,015–km² (8,500–mi²) "dead zone" in the waters of the Gulf of Mexico near the mouths of the Mississippi and Atchafalaya Rivers off the Louisiana coast. The so-called dead zone was a region of water so depleted of oxygen that marine organisms were being killed or driven away. These organisms included commercially important fish and shellfish, such as menhaden and white and brown shrimp. The dead zone was not new—scientists had tracked it since at least 1985—but it had grown bigger than ever, covering an area larger than the

state of Massachusetts. And it was posing a grave new threat to the region's environment and to its economy.

The extremely low dissolved oxygen concentrations in the bottom waters of this region represent a condition called **hypoxia** (see The Science behind the Story on page 161 and Figure 6.5 on page 180). Aquatic animals obtain oxygen by respiring through their gills, as we do through our lungs, and like us, these animals will suffocate if deprived of it. When oxygen concentrations drop below 5 parts per million (ppm), marine organism growth is slowed, and when concentrations drop below 2 ppm, the organisms that can leave an affected area will do so. At concentrations below 0.2 ppm, sediments will turn black and become covered by thick layers of sulfur-oxidizing bacteria. In the hypoxic zone off the Louisiana coast, oxygen concentrations frequently drop below 2 ppm.

The spread of the Gulf of Mexico's hypoxic zone threatens the fishing industry, which has long been a cornerstone of the Gulf Coast economy. One of the most productive fisheries in the United States, Gulf

Coast waters were estimated in 1999 to yield $2.8 billion in seafood production per year. But in the summer of 2002, shrimp boat captains were coming up with nets nearly empty. One shrimper derided his meager catch as "cat food" when talking to a newspaper reporter, while another said he hoped a hurricane would strike and stir some oxygen into the Gulf's stagnant waters.

Environmental advocates and fishermen across the country joined local shrimpers in expressing anxiety about the dead zone. Their concerns were not just speculation or sensationalism; they were based on a well-documented environmental problem with surprising origins. Scientists studying the dead zone, its formation, and its origins have fingered three unlikely culprits: the invention of synthetic ammonia by a German chemist, modern Midwestern farm practices, and a global nitrogen cycle thrown out of balance. The story of how scientists discovered these connections involves understanding environmental systems and the often-unexpected behavior they exhibit.

Earth's Environmental Systems

There is nothing static and simple about our planet. Earth's environment is made up of a complex network of interlinked systems. On the community level, these systems include the ecological webs of relationships among species; on the ecosystem level, they include the interaction of living species with the abiotic elements around them. Earth's systems also include a variety of cycles that determine the flow of key chemical elements and compounds that support life and that regulate climate and other aspects of Earth's functioning. We depend on these systems for our survival, and if they are subjected to severe shock or alteration they can cause loss of life, economic turmoil, and other unpredictable and undesirable environmental changes. Scientists are increasingly taking a "systems-level approach" in researching questions in environmental science, and such a holistic approach is also ideal for designing solutions to environmental problems.

Systems show several defining properties

Before we investigate environmental systems in particular, it will help to define what exactly we mean by *system* in general. A **system** is a network of relationships among a group of parts, elements, or components that interact with and influence one another through the exchange of energy, matter, and/or information.

Systems receive inputs of energy, matter, and/or information, process these inputs, and produce outputs. For example, the Gulf of Mexico receives inputs of fresh water, dissolved oxygen, nutrients, and pollutants from the Mississippi and other rivers. When shrimpers harvest shrimp from the Gulf, its system is providing an output of matter and food energy. This output subsequently becomes an input to the human economic system and to the digestive systems of the individual people who consume the shrimp. Widespread energy inputs to Earth's environmental systems include solar radiation and the heat released by industrial activities, metabolism, geothermal activity, and fossil-fuel combustion. Matter inputs include the flow of nutrients from neighboring systems. Information inputs can come in the form of sensory cues from visual, olfactory (chemical), magnetic, thermal, and other types of signals.

Sometimes a system's output can serve as input to that same system, a circular process described as a **feedback loop.** Feedback loops can be classified into two types, negative and positive. In a **negative feedback loop** (Figure 6.1a), output of one type acts as input that moves the system in the other direction. The input and output essentially neutralize one another's effects, stabilizing the system. A room with a thermostat, for instance, stabilizes temperature through negative feedback, turning the furnace on when the room gets too cold and shutting the furnace off when the room gets too hot. Similarly, a negative feedback

Figure 6.1 (a) Negative feedback loops exert a stabilizing influence on systems and are common in nature. The human body's response to heat and cold involves a negative feedback loop.

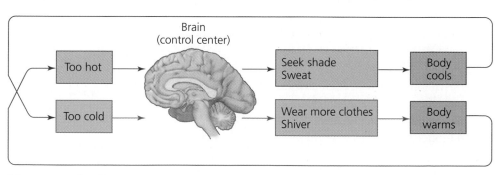

(a) Negative feedback

Figure 6.1 (b) Positive feedback loops have a destabilizing effect on systems and push them toward extremes. Rare in nature, they are common among human impacts on natural systems.

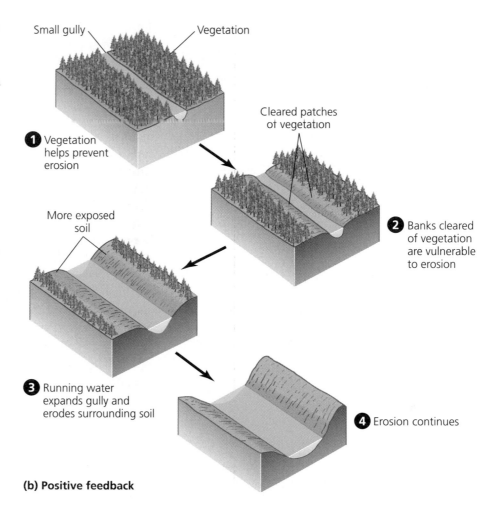

Small gully — Vegetation

1 Vegetation helps prevent erosion

Cleared patches of vegetation

2 Banks cleared of vegetation are vulnerable to erosion

More exposed soil

3 Running water expands gully and erodes surrounding soil

4 Erosion continues

(b) Positive feedback

loop regulates our body temperature. If we've had too much sun, our brains tell us to move into the shade, and if we're overheated, our sweat glands pump out moisture to cool us down; if we're cold, we shiver and our brains tell us to move into the sun or put on more clothing to warm up. Most systems in nature involve negative feedback loops. Negative feedback loops enhance stability, and, in the long run, only those systems that are stable will persist.

Positive feedback loops have the opposite effect. Rather than stabilizing a system, they drive it further toward one extreme or another. This can occur with the process of erosion, the weathering away of soil by water or wind. Once soil has been exposed by removal of vegetation, erosion may become progressively more severe if the forces of water or wind surpass the rate of vegetative regrowth. Water flowing through an eroded gully, for instance, may expand the gully and lead to further erosion (Figure 6.1b). Alternatively, if vegetative regrowth is rapid enough, then erosion may progressively decrease as plants secure the soil. Positive feedback can alter a system substantially. Positive feedback loops are relatively rare in nature, but are common in natural systems altered by human action.

The inputs and outputs of a complex natural system often occur simultaneously, keeping the system constantly active; Earth's climate system, for instance, or the nutrient cycles we will soon discuss, do not ever stop. When processes within a system are moving in opposing directions at equivalent rates so that their effects balance out, the process is said to be in **dynamic equilibrium.** Processes in dynamic equilibrium can contribute to **homeostasis,** the tendency of a system to maintain constant or stable internal conditions. When homeostasis exists, organisms and other systems can keep their internal conditions within the range of tolerance that allows them to function. If there is an ongoing net buildup or depletion of material or energy, then a system is not considered homeostatic, since homeostatic systems are often thought of as being in a steady state. However, the steady state itself may change slowly over time, while a system maintains its ability to stabilize conditions internally. For instance, Earth has experienced a very slow increase in oxygen concentration in its atmosphere over its history (Chapter 4), yet life has adapted, and Earth remains by most definitions a homeostatic system.

Often it is difficult to understand systems fully by focusing on their individual components, because systems can show **emergent properties**, characteristics that are not evident in the system's components. Stating that systems possess emergent properties is a lot like saying "The whole is more than the sum of its parts." For example, if you were to reduce a tree to its component parts (leaves, branches, trunk, bark, roots, fruit, and so on) you would not be able to predict the whole tree's emergent properties, such as the role the tree plays as habitat for birds, insects, parasitic vines, and other organisms (Figure 6.2). You could chemically analyze the tree's chloroplasts (photosynthetic cell organelles), diagram its branch structure, and evaluate its fruit's nutritional content, but you would still be unable to understand the tree as habitat, as part of the forest landscape, or as a reservoir or "sink" for carbon storage. Emergent properties make it difficult to predict system behavior, but if we try to understand systems solely by breaking them into component parts, we will miss much of what makes them work and much of what makes them important.

Weighing the Issues:
Emergent Properties and the Mississippi River

List and describe the components of the Mississippi River system. Now list and describe at least four emergent properties of this system. What are the advantages and disadvantages of thinking in terms of whole systems and emergent properties rather than in terms of components and parts?

Finally, systems seldom have well-defined boundaries, so deciding where one system ends and another begins can be difficult. Consider a desktop computer system. It is certainly a network of parts that interact and exchange energy and information, but what are its boundaries? Is the system that which arrives in a packing crate and sits on top of your desk? Or does it include the network you connect it to at school, home, or work? What about the energy grid you plug it into, with its distant power plants and distribution and transmission

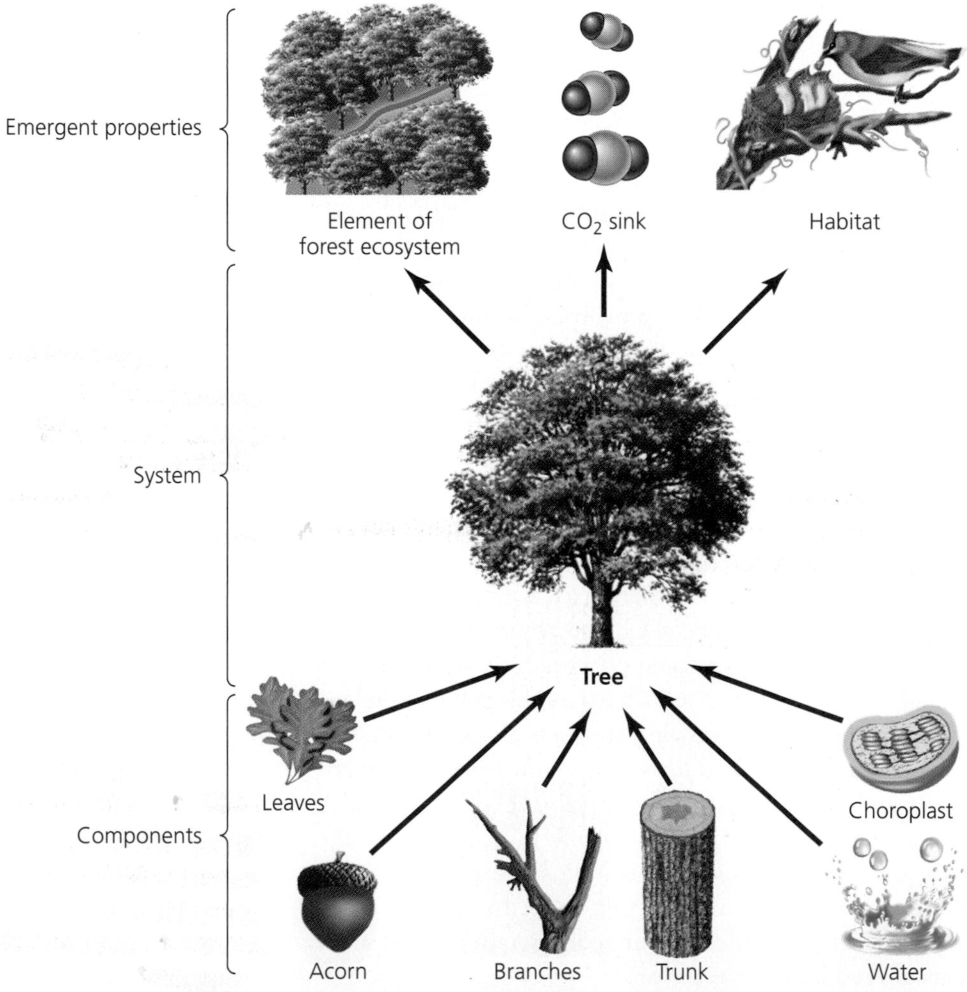

Figure 6.2 A system's emergent properties are not evident when we break the system down into its component parts. For example, a tree serves as wildlife habitat and plays roles in forest ecology and global climate systems, but you wouldn't know that from considering the tree as a collection of leaves, branches, and chloroplasts.

Emergent properties

Element of forest ecosystem   CO₂ sink   Habitat

System

Tree

Components

Leaves   Choroplast

Acorn   Branches   Trunk   Water

lines? And what of the Internet? While you're browsing the Web, you are constantly drawing in digitized text, light, color, and sound from around the world. No matter how we attempt to isolate or define a system, we soon see that it has many connections to systems both larger and smaller than itself. Scientists will often treat a system as if it is a **closed system,** one that is isolated and self-contained, for the purpose of simplifying some problem with which they are grappling. However, no matter how closed a system might seem, if we look closely enough or wait long enough, we will see connections between the system and the universe beyond. Thus, when viewed in context, all systems are **open systems,** exchanging energy, matter, and information with other systems.

Weighing the Issues:
Earth as a System

Do you think Earth is a closed system or an open system? In what ways? What implications might your answer have for how we as a society should use resources and develop energy policy?

Figure 6.3 The Mississippi River watershed is the biggest in North America, encompassing 41% of the lower 48 states—an area of 3.2 million km^2 (1.2 million mi^2). Water, sediments, and pollutants from a variety of sources are carried downriver to the Gulf of Mexico.

The Mississippi River is an environmental system

Let's now take a look at a natural system to see whether its boundaries are any easier to understand than those of a computer system. The Mississippi River is a system because it is an interacting collection of components that receives input, produces output, and shows emergent properties; one could hardly predict steamboats, barges, hundred-year floods, and Huckleberry Finn solely from a study of water molecules.

On a map, the Mississippi River system stands out as a branched and braided network of water channels lined by farms, cities, and forests (Figure 6.3). But where are this system's boundaries? You might argue that the Mississippi consists primarily of water, originates in Minnesota, and ends in the Gulf of Mexico near New Orleans. But what about the rivers that feed it and the farms that line its banks? Major rivers such as the Missouri, Arkansas, and Ohio flow into the Mississippi. Hundreds of smaller rivers and their tributaries flow through vast expanses of farmland, woodland, fields, cities, towns, and industrial areas before their water

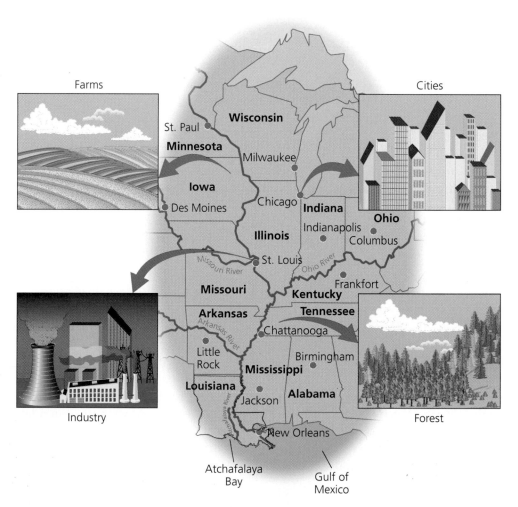

joins the Mississippi's. As these waterways flow, they carry with them millions of tons of sediment, hundreds of species of plants and animals, and a sizeable variety of pollutants. And what about the Mississippi River's output? Its waters flow into, and influence, the system of the Gulf of Mexico.

For an environmental scientist interested in runoff and the flow of water, sediments, pollutants, or anything else the Mississippi River carries along, it may make the most sense to view the river's watershed as a system. As with the Tijuana River in Chapter 3, considering the entire area of land a river drains, as a whole, may be most helpful in understanding how to solve problems of river pollution. For a scientist interested in the Gulf of Mexico's dead zone, however, it may make more sense to view the Mississippi River watershed together with the Gulf as the system of interest, because their interaction is so central to the problem. In environmental science, one's delineation of a system can and should depend upon the questions one is seeking to address.

Understanding the dead zone requires considering River and Gulf together as a system

Concerns about the Gulf of Mexico's ecosystem and fishery have centered on hypoxia, but the reason for these dangerously low levels of oxygen, scientists soon learned, was abnormally high levels of nitrogen. The excess nitrogen originates with various sources in the Mississippi River watershed, particularly nitrogen-rich fertilizers applied to crops. Additionally, about 90% of the nitrate that flows into the Gulf each year comes from agriculture, leaky septic fields, and runoff from city streets. Much of this nitrate originates from farms in Iowa, Illinois, Indiana, Minnesota, and Ohio (Figure 6.4a). Nitrogen input to farmland in the Mississippi River system has increased dramatically since 1950 (Figure 6.4b). From 1980 to the present, the Mississippi River and the Atchafalaya River (which drains a third of the Mississippi's diverted water) have pumped an average of 1.6 million metric tons (1.76 million tons) of nitrogen into the Gulf of Mexico each year, and this amount is still increasing.

The enhanced nitrogen input to the Gulf has boosted the growth of phytoplankton, microscopic photosynthetic organisms that live near the ocean surface, which ordinarily are limited by nitrogen scarcity. As more phytoplankton flourish at the surface, more of their waste products and more dead phytoplankton drift to the bottom, providing food for organisms (mainly

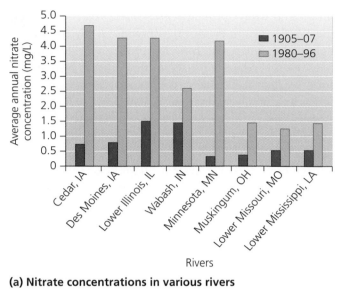

(a) Nitrate concentrations in various rivers

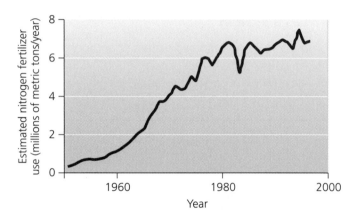

(b) Nitrogen use in the Mississippi-Atchafalaya River basin

Figure 6.4 **(a)** Concentrations of nitrates in the Mississippi and its tributaries rose during the 20th century. Blue bars show average concentrations from 1905 to 1907 (based on analysis of sediments), and green bars show average concentrations from 1980 to 1996. **(b)** Nitrogen fertilizer use in the Mississippi-Atchafalaya River basin skyrocketed after 1950. Data from National Science and Technology Council: Committee on Environment and Natural Resources, Hypoxia: An integrated assessment in the Northern Gulf of Mexico, May, 2000.

bacteria) that decompose them. The result is a population explosion of bacteria. These decomposers consume enough oxygen to cause oxygen concentrations in bottom waters to plummet, suffocating shrimp and fish that live at the bottom and giving rise to the dead zone. Because the fresh water from the river remains naturally stratified in a layer at the surface that mixes only very slowly with the salty ocean water, oxygenated surface water does not make its way down to the bottom-dwelling life that needs it. This process of nutrient

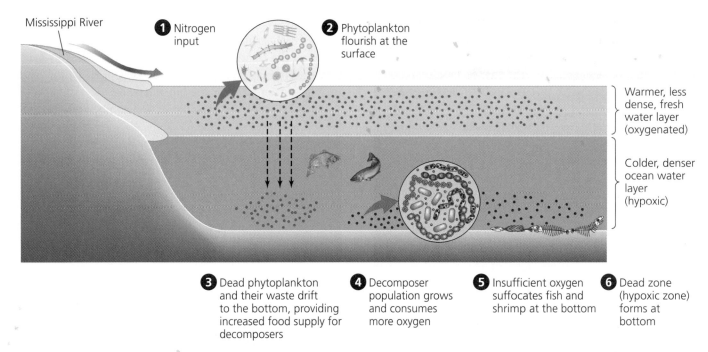

1 Nitrogen input

2 Phytoplankton flourish at the surface

Warmer, less dense, fresh water layer (oxygenated)

Colder, denser ocean water layer (hypoxic)

Mississippi River

3 Dead phytoplankton and their waste drift to the bottom, providing increased food supply for decomposers

4 Decomposer population grows and consumes more oxygen

5 Insufficient oxygen suffocates fish and shrimp at the bottom

6 Dead zone (hypoxic zone) forms at bottom

Figure 6.5 Excess nitrogen causes eutrophication in coastal marine ecosystems such as that in the Gulf of Mexico. Coupled with stratification (layering) of marine waters, eutrophication can severely deplete dissolved oxygen. Nitrogen from river water boosts phytoplankton growth, while stable surface waters prevent deeper water from absorbing oxygen to replace that consumed by decomposers. This systemic interaction gives rise to hypoxic zones like that of the Gulf.

enrichment, increased production of organic matter, and subsequent ecosystem degradation is known as **eutrophication** (Figure 6.5).

Organisms that can leave a hypoxic zone do, but those that are not mobile enough to escape are left to asphyxiate. Shrimp are mobile, but find escape difficult when hypoxia occurs rapidly over a wide area. Although additional nitrogen may increase the productivity of fisheries at first, at higher concentrations this fertilizing effect is offset by hypoxia, and fishery yields decline (Figure 6.6).

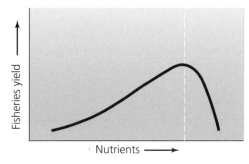

Figure 6.6 It is possible to have too much of a good thing. As nutrient loading increases, production and fishery yields may increase for many species but will drop off sharply once eutrophication begins.

Environmental systems may be perceived in different ways

Thinking in terms of watersheds is one way to perceive and delineate environmental systems, and combining two interacting systems such as the Mississippi River and the Gulf of Mexico is another. There are many other ways to delimit systems, and your choice will depend on the issues in which you are interested. We will now touch on several traditional ways of categorizing environmental systems, including structural spheres, biomes, ecosystems, landscapes, and cycles. These are each ways we can break down Earth's complex intertwined diversity of systems to make this complexity comprehensible to the human brain—and accessible to problem solving.

Earth can be divided into several structural spheres

One way to comprehend Earth's systems is to arrange them into major structural categories: the lithosphere (Chapter 8 and Chapter 16), hydrosphere (Chapter 13 and Chapter 14), atmosphere (Chapter 11 and Chapter 12), and biosphere (Chapter 5 and Chapter 15). The **lithosphere** is everything that is solid earth beneath our feet, including the rocks, sediment, and soil at the surface and extending down many miles underground. The **hydrosphere** encompasses all water—salt or fresh, liquid, ice, or vapor—in surface bodies, underground, and in the atmosphere. The **atmosphere** is comprised of the air surrounding our planet. The **biosphere** consists of the sum total of all the planet's living organisms and the abiotic portions of the environment with which they interact.

Although this division is useful, it is important to realize that the boundaries of categories even so fundamental as these do overlap, allowing the systems to interact. Picture the simple image of a robin plucking an earthworm from the ground after a rain. You are witnessing a component of the biosphere (the robin) consuming another component of the biosphere (the earthworm) by removing it from part of the lithosphere (soil) that the earthworm had been modifying—all this made possible because the hydrosphere (rain) recently wet the ground. The robin might then fly through the air (the atmosphere) to a tree branch (the biosphere), in the process respiring (combining oxygen from the atmosphere with glucose from the biosphere, and adding water to the hydrosphere and carbon dioxide and heat into the atmosphere). Finally, the bird might defecate, adding nutrients from the biosphere to the lithosphere below. The study of such interactions among living and nonliving things is part of ecology. As scientists become more inclined to approach whole systems in their work, ecology at and above the community and ecosystem levels is increasingly in demand. We will now look at several ecological systems, beginning with biomes.

Biomes are fundamental groupings of plant communities that cover large geographic areas

One way to categorize environmental systems is to view the world through the lens of biomes. A **biome** is a major regional complex of similar plant communities, a large ecological unit defined by its dominant plant type and vegetation structure. The world contains a number

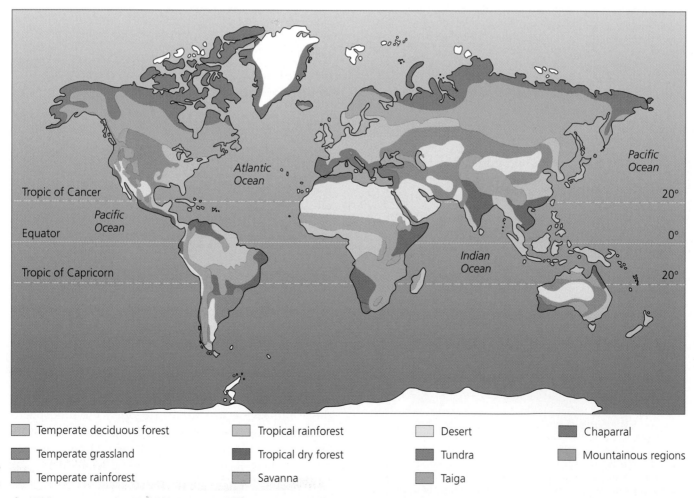

☐ Temperate deciduous forest	☐ Tropical rainforest	☐ Desert	☐ Chaparral
☐ Temperate grassland	☐ Tropical dry forest	☐ Tundra	☐ Mountainous regions
☐ Temperate rainforest	☐ Savanna	☐ Taiga	

Figure 6.7 Biomes are distributed around the world according to temperature, precipitation, and other factors.

of different biomes, each covering large contiguous geographic areas (Figure 6.7).

Which biome covers any particular portion of the planet depends on a variety of abiotic factors, including temperature, precipitation, atmospheric circulation, and soil characteristics, of which temperature and precipitation exert the greatest influence (Figure 6.8). Because biome type is largely a function of climate, and average monthly temperature and precipitation are among the best indicators of an area's climate, scientists often use **climate diagrams**, or **climatographs**, to depict such information. Global climate patterns may cause biomes to occur in large discrete patches in different parts of the world. For instance, temperate deciduous forest occurs in eastern North America, north-central Europe, and much of eastern China. Note in Figure 6.7 how patches representing the same biome tend to occur

along similar latitudes. This phenomenon is due to the north-south gradient in temperature and to atmospheric circulation patterns we will discuss in Chapter 11.

Biomes and ecosystems represent different ways of viewing and categorizing the same natural reality. As one example, vegetation across the eastern third of the United States is part of the temperate deciduous forest biome. From New Hampshire to eastern Texas, precipitation and temperature are similar enough that most of the region's natural plant cover consists of broad-leafed hardwood trees that lose their leaves in winter. Within this large region, however, there exist many different types of temperate deciduous forests, such as oak-hickory, beech-maple, and pine-oak forests, each of these sufficiently different to be called a separate community. When we expand our view to include the abiotic factors with which each community interacts, we are considering ecosystems. Thus a

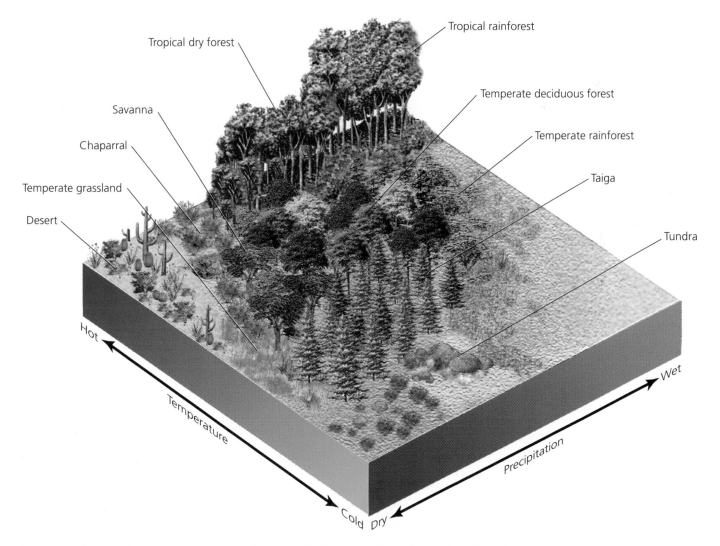

Figure 6.8 As precipitation increases, vegetation generally becomes taller and more luxuriant. As temperature increases, types of plant communities change. Together, temperature and precipitation are the main factors determining what biome occurs in a given area. Deserts occur in hot dry regions, for instance, tropical rainforests occur in hot wet regions, and tundra occurs in the coldest regions.

biome consists of similar plant communities growing over a more or less contiguous geographic area, whereas each of the communities and its accompanying abiotic factors comprise an ecosystem. Because of the similarities in plant communities and climatic conditions within biomes, animal communities tend to be more similar within biomes than among biomes.

We can divide the world into roughly ten biomes

Temperate deciduous forest The Mississippi River is fed by tributaries that drain portions of two biomes. The **temperate deciduous forest** (Figure 6.9) to the river's east is characterized by broad-leafed trees that lose their leaves each fall so that they remain dormant during winter, when hard freezes would endanger leaves. These mid-latitude forests occur in much of Europe and eastern China as well as in eastern North America, all areas where precipitation is spread relatively evenly throughout the year. Although soils of the temperate deciduous forest are relatively fertile, the biome generally consists of far fewer tree species than are found in tropical rainforests. Oaks, beeches, and maples are a few of the most dominant types of trees in these forests. A sampling of typical types of animals in the temperate deciduous forest of eastern North America is shown in Chapter 5, Figure 5.14.

Temperate grassland As one moves westward across the Mississippi River, temperature differences between winter and summer become more extreme, and rainfall diminishes. The amount of precipitation in this region can support grasses more easily than trees. Thus west of the Mississippi we find **temperate grasslands** (Figure 6.10). Also known as steppe or prairie, temperate grasslands were once widespread throughout regions of North and South America and much of central Asia. Today most of the world's grasslands have been converted to farmland or rangeland; the result has been a great reduction in the abundance of native grasses, the grazing animals that depend on them, and the predators of these grazers. Characteristic vertebrate animals of the wide-open spaces of the North American grasslands include American bison, prairie dogs, pronghorn antelope, and ground-nesting birds like meadowlarks.

Temperate rainforest Moving further west in North America, the topography becomes more varied, and biome types are intermixed. The Pacific Northwest region, with its heavy rainfall, features **temperate rainforest**

(a) Temperate deciduous forest

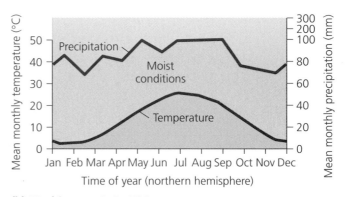

(b) Washington, D.C., USA

Figure 6.9 Temperate deciduous forests experience relatively stable seasonal precipitation and stronger variation in seasonal temperatures. Scientists use climate diagrams to illustrate an area's average monthly precipitation and temperature. Typically in these diagrams, the X-axis marks months of the year (beginning in January for regions in the northern hemisphere and in July for regions in the southern hemisphere) and paired Y-axes denote average monthly temperature and average monthly precipitation. The twin curves plotted on a climate diagram indicate the trends in precipitation and in temperature from month to month. When the precipitation curve lies above the temperature curve, as is the case throughout the year in the temperate deciduous forest biome around Washington, D.C., the region experiences relatively "moist" conditions. *Climatograph adapted from Siegmar-Walter Breckle.* Walter's Vegetation of the Earth: The Ecological Systems of the Geo-Biosphere, fourth edition. *Springer-Verlag Berlin Heidleberg, New York, 1999.*

(Figure 6.11), a forest-type known for its potential to produce large volumes of commercially important forest products, such as lumber and paper. Coniferous trees like cedars, hemlocks, and Douglas fir grow very tall in the temperate rainforest, and the forest interior is darkly shaded and damp. In the Pacific Northwest, moisture-loving animals like bright yellow banana slugs are common, and old-growth stands hold the endangered spotted

(a) Temperate grassland

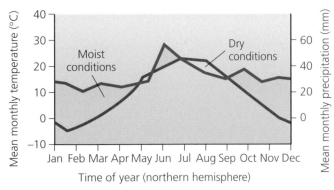

(b) Odessa, Ukraine

Figure 6.10 Temperate grasslands experience temperature variations throughout the year and too little precipitation for many trees to grow. Constructed for Odessa, Ukraine, this climatograph indicates both "moist" and "dry" climate conditions. When the temperature curve is above the precipitation curve, as is the case in May and mid-June through September, the climate conditions are "dry." *Climatograph adapted from Siegmar-Walter Breckle, 1999.*

(a) Temperate rainforest

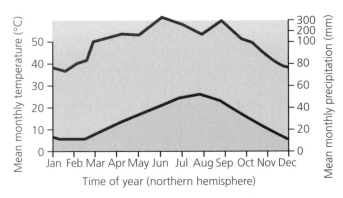

(b) Nagasaki, Japan

Figure 6.11 Temperate rainforests receive a great deal of precipitation, and include moist mossy interiors. *Climatograph adapted from Siegmar-Walter Breckle, 1999.*

owl. The soils of temperate rainforests are usually quite fertile but are susceptible to landslides and erosion after forest clearing. Temperate rainforests have been the focus of much controversy in the Pacific Northwest, where a century of overharvesting has driven certain species toward extinction and pushed many forest-dependent human communities toward economic ruin.

Tropical rainforest In tropical regions we see the same pattern found in temperate regions: Areas of high rainfall grow rainforests, areas of intermediate rainfall host dry or deciduous forests, and areas of lower rainfall become dominated by grasses. However, tropical biomes differ from their temperate counterparts in other

ways because they are closer to the equator and therefore warmer. For one thing, they hold far greater biodiversity. **Tropical rainforest** (Figure 6.12) is found in Central America, South America, southeast Asia, west Africa, and other tropical regions and is characterized by year-round rain and uniformly warm temperatures. Tropical rainforests have dark damp interiors, lush vegetation, and highly diverse biotic communities, with greater numbers of species of insects, birds, amphibians, and various other animals than any other biome. These forests are not dominated by single species of trees as in forests closer to the poles, but instead consist of very high numbers of tree species intermixed, with each at a low density. Any given tree may be draped with vines, enveloped by strangler figs, and its branches loaded with epiphytes (orchids and other plants that can grow without soil), such that trees occasionally collapse under the weight of all the life they support. Despite this profusion

(a) Tropical rainforest

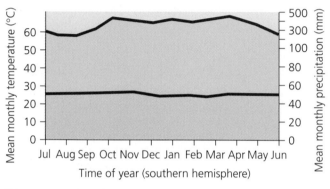

(b) Bogor, Java, Indonesia

Figure 6.12 Tropical rainforests, famed for their biodiversity, are defined by constant, warm temperatures and a great deal of rain. *Climatograph adapted from Siegmar-Walter Breckle, 1999.*

(a) Tropical dry forest

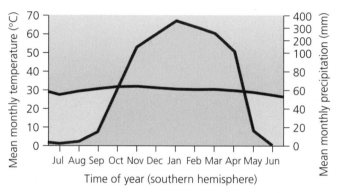

(b) Darwin, Australia

Figure 6.13 Tropical dry forests experience significant seasonal variations in precipitation and relatively stable warm temperatures. *Climatograph adapted from Siegmar-Walter Breckle, 1999.*

of life, however, tropical rainforests have very poor, acidic soils low in organic matter (matter derived from living things). Nearly all nutrients present in this biome are contained in the trees, vines, and other plants—not in the soil. An unfortunate consequence is that once tropical rainforests are cleared, the nutrient-poor soil can support agriculture for only a very short time. As a result, farmed areas are abandoned quickly, and the soil and forest vegetation recovers slowly.

Tropical dry forest Tropical areas that are warm year-round but where rainfall is lower overall and highly seasonal give rise to **tropical dry forest**, or tropical deciduous forest (Figure 6.13), a biome widespread in India, Africa, South America, and northern Australia. Wet and dry seasons each span about half a year in tropical dry forest. Rains during the wet season can be extremely heavy and, coupled with erosion-prone soils,

can lead to severe problems of soil loss when forest clearing occurs over large areas. Indeed, across the globe, much land originally covered in tropical dry forest has been converted to agriculture. Clearing for farming is made easier by the fact that vegetation heights are much lower and the canopies less dense than in tropical rainforest. Organisms that inhabit tropical dry forest have adapted to seasonal fluctuations in precipitation and temperature. For instance, plants often leaf out and grow profusely with the rains, then drop their leaves during the driest part of year.

Savannas Drier tropical regions may give rise to **savannas** (Figure 6.14), regions of grasslands interspersed with clusters of acacias and other trees. The savanna biome is found today across large stretches of Africa (the ancestral home of our species), South America, Australia, India, and other dry tropical regions.

(a) Savanna

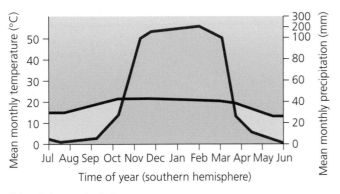

(b) Salisbury, Zimbabwe

Figure 6.14 Savannas, grasslands with clusters of trees, experience slight seasonal variation in temperature but significant variation in rainfall. *Climatograph adapted from Siegmar-Walter Breckle, 1999.*

Precipitation in savannas usually arrives during distinct rainy seasons and brings with it the periodic gathering of many grazing animals near widely spaced water holes. Common herbivores on the African savanna include zebras, gazelles, and giraffes, and the predators of these grazers include lions, hyenas, and other highly mobile carnivores.

Desert Where rainfall is very sparse, **desert** (Figure 6.15) forms. This is the driest biome on Earth; most deserts receive less than 25 cm (9.8 in) of precipitation per year, much of it during isolated storms that may occur months or years apart. Depending on rainfall, deserts vary greatly in the amount of vegetation they support. Some, like the Sahara and Namib Deserts of Africa, are mostly bare sand dunes; others, like the Sonoran Desert of Arizona and northwest Mexico, are quite heavily vegetated. Contrary to popular belief,

(a) Desert

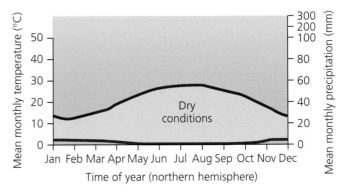

(b) Cairo, Egypt

Figure 6.15 Deserts are dry year-round but they are not always hot. Precipitation can arrive in intense, widely spaced storm events. The temperature curve is consistently above the precipitation curve in this climatograph of Cairo, Egypt, indicating that the region experiences "dry" conditions all through the year. The photograph, from the Sonoran Desert in Arizona, shows the maximum amount of vegetation a desert would support. *Climatograph adapted from Siegmar-Walter Breckle, 1999.*

deserts are not always hot; the high desert of the western United States is an example. Because deserts have relatively little vegetation to insulate them from temperature extremes, sunlight readily heats them in the daytime, but daytime heat is quickly lost at night; so, temperatures vary widely from day to night and across seasons of the year. Deserts are also frequently subjected to high winds and, during the infrequent rainy spells, widespread flooding. Desert soils can often be quite saline and are sometimes known as lithosols, or stone soils, for their high mineral and low organic-matter content. Desert animals and plants show many fascinating adaptations to deal with a harsh climate; most reptiles and mammals, such as rattlesnakes and kangaroo mice,

(a) Tundra

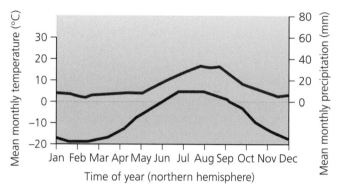

(b) Vaigach, Russia

Figure 6.16 Tundra is a cold, dry biome found near the poles and in the high mountains at lower latitudes. *Climatograph adapted from Siegmar-Walter Breckle, 1999.*

(a) Taiga

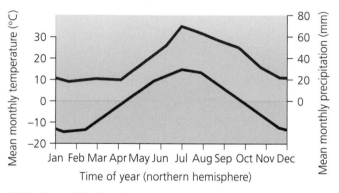

(b) Archangelsk, Russia

Figure 6.17 Taiga is defined by long, cold winters, relatively cool summers, and moderate precipitation. *Climatograph adapted from Siegmar-Walter Breckle, 1999.*

are active in the cool of night, and many Australian desert birds are nomadic, wandering long distances to find areas of recent rainfall and plant growth. Many desert plants have thick leathery leaves to reduce water loss, and some have green trunks so the plant can photosynthesize while foregoing leaves that would lose water. The spines of cacti and many other desert plants guard those plants from being eaten by herbivores desperate for the precious water they hold.

Tundra Tundra (Figure 6.16) is nearly as dry as desert, but is located at very high latitudes along the northern edges of Russia, Canada, and Scandinavia. Extremely cold winters with little daylight and moderately cool summers with lengthy days characterize this landscape of lichens and low scrubby vegetation. The great seasonal variation in temperature and daylength results from this biome's position close to the poles, which are angled toward the sun in the summer and away from the sun in

the winter. Because of the cold climate, underground rock and soil remains more or less permanently frozen; this material is called *permafrost*. During the long cold winters, the surface soils freeze as well; then, when the weather warms, they melt and produce seasonal accumulations of surface water that make ideal habitat for mosquitoes and other biting insects. The swarms of insects for which areas like Alaska are notorious may deter us, but they benefit bird species that migrate long distances to breed during the brief but productive summer. Caribou also migrate to the tundra to breed, then leave for the winter. Only a few animals, like polar bears and musk oxen, can survive year-round in this extreme climate.

Taiga The northern coniferous, or boreal, forest—often called **taiga** (Figure 6.17)—stretches in a broad band across much of Canada, Alaska, Russia, and Scandinavia. It consists of a limited number of species of evergreen trees, such as black spruce, that dominate large stretches

(a) Chaparral

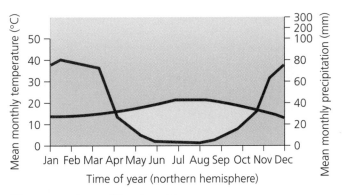

(b) Los Angeles, California, USA

Figure 6.18 Chaparral is a highly seasonal biome dominated by shrubs, influenced by marine weather, and dependent on fire. *Climatograph adapted from Siegmar-Walter Breckle, 1999.*

of forests interspersed with occasional bogs and lakes. Taiga's uniformity over huge areas reflects the climate common to this latitudinal band of the globe: these forests develop in cooler, drier regions than do temperate rainforests, and they experience long, cold winters and short, cool summers. Soils are typically nutrient-poor and somewhat acidic. As a result of the strong seasonal variation in day length, temperature, and precipitation, many organisms compress a year's worth of feeding, breeding, and rearing of young into a few warm, wet months. Year-round residents of taiga include mammals such as moose, wolves, bears, lynx, and many burrowing rodents. Taiga is also home to many insect-eating birds that migrate from the tropics to breed during the short intensely productive summer season.

Chaparral In contrast to taiga's broad continuous distribution, **chaparral** (Figure 6.18) is limited to fairly small patches widely flung around the globe. Chaparral consists mostly of evergreen shrubs and is densely thicketed. This biome is also highly seasonal, with mild wet winters and warm dry summers. This type of climate is induced by oceanic influences and is often termed "Mediterranean." In addition to ringing the Mediterranean Sea, chaparral occurs along the coasts of California, Chile, and southern Australia. Chaparral communities experience frequent fire, and their plant species are adapted to resist fire or even to depend on it for germination of their seeds.

Altitude creates patterns analogous to latitude

As any hiker or skier knows, climbing in elevation causes a much more rapid change in climate than moving the same distance toward the poles. Vegetative communities change along mountain slopes in correspondence with this small-scale climate variation (Figure 6.19). This is why it is often said that hiking up a mountain in the southwestern United States can be like walking from Mexico to Canada. A hiker ascending one of southern Arizona's higher mountains would start in Sonoran Desert or desert grassland, proceed through oak woodland, enter pine forest, and finally arrive in spruce-fir forest—the equivalent of passing through several biomes. A hiker scaling one of the great peaks of the Andes in Ecuador could begin in tropical rainforest and end in alpine tundra.

Because of the way vegetation—and the animals associated with particular plant communities—changes with altitude, mountains often have the effect of isolating communities in high-elevation patches. Thus mountains can be considered terrestrial equivalents of oceanic islands, as we saw with the Monteverde cloud forest in Chapter 5. Such isolation can help drive the process of speciation in the long term but can also make isolated species vulnerable to extinction in the short term if climate changes too quickly.

Aquatic systems also show biome-like patterns of variation and similarity

You may have noticed that in our discussion of biomes we have focused exclusively on terrestrial (land-based) systems. This is because the biome concept, as traditionally developed and applied, has been limited to terrestrial systems. However, this limit may be largely a matter of historical convention, because there seems little obvious reason not to consider aquatic systems in the same way. One could, for example, consider all large freshwater lakes of the world collectively as a biome, albeit a very dispersed one. Similarly, areas equivalent to biomes

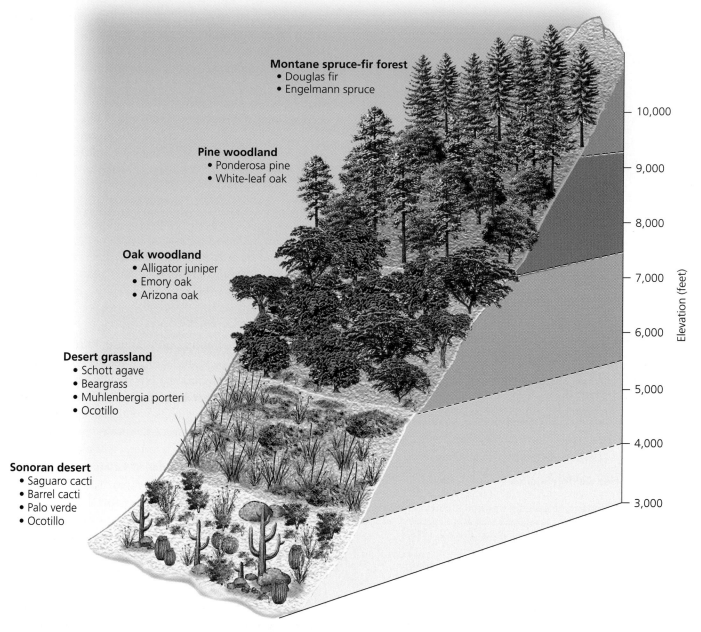

Montane spruce-fir forest
• Douglas fir
• Engelmann spruce

Pine woodland
• Ponderosa pine
• White-leaf oak

Oak woodland
• Alligator juniper
• Emory oak
• Arizona oak

Desert grassland
• Schott agave
• Beargrass
• Muhlenbergia porteri
• Ocotillo

Sonoran desert
• Saguaro cacti
• Barrel cacti
• Palo verde
• Ocotillo

Elevation (feet): 10,000 — 9,000 — 8,000 — 7,000 — 6,000 — 5,000 — 4,000 — 3,000

Figure 6.19 Vegetation changes as one climbs in altitude in ways similar to how it changes as one moves toward the poles. Climbing a mountain in southern Arizona, as pictured here, can be likened to traveling from Mexico to Canada, taking the hiker through the local equivalent of several biomes.

certainly exist in the oceans, but their geographic shapes would look very different from those of terrestrial biomes if plotted on a world map. One might consider the thin strips along the world's coastlines to represent one aquatic system, the continental shelves another, and the open ocean, the deep sea, and the coral reefs still other distinct sets of communities. There are in addition many coastal systems that straddle the line between terrestrial and aquatic, such as salt marshes, rocky intertidal communities, mangrove forests, and estuaries.

Unlike terrestrial biomes, however, aquatic systems are shaped not by air temperature and precipitation, but instead by factors such as water temperature, salinity, dissolved nutrients, wave action, currents, depth, and type of substrate (sandy, muddy, rocky bottom, and so on). In addition, marine communities in many cases may be more clearly delineated by their animal life than by their plant life. We will examine freshwater and marine ecosystems in the greater detail they deserve in Chapter 13 and Chapter 14.

Weighing the Issues:
Human Impacts on Biomes

Temperature, precipitation, and soil type affect biome distribution in terrestrial systems, whereas temperature, salinity, dissolved nutrients, wave action, currents, depth, and the type of substrate affect biome distribution in aquatic systems. Describe some human activities that could affect these biome-shaping factors. How might these effects give rise to undesirable environmental changes?

Ecosystems are key environmental systems

Whereas biomes are associations of vegetation critically influenced by abiotic factors, an ecosystem describes all the interacting organisms and abiotic factors that occur in a particular place at the same time. An ecosystem might be as small as an aquarium containing a few goldfish, some aquatic plants, gravel, and water, or it might be as large as a continent. For some purposes, scientists view the entire planet as a single all-encompassing ecosystem. But the term is most often used to refer to interactive systems limited to a somewhat small geographic area with somewhat discrete boundaries. For example, the Monteverde cloud forest of the previous chapter is easily classified as an ecosystem. It includes a diversity of biotic members interacting with abiotic factors, such as moisture from clouds. It is somewhat delineated in that it is surrounded by drier forests on the mountains' lower slopes and by cleared agricultural land below that. The ecosystem of the Gulf of Mexico's continental shelf is not quite as neatly demarcated, but still is a distinctive entity.

A key distinction between ecosystems and communities is that ecosystems include abiotic elements and involve the flow of energy and nutrients among both living and nonliving parts of the system. Input of moisture from clouds played a key role in the Monteverde ecosystem, for example, and inputs of water, sediment, and nutrients from the Mississippi River play a key role in the Gulf of Mexico ecosystem.

Energy for most ecosystems comes from the sun (Chapter 4) and is converted to **biomass** (matter contained in living organisms) by producers through photosynthesis. Ecosystems vary in the rate at which plant biomass is produced. Those that convert solar energy to biomass rapidly are said to have high **primary productivity.** Coral reefs and certain other marine systems tend to have the highest primary productivities of the world's ecosystems. Producers—such as ocean phytoplankton or the microbial photosynthesizers living symbiotically with corals on reefs—need to use much of this energy conversion for themselves, however. Only some of the energy is left for consumers—such as fish, shrimp, and starfish—to use. The rate at which biomass available to consumers is produced is termed **net primary productivity.** Among the world's ecosystems, wetlands and tropical rainforests tend to have the highest net primary productivities. Soon we will explore the flow of matter—specifically, key nutritive chemicals—through ecosystems in more detail, but first let's look at one more way of viewing environmental systems.

Landscape ecology studies geographical areas that include multiple ecosystems

Ecosystems are open systems, and those that physically abut one another will interact extensively, no matter how distinctly different they may appear. For instance, a pond ecosystem is very different from the forest ecosystem that may surround it, but the two interact. Salamanders that breed and develop in the pond may live their adult lives under logs on the forest floor. Rainwater that nourishes forest plants may eventually make its way to the pond, carrying with it nutrients from the forest's leaf litter. Similarly, coastal dunes, the ocean, and the lagoon or salt marsh in between them all interact, as do forests and prairies in areas where they meet. Areas where ecosystems meet may consist of transitional zones called **ecotones,** in which biotic and abiotic elements of the ecosystems mix.

Because of this mixing, sometimes ecologists find it useful to view these systems at a larger, landscape scale that focuses on geographical areas that include multiple ecosystems. For instance, if you are interested in large mammals, such as black bears, that move seasonally from mountains to valleys or can move between mountain ranges, you had better consider the landscape scale that includes all these habitats. Such a broad-scale approach, often called **landscape ecology,** is important in studying birds that migrate between continents, or fish like salmon that move between marine and freshwater ecosystems. Some conservation groups, such as the Nature Conservancy, have begun applying the landscape ecology approach widely in their land management strategies (Chapter 16).

How Environmental Systems Work

We now have some idea of ways to classify environmental systems. Let's now turn to how they actually work. The flow of matter through Earth and its ecosystems involves several crucial cycles of rock, water, and key nutrients.

The rock cycle is a slow, but important, environmental system

We tend to think of rock as pretty solid stuff. This is clear when we say something is "hard as rock." Yet in the long run, over geological time, rocks do change. Rocks and the minerals that make them up are heated, melted, cooled, broken, and reassembled (Figure 6.20) in a very slow process called the **rock cycle.**

Igneous rock All rocks can melt. At temperatures that are high enough, rocks will turn to a molten, liquid state

called **magma.** If magma is released from the lithosphere (as in a volcanic eruption), it may flow or spatter across Earth's surface as **lava.** Rocks that form when magma cools are called **igneous** (from the Latin *ignis,* meaning "fire") **rocks** (Figure 6.20a). Igneous rock comes in several different types, because there are different ways in which magma can solidify. Magma may sometimes cool slowly while it is still well below Earth's surface, insulated by a layer of overlying lithosphere. In this case it is known as **intrusive** igneous rock. Half Dome and many other famous rock formations at Yosemite National Park in California were formed in this way and later uncovered. Granite is the best-known type of intrusive rock. A slow cooling process allows minerals of different types to segregate from one another and aggregate with minerals of their own type, forming the crystals that give granite its multicolored, coarse-grained appearance. In contrast, when magma is spewed from a volcano, it cools relatively quickly, such that minerals have little time to consolidate into different types. This kind of igneous rock is called **extrusive** igneous rock, and its most common representative is basalt.

Figure 6.20 The rock cycle is a slow but essential process that shapes Earth's crust and the various nutrient cycles. Rocks occur in three main types: (**a**) igneous, (**b**) sedimentary, and (**c**) metamorphic.

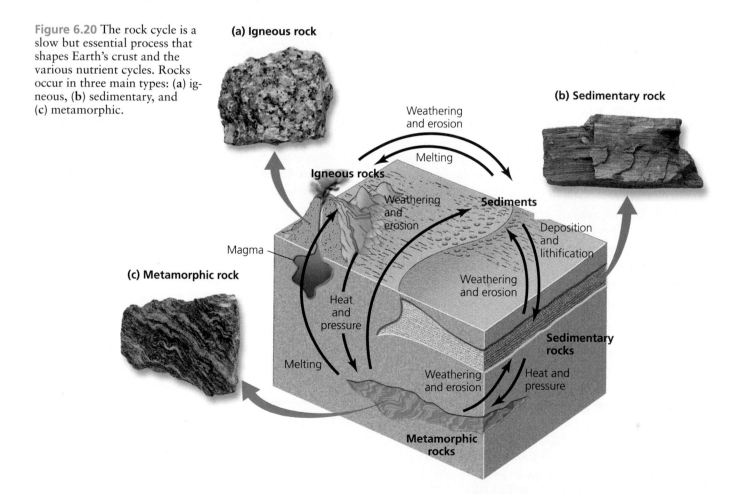

(a) Igneous rock

(b) Sedimentary rock

(c) Metamorphic rock

Weathering and erosion

Melting

Igneous rocks

Weathering and erosion

Sediments

Magma

Deposition and lithification

Weathering and erosion

Heat and pressure

Sedimentary rocks

Melting

Weathering and erosion

Heat and pressure

Metamorphic rocks

Sedimentary rock All rocks weather away with time. The relentless forces of wind and water, and freezing and thawing, gradually eat away at rocks, stripping away one tiny grain (or large chunk) at a time. Particles of rock get blown by wind or washed away by water, finally coming to rest somewhere, where they help to form **sediments.** These eroded remains of rocks usually are deposited very slowly, but floods—like those that sweep through the Mississippi and the Tijuana Rivers—can accelerate the process. Human activity, such as mining, farming, road-building, and industrial development, can also speed the process of erosion. Sediments collect downhill, downstream, or downwind from their sources and are deposited in layers. These layers accumulate over time, causing the weight and pressure of overlying layers to increase. **Sedimentary rock** is formed when dissolved minerals seep through sediment layers and act as a kind of glue, crystallizing and binding sediment particles together (Figure 6.20b). The fossils of organisms, and the fossil fuels we utilize for energy, are formed along with sediments during this process.

Like igneous rock, the several types of sedimentary rock are classified by the way they form and the size of the particles they contain. One major type of sedimentary rock forms when rocks dissolve and their components crystallize to form limestone, rock salt, and other familiar chemical sedimentary rocks. A second type of sedimentary rock forms when layers of sediment are compressed and physically bonded to one another. Conglomerate is one example made up of large particles that can give it the appearance of nougat. Sandstone is rock made of cemented sand particles. Shale is comprised of even smaller particles—mud particles—compacted together. Shale has been proposed as a source of fuel to turn to once existing oil and gas deposits start to run dry.

Metamorphic rock Rock does not lie around untouched forever. Geological processes may uplift it, bend its layers, compress it, or stretch it. When great heat or pressure is exerted on rock, the rock may change its form, becoming **metamorphic** (from the Greek for "changed form" or "changed shape") **rock** (Figure 6.20c). The forces that act to transform rock into metamorphic rock occur at temperatures lower than the melting point of the rock but high enough to reshape the crystals within the rock and change its appearance and physical properties. Metamorphic rock may resemble the rock it was created from but may be different in many ways. Common types of metamorphic rock include marble, formed when limestone is heated and pressurized, making its

structure stronger; and slate, formed when shale is heated and metamorphosed.

The changes that occur as rocks are altered from one type to another may proceed in any direction. Understanding the transition of rocks from one stage in the rock cycle to another will enable you to appreciate more clearly the formation and conservation of mineral resources, soils, fossil fuels, and other resources.

Plate tectonics shapes Earth's geography

The rock cycle takes place within the larger context of **plate tectonics**, a process that underlies earthquakes and volcanoes and that determines the geography of the Earth's surface. Like rocks, in the long run even continents are altered. Earth's surface consists of a lightweight thin **crust** of rock floating atop a malleable **mantle**, which in turn surrounds a molten heavy **core** made mostly of iron. Earth's internal heat drives convection currents that flow in loops in the mantle, pushing the mantle's soft rock cyclically upward (as it warms) and downward (as it cools), like a gigantic conveyor belt. As the mantle material moves, it drags along its surface edge large plates of crust. Earth's surface consists of about 15 major tectonic plates, each including some combination of ocean and continent (Figure 6.21). Imagine peeling an orange and then putting the pieces of peel back onto the spherical fruit; the ragged pieces of peel are like the plates of crust riding atop the Earth's surface.

Magma surging upward to the surface where plates divide pushes plates apart and creates new crust as it cools and spreads (Figure 6.22a). The mid-Atlantic ridge is part of a 74,000-km (46,000-mi) system of magmatic extrusion cutting across the seafloor. Plates moving outward from these divergent plate boundaries at the mid-ocean ridges bump and grind against one another, resulting in long-term continental movement, at rates of roughly 2–15 cm (1–6 in) per year. This movement has influenced Earth's climate and life's evolution throughout our planet's history, as the continents combined, separated, and recombined in different configurations. At least once, all landmasses were joined together in a supercontinent scientists have dubbed "Pangaea."

When two plates meet they may slip and grind alongside one another (Figure 6.22b). This is the case with the Pacific Plate and the North American Plate, which rub against each other along California's San Andreas Fault, famed for the earthquakes the plates' friction engenders. Southern California is slowly inching its way toward northern California along this transform fault, and Los Angeles will eventually reach San Francisco.

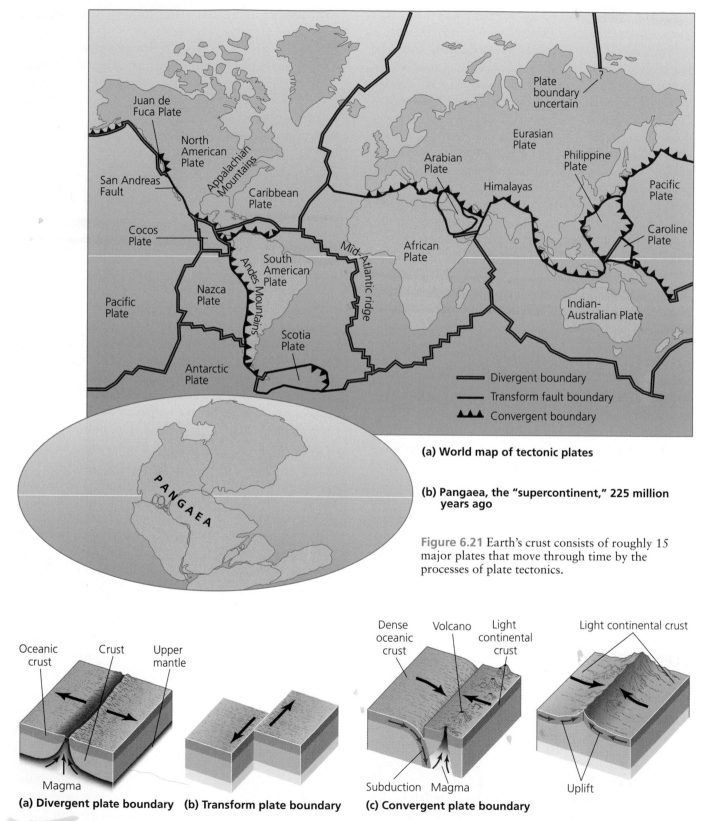

(a) World map of tectonic plates

(b) Pangaea, the "supercontinent," 225 million years ago

Figure 6.21 Earth's crust consists of roughly 15 major plates that move through time by the processes of plate tectonics.

(a) Divergent plate boundary **(b) Transform plate boundary** **(c) Convergent plate boundary**

Figure 6.22 Different types of boundaries between tectonic plates result in different geologic processes. At a divergent plate boundary (**a**) such as a mid-ocean ridge on the seafloor, magma extrudes from beneath the crust, and the material of the two plates moves gradually away from the boundary in the manner of conveyor belts. At a transform boundary (**b**), two plates slide alongside one another, creating friction that leads to earthquakes. Where plates collide (**c**), one plate may be subducted beneath another, leading to volcanism, or both plates may be uplifted, leading to the formation of mountain ranges.

When plates collide, either of two consequences may result (Figure 6.22c). Denser ocean crust will slide beneath lighter continental crust in a process called **subduction.** The subducted crust gets heated as it dives into the mantle, and often sends up magma that erupts through the surface in volcanoes. South America's Andes Mountains were formed by subduction, and their volcanoes are still active as the Nazca Plate continues to slide below the South American Plate. The 1980 eruption of Mt. St. Helens in Washington was likewise fueled by magma from subduction. Alternatively, two colliding plates of continental crust may slowly lift up material. The Himalayas, the world's highest mountains, are the result of the Indian-Australian Plate's collision with the Eurasian Plate 40–50 million years ago, and these mountains are still being uplifted today. The Appalachian Mountains of the eastern United States, once the world's highest mountains themselves, resulted from a much earlier collision with the edge of what is today Africa.

Amazingly, this environmental system of such fundamental importance was unknown to us just half a century ago. Our civilization was sending humans to the moon at the same time it was confirming the theory of plate tectonics to explain the movement of land under our very feet. It is humbling to reflect on this; what other fundamental systems on Earth might we not yet appreciate or understand while our technology—and our ability to affect Earth's processes—continues racing ahead?

The hydrologic cycle influences all other cycles

Water is so integral to life that we frequently take it for granted. Its supply is finite, but for most of human history people were unaware of this limitation. Water is the essential medium for all manner of biochemical reactions, as discussed in Chapter 4, and it plays a role in nearly every environmental system, including each of the nutrient cycles we will discuss shortly. Along with nutrients, water carries sediments from the continents to the oceans via rivers, streams, and surface runoff, and it distributes sediments onward in ocean currents. Increasingly, water also distributes artificial pollutants. The water cycle, or **hydrologic cycle,** (Figure 6.23) summarizes how water—in liquid, gaseous, and solid forms—flows through our biotic and abiotic environment.

The oceans are the main reservoir of water in the hydrologic cycle, holding over 97% of all water on Earth. The freshwater we depend on for our survival accounts for less than 3%, and two-thirds of this already small amount is tied up in glaciers, permanent snowfields, and ice caps. Thus considerably less than 1% of the planet's water is in a form in which we can readily use it—groundwater, surface fresh water, or rain from water vapor in the atmosphere.

Evaporation and transpiration Let's step into the hydrologic cycle at the stage of surface water—water in

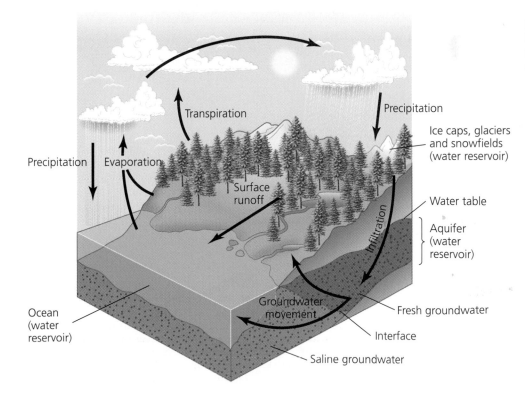

Figure 6.23 The hydrologic cycle is a system unto itself but also plays a key role in other nutrient cycles.

oceans, lakes, ponds, rivers, and moist soil. Water moves from these reservoirs into the atmosphere via **evaporation,** conversion from a liquid to a gaseous form. Warmer temperatures and stronger winds speed rates of evaporation. A greater degree of exposure has the same effect; an area logged of its forest or converted to agriculture or residential use will lose water more readily than a comparable area that remains heavily vegetated, for example. Water also enters the atmosphere via **transpiration**. This is the release of water vapor by plants through their leaves. Transpiration and evaporation act as natural processes of distillation, effectively creating pure water by separating it from any minerals carried with it in solution.

Precipitation, runoff, and surface water Water returns from the atmosphere to Earth's surface as **precipitation** when water vapor condenses and falls in droplet or crystal form. Rain or snow that falls on land may be taken up by plants and used by animals, but much of it flows into streams, rivers, lakes, ponds, and in many cases eventually to the ocean as **runoff.** Amounts of precipitation vary greatly from region to region globally, helping give rise to the variety of biomes we investigated earlier. Precipitation amounts can also vary greatly on small scales, especially due to the effects of mountains. Air currents forced upward by mountain ranges cause water vapor to condense and fall as rain on the windward side of a range, leaving the leeward side, often called the **rainshadow,** to receive dry air depleted of its moisture. Weather systems from the Pacific Ocean drop rain on the western slopes of major ranges in the western United States, for example, as they move east. This is why areas to the west of the Cascades in Washington and Oregon, and California's coast ranges and Sierra Nevada, have rainy climates, whereas east of these ranges the land gives way to the desert of the Great Basin and other regions.

Groundwater Some precipitation and surface water soaks down through soil and rock to recharge underground water reservoirs known as **aquifers.** Aquifers are spongelike regions of rock and soil that hold **groundwater,** water found underground beneath layers of soil. The upper limit of groundwater held in an aquifer is referred to as the **water table.** Aquifers may hold groundwater for long periods of time, so the water may often be quite ancient. In some cases groundwater can take hundreds or even thousands of years to fully recharge after being depleted (Figure 6.24).

Human impacts on the hydrologic cycle Human activity has affected every aspect of the water cycle. By damming rivers to create reservoirs, we have increased evaporation and, in some cases, infiltration of surface water into aquifers. By altering Earth's surface and its vegetation, we have increased surface runoff and the erosion that results from it. By drawing groundwater to the surface for drinking, irrigation, and industrial uses, we have begun to deplete groundwater resources (Figure 6.24). By spreading water on agricultural fields, we have depleted rivers, lakes, and streams and also have increased evaporation. By removing forests and other vegetation, we have reduced transpiration in some areas. By emitting into the atmosphere certain pollutants that dissolve in water droplets, we have changed the chemical nature of precipitation, in effect sabotaging the natural distillation process that evaporation and transpiration provide.

Figure 6.24 This villager in Kenya dishes drinking water from a well that must be dug deeper every year because of depletion of groundwater. Water tables are falling worldwide as humans extract water from aquifers faster than they can be recharged.

In recent years many political scientists and economists have claimed that shortages of fresh water will be a widespread cause of wars in the near future. Water shortages have already given rise to numerous conflicts worldwide, from the Middle East to the American West (Chapter 14).

Biogeochemical cycles are essential to understanding ecosystems

Some of the most important environmental systems are those that move essential chemicals known as nutrients through the environment. **Nutrients** are elements and compounds that organisms consume and require for survival, and these substances move through the environment in cycles called **nutrient cycles** or **biogeochemical cycles.** In the process, they travel within and among ecosystems, pass through the atmosphere, hydrosphere, lithosphere, and biosphere, and pass from one organism into another. A carbon atom that nourishes you today might have been incorporated into the body of a cow a month earlier, may have been in a blade of grass a month before that, and may have been part of a dinosaur's body 100 million years ago. After our deaths, the nutrients in our bodies will return to the environment and spread widely throughout it, eventually being incorporated by an untold number of organisms.

Biogeochemical cycles consist of systems within systems. They support life on Earth and are altered by it. These cycles shape many environmental problems, not least the presence of hypoxic waters in the Gulf of Mexico. As we discuss these cycles, think about how negative feedback loops are involved in them, and how positive feedback loops may be created by some human actions.

All organisms require 24 of the naturally occurring chemical elements to survive. Elements and compounds required in relatively large amounts are called **macronutrients.** In contrast, **micronutrients** are nutrients needed in smaller doses. Nitrogen, carbon, and phosphorus are macronutrients whose cycles tell us a great deal about the nature of environmental systems. They can also remind us that it is possible to have "too much of a good thing."

The nitrogen cycle has changed dramatically in recent decades

Nitrogen makes up 78% of our atmosphere, is the sixth most abundant element on Earth, and is an essential ingredient in proteins, DNA, and RNA. Thus the **nitrogen cycle** (Figure 6.25) is of great importance to us and to all other organisms. Despite its abundance in the air, nitrogen gas (N_2) is chemically inert and cannot cycle out

Figure 6.25 The nitrogen cycle is one of Earth's critical biogeochemical cycles. Key human impacts occur via fossil fuel combustion, nitrogen oxide emissions from industry, and application of nitrogen fertilizer produced by the Haber-Bosch process. Humans have doubled the amount of nitrogen fixed in terrestrial ecosystems since the early 1900s. In the past half-century, human inputs of nitrogen into the environment have greatly increased, such that today fully half of the nitrogen entering the environment is of human origin. Agricultural fertilizer has for the past 35 years been the leading source of all nitrogen inputs, natural and artificial.

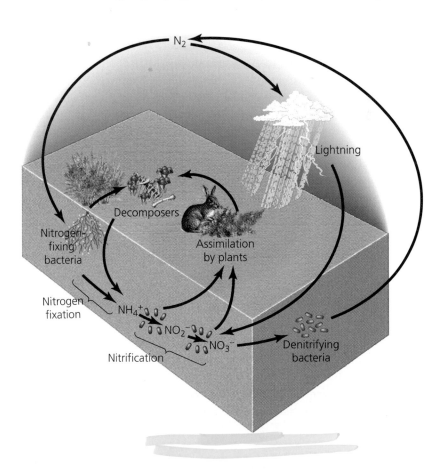

of the atmosphere and into living organisms without assistance from lightning, highly specialized bacteria, or human intervention. For this reason, the element is fairly scarce in the lithosphere, hydrosphere, and biosphere. However, once nitrogen undergoes the right kind of chemical change, it becomes available to the organisms that need it and can then act as a potent fertilizer. Because of its scarcity, biologically active nitrogen is a limiting factor for plant growth. By limiting production of agricultural crops, nitrogen has long placed an effective cap on human population growth.

Nitrogen fixation and nitrification To become biologically available, inert nitrogen gas (N_2) must become "fixed," or combined with hydrogen in nature to form ammonium ions (NH_4^+), which are chemically and biologically active and can be taken up by plants. This conversion process, called **nitrogen fixation,** can be accomplished in two ways: by the intense energy of lightning strikes, or when air in the top layer of soil comes in contact with particular types of bacteria known as **nitrogen-fixing bacteria.** These bacteria live in a mutualistic relationship (Chapter 5) with many types of plants, including soybeans and other legumes, providing them nutrients by converting nitrogen to a usable form. Another type of bacteria then performs a process known as **nitrification,** turning ammonium ions first into nitrite ions (NO_2^-) and then into nitrate ions (NO_3^-). Ammonium and nitrate ions are the only forms in which plants can absorb nitrogen. These can enter plants with the assistance of nitrogen-fixing bacteria, by atmospheric deposition on soils or in water, or by the application of nitrate-based fertilizer.

Consumption, decomposition, and denitrifying bacteria Animals obtain the nitrogen they need by consuming plants or other animals. Recall from Chapter 4 (The Science behind the Story on page 180) how scientists use stable isotopes of nitrogen to study the trophic level and nutritional condition of animals. Decomposers obtain nitrogen via dead and decaying plant and animal matter and their waste products. Once decomposers process the nitrogen-rich compounds they take in, they release more ammonium, making this available to nitrifying bacteria to convert again to nitrites and nitrates that continue through the cycle. Another important step in the nitrogen cycle occurs when a third category of bacteria, **denitrifying bacteria,** converts nitrates in soil or water to gaseous nitrogen. Thus denitrifiers complete the cycle by releasing nitrogen back into the atmosphere as a gas.

Decomposers act as a kind of shortcut, setting up a smaller cycle within the nitrogen cycle.

Human impacts on the nitrogen cycle The impacts of excess nitrogen from agriculture in the Mississippi River watershed have become painfully evident to shrimpers and scientists with an interest in the Gulf of Mexico and its dead zone (see The Science behind the Story on page 180). But hypoxia in the Gulf is hardly the only problem resulting from human manipulation of the nitrogen cycle.

Historically, nitrogen fixation was a bottleneck, a step that limited the flow of nitrogen through the environment. This changed when the research of two German chemists enabled humans to fix nitrogen on an industrial scale. Their work has given our species the ability to double the amount of nitrogen fixed on Earth, essentially blowing the bottleneck out of the nitrogen cycle. Fritz Haber (1868–1934) was a chemist who supported the German army's chemical weapons program during World War I. Shortly before the war Haber found a way to combine nitrogen and hydrogen gases to synthesize ammonia (NH_3), a key ingredient in modern explosives and agricultural fertilizers. Several years later Karl Bosch, another German chemist, built on Haber's work and found a way to produce ammonia on an industrial scale. The work of these two scientists enabled people to overcome the limits on productivity long imposed by nitrogen scarcity in nature. Their discovery and its widespread application has led to a dramatic alteration of the nitrogen cycle and, through the enhancement of agriculture, has contributed to the enormous increase in the human population over the past 90 years. Today, using the Haber-Bosch process, we are fixing at least as much nitrogen artificially as is occurring naturally; that is, we have effectively doubled the natural rate of nitrogen fixation. Not only farmers, but also golf course managers, homeowners, sewage-treatment plants, and others all release nitrogen-rich compounds into surface waterways.

By fixing nitrogen, we accelerate its movement into other reservoirs within the cycle. When we burn forests and fields, we force nitrogen out of soils and vegetation and into the atmosphere. When we burn fossil fuels, we increase the rate at which nitric oxide (NO) enters the atmosphere and reacts to form nitrogen dioxide (NO_2). This compound is a precursor to nitric acid (HNO_3), a key component of acid rain, which damages both the buildings and statues of the built environment and the structure and function of ecosystems (Chapter 11). We also introduce another nitrogen-containing gas, nitrous oxide (N_2O), into the mix by allowing anaerobic bacteria

Figure 6.26 In the past half-century, human inputs of nitrogen into the environment have greatly increased, such that today fully half of the nitrogen entering the environment is of human origin. Agricultural fertilizer has for the past 35 years been the leading source of all nitrogen inputs, natural and artificial. Data from National Science and Technology Council: Committee on Environment and Natural Resources, Hypoxia: An integrated assessment in Northern Gulf of Mexico, May, 2000.

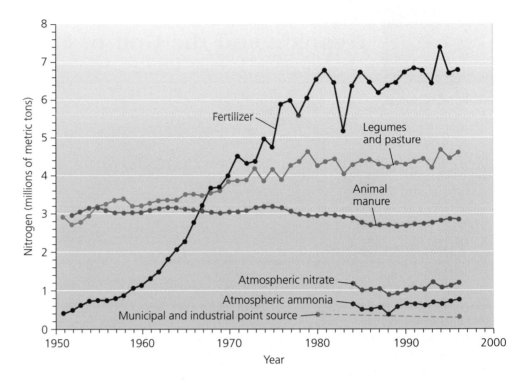

to break down the tremendous volume of animal waste produced by large-scale feedlots (Chapter 9). And we have accelerated the introduction of nitrogen-rich compounds into terrestrial and aquatic systems through the destruction of wetlands and the increased cultivation of legume crops that host nitrogen-fixing bacteria in their roots (Figure 6.26).

All these activities increase amounts of nitrogen available to aquatic plants, producing an effect much like that of intentionally fertilizing agricultural crops: a boom in aquatic plant growth. When this happens, plant populations soon outstrip the availability of other required nutrients and begin to die and decompose. As in the Gulf of Mexico, this large-scale decomposition can rob other aquatic organisms of oxygen, leading to massive fish die-offs and other significant impacts on ecosystems. In a 1997 report, a team of scientists led by Peter Vitousek of Stanford University summarized the dramatic changes humans have caused in the global nitrogen cycle. Each of these alterations can reverberate, through positive and negative feedback loops, to produce unintended outcomes and unpredictable results. Human alterations of the nitrogen cycle, according to the Vitousek team's report, have:

- doubled the rate that fixed nitrogen enters terrestrial ecosystems, and the rate is still increasing.
- increased atmospheric concentrations of the greenhouse gas N_2O and of other oxides of nitrogen that produce smog.

- depleted essential nutrients, such as calcium and potassium, from soils.
- acidified surface water and soils.
- greatly increased transfer of nitrogen from rivers to oceans.
- caused more carbon to be stored within terrestrial ecosystems.
- reduced biological diversity, especially plants adapted to low nitrogen concentrations.
- changed the composition and function of estuaries and coastal ecosystems.
- harmed many coastal marine fisheries.

Weighing the Issues:
Nitrogen Pollution and Its Financial Impacts

Most of the nitrate that enters the Gulf of Mexico originates on farms and other sources in the middle of the United States, yet many of its negative impacts are borne by downstream users, including fishermen and shrimpers along the Gulf Coast. Who do you believe should be responsible for solving this problem? Should environmental policies on this particular problem be developed and enforced by state governments, the federal government, or both? Explain the reasons for your answer.

Hypoxia and the Gulf of Mexico "Dead Zone"

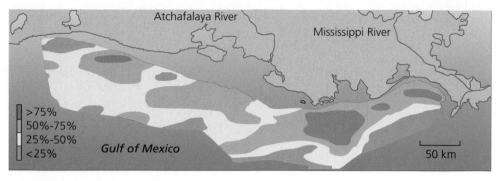

(a) Frequency of hypoxia, 1985-1999

(a) Some parts of the Gulf of Mexico suffer from hypoxia more frequently than others. Areas in red experience oxygen concentrations below 2 ppm more than 75% of the time; those in orange experience hypoxia 50–75% of the time; those in yellow are affected 25–50% of the time; and those in green are affected less than 25% of the time. Data from National Science and Technology Council: Committee on Environment and Natural Resources, Hypoxia: An integrated assessment in the Northern Gulf of Mexico, May, 2000.

Leaning over the side of an open boat idling miles from shore in the Gulf of Mexico, a young scientist hauled a water sample aboard—and thereby launched the long effort to breathe life back into the Gulf's "dead zone."

The researcher was Nancy Rabalais, and that expedition in 1985 marked the first major inquiry into the forces draining oxygen from stretches of the northern Gulf.

Other scientists, studying separate issues such as offshore oil drilling, had noticed very low levels of oxygen, or hypoxia, in some parts of the Gulf in the 1970s. A few subsequent studies found summertime hypoxia in spots relatively near the coast. Those observations gained new urgency in the 1980s, as fishing trawlers had to head further and further offshore to find anything to catch.

Such scattered signs of trouble led to bigger questions for Rabalais and fellow scientists at the Louisiana Universities Marine Consortium (LUMCON), an institute researching the health of the Gulf. How widespread was the hypoxia? Did this oxygen loss add up to a major environmental problem?

As Rabalais started sampling, clues came together. Even at that early stage of research, the known dead spots clearly occurred closer to the Louisiana shore, where the Mississippi River flows into the ocean. The LUMCON scientists hypothesized that the hypoxia was

neither isolated nor rare, and they started painstaking experiments.

Rabalais and other LUMCON researchers tracked oxygen levels at nine sites in the Gulf every month, and kept making those measurements for 5 years. At dozens of other spots near the shore and in deep water, they took regular, although less frequent, oxygen readings. To do some of the work, the researchers have relied on mobile oxygen probes. Sensors, as they are lowered into the water, measure oxygen levels and send continuous readings back to a shipboard computer. Further findings have come from fixed, submerged oxygen meters that continuously measure oxygen in the surrounding water and store the retrieved data.

The team also collected hundreds of coastal and Gulf water samples, using lab tests to scan the samples for elements other than oxygen such as nitrogen. Other tests measured the water's levels of salt, levels of bacteria, and amounts

of microscopic aquatic plants, or phytoplankton. Together, these factors painted a picture of the water's contents and quality.

The LUMCON scientists logged hundreds of miles in their boats, regularly monitoring more than 70 sites in the Gulf. They also donned scuba gear to view the condition of shrimp, fish, and other sea life. Such a wide range of long-term data allowed the scientists to build a "map" of the dead zone, tracking its location and effects.

In 1991, Rabalais made that map public and earned immediate headlines. Her findings showed that the Gulf's dead zone was neither small nor sporadic. In that year alone, her group mapped the size of the zone at more than 10,000 km^2 (about 4,000 mi^2). Bottom-dwelling shrimp were stretching out of their burrows, straining for oxygen. Many fish had fled. The bottom waters smelled of rotten eggs.

The source of the problem, Rabalais said, lay back on land. The

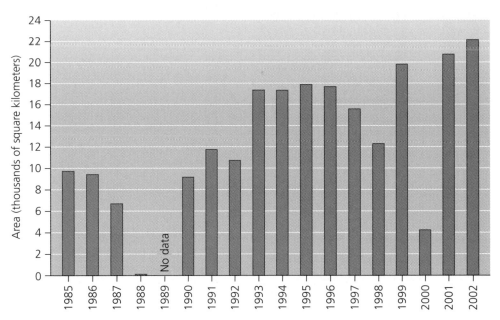

(b) Area of hypoxic zone in the Northern Gulf of Mexico

(b) The size of the Gulf's hypoxic zone varies as a result of a number of factors, including the presence or absence of flooding in the Mississippi River basin. Between 1985 and 2002 its extent varied from 4,000 km^2 to 22,000 km^2. Data from National Science and Technology Council: Committee on Environment and Natural Resources, Hypoxia: An integrated assessment in the Northern Gulf of Mexico, May, 2000.

major rivers draining into the Gulf —the Mississippi and the Atchafalaya—were polluted from runoff, and that pollution snuffed out oxygen in wide stretches of ocean water for at least a few months each year. The rivers carried high levels of nitrogen and other chemicals from farm fertilizer and other sources. As previously outlined in this chapter, nitrogen sets off a chain reaction in ocean waters that eventually drains oxygen at deeper levels.

The group's years of continuous tracking also explained the dead zone's predictable emergence. As the rivers rose each spring, oxygen would start to disappear in the northern Gulf. The hypoxia would last at least through the summer,

until seasonal storms mixed oxygen into the dead areas.

Over time, further water monitoring linked the size of the dead zone to the volume of river flow; the 1993 flooding of the Mississippi created a zone that almost doubled in size from the year before. Conversely, a drought in 2000 brought smaller river flows and a smaller dead zone.

Some scientists, especially those from farming states, challenge LUMCON's findings, saying the high levels of nitrogen flowing in the Gulf could come from sources other than agriculture. But an independent federal study of nitrogen in the Mississippi, published in 2000, revealed that about 7 million metric tons of nitrogen from

fertilizer enter the river basin each year, and that amount has risen more than sixfold over the last 50 years.

Rabalais's research has led to broader understanding of dead zones in oceans around the world. Using LUMCO's work, the United States is implementing a far-reaching plan to reduce farm runoff, clean up the Mississippi River, and try to shrink the dead zone.

"What people do 800 miles away from the Gulf of Mexico directly affects the Gulf of Mexico," Rabalais said when she accepted a major environmental award in 1999. "It's hard for many people to realize that."

The carbon cycle moves organic nutrients through the environment

As the definitive component of **organic molecules**, carbon is an ingredient in the carbohydrates, fats, and proteins, and the bones, cartilage, and shells of all living things. From fossil fuels to DNA, from plastics to pharmaceuticals, carbon atoms are everywhere. The **carbon cycle** describes the routes that carbon atoms take through the nested networks of environmental systems (Figure 6.27). The atmosphere is a convenient conceptual starting point; as we will see in Chapter 12, atmospheric carbon dioxide plays a large role in regulating climate.

Photosynthesis and food webs Producers, including terrestrial and aquatic plants, algae, and cyanobacteria, pull carbon dioxide (CO_2) out of the atmosphere and out of surface water (where a portion of it from the atmosphere dissolves) to use in photosynthesis. As we noted in Chapter 4, photosynthesis breaks the bonds in carbon dioxide and water and produces oxygen (O_2) and carbohydrates ($C_6H_{12}O_6$). Producers use some of the carbohydrates to fuel their own respiratory processes, thereby releasing some of the carbon back into the atmosphere and oceans as carbon dioxide. When producers are eaten by consumers, who are in turn eaten by other consumers, more carbohydrates are broken down by respiration to produce carbon dioxide and water. The same process occurs when decomposers consume waste materials and dead organic matter. Respiration from organisms releases carbon back into the atmosphere and oceans. Food webs, therefore, serve as part of a carbon atom's journey through the carbon cycle.

As the key building block of organic matter, carbon is utilized by all organisms for structural growth, so a portion of the carbon an organism takes in is incorporated into its tissues. The abundance of plants and the fact that they take in so much carbon dioxide for photosynthesis makes plants a major reservoir, or sink, for CO_2, and thus for carbon. Because CO_2 is a greenhouse gas of primary concern, much research on global climate change has been directed toward measuring the amount of CO_2 that plants tie up. So far it remains uncertain exactly how much this portion of the carbon cycle influences Earth's climate (Chapter 12).

Figure 6.27 The carbon cycle moves this essential element from the atmosphere into the biosphere via photosynthesis and back again via respiration, decomposition, and biomass burning. Fossil-fuel combustion releases carbon stored for many millions of years into the atmosphere, and the oceans sequester carbon in deep sediments.

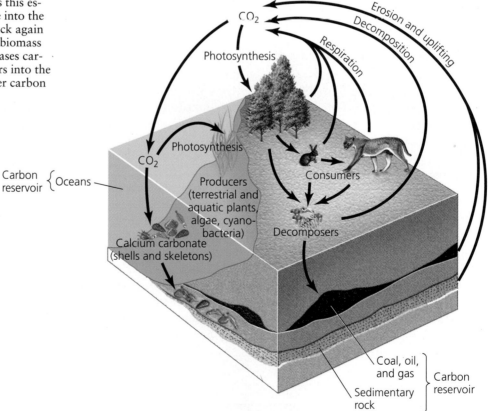

Sediment storage of carbon atoms As producers and consumers die, some of their remains may settle as sediments in ocean basins or in freshwater lakes, ponds, streams, and wetlands. As layers of sediment accumulate, older layers are buried more and more deeply, experiencing high pressure over long periods of time. These are the conditions that convert soft tissues into fossil fuels, such as coal and oil, and shells and skeletons into sedimentary rock, such as limestone. Organic matter converted into fossil fuel and sedimentary rock comprises the largest single reservoir, or sink, in the carbon cycle. While any given carbon atom spends a relatively short time in the atmosphere, carbon trapped in sedimentary rock may reside there for hundreds of millions of years.

Carbon trapped in sediments and fossil fuel deposits may eventually be released into the oceans or atmosphere by geological processes such as uplift and erosion. It also may reenter the atmosphere when we extract and burn fossil fuels. However, many of these sediments descend into the Earth's mantle at subduction zones, and when this happens the carbon may not find its way back to the oceans or atmosphere for hundreds of millions of years. When it does return it is likely to be through the agency of deep-sea spreading zones or volcanoes.

The oceans and the carbon cycle The world's oceans are the second-largest reservoir in the carbon cycle. They absorb carbon-containing compounds from the atmosphere and from terrestrial runoff, from undersea volcanic eruptions, and from the waste products and detritus of marine organisms. Some of the carbon atoms absorbed by the oceans—in the form of carbon dioxide, carbonate ions (CO_3^{2-}), and bicarbonate ions (HCO_3^-)—combine with calcium ions (Ca^{2+}) to form **calcium carbonate** ($CaCO_3$), an essential ingredient in the skeletons and shells of microscopic marine organisms. As these organisms die, their calcium carbonate shells sink to the ocean floor and begin to form the sedimentary rocks mentioned above. The rate at which the oceans take up and release carbon depends on many factors, including temperature and the numbers of marine producer organisms converting carbon dioxide into carbohydrates and carbonates.

Human impacts on the carbon cycle When the amount of time a chemical stays in a part of the cycle changes, or when the relative concentrations of carbon in each reservoir change significantly, we may see major consequences throughout the cycle. Humans have altered in several ways the relative rates at which steps in the cycle occur. Two activities in particular—burning fossil-fuel deposits and burning forests and fields—have pushed much more carbon dioxide into the atmosphere than in the past. Since the mid-18th century, our fossil-fuel combustion has added 245.9 billion metric tons (271 billion tons) of carbon to the atmosphere. The movement of carbon dioxide from the atmosphere to the hydrosphere, lithosphere, and biosphere has not kept pace. Some scientists calculate that the atmospheric carbon dioxide reservoir in the year 2004 is the largest Earth has experienced in the last 420,000 years, and perhaps in the last 20 million years. This ongoing movement of carbon out of the fossil-fuel reservoir and into the atmosphere is one of the driving forces behind global warming and related climate-change processes (Chapter 12).

The phosphorus cycle plays a key role in the conversion of energy via metabolism

The element phosphorus is a key component of DNA and RNA (Chapter 4). Organisms also use phosphorus to build two important biochemical compounds, adenosine diphosphate (ADP) and adenosine triphosphate (ATP). Cells use ADP and ATP to transfer and convert energy from one form to another during metabolism, and these compounds play a role in processes involving DNA and RNA. Phosphorus, however, is most abundant in rocks. The weathering of rocks releases phosphate (PO_4^{3-}) ions into water, which delivers these ions into plants through their roots.

Food webs and phosphorus Primary consumers acquire phosphorus from water and from plants, and in turn pass it on to secondary and tertiary consumers. Consumers also pass phosphorus to the soil through the excretion of waste. Decomposers break down these phosphorus-rich producers, consumers, and their wastes, and in so doing return phosphorus to the soil. Phosphates dissolved in deep lakes or in the oceans precipitate out into solid form, settle to the bottom, and reenter the lithosphere's phosphorus reservoir as sediments (Figure 6.28, page 186).

Geology and phosphorus availability Because most of Earth's phosphorus is contained within rocks and is released only by weathering, an extremely slow process, concentrations of available phosphorus in the environment tend to be very low. Because of its relative rarity, phosphorus is frequently a limiting factor in the growth of plants and other organisms. For this reason, an influx

The Science behind the Story

Biosphere 2

In September 1991, eight people and nearly 3,800 species of plants and animals were sealed within Biosphere 2, a collection of airtight, interconnected domes spanning more than 1.2 hectares (3.0 acres) in the Arizona desert. The goal of the biospherians, as they were known, was to survive for 2 years within a self-contained ecosystem that could someday be used to colonize other planets, while also learning about environmental processes on Earth. Only 9 months later, however, oxygen levels in Biosphere 2's artificial atmosphere began to drop at an alarming rate. Within 18 months, the biospherians were literally gasping for breath.

The near-failure of Biosphere 2's life-support system was not due to a lack of data. Scattered throughout Biosphere 2's ocean, rainforest, savanna, and desert biomes were more than 1,000 sensors that tracked day-to-day changes in oxygen, carbon dioxide, temperature, pH, and other environmental variables.

Within 18 months of being sealed off from the Earth's atmosphere, Biosphere 2 had oxygen levels of less than 14%—close to the lowest levels at which humans can survive.

Nonetheless, despite a sophisticated computer monitoring system and the help of external advisers, the biospherians, only some of whom were scientists, were unable to locate the missing oxygen. In desperation, they called on Wallace S. Broecker, a geochemist at Columbia University's Lamont-Doherty Earth Observatory.

Broecker and graduate student Jeff Severinghaus quickly ruled out the possibility that the biospherians themselves were consuming the oxygen. The amount disappearing—on the order of 450 kg (1,000 pounds) of pure oxygen per month—was far larger than they, or any of the project's large animals, could have used. In-

---

of phosphorus due to human activity can produce much more immediate and dramatic effects than that of more commonly available nutrients.

Human impacts on the phosphorus cycle Humans influence the phosphorus cycle in several ways. Effluents from sewage treatment plants tend to be phosphate-rich. The introduction of these extra phosphates to aquatic ecosystems can boost algal growth, leading to murkier waters and changing the structure and functioning of ecosystems. Phosphates are also common ingredients in fertilizers and thereby find their way into streams, rivers, ponds, and lakes, where they help create eutrophic waters.

The Gaia hypothesis portrays Earth as a self-regulating system

Whereas many scientists have viewed biogeochemical cycles as systems, one scientist has received a great deal of attention for his controversial belief that the entire planet behaves like a single self-regulating system. In the 1970s British scientist James Lovelock described Earth as a superorganism, an integrated system that is, in some sense, alive. Lovelock was not suggesting Earth could make conscious, rational decisions, but he was arguing that it is a homeostatic system that keeps itself stable with negative feedback loops. This notion of a living Earth is at least as old as the aboriginal dreaming track stories we discussed in Chapter 2, but Lovelock

stead, they focused their investigation on Biosphere 2's 30,000 tons of soil, which had an extraordinarily high percentage—nearly 30%—of organic matter. They quickly determined that microbes in the soil were converting unexpectedly large amounts of oxygen to carbon dioxide. Still, one question remained: If the microbes were responsible for the severe decrease in oxygen, then carbon dioxide levels in the atmosphere would have skyrocketed, when instead the opposite had occurred. Broecker and Severinghaus suspected that the missing carbon dioxide was being stored in the soil itself.

On close examination, however, that proved not to be the case, so the two researchers turned their attention to another possible culprit: the exposed concrete supporting the building's glass and metal shell. In Earth's atmosphere, they knew, concrete can react with carbon dioxide to form a solid substance called calcium carbonate. Usually, the reaction takes place only in a thin outer layer of exposed concrete, which is what the designers of Biosphere 2 had expected. Concentrations of carbon dioxide in Biosphere 2's atmosphere, however, ranged from 3 to 10 times higher than Earth's, and the investigators found that carbon dioxide, in the form of calcium carbonate, had been deposited up to 6 inches deep in the concrete walls.

The most effective solution would have been to replace all 30,000 tons of soil, but that would have cost millions of dollars and set the project back several years. Instead, Biosphere 2's management team settled on two stopgap measures. First, to address the urgent need for oxygen, they injected more than 23 tons of pure oxygen gas. Then, to minimize calcium carbonate deposits, they covered the exposed concrete with a layer of paint. With the most pressing problem solved, at least temporarily, the biospherians were able to focus on other issues, such as invasive species. Aquatic fire worms were preying on coral, hundred-foot-long morning glory vines were overwhelming the rainforest, and crazy ants—an aggressive species unintentionally included in the project—had spread throughout Biosphere 2 and driven most of its other insect species extinct. Nonscientific problems also plagued the project. By the time the biospherians emerged from their 2-year seclusion in September 1993, the project had become embroiled in accusations of mismanagement and scientific fraud. In 1994, its financial backer, Texas oil billionaire Edward P. Bass, dismissed the project's management team and invited a panel of independent scientists to assess its future.

In 1996, Columbia University began the difficult process of transforming Biosphere 2 into a center for research and education on global climate change. Despite some successes, the project remains troubled. Whatever the future of Biosphere 2, however, the lesson of its first mission is clear: Creating a self-contained ecosystem from scratch is no easy task. We're better off preserving the one we have.

was among the first to express such a view in scientific terms. His **Gaia hypothesis**, named for the Greek goddess of Earth, states that living things affect the environment in ways that stabilize the climate and make it possible for life to persist and flourish. For example, by converting carbon dioxide and water to oxygen and carbohydrates, green plants and cyanobacteria influence atmospheric composition and make life possible for other organisms.

Lovelock developed his ideas after concluding that Earth's atmosphere was highly unusual, composed of molecules that would not be expected in the absence of life. Mars and Venus have atmospheres dominated by carbon dioxide, yet Earth's atmosphere has low CO_2 levels and instead contains abundant oxygen and nitrogen and biologically produced compounds like methane. Something was keeping our planet's atmosphere from being dominated by CO_2, Lovelock figured, and it must be what Mars and Venus appeared not to have—life. Lovelock was further struck by the fact that life had survived on Earth for 3.5 billion years despite the fact that the sun had grown 30% brighter over this time period. Something on Earth had to be compensating for this change in solar energy in order to keep conditions constant enough for life to survive, he surmised.

The Gaia hypothesis is less a hypothesis than a way of viewing the world and a framework for approaching research questions. Although it has generated immense controversy, it has also stimulated new research, much

Figure 6.28 The phosphorus cycle moves this essential element from Earth's crust into its soil, freshwater ecosystems, terrestrial ecosystems, and oceans. This process, especially the portion involving the sea floor and Earth's crust, typically takes many millions of years. Humans have accelerated this process by mining phosphorus for fertilizer and by clearing land for farms and forests. Both activities dramatically accelerate the rate at which phosphorus enters the biosphere.

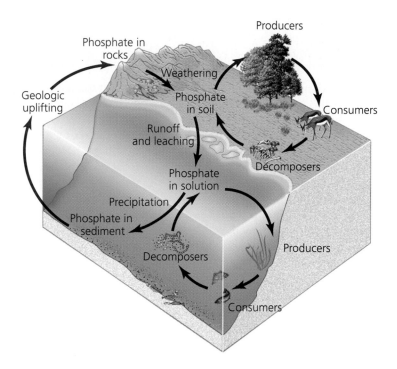

of which has born out the idea that organisms have influenced and stabilized the planet's climate. For instance, work on phytoplankton in the 1980s showed that aromatic sulfur compounds these organisms release were serving as nuclei for water condensation and cloud formation in the atmosphere. By helping clouds form over the ocean, phytoplankton were apparently helping keep Earth cooler than it would otherwise be.

The Gaia hypothesis has attracted much criticism from scientists, due in part to the mystical, almost religious, way some of its supporters have expressed its ideas. Among other things, critics of Gaia stress that any stabilization of Earth's environment for life could not have been guided by natural selection, since *individual* organisms would not benefit from influencing the environment in this way. Rather, the remarkable stability of Earth as a planet habitable to life, critics say, is probably an accident. We are witnessing this extraordinary accident, they explain, because the only type of planet that could possibly have observers like ourselves is a planet that has been stable enough to support several billion years of evolution. In other words, there could well be plenty of other planets that supported life for only a brief time and never evolved creatures intelligent enough to observe and understand their planet's operation as a system. But only those rare planets that manage to support life for billions of years evolve observers able to reflect on their own planet. This phenomenon is called the **anthropic principle.** If it is true, then we may be a rare exception to the rule, and our planet's apparently homeostatic system may be a rarity in the universe.

Weighing the Issues:
Gaia and the Anthropic Principle

James Lovelock once wrote that the Gaia hypothesis "is an alternative to that pessimistic view which sees nature as a primitive force to be subdued and conquered. It is also an alternative to that equally depressing picture of our planet as a demented spaceship, forever travelling driverless and purposeless around an inner circle of the sun." Contrast Lovelock's view with that of his critics who maintain that Earth's homeostasis results from the anthropic principle. Only those planets that evolve intelligent observers can be observed, they contend, and therefore Earth may not be representative. Life's apparent tendency to promote conditions conducive to life may simply be a happy accident that we are witnessing simply because those conditions had to exist in order for us to be here. Does it matter whether Lovelock or his critics are right—whether Earth's apparent homeostasis as a home for life reflects real universal tendencies, or is rare and accidental? What implications for human conduct on Earth does each view suggest?

The Dead Zone
Recent scientific research has indicated that nitrogen fertilizers in the Mississippi River Watershed are contributing to water pollution problems in the Gulf of Mexico. Do you agree with the researchers' conclusions? Why or why not? If this is occurring, what steps should be taken to solve the problem?

Evidence Not Conclusive
Scientific evidence that nitrogen (N) fertilizer is polluting the Gulf of Mexico is not conclusive. Hypoxia in the Gulf has been recognized since 1935 and has likely been present long before commercial N fertilizer use became widespread in the U.S. in the 1960s. The areal extent of the hypoxic zone has varied dramatically from year to year and in 2000 was smaller than any year since 1985, except for 1988, a year when flow from the Mississippi River was low. The large decline in 2000 occurred with no appreciable change in N fertilizer use.

According to the U.S. Geological Survey, the annual discharge of N from the Mississippi River has tripled in the last 30 years, with most of the increase occurring from 1970 to 1983. However, since 1980 it has changed very little, while N fertilizer use has grown almost 10%. There are no definitive data to identify the sources of N entering the Mississippi River. Scientists have reported a strong correlation between long-term (1930s to 1988) annual N fertilizer consumption and the nitrate-N concentration in the lower Mississippi River, but this does not mean there is a cause and effect relationship between N fertilizer consumption and the total amount of nitrate-N delivered to the Gulf.

Numerous N sources contribute to Gulf loading. Fertilizer is essential to crop production in the Mississippi River Basin, but its contribution to the Mississippi River in runoff and subsurface drainage is no more likely than N from atmospheric deposition, crop residue and soil organic matter decomposition, legumes, animal manure, municipal sewage sludge and effluent, or composted household wastes.

Hypoxia results from a complex interaction of chemical, biological, and physical factors. Fertilizer is a potential pollutant if used improperly, but used correctly, it increases food production and helps protect the environment.

Dr. Terry L. Roberts is President of the Foundation for Agronomic Research (FAR), Norcross, Georgia. At FAR he directs the Foundation's research and education programs. Dr. Roberts is a Certified Crop Adviser and a Fellow of the American Society of Agronomy.

Act Now to Save These Resources
The springtime area of low oxygen (anoxic) water in the Gulf of Mexico, known as the Dead Zone, is driven by a massive influx of nutrients into a system no longer able to process them. Eutrophication begins when nutrients from farmlands in the floodplain states wash to the sea. These nutrients (nitrogen fertilizers) now present in the water, lead to plankton blooms, which in turn reduce dissolved oxygen in the water and eventually, kill fish.

Taking a system view is slightly more complicated, but understanding the system is important for the most effective long-term management. Before man built levees all along the delta, the Mississippi River flooded each spring and the waters of the river covered the extensive wetlands. This important renewal process deposited sediment on the wetlands to build up the soil base while the plants of the wetlands made use of the nutrients (most produced though nitrogen fixation) in the water. The result was cleaner water, richer wetlands, and a sustained environment. Levees now prevent the flooding, the Dead Zone emerges, and the wetlands are lost as they sink below sea level. Sea level rise speeds the loss.

Loss of wetlands is serious. The wetlands are the base of the fisheries of the Gulf of Mexico and their loss is irreversible. Saving the wetlands and reversing the Dead Zone requires a two-fold approach; first reduce the amount of fertilizer use so it's used more efficiently by plants and less enters streams. This has the added benefit of saving money and reducing energy consumption (making fertilizers is energy intensive). Second, reinstate the flooding of the wetlands. Should we wait to act? No. We know enough now to design strategies that can sustain these resources and new information is unlikely to change what we know. The Precautionary Principle, which environmental managers use, says that even if information is imperfect its important to act before the resource is lost entirely and while any possible cost of error is small and manageable. We need to act now to save these resources.

Dr. Paul Templet is a Professor at the Institute for Environmental Studies at Louisiana State University. He organized the first Earth Day at LSU in 1970 and served as the Secretary of the Louisiana Department of Environmental Quality from 1988–92.

Conclusion

The Gaia hypothesis presents the most radically expansive notion of an environmental system, but as we have seen, there are many interacting systems involved in Earth's functioning, and the way one perceives them depends on the questions one is interested in. Approaching questions holistically by taking a systems approach is helpful in environmental science, in which so many issues are multifaceted and complex. This approach, however, does pose a challenge, because systems by their nature show behavior that is difficult to understand and predict. Furthermore, the scientific method is geared toward operating in a largely reductionist manner, and many of science's successes have come by researching components in depth rather than studying systems broadly. However environmental science benefits by incorporating slightly different skills and approaches. Thinking in terms of systems is important to understand how Earth works, so that we may learn how to avoid disrupting its processes and how to mitigate any disruptions we cause.

The case of the Gulf of Mexico's hypoxic zone provides evidence that systems thinking can lead the way to solutions. On November 13, 1998, President Bill Clinton signed into law the Harmful Algal Bloom and Hypoxia Research and Control Act, which called for an "integrated assessment" of hypoxia in the northern Gulf to address the extent, nature, and causes of the dead zone as well as its ecological and economic impacts. The report was also to document methods for solving the problem, along with the social and economic costs associated with these solutions. A 1999 report by the National Oceanic and Atmospheric Administration's Coastal Ocean Program outlined potential solutions, as follows:

- Reduce nitrogen fertilizer use on Midwestern farms.
- Change the timing of fertilizer application to minimize rainy-season runoff.
- Use alternative crops.
- Manage nitrogen-rich livestock manure more effectively.
- Restore nitrogen-absorbing wetlands in the Mississippi River basin.
- Use artificial wetlands to filter farm runoff.
- Install more efficient nitrogen-removing technologies in sewage treatment plants.
- Restore frequently flooded lands to reduce runoff.
- Restore land near the Mississippi River's mouth to enhance its nitrogen-absorbing ability.
- Continue evaluating which of these approaches work and which do not.

By studying the environment from a systems perspective and by integrating scientific findings with the policy process, people who care about the Gulf of Mexico are working today to solve its pressing problem. Their model is one that we can adapt to many other issues in environmental science.

We might also consider adopting other models more generally. Think again about unperturbed ecosystems, their use of renewable solar energy, their recycling of nutrients, and the extent to which they exhibit homeostasis and involve negative feedback loops. The ecosystems and other environmental systems we see on Earth today are those that have survived the test of time; our industrialized civilization is young in comparison. Might we be able to take a few lessons about sustainability from a careful look at the natural systems of our planet?

REVIEW QUESTIONS

1. How can hypoxic conditions develop in marine ecosystems? Describe the environmental systems that play a part in this process.
2. What is the difference between a cycle and a system?
3. Give an example of a system and two of its emergent properties. Do not use any of the examples described in this chapter.
4. What is the difference between dynamic equilibrium and homeostasis?
5. How might the emergence of a positive feedback loop affect a system in homeostasis?
6. What is the difference between an ecosystem, a community, and a biome?
7. What factors exert the strongest influence over the type of biome that forms in a particular place on land? What factors determine the type of aquatic system that may form in a given location?
8. Draw climate diagrams for a tropical rainforest and for a desert. Label all parts of the diagram and describe all of the types of information an ecologist could glean from such a diagram.
9. Which are hotter: tropical rainforests or deserts? Explain your answer.

10. Name the three main types of rocks, and describe how each type may be converted to the others via the rock cycle.
11. Name the three types of plate tectonic boundaries and describe what typically happens at each.
12. Identify three ways in which humans have influenced the hydrologic cycle.
13. Explain the connection between the work of the German chemists Fritz Haber and Karl Bosch and the hypoxic zone in the Gulf of Mexico.
14. What is the difference between the function performed by nitrogen-fixing bacteria and that performed by denitrifying bacteria?

15. How has human activity altered the carbon cycle? What environmental problems has this given rise to?
16. What role do each of the following play in the carbon cycle?
    • cars
    • photosynthesis
    • the oceans
    • Earth's crust
17. How have humans altered the phosphorus cycle and what environmental effects has this had?

DISCUSSION QUESTIONS

1. In this chapter you learned that system boundaries can be difficult to determine. Can you think of a truly closed system whose boundaries are easily defined? If so, try to describe such a system and its boundaries.
2. From year to year, biomes are stable entities, and our map of world biomes appears to be a permanent record of the patterns of biomes across Earth. But are the location of biomes permanent, or could they move over long periods of time? Is the identity of biomes permanent, or could they change over long periods of time? Provide several reasons that biomes might change in their composition or location over time.
3. How does plate tectonics account for mountains, volcanoes, and earthquakes?
4. Explain how human alteration of the nitrogen cycle shows us that it is possible to have "too much of a good thing."

5. How do you think we might solve the problem of eutrophication in the Gulf of Mexico? List several possible solutions, your reasons for believing they might work, and the likely hurdles we might face. Explain who should be responsible for implementing solutions and why.
6. Explain the Gaia hypothesis and its relevance (or lack thereof) to systems thinking.

Media Resources *For further review, go to the website* **www.envscienceplace.com** *or student CD-ROM, where you will find quizzes, flashcards, a glossary, additional interactive exercises, and links to relevant news and research sources. Also, on the website and CD-ROM is* **GRAPH IT**, *a series of interactive graphing tutorials to help you interpret graphs and plot data.*

PART TWO

Environmental Problems and the Search for Solutions

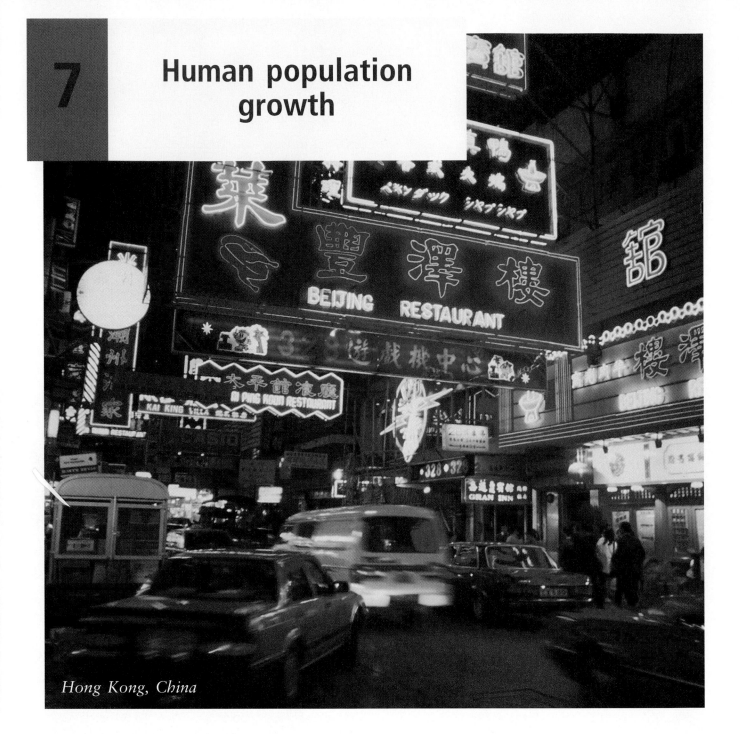

7

Human population growth

Hong Kong, China

This chapter will help you understand:

- The scope of human population growth

- The fundamentals of demography

- How human population, affluence, and technology affect the environment

- How wealth and poverty, the status of women, and other factors affect population growth

- Programs effective in controlling human population growth

- Demographic transition theory

- Consumption and the ecological footprint

- Dimensions of the HIV/AIDS crisis

Billboard in Chengdu, China Promoting One-Child Policy

Central Case: One-Child Policy

"Population growth is analogous to a plague of locusts. What we have on this earth today is a plague of people."
—*Ted Turner, media magnate and supporter of the United Nations Population Fund*

"There is no population problem."
—*Sheldon Richman, senior editor, Cato Institute*

The People's Republic of China is the world's most populous nation, home to over one-fifth of the 6.3 billion people living on Earth at the start of 2004. When Mao Zedong founded the country's current regime 55 years earlier, roughly 540 million people lived in a mostly rural, war-torn, impoverished nation. Mao believed population growth was desirable, and under his leadership China grew and changed. By 1970, improvements in food production, food distribution, and medical care allowed China's population to swell to approximately 790 million people, and the average Chinese woman was giving birth to 5.8 children in her lifetime. Unfortunately, the country's burgeoning population and its industrial and agricultural development were also eroding the nation's

soils, depleting its water, leveling its forests, and polluting its air. Chinese leaders realized that the nation might not be able to feed its people if their numbers grew much larger and that continued population growth could use up resources and threaten the stability of Chinese society. The government decided to institute a population-control program that precluded large numbers of Chinese couples from having more than one child.

The program began with education and outreach efforts encouraging people to marry later and have fewer children. Along with these efforts, the Chinese government increased the accessibility of contraceptives and abortion. By 1975 China's annual population growth rate had dropped from 2.8% to 1.8%. Despite this success in decreasing births, in 1979 the government decided to take the more drastic step of instituting a system of rewards and punishments to enforce the one-child limit. For example, one-child families received better access to schools, medical care, housing, and government jobs, and mothers with only one child were given longer maternity leaves. Families that had more than one child, meanwhile, were subjected to social scorn and ridicule, employment discrimination, and monetary fines. In some cases, the fines exceeded half the offending couple's annual income.

In enforcing these policies, China has, in effect, been conducting one of the largest and most controversial social experiments in history. In purely quantitative terms, the experiment has been a major success; the nation's growth rate is now down to 0.7%, making it easier for the country to deal with its many social, economic, and environmental challenges. However, China's population control policies have also produced some unintended consequences, such as widespread killing of female infants and an unbalanced sex ratio. Furthermore, the policies have elicited intense criticism from those opposing government intrusion into personal reproductive choices.

China embarked on its policy because its leaders felt it necessary. As other nations become more and more crowded, might their governments also feel forced to turn to draconian policies that restrict individual freedoms? For every society today, it is worth taking a hard look at population growth patterns, their causes, and their consequences. In this chapter, we examine the phenomenon of human population growth worldwide and consider its effects on our environment and our society.

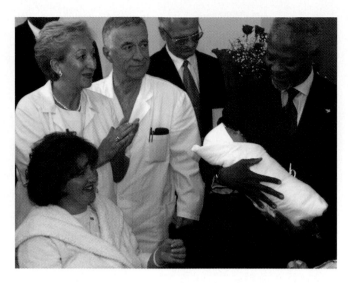

Figure 7.1 U.N. Secretary-General Kofi Annan recognized the 3.55-kg (8-lb) son of 29-year-old Fatima Nevic and her husband, Jasminko, as our six-billionth neighbor. Although it is impossible to know the precise moment—or day, week, or even month—the world's population reached 6 billion, U.N. population experts pinpointed October 12, 1999, as the best approximation to make the symbolic declaration. Many observers interpreted the selection of a child born in war-ravaged Sarajevo as a harbinger of the hard times that could face future generations as population grows and competition for scarce resources increases.

Human Population Growth: Baby 6 Billion and Beyond

While China was working to slow its population growth and speed its economic growth, in 1999 on the other side of the Eurasian continent a milestone was reached. On the morning of October 12 of that year, the first cries of a newborn baby in Sarajevo, Bosnia-Herzegovina marked the arrival of the six-billionth human being on our planet (Figure 7.1). At least that was how the milestone was symbolically marked by the United Nations, which today monitors human population growth and an increasing number of other environmental trends.

The human population is growing nearly as fast as ever

As we saw in Chapter 1, the human population has been growing at an unprecedented rate. Our population has doubled just since 1963, and we are adding about 79 million people annually (2.5 people every *second*), the equivalent of adding one country the size of Germany to the world each year. It took until 1800, virtually all of

human history, for our population to reach 1 billion, but we expanded from 5 billion to 6 billion in just 12 years (Figure 7.2).

We saw in Chapter 5 how exponential growth—the increase in a quantity by a *fixed percentage* per unit time—accelerates the absolute increase of population size over time, just as compound interest accrues in a

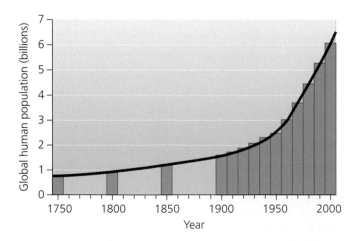

Figure 7.2 The global human population has grown more than exponentially in recent decades, rising from less than 1 billion in 1800 to over 6 billion today. Data from U.S. Bureau of the Census, 2002.

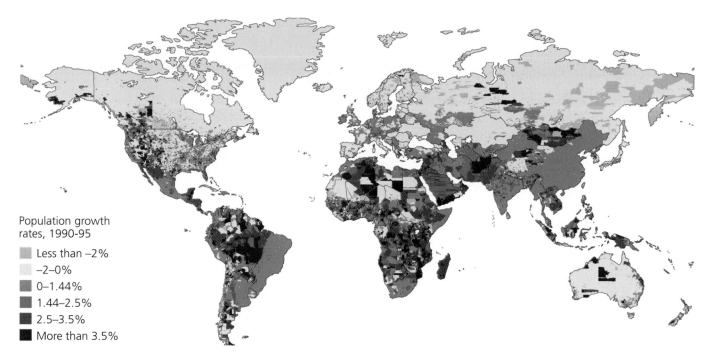

Population growth
rates, 1990-95

- Less than −2%
- −2–0%
- 0–1.44%
- 1.44–2.5%
- 2.5–3.5%
- More than 3.5%

Figure 7.3 A map of population growth rates from the period 1990-1995 shows great variation from place to place. Population is growing fastest in tropical regions and in some desert and rain-forest areas that have been sparsely populated. Data from American Association for the Advancement of Science (AAAS) *Atlas of Population & Environment,* University of California Press, 2000.

savings account. The reason, you will recall, is that a given percentage of a large number is a greater quantity than the same percentage of a small number. Thus, even if the growth rate remains steady, population size will accelerate with each successive generation. Although ecologists expect temporary exponential growth from populations of organisms that have not yet run up against limiting factors in their environment, for much of the 20th century human population actually grew at a *greater-than-exponential* rate—that is, the annual percentage increase in population actually rose from year to year. It peaked at 2.1% during the 1960s and has declined slightly since then. In today's world, rates of annual growth vary greatly from region to region; Figure 7.3 maps this variation.

At a 2.1% annual growth rate, the human population would take only 33 years to double in size. For low rates of increase, we can calculate doubling rates via a handy formula: Just take the number 70, and divide it by the annual growth rate: 70 ÷ 2.1 = 33.3. Had China not instituted its one-child policy—that is, had its growth rate remained unchecked at 2.8%—it would have taken only 25 years to double in size. Had population growth continued at this rate, in 2004 China's population would have surpassed 2 billion people.

Is population growth really a "problem"?

Our ongoing burst of population growth has resulted largely from technological innovations, improved sanitation, better medical care, increased agricultural output, and other factors that have led to a decline in death rates, particularly a drop in rates of infant mortality. Meanwhile, birth rates have not declined as much, so that births have outpaced deaths for many years now. Thus the so-called population problem actually arises from a very good thing—our ability to keep more of our fellow humans alive longer.

Indeed, just as the mainstream view in Malthus's day held that population increase was a good thing, there are many people today who deny that population growth is a problem at all. Under the Cornucopian view held by many economists, resource depletion due to population increases is not a problem if new resources can be found to replace depleted resources (Chapter 2). As libertarian writer Sheldon Richman expressed it at the time the six-billionth baby was born,

> The idea of carrying capacity doesn't apply to the human world because humans aren't passive with respect to their environment. Human beings create resources. We find potential stuff and human

intelligence turns it into resources. The computer revolution is based on sand; human intelligence turned that common stuff into the main component [silicon] of an amazing technology.

In contrast to Richman's point of view, environmental scientists argue that not all resources can be replaced once they have become depleted and that, moreover, few resources are actually created by humans. Some resources associated with living organisms, for example, cannot be replaced: Once species have gone extinct, we cannot replicate their exact function in ecosystems, or know what medicines or other practical applications we might have obtained from them, or regain the educational and aesthetic value of seeing them. Another irreplaceable resource is land, that is, space in which to live; we cannot expand Earth like a balloon to increase its surface area.

Furthermore, even if resource substitution could hypothetically enable population growth to continue indefinitely, could we maintain the *quality* of life that we would desire for ourselves and our descendants? Surely some resources will be easier or cheaper to use, and less environmentally destructive to harvest or mine, than any resources that can replace them. Replacing such resources might make our lives more difficult or less pleasant. In any case, unless resource availability keeps pace with population growth forever, eventually the average person on Earth will have less space to live in, less food to eat, and less material wealth than the average person does today. Thus population increases are indeed a problem if they create so much stress on resources, social systems, or the natural environment that the quality of life for humans declines significantly.

Despite these considerations—and despite the fact that in today's world population growth is correlated with poverty, not wealth—many governments have found it difficult to let go of the notion that population growth increases a nation's economic, political, and military strength. Many national governments, even those that view global population increase as a problem, still offer financial and social incentives that encourage their own citizens to produce more children. Governments of countries currently experiencing population declines feel especially uneasy; the majority of European national governments now take the view that their birth rates are too low, for instance. However, according to the Population Reference Bureau in Washington, D.C., outside Europe, 57% of national governments feel their birth rates are too high, and only 7% feel they are too low.

Population is one of several major factors that affect the environment

The extent to which population increase can be considered a problem involves more than just numbers of people. One widely used formula gives us a handy way to think about factors that affect our environment. Nicknamed the **IPAT model**, it is a variation of a formula proposed in 1974 by Paul Ehrlich (Chapter 1) and John Holdren, professor of environmental policy at Harvard University. The IPAT model represents how humans' total impact (I) on the environment results from the interaction among three factors, population (P), affluence (A), and technology (T):

$$I = P \times A \times T$$

Increased population will intensify impact on the environment as more individuals take up space, tap into natural resources, and generate waste. Likewise, increased affluence will magnify environmental impact through the greater per capita resource consumption that generally accompanies enhanced wealth. Changes in technology are potentially capable of either decreasing or increasing human impact on the environment. Technology that enables us to exploit minerals, fossil fuels, old-growth forests, or ocean fisheries more thoroughly might increase impact, but technology to reduce smokestack emissions, for instance, might decrease it. In addition to these three elements of the model, we might add a sensitivity factor (S) to the equation to denote how sensitive a given environment is to these human pressures:

$$I = P \times A \times T \times S$$

We could also refine the model by adding terms for the effects of social institutions. Education, laws and their enforcement, stable and cohesive societies concerned with environmental well-being—such factors also affect how population, affluence, and technology translate into environmental impact.

The IPAT equation was developed as a conceptual model and was not necessarily meant to receive exact numerical values. As such it is useful to help us grasp the major influence of population growth on the environment and to recognize that population is only one of several important factors that determine how we affect our environment. These factors are intertwined and can affect each other. They all touch on resource use, and it is the depletion of resources by larger and hungrier populations that has been the focus of numerous scientists and philosophers since the time of Thomas Malthus (Chapter 1 and Chapter 2). We saw with the story of Easter Island how islanders brought down their own

Figure 7.4 In this model, as resources are depleted, global food production and industrial output falls, leading to a drop in population after 2030. Data from D. Meadows et al., *Beyond the Limits,* Earthscan Publications, 1992.

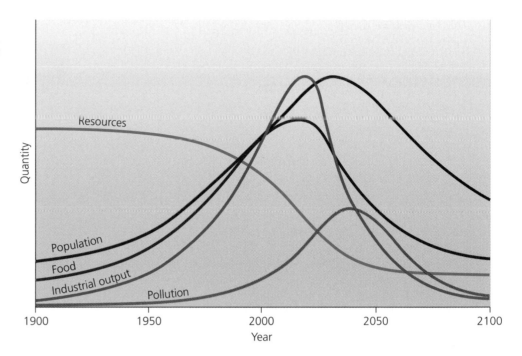

civilization by depleting their most important limited resource, trees. History appears to offer other cases in which resource depletion helped end a civilization, from the Maya to the Mesopotamians. In predicting our own society's future, some environmental scientists have used computer simulations to generate projections of trends in population increase, availability of resources, food production, industrial output, and pollution (Figure 7.4).

However, as we noted in Chapter 1, Malthus and his "neo-Malthusian" followers have not yet seen their direst predictions come true. This is because we have developed technology—the T in the IPAT equation—time and again to alleviate our strain on resources and allow us to further expand our population. For instance, world food production has risen faster than has population, so that the world's widespread malnutrition is more a result of wealth inequality and failures in food distribution on the part of economies and governments than of inadequate production. These are important points to keep in mind as we focus on the population aspect of the IPAT formula in much of this chapter.

Modern-day China is an instructive example of how all elements of the IPAT formula can combine to result in tremendous environmental impact in very little time. The world's fastest-growing economy over the past two decades, China "is telescoping history," in the words of Earth Policy Institute president Lester Brown, "demonstrating what happens when large numbers of poor people rapidly become more affluent." While many millions of Chinese are increasing their material wealth and their consumption of resources, the country is also battling un-

precedented environmental challenges on multiple fronts, brought about by its pell-mell economic development. For example, intensive agriculture has expanded westward out of the country's historic moist rice-growing areas, causing farmland to erode and literally blow away, much like the Dust Bowl tragedy that befell many areas in the U.S. agricultural heartland in the 1930s. China has overpumped many of its aquifers and has drawn so much water for irrigation from the Yellow River that the once-mighty waterway now dries up in many sections. Although China has been reducing its air pollution from industry and charcoal-burning homes, the country faces new urban pollution and congestion threats from rapidly increasing numbers of automobiles. As the world's developing countries—those that are not yet fully industrialized—try to attain the level of material prosperity enjoyed by the developed, or industrialized, nations, China is a window on what much of the rest of the world could become.

The principles of population ecology apply to humans

As we have seen, it is a fallacy to think of humans as being somehow outside nature. Humans exist within their environment as one species of the many that are part of the broader natural world. As such, all the principles of population ecology we outlined in Chapter 5 that apply to toads, frogs, and passenger pigeons apply to humans as well. Like other organisms, humans have a carrying capacity set by environmental limitations on our population growth.

Figure 7.5 Tool-making, the advent of agriculture (the Neolithic revolution), and industrialization each allowed humans to raise their global carrying capacity. The logarithmic scale of the axes makes it easier to see this pattern. Data from Andrew Goudie, *The Human Impact*, MIT Press, 2000.

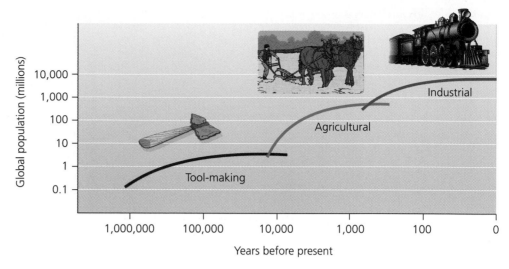

We happen to be a particularly successful animal, however—one that has repeatedly pushed out the natural limits on its growth through technology, and raised its carrying capacity. We did so with the agricultural and the industrial revolutions (Chapter 1), and likely before that with the invention of tools (Figure 7.5).

Environmental scientists who have tried to pin a number to the human carrying capacity have come up with wildly differing estimates. The most rigorous human carrying capacity estimates range from 1–2 billion people living prosperously in a healthy environment to 33 billion living in poverty in an environment of intensive cultivation without natural areas. As our population climbs toward 7 billion and beyond, we may yet continue to find ways to raise our carrying capacity. Given our knowledge of population ecology, however, we have no reason to assume that human numbers can go on growing indefinitely.

Demography is the science of human population

The application of the principles of population ecology to the study of statistical change in human populations is the focus of the social science of **demography.** In fact, however, the field of demography developed along with and partly preceding population ecology, so that both disciplines have influenced and borrowed from one another. Data gathered by demographers help us understand how differences in population characteristics and related life-cycle experiences (for instance, decisions about reproduction) can affect human communities and their environments. Demographers study population size, density, distribution, age structure, sex ratio, and rates of birth, death, immigration, and emigration of humans, just as population ecologists study these characteristics

Figure 7.6 The world's nations range in human population from several thousand (on some South Pacific islands) up to China's 1.3 billion. Shown here are the 2002 populations for the world's most populous 15 countries, followed by various other countries. Data from *2002 World Population Data Sheet,* Population Reference Bureau.

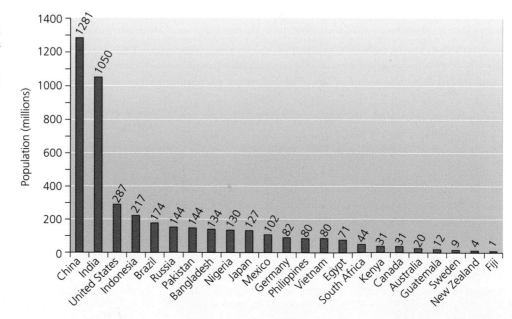

for other organisms. Each of these population characteristics is useful both for predicting population dynamics and for predicting potential human population impacts on the environment.

Population size The global human population of over 6.3 billion consists of well over 200 nations with populations ranging from China's 1.3 billion, India's 1.1 billion, and the United States's 290 million down to a number of island nations with populations below 100,000 (Figure 7.6). What size our population will eventually reach is uncertain (Figure 7.7). However, population size alone—the absolute number of individuals—doesn't tell the whole story. Rather, a population's environmental impact depends on its density, distribution, and composition (as well as on affluence, technology, and other factors outlined earlier).

Population density and distribution People are very unevenly distributed over the globe. In ecological terms, the distribution of *Homo sapiens* is clumped (Chapter 5) at all spatial scales. At the largest scales (Figure 7.8),

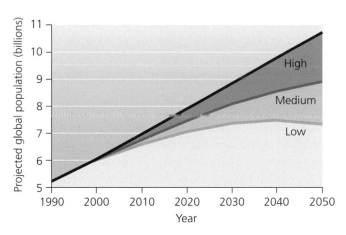

Figure 7.7 The United Nations predicts trajectories of world population growth from time to time, presenting its estimates in several scenarios based on different assumptions of fertility rates. In this 2002 projection, population is estimated to reach 12.8 billion in the year 2050 if fertility rates remain constant at 2002 levels. However, U.N. demographers expect fertility rates to continue falling, so they arrived at a best guess (*medium* scenario) of 8.9 billion for the human population in 2050. In the *high* scenario, if women on average have one-half a child more than in the medium scenario, population will reach 10.6 billion in 2050. In the *low* scenario, if women have one-half a child less than in the medium scenario, the world will host 7.4 billion people in 2050. Data from *World Population Prospects: The 2002 Revision*, United Nations Population Division.

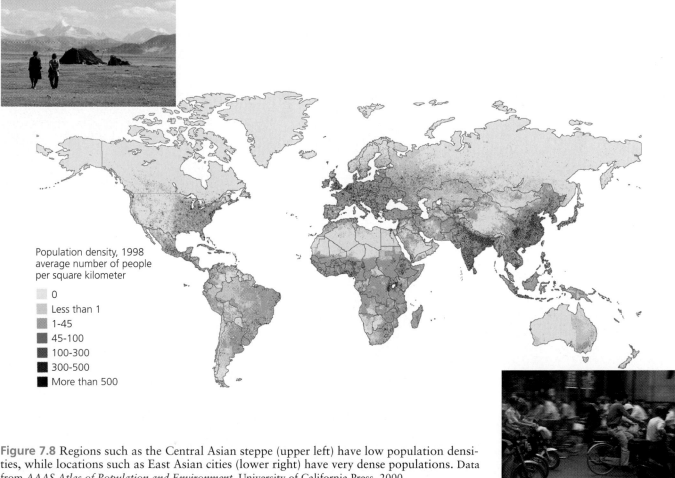

Population density, 1998
average number of people
per square kilometer

- 0
- Less than 1
- 1-45
- 45-100
- 100-300
- 300-500
- More than 500

Figure 7.8 Regions such as the Central Asian steppe (upper left) have low population densities, while locations such as East Asian cities (lower right) have very dense populations. Data from *AAAS Atlas of Population and Environment*, University of California Press, 2000.

population density is high in areas with temperate, subtropical, and tropical climates, from Europe to China to Mexico to southern Africa to India. Population density is lower in areas of extreme-climate biomes, such as desert, deep rainforest, and tundra. Dense along seacoasts and rivers, human population is less dense at inland locations lacking surface water. At intermediate scales, we clump ourselves into cities and suburbs, while we are spread more sparsely across rural areas. At small scales, we cluster in certain neighborhoods and in individual households.

This uneven distribution means that certain areas bear far more environmental impact than others. Just as the Yellow River has been subject to great pressure from millions of Chinese farmers, the world's other major rivers, from the Nile to the Danube to the Ganges to the Mississippi, have all seen more than their share of human impact. The concentration of people in cities may relieve pressures on more rural ecological systems by releasing some of them from human land use, but the urban way of life, along with the packaging and transport of goods it entails, also makes for more intensive energy use, fossil fuel consumption, and production of pollution and waste material.

At the same time, areas with low population density are often vulnerable to environmental impacts as well, because the reason they have low populations in the first place is that they are sensitive and cannot support many people (a high S value in our revised IPAT model). Deserts, for instance, are easily affected by development that commandeers a large share of available water. Grasslands can too easily be turned to desert if they are farmed too intensively, as has happened across vast stretches of the Sahel region bordering Africa's Sahara Desert, in the Middle East, in parts of China and the United States, and in central Asian nations such as Kazakhstan.

Age structure Data on the age structure or age distribution of human populations is especially valuable to demographers trying to predict future dynamics of a population. Why do the relative sizes of each age group in a population affect the overall size of a population over time? Consider this: All else being equal, where would you expect to see a larger and more rapid population increase—on a desert island populated entirely by senior citizens or one populated entirely by college freshmen (Figure 7.9)?

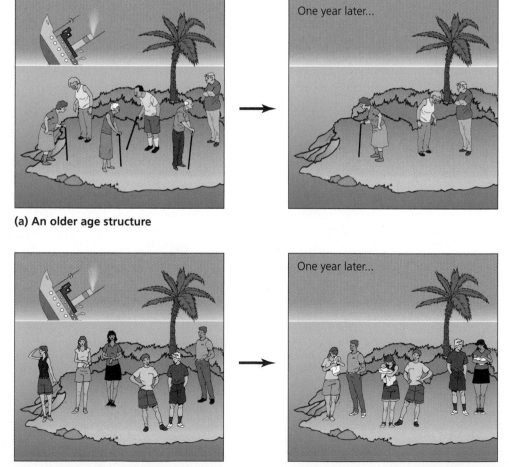

Figure 7.9 Age structure can influence population growth rates. Here, **(a)** one desert island is populated by a group of shipwrecked senior citizens, and **(b)** another desert island is populated by a group of shipwrecked college students. Assuming rescue were not an option and food and shelter were adequate for both groups of castaways, what population changes would you predict for each island over time?

(a) An older age structure

(b) A younger age structure

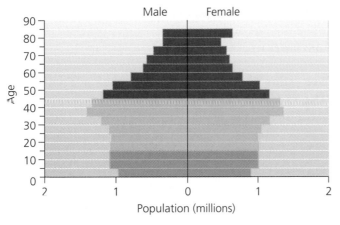

(a) Age pyramid of Canada in 2000

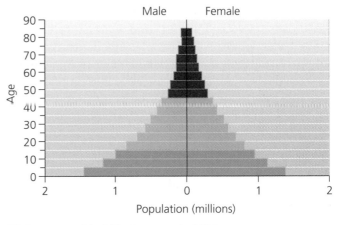

(b) Age pyramid of Madagascar in 2000

Figure 7.10 (**a**) Canada shows a balanced age structure, with relatively even numbers of individuals in various age classes. (**b**) Madagascar shows an age distribution heavily weighted toward young people. Madagascar's population growth rate is 10 times that of Canada's. Go to **GRAPH IT** on the website or CD–ROM. Data from U.S. Bureau of the Census, International Data Base, 2002.

With this in mind, now look at age pyramids for Canada and Madagascar (Figure 7.10). Not surprisingly, it is Madagascar that has the greater population growth rate. In fact, its annual growth rate, 3.0%, is 10 times that of Canada's 0.3%.

By causing dramatic reductions in the number of children born since 1970, China virtually guaranteed that its population age structure would change. In 2000 there were 87 million people older than 65 in China. By 2025 there will be 198 million people older than 65 (Figure 7.11). In 1997 there were 125 children under age 5 for every 100 people 65 or older in China; by 2025 there will be only 40. In 1995 the median age in China was 27; by 2025 it will be 39. This dramatic

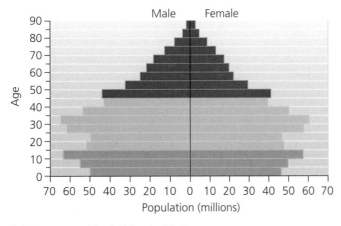

(a) Age pyramid of China in 2000

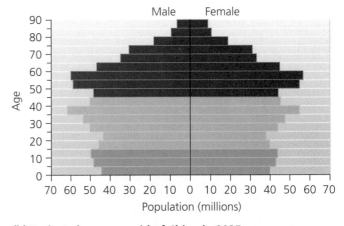

(b) Projected age pyramid of China in 2025

Figure 7.11 As China's population ages, older people will outnumber the young. Age pyramids show the predicted graying of the Chinese population between (**a**) 2000 and (**b**) 2025. Today's children may, as working-age adults (**c**), face pressures to support greater numbers of older citizens than any previous generation. Data from U.S. Bureau of the Census, International Data Base, 2002.

(c) Young women factory workers in Hong Kong

shift in age structure will challenge China's economy, healthcare systems, families, and military forces because fewer working-age people will be available to support social programs that assist the increasing number of older people. On the other hand, the shift in age structure also involves a reduction in the proportion of dependent children; the reduced number of young adults may mean a decrease in the crime rate; and elderly people are often productive members of society, contributing through volunteer activities and services to their children and grandchildren. Clearly, in terms of both benefits and drawbacks, life in China has been and will continue to be profoundly affected by the particular approach its government has taken to population control.

Weighing the Issues:
China's Reproductive Policy

Using China's reproductive policy as a model, consider the benefits as well as the problems associated with such a policy. Do you think that a government should be able to enforce such strict penalties for people who fail to abide by the policy? If you disagree with China's policy, what might be some alternatives for dealing with the resource demands of a quickly growing population?

This pattern of aging in the population is occurring in many countries worldwide, including the United States (Figure 7.12). Older populations will present new challenges for the United States, Canada, and many other

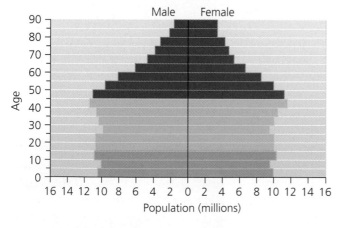

Figure 7.12 The "baby boom" is visible in the age pyramid for the United States, in the age brackets between 40 and 50. In future years the nation will experience an aging population as baby-boomers grow older. Data from U.S. Bureau of the Census, International Data Base, 2002.

developed countries as more and more older people require the care and financial assistance of relatively fewer and fewer working-age citizens.

Sex ratios The ratio of males to females in a population also can affect population dynamics. Imagine two isolated islands, one populated by 99 males and one female and the other by 50 males and 50 females (Figure 7.13). Where would we be likely to see the most significant population increase over time? Of course, the island with an equal number of males and females would have a greater number of potential mothers and, as a result, a greater potential for population growth.

The naturally occurring sex ratio in human populations at birth features a slight preponderance of males; for every 100 female infants born, 105 to 106 male infants are born. This phenomenon may be an evolutionary adaptation to the fact that males are often more prone to death during any given year of life. It usually ensures that the ratio of males to females is approximately equal by the time people reach reproductive age. Thus, a slightly uneven sex ratio at birth may be beneficial. However, a greatly distorted ratio can lead to problems.

In 2002, researchers reported an unsettling trend in China's Hainan province: The ratio of newborn males to females was wildly out of alignment. In mid-2002, the sex ratio at birth in Hainan province was 135.64 males for every 100 females. The sex ratio at birth for China as a whole was also skewed that year: 116.86 boys were born for every 100 girls. The reported reason for these unusual sex ratios was that many pregnant women, having learned the sex of their fetuses by ultrasound, were selectively aborting female fetuses.

Many observers interpret these selective abortions and the resulting distortion of newborn sex ratios as unintended consequences of the country's one-child policy. Traditionally, Chinese culture has valued male children because they will carry on the family name, assist with farm work in rural areas, and care for parents as they grow older. Girls, in contrast, will most likely grow up to marry and leave their parents, as the culture dictates. As a result, they will not provide the same benefits to their parents as male children. Some sociologists maintain that the combination of cultural gender preference and the government's one-child policy has led some couples to selectively abort female fetuses, to abandon female infants, and even to kill female infants. Similar cultural biases have led populations in parts of India to have distorted sex ratios as well, where up to as many as 160 boys are born for every 100 girls.

Figure 7.13 Sex ratios can affect population growth rates in the real world, just as they can on imaginary islands like these. Populations with even sex ratios will show greater capacity for reproduction and growth than populations with highly skewed ratios.

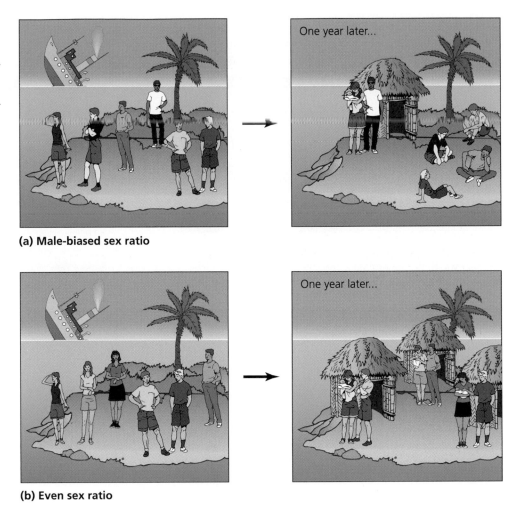

(a) Male-biased sex ratio

(b) Even sex ratio

China's skewed sex ratio may have the effect of further lowering population growth rates. However, it has entailed tragic consequences for those female infants lost and will also have the unfortunate social effect of leaving many Chinese men single.

Population growth depends on rates of birth, death, immigration, and emigration

Just as they do for other organisms, rates of birth, death, immigration, and emigration help determine whether a human population grows, shrinks, or remains stable. The formula for measuring population growth rate that we used in Chapter 5 pertains to humans: Crude birth rate and immigration rate add individuals to a population, while crude death rate and emigration rate remove individuals. As we have said, technological advances have led to a dramatic decline in human death rates, widening the gap between birth rates and death rates and resulting in the global human population expansion.

In today's ever-more-crowded world, immigration and emigration are playing increasingly large roles because of refugees. Refugees, people who flee their home country or region for economic or political reasons, have become more numerous in recent decades as a result of wars, civil strife, and environmental degradation. The United Nations puts the number of refugees who flee to escape poor environmental conditions at 25 million per year and possibly many more. Often refugee movement causes environmental problems in the receiving region as people try to eke out an existence without a livelihood and with no cultural or economic attachment to the land or incentive to conserve its resources. The millions who fled Rwanda following the genocide there in the mid-1990s, for example, destroyed large areas of forest while trying to obtain fuelwood, food, and shelter when they reached the Democratic Republic of Congo (Figure 7.14).

Figure 7.14 The flight of refugees from Rwanda into the Democratic Republic of Congo in 1994 following the Rwandan genocide caused tremendous hardship for the refugees and tremendous stress on the environment into which they moved.

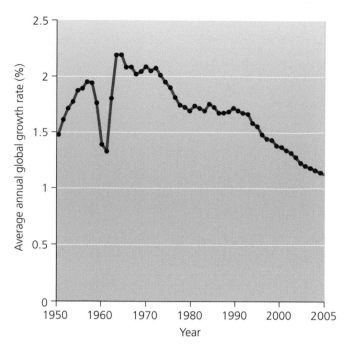

Figure 7.15 The annual growth rate of the global human population has been declining since the 1960s. Data from U.S. Bureau of the Census, International Data Base, 2002.

Population growth rates in any given region change over time because birth, death, and migration rates change. For most of the past 2,000 years, China's population has been relatively stable. The first significant increases began as a result of increased agricultural production and a powerful government during the Qing, or Manchu, Dynasty in the 1800s. Unfortunately, population growth began to outstrip food supplies by the mid-1850s, and the quality of life of the average Chinese peasant began to decline. From the mid-1800s, an era of increased European involvement in China, until 1949, China's population grew very slowly, at about 0.3% per year. This slow population growth was due, in part, to food shortages and political instability in the nation. As we have seen, population growth rates rose again following Mao's establishment of the People's Republic and have declined since the establishment of the one-child policy (Table 7.1).

Since 1970, growth rates in many countries have been declining, even without population control policies, and the global growth rate has declined (Figure 7.15). This decline has come about in part from a steep drop in birth rates, for which there are several apparent causes.

Weighing The Issues:
Malaria and Africa's Dynamics

More than one million people die from malaria every year and about half a billion others are debilitated by the disease in some way. Nine of every ten malaria-related deaths occur in Africa. Pregnant women and children are especially vulnerable to malaria, and in 2003, the World Health Organization estimated that the disease was causing 3,000 child deaths a day. Beyond its physical effects, widespread malaria also contributes significantly to sustained poverty. What effects on population dynamics would you expect this disease to have on African populations, and why?

A population's total fertility rate can shape population growth rate

One key statistic demographers calculate to examine a population's potential for growth is the **total fertility rate (TFR),** or the average number of children born per

Table 7.1   Recent Trends in China's Population Growth

	1970	1993	2002
Total fertility rate	5.8	2.0	1.8
Rate of natural population increase (% per year)	2.6	1.2	0.7
Doubling time (years)	26.9	58.3	100.0

Data from *2002 World Population Data Sheet*, Population Reference Bureau; Peng Peiyun, One family, one child, *Harvard International Review*, 1994; United Nations Economic and Social Commission for Asia and the Pacific, Population and development indicators for Asia and the Pacific, 2002; Penny Kane and Ching Y. Choi, China's one child family policy, *British Medical Journal*, 1999.

female member of a population during her lifetime. The **replacement fertility** is the TFR that keeps the size of a population stable. For humans, replacement fertility is equal to a TFR of 2.1. When the TFR drops below 2.1, population size, in the absence of immigration, will shrink.

Various factors influence TFR and have acted to drive it downward in many countries in recent years. A lower infant mortality rate has alleviated people's tendency to conceive many children in order to ensure that at least some survive. Increasing urbanization has also driven TFR down; whereas populations in rural areas need children to contribute to the labor pool, in urban areas children are usually excluded from the labor market, are required to go to school, and represent more economic costs to their families. If a government provides some form of social security, as most do these days, parents need fewer children to support them in their old age when they can no longer work. Furthermore, with greater education and changing roles in society, women tend to shift into the labor force, putting less emphasis on child-rearing.

All of these factors have come together in Europe, where TFR has dropped from 2.6 to 1.4 in the past half-century. Fertilities below replacement level are already resulting in slightly declining populations in 17 European nations. In 2002, Europe's overall annual **natural rate of population change** (change due to birth and death rates alone, excluding migration), was −0.1%. Worldwide by 2002, a total of 70 countries had fallen below the replacement fertility of 2.1. These countries made up more than 44% of the world's population and included China (with a TFR of 1.8). Table 7.2 shows the TFRs of major continental regions.

Weighing the Issues:
Economic Consequences of Low Fertility Rates?

In many European nations, the total fertility rate has dipped below the replacement fertility rate. What might be some economic and social consequence of below-replacement fertility rates?

Women's empowerment and family planning greatly affect population growth rate

Many demographers had long assumed that cultural norms and levels of wealth or poverty alone dictated TFR, but most data now are highlighting factors

Table 7.2 Total Fertility Rates for Major Continental Regions

Region	Total fertility rate (TFR)
Africa	5.2
Latin America and Caribbean	2.7
Asia	2.6
Oceania	2.5
North America	2.1
Europe	1.4

Data from *2002 World Population Data Sheet*, Population Reference Bureau.

pertaining to women and their social empowerment. Certainly it is not just wealth that determines TFR; several developing countries now have TFRs lower than that of the United States. Drops in TFR have been most noticeable in countries where women have gained improved access to contraceptives and education, particularly family-planning education (see The Science behind the Story, Figure 7.16, and Figure 7.17). In 2002, 55% of married women worldwide (ages 15–49) reported using some type of modern contraceptive to plan or prevent pregnancy. China, at 83%, had the highest rate of contraceptive use of any nation. Seven western European nations had rates of contraceptive use above 70%, as did Costa Rica, Cuba, New Zealand, the United States, Brazil, and Thailand. At the other end of the spectrum, 24 African nations had rates below 10%. These low

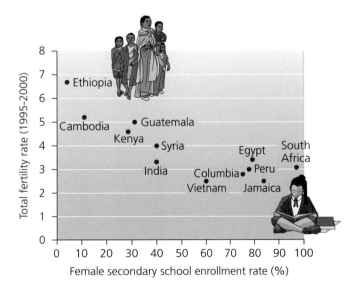

Figure 7.16 Increasing female literacy is strongly associated with reduced birth rates in many nations. Data from M. McDonald and D. Nierenberg, Linking population, women, and biodiversity, *State of the World*, Worldwatch Institute, 2003.

Figure 7.17 (a) Four pairs of neighboring countries demonstrate the effectiveness of family planning in reducing fertility rates. In each case, the nation that invested in family planning and (in some cases) made other reproductive rights, education, and health care more available to women (green lines) reduced its total fertility rate (TFR) far more dramatically than its neighbor (red lines). **(b)** A family planning counselor advises African women on health care and reproductive rights. Data from U.N. Population Division and *AAAS Atlas of Population & Environment,* University of California Press, 2000.

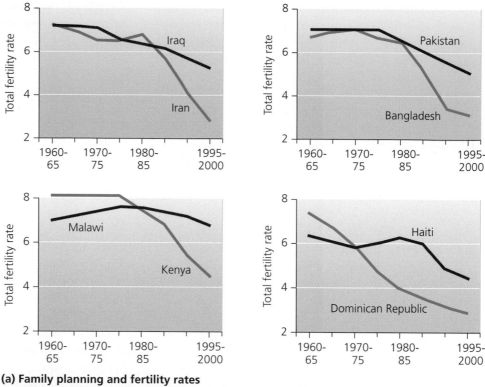

(a) Family planning and fertility rates

(b) Counseling and outreach

rates of contraceptive use contribute to sub-Saharan Africa's high fertility rates, although the region's TFR has fallen slightly from 6.5 births per woman in 1950 to 5.6 in 2002. By comparison, in 1950 the TFR in Asia was 5.9 children per woman, whereas today it is 2.6—in part a legacy of the population control policies of China and of Thailand. (Thailand's population control policies, which are very different from those of China, are discussed later in the chapter.) These patterns of data clearly suggest that in societies in which women have

little power, substantial proportions of pregnancies are unintended.

Unfortunately, many women still lack the information and personal freedom of choice that would allow them to make their own decisions about when to have children and how many to have. The United States has often declined to fund family-planning efforts by the United Nations, a stance for which it has been criticized. Canceling this funding, for example, was one of George W. Bush's first acts on becoming U.S. president in 2001.

Figure 7.18 The demographic transition is an idealized process that has taken some populations from a pre-industrial state of high birth rates and high death rates to a post-industrial state of low birth rates and low death rates. In this diagram the wide colored area between the two curves illustrates the gap between birth and death rates that causes rapid population growth during the middle of this process. Data from M.M. Kent and K.A. Crews, *World Population: Fundamentals of Growth*, Population Reference Bureau, 1990.

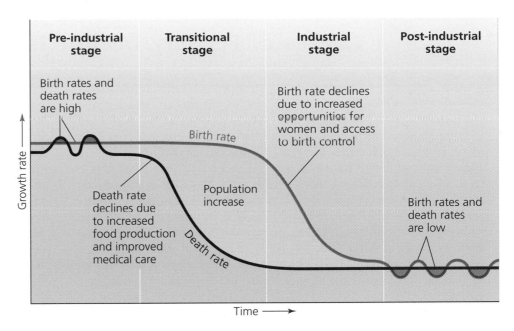

Weighing the Issues:
Population Growth and
Reproductive Freedom

It has been suggested that if human population growth remains unchecked, everyone will eventually begin to suffer a poorer quality of life. Would you be willing to make this sacrifice if it meant that people in other countries (such as China) could avoid government-imposed limitations on their reproductive freedom? If it came to the point that the U.S. government implemented a strict reproductive policy, how would you feel? Would you rather have your reproductive freedom limited by the government or your consumption limited by the government?

Some nations have experienced a change called the demographic transition

Many nations that have lowered their birth rates and TFRs have been going through a similar set of interrelated changes. In countries with good sanitation, good health care, and reliable food supplies, more people than ever before are living long lives. As a result, over the past 50 years the life expectancy for the average human has increased from 46 to 66 years, as the global crude death rate has dropped from 20 deaths per 1,000 people to fewer than 10 deaths per 1,000 people. Strictly speaking, **life expectancy** is the average number of years that

individuals in particular age groups are likely to continue to live, but often people use this term to refer to the average number of years a person can expect to live from birth. Much of the increase in life expectancy is due to reduced infant mortality rates. Societies going through these changes are mostly the ones that have urbanized, have industrialized, and have been able to generate personal wealth for their citizens.

To make sense of these trends, demographers developed a concept called the **demographic transition.** This is a theoretical model of economic and cultural change proposed in the 1940s and 1950s by demographer Frank Notestein and elaborated on by others to explain the declining death rates and birth rates that occurred in Western nations as they became industrialized. Notestein believed nations moved from a stable pre-industrial state of high birth and death rates to a stable post-industrial state of low birth and death rates. Industrialization, he proposed, caused these rates to fall naturally by decreasing mortality and by lessening the need for large families. Parents would thereafter choose to invest in quality of life rather than quantity of children.

The pre-industrial stage Notestein's demographic model describing the population impacts of industrialization proceeds in several stages (Figure 7.18). The first is the **pre-industrial stage,** characterized by conditions that defined most of human history. In pre-industrial societies, both death rates and birth rates are high. Death rates are high because disease is widespread, medical care

Causes of Fertility Decline in Bangladesh

Research in developing countries suggests that poverty and overpopulation can create a vicious cycle, in which poverty encourages high fertility and high fertility acts as an obstacle to economic development. Are there certain policy steps that such countries can take to bring down fertility rates? Scientific analysis of family planning programs in the South Asian nation of Bangladesh suggests the answer is yes.

Bangladesh is one of the poorest, most densely populated countries on the planet. Its 120 million people live in an area about the size of Wisconsin, and 45% of them live below the poverty line. With few natural resources and more than 850 people per km^2 (2,200 per mi^2), limiting population growth is critically important. As Bangladeshi president Ziaur Rahman declared in 1976, "If we cannot do something about population, nothing else that we accomplish will matter much."

Fortunately, Bangladesh has made striking progress in controlling population growth in the past three decades. Despite stagnant economic development, low literacy rates, poor health care, and limited rights for women, the nation's total fertility rate (TFR) has

dropped markedly. In the early 1970s, the average woman in Bangladesh had more than 6 children over the course of her life. Today, the TFR is about 3.3.

Researchers interested in population issues have hypothesized that family planning programs were responsible for Bangladesh's rapid reduction in TFR. Because conducting an experiment to test such a hypothesis is difficult, some researchers have taken advantage of a natural experiment. By comparing Bangladesh to countries such as Pakistan that are socioeconomically similar but have had less success in lowering TFR, researchers have concluded that Bangladesh succeeded because of aggressive, well-funded outreach efforts that were sensitive to the values of its traditional society.

However, because no two countries are identical, it is difficult to draw firm conclusions from such broad-scale studies, which is why the Matlab Family Planning and Health Services Project, in an isolated rural area of Bangladesh, has become one of the best-known experiments in family planning in developing countries. The Matlab Project was an intensive outreach program run collaboratively by the Bangladeshi government and international aid organizations. Each

household in the project area received biweekly visits from local women offering counseling, education, and free contraceptives. Compared to a similar government-run program in a nearby area, the Matlab Project featured more training, more services, and more frequent visits. In both areas, however, a highly organized health surveillance system gave researchers detailed information about births, deaths, and health-related behaviors such as contraceptive use. The result was an experiment in which the Matlab Project area was the "treatment" condition and the government-run area was the "control."

When Matlab Project director James Phillips and his colleagues reviewed a decade's worth of data in 1988, they found that fertility rates had declined in both treatment and control areas. The decline appeared to be due almost entirely to a rise in contraceptive use, because other factors—such as the average age of marriage—remained the same. Phillips and his colleagues also found that the declines had been significantly larger in the treatment area than in the control area. These findings suggested that high-intensity outreach efforts can affect fertility even in the absence of significant improve-

rudimentary, and food supplies unreliable and difficult to obtain. Birth rates are high because people must compensate for high mortality rates in infants and young children by having several children. In this stage, children are valuable as additional workers who can help meet a family's basic needs. Populations within the pre-industrial stage are not likely to experience much growth, and this is one reason why the human population was relatively stable from Neolithic times until the industrial revolution.

Industrialization and falling death rates Industrialization initiates the second stage of the demographic transition, known as the **transitional stage.** This transition from the pre-industrial stage to the industrial stage is generally characterized by declining death rates due to increased food production and improved medical care. Birth rates in the transitional stage, however, remain high because citizens have not yet grown used to the new economic and social conditions. As a result, population growth surges.

Total Fertility Rate, Bangladesh 1963–1988

Year	TFR	Source
1963–65	7.1	Population Growth estimate, 1962–65
1967–68	6.0	National Impact Survey, 1966–68
1974	7.1	BRSFM,* 1974
1975	6.4	Bangladesh Fertility Survey, 1975
1979	6.3	Contraceptive Prevalence Survey, 1979
1979	6.9	ICDDR,[†] B Matlab
1981	6.3	ICDDR,[†] B Matlab
1984	6.4	ICDDR,[†] B Sirajgonj
1985	6.1	ICDDR,[†] B Matlab Sirajgonj
1985	5.9	ICDDR,[†] B Matlab Comparison Area
1986	5.5	ICDDR,[†] B Matlab Comparison Area
1986	5.6	USAID,[‡] Dhaka
1986	5.4	ICDDR,[†] D Sirajgonj
1986	5.6	National Family Planning and Fertility Survey.
1987	4.8	ICDDR,[†] B Sirajgonj
1987	4.8	Bangladesh Fertility Survey, 1989
1988	4.6	Bangladesh Fertility Survey, 1989

*Bangladesh Retrospective Survey of Fertility and Mortality
Data from The Global Reproductive Health Forum @ Harvard, 2000.
[†]The International Centre for Diarrhoeal Disease Research, Bangladesh
[‡]The United States Agency for International Development

ments in women's status, education, or economic development.

But why exactly was the outreach program successful? one hypothesis was that visits from health care workers had helped convince local women that small families are desirable. However, in 1999, Mary Arends-Kuenning, a graduate student in economics at the University of Michigan, and her colleagues reported that there was no relationship between ideal family size and the number of visits made by outreach workers, either in Matlab or nearby comparison areas. In fact, the ideal family size declined equally in all areas. Instead of creating new demand for birth control, the Matlab Project appears to have helped women convert an already-existing desire for fewer children into behaviors, such as contraceptive use, that reduce fertility.

Bangladesh's ability to rein in fertility rates despite unfavorable social and economic conditions bodes well not only for that country, but also for many other impoverished nations now facing explosive population growth. However, significant challenges remain. In the 1990s, Bangladesh's TFR showed signs of having leveled off at slightly more than 3 children per woman. If rates fail to decline further, the country's population could double to 250 million within 30 years. Although scientific research has been helpful in understanding the impact of family planning programs on fertility, further reductions may require fundamental social, political, and economic changes that are difficult to implement in traditional, resource-strapped countries such as Bangladesh. Nonetheless, the scientific evidence collected at Matlab since the 1970s has played an important role in informing population control efforts in Bangladesh and elsewhere.

The industrial stage and falling birth rates The third stage in the demographic transition is the **industrial stage.** Widespread industrialization creates increased opportunities for employment outside the home, in particular for women. Children become less valuable, in economic terms, because they do not help meet family food needs as they did in the pre-industrial stage. If couples are aware of this, and if they have access to birth control, they may choose to have fewer children. Thus birth rates fall, closing the gap with death rates and reducing the rate of population growth.

The post-industrial stage In the final stage, the **post-industrial stage,** both birth and death rates have fallen to a low level and remain stable there, and populations may even decline slightly. The society enjoys the fruits of industrialization without the threat of runaway population growth.

Is the demographic transition a universal process?

The demographic transition has occurred in many European countries, the United States, Canada, Japan, and several other developed nations over the past 200 to 300 years. Nonetheless, it is a model that may not apply to all of the developing countries as they industrialize now and in the future. Some social scientists doubt that it will apply. They point out that population dynamics may be different for developing nations that adopt the Western world's industrial model rather than devising their own. Some social scientists assert that the transition will fail in cultures that place greater value on childbirth or grant women fewer freedoms. Moreover, some natural scientists warn that there are not enough resources in the world to enable all countries to attain the standard of living now enjoyed by the developed countries. It is often said that for all nations to enjoy the quality of life that United States citizens enjoy, we would need the natural resources of three more planet Earths. Whether or not developing nations, which include the vast majority of the planet's people, pass through the demographic transition as developed nations have is one of the most important and far-reaching questions for the future of our civilization and for Earth's environment.

Weighing the Issues:
The Demographic Transition in Developed Nations

Do you think that all developed nations will necessarily complete the demographic transition as the model predicts and end in a permanent state of low birth and death rates? Or could the future yet take another path? Think about developed nations like the United States and Canada; can you think of reasons why these nations might or might not continue to lower and stabilize their birth and death rates in a state of prosperity? What factors might affect the likelihood of their doing so? How complete and realistic a model would you consider the demographic transition to be?

Population policies and family-planning programs are working around the globe

Regardless of whether industrialization allows developing countries to pass through all stages of the demographic transition, available data show that funding and policies that encourage family planning have been effective in lowering population growth rates in all types of nations, even those that are least industrialized. No nation has pursued a population control program as extreme as China's, but other nations with fast-growing populations have implemented less restrictive programs.

The government of Thailand has relied on an education-based approach to family planning that has reduced birth rates and slowed population growth. In the 1960s Thailand's growth rate was 2.3%, but by 2002 it had declined to 0.8%. This decline has been achieved without a one-child policy. It has been due, in large part, to government-sponsored programs devoted to family-planning education and increased contraceptive availability.

India has had long-standing policies, but many observers think they need to be strengthened because India—although it has less land area—seems set to overtake China in population quite soon. Brazil, Mexico, Iran, Cuba, and many other developing countries have instituted active programs to reduce their population growth. These programs entail setting targets and providing incentives, education, contraception, and reproductive health care. One study in 2000 looked at four pairs of nations located in the same parts of the world, but with one country in each pair having a stronger program than the other: Thailand and the Philippines, Pakistan and Bangladesh, Tunisia and Algeria, and Zimbabwe and Zambia. The demographers concluded that in three of four cases, the country with the stronger program (Bangladesh, Tunisia, and Zimbabwe) initiated a decline in fertility with its policies; in the case of Thailand, whose program is stronger than that of the Philippines, the fertility decline began before the policies but was accelerated by them. The researchers also found that the factor driving implementation of population control policies in these nations was not public demand, but rather the political ideology of the ruling elite. In the case of Thailand and the Philippines, religion also played a role; the Catholic Church's strong presence in the Philippines held back the success of family planning there, the researchers concluded.

In 1994 the United Nations hosted a milestone conference on population and development in Cairo, Egypt, at which 179 nations endorsed a platform calling for all governments to offer universal access to reproductive health care within 20 years. The conference marked a turn away from older notions of command-and-control population policy geared toward pushing contraception and lowering population to preset targets, and toward a newer recognition that governments should offer better education and health care and address social needs that indirectly bear on population (such as alleviating poverty,

Table 7.3 Targets and Timetables Set by the United Nations Conference on Population and Development

Targets set by 1994 International Conference on Population and Development, Cairo, Egypt	Year to achieve
1. Provide universal access to reproductive health services, including family planning and sexual health care	2015
2. Provide universal primary education for all	2015
3. Reduce infant mortality rate to below 35 per 1,000 live births, and under-5 child mortality rate to below 45 per 1,000 live births in all countries	2015
4. Reduce maternal mortality to 50% below 1990 levels To 75% below 1990 levels	2000 2015
5. Increase life expectancy at birth to 75 years or more	2015

Targets set by 1999 progress session to follow up Cairo conference	Year to achieve
6. Reduce illiteracy rate for women and girls to 50% below 1990 levels	2005
7. Enroll 90% of boys and girls in primary school	2010
8. Ensure that 60% of primary health care and family planning facilities offer a wide range of services, including family planning, obstetric care, and prevention and treatment of reproductive tract infections, including sexually transmitted diseases Raise level to 80%	2005 2010
9. Provide skilled birth attendants at 80% of all births globally, and at least 40% of all births where maternal mortality rate is very high	2005
10. Reduce unmet need for contraceptives by 50% By 75% By 100%	2005 2010 2015
11. Guarantee that 90% of 15 to 24-year-olds have access to information and services to help them avoid HIV infection, including condoms, voluntary testing, counseling, and follow-up	2015

Data from United Nations Population Fund (UNFPA), 1999.

disease, and sexism). This conference and a follow-up meeting five years later set a number of goals for nations to meet (Table 7.3).

Weighing the Issues:
U.S. Involvement in International Family Planning

In 2002 the U.S. government announced that it was with-holding $34 million in funding that had been designated for the U.N. Population Fund (UNFPA), the agency responsible for promoting family planning programs in many nations, including China. UNFPA programs provide education in family planning, HIV/AIDS prevention, and teen pregnancy prevention. According to U.S. law, the federal government is not allowed to fund any organization that "supports or participates in the management of a program of coercive abortion or involuntary sterilization." Because the Chinese government has been implicated in both of these activities and the UNFPA is active in China, the U.S. government decided not to fund UNFPA activities in 2002, even though it did provide $46.5 million to the UNFPA between 1998 and 2001. Many nations and non-governmental organizations criticized the U.S. decision and alleged that it was motivated more by politics than by scientific or legal concerns. In response to the new policy, the European Union offered additional funding to UNFPA to offset the loss of the U.S. contributions. What do you think of the U.S. decision? Should the U.S. fund family planning efforts in other nations? What conditions, if any, should the U.S. place on the use of such funds?

Women still need more empowerment

Improving the treatment of women, through health care and education, was one goal of the Cairo conference. More social scientists and policymakers today recognize that for population growth to slow and finally stabilize, and for family-planning programs to fulfill their potential, women need to be granted equal power with men in societies worldwide. Clearly we are still a substantial way from achieving this goal. Over two-thirds of the world's people who cannot read, and 60% of those living in poverty, are female. In many societies by tradition men restrict women's decision-making abilities regarding

Figure 7.19 Although women make up more than half of the world's population, they are vastly underrepresented in positions of political power. Measured by the percentages of seats held by women in national legislatures, women in some regions fare better than those in others. Note that the United States may not rank quite as highly as you might guess. Data from Inter-Parliamentary Union, Women in National Parliaments, 2003.

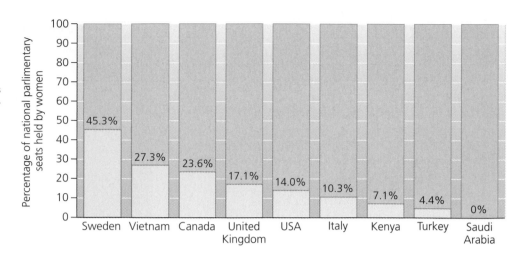

how many children they will have. Studies show that in those societies in which women are freer to decide whether and when to have children, fertility rates have fallen, and the resulting children are better taken care of, healthier, and better educated.

Proponents of the view emerging from the Cairo conference also maintain that gender equality promotes more efficient resource use, because it is women who in much of the world are most attuned to the availability of resources such as water and fuelwood. Proponents of this new view see a number of important environmental and social issues as being inextricably tied together—poverty alleviation, gender equality, resource conservation, population control, biodiversity conservation, and better education and health for women and children.

Although women at the level of the domestic household generally hold less decision-making power than men, and although violence against women remains shockingly common in most societies, the gap between the power (economic and political) held by men and the power held by women is just as obvious at the highest levels of government. Worldwide, only about 13% of elected government officials in national legislatures are women (Figure 7.19). The United States lags behind not only Europe but also many developing nations in the proportion of women in positions of power in its government. As more women come into positions of power, perhaps gender equality will become a more tangible reality. Such equality would undoubtedly have environmental effects, for when women have economic and political power and access to education, they gain the option, and often the motivation, to limit the number of children they bear.

Poverty is strongly correlated with population growth

As we noted above, the alleviation of poverty is one factor in the web of social and environmental issues that has been linked to population, and it was one of the targets of the Cairo conference. With some exceptions, wealthier societies tend to have lower population growth rates whereas poorer societies tend to have higher population growth rates. This phenomenon is in line with demographic transition theory. Note in Table 7.4 how well per capita gross national income correlates inversely with measures of fertility and population growth rate; the poorer nations tend strongly to have higher fertility and growth rates. These rates are a result of the higher birth and infant mortality rates shown by poorer nations. Also, in poorer countries, fewer couples tend to use contraception.

Trends such as these have affected the distribution of people on the planet. In 1960, 70% of all people lived in developing nations. By late 1999, 80% of the world's population was living in these countries. Moreover, fully 98% of the next billion people to be added to the global population will be born in these poor, less developed regions (Figure 7.20). This is unfortunate from a social standpoint, because these people will be added to the countries that are least able to provide for them. It is also unfortunate from an environmental standpoint, because poverty often results in environmental degradation. People dependent on agriculture in an area of poor farmland, for instance, may need to try to farm even if doing so destroys the soil and is not sustainable. This is largely why Africa's once-productive Sahel region and many regions of western China are turning to desert

Table 7.4 Per Capita Wealth, with Rates of Fertility, Population Growth, and Contraceptive Use, for Selected Nations

Country	Per capita GNI PPP (US$)*	Population increase (%/year)	Children born per woman	Population density (per mile)^2	Infant mortality (per 1000)	Percentage of couples using birth control
Ethiopia	660	2.5	5.9	159	97	6
Niger	740	3.5	8.0	24	123	4
Haiti	1,470	1.7	4.7	659	80	22
Cameroon	1,590	2.5	4.9	88	77	8
Pakistan	1,860	2.1	4.8	467	86	20
Nicaragua	2,080	2.8	4.1	107	40	57
India	2,340	1.7	3.2	827	68	43
Syria	3,340	2.6	4.1	245	24	28
China	3,920	0.7	1.8	347	31	83
Brazil	7,300	1.3	2.2	53	33	70
Mexico	8,790	2.1	2.9	135	25	60
Czech Republic	13,780	−0.2	1.1	337	4	58
Spain	19,260	0.1	1.2	211	5	67
United Kingdom	23,550	0.1	1.6	637	6	71
Japan	27,080	0.2	1.3	873	3	48
Canada	27,170	0.3	1.5	8	5	68
United States	34,100	0.6	2.1	77	7	72

*GNI PPP is gross national income in purchasing power parity, a measure in which income is converted to "international" dollars using a purchasing power parity conversion factor. International dollars indicate the amount of goods and services one could buy in the United States with a given amount of money. Data from *2002 World Population Data Sheet*, Population Reference Bureau.

(Figure 7.21). Poverty also drives the hunting of many large mammals in Africa's forests, including the great apes that are now disappearing as local settlers and miners kill them for their "bush meat."

The consumption that comes with affluence creates a large environmental impact

Poverty can lead people into environmentally destructive behavior, but wealth can produce even more severe and far-reaching environmental impacts. The affluence that characterizes a society such as the United States,

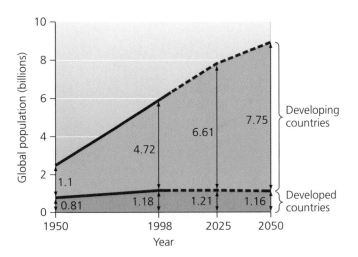

Figure 7.20 Nearly 98% of the next one billion people added to Earth's human population will reside in the less developed, poorer parts of the world. Data from U.N. Population Division and *AAAS Atlas of Population & Environment*, University of California Press, 2000.

Figure 7.21 Dependence on grazing agriculture by poor people in an area where population is increasing beyond the land's ability to handle it has led to environmental degradation of the semi-arid Sahel region of Africa.

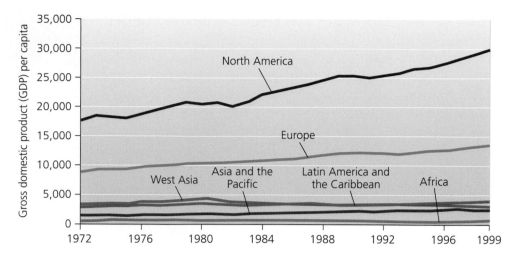

Figure 7.22 Affluence, as measured by GDP (Chapter 2), shows marked variation among different regions of the world, with North America and Europe standing head and shoulders above other continents comprised mostly of developing nations. Data from World Bank, 2001.

Japan, or the Netherlands is built on massive and unprecedented levels of resource consumption. Much of this chapter has dealt with numbers of people rather than on the amount of resources each member of the population consumes or the amount of waste each member produces. The environmental impact of human activities, however, depends not only on the number of people involved but also on the way those people live. Patterns of affluence and consumption are spread unevenly across the world (Figure 7.22).

Individuals from affluent societies leave a larger "ecological footprint"

Recall that affluence was one element of the formula we discussed earlier in this chapter, $I = P \times A \times T$. We have addressed population (P) at some length and have touched on how technology (T) can either reduce or exacerbate overall impact. Affluence (A), too, is an important factor, increasing environmental impact just as population growth does. Interpretation of the IPAT formula leads us to understand that we can change our population's total environmental impact not only by altering population size but also when individuals change their behavior. The environmental impact of an individual or of a population can be expressed in terms of an **ecological footprint**, the cumulative amount of Earth's surface area required to provide the raw materials a person or population consumes and to dispose of or recycle the waste that is produced. The ecological footprint of the average U.S. citizen is considerably larger than that of the average resident of a developing country (Figure 7.23)—a fact

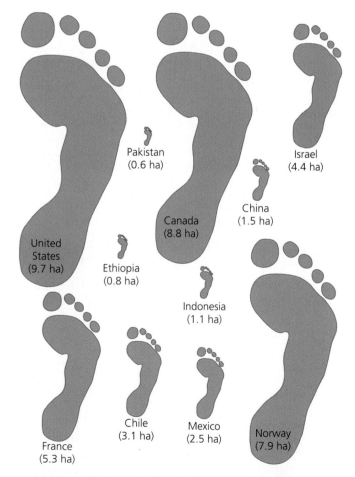

Figure 7.23 The citizens of some nations have larger ecological footprints than the citizens of others. U.S residents consume more resources—and thus use more land—than residents of any other nation. Shown here are ecological footprints for the average citizens of several nations, as of 1999. Data from Redefining Progress, 1999.

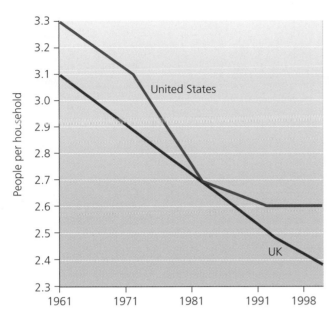

(a) A family living in the United States

Figure 7.24 The average number of people per household has dropped greatly in many developed countries, including Great Britain and the United States. This trend has ramifications for the environment: Because many goods are purchased by the household rather than by the individual, more households mean more consumption. Data from *AAAS Atlas of Population & Environment,* University of California Press, 2000.

(b) A family living in Egypt

Figure 7.25 A typical U.S. family (**a**) may own a large house, keep numerous material possessions, and have enough money to afford luxuries such as vacation travel. A typical family in a developing nation such as this family in Egypt (**b**) may live in a small sparsely furnished dwelling with few material possessions and little money or time for luxuries. The ubiquity of television sets, even among poor families of the developing world, means that the world's poor see representations (both real and exaggerated) of wealth in the United States as depicted on American TV shows. Many sociologists hold that this has increased the poor's awareness of the global wealth gap and has spurred aspirations toward consumerism among the poor of developing nations.

that should remind us that the "population problem" does not lie entirely with the developing world.

In much of the developed world, consumption is rising faster than population. Some scientists have suggested that increasing consumption poses a larger environmental problem than increasing population because whereas sooner or later an expanding population will run into limits to its growth, there is no theoretical limit to the desire to consume. In the face of demand for luxury products and the all-too-human desire not only to use these products but also to flaunt them as status symbols, the desire to consume could conceivably keep rising without limit.

One factor contributing to increased consumption in affluent societies is that fewer people are living in each household; that is, household size is decreasing (Figure 7.24). For this reason, the number of households is rising faster than is population. This statistic is important because consumption of goods often reflects numbers of households rather than number of individuals.

The wealth gap and population growth contribute to violent conflict

The stark contrast between affluent and poor societies in today's world is, of course, the cause of social as well as environmental stresses. In 1999, the richest one-fifth of the world's people possessed 82 times the income of the poorest one-fifth (Figure 7.25). The richest 20% also used 86% of the world's resources. That left only 14% of global resources—energy, food, water, and other essentials—for the remaining 80% of the world's population to share. The impact of this inequitable distribution of wealth is one of the key factors the U.S. Departments of Defense and State take into account when assessing the potential for armed conflict around the world, whether it be conventional warfare or terrorism.

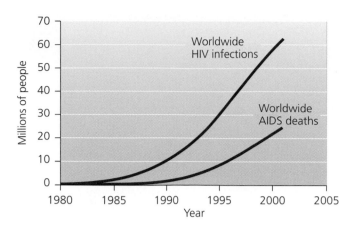

Figure 7.26 AIDS cases are increasing rapidly in much of the world. Data from *Vital Signs 2002,* Worldwatch Institute.

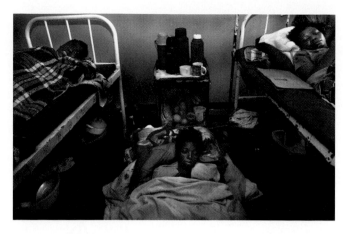

Figure 7.27 In 1999, AIDS patients occupied 60% of South Africa's hospital beds. By 2010, AIDS in Africa will orphan an estimated 40 million children.

As the gap between rich and poor grows wider and as the sheer numbers of those living in poverty continue to increase, it is reasonable to predict increasing tensions between the haves and the have-nots.

HIV/AIDS is a major factor affecting populations in some parts of the world

The rising material wealth and falling fertilities of many industrialized nations today is slowing population growth in accordance with demographic transition theory. Some other nations, however, are not following Notestein's script. Instead, in these countries, mortality is beginning to increase, presenting a scenario more akin to Malthus's fears. This is especially the case in countries where the HIV/AIDS epidemic has taken hold and is now raging out of control (Figure 7.26). African nations are being hardest hit. Of the 40 million people around the world infected with HIV/AIDS as of 2002, 28.5 million live in the nations of sub-Saharan Africa. During 2001, 5 million people became infected with the disease worldwide, 3.5 million of them in sub-Saharan Africa. The low rate of use of contraceptives, specifically those that diminish risk of infection, not only contributes to this region's high fertility rate (5.6 children per woman) but also fuels the expansion of AIDS. One in every 11 people aged 15 to 49 in sub-Saharan Africa is infected with HIV; for southern African nations alone, the figure is more than one in five. One of every four students at the University of Durban-Westville in South Africa is HIV-positive, and AIDS patients occupy 60% of South Africa's hospital beds (Figure 7.27).

Although other environmental problems (such as groundwater depletion) may eventually lead to greater global death rates, the AIDS epidemic for now illustrates most clearly the dramatic effect of a single cause of death on population dynamics and social cohesion. It is having the biggest impact on human populations of any disease since the Black Plague (which killed roughly one of three people in 14th-century Europe) and the smallpox epidemics brought by Europeans to the New World (which wiped out perhaps millions of native people) (see The Science behind the Story).

Besides raising death rates, the epidemic is causing a variety of demographic changes. In 2001 AIDS took 6,030 lives in Africa every day, a figure expected to double by 2010. It is estimated that between 2000 and 2020 the scourge will cause 68 million premature deaths in the 45 countries most affected. Of this total, 55 million will strike sub-Saharan Africa—39% more than would occur in the absence of AIDS. Children between the ages of one and five are particularly susceptible to the disease's lethal effects, so infant mortality is also elevated in this region (91 deaths out of 1000 live births, 13 times the rate in the developed world). In Botswana, one of the hardest-hit nations, 44.9% of pregnant women were found to be infected in 2001.

The high numbers of premature deaths are slashing average life expectancy. In parts of southern Africa, life expectancy has fallen from a high of close to 59 years in the early 1990s back to nearly 44 years, where it stood in the early 1950s. In some African nations, average life expectancy could drop to as low as 30 years in the near future. AIDS not only is killing millions of people, but also is leaving behind millions of orphans. As of 2002, 14 million children under the age of 15 worldwide had lost one or both parents to the disease. Current estimates indicate that the disease will orphan 40 million African children by 2010.

The Science behind the Story

AIDS Resistance Genes and the Black Death: An Unexpected Connection

When HIV was first identified as the cause of AIDS in the early 1980s, some scientists predicted that a cure for the disease would be found within a few years. In the two decades that followed, however, progress toward a cure or vaccine remained slow, even as the number infected grew to more that 60 million.

Thus the discovery in 1996 of a genetic mutation that appeared to confer resistance to AIDS was an important breakthrough. Still more surprising was a possible connection between AIDS and the Black Death, a plague that decimated Europe in the mid-14th century.

Beginning in 1984, geneticists Stephen J. O'Brien, Michael Dean, and colleagues at the National Cancer Institute (NCI) began searching for individuals who were naturally resistant to HIV. Over the next decade, they gathered genetic samples from people with and without the virus, trying to identify mutations unique to people who had been highly exposed to HIV but remained uninfected. Such mutations, they hoped, would provide clues to preventing or curing AIDS.

Until the mid-1990s, the search was largely unsuccessful. Then in 1995, researchers at the NCI identified molecules called chemokines that carry signals from one part of the immune system to another. Other researchers discovered that HIV uses the immune system's own chemokine receptors to penetrate the cell membrane. Together, these advances provided a window on how HIV gains entry into human hosts. They also gave O'Brien, Dean, and other researchers a promising place to look for resistance genes.

The search quickly produced exciting results. In 1996, three separate groups reported the discovery of a mutation conferring resistance to HIV. The mutation—a deletion of 32 base-pairs of DNA—affected the chemokine receptor CCR5, which is expressed on the surface of macrophages, the first cells HIV attacks when it enters a new host. People with two copies of the mutation appeared to be immune to HIV; people with one copy developed AIDS, but slowly.

Surprisingly, the mutation was unevenly distributed across human populations. One of the teams that discovered the mutation, a group of Belgian researchers led by Michael Samson and Marc Parmentier, found that about 9% of the Caucasians they tested had at least one copy of the mutation, but none of the Japanese or Africans they tested had the mutation at all. Other groups found similar results, and it soon became clear that the mutation was present only in Europeans and their descendants.

A key step toward answering why the mutation was unique to Europeans was taken when O'Brien and Dean determined the date the mutation had first arisen. To do so, they used a technique called haplotype analysis, which analyzes a mutation's association with distinctive genetic variants, or alleles, on the chromosome. In each generation, the linkage between the mutation and its original neighbors is weakened by recombination, the chromosomal reshuffling that takes place during reproduction. Scientists can thus use the strength of the linkage to identify the mutation's approximate age. When O'Brien and Dean used this technique, they found that the CCR5 mutation was about 700 years old.

That put the mutation's origin at about the time when the Black Death, which most scientists have identified as the bubonic plague, was ravaging Europe. Like HIV, the bacterium responsible for the bubonic plague *(Yersinia pestis)* initially attacks macrophages, using them as a refuge before spreading to the rest of the body. Thus a mutation in CCR5 could potentially have conferred resistance to bubonic plague, just as it now confers resistance to AIDS. Under the strong selection pressure of the Black Death, which killed a third of Europe's population, the prevalence of the mutation could have risen greatly. The connection between AIDS and the Black Death may never be proven beyond a doubt, but it suggest that diseases having a significant impact on human populations also can affect human evolution.

Population Control

The debate regarding human population growth and environmental problems is often contentious and divisive. Do you believe that national governments should implement policies, subsidies, or other programs to reduce birth rates?

Implement ICPD Program of Action

Access to reproductive health care, including family planning, is a basic human right. In order to exercise this right, men and women need to be informed about and have access to safe, effective, affordable, and acceptable methods of family planning of their choice.

All national governments should adopt policies, subsidies, and other programs to help implement the Program of Action, which the United Nations agreed to at the International Conference on Population and Development (ICPD), held in Cairo in 1994.

The ICPD is based on principles virtually every country in the world agreed to, one of which is as follows:

> . . . States should take all appropriate measures to ensure, on a basis of equality of men and women, universal access to health-care services, including those related to reproductive health care, which includes family planning and sexual health. Reproductive health-care programs should provide the widest range of services without any form of coercion. All couples and individuals have the basic right to decide freely and responsibly the number and spacing of their children and to have the information, education and means to do so.

In the same way that democratic nations are obligated to assist emerging democracies in holding fair and free elections, developed nations ought to assist developing nations in implementing the ICPD Program of Action. In every society where the principles of the ICPD have been implemented, birth rates have gone down, infant survival rates have gone up, maternal mortality have declined, and the quality of life has improved.

Timothy Cline is Director of Communication at Population Connection, where he has held a variety of positions including Publications Manager. He has also served as Chief of Advocacy and Policy Research at the Johns Hopkins University Center for Communication Programs. Before joining Population Connection, Mr. Cline served as Operations Manager for the Washington Regional Alliance and was a Senior Editor at Ecomedia.

Fertility Is Not an Issue for Government

To address this issue, we must ask two interrelated questions. First, do we know that population growth is responsible for serious environmental problems—in other words, does the world have a legitimate *interest* in reducing population growth? And, second, do governments possess the rightful *authority* to engage in population control programs?

According to the UN Population Division, "Even for those environmental problems that are concentrated in countries with rapid population growth, it is not necessarily the case that population increase is the main root cause, nor that slowing population growth would make an important contribution to resolving the problem." The cause of environmental degradation does not appear to be the base number of people in a country, but how those people produce and consume goods, as well as how they are organized politically. Thus, we should seek economic, technological, and political solutions to environmental problems, not demographic ones.

Paradoxically, fertility *decline,* not fertility *growth,* is now emerging as the most serious population problem. About half of the nations on Earth now have below-replacement rate fertility. In Russia, for instance, there were one million fewer births than deaths during last year alone. This worldwide fertility decline is historically unprecedented.

Governments do not have an interest in further reducing fertility. Nor should they have the authority to do so. The very concept of population is simply an abstraction of millions of families making the most intimate decisions about their lives. Do you want to be told how many children you can have? To meet population targets, governments have deceived, bribed, and even coerced their own people. In China, widely considered the most successful population control program on earth, over 100 million women have been forced to abort unborn babies or to be sterilized. What societal or environmental problems—however severe—could ever justify such barbarity?

Douglas A. Sylva is the director of research at the Catholic Family and Human Rights Institute, a pro-life, pro-family think-tank and lobbying group working at the United Nations. He received a doctorate in political theory from Columbia University.

Weighing the Issues:
HIV/AIDS and Population Growth

Malthus predicted that under certain conditions, nature would take over and limit population growth. Some people feel that the current HIV/AIDS epidemic in Africa is nature's way of limiting population growth. Would you agree with this idea? Why or why not? What sorts of problems would you predict to occur in the surviving population after a major disease such as AIDS kills a high percentage of the population?

Severe demographic changes have social, political, and economic repercussions

In the long run, it remains to be seen whether AIDS will leave barely a blip in our global population growth or whether it is a dark harbinger of the checks on growth that Malthus predicted. The Black Plague caused a noticeable drop in world population but did little to slow our long-term growth. Beyond the demographic effects of AIDS, however, the disease is clearly having immediate social, economic, and political consequences. Everywhere in sub-Saharan Africa, AIDS is undermining the ability of developing countries to make the transition to modern technologies because it is removing many of the youngest, most productive members of society. For example, in 1999 Zambia lost 600 teachers to AIDS, while only 300 new teachers graduated to replace them. In Rwanda, more than one in three college-educated residents of the city of Kigali are infected with the virus. The loss of productive household members to AIDS causes families and communities to break down as income drops and food production declines while medical expenses and debt skyrocket.

These problems are hitting many AIDS-infected countries at a time when their governments are already experiencing what has been called *demographic fatigue*. Demographically fatigued governments face overwhelming challenges related to population growth, including educating and finding jobs for their swelled ranks of young people. With the added stress of HIV/AIDS, these governments face so many demands that they are stretched beyond their capabilities to address problems, so that the problems grow worse and citizens lose faith in their governments' abilities to help them. South Africa's government has exacerbated its HIV/AIDS crisis by distrusting and refusing to use Western medicines designed to combat the disease. In contrast, the government of Uganda is a positive role model for Africa. This is the country where

the disease first broke out, but its government has been proactive in educating its people of the dangers. Despite being in the thick of a region of Africa with high rates of infection, Uganda has managed to hold its infection rate down to only 5% of adults of childbearing age.

If nations in sub-Saharan Africa—and other regions where the disease is spreading fast, such as India and southeast Asia—do not take aggressive steps soon, and if the rest of the world does not step in to offer considerable help, many people fear that these countries may fail to advance through the entire demographic transition. Instead, their rising death rates may push birth rates back up, potentially causing these countries to fall back to the pre-industrial stage of the demographic transition model. Such an outcome would lead to greater population growth while economic and social conditions worsen; it would be a profoundly negative outcome both for human well-being and for the well-being of the nonhuman environment.

Conclusion

Today, several years after welcoming its six-billionth member, the human population is more urban, much larger, and nearly as polarized between rich and poor, as at any time in the past. These changes are affecting the environment and our ability to meet the needs of all the world's people. Approximately 90% of children born on October 12, 1999, are likely to live their lives in conditions far less healthy and prosperous than most of us in the industrialized world are accustomed to.

However, there are at least two major reasons to be encouraged. First, although global population is still growing, the rate of growth has decreased nearly everywhere and is still decreasing. Some countries have even seen the beginning of a gentle population decline. Most developed nations have passed through the demographic transition, showing that it is possible to lower death rates while also stabilizing population and creating more prosperous societies. These countries are now in a position to make another equally important transition: from consumption-oriented and growth-oriented economies to ecologically sustainable ones. Among the developing nations, many are making their way through the demographic transition, and, we can hope, will arrive at their post-industrial stage soon. Their equitable treatment by the developed nations, together with greatly improved efficiency for using limited resources, will be vital in helping them succeed.

A second reason to feel encouraged is the progress in expanding rights for women worldwide. Although there is still a long way to go, women are slowly being treated more fairly, receiving better education, obtaining more economic independence, and gaining the ability to control more of their reproductive decisions. Besides the obvious ethical progress these developments entail, they also are helping slow population growth and, some argue, are helping us use resources more sustainably.

Whether one focuses on the progress being made or on the hurdles to be overcome, it is obvious that population growth underlies many of the environmental problems we face. As you proceed through this book, and as you look around you in the world, you will notice population issues coming up again and again, seemingly attached to everything. Some say that population growth is at the root of all our environmental problems and that if we can only stop it, all our problems would be solved, or at least solvable. Others contend that the challenges are more complex. Regardless, human population cannot continue to rise forever. The question is how it will stop rising: whether through the gentle and benign process of the demographic transition, through the imposition of severe governmental intervention such as China's one-child policy, or through the miserable Malthusian checks imposed by disease and social conflict caused by overcrowding and competition for scarce resources.

In addition, true sustainability demands a further challenge—not just that we stabilize our population size fairly soon, but that we do so in time to avoid destroying the natural systems that support our economies and societies. Some success stories give us hope: China is mitigating its air pollution and Kenya its soil erosion even while their populations continue to rise, for instance. The technology portion of the IPAT equation may provide part of the answer, but only if we are creative and innovative and use technology wisely. We are indeed a special species. We are the only one to come to such dominance on Earth as to fundamentally change so much of its landscape and even its climate system. We are also the only species with the intelligence needed to turn around an increase in our own numbers before we destroy the very systems on which we depend.

REVIEW QUESTIONS

1. What is the approximate current human global population? How many people are being added to the population each day? Does the model of exponential growth apply to this increase?

2. Why has the human population continued to grow in spite of environmental limitations?

3. Contrast the views of environmental scientists with those of the libertarian writer Sheldon Richman and similar-thinking economists over whether population growth is a problem. Why does Richman think the concept of carrying capacity does not apply to human populations?

4. Explain the IPAT model and how technology helps to increase or decrease environmental impact. How does affluence affect environmental impact? Provide at least two examples.

5. What statistics do demographers use to study human populations? Which of these factors help determine the impact of human population on the environment?

6. What is the total fertility rate (TFR)? Can you explain why the replacement fertility for humans is approximately 2.1? How is Europe's TFR affecting its natural rate of population change?

7. Why have fertility rates fallen in many countries?

8. According to the demographic transition model, societies pass through stages based on their levels of industrialization. Why is the pre-industrial stage characterized by high birth and death rates and the industrial stage by falling birth and death rates? Why might this model of transition apply to some developing nations and not others?

9. See Table 7.3. What are the educational targets and dates set by the 1994 United Nations Conference on Population and Development in Cairo and its followup conference? How might these targets affect populations and the environment in developing countries?

10. How does poverty affect population growth? Why do poorer societies have higher population growth rates than wealthier societies? How does poverty affect the environment?

11. Why may affluence be worse for the environment than poverty? What is an ecological footprint?

12. Why is the number of households rising faster than population in affluent societies? How does this affect levels of consumption?

13. Review the statistics dealing with AIDS. About how many more deaths will occur between 2000–2020 in Sub-Saharan Africa due to AIDS alone? During this time, about how many deaths will occur from AIDS globally? How many African children are ex-pected to be orphans by 2010? How is this disease preventing demographic transition? What related factors are causing the spread of the disease?

14. What is *demographic fatigue*, and why is it likely to worsen the crisis in countries affected by AIDS?

DISCUSSION QUESTIONS

1. China's recent reduction in birth rates will lead to a significant change in the nation's age structure. Review Figure 7.11, which portrays the projected changes to China's age structure. You can see that the population is growing older based on the top-heavy age pyramid. What sorts of effects might this ultimately have on Chinese society? Explain your answer.

2. In 2001, approximately 1.2 billion people were so poor that they were forced to survive on less than the equivalent of one dollar per day. What effect would you expect this situation to have on the political stability of the world? Explain your answer.

3. Apply the IPAT model to the example of China provided in the chapter. How do population, affluence, technology, and ecological sensitivity affect China's environment? Given that government policies to promote economic development and stability caused both increases and decreases in population growth in China, do you think it is also possible or desirable for the Chinese government to regulate the relationship between population and its effects on the environment?

4. Examine the demographic transition theory. What major economic assumption does the theory make? Evaluate this assumption based on what you have learned from this chapter.

5. How has contraception influenced fertility rates and population dynamics in different nations? What factors determine the effectiveness of contraception as a method of population control? Use examples from this chapter to answer these questions.

6. Review the data presented in Table 7.4 for Nicaragua and Syria. The percentage of couples using birth control in Nicaragua is more than double the percentage in Syria. You might therefore expect that the number of children born per woman to be less in Nicaragua than in Syria because more people in Nicaragua use birth control. However, the number of children born per woman is the same in both countries. Propose an explanation for why this might be the case.

Media Resources *For further review go to the website* **www.envscienceplace.com** *or student CD-ROM, where you will find quizzes, flashcards, a glossary, additional interactive exercises, and links to relevant news and research sources. Also, on the website and CD-ROM is* **GRAPH IT**, *a series of interactive graphing tutorials to help you interpret graphs and plot data.*

8 Agriculture and soil formation, degradation, and conservation

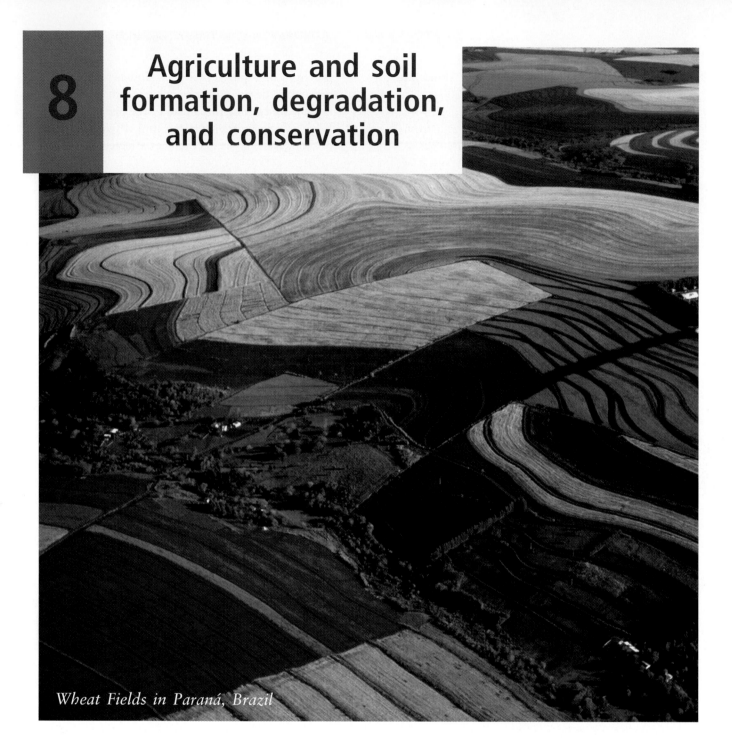

Wheat Fields in Paraná, Brazil

This chapter will help you understand:

- The importance of soils to agriculture and the impact of agriculture on soils

- A brief history of agriculture

- The fundamentals of soil science, including soil formation and the properties of soil

- The causes and consequences of soil erosion and degradation

- The history and principles of soil conservation

- U.S. and international soil conservation policies and practices

Atlantic
Ocean

Brazil

Pacific
Ocean

Paraná

Santa
Catarina

Rio Grande
do Sul

No-Till Farm in Ceara, Brazil

Central Case: No-Till Agriculture in Southern Brazil

"The nation that destroys its soil destroys itself."
— *U.S. President Franklin D. Roosevelt*

"There are two spiritual dangers in not owning a farm. One is the danger of supposing that breakfast comes from the grocery, and the other that heat comes from the furnace."
—*Aldo Leopold*

In southernmost Brazil, hundreds of thousands of people make their living farming. The warm climate and rich soils of this region's rolling highlands and coastal plain have historically made for bountiful harvests. However, repeated cycles of plowing and planting over many decades had diminished the productivity of the soil. More and more topsoil—the valuable surface layer of soil richest in organic matter and nutrients—was being eroded away by water and wind. Meanwhile, the synthetic fertilizers used to restore nutrients were polluting area waterways. Yields were falling, and by 1990 farmers were looking for help.

As a result, many of southern Brazil's farmers started turning to "no-tillage" farming, otherwise known as "zero-tillage," "no-till," or "zero-till"—or in Brazil, *plantio direto*. They abandoned the conventional practice of plowing, or tilling, the soil after harvests—a practice that farmers around the world have followed for centuries. Turning the earth with a plow aerates the soil and works weeds and old crop residue into the soil to nourish it. However, tilling also leaves the surface bare of vegetation for a period of time, during which erosion by wind and water can remove precious topsoil.

Working with agricultural scientists and government extension agents, southern Brazil's farmers started leaving crop residues on their fields after harvesting and began planting "cover crops" to keep their soil protected during periods when they weren't raising a commercial crop. When they went to plant the next crop, they merely cut a thin shallow groove into the soil surface, dropped in seeds, and covered them. They did not invert their soil as they had with plowing, and the soil stayed covered with plants or their residues at all times, reducing erosion by 90%.

With less soil eroding away, and more organic material being added to it, the soil could hold water better and was better able to support crops. The improved soil quality meant better plant growth and, over several years, greater crop production. In the state of Santa Catarina, maize yields per hectare increased by 47% between 1991 and 1999, wheat yields rose by 82%, and soybean yields by 83%, according to local farmers, extension agents, and international scientists. In the states of Paraná and Rio Grande do Sul, maize yields were up 67% over 10 years, and soybean yields were up 68%.

Besides boosting yields, the no-till farming methods reduced farmers' costs, because farmers now used less labor and less fuel. No-till agriculture spread astonishingly quickly in the region, as farmers saw their neighbors' successes and traded information in groups called "Friends of the Land" clubs that were organized on local, municipal, regional, and statewide levels. In Paraná and Rio Grande do Sul, the area being farmed with no-till methods shot up from 700,000 ha (1.7 million acres) in 1990 to 10.5 million ha (25.9 million acres) in 1999, when it involved 200,000 farmers. In Santa Catarina, where farms are generally smaller, over 100,000 farmers now apply no-till methods to 880,000 ha (2.2 million acres) of farmland. No-till farming is now spreading northward into Brazil's tropical regions, and to other parts of Latin America.

By enhancing soil conditions and reducing erosion, no-till techniques have benefited southern Brazil's society and environment as well; its air, waterways, and ecosystems are less polluted. Similar effects are being felt elsewhere in the world where no-till and reduced-tillage methods (together often called *conservation tillage*) are being applied, although the benefits and drawbacks of these approaches vary with location, soil characteristics, and type of crop. Proponents say locally based sustainable agriculture like southern Brazil's is the type of win-win solution we will need if we are to feed the world's human population while protecting the natural environment, including the soils that vitally support our production of food.

Soils: The Foundation for Feeding a Growing Human Population

The switch that many Brazilian farmers have made from conventional plowing methods to no-till agriculture is one of many ways that people have devised to increase food production in a sustainable way. Increasing food production sustainably is necessary if we are to feed the world's rising human population. Most of our food comes from agriculture, and less and less is harvested directly from the wild. We can define **agriculture** as the practice of cultivating soil, producing crops, and raising livestock for human use and consumption. We obtain much of our food from **croplands**, lands used to raise plants for human food, and **rangelands** or pastures, lands used for grazing livestock. In this chapter we focus on soils and the role of soil conservation in preserving the productivity of croplands and rangelands, a crucial effort in a time of rapid population growth.

As the human population increases, so does the amount of land and other resources devoted to agriculture, which currently covers 38% of Earth's land surface. This means it is vital to find ways to improve the *efficiency* of food production in areas that are already cultivated, because land suitable and available for farming is decreasing. Already many lands are being farmed that are unsuitable, causing great environmental damage. Mismanaged agriculture has turned grasslands into deserts; removed ecologically precious forests for minimal human gain; extracted nutrients from soils and added them to water bodies, harming both systems; allowed countless tons of fertile soil to be blown and washed away; diminished biodiversity; encouraged invasive species; and polluted soil, air, and water with toxic chemicals. It is imperative for our future and the future of our environment that we learn to farm in sustainable ways that are gentler on the land and that maintain the integrity of soil.

Healthy soil is vital for agriculture, for forestry (Chapter 16), and for the preservation of Earth's natural systems. Soil quality is determined largely by its physical, chemical, and biological content, for soil is not merely lifeless dirt. Rather, **agronomists**, the scientists who combine soil and plant science to manage agricultural systems, define **soil** as a complex plant-supporting system consisting of disintegrated rock, organic matter, air, water, nutrients, and microorganisms. Each of these components can be altered by our treatment of soil. Productive soil is a renewable resource, but if we abuse it through careless or uninformed agricultural practices, we can greatly reduce its renewability.

As population increases, soils are being degraded

As our planet gains nearly 80 million people each year, we are losing 5–7 million ha (12–17 million acres) of productive cropland annually. Throughout the world, especially in drier regions, it has gotten more difficult to

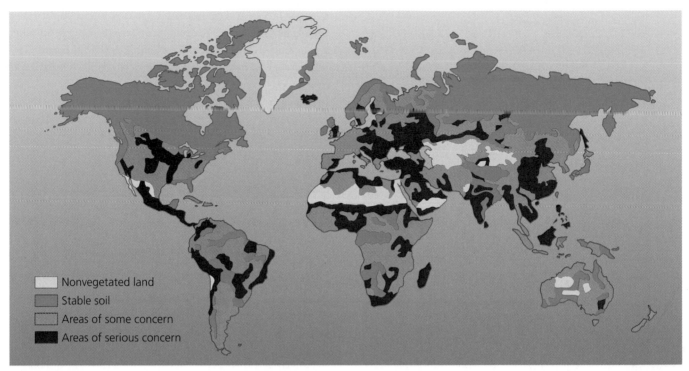

(a) World soil conditions

Legend:
- Nonvegetated land
- Stable soil
- Areas of some concern
- Areas of serious concern

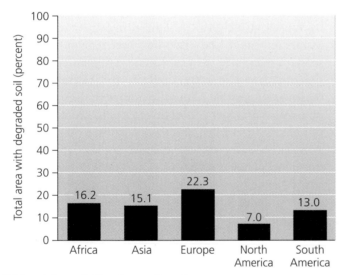

(b) Land degradation by continent

y-axis: Total area with degraded soil (percent)

Continent	Value
Africa	16.2
Asia	15.1
Europe	22.3
North America	7.0
South America	13.0

Figure 8.1 (a) Soils are becoming degraded in many areas worldwide. **(b)** Europe currently has a higher proportion of degraded land than other continents because of its long history of intensive agriculture, but degradation is rising quickly in developing countries in Africa and Asia. Go to **GRAPH IT** on the website or CD–ROM. Data from International Soil Reference and Information Centre (ISRIC) and United Nations Environment Programme (UNEP), "Human–induced soil degradation," 1996; UNEP, "Global Environmental Outlook 3," 2000.

raise crops and graze livestock as soils have become eroded and degraded (Figure 8.1). Soil degradation around the globe has resulted from roughly equal parts forest removal, cropland agriculture, and overgrazing of livestock (Figure 8.2).

In this chapter we focus particularly on cropland agriculture, since soil degradation has direct and immediate impacts on agricultural production. It is estimated that such degradation over the past 50 years has reduced rates of global grain production by 13% on cropland and 4% on rangeland. World food production had long been growing faster than global population, but in 1983 the amount of grain produced per capita globally leveled off and began to decline (Figure 8.3). When our six-billionth

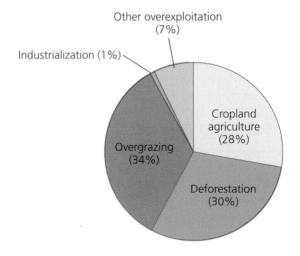

Pie chart labels:
- Other overexploitation (7%)
- Industrialization (1%)
- Cropland agriculture (28%)
- Overgrazing (34%)
- Deforestation (30%)

Figure 8.2 The great majority of the world's soil degradation results from cropland agriculture, overgrazing by livestock, and deforestation. Data from M.K. Wali et al., "Assessing terrestrial ecosystem sustainability: usefulness of regional carbon and nitrogen models," *Nature and Resources, 1999, 3.*

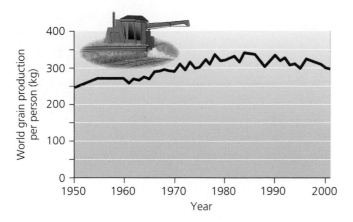

Figure 8.3 The amount of grain produced per person worldwide was rising but is now falling. Per capita yields peaked in 1984 at 343 kg (756 lbs), and have since declined 12% to 302 kg (666 lbs) in 2000. Total grain production may possibly also now have peaked; yields have fallen from 1.880 billion tons in 1997 to 1.836 billion tons in 2000. Data from *U.S. Department of Agriculture, 2001.*

global citizen was born in 1999, 841 million people were already chronically malnourished. What will it take to feed the nine billion people who will likely populate the planet by the middle of the 21st century?

Human agriculture began to appear around 10,000 years ago

Agriculture is a relatively new approach to meeting our nutritional needs; on the scale of human history, there was no such thing as a farm until very recently. During most of our species' 100,000-year existence, we were hunter-gatherers, depending solely on wild plants and animals for our nutritional needs. Then about 10,000 years ago, humans began to raise plants from seed. The earliest widely accepted archaeological evidence for plant domestication is from the Middle East about 8,500 B.C., and the earliest evidence for animal domestication is from that region about 500 years later. Crop remains have been dated using radiocarbon dating (Chapter 4) and similar methods.

But how and why did agriculture begin? The most plausible hypothesis is that it began as hunter-gatherers collected and brought back to their encampments wild fruits, grains, and nuts. Some fell accidentally to the ground, some were thrown away, and others were eaten and survived passage through the digestive system. Thus, seeds ended up in the nutrient-rich repositories of waste that archaeologists often find near human encampments.

The plants that grew from refuse piles near inhabited areas likely would have produced fruits that were on average larger and tastier than those in the wild, because they sprang from seeds that were selected by people because they were especially sizeable and delicious. The plants that sprouted near settlements thus would have inherited the very qualities that appealed to the hunter-gatherers who had collected their progenitors. As these plants pollinated others nearby that shared their characteristics, they gave rise to subsequent generations of plants with large and flavorful fruits. As time passed, people must have realized that the accidental selection occurring in their waste piles could also occur through conscious effort. In this hypothesized scenario, our ancestors began intentionally planting seeds from the plants whose produce was most desirable. This is, of course, artificial selection—the human-induced version of natural selection—at work (Chapter 4). This practice of selective breeding continues to the present day and has produced our modern-day crops, all of which are artificially selected versions of natural plants.

Once our ancestors learned to cultivate crops, they began to settle in more permanent camps and villages often near water sources. Agriculture and a sedentary lifestyle likely reinforced one another in a positive feedback cycle (Chapter 6); the need to harvest crops kept people sedentary, and once they were sedentary, it made sense to maximize the crops that were planted. Population increase also resulted from and further spurred these developments. Thus, agriculture enabled the development of a modern lifestyle revolving around densely populated urban centers. Furthermore, the ability to grow excess farm produce enabled some people to leave farming and live off the food that others produced, leading to the development of professional specialties, commerce, technology, social stratification, and politically powerful elites. For better or worse, the advent of agriculture eventually brought us the civilization we have today.

Archaeological and paleoecological evidence suggests that agriculture was invented independently multiple times by different cultures in different areas of the world. Syntheses of known data suggest that agriculture originated independently in at least five different places, and possibly ten or more (Figure 8.4). In the "Fertile Crescent" region of the Middle East, wheat and barley originated, as did rye, peas, lentils, onions, garlic, carrots, grapes, and other food plants familiar to us today. The people of this region also domesticated goats and sheep. Domestication began in China as early as 7,500 B.C., leading eventually to the rice, millet, and pigs we know today. Agriculture in Africa (coffee, yams, sorghum, and more) and the Americas (corn, beans, squash, potatoes,

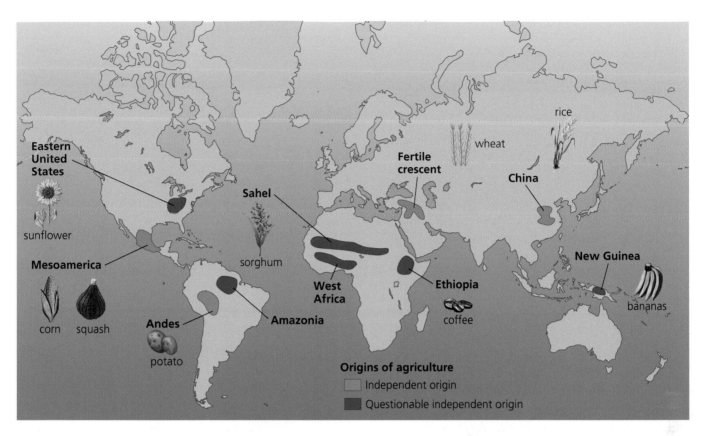

Figure 8.4 Agriculture appears to have originated independently in multiple locations throughout the world, as different cultures domesticated certain plants and animals from wild species living in their environments. In this depiction summarizing conclusions from diverse sources of research on evidence for early agriculture, areas where people are thought to have independently invented agriculture are colored in green. (China may represent two independent inventions.) Those areas colored in blue represent regions where people may have invented agriculture independently, or may have obtained the idea from cultures of other regions. A few of the many crop plants domesticated in each region are shown. Data from syntheses in Diamond, J. *Guns, Germs, and Steel;* A. Goudie, *The Human Impact, 5th ed., 2000.*

llamas, and more) developed later in several areas, 4,500–7,000 years ago.

For most of the thousands of years that humans have practiced agriculture, human and animal muscle power, along with hand tools and simple machines, performed the work of cultivating, harvesting, storing, and distributing crops (Figure 8.5). This biologically powered agriculture is known as **traditional agriculture.** Scholars have divided traditional agriculture into two types. In the oldest form, known as **subsistence agriculture,** the members of a farming family produce only enough food for themselves and do not make use of large-scale irrigation, fertilizer, or large teams of laboring animals. **Intensive traditional agriculture,** in contrast, sometimes uses draft animals and also employs significant quantities of irrigation water and fertilizer, but stops short of using fossil fuels. This type of agriculture aims to produce not only enough food for the farming family, but also excess food to sell in the market.

Industrialized agriculture is newer still

The industrial revolution introduced large-scale fossil fuel combustion and mechanization to agriculture just as it did to industry, enabling farmers to replace horses and oxen with faster and more powerful means of cultivating, harvesting, transporting, and processing crops. Other advances facilitated irrigation and fertilizing, while the invention of chemical herbicides and pesticides reduced competition from weeds and herbivory by insects. To be efficient, however, this **industrialized agriculture** demands vast fields planted with a single type of crop. The uniform planting of a single crop is termed **monoculture,** and is distinct from the **polyculture** approach of much traditional agriculture, such as Native American farming systems that mixed maize, beans, squash, and peppers in the same fields. Today, about 25% of the world's croplands are cultivated with some form of industrialized agriculture.

Figure 8.5 Hunting and gathering was the predominant human lifestyle until the onset of agriculture and sedentary living, which centered around farms, villages, and cities, beginning close to 10,000 years ago. Over the millennia, hunter-gatherer cultures gradually were replaced by those practicing traditional agriculture. Only within the past century has industrialized agriculture spread, replacing much traditional agriculture.

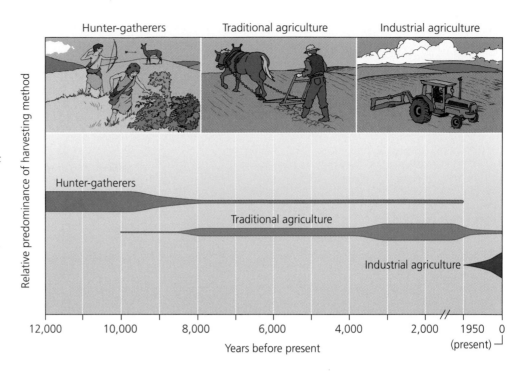

Industrialized agriculture is widespread today because not everyone is a farmer. In developed countries, relatively few people produce food for many, so these societies need their few farmers to produce massive amounts of produce. In developing nations, a great number of people still work the land as farmers, but these nations also are urbanizing as young people move from farms to cities. Increasingly, developing nations are taking up the industrialized agricultural model.

The Green Revolution applied technology to boost crop yields

Despite historical advances in agricultural production, some critics—from Thomas Malthus in the 18th century to Paul Ehrlich in our own time (Chapter 1)—have argued that our growing population would overwhelm our best efforts to keep pace with food production. In 1968, for instance, Ehrlich predicted that global agricultural systems would fail utterly by 1980, leading to widespread starvation. Although malnutrition and even starvation are constant specters in some parts of the world, the global catastrophe Ehrlich predicted has not yet occurred. One reason is that Ehrlich underestimated the longevity and power of an agricultural phenomenon known as the Green Revolution.

The **Green Revolution** refers to an intensification of the industrialization of agriculture, a change in agricultural practices that dramatically increased the crops

produced per acre of farmland between 1950 and the start of the 21st century. New practices have involved:

1. Devoting large areas to identical crops specially bred for high yields and rapid growth
2. Heavy use of fertilizers, pesticides, and irrigation water
3. Sowing and harvesting on the same piece of land more than once per year or per season

By increasing food production on existing farmlands, the Green Revolution has helped us avoid, or at least postpone, Ehrlich's apocalypse.

Despite its successes, however, the Green Revolution is exacting a high price. The intensive cultivation of farmland is creating new problems and exacerbating old ones. Many of these pertain to the health of the soil, which is the very foundation of our terrestrial food supply. We will revisit the Green Revolution in more detail in the next chapter, but now let's examine the nature of this vital yet little-understood resource, soil.

Soil as a System

Because most of us spend little time gardening, or even keeping potted plants alive, we generally overlook the startling complexity of soils. English-speaking people

Figure 8.6 Soil is not merely lifeless dirt; it is a complex mixture of organic and non-organic components and is full of living organisms whose actions help keep it fertile. In fact, entire ecosystems exist in the soil. Most soil organisms, from bacteria to fungi to insects to earthworms, decompose organic matter. Many, such as earthworms, also help to aerate the soil.

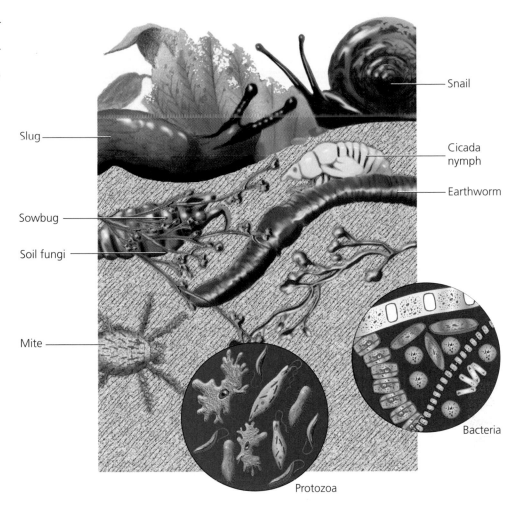

Slug

Sowbug

Soil fungi

Mite

Snail

Cicada nymph

Earthworm

Bacteria

Protozoa

tend to equate the word *soil* with the word *dirt,* which connotes something useless or undesirable. Soil, however, is much more. It is not merely unconsolidated material derived from rock; such material is termed **regolith.** In contrast, soil contains a large biotic component, is molded by life, and is capable of supporting plant growth (Figure 8.6).

By volume, most soil consists very roughly of half mineral matter, half pore space variably taken up by air or water, and up to 5% organic matter. The organic matter in soil includes both living and dead microorganisms and decaying material derived from plants and animals. Most of us tend to think of soil as an inert, lifeless substance, but the opposite is true. A single gram of soil can contain 100 million bacteria, 500,000 fungi, 100,000 algae, and 50,000 protozoa. Soil also contains earthworms, insects, mites, millipedes, centipedes, nematodes, sow bugs, and even burrowing mammals, amphibians, and reptiles. The composition and quality of a region's soil can have as much influence on the region's ecosystems as do the climate, latitude, and elevation. In fact, because soil is composed of living and nonliving components that interact in complex ways, soil itself meets the definition of an ecosystem (Chapter 6).

Soil formation is slow and complex

The formation of soil also plays a key role in the ecological process of succession (Chapter 5). Terrestrial primary succession begins when some of the lithosphere's parent material is exposed to the effects of the atmosphere, hydrosphere, and biosphere. **Parent material** is the base geological material in a particular location. It can include lava or volcanic ash; sediment and chunks of rock deposited by glaciers; wind-blown dunes; sediments deposited from rivers or in the ocean; or **bedrock,** the continuous mass of solid rock that makes up Earth's crust. Soil formation is an early step in the process of succession that follows the pioneering of bare mineral substrate by lichens and plants. The processes that are most directly responsible for soil formation are weathering, erosion, and the deposition and decomposition of organic matter.

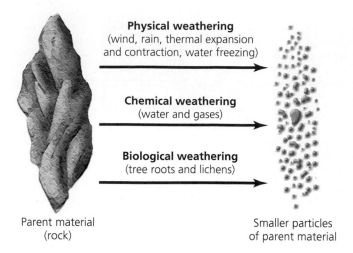

Physical weathering
(wind, rain, thermal expansion
and contraction, water freezing)

Chemical weathering
(water and gases)

Biological weathering
(tree roots and lichens)

Parent material
(rock)

Smaller particles
of parent material

Figure 8.7 The weathering of parent material is the first step in the formation of soil. Rock and minerals may be broken down into finer particles physically, chemically, or biologically.

Weathering is the term scientists use to describe the physical, chemical, and biological processes that break down rocks and minerals, turning large particles into smaller particles. The three types of weathering are shown in Figure 8.7. **Physical** or **mechanical weathering** breaks rocks down into smaller particles without triggering a chemical change in the parent material. Wind and rain are two main forces of physical weathering. Their action is aided by daily and seasonal temperature variation that causes the thermal expansion and contraction of parent material. Areas with great temperature extremes may see high rates of physical weathering. Water freezing (and thus expanding) in the cracks of rock is also a major cause of physical weathering.

Chemical weathering results from the chemical interaction of water, atmospheric gases, and other substances with parent material. Climate, which plays a role in physical weathering, also affects chemical weathering. For example, warm rainy conditions usually accelerate chemical weathering and may also trigger positive feedback mechanisms that further increase the rates of both chemical and physical weathering.

Biological weathering is the breakdown of parent material into smaller particles through the activities of living things. Organisms can weather rocks by either physical or chemical means and can add to ongoing physical or chemical processes. For example, the lichens we described as the initiators of primary terrestrial succession produce acid, which weathers rock. A tree may accelerate weathering through the physical action of its roots as they grow and rub against rock. It may also accelerate weathering through the chemical action resulting

from the decomposition of its leaves and branches or from chemicals it releases from its roots as it grows.

Weathering produces fine particles and thus is the first step toward the creation of soil. It is only one of many processes that may be involved, however. Erosion, the process of moving soil from one area to another, may contribute to the formation of soil in one locality even as it depletes topsoil from another. Erosion is particularly prevalent when soil has been denuded of vegetation, leaving the surface exposed to water and wind that may wash or blow soil away. On the timescale of human lifetimes and for the natural systems on which we depend, erosion is generally perceived as a destructive process, reducing the amount of life that a given area of land can support. Over longer time periods, however, material eroded from one area is deposited in others, where it contributes to the building of new soil.

Biological activity also contributes to soil formation through the deposition, decomposition, and accumulation of organic matter. As plants, animals, and microbes die or leave their waste, this material is incorporated into the substrate, mixing with minerals. The deciduous trees of temperate forests, for example, drop their leaves each winter, making leaf litter available to the detritivores and decomposers that can break it down and incorporate its nutrients into the soil. In decomposition, complex organic molecules are broken down into simpler ones, including those that plants can take up through their roots. One process involved is mineralization, in which minerals are released from organic matter decomposition. Another is humification, the generation of humus from organic material. **Humus** is the result of the partial decomposition of organic matter; it is a dark, spongy, crumbly mass of undifferentiated material made up of complex organic compounds. High humus content can allow soils to hold moisture and become more productive for plant life.

Soil formation is influenced by five main factors

Weathering, erosion, the accumulation and transformation of organic matter, and various other processes contribute to soil formation. All these processes are influenced by outside factors that affect the rate or nature of the processes. Soil scientists cite five primary factors that influence the formation of soil (Table 8.1). These factors were proposed in the 19th century by Russian soil scientist Vasily Dokuchaev and elaborated on in a 1941 book by Swiss-born American soil scientist Hans Jenny. While these five factors—climate, organisms, relief, parent

Table 8.1 Five Factors That Influence Soil Formation

Factor	Effects
Climate	The two key elements of climate—temperature and moisture—each play roles at site-specific and regionwide scales. Warmer temperatures speed up rates of weathering, decomposition, and biological growth. Because chemical reactions take place more quickly at warmer temperatures, many of the processes that contribute to soil formation are sped up. Moisture also is necessary for many biological processes and can speed physical and chemical weathering, so in general soils develop more quickly in wetter climates.
Organisms	As soil develops, earthworms and other burrowing animals mix soil, aerate soil, add organic matter, and facilitate microbial decomposition. The type of vegetation that grows in a developing soil (and adds its organic matter to it) affects the composition and structure of the soil.
Relief	Features of topographical relief such as hills and valleys affect exposure to sun, wind, and water, and influence where and how soil moves. Steeper slopes, for instance, result in greater erosion, less accumulation of organic matter, reduced amounts of leaching, and less differentiation of soil layers.
Parent Material	The chemical and physical attributes of the parent material exert a major influence on the attributes of the soil that results. Soils that originate from riverine sediments, for example, will be very different in chemistry, texture, and other properties than those that originate from volcanic rock.
Time	Soil formation is slow and takes decades, centuries, or millennia. Over time, factors influencing soil formation may change, so that the soil we see today may be the result of multiple sets of factors.

material, and time—are thought to be most important in influencing how soil is formed, many other factors may also come into play. The process of soil formation is slow and complex.

Weighing the Issues: Earth's Soil Resources

It can take 500–1,000 years to produce 1 inch of topsoil. Is soil a renewable resource? How should soil's long renewal time influence its management? What types of practices encourage the formation of new topsoil?

A soil "profile" consists of distinct layers known as "horizons"

The processes of soil formation lead to characteristic aspects of soil morphology. Once physical, chemical, and biological weathering have produced a layer of smaller particles between the parent material and the atmosphere, wind and water begin to move and sort them. Organisms migrate in and move them as well. Eventually, distinct zones and patterns appear. Some can be recognized by differences in color or texture; others require closer analysis. Each layer of soil is known as a **horizon,** and the cross-section as a whole, from the surface to the bedrock, is known as a **soil profile.** Soils from different locations differ due to factors influencing their formation, but soil from any given location can nonetheless be divided into recognizable horizons. The simplest way to

categorize soil horizons is to recognize A, B, and C horizons corresponding to topsoil, subsoil, and parent material. However, it is usually useful to subdivide the layers more finely, by their characteristics and the processes that take place within them. While soil researchers may divide soil profiles into many fine layers, for our purposes we will discuss six major horizons found in a typical soil profile. These are known as the O, A, E, B, C, and R horizons (Figure 8.8). Generally, the degree of weathering and the concentration of organic matter decrease as one moves downward in the soil profile.

Many soil profiles contain an uppermost layer consisting mostly of organic matter, such as decomposing branches, leaves, crop residue, and animal waste. This layer is designated the **O horizon** (O for *organic*) or litter layer. Just below the O horizon in a typical soil profile lies the **A horizon,** which consists of mostly inorganic mineral components such as weathered substrate, with some organic matter and humus from above mixed in. The A horizon is often referred to as **topsoil,** that portion of the soil that is most nutritive for plants and is thus of the most direct importance to ecosystems and to agriculture. Topsoil takes its loose texture and dark coloration from its humus content. The O and A horizons are home to most of the trillions upon trillions of organisms that give life to soil; from earthworms and insects to fungi and bacteria, these horizons harbor a vast array of organisms essential to the productivity of our soils.

Beneath the A horizon lies the **E horizon,** also known as the *eluviation* horizon. *Eluviation* refers to loss, and the E horizon is characterized by the loss of certain minerals

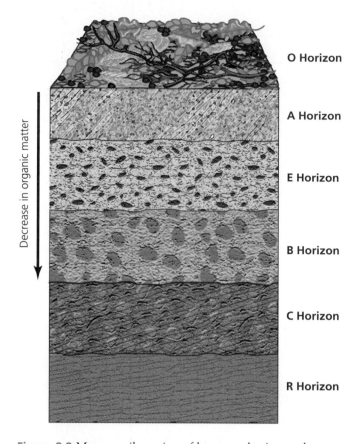

Decrease in organic matter

O Horizon

A Horizon

E Horizon

B Horizon

C Horizon

R Horizon

Figure 8.8 Mature soil consists of layers, or horizons, that have different compositions and characteristics. The number and depth of these horizons varies from place to place and from soil type to soil type, producing different soil profiles. In general organic matter and the degree of weathering decrease as one moves downward in a soil profile. The O horizon consists mostly of organic matter deposited by organisms, while the A horizon, or topsoil, consists of some organic material mixed with mineral components. Minerals tend to leach out of the E horizon down into the B horizon. The C horizon consists largely of parent material, which may overlie an R horizon of bedrock.

through leaching. **Leaching** is the process whereby solid materials such as minerals are dissolved in a liquid and transported to another location. Generally in soils the solvent is water and leaching carries minerals downward in the soil profile. Soil that undergoes leaching is a bit like coffee grounds in a drip filter; when it rains, water infiltrates the soil (like it infiltrates coffee grounds), dissolves some of its components, and carries them downward into the deeper horizons. Iron, aluminum, and silicate clay are examples of minerals that are commonly leached from the E horizon. Some soils are more prone to leaching of minerals than others, and this measure is important for at least two reasons. If minerals are leached from soils too rapidly, plants may be deprived of

the nutrients they require to grow properly. In addition, minerals that leach rapidly from soils may be carried into groundwater and pose human health threats there.

The minerals that leach out of the E horizon are carried down into the layer beneath it, the **B horizon**, or subsoil. This horizon, also called illuviation, collects and accumulates the minerals deposited from above. Thus while the E horizon is sometimes called the "zone of leaching," the B horizon is called the "zone of accumulation" or "zone of deposition." It contains a greater concentration of minerals, and also of organic acids leached from above, than does the E horizon.

The **C horizon**, located below the B horizon, contains rock particles that are larger and less-weathered than the layers above. It consists of parent material that has been unaltered or only slightly altered by the processes of soil formation. The C horizon sits directly above the **R horizon**, which is also known as *bedrock*.

Soil can be characterized by its color, texture, structure, and pH

The six horizons presented above depict an idealized model of a "typical" soil. Real soils display great variety, and scientists have developed a wide array of categories for classifying them. U.S. soil scientists have classified soils into 11 major groups, based largely on the processes thought to lead to their formation. Within these 11 "orders" there are dozens of "suborders," hundreds of "great groups," and thousands of soils belonging to lower categories, all arranged in a hierarchical system. A wide variety of factors are used to classify soils, including properties such as color, texture, structure, and pH.

Soil color The distinctive colors of soil in certain parts of the world (Figure 8.9) are indicators of soil composition and sometimes soil fertility. Rust-red soils result from the presence of oxidized iron, for example. Black or dark brown soils are usually rich in organic matter, whereas pale gray to white color often indicates leaching or low organic content. This color variation occurs among soil horizons in any given location and also among the topsoil of different geographical locations. Long before modern analytical tests of soil content were developed, the color of topsoil provided farmers and ranchers with information about a region's potential to support crops and provide forage for livestock.

Soil texture Soil texture is determined by the size of particles and is the basis on which the United States Department of Agriculture (USDA) assigns soils to one

Figure 8.9 The color of soil may vary drastically from one location to another. The soil's composition affects its color. For instance, soils high in organic matter tend to be dark brown or black.

Clay

Peat

Sand

Chalk

Silt

of three general categories (Figure 8.10). **Clay** consists of particles less than 0.002 mm in diameter, **silt** of particles 0.002–0.05 mm, and **sand** of particles 0.05–2.0 mm. Sand grains, as any beachgoer knows, are large enough to see individually, and do not adhere to one another. Clay particles, in contrast, readily adhere to one another and give clay a sticky feeling when moist. Soil with a relatively even mixture of all three particle sizes is known as **loam.**

Soil texture is important to plant growth for several reasons. Soil texture influences **soil porosity,** which is a measure of the size of spaces between soil particles. In general, the finer the particles, the smaller the spaces between them. The smaller the spaces, the harder it is for water and air to travel through the soil, slowing infiltration and reducing the amount of oxygen available to biotic soil components. Conversely, soils that consist of larger particles allow water to pass through them (and beyond the reach of plants' roots) too quickly. Thus, crops planted in sandy soils require frequent and extensive irrigation. For this reason, silty soils with medium-sized pores or loamy soils with mixtures of pore sizes are generally best for plant growth and crop agriculture. In addition, texture influences the "work-

ability" of a soil; that is, the relative ease or difficulty of cultivation. A soil with a high clay content can be sticky and difficult to manipulate when wet and turns hard and brittle when dry.

Soil structure Soil structure is a measure of the arrangement of sand, silt, or clay particles into clumps or aggregates. Some degree of structure encourages soil productivity, and biological activity in the soil helps promote this structure. However, soil aggregates that are too large can discourage plant root establishment if the soil particles are compacted too tightly together. Repeated plowing can cause soil to become compacted and less capable of absorbing water. Thus when farmers continually plow the same field at the same depth, they may end up forming **plowpan,** a hard layer that resists both the infiltration of water and the penetration of roots and can make cultivation more difficult.

Soil pH The degree of acidity or alkalinity (Chapter 4) is another characteristic that differentiates soils and influences their ability to support plant growth. Plants can die in soils that are extremely acidic or alkaline, but even more moderate extremes can influence the availability of

Figure 8.10 The texture of soil depends on its particular mixture of particle sizes. Using a triangular diagram such as this, scientists classify soil texture according to the relative proportions of sand, silt, and clay. After measuring the percentage of each type of particle size in a soil sample, a scientist can trace the appropriate white lines extending inward from each side of the triangle to determine what type of soil texture that particular combination of values creates. Loam and silty soils are generally the best for plant growth.

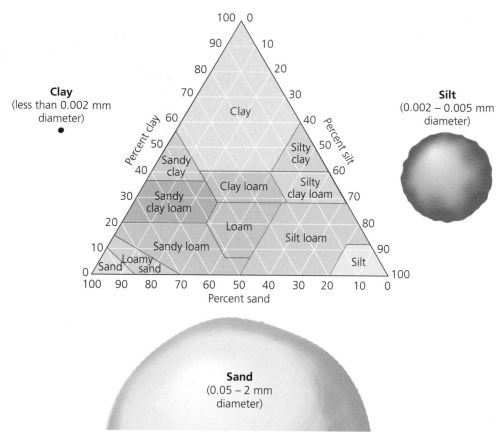

nutrients for plants' roots. During the process of leaching, for instance, acids from organic matter remove nutrient cations from the sites of exchange between plant roots and soil particles, and water carries the cations deeper, making the uppermost layers of the soil more acidic.

Regional differences in soil traits can affect agriculture

The characteristics of soil and soil profiles can vary from place to place. One example that bears on agriculture is the difference between soils of tropical rainforests and those of temperate grasslands. Although rainforest ecosystems appear to be rich and productive, most of their nutrients are tied up in plant tissues and not in the soil. The soil in an area of Amazonian rainforest in northern Brazil is in fact much less productive than the soil in a plot of grassland in Kansas. To understand how this can be, consider the main differences between the two regions: temperature and rainfall. The enormous amount of rain that falls in the Amazon readily leaches minerals and nutrients out of the topsoil and E horizon. Those that are not captured by plants are taken quickly down to the B horizon and the water table, out of reach of most plants' roots. High temperatures speed the

decomposition of leaf litter, but amounts of humus remain small and the topsoil layer remains thin. Thus when forest is cleared for farming, cultivation quickly depletes the soil's fertility. This is why the traditional form of agriculture in tropical forested areas is swidden agriculture, in which a plot is farmed for one to a few years, and the farmer then moves on to clear another plot, leaving the first to grow back to forest. This method may work well at low population densities, but with today's high human populations, soils may not be allowed enough time to regenerate, and intensive agriculture has ruined the soils and forests of many areas.

In temperate grassland areas like the Kansas prairie, in contrast, rainfall is low enough that leaching is reduced and nutrients remain high in the soil profile, within reach of plants' roots. Plants take up nutrients, then return them to the topsoil when they die, and the soil's fertility is maintained in this cycle. The thick, rich topsoil of temperate grasslands can be farmed repeatedly with minimal loss of fertility if proper farming techniques are used. However, growing crops and harvesting them without returning much organic matter to the soil gradually depletes soil of organic material, and leaving soil exposed to the elements increases the outright loss of topsoil through erosion. It is such consequences which farmers in the temperate grassland areas of southern Brazil have sought to forestall through the use of conservation tillage, and by which farmers the world over are constantly challenged.

Soil Degradation: Problems and Solutions

Scientists' studies of soil and the practical experience of farmers have shown that the most desirable soil for agriculture is a loamy mixture with a pH close to neutral that is workable and capable of holding nutrients. Many soils deviate from this ideal naturally and thus prevent land from being arable or limit the productivity of arable land. Increasingly, however, limits to productivity are being set by human impact, as many once-excellent soils have been degraded by our actions. The most common problems affecting soil productivity include erosion, desertification, salinization, waterlogging, nutrient depletion, structural breakdown, and pollution.

Erosion is the movement of soil from one place to another

Erosion is the removal of material from one place and its transport toward another by the action of wind or water. **Deposition** is the arrival of eroded material at another location. Erosion is a problem when it removes fertile topsoil, whereas deposition is a problem when it deposits materials in inconvenient places, clogging streams, lakes, or reservoirs. Although they are frequently vilified as threats to agriculture, erosion and deposition are natural processes that actually helped create many of the most productive soils throughout the world. Water tends to deposit large amounts of eroded sediment in river valleys and deltas, producing rich and productive soils. This is why many floodplains are excellent for farming and why flood-control measures can decrease the productivity of agriculture in the long run.

While erosion is a natural part of the soil formation process, it can often become a problem for ecosystems and agriculture locally when and where it occurs. This is because it nearly always takes place at much faster rates than the (extremely slow) rates of soil formation, so that soil is lost much faster than it is formed. Furthermore, erosion takes place on the land surface and thus tends to remove topsoil, the most valuable soil

layer for living things. Humans have increased the vulnerability of fertile lands to erosion through three widespread practices:

1. Overcultivating fields with excessive plowing or poor planning leads to the erosion of croplands.
2. Overgrazing rangelands with more livestock than the land can support degrades grassland communities and thus leads to erosion.
3. Deforestation is also a major cause of erosion, whether trees are cut for timber or to open areas for farming. Harvesting trees in large clear-cuts or on steep slopes is especially likely to lead to erosion that can seriously degrade both the land and the waterways into which soil washes.

Erosion can be gradual and hard to detect, even for a farmer who knows his or her land. For example, an erosion rate of 12 tons/ha (5 tons/acre) removes only a penny's thickness of soil. In many parts of the world, scientists, farmers, and extension agents are measuring erosion rates in hopes of catching areas in danger of serious degradation before they are too badly damaged (see The Science behind the Story).

Soil may erode by various mechanisms

Healthy and well-developed grasslands, forests, and other plant communities can protect soils from wind and water erosion. Vegetation breaks the wind and slows water movement, while plant roots hold soils in place and take up water. Removing long-established plant communities will, in almost every instance, accelerate erosion. When vegetation is removed or disturbed, several types of erosion can occur. These include wind erosion and four principal kinds of water erosion: splash, sheet, gully, and rill erosion (Figure 8.11).

Splash erosion occurs when rain striking the soil surface breaks aggregates into smaller sizes. Soil particles are released in the process and fill in gaps between the remaining clumps, decreasing a soil's ability to absorb water. In **sheet erosion,** or **overland flow** (Figure 8.11b), surface water—whether from precipitation or irrigation—flows downhill, washing topsoil away in relatively uniform layers. **Rill erosion** (Figure 8.11b) takes place when surface water runs along small contours on the surface of the topsoil, gradually deepening and widening the contours into rills, or small channels. Rills can merge to form larger and larger channels and eventually gullies. **Gully erosion** (Figure 8.11c) causes the most dramatic and visible changes in the landscape.

(a) Splash erosion

(c) Rill erosion

(d) Gully erosion

(b) Sheet erosion

Figure 8.11 The erosion of soil by water can be classified by type and scale into at least four categories. Splash erosion (**a**) occurs as raindrops strike soil with enough force to dislodge small amounts of soil. Sheet erosion (**b**) results when thin layers of water traverse broad expanses of sloping land. Rill erosion (**c**) leaves small pathways along the surface where water has carried topsoil away. Gully erosion (**d**) cuts deep into soil, leaving large gullies that can expand as erosion proceeds.

Research indicates that rill erosion has the greatest potential to move topsoil, followed by sheet erosion and splash erosion, respectively. All types of water erosion are more likely to occur where slopes are steeper, but the tendency is especially pronounced in the case of gully erosion. In general, steeper slopes, greater precipitation intensities, and sparser vegetative cover all lead to greater potential for water erosion. Wind erosion is more difficult to quantify, and its ability to decrease soil productivity varies widely according to a number of factors.

One study conducted in the early 1990s determined that at erosion rates typical for the United States, U.S. croplands would lose about 2.5 cm (1 in) of topsoil every 15–30 years. According to the study, this amount of loss would reduce corn yields by 4.7–8.7% and

Figure 8.12 Industrial cropland agriculture promoted by the Soviet Union (as seen on this Soviet collective farm) made Kazakhstan into Asia's breadbasket. However, it also resulted in widespread erosion and soil degradation, because northern Kazakhstan's soils were better suited for rangeland agriculture.

wheat yields by 2.2–9.5%. More severe erosion in Indiana has been shown to cause reductions of 17% for corn and 30% for soybeans.

Soil erosion is a massive problem globally

Erosion has become a major problem in many areas of the world, including Australia, sub-Saharan Africa, central Asia, India, the Middle East, and parts of South America, Central America, Europe, and the United States. In total, more than 19 billion ha (47 billion acres) of the world's croplands suffer from erosion and other forms of soil degradation stemming from human activities. Between 1957 and 1990, China lost as much arable farmland as exists in all of Denmark, France, Germany, and the Netherlands combined. For Africa, projections indicate that soil degradation over the next 40 years could reduce crop yields by half. Couple these declines in soil quality and crop yields with the rapid population growth occurring in many of these areas, and you will begin to see why some people describe the future of agriculture as a crisis situation.

Let's take a closer look at just one country, Kazakhstan, the largest nation in central Asia (Figure 8.12). The temperate grassland in Kazakhstan has made for excellent rangeland, which is why people in the northern region of the country traditionally lived as nomadic herders, moving from place to place across the landscape with their livestock. However, when the Soviet Union brought Kazakhstan under its control, it introduced industrial cropland agriculture on a massive scale.

It turned Kazakhstan into the region's breadbasket but degraded the land severely in the process. Statistics from 1993 classify 26 million ha (64 million acres) as severely eroded by wind and another 72 million ha (178 million acres) as at-risk from wind erosion. An additional 413,000–465,000 ha (1.0–1.1 million acres) were threatened by water erosion. In 2001, the U.N. Environmental Programme (UNEP) reported that humus content of the soil in northern Kazakhstan had decreased by 5–20%. Since the dissolution of the Soviet Union and Kazakhstan's independence, many Soviet collective farms have been abandoned, such that today fully half the former cropland lies fallow. Unfortunately, this has contributed to massive outbreaks of locusts that in recent years have greatly damaged the remaining grain crops. The locust outbreaks suggest that the overextension of industrial agriculture can have indirect and unintended consequences even after agriculture has been reduced in scale.

Arid land may lose productivity in the process of desertification

Much of the world's population lives and farms in arid environments, where **desertification** is a source of concern. This term describes a loss of more than 10% productivity due to erosion, soil compaction, forest removal, overgrazing, drought, salinization, climate change, depletion of water sources, and an array of other factors. Severe desertification can result in the actual expansion of desert areas or creation of new ones in areas that once supported fertile land. As we have noted, this process has occurred in many areas of the Middle East that have been inhabited for long periods of time. To appreciate the cumulative impact of centuries of traditional agriculture, we need only look at the present desertified state of that portion of the Middle East where agriculture originated, nicknamed the "Fertile Crescent." These arid lands—in present-day Iraq, Syria, Turkey, Lebanon, and Israel—are not so fertile anymore.

Arid and semiarid lands are particularly prone to desertification because their precipitation is too meager to meet the demand for water to support the needs of growing human populations. According to UNEP, 40% of Earth's land surface can be classified as drylands, arid areas especially subject to erosion and other forms of degradation. The breakdown of soil quality in these areas has endangered the food supply or well-being of more than 1 billion people around the world. Of the affected lands, most degradation comes from wind and water erosion (Figure 8.13).

The Science behind the Story

Measuring Erosion

Can a row of scrubby grass help stop erosion? With the help of soil study techniques that range from simple measuring pins to complex radiation detectors, federal agricultural scientist Jerry Ritchie was able to find out.

Grass hedges are widely used, especially in the tropics, to trap eroding soil by slowing runoff from rain. But Ritchie, a researcher with the U.S. Department of Agriculture, wanted to measure how well they actually work. He went beyond the visible signs of soil loss, using methods and technology that document soil erosion and calculate just how much soil has moved.

In the 1990s, Ritchie and his team of researchers began by making physical measurements of erosion around hedges planted near a set of gullies in Maryland. Such calculations rely on cheap, low-technology tools developed decades ago. In the 1960s and 1970s, scientists working for the United Nations Food and Agriculture

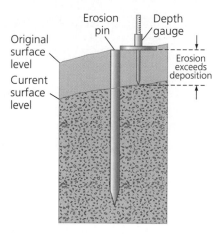

The degree to which an erosion pin is exposed indicates the extent of erosion.

Organization found they could quantify soil loss simply by shoving spikes in the ground, digging big holes nearby, and then watching the evidence literally pile up.

The spikes used in such research, known as erosion pins, can be made from almost anything, including bamboo stakes or pieces of plastic pipe. The pins, each cut to a uniform length, are driven into the soil until their tops are level

with the ground's surface. Over time, if the area is eroding, the soil surface will recede, and more and more of the erosion pins will be exposed. By using a large number of pins over a wide area and multiplying the average change in soil levels shown on the pins by the area of the plot being studied, scientists can determine an approximate erosion rate for that spot.

At the same time, researchers often dig deep and wide holes, called catchpits, nearby. The pits are lined with plastic, and if erosion is occurring, the holes turn into collection sites for eroding soil. Researchers measure the volume of accumulating soil and compare that volume with the extent of exposure on erosion pins. In the Maryland experiment, Ritchie used such techniques to find that between 1 and 2 cm (between 0.4–0.8 in) of soil was accumulating upslope from the hedges per year, indicating that the grass was trapping soil as it moved.

Erosion pins and catchpits work well in one concentrated spot but

It has been estimated that fully one-third of the planet's land area suffers from desertification, affecting people in 110 countries. Desertification cost the world's people at least $300–$600 billion in income just in the period 1978–1991, UNEP estimates. China alone loses $6.5 billion annually from desertification; in its western reaches once-isolated desert areas are expanding and joining one another, due to overgrazing from over 400 million goats, sheep, and cattle. In the Sistan Basin along the border of Iran and Afghanistan, an oasis that supported a million livestock just five years ago has turned barren, and windblown sand has buried over 100 villages. In Africa, the continent's most populous nation, Nigeria, loses an amount of land equal to half the state of Delaware each year to the expanding Sahara Desert. Kenya's rapid population growth and

resultant overgrazing and deforestation has left 80% of its land vulnerable to desertification. In a positive feedback cycle, ranchers are crowded onto more marginal land, while farmers are forced to reduce fallow periods, both of which exacerbate soil degradation.

Problems from desertification, moreover, do not always stay localized; in recent years gigantic dust storms from denuded land in China have blown across the Pacific Ocean to North America, and dust storms from Africa's Sahara Desert have blown across the Atlantic Ocean to the Caribbean Sea. While the scale of these problems is now global, none are brand new; many such tragedies occurred in the United States during the Dust Bowl days of the early 20th century, when desertification shook American agriculture and society to their very roots.

are impractical over a large region. To get the best evidence of erosion on a wide scale, scientists have turned to measuring a modern-day leftover rarely considered an environmental benefit—nuclear fallout. Such fallout comes largely from nuclear weapons testing. In 1945, the United States exploded the world's first nuclear bomb, and since then more than 2,000 nuclear devices have been tested by the United States, the former Soviet Union, and other countries. Nuclear weapons testing has scattered radioactive material worldwide, covering Earth's surface with fine layers of nuclear debris.

Fallout includes cesium-137, a radioactive isotope of the element cesium. As discussed in Chapter 4, isotopes of chemical elements are often tracked and measured by environmental scientists. Cesium-137, a product of nuclear fuel and weapons reactions, has a half-life of 30 years—a duration that enables soil scientists to turn the sometimes hazardous isotope into a universal environmental

tracer for erosion and sediment deposits.

Scientists have discovered that soil tends to absorb cesium-137 quickly and evenly. Hence, if soils haven't moved or been heavily disturbed, testing will show fairly uniform levels of the isotope in the area. But if such measurements are not uniform—with some areas showing lower concentrations of cesium-137 and others showing higher concentrations—then erosion may be at work. Ritchie decided to use cesium-137 tests on the area around the hedges and compare the results against his physical soil measurements.

In studies involving cesium-137, soil samples are tested using a device called a gamma spectrometer, which measures gamma rays, unique signatures of energy emitted by chemical elements in soil or rock. As they are emitted, gamma rays show up as sharp emission lines on a spectrum. The energy represented in these emissions determines which elements are present, and the intensity of the lines

reveals the concentration of the element. Thus it is possible to calculate the content of an isotope such as cesium-137 in a test sample. In erosion tests, each sample from the study area is measured for cesium-137, and those levels are compared to baseline levels for the region. By pinpointing places in the study area with lower or higher levels of accumulated cesium, scientists can often detect where soil has moved and how much has shifted.

In Maryland, the radioactive testing helped Ritchie determine that hedges may offer only short-term help against erosion. Although his team's physical measurements of soil accumulation showed soil being deposited near the hedges, the cesium-137 tests revealed that the area had nonetheless undergone a net loss of soil around the hedges over a period of four decades. Grass hedges can help, Ritchie wrote when releasing his findings for the Federal Agricultural Research Service in 2000, but they "should not be seen as a panacea."

The Dust Bowl was a monumental event in the United States

Prior to large-scale cultivation of the southern Great Plains of the United States, native prairie grasses of this temperate grassland region held erosion-prone soils in place. In the late 19th century and first 15 years of the 20th century, however, many homesteading settlers arrived in Oklahoma, Texas, Kansas, New Mexico, and Colorado with hopes of making a living there as farmers. Between 1879 and 1929, cultivated area in the region soared from around 5 million ha (12 million acres) to 40 million ha (100 million acres). Farmers grew abundant wheat, and ranchers grazed many thousands of cattle. Both types of agriculture contributed to soil erosion by removing native grasses and breaking down soil structure.

During the early 1930s a drought in the region exacerbated the ongoing human impacts on the soils. The area's strong winds began to carry away millions of tons of topsoil. Dust storms traveled up to 2,000 km (1,250 mi) before dying down, blackening rain and snow as far away as New York and Vermont. Some areas in the affected states lost as much as 10 cm (4 in) of topsoil in a few short years (Figure 8.14). The affected region in the Great Plains became known as the **Dust Bowl**, a term now also used for the historical event itself and generally to describe any arid area losing huge amounts of topsoil to wind as a result of drought and/or human impact. The so-called black blizzards of the American Dust Bowl destroyed livelihoods and also caused many people to suffer a type of chronic lung irritation and degradation known as dust pneumonia, which is similar to

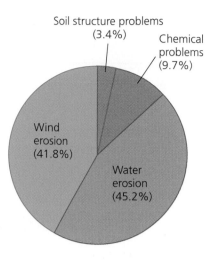

Soil structure problems
(3.4%)

Chemical problems
(9.7%)

Wind erosion
(41.8%)

Water erosion
(45.2%)

Figure 8.13 Soil degradation on drylands is primarily due to erosion by wind and water. Data from UNEP, "Tackling Land Degradation and Desertification," 2002.

Figure 8.14 Drought combined with poor agricultural practices brought devastation and despair to millions of U.S. farmers in the 1930s, especially in the Dust Bowl region of the southern Great Plains. The tragedy spurred the development of many soil conservation practices that have since been put into place in the United States and around the world.

the silicosis that afflicts coal miners exposed to high concentrations of coal dust. Large numbers of farmers were forced off their land, and many who remained were unable to support themselves, having to rely on government assistance programs to survive.

In response to the devastation occurring in the Dust Bowl, the U.S. government, along with state and local governments, increased its support of research into soil conservation measures.

The U.S. Soil Conservation Service pioneered many measures to slow soil degradation

As part of the effort to address the problem of soil degradation, the U.S. Congress passed the Soil Conservation Act of 1935. This act described soil erosion as a threat to the nation's well-being and established the Soil Conservation Service (SCS) to try to solve the problem. The new agency worked closely with farmers to develop conservation plans for individual farms, following several aims and principles:

1. Assess the land's resources, its problems, and the opportunities for conservation.
2. Draw on various scientific disciplines to prepare an integrated plan for the property.
3. Work closely with land users to ensure that the conservation plans harmonize with the users' objectives.
4. Implement conservation measures on individual properties to contribute to the overall quality of life in the watershed or region.

The early teams that the SCS formed to combat erosion typically included soil scientists, forestry experts, engineers, economists, and biologists. These teams were among the earliest examples of interdisciplinary approaches to environmental problem solving. The first director of the SCS, Hugh Hammond Bennett, was an innovator and evangelist for soil conservation. Under his leadership the agency promoted soil-conservation practices through county-based entities known as **conservation districts.** These districts are operated with federal direction, authorization, and funding, but they are organized by state law, while the actual implementation of soil conservation programs is carried out locally by the districts. The districts aim to empower local residents to plan and set priorities in their home areas. In 1994 the SCS was renamed the Natural Resources Conservation Service, and its responsibilities expanded to include water quality protection and pollution control, as well as soil conservation.

The SCS has served as a model for similar efforts elsewhere in the world (Figure 8.15). Southern Brazil's no-till movement came about through local grass-roots organization by farmers, with the help of agronomists and government extension agents who provided them information and resources. In this model of collaboration between local farmers and trained experts, 8,000 Friends of the Land clubs now exist in Paraná and Rio Grande do Sul, and 7,700 in Santa Catarina. Many of

Figure 8.15 Government agricultural extension agents assist farmers by providing information on the newest research and techniques that can help them farm productively while minimizing damage to the land. Such specialists have helped U.S. farmers since the Dust Bowl and are currently assisting farmers worldwide. Here, an extension agent from Colombia's Instituto Colombiano Agropecuario inspects yuca plants grown by farmer Pedro Gomez on a farm in Valle del Cauca.

these groups are delineated by the boundaries of the more than 3,000 small-scale watersheds *(microbacias)* in which they farm.

No-till agriculture as practiced in southern Brazil is merely one of many approaches to soil conservation, however. SCS director Hugh Hammond Bennett advocated a complex approach, combining techniques known as contour farming, strip-cropping, crop rotation, terracing, grazing management, and reforestation, as well as wildlife management. The SCS implemented such soil conservation measures, and they have by now been widely applied in the United States and in many places around the world. We will now briefly examine several of the most common and effective methods of controlling soil degradation.

Farmers can protect soils against degradation in various ways

Several farming techniques can reduce the impacts of conventional cultivation on soils (Figure 8.17). Some of these have been promoted by the SCS since the Dust Bowl; some, like no-till farming in Brazil, are finding popularity more recently; and others have been practiced by certain cultures for centuries.

Crop rotation The practice of alternating the kind of crop grown in a particular field from one season or year to the next is **crop rotation** (Figure 8.16a). Rotating crops can return nutrients to the soil, can break cycles of disease associated with continuous cropping, and can minimize the erosion that can come from letting fields lie fallow. U.S. farmers rotate many of their fields between wheat or corn and soybeans from one year to the next. Soybeans are legumes, plants that have specialized bacteria on their roots that can fix nitrogen in the soil; thus, a soybean crop revitalizes soil that the previous crop had partially depleted of nutrients. Crop rotation also can reduce numbers of insect pests; if an insect is adapted to feed and lay eggs on one particular crop, planting a different crop will leave offspring that may hatch with nothing to eat.

Contour farming Water running down a hillside can easily carry soil away, particularly if there is inadequate vegetative cover to hold the soil in place. Thus, slopes on agricultural land heighten the threat of erosion. Several methods have been developed for farming on slopes. **Contour farming** (Figure 8.16b) consists of plowing furrows sideways across a hillside, perpendicular to its slope, to help prevent the formation of rills and gullies. The technique is so named because the furrows follow the natural contours of the land. In contour farming, the downhill side of each furrow acts as a small dam that slows the runoff of water and catches soil before it is carried away. Contour farming is most effective on gradually sloping land planted with crops that grow well in rows. The extension agents from Santa Catarina in Brazil have helped those farmers who still plow to switch to contour plowing and to use barriers of grass along their contours.

Intercropping The contour farmer may gain further protection against erosion by planting alternating bands of different types of crops across a slope, a method called **strip cropping** (Figure 8.16c). Strip cropping provides more complete ground cover than does a single crop and thus helps to slow erosion. Like crop rotation,

(a) Crop rotation　　　**(b) Contour farming**　　　**(c) Intercropping**

(d) Terracing　　　**(e) Shelterbelts**　　　**(f) No–till farming cover crop**

Figure 8.16 The world's farmers have adopted a number of strategies to conserve their soil and safeguard their crop yields. Rotating the type of crop planted in a particular field, such as between soybeans and corn (**a**), can help restore nutrients to the soil and reduce impacts of crop pests and diseases. Contour farming (**b**), or planting crops along contour lines of elevation across slopes, is a method of reducing erosion on hillsides. Intercropping certain combinations of crops, shown by this agroforestry plantation in the Philippines (**c**), can also reduce erosion while providing plants the nutrients they need. Terracing is a timeworn method for reducing erosion in mountainous areas, such as in the Himalayas, the Andes, or in regions of China, as shown (**d**). This technique makes farming possible on extremely steep slopes. In open landscapes, shelterbelts of trees or shrubs protect against wind erosion; the young trees in this orchard in Argentina (**e**) are being sheltered by rows of tall, fast-growing poplars. No-till and reduced-tillage agriculture involve leaving vegetation on the surface to prevent erosion, rather than plowing crop residue into the soil. The new crop here, corn, grows up from amid the remnants of a "cover crop" (**f**). Often special machinery is used that cuts a furrow into the surface, deposits a seed, adds fertilizer, and covers the seed.

strip cropping offers the additional benefits of reducing an area's vulnerability to insect and disease incidence, and, when a nitrogen-fixing legume is one of the crops, of replenishing the soil. Strip cropping is one type of *intercropping,* a term that denotes any spatial mixing of crops. Southern Brazil's farmers have had success with intercropping food crops and "cover crops" designed to prevent erosion and forestall nitrogen loss from leaching. Santa Catarina's extension agents have worked with farmers to test over 60 species as cover crops, including both legumes and other plants, such as oats and turnips. Cover crops are planted in ways that physically mix them with primary crops, which include maize, soybeans, wheat, onions, cassava, grapes, tomatoes, tobacco, and orchard fruit.

Terracing On extremely steep terrain, **terracing** (Figure 8.16d) is often the most effective method for preventing

erosion. Terraces are level platforms, sometimes with raised edges, that are cut into steep hillsides to contain water from irrigation and precipitation. As it transforms slopes into series of steps like a staircase, terracing enables farmers to cultivate hilly land without losing as much soil to water erosion as would occur otherwise. Terracing is common in ruggedly mountainous regions, such as the foothills of the Himalayas and the Andes, and has been used for centuries by farming communities in such areas. It is labor-intensive to establish, so some farmers colonizing new areas do not or cannot bother to terrace their crops. In the long term, however, terracing is likely the only sustainable way to farm in mountainous terrain.

Shelterbelts Although many of the methods we've discussed protect against erosion by water, they do not help alleviate erosion by wind, the type of threat that leads to dust bowls. A widespread technique to reduce wind erosion is to establish **shelterbelts,** also known as *wind-breaks* (Figure 8.16e). These are rows of trees or other tall, perennial plants that are planted along the edges of farm fields in order to break the wind. Shelterbelts have been widely planted across the U.S. Great Plains. Often fast-growing species, such as poplars, are used. Shelterbelts have also been combined with strip cropping in a practice known as **agroforestry,** or alley cropping. In this approach, fields planted in rows of mixed crops are surrounded by or interspersed with rows of trees that may provide fruit, wood, or protection from wind. Such methods have been used in India, Africa, and Brazil, where coffee growers near a national conservation area have established farming systems combining farming and forestry.

Conservation tillage

As we have seen, agriculture that bypasses plowing is an approach to soil conservation that has gained great popularity in Brazil in recent years (Figure 8.16f). To plant using this method, a tractor pulls a no-till device (sometimes known as a no-till drill) through the field that (1) cuts long furrows through the O horizon of dead weeds and crop residue and the upper levels of the A horizon, (2) drops seeds into the furrow, and (3) closes the furrow over the seeds. Often a localized dose of fertilizer is added to the soil along with the seed.

These methods were actually pioneered in the United States and United Kingdom, where no-till is still rare but where limited tilling has been slowly spreading for decades. Minimum-tillage or reduced-tillage agriculture requires somewhat greater disturbance of the soil surface than no-tillage, but significantly less than conventional cultivation. Today nearly half of U.S. acreage is farmed with reduced-tillage methods.

Critics of no-till and reduced-tillage farming in the United States have noted that these techniques often require substantial use of chemical herbicides (because weeds are not physically removed from fields) and synthetic fertilizer (because other plants take up a fair portion of the soil's nutrients). In many industrialized countries, this has indeed been the case. Proponents of the Brazilian program, however, assert that it does not always need to be so. Southern Brazil's farmers have departed somewhat from the industrialized model in relying more heavily on green manures (dead plants as fertilizer) and rotating fields with cover crops, including nitrogen-fixing legumes. The manures and legumes nourish the soil, and cover crops also reduce weeds by taking up space the weeds might occupy. Critics maintain, however, that green manures are generally not practical for

large-scale intensive agriculture. What is certain is that conservation tillage methods work well in some areas but not in others, and work better with some crops than with others. Farmers will do best by educating themselves on the options and doing what is best for their particular crops on their own land.

Although no-till farming may seem to be based on a simple physical principle (avoid plowing), it can have substantial repercussions for the biological components of soil. By increasing organic matter and the soil biota, no-till farming not only conserves soil but also actually builds it up, restores it, and improves it. One researcher in Brazil has called no-till farming "the only macro-economic solution which can respond to the conflicting demands of more food at lower prices while ensuring sustainability—in fact land quality is continually increasing." A leaflet prepared for an international environmental conference in 1997 proclaimed a number of benefits for no-till farming based on the experience in Southern Brazil (Table 8.2).

At the same time the appeal of no-till farming was spreading among farmers in Brazil, it was also spreading in neighboring countries, Argentina and Paraguay. In Argentina, the area under no-till farming exploded from 100,000 ha (247,000 acres) in 1990 to 7.3 million ha (18.0 million acres) in 1999, covering 30% of all arable land in the country. The results there parallel those in Brazil: increased crop yields, reduced erosion, enhanced soils, and a healthier environment. Maize yields grew by 37% and soybean yields by 11%, while costs to farmers fell by 40–57%, and erosion, pesticide usage, and water pollution declined. As in Brazil, the techniques spread largely because of the actions of farmers themselves and their national no-till farmers' organization.

Of course, these various methods to halt soil degradation can be used in combination. When they have been, the results have sometimes been dramatic. One town in the Guatemalan highlands that established shelterbelts, crop rotation, and cover crops improved its corn production from 0.4 tons/ha (1.0 tons/acre) in 1972 to 2.5 tons/ha (6.2 tons/acre) in 1979 with the help of a U.S.-based nonprofit organization, and went on to further improve production to 4.5 tons/ha (11.1 tons/acre) by 1994.

Weighing the Issues:
How Would You Farm?

You are a farmer owning land on both sides of a steep ridge. You want to plant a sun-loving crop on the sunny but very windy south slope of the ridge and a crop that needs a great deal of irrigation water on the north slope.

Table 8.2   No-Till Farming in Brazil

Benefits of no-till farming

Conserves biodiversity in soil and in terrestrial and aquatic ecosystems

Produces sustainable, high crop yields

Heightens environmental awareness among farmers

Provides shelters and winter food for fauna

Reduces irrigation demands by 10–20%

Crop residues act as sink for carbon (1 metric ton/ha)

Reduces fossil fuel use by 40–70%

Enhances food security by increasing drought resistance

Reduces erosion by 90%

Other benefits arising from the reduction in erosion

Reduces silt deposition in reservoirs

Virtually eliminates water pollution from chemicals

Increases groundwater recharge and lessens flooding

Increases sustained crop yields and lowers prices of food

Lowers costs for treating drinking water

Reduces costs for maintaining dirt roads

Eliminates dust storms in towns and cities

Increases efficiency in use of fertilizer and machinery

Modified from T. F. Shaxson, *The Roots of Sustainability, Concepts and Practice: Zero Tillage in Brazil*, ABLH Newsletter ENABLE; World Association for Soil and Water Conservation (WASWC) Newsletter 1999.

What type of farming techniques might you choose in order to maximize conservation of your soil? What other factors might you want to know about before you decide to commit to certain methods?

Protecting and restoring plant cover is the theme of most erosion-control practices

Farming methods to control erosion make use of the general principle that maximizing vegetative cover will protect soils, and this principle has been applied widely beyond farming. It is common throughout the developed world to stabilize eroding banks along creeks and roadsides by planting plants to anchor the soil. In areas with severe and widespread erosion, some nations have planted vast plantations of fast-growing trees. China in particular has embarked on the world's largest tree-planting program to slow its soil loss. While such "reforestation" efforts do help slow erosion, they do not at the same time produce an ecologically functioning forest, since tree species are selected only for their fast growth and are planted in monocultures.

Irrigation has boosted productivity but has also caused long-term soil problems

The numerous methods developed to combat erosion have been widely successful in reducing the loss of topsoil in many countries. However, erosion is not the only threat to the health and integrity of soils. Soil degradation can result from other factors as well, some of which are caused by our application of water and synthetic chemicals to our crops.

Water is an integral part of any agricultural system. The amount of water needed to support a specific crop in a particular place depends on several factors. The type of crop and its metabolic needs is one primary factor; some crops, such as rice and cotton, require large amounts of water, whereas others, such as beans and wheat, require relatively little. Other factors influencing the amount of water required for growth include the rate of evaporation, as influenced by climate, and the soil's ability to hold water and make it available to plant roots. If too much water evaporates or runs off before it can be absorbed into the soil, crops may require the delivery of additional amounts of water. The artificial provision of water to support agriculture is known as **irrigation.** By irrigating crops, humans have managed to turn previously dry and unproductive regions into fertile farmland. Seventy percent of all fresh water withdrawn by humans is used for irrigation, and irrigated acreage has increased dramatically around the world, reaching 274 million ha (677 million acres) by 1999, greater than the entire area of Mexico and Central America. Irrigation is involved in many environmental issues, and we will examine it further in Chapter 14.

If some water is good for plants and soil, it might seem reasonable that more must be better. But this is not necessarily the case; there is indeed such a thing as too much water. Overirrigation in poorly drained areas can cause or exacerbate certain types of soil problems. Soils too saturated with water may become **waterlogged.** When waterlogging occurs, the water table is raised to the point that water bathes plant roots. This deprives roots of access to gases, essentially suffocating them; if it lasts long enough, waterlogging can damage or kill the plants.

An even more frequent problem is **salinization**, the buildup of salts in surface soil layers. In dryland areas where precipitation is minimal and evaporation rates are high, water evaporates from the A horizon of the soil. This may pull additional water from lower horizons up through the soil by capillary action. As this water rises through the soil it carries dissolved salts, and when it evaporates at the surface those salts precipitate and are left at the surface. Irrigation in arid areas generally hastens salinization, because it provides repeated

VIEWPOINTS

Soil Conservation
Productive farming depends on fertile soil, but our farming practices have all too often eroded and degraded soil. How much erosion constitutes a problem? How can we best protect the condition of soil while we farm?

Land Policy Is Necessary for Soil Conservation

Claiming that soil erosion is a major problem in the more-developed parts of the world is environmental nonsense. Pierre Crosson of Resources for the Future, Washington, D.C., has looked carefully at the U.S. data over many years and concluded that soil erosion does *not* constitute a significant problem for contemporary agriculture and its sustainability. There are surely many cases of local erosive effects but these usually amount to small quantities of soil shifting around the agricultural landscape. Erosion can be beneficial in some cases, such as in the creation of new farmland in Bangladesh, when materials washed down from Nepal and other parts of the Himalayas.

Soil conservation is a problem that must be approached from all sides. Most importantly, will be the matter of securing legally protected property rights for resource users. Farmers must have the right to own their own land resource. Ethiopia is usually regarded as one of the most eroded countries in the world. The State owns all farmland, and there is a history of moving people against their will to different areas. If a farmer thinks he will not be able to continue managing his farm for the next several years, he will not invest his scarce resources in erosion-reducing practices such as terracing, stone building, maintaining vegetative contour strips, and reducing grazing pressure. For resource custody, and hence the treatment of soil, land policy really matters.

Land policy also affects public investment in infrastructure such as rural roads, which reduce the cost of farmers' transactions. Farmers must haul supplies such as fertilizer to their farms and must transport their product to market. Inadequate access to markets can significantly influence the incentives to employ resource-consuming soil conservation practices. Thus, social science has key roles to play along with physical and biological sciences in addressing soil management.

Jock R. Anderson joined the World Bank in its Agriculture and Natural Resources Department, where he served variously as adviser of Agricultural Technology Policy and more recently adviser of Strategy and Policy for Agriculture and Rural Development. He is a Fellow of the Australian Institute of Agricultural Science and of the American Agricultural Economics Association.

Sustainable Soil Productivity Requires Good Land Stewards

How much soil erosion is tolerable? The standard answer to this question ranges from 1 to 5 tons of soil per acre annually, depending on soil depth and other characteristics that affect sustainability. However, this answer is admittedly subjective and debatable. Some say these values are too low. They argue that land has been used for centuries with higher rates of erosion, and it is difficult or impossible to reduce erosion on some land without drastic reductions in high value crops. Others say the standard tolerable erosion values are too high. They argue that the average annual rate of soil formation is less than 1 ton per acre, so a greater erosion rate will inevitably make the soil shallower and less productive. Most agree that excessive erosion is undesirable and a major cause of land abandonment.

Tillage was fundamental to the development of civilization, but it also exposed the soil to erosion—a problem that became more serious as population increased. Tillage can now be reduced or even eliminated by using modern no-till equipment that can cut through crop residues and plant seeds in a narrow slot. Weeds, insects, and plant diseases are controlled by integrated pest management using crop rotations, biological methods, and chemicals. We can now grow crops with much less erosion and still produce good yields.

Conserving water and planting drought-tolerant crops in semi-arid climates is another important advance in areas where erosion due to summerfallow has been common. These practices are effective ways to conserve soil.

The most important requirement for sustainable productivity is a fervent desire to be good land stewards—people who use the best soil conservation practices and are alert for detecting problem spots.

Frederick R. Troeh grew up in Idaho, surveyed soils for the Soil Conservation Service, earned his Ph.D. from Cornell University, and taught soil science and researched for Iowa State University. He has worked internationally in Uruguay, Argentina, and Morocco.

doses of moderate amounts of water, which dissolve salts in the soil and gradually raise them to the surface. Furthermore, because irrigation water often contains some small amount of dissolved salt in the first place, irrigation introduces new sources of salt to the soil. Overirrigation and resultant waterlogging can worsen salinization problems. The salinization process accounts for natural salt playas or salt deserts, such as are found in parts of the American West, and is responsible for the increasing areas of farmland where soil is turning whitish with encrusted salt. Salinization now inhibits agricultural production on one-fifth of all irrigated cropland globally, costing more than $11 billion.

Salinization is easier to prevent than to correct

The remedies for mitigating salinization once it has occurred are more expensive and difficult to implement than the techniques for preventing it in the first place. The best way to prevent salinization is to avoid planting crops that require a great deal of water in areas that are especially prone to the problem. A second way is to irrigate with water that is as low as possible in salt content. Another way to prevent salinization is to irrigate efficiently, supplying no more water than the crop requires, thus minimizing the amount of water that evaporates and hence the amount of salt that accumulates in the topsoil. Currently, irrigation efficiency worldwide is low; only 43% of water applied actually gets used by plants. Drip irrigation systems (Figure 8.17) that target their water directly to plants are helping, giving more control over how much water is used and wasting far less. Once considered expensive to install, these systems are becoming cheaper such that more farmers in developing countries will be able to afford them.

If salinization has already occurred, one potential way to mitigate it would be to cease irrigation and wait for precipitation to flush salts from the soil. However, this solution is unrealistic because salinization generally becomes a problem in those dryland areas in which precipitation is never adequate to flush soils. A better solution is to plant salt-tolerant plants such as barley that can be used as food or pasture. A third option is to bring in large quantities of less-saline water with which to flush the soil. However, using too much water may cause waterlogging. As is the case with many environmental problems, preventing salinization is easier than correcting it after the fact.

(a) Conventional irrigation

(b) Drip irrigation

Figure 8.17 Irrigation methods are being improved to increase efficiency, because currently less than half the water applied for irrigation actually is taken up by plants. Conventional methods that lose a great deal of water to evaporation **(a)** are being replaced by more efficient ones in which water is more precisely targeted to the plants. In drip irrigation systems, such as this one watering grape vines in California **(b)**, hoses are arranged so that water drips from holes in the hoses directly onto the plants that need the water.

Weighing the Issues:
Can We Measure Soil Quality?

The U.S. EPA has adopted measures of air quality and water quality and has set legal standards for allowable levels of various pollutants in air and water. Can and should such standards be developed for soil quality? What makes it more difficult to assess soil quality? How might it be best to go about using measurements of soil properties to determine quality?

Agricultural fertilizers boost crop yields but can be over-applied

Salinization is not the only source of chemical damage to soil. The fertilizers we use to boost the growth of our crops can also chemically damage soils. Plants grow through photosynthesis, which requires sunlight, water, and carbon dioxide; but many plants also require substantial amounts of nitrogen, phosphorus, and potassium, as well as smaller amounts of over a dozen other nutrients. These **macronutrients** (nutrients required in relatively large amounts) and **micronutrients** (nutrients required in much smaller amounts) occur in suitable quantities for wild plants in healthy soils. However, crop plants often need more of these nutrients, and if soils contain too little of them, crop yields will decline. Therefore, a great deal of research has gone into efforts to increase the concentrations of nutrients in nutrient-limited soils by adding **fertilizer,** any of various substances that contain essential nutrients. The overapplication of fertilizer, however, has been the source of many environmental problems.

Nutrient depletion creates a need for fertilizers

Soils can become depleted of nutrients for several reasons. All plants remove nutrients from soil as they grow. In addition, leaching may remove essential nutrients, especially ions of phosphate, potassium, and calcium. If both of these processes continue unchecked, soils lose their ability to support subsequent crops, and fertilizer application becomes necessary (Figure 8.18).

There are two main types of fertilizers. **Organic fertilizers** consist of natural materials (largely the remains or wastes of organisms) and include animal manure; crop residues; fresh vegetation, known as **green manure;** and **compost,** a mixture produced when decomposers break down organic matter, including food and crop waste, in a controlled environment. **Inorganic fertilizers** are mined or synthetically manufactured mineral supplements. They are generally more susceptible than are organic fertilizers to leaching and runoff, and may be more likely to cause unintended off-site impacts. Inorganic and organic fertilizer use is widespread and its mismanagement is causing increasingly severe problems around the world (Figure 8.19).

Excess nitrogen impacts the environment

We have already seen in Chapter 6 one impact of excess fertilizer use. The nitrogen and phosphorus runoff from farms and other sources in the Mississippi River basin

Figure 8.18 Farmers often find it necessary to add or restore nutrients to soils with fertilizers. Organic fertilizers such as manure and vegetation may be used, or synthetically manufactured chemicals may be applied to supply nitrogen, phosphorus, and other nutrients, as this North Dakota farmer is doing.

each year spurs phytoplankton blooms in the Gulf of Mexico and creates an oxygen-depleted "dead zone" that kills fish and shrimp off the Louisiana coast. Such eutrophication occurs to one degree or another in aquatic ecosystems at river mouths and in inland lakes and ponds throughout the world, and we will revisit this process in Chapter 14. Nitrogen, which plays a key role in photosynthesis and is an essential component of proteins, is the macronutrient most likely to be a limiting factor (Chapter 5) to a plant's growth. Even though every hectare of arable land in the world sits beneath 75,000 metric tons of atmospheric nitrogen gas (N_2), that gas is extremely stable and thus unavailable to satisfy the metabolic needs of plants. The bond within the N_2 molecule must be broken for plants to make use of this element, and the only naturally occurring phenomena that cause this to happen are lightning bolts and the metabolic activities of certain bacteria, such as those associated with the roots of legumes. Because these processes alone cannot produce enough nitrogen to meet human agricultural needs, farmers have depended on nitrogen-rich organic and inorganic fertilizers for many years.

As you'll recall from Chapter 6, the nitrogen cycle is a system whose efficacy may change when humans tinker with it—and tinker we have. In the early 19th century, merchants from Chile recognized the global need for nitrogen-rich fertilizers and began mining and exporting the nitrates (NO_3^-) so abundant in their country's Atacama Desert. Then in 1913, construction was completed on the world's first commercial plant

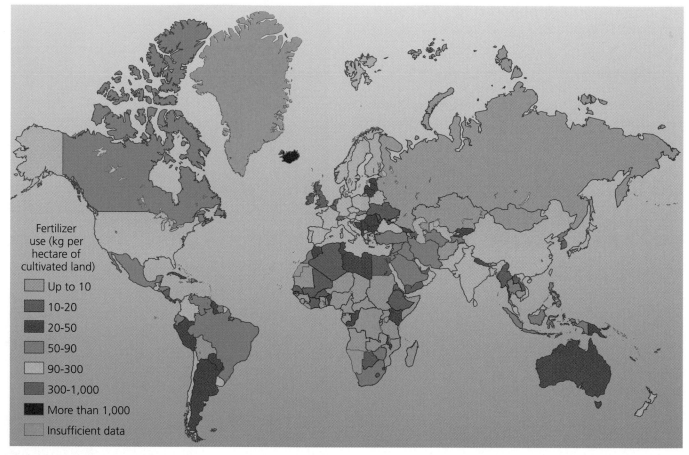

Fertilizer use (kg per hectare of cultivated land)

Up to 10
10-20
20-50
50-90
90-300
300-1,000
More than 1,000
Insufficient data

(a) World fertilizer use, 1998

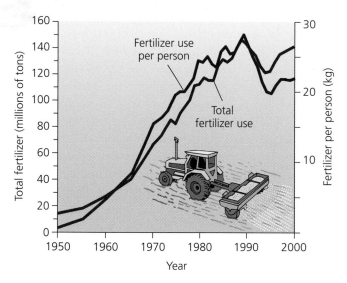

(b) World and per person fertilizer use

Figure 8.19 The use of synthetic fertilizers is widespread across the world (**a**), and fertilizer use has risen sharply over the past half-century (**b**). The temporary drop during the early 1990s was due to the economic decline of agriculture in countries of the former Soviet Union following that nation's dissolution. Data from AAAS Atlas of Population and Environment, "World Fertilizer Use, 1998," The University of California Press, 2000; Vital Signs, "World Fertilizer Use, 1950–2000," Worldwatch Institute, 2001.

producing ammonia (NH_3), a reactive compound used as fertilizer. Global application of synthetic nitrogen-rich fertilizers began increasing dramatically in 1950. From 1960 to 1995, nitrogen, phosphorus, and potassium fertilizer usage increased by 300–800%. In the 1990s, about 100 metric tons (110 million tons) of fertilizer were added to farm fields annually, and since 2000 humans have been adding more nitrogen to Earth's systems than all of the natural inputs combined. Without synthetic nitrogen-rich fertilizers, it has been estimated that we would have to reduce Earth's human population to approximately 2–3 billion in order to maintain our current levels of food consumption.

Applying such significant amounts of fertilizer to croplands has impacts far beyond the boundaries of the fields (Figure 8.20). Nitrates readily leach through the soil and contaminate groundwater. Phosphates readily are carried away with sediment by runoff into surface waterways. And components of nitrogen fertilizers can even volatilize (evaporate) and enter the air. Through these processes, unnatural amounts of nitrates and phosphates spread through terrestrial ecosystems

Figure 8.20 The overapplication of inorganic (and organic) fertilizers can have effects beyond the farm field, because nutrients that are not taken up by plants may end up in other places. Nitrates can leach into groundwater, for instance, where they can pose a threat to human health in drinking water. Phosphates and some nitrogen compounds can run off into surface waterways and alter the ecology of streams, rivers, ponds, and lakes through eutrophication. Some compounds like nitrogen oxides can even enter and pollute the air. Anthropogenic inputs of nitrogen have greatly modified the natural nitrogen cycle (Chapter 6) and now account for one-half the total nitrogen flux on Earth.

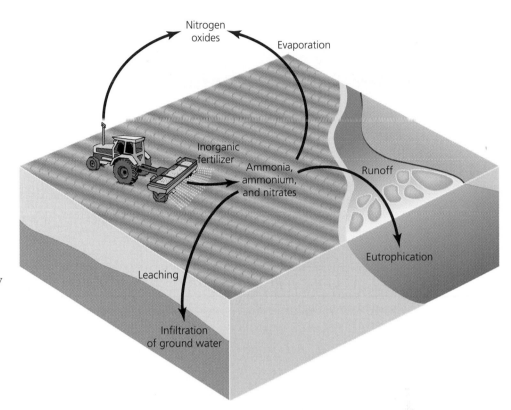

and cause eutrophication of aquatic ones. Boosting nitrogen inputs to the environment can also affect human health. Nitrate-contaminated drinking water can cause methemoglobinemia, or blue-baby disease, which asphyxiates and sometimes kills infants. Nitrates are also suspected of causing esophageal and stomach cancer in adults. The U.S. EPA has determined that nitrate concentrations in excess of 10 mg/l for adults and 5 mg/l for infants in drinking water are unsafe, yet many sources around the world exceed even the higher standard of 50 mg/l set by the World Health Organization.

Organic fertilizers have benefits synthetic ones do not

Although abnormally high inputs of nutrients can pose problems for ecosystems, crops, soils, and human health regardless of the source from which they originate, scientists have realized that organic fertilizers, including compost and animal manure, can provide many benefits that inorganic fertilizers cannot. The proper use of compost, for example, not only delivers essential nutrients to the soil but also improves soil structure, nutrient retention, and water-retaining capacity, thus helping to prevent erosion. In addition, as a form of recycling, composting reduces the amount of waste

consigned to landfills and incinerators. However, organic fertilizers are no panacea. For instance, manure, when applied in amounts needed to supply sufficient nitrogen for a crop, may introduce excess phosphorus that may then run off into area waterways.

Grazing practices and policies can contribute to soil degradation

We have focused in this chapter largely on the cultivation of crops as a source of impact on soils and ecosystems, but our raising of livestock also has such impacts. When sheep, goats, cattle, or other livestock are allowed to graze on the open range, they feed primarily on grasses. As long as livestock populations do not exceed a range's carrying capacity (Chapter 5) and do not consume grasses faster than grasses can be replaced, grazing may be sustainable. However, when too many animals eat too much of the plant cover, impeding plant regrowth and preventing the replacement of biomass, the result is **overgrazing.**

Overgrazing has been shown by rangeland scientists to cause a number of impacts, some of which give rise to positive feedback cycles that exacerbate damage to soils, natural communities, and the land's productivity for grazing (Figure 8.21). When livestock remove or harm too much of an area's native plant cover, more of the soil surface is exposed and made vulnerable to erosion by wind and

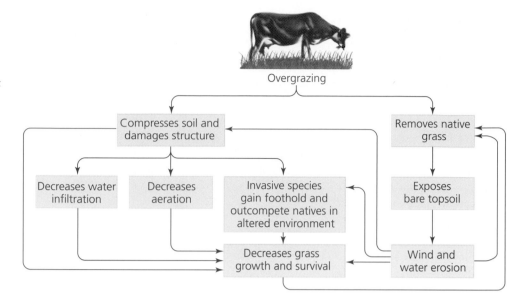

Figure 8.21 When grazing by livestock exceeds the carrying capacity of rangelands and their soil, overgrazing can set in motion a series of consequences and positive feedback loops that degrade soils and grassland ecosystems.

water. Soil erosion makes it difficult for native vegetation to regrow, thus perpetuating the lack of cover and giving rise to more erosion. Bare, eroded soils also may be invaded by non-native weedy plants (Figure 8.22). These invasive plants are usually less palatable to livestock and can outcompete native vegetation in the new, modified environment, further decreasing native plant cover.

In addition, overgrazing can also compact soils and alter their structure. Soil compaction makes it harder for water to infiltrate, harder for soils to be aerated, harder for plants' roots to expand, and harder for roots to conduct cellular respiration (Chapter 4). All of these effects

Figure 8.22 The effects of overgrazing can be striking, as shown in this photo along a fenceline separating a grazed plot (right) from an ungrazed plot (left) in the Konza Prairie Reserve in Kansas. The overgrazed plot has lost much of its native grass and has been invaded by weedy plants that compete more effectively in conditions of degraded soil.

cause further decreases in the growth and survival of native plants.

Overgrazing is a serious problem worldwide, a cause of soil degradation equal to that of cropland agriculture and a greater cause of desertification. Humans keep a total of 3.3 billion cattle, sheep, and goats. Rangeland classified as degraded now adds up to 680 million ha (1.7 billion acres), five times the area of U.S. cropland, although some estimates put the number as high as 2.4 billion ha (5.9 billion acres), fully 70% of the world's rangeland area. Rangeland degradation is estimated to cost humans $23.3 billion per year. Grazing exceeds the sustainable supply of grass in India by 30% and in parts of China by up to 50%. Both nations are now beginning to feed livestock crop residues in order to relieve pressure on rangelands.

Range managers in the United States do their best to assess the carrying capacity of rangelands and inform livestock owners of these limits, so that herds are rotated from site to site as needed to conserve grass cover and soil integrity. Managers also can establish and enforce limits on grazing on publicly owned land when necessary. U.S. ranchers have traditionally had little incentive to conserve rangelands because most grazing has taken place on public lands leased from the government, not on lands privately owned by ranchers. In addition, the U.S. government has heavily subsidized grazing. As a result, overgrazing has been extensive and has caused many environmental problems in the American West. Today increasing numbers of ranchers are working cooperatively with government agencies, environmental scientists, and even environmental advocates to find ways to ranch more sustainably and safeguard the health of the land (see The Science behind the Story on page 248).

Forestry, too, has impacts on soils

Farming and grazing are agricultural practices that help feed human populations, that depend upon healthy soils, and that affect the conditions of those soils. Forestry, the cultivation of trees, is a similar practice that we will cover in detail in Chapter 16. Forestry practices can have substantial impacts on soils just as can farming and ranching. As with farming and ranching techniques, forestry practices have been modified over the years to try to minimize damage to soils. Some practices such as clear-cutting, the removal of all trees from an area at once, can lead to severe erosion due to the sudden exposure of the soil. This is particularly the case on steep slopes (Figure 8.23). Alternative timbering methods have been devised, and those that remove fewer trees over longer periods of time are more successful in minimizing erosion.

In addition, the maintenance of ecosystems and their biodiversity is becoming a larger issue in forestry, one that often conflicts with approaches to maximize short-term timber harvests. Although most farmlands under the industrialized agriculture model are quite devoid of wildlife, managed forests can shelter biodiversity and act as functional ecosystems even while they produce timber for our consumption. Many managed forests, however, resemble industrialized cropland in that they are monocultures of a particular tree species, with little biotic diversity. Monocultures of fast-growing species may maximize timber growth in the short run, but in the long run they are susceptible to insect outbreaks and can degrade soil quality by reducing biodiversity.

Recent U.S. laws have promoted soil conservation

In recent years, the U.S. Congress has enacted a number of laws promoting soil conservation. The Food Security Act of 1985 requires farmers to adopt soil conservation plans and practices as a prerequisite for receiving price supports and other government benefits. The Conservation Reserve Program, also enacted in 1985, pays farmers to stop cultivating highly erodible cropland and instead place it in conservation reserves planted with grasses and trees. The U.S. Department of Agriculture (USDA) estimates that for an annual cost of $1 billion,

Figure 8.23 Deforestation, discussed further in Chapter 16, can be a major cause of erosion, especially when trees are clear-cut for timber on steep slopes.

this program saves 700 million metric tons (771 million tons) of topsoil each year. In addition to reducing erosion, the Conservation Reserve Program has generated income for farmers and provided habitat for native wildlife. In 1996, Congress extended the program by passing the Federal Agricultural Improvement and Reform Act. Also known as the "Freedom to Farm Act," this law aimed to reduce subsidies and government influence over many farm products. It also created the Environmental Quality Incentive Program and the Natural Resources Conservation Foundation to promote and pay for the adoption of conservation practices in agriculture. In 1998 the USDA initiated the Low-Input Sustainable Agriculture Program to provide funding for individual farmers to develop and practice sustainable agriculture.

Overgrazing and Fire Suppression in the Malpai Borderlands

In the high desert of southern Arizona and New Mexico, scientists trying to heal the scars left by decades of cattle ranching and overgrazing found they had to contend with a creature even more damaging than a hungry steer: Smokey the Bear.

Wildfires might seem a natural enemy of grasslands, but here in the Malpai Borderlands, researchers realized it was people's efforts to suppress fire that had done far more harm. Before large numbers of Anglo settlers and ranchers arrived more than a century ago, this rocky stretch of arid Western landscape thrived under an ecological cycle common to many grasslands. Scrubby trees such as mesquite grew near creeks. Hardy grasses, such as black grama, covered the drier plains. Periodic wildfires, usually sparked by lightning, burned back shrubs and trees and kept grasslands open. Native grazing animals, such as deer, rabbits, and bighorn sheep, ate the grasses, but rarely in large enough numbers to deplete the range. Fed by seasonal rains, new grasses sprouted without being overeaten or crowded out by larger plants.

By the early 1990s, however, those rangelands were increasingly

To restore native grasslands and savanna habitat to the Malpai Borderlands, the Malpai Borderlands Group reinstated fire as a natural landscape process, conducting controlled burns on over a hundred thousand ha since 1994. Monitoring indicates that restoring fire has improved ecological conditions in the region.

scarce. Ranchers had brought large cattle herds to the area in the late 1800s. The cows chewed through vast expanses of grass, trampled soil, and scattered mesquite seed into areas grasses had dominated. Ranchers fought wildfires to keep their herds safe, and the federal government joined in the firefighting efforts. Gradually, the Malpai's ranching families found themselves struggling to feed their cattle and make a living from the land.

Decades of photos taken by ranchers showed how the area's soil had eroded, with trees and brush overgrowing grass. The ranchers knew their cattle were part of the problem, but they also suspected that firefighting efforts might be to blame.

In 1993, a group of ranchers launched an innovative plan. They formed the Malpai Borderlands Group, designating about 325,000 ha (800,000 acres) of

Several key international programs also promote soil conservation

The United Nations is actively involved in promoting soil conservation and sustainable agriculture through a variety of programs of its Food and Agriculture Organization (FAO). The FAO's Farmer-Centered Agricultural Resource Management Program (FAR) is a project undertaken in partnership by China, Thailand, Vietnam, Indonesia, Sri Lanka, Nepal, the Philippines, and India

to support innovative approaches to resource management and sustainable agriculture. This program studies agricultural success stories and tries to help other farmers duplicate the successful efforts. Rather than following a top-down, government-controlled approach, the FAR program makes a point of utilizing the creativity of preexisting communities in order to educate and encourage farmers throughout Asia to conserve soils and secure their food supply.

land as an area for protection and study. Through study of the region, ranchers joined government agencies, environmentalists, and scientists to bring back grasses, restore native animal species, and return periodic fires to the landscape.

The group's research efforts have centered on the Gray Ranch, a 121,000-ha (300,000-acre) parcel in the heart of the borderlands. Scientists led by researcher Charles Curtin have created study sites by dividing pastureland on the ranch into four study areas of about 890 ha (2,200 acres) each. Each area is further divided into four different "treatments," or areas with varying land management techniques:

- In Treatment 1, fire is not suppressed, and grazing by cattle and native animals is permitted.
- In Treatment 2, fire is not suppressed, and grazing is not allowed.
- In Treatment 3, animals can graze, but fire is suppressed.
- In Treatment 4, fire is suppressed, and grazing is not permitted.

Treatments 1 and 3, which allow grazing, also include small areas that block most animals from eating grass. These exclosures allow scientists to make precise side-by-side comparisons of how grazing affects grasses. Rainfall is measured in each area, and soils are monitored for degradation and erosion. Teams of researchers, including wildlife and vegetation specialists, monitor each treatment for the distribution and abundance of birds, insects, animals, and vegetation.

By comparing findings from each area, scientists have been able to take important steps toward restoring the area's rangelands. Bringing back grass, they have found, takes far more than getting rid of hungry cows. Some of the most important findings have come from comparing areas where fire is suppressed to those where fires can burn. Such research has documented how the suppression of fire leads to more brush and trees, with grass getting crowded out. When fire burns through an area, however, woody plants such as mesquite are damaged and their seed production disrupted, and the flames are often followed by a return of grass. Such benefits follow both natural fires and carefully monitored, deliberately set controlled burns.

Cattle don't do heavy damage to grass if they are managed carefully, the researchers have found.

Rather, it is interference with the natural cycles of the land that does the greatest harm. Researchers have helped ranchers develop a cycle of grassbanking, in which herds are allowed to graze on shared plots of land while other areas recover. And ranchers must work with the weather. Scientists have found that controlled burns or grassbanking should track with cycles of rain and drought to keep the land healthy and bring back grass.

Because of such research, controlled burns are now a regular part of the Malpai landscape. More than 100,000 ha (250,000 acres) have been burned since 1994, with 19,000 ha (48,000 acres) set ablaze in the summer of 2003 alone. Natural fires are often allowed to run their course with little to no intervention. Damaged areas have been reseeded with native grasses. Scientists increasingly believe that ranching, if managed properly, can help bring damaged grazing areas back to life. "We cannot assume rangelands will recover on their own," Curtin wrote in a recent study on the Malpai Borderlands. "Conservation of grazed lands requires restoring and sustaining natural processes."

Conclusion

Many of the numerous policies enacted and practices developed to combat soil degradation in the United States and worldwide have been quite successful, particularly in reducing the erosion of topsoil. Despite these successes, however, soil is still being lost at a rate that calls into question the sustainability of industrial agriculture. For instance, erosion rates in the United States have declined from 2.65 tons/ha (6.55 tons/acre) in 1982 to 1.8 tons/ha 4.4 tons/acre in 1997, thanks to soil conservation measures. Yet in spite of these measures, U.S. farmlands still lose 6 tons of soil for every ton of grain harvested.

Our species has enjoyed a 10,000-year history with agriculture, yet despite all we have learned about soil degradation and conservation, many challenges remain. It is clear that even the best-conceived soil conservation programs require research, education, funding, and commitment from both farmers and governments if they are to fulfill their potential. Most agronomists would agree that in light of continued population growth, we will need better technology and wider adoption of soil conservation techniques to avoid an eventual food crisis. Increasingly, it seems relevant to consider whether the universal adoption of Aldo Leopold's land ethic (Chapter 2) will also be required if we are to feed the 9 billion people expected to crowd our planet in mid-century.

REVIEW QUESTIONS

1. How do agronomists define soil? What components of soil determine its value? Why is soil such a valuable resource?

2. How did the practices of selective breeding, or artificial selection, and human agriculture begin roughly 10,000 years ago? Summarize the influence of agriculture on the development and organization of human communities.

3. Describe the methods used in traditional agriculture, and contrast subsistence agriculture with intensive traditional agriculture. What makes industrialized agriculture different from traditional agriculture and why has industrialization occurred? Roughly how much of the world's croplands are cultivated through industrialized agriculture methods?

4. How has the Green Revolution used technology to boost crop yields in the developing world?

5. What processes are most responsible for the formation of soil? Describe the three types of weathering that may contribute to the process of soil formation.

6. How can soil formation be a positive feedback process? What are the ecological functions of mature soils?

7. Name the five primary factors thought to influence soil formation, and describe one effect of each.

8. How are soil horizons created? What is the general pattern of distribution of organic matter in a typical soil profile?

9. What does the color of a soil indicate? Based on what you know about soil composition and soil profiles, what color do you think would indicate a soil's strong potential for supporting crops and forage for livestock?

10. Explain one trait that texture gives to soil that helps determine its usefulness for agriculture. Why are loam and silt considered productive soils for plant growth?

11. How does erosion aid in soil creation? Why is erosion generally considered a destructive process?

12. Name three human activities that can promote soil erosion. Describe four kinds of soil erosion by water. What factors affect the intensity of water erosion?

13. How has erosion become a global problem for agriculture? What methods might be used to detect soil erosion?

14. Define desertification. What kinds of soils are prone to desertification?

15. List innovations in soil conservation introduced by Hugh Hammond Bennett, first director of the SCS. What other farming techniques can help reduce the risk of erosion due to conventional agricultural cultivation methods?

16. How does terracing effectively turn very steep and mountainous areas into arable land?

17. What is agroforestry or alley cropping, and why is it effective against both water and wind erosion?

18. Explain the method of no-till farming. Why does this method reduce soil erosion?

19. How can irrigation lead to salinization? How might salinization be prevented or reduced?

20. Describe the process by which fertilizers boost crop growth. How do large amounts of fertilizer added to soil also end up in water supplies and the atmosphere?

21. Name one benefit of organic fertilizers over synthetic fertilizers.

22. What policies can be linked to the practice of overgrazing? Describe the effects of overgrazing on soil. With regard to soil conservation, what conditions characterize sustainable grazing practices?

23. Name two forestry practices that have had negative impacts on soil.

24. What approaches have the United States and the United Nations taken to reconciling the needs and practices of farmers with the need for soil conservation?

DISCUSSION QUESTIONS

1. Since 1983, per capita grain production has leveled off and begun to decline. This has happened in spite of the spread of industrialized agriculture and its many strategies for increasing crop yields and despite the spread of these techniques to the developing world in the Green Revolution. Why do you think per capita grain production has declined? What might we do about it?

2. How might actual soils differ from the idealized six-horizon soil profile presented in the chapter? How might departures from the idealized profile indicate the impact of human activities? Provide at least three examples. What do you think happened in the Fertile Crescent?

3. What method of farming would you employ on a gradual slope with the threat of natural erosion? What kinds of plants might you use to prevent erosion and why?

4. Discuss how the method of no-till or zero-tillage farming, as described in this chapter, can contribute to soil quality. What drawbacks or negative effects might no-till or reduced-tillage practices have on soil quality, and how might these be prevented?

5. Discuss how methods of locally-based sustainable agriculture described in this chapter are promoting the science of soil conservation. In reference to Aldo Leopold's land ethic (Chapter 2), how are humans and the soil members of the same community?

Media Resources *For further review go to the website* **www.envscienceplace.com** *or student CD-ROM, where you will find quizzes, flashcards, a glossary, additional interactive exercises, and links to relevant news and research sources. Also on the website and CD-ROM is* **GRAPH IT**, *a series of interactive graphing tutorials to help you interpret graphs and plot data.*

9 Agriculture, biotechnology, and the future of food

Organic Vegetable Farm in Whatcom County, Washington

This chapter will help you understand:

- The challenge of feeding a rapidly growing human population

- The environmental impacts of food production

- Human nutritional needs and challenges

- The "green revolution"

- Pest management methods

- The importance of pollination

- The science behind genetically modified food

- Controversies and debate over genetically modified food

- Approaches for preserving crop diversity

- Feedlot agriculture for livestock and poultry

- Aquaculture

- Organic agriculture

Farmer among Dry Stalks of Corn in Field of Oaxaca, Mexico

Central Case: Possible Transgenic Maize in Oaxaca, Mexico

"Worrying about starving future generations won't feed them. Food biotechnology will. . . . At Monsanto, we now believe food biotechnology is a better way forward."
—*Ad campaign of the Monsanto Company*

"Industrial agriculture has not produced more food. It has destroyed diverse sources of food, and it has stolen food from other species . . . using huge quantities of fossil fuels and water and toxic chemicals in the process "
—*Vandana Shiva, director of the Research Foundation for Science, Technology, and Natural Resource Policy, Dehra Dun, India*

Corn is a staple grain of the world's food supply. We can trace its ancestry back roughly 5,500 years, when people in the highland valleys of what is now the state of Oaxaca in southern Mexico first domesticated that region's wild maize plants. The corn we eat today arose from some of the many varieties that evolved from the early selective crop breeding conducted by the people of this region. Today Oaxaca remains a world center of biodiversity for maize, with numerous native varieties, or cultivars, growing in the rich, well-watered soil. Preserving such varieties of major crops in their ancestral homelands is important for securing the future of our food supply, scientists maintain, because these varieties serve as reservoirs of genetic diversity—reservoirs we may need to draw on to sustain or further advance our agriculture.

Thus it caused global consternation when researchers announced in 2001 that they had found evidence for what they contended was a threat to the region's ancestral crop diversity. Mexican scientists conducting routine genetic tests of Oaxacan farmers' maize had turned up suspect DNA that seemed to match genes from genetically modified corn. Such corn was widely grown in the United States, but Mexico had banned its cultivation in 1998. Corn is one of many crops that scientists have genetically engineered to express certain desirable traits, such as large size, fast growth, or resistance to insect pests. To genetically engineer crops, scientists extract genes that encode specific traits from the DNA of one plant and then transfer these genes into the DNA of another. The aim is to improve crop performance and feed the world's hungry, but many activists protest that the transgenes from these transgenic plants might cause unpredictable harm. One concern is that transgenic crops might crossbreed with native crops and thereby "contaminate" the genetic makeup of native crops.

Two researchers at the University of California at Berkeley, David Quist and Ignacio Chapela, shared these concerns, and ventured to Oaxaca to collect samples of native maize to see whether they could confirm the Mexican scientists' findings. Their lab analyses revealed what they argued were traces of DNA from genetically engineered corn in the genes of native maize plants. They also suggested that the invading genes had split up and spread throughout the maize genome. Quist and Chapela published their findings in the scientific journal *Nature* in November 2001. Activists opposed to genetically modified food trumpeted the news and urged a ban on all importation of transgenic crops from producer countries such as the United States into developing nations. Agrobiotech industries, however, defended the safety of their crops and questioned the validity of the research—as did many of Quist and Chapela's peers. Responding to criticisms from researchers, *Nature* took the unprecedented step of stating that Quist and Chapela's paper should never have been published, a move that created a firestorm of controversy (See "The Science behind the Story").

Had transgenes actually invaded Mexican maize? No one yet knows for sure, but even the harshest critics of Quist and Chapela's work say that transgenic invasion is likely to have happened already or will likely happen soon.

An even harder question to answer is how genetically modified crops may affect people and the environment. In this chapter, we take a wide-ranging view of the ways people have devised to increase agricultural output to keep pace with the growing human population, what the environmental effects of these efforts have been, and what the food of the future has in store for us.

The Race to Feed the World

Although human population growth has slowed, we can still expect our numbers to swell to 9 billion by the middle of this century. For every two people living today, there will be three in 2050. The kind of world we live in then will depend on the choices we make now, and some of the most important choices involve our means of producing food. Perhaps the genetically modified crops eliciting so much controversy in Oaxaca and elsewhere may be the solution to feeding 50% more mouths half a century from now while protecting the integrity of our soil, water, and ecosystems. Or perhaps the solution will lie in organic farming and sustainable agriculture.

Agricultural production has so far outpaced population increase

Over the past half-century, our ability to produce food has grown even faster than has global population (Figure 9.1). However, largely due to political obstacles and

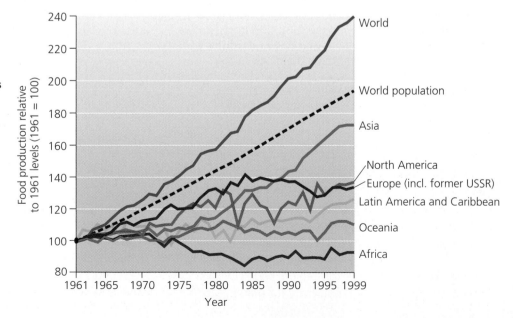

Figure 9.1 Global agricultural food production rose by nearly two-and-a-half times in the past four decades, growing at a faster rate than world population. Food production increased in all regions except Africa during this time period and particularly rose in Asia, where effects of the green revolution were greatest. Data from *AAAS Atlas of Population & Environment,* University of California Press, 2000.

Transgenic Contamination of Native Maize?

The Science behind the Story

David Quist and Ignacio Chapela's *Nature* paper claiming to find DNA from genetically engineered corn in native Mexican maize was controversial from the moment it was published. One claim in particular came under fire: the idea that a transgenic promoter, which determines where to begin transcribing a gene, had not only been found in maize from remote areas of Oaxaca but also had been jumping at random throughout the genome. The possibility that transgenic corn might be cross-pollinating with Mexico's native maize was worrisome; the possibility that a promoter might be injected into maize DNA, with unknown effects for agriculture, was terrifying.

Soon after publication, a number of geneticists pointed out flaws in Quist and Chapela's methods, particularly those used to determine the location of the promoter. Using the results of an experimental technique known as inverse polymerase chain reaction (i-PCR), Quist and Chapela had reported that the transgenic promoter was surrounded by essentially random sequences of DNA. However, critics argued that i-PCR was unreliable and that Quist and Chapela had used insufficient controls. Their findings, the critics suggested, could have arisen from similarities between the promoter and segments of maize DNA.

Several geneticists also attacked Quist and Chapela's more fundamental claim that transgenes had been integrated into the genome of native maize. Matthew Metz of the University of Washington and Johannes Fütterer of the Institute of Plant Sciences in Zürich, Switzerland, argued that the low levels of transgenic DNA discovered by Quist and Chapela corresponded,

Ignacio Chapela (left) and David Quist (right) ignited a firestorm of controversy with their scientific paper on finding transgenic corn DNA in Mexican maize.

at most, to first-generation hybrids between local landraces and transgenic varieties. Other critics, such as Paul Christou, editor of the journal *Transgenic Research,* noted that PCR was a highly sensitive technique and that all of Quist and Chapela's findings could have been unreliable if laboratory practices had been careless.

On April 11, 2002, *Nature* published two letters from scientists critical of the study, along with an editorial note concluding that "the evidence available is not sufficient to justify the publication of the original paper." Quist and Chapela responded by acknowledging that some of their initial findings, particularly those based on the i-PCR technique, were probably invalid. They also presented new analyses to support their fundamental claim and pointed to a recent press release on a Mexican government study that had also found high rates of transgenic contamination. The new data, however, did little to convince skeptics, especially because the Mexican government's results remained unpublished.

What was published in *Nature* was only a small part of a broader debate that took place on the editorial pages of journals such as *Nature Biotechnology* and *Transgenic Research*, in newspapers and magazines, over the Internet, and

within academic institutions. The debate often turned away from the science and toward the personal. GM supporters pointed out that Quist and Chapela had a long history of opposing transgenic maize and biotechnology corporations. Chapela had spoken out against a proposal for University of California at Berkeley to partner with a biotechnology firm. GM opponents countered that most vocal critics of Quist and Chapela's work received funding from biotechnology corporations. As the controversy progressed, it became clear that few of the participants in the debate could be entirely objective. Even *Nature*, whose impartiality is critical to its reputation as a first-tier science journal, was engaged in commercial partnerships with biotechnology corporations.

The twists and turns of the still-unresolved debate illustrate that there can be more to the scientific process than merely the scientific method. Almost overlooked, however, was the fact that nearly all researchers agree that gene flow between transgenic corn and Mexico's native landraces will likely happen soon, if it has not taken place already. Researchers who are anxious about transgenic crops emphasize that such gene flow could decrease genetic diversity and negatively affect agriculture and the environment. Those in favor of transgenic foods have argued that gene flow is natural, even if the genes being transmitted are not, and that the genetic diversity of Mexican maize has persisted despite thousands of years of cross-pollination. The paucity of adequate scientific data, and the uncertain but potentially enormous economic and environmental consequences of Mexico's ban on transgenic corn, go a long way toward explaining why the debate has been so heated.

inefficiencies in distribution, almost 800 million people in developing countries do not have enough to eat. Agricultural scientists and policymakers pursue a goal of **food security,** an adequate, reliable, and available food supply to all people at all times. Whether a food supply is sustainable depends largely on maintaining healthy soil, water, and biodiversity. As we saw in Chapter 8, careless expansion of agriculture can have devastating effects on the environment and the long-term ability of the world's soils to continue supporting crops and livestock.

Starting in the 1960s, a number of doomsdayers including Paul Ehrlich (Chapter 1) predicted widespread starvation and a catastrophic failure of agricultural systems, arguing that the human population could not continue to grow without outstripping its food supply. The human population, however, has continued to increase well past their predictions. Although it is tragic that today we have 800 million hungry people, this number is smaller than the 830 million who did not have enough to eat a decade earlier, or the 960 million who lacked a reliable and sufficient food supply between 1969 and 1971.

We have achieved these dramatic increases in our carrying capacity in part by increasing our ability to produce food (Figure 9.1). We have increased food production by devoting more energy (including that of humans, draft animals, and fossil fuels) to agriculture; by planting and harvesting more frequently; by increasing the use of irrigation, fertilizer, and pesticides; by increasing the amount of cultivated land; and by developing (through crossbreeding and genetic engineering) more productive crop and livestock varieties. Our ability to raise agricultural production to keep pace with rising population over the past few decades has caused many people to believe that all is well and that we won't face any serious food shortages in the foreseeable future. However, the world's soils are in decline and nearly all the planet's arable land has been claimed. The fact that up to this point in time agricultural production has outpaced population growth does not guarantee that it will continue to do so.

Undernourishment, overnutrition, and malnourishment are all common dietary problems

Although some people do not have access to enough food to stay healthy, others are affluent enough to consume far more than is healthy. Those who are **undernourished,** receiving less than 90% of their daily caloric needs, live mostly in the developing world, whereas those who suffer from **overnutrition,** receiving too many calories each day, live largely in the developed world. A report issued in 1999 by the Washington, D.C.-based Worldwatch Institute indicated that for the first time in human history, the overfed and undernourished portions of the human population were roughly equal; the world held 1.1 billion of each. In the developed world, food is available in abundance, and people tend to lead sedentary lives with too little exercise. Sixty-one percent of U.S. adults are technically overweight, and 27% are obese. The trend extends to U.S. children, threatening a future of severe health problems and extensive health care costs.

While the *quantity* of food a person eats is important for health, so is the *quality* of food. **Malnutrition,** the lack of nutritional elements the body needs, including a complete complement of vitamins and minerals, can occur in both undernourished and overnourished individuals. In fact, malnutrition combined with obesity is becoming a common condition in the United States and other developed nations.

It is not difficult to understand why so many people have trouble meeting their nutritional needs. One-fifth of the world's people live on less than $1 per day, and almost half live on less than $2 per day, the World Bank estimates. Without adequate income, it can be impossible to purchase adequate food. Hunger is a problem even in the United States, which is by most measures one of the wealthiest nations on Earth. The U.S. Department of Agriculture (USDA) has classified 31 million Americans as "food insecure," lacking the income required to procure sufficient food at all times. In 1999, 10% of U.S. households were not food secure and people in 3%—3.1 million households—at times went hungry because of a lack of money.

Malnutrition causes various diseases

Malnutrition can lead to disease (Figure 9.2). When people eat a high-starch diet but not enough protein, or protein that does not contain all the essential amino acids (Chapter 4) in sufficient quantities, then kwashiorkor results. Children who have recently stopped breastfeeding are particularly at risk for developing kwashiorkor and, unless treated, will suffer bloating of the abdomen, deterioration and discoloration of their hair, mental disability, immune suppression, developmental delays, anemia, and reduced growth. Protein deficiency and a lack of calories can lead to marasmus, which causes wasting or shriveling among millions of children in the developing world. Several disorders are caused by dietary deficiencies in protein, carbohydrates, and vitamins—all organic molecules—but others result from

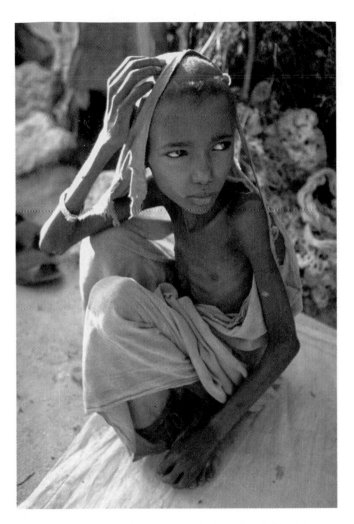

Figure 9.2 Millions of children, including this child of Somalia, suffer from forms of malnutrition, such as kwashiorkor and marasmus.

deficiencies in the inorganic nutrients known as minerals. Anemia, caused by an inability to absorb dietary iron, afflicts hundreds of millions of people, especially women of childbearing age. Hypothyroidism, caused by a lack of dietary iodine, can give rise to goiter, an enlargement of the thyroid gland, severe mental disability, and lethargy. Prevalence of hypothyroidism has been reduced in parts of the world by the addition of iodine to table salt.

The "green revolution" led to dramatic increases in agricultural production

The need for higher quantity and better quality of food for the growing human population led in the mid and late 20th century to a phenomenon called the **green revolution.** Realizing that farmers could not go on indefinitely increasing crop output by cultivating more and more land, agricultural scientists created methods and technology to increase crop output per unit area of existing cultivated land. The United States and other countries had been dramatically increasing their yields per unit of cultivated land; the average hectare of U.S. cornfield during the 20th century, for instance, upped its corn output fivefold. Many people saw such growth in production and efficiency as key to ending starvation in developing nations.

The transfer of technology to the developing world that marked the green revolution began in the 1940s, when U.S. agricultural scientist Norman Borlaug introduced Mexico's farmers to a specially bred type of wheat that produced large seed heads, was short in stature to resist wind, was resistant to diseases, and produced high yields. Within two decades of planting and harvesting this specially bred crop, Mexico tripled its wheat production and began exporting wheat. The stunning success of this program inspired others. Borlaug—who won the Nobel Peace Prize for his work—took his wheat to India and Pakistan and helped transform agriculture there. Soon many developing countries were increasing their crop yields using selectively bred grains of wheat, rice, corn, and other crops from developed nations. Some varieties of rice and wheat yielded three or four times as much per acre as did their predecessors.

The green revolution has caused the environment both benefit and harm

Along with the new grains, developing countries imported the methods of industrialized agriculture that had proved so successful at expanding agricultural production in developed nations. These countries began applying large amounts of synthetic fertilizers and chemical pesticides on their fields, irrigating crops with generous amounts of water, and using heavy equipment powered by fossil fuels. We can get a quantitative idea of the increase of these inputs by considering that from 1900 to 2000, humans expanded the world's total cultivated area by 33% and increased energy inputs into agriculture by 80 times.

This high-input agriculture was dramatically successful at allowing farmers to harvest more corn, wheat, rice, and soybeans from each hectare of land. Intensive agriculture saved millions in India from starvation in the 1970s and eventually turned the country into a net exporter of grain (Figure 9.3). However, it had mixed effects on the environment. On the positive side, the intensified use of already-cultivated land made it less necessary to convert additional natural

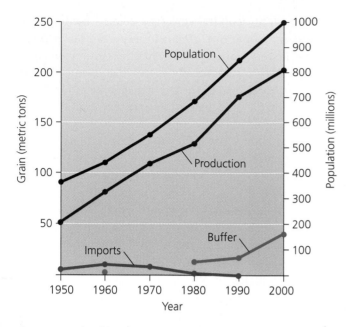

Figure 9.3 Thanks to green revolution technology, India's grain production has kept pace with its rapid population growth. This has enabled India to stop importing grain and to maintain extra grain in reserve as a buffer against food shortages.

lands for new cultivation. For this reason, the green revolution greatly slowed rates of deforestation and habitat conversion in many countries at the very time those countries were experiencing their fastest population growth rates. In this sense, the green revolution was a boon for the preservation of biodiversity and natural ecosystems. However, because of the intensive use of water, chemical fertilizers and pesticides, and fossil fuels, green revolution agriculture had negative impacts on the environment in terms of pollution, salinization, and desertification (Chapter 8).

Weighing the Issues:
The Green Revolution and Population

In the 1960s India's population was growing at an unprecedented rate, and its traditional agriculture was not producing enough food to support the growth. The future looked grim, and many predicted mass starvation. By adopting green revolution agriculture, India largely sidestepped this disaster. Producing more food enabled India to feed its people, but it also permitted India's population to keep growing. In the years since intensifying its agriculture, India has added several hundred million more people, and continues to suffer widespread poverty. Nor-

man Borlaug called his green revolution methods "a temporary success in man's war against hunger and deprivation," something to give us breathing room in which to deal with what he called "the Population Monster." Do you think we can call the green revolution a success? Do you think the green revolution has solved problems, or delayed our resolution of problems, or created new ones? How sustainable are green revolution approaches? Consider our discussion of demographic transition theory from Chapter 7. Have we been dealing with the "Population Monster" during the breathing room that the green revolution has bought for us?

One key aspect of green revolution techniques has had negative consequences for biodiversity and mixed consequences for crop yields. The planting of crops in monocultures, large expanses of single crop types (Chapter 8; Figure 9.4), has made for great efficiency in planting and harvesting and has thereby increased output overall. However, monocultural planting has reduced biodiversity over large areas; many fewer wild organisms are able to live in monocultures than in traditional small-scale polycultures. Furthermore, when all plants in a particular field are nearly identical genetically, as is the case in monocultures, all will be equally susceptible to particular viral diseases, fungal pathogens, or insect pests that can spread quickly from plant to plant. For this reason, monocultures bring significant risks of catastrophic failure.

In addition, monocultures have contributed to a narrowing of the human diet. Globally, 90% of the food humans consume now comes from only 15 crop species and eight livestock species, a drastic reduction in diversity from earlier times. Fortunately, the nutritional dangers of such dietary restriction have been alleviated by the fact that expanded global trade has provided many people (at least wealthy ones) access to a wider diversity of foods from around the world.

Scientists manage the risks posed by monocultures by searching for the genetically diverse ancestors of today's most popular crops and protecting their seeds in **seed banks** or **gene banks,** storehouses for samples of the world's crop diversity. We will examine such efforts to improve food security and conserve diversity later in this chapter. One of the reasons scientists were so concerned about the potential contamination of Oaxaca's native maize fields with DNA from transgenic corn is that Oaxaca's many types of maize served as an important source for these seed banks.

Per capita grain production has begun to decline

Due to the green revolution, increases in overall food production have been outpacing human population growth, such that there has been more food per person each year. However, this is no longer the case with grain crops, the world's staple foods. As we saw in Chapter 8, world grain production per person peaked in the mid-1980s and has since slowly fallen.

Most environmental scientists attribute these declining per capita grain yields to the declining health of agricultural lands due to the environmental effects of industrialized agriculture. As Chapter 8 made clear, problems such as desertification are reducing the productivity of many lands and making it impossible to farm some once-fertile areas. During the 1990s, per capita cropland declined by 20% worldwide. This decline in turn caused per capita declines during that decade in irrigation (down 15%) and fertilizer use (down 23%). Although today's food security problems are largely matters of distribution, not inadequate production, they could become driven by production if these trends of declining health in agricultural lands continue. As croplands, rangelands, and fisheries experience increasing stress from rising human population and environmental degradation, scientists have turned to solutions as diverse as genetic engineering and organic agriculture, in an attempt to obtain food security while protecting environmental quality.

Pest Management

Before we discuss genetically modified crops and organic agriculture, let's examine the strategies developed for dealing with the insects, fungi, viruses, and weeds that eat or compete with our crop plants. The strategies and chemical products involved in pest control efforts have far-reaching impacts for agriculture and for the health of wild plants, animals, ecosystems, and humans. In this chapter we will investigate the role of pest management in agriculture. We will address its health consequences for humans and other organisms in Chapter 10.

What humans term a *pest* is any organism that damages crops that are valuable to us. Similarly, what we term a *weed* is any plant that competes with our crops. There is nothing inherently malevolent in the behavior of a pest or a weed; these are subjective categories defined entirely by our own economic interests. The organisms themselves are simply trying to survive and reproduce. From the viewpoint of an insect that happens to specialize on corn, rice, grapes, or apples, a grain field, vineyard, or orchard represents a smorgasbord. The presence of so many similar plants adjacent to one another means abundant food for the insect and its offspring.

(a) Wheat monoculture

(b) Armyworm

Figure 9.4 Most agricultural production in developed countries comes from monocultures—large stands of single species of crop plant, such as this wheat field in Washington (**a**). Clustering crop types in uniform fields on large scales greatly improves the efficiency of planting and harvesting but also causes losses in biodiversity and makes crops susceptible to outbreaks of pests that specialize on particular crops. Armyworms (**b**) are major agricultural pests whose outbreaks can lead to substantial reductions in crop yields. The caterpillars of a number of species of these moths defoliate a wide variety of grains, vegetables, and other crops, including wheat, corn, cotton, alfalfa, and beets.

Throughout the history of agriculture, insects, fungi, viruses, rats, and other organisms have taken advantage of our clustering food plants into agricultural fields. These organisms, in making a living for themselves, cut crop yields and make it harder for farmers to make a living. As just one example of thousands, various species of moth caterpillars known as armyworms stage outbreaks in fields of a wide variety of crops worldwide, lowering yields of everything from beets to sorghum to millet to canola to pasture grasses. While pests and weeds have always posed problems for traditional agriculture, they can pose even more of a threat for monocultures, where a pest specializing on the crop can easily move from one individual plant to many others of the same type.

Many thousands of chemical pesticides have been developed and applied

To prevent pest outbreaks and to limit competition with weeds, people have developed thousands of artificial chemicals to kill insects (insecticides), plants (herbicides), and fungi (fungicides). Together, along with poisons targeting other organisms, insecticides, herbicides, and fungicides are termed **pesticides.** Table 9.1 shows the top ten most widely used pesticides on U.S. crops, along with the estimated amounts of each active ingredient applied each year. All told, close to 1 billion kg (over 2 billion lbs) of pesticides are applied in the United States each year, most of this total to agricultural land. Pesticide

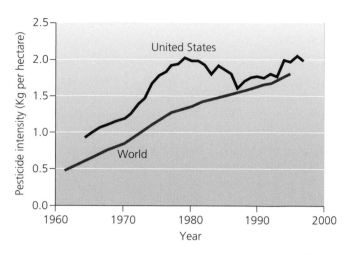

Figure 9.5 Pesticide use in developed nations such as the United States has leveled off in recent years but continues to grow across the world as a whole. Data from Food and Agriculture Organization of the United Nations (FAO), 1999.

use has increased greatly worldwide since the middle of the 20th century, rising fourfold since 1960. Usage in the United States and the rest of the developed world has more or less leveled off in the past two decades, but the developing world continues to use more and more (Figure 9.5). The monetary value of pesticide imports and exports has risen almost ninefold since 1960, and today more than $33.5 billion is expended annually on pesticides, with one-third of that total being spent in the United States.

Pests evolve resistance to pesticides

Despite the high toxicity of many of these chemicals, their usefulness tends to decline with time, as pests evolve resistance to them. Recall from our discussion of natural selection in Chapter 4 that organisms within a population vary in their traits. Because most insects and microbes can occur in huge numbers, it is likely that a small fraction of individuals may by chance have genes that confer some degree of immunity to a given pesticide. Even if a pesticide application kills 99.99% of all the insects in a field, still 1 in 10,000 survives. If an insect survives by virtue of being resistant to a pesticide, and if it mates with other insects that are similarly resistant, the insect population may build up. The new population to which these insects give rise will consist of individuals that are genetically resistant to the pesticide. As a result, pesticide applications will cease to be effective (Figure 9.6). This is evolution by natural selection, and it has had major effects on agriculture throughout

Table 9.1 Most commonly used pesticides in agriculture in the United States		
Active ingredient	Type of pesticide*	Millions of kilograms applied per year*
Atrazine	herbicide	34–36
Glyphosate	herbicide	30–33
Metam Sodium	fumigant	27–29
Acetochlor	herbicide	14–16
Methyl Bromide	fumigant	13–15
2,4-D	herbicide	13–15
Malathion	insecticide	13–15
Metolachlor	herbicide	12–14
Trifluralin	herbicide	8–10
Pendamethalin	herbicide	8–10

*Includes only *conventional* pesticide active ingredients and not ingredients such as oil, sulfur, and sulfuric acid and only those used in agriculture (72% of total usage). Data from D. Donaldson, T. Kield, and A. Grube, *U.S. EPA, Pesticides Industry Sales and Usage: 1998 and 1999 Market Estimates*, 1999.

Figure 9.6 Crop pests frequently evolve resistance to the poisons we apply to kill them, through the process of natural selection (Chapter 4). When a pesticide is applied to an outbreak of insect pests, it may kill virtually all individuals except those few with an innate immunity to the poison. Those surviving individuals may then serve as the founders of a population with genes for resistance to the poison. In such a case, future applications of the pesticide may be ineffective, forcing us to develop a more potent poison or an alternative means of pest control.

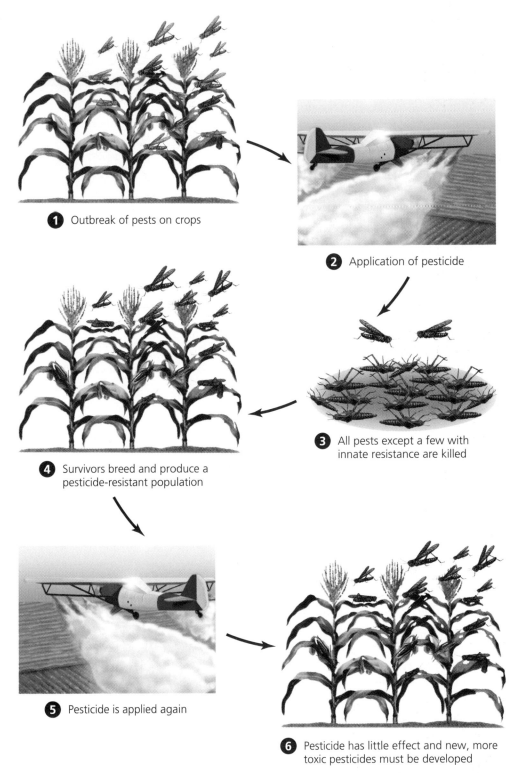

1 Outbreak of pests on crops

2 Application of pesticide

3 All pests except a few with innate resistance are killed

4 Survivors breed and produce a pesticide-resistant population

5 Pesticide is applied again

6 Pesticide has little effect and new, more toxic pesticides must be developed

recent history. In many cases, industrial chemists are caught up in an "evolutionary arms race" with the pests they battle, racing to increase or retarget the toxicity of their chemicals while the armies of pests evolve ever-stronger resistance to their efforts.

A number of insects like the green peach aphid, the Colorado potato beetle, and the diamondback moth, have evolved resistance to multiple insecticides. Resistant pests can take a significant economic toll on crops. As just one example, gummy stem blight destroyed two-thirds of Texas's melon crop in 1997 after blight evolved resistance to the pesticide Benlate.

Biological control pits one organism against another

Because of pesticide resistance, and because of the potential health dangers of synthetic chemicals in the environment, agricultural scientists have increasingly attempted to battle pests and weeds with other organisms that eat or infect them. This strategy, called **biological control**, or **biocontrol** for short, operates on the principle that "the enemy of one's enemy is one's friend." For example, parasitic wasps that lay eggs on particular species of leaf-munching caterpillars are natural enemies of caterpillars that attack crops. The eggs hatch, and the young wasp larvae feed on, and eventually kill, the caterpillar. Such wasps have been used as biocontrol agents in many situations. Such efforts have sometimes succeeded at pest control and have led to steep reductions in chemical pesticide use where they have been tried. One classic case of successful biological control is the introduction of the cactus moth, *Cactoblastis cactorum*, from Argentina to Australia in the 1920s to control invasive prickly pear cactus that was overrunning rangeland (Figure 9.7). Within just a few years, the moth managed to free millions of hectares of rangeland from the cactus.

One of the most widespread biocontrol efforts has been the use of *Bacillus thuringiensis* **(Bt)**, a naturally occurring soil bacterium that produces a protein that kills many pests, including caterpillars and the larvae of some flies and beetles. Many farmers have used the natural pesticide-producing abilities of this bacterium to their advantage by spraying the bacteria on their crops. If used correctly, Bt can protect crops from pest-related losses when applied periodically. Some evidence that Bt poses risks to humans and non-target species exists, but such evidence is limited.

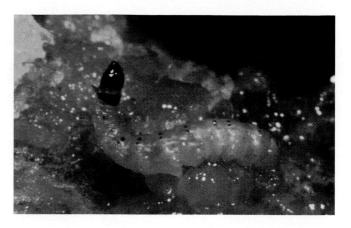

Figure 9.7 In one of the classic cases of biocontrol, larvae of the cactus moth, *Cactoblastis cactorum*, were used to clear non-native prickly pear cactus from millions of hectares of Australian rangeland.

Biological control agents themselves may become pests

In most cases, biological control involves introducing an animal or microbe to a foreign ecosystem. Frequently, this means moving the organism from one continent to another for its use as a biocontrol agent. Although such relocation helps ensure that the target pest has not already evolved ways to deal with the agent, it also introduces a number of risks. Because the biocontrol agent's presence in the new system is unprecedented, no one can know for certain in advance what effect it might have. In many cases, biocontrol has produced unintended consequences as the biocontrol agent became invasive and affected non-target organisms. Following the cactus moth's success in Australia, for example, the moth was introduced in other countries to control prickly pear. Moths introduced to Caribbean islands spread to Florida on their own and are now eating their way through rare native cacti in Florida and spreading to other states. If these moths reach Mexico and the southwestern United States, they could do great damage to many native and economically important species of prickly pear there.

One recent study examined the extent to which some biocontrol agents have missed their targets in Hawaii. Wasps and flies have been introduced for control of agricultural pests at least 122 times in Hawaii over the past century, and biologists Laurie Henneman and Jane Memmott suspected that some of these might be adversely affecting native Hawaiian caterpillars that were not pests. They sampled parasitoid wasp larvae from

2,000 caterpillars of various species in a remote mountain swamp far from farmland. In this area, designated as a wilderness preserve, they found that 83% of the parasitoids were biocontrol agents that had been intended to combat lowland agricultural pests. If biocontrol agents were exerting such effects on non-target organisms so far from farmland, most likely their effects were strong and widespread throughout the islands.

Scientists argue over the relative benefits and risks of biocontrol measures. If biocontrol works as planned, it can be a permanent solution that requires no further maintenance and is environmentally benign. However, if the agent has non-target effects, the harm done may also become permanent, because removing the agent from the system once it is established is far more difficult than simply stopping a chemical pesticide application. A noted skeptic of biocontrol, ecologist Daniel Simberloff (see Chapter 15), has remarked that biocontrol "should be used with our eyes wide open—and as a last resort." However, two British scientists reviewing cases as of 2000 concluded that only a small percentage of efforts resulted in demonstrable non-target effects, and perhaps less than 10% of these effects were substantial. Because of concerns about unintended effects, researchers now study biocontrol proposals carefully before putting them into action, and these efforts must be approved by government regulators. However, there is no sure-fire way of knowing whether biocontrol will work before it is implemented.

Integrated pest management combines biocontrol and chemical methods

As it became clear that both chemical and biocontrol approaches have their drawbacks, many agricultural scientists and farmers developed a more sophisticated strategy, trying to combine the best attributes of the two approaches. In **integrated pest management (IPM)**, numerous techniques are integrated to achieve long-term suppression of pests, including biocontrol, use of chemicals, close monitoring of populations, habitat alteration, crop rotation, transgenic crops, alternative tillage methods, and mechanical pest removal. IPM is broadly enough defined that it encompasses a wide variety of strategies.

The state of Texas can boast a number of success stories with IPM. Texas farmers reduced the impact of a virus on peanut yields by delaying planting until the ground warmed and by reducing use of insecticides that killed predators and parasites of peanut pests. Farmers of cotton and other crops reduced whitefly infestations by

following nine measures, including plowing under crop residues that had helped nourish the insects between plantings. In pecan orchards, surveys and computer models of the population dynamics of pests and their natural enemies helped farmers reduce unnecessary insecticide spraying. In addition, stink bugs that ruin pecans were successfully lured away from the nuts by planting peas in the pecan orchards, since stinkbugs prefer peas to pecans. All in all, Texas agricultural agencies have estimated that the state economy benefits by $340 million per year from farmers who use IPM methods to reduce their pesticide costs and/or increase their yields.

In recent decades, IPM has seen great popularity in many parts of the world. Indonesia stands as an exemplary case (Figure 9.8). The nation had subsidized pesticide use heavily for years, but its scientists came to understand that pesticides were actually making pest problems worse. They were killing the natural enemies of the brown planthopper, which began to devastate rice fields as its populations exploded. Concluding that pesticide subsidies were costing money, causing pollution, and apparently decreasing yields, the Indonesian government in 1986 banned the importation of 57 pesticides, slashed pesticide subsidies, and encouraged IPM. Within four years, pesticide production fell to below

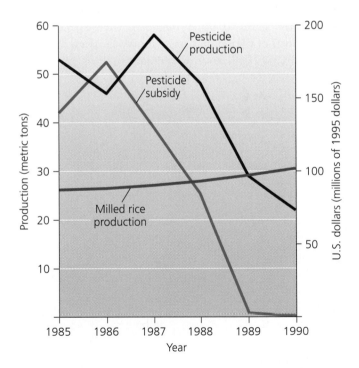

Figure 9.8 The Indonesian government threw its weight behind integrated pest management starting in 1986. Within just a few years, pesticide production and pesticide imports were down drastically, pesticide subsidies were phased out, and yields of rice increased slightly.

half its 1986 level, imports fell to one-third, subsidies were phased out (saving $179 million annually), and rice yields rose 13%.

Pollination: "Good" Insects, Unsung Heroes

The management of pests is such a major issue in agriculture that many people fall into a habit of thinking of all insects as somehow bad or dangerous. In fact, not only are most insects harmless to agriculture, but some are absolutely essential. The insects that pollinate agricultural crops are one of the most vital, yet least understood and least appreciated, factors in cropland agriculture.

Pollination is the process by which male sex cells of a plant fertilize female sex cells of a plant; it is the botanical version of sexual intercourse. Without pollination, no plants could reproduce sexually, and no plant species would persist for long. Pollination occurs when pollen, the male sex cells of a plant, reaches the female sex organs of a plant of the same species. Many plants, such as ferns, conifer trees, and grasses, achieve pollination by the wind; millions of miniscule pollen grains are blown long distances, and a small number by chance land on the female parts of other plants of their species. The many kinds of plants that sport showy flowers, however, are as a rule pollinated by animals such as humming-birds, bats, and insects (Figure 9.9). Flowers are, in fact, evolutionary adaptations that function specifically to attract pollinators. The sugary nectar and protein-rich pollen in flowers serve as rewards to lure these sexual intermediaries, and the sweet smells and bright colors of flowers are signals to advertise their rewards. To serve as further advertisement, many flowers even sport distinctive patterns visible only in ultraviolet light (Chapter 4) that insects and birds can see.

We depend on insects to pollinate many of our crops

Although our staple grain crops are derived from grasses and therefore are wind-pollinated, many of our other crops depend on insects for pollination (Table 9.2). The most complete survey to date, by tropical bee biologist Dave Roubik, has documented fully 800 species of cultivated plants that rely on bees and other insects for pollination. An estimated 73% of cultivars are pollinated, at least in part, by bees; 19% by flies; 5% by wasps; 5% by beetles; and 4% by moths and butterflies. In addition, the study estimates 6.5% to be pollinated by bats, and 4% by birds. As one of many examples, alfalfa has long been a major cash crop in the Great Basin states of the United States and is pollinated mostly by native alkali bees that live in the soil as larvae. For decades farmers were unaware of the bees' services, however, and

Figure 9.9 Many agricultural crops depend on insects or other animals to pollinate them. Our food supply thus is partly dependent on conservation of these important creatures. These apple blossoms are being visited by a European honeybee. Flowers use colors and sweet smells to advertise nectar and pollen, enticements that attract pollinators.

Table 9.2 Pollinators of Some Major Crop Plants

Crop	Pollinator(s)
Potato	Bumblebees
Yams	Bees, beetles, flies
Sugarcane	Bees, thrips
Soybean	Bees
Beans	Bees, thrips
Sunflower	bees, flies
Cotton	Bees
Sesame	Bees, flies, wasps
Almond	Bees
Tomato	Bees
Cabbage	Honeybees
Onions	Honeybees, flies
Carrot	Flies, honeybees
Eggplant	Bees
Banana	Birds, bats

Data adapted from Stephen L. Buchmann and Gary Paul Nabhan, *The Forgotten Pollinators*, Shearwater Books, 1996.

in the 1940s to 1960s they began plowing the land and increasing pesticide usage in an effort to boost yields. These measures killed vast numbers of the soil-dwelling bees, and alfalfa production declined. By 1990, only 15% of U.S. alfalfa-growing lands were inhabited by alkali bees and other native bees called leaf-cutter bees. Despite their decline in numbers, bees and their pollination services help these lands produce fully 85% of the U.S. alfalfa crop.

Preserving the natural biodiversity of native pollinators is especially important today because the domesticated workhorse of pollination, the European honeybee, is being devastated by parasites. North American farmers regularly hire beekeepers to bring colonies of this introduced honeybee to their fields when it is time to pollinate crops (Figure 9.10), but in recent years certain parasitic mites have swept through honeybee populations, decimating many hives and pushing many beekeepers toward financial ruin. Furthermore, research indicates that honeybees are less effective pollinators than many native species and in fact often outcompete them, keeping the native species away from the plants. Recognizing our dependence on a single species for pollinating so many crops, the USDA has recently begun acting on advice from ecologists and is funding research to develop ways to protect and encourage native species of pollinators.

All insect pollinators, including honeybees, are vulnerable to the vast arsenal of insecticides that are applied to crops in modern industrial agriculture. Farmers want to control costs and maximize profits and also want to keep the land on which they and their families live unpolluted, so they try to minimize pesticide use whenever possible. However, farmers do not often have full and detailed information on the effects of pesticides on pest populations inhabiting their crops, so in their efforts to control the "bad" bugs that threaten their crops, farmers too often kill the "good" insects as well. Some pesticides are designed to be somewhat specific, targeting certain types of insects, but many are not. We will examine chemical effects on nontarget organisms more in Chapter 10, but it is clear that by killing pollinators, excess or careless use of insecticides can sometimes cause greater harm than good to crop yields.

Luckily, there are ways that farmers—and in fact anyone with a backyard—can help maintain populations of pollinating insects. Reducing or eliminating pesticide use on fields and lawns is the best way. In addition, planting gardens of flowering plants that nourish pollinating insects, even in the middle of a large city, can encourage populations of these valuable creatures. Homeowners and farmers alike can also provide small contraptions of wood and plastic straws that provide holes to serve as nesting sites for bees. And when farmers allow certain noncrop flowering plants (such as clover) that nourish pollinators to grow around the edges of their fields, they can maintain a diverse community of insects, some of which will pollinate their crops.

Genetic Modification of Food

The green revolution enabled us to feed a greater number and proportion of the world's people. Relentless population growth demanded more, however. A new set of potential solutions began to arise in the 1980s and 1990s as advances in the field of genetics enabled scientists to directly alter the genes of organisms, including crop plants and livestock. The genetic modification of organisms that provide us food holds promise for increasing the efficiency of agriculture and the nutrition of foods while lessening the impacts of agriculture on the planet's environmental systems. However, genetic modification may also pose risks that are not yet well understood, which has given rise to protest around the globe from consumer advocates, small farmers, opponents of big business, and environmental activists. The techniques have also caused political divisiveness among nations, as some have refused the genetically altered foods exported by others.

Figure 9.10 European honeybees are widely used as pollinators of crop plants, although they sometimes compete with native pollinators and have recently suffered devastating epidemics of parasitism. Beekeepers transport hives of bees to crops when it is time for flowers to be pollinated.

Because genetically modified foods have generated so much emotion and controversy, it is important at the outset to clear up the terminology and clarify exactly what the techniques involve.

Genetic modification of organisms depends on recombinant DNA

The genetic modification of crops and livestock is one type of genetic engineering. **Genetic engineering** is any process whereby scientists directly manipulate an organism's genetic material in the lab, by adding, deleting, or changing segments of its DNA. Such human intervention aims to create genetic variants that will produce organisms with traits desired by the researcher. **Genetically modified (GM) organisms** are organisms that have been genetically engineered using a technique called recombinant DNA technology. **Recombinant DNA** is DNA that has been patched together from the DNA of multiple organisms. In this process, scientists break up DNA from multiple organisms and then splice segments together, trying to place genes that produce certain proteins and code for certain desirable traits (such as rapid growth, disease and pest resistance, or higher nutritional content) into the genomes of organisms lacking those traits.

Recombinant DNA technology was developed during the 1970s by scientists studying the *Escherichia coli* (*E. coli*) bacterium. As shown in Figure 9.11, scientists first isolate plasmids, small, circular DNA molecules, from a bacterial culture. At the same time, DNA containing a gene of interest is removed from the cells of another organism. Scientists insert the gene of interest into the plasmid to form recombinant DNA. This recombinant DNA enters new bacteria, which then reproduce, generating many copies of the desired gene.

When scientists use recombinant DNA technology to develop new varieties of crops, they can often introduce the recombinant DNA directly into a plant cell and then regenerate an entire plant from that single cell. Some plants, including many grains, are not receptive to plasmids, in which case scientists often use a "gene gun" to shoot the DNA directly into plant cells. An organism that contains DNA from another species is called a **transgenic** organism, and the genes that have moved between them are called transgenes. The creation of transgenic organisms is one type of **biotechnology,** the material application of biological science to create products derived from organisms. Recombinant DNA and other types of biotechnology have helped us develop medicines, clean up pollution, understand the causes of cancer and many other diseases, dissolve

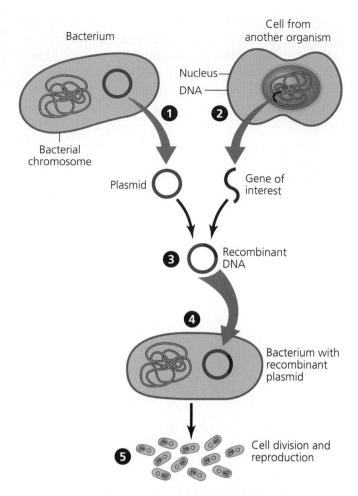

Figure 9.11 The creation of recombinant DNA is a key to genetically modifying organisms. In this process, a gene of interest is excised from the DNA of one type of organism and is inserted into a stretch of bacterial DNA called a plasmid. The plasmid is then introduced into cells of the organism to be modified. If all goes as planned, the new gene will be expressed in the GM organism as a desirable trait, such as rapid growth or high nutritional content in a food crop.

blood clots after heart attacks, make better beer and cheese, and even enable snow to form at higher temperatures. Figure 9.12 details several of the most notable developments in GM foods. These examples and the stories behind them illustrate both the promises and pitfalls of food biotechnology.

Genetic engineering is like, and unlike, traditional agricultural breeding

The genetic alteration of plants and animals by humans is nothing new; we have been influencing the genetic makeup of our livestock and crop plants for thousands of years, ever since we invented agriculture. As we saw in Chapter 8, our ancestors altered the gene pools of our domesticated plants and animals through selective

Genetically Modified Foods

Proponents of genetically modified foods say these products can alleviate hunger and malnutrition and pose no known threats to human or environmental health. Opponents say they help only the large corporations that sell them, while posing unknown risks to human health and threatening to genetically alter wild organisms and affect ecosystems. What do you think? Should we encourage the continued development of GM foods or not, and what, if any, restrictions should we put on their dissemination?

The United States Should Begin a Phased Deregulation of Biotech Crops

During the past two decades the international scientific community, biotechnology industry, and regulatory agencies in many countries have accumulated and critically evaluated a wealth of information about the production and use of biotech crops and products. Biotech crops have been planted since 1996 on more than 700 million acres of farmland in nearly 20 countries. Biotech foods and products have been consumed by more than a billion humans and hundreds of millions of farm animals. Yet there is not a single instance in which biotech crops and foods have been shown to cause illness in humans or animals or to damage the environment.

In spite of this exemplary safety record, a small but well-organized, well-financed, and vocal anti-biotechnology lobby has alleged that biotech crops and products are unsafe for humans and a danger to the environment, demanding a moratorium or outright ban on biotech crops. The rhetoric of the anti-biotechnology groups is alarming, confusing, and frightening to the public, but it is devoid of any substance, as they have never provided any credible scientific evidence to support their allegations.

Any further delay in combining the power of biotechnology will seriously endanger future food security, political and economic stability, and the environment. Plant biotechnology is still the best hope for meeting the food needs of the ever-growing world population. Biotech crops are already helping to conserve valuable natural resources, reduce the use of harmful agro-chemicals, produce more nutritious foods, and promote economic development.

Twenty years ago, the United States set the precedent by developing regulations for the development and use of biotech crops. Now, as the world-leader in plant biotechnology, it is imperative that it lead again by phasing out these redundant regulations in an organized and responsible manner.

Indra K. Vasil is Graduate Research Professor Emeritus at the University of Florida (Gainesville, FL). His research focus is on the biotechnology of cereal crops, and he has authored numerous papers on the plant and animal sciences.

A Global Experiment Without Controls

Genetic engineering, specifically transgenesis, gives us the unprecedented capacity to move DNA. In so doing, this technology breaches boundaries established through millions of years of evolution. As such, we should expect fundamental alterations in ecosystems with the release of transgenic crops, fish, insects, microbes, and so forth into uncontrolled areas. These alterations are similar in nature to those caused by the introduction of exotic species into new environments. Both processes are unpredictable and could have serious consequences.

Yet because of political and short-term economic imperatives, releases of transgenic organisms have continued unabated for at least a decade. Science has barely started to imagine the ecological and evolutionary consequences of releasing transgenic crops. Not only are we experiencing a global experiment without controls, we don't have the tools to document it. Serious research, although extremely scarce, has already confirmed some of the theoretical fears concerning transgenesis aired by scientists a quarter-century ago.

Today, we have cataclysmic world hunger paired with food surpluses. The claim that transgenesis can solve this problem is merely a diversion tailored to conceal how transgenesis manipulates the biosphere. Molecular biology might one day become part of the solution to world hunger, but it is certainly not the science most relevant to address the problem today.

What checks should be placed on the release of transgenic organisms? Every check. Through the unaccountable releases so far, we have seen enough, and possibly caused enough, environmental insult for me to say today that we should stop. We need to take stock of the consequences of transgenesis and continue researching under strictly regulated conditions.

Ignacio H. Chapela is Assistant Professor (Microbial Ecology), in the Department of Environmental Science, Policy, and Management at the University of California, Berkeley. He helped found the Mycological Facility: Oaxaca, Mexico, where he also serves as Scientific Director.

	Several Notable Examples of Genetically Modified Food Technology
Food	**Development**
Golden rice 	Millions of people in the developing world get too little vitamin A in their diets, causing diarrhea, blindness, immune suppression, and even death. The problem is worst with children in east Asia, where the staple grain, white rice, contains no vitamin A. Researchers took genes from plants that produce vitamin A and spliced the genes into rice DNA to create more-nutritious "golden rice" (the vitamin precursor gives it a golden color). Critics charged that biotech companies over-hyped their product, which contains only small amounts of the nutrient and may not be the best way to combat vitamin A deficiency. India's foremost critic of GM food, Vandana Shiva, charged that "vitamin A rice is a hoax… a very effective strategy for corporate takeover of rice production, using the public sector as a Trojan horse." Backers of the technology counter that the nutritive value can be further improved and could enhance the health of millions of people.
FlavrSavr tomato 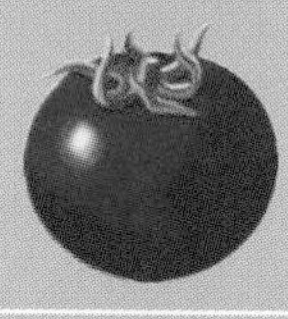	By reversing the function of a normal tomato gene, the Calgene Corporation created the FlavrSavr tomato, which Calgene maintained would ripen longer on the vine, taste better, stay firm during shipping, and last longer in the produce department. The U.S. Food and Drug Administration approved the FlavrSavr tomato for sale in the United States in 1994. Calgene stopped selling the FlavrSavr in 1996, however, for several reasons, including problems with the technique and public safety concerns.
Ice-minus strawberries 	University of California-Berkeley researcher Steven Lindow removed a gene that facilitated the formation of ice crystals from the DNA of a particular bacterium, *Pseudomonas syringae*. The modified, frost-resistant bacteria could then serve as a kind of anti-freeze when sprayed on the surface of frost-sensitive crops such as strawberries. The multiplying bacteria would coat the berries, protecting them from frost damage. However, early news coverage of this technique showed scientists spraying plants while wearing face masks and protective clothing, an image that caused public alarm.
Bt crops 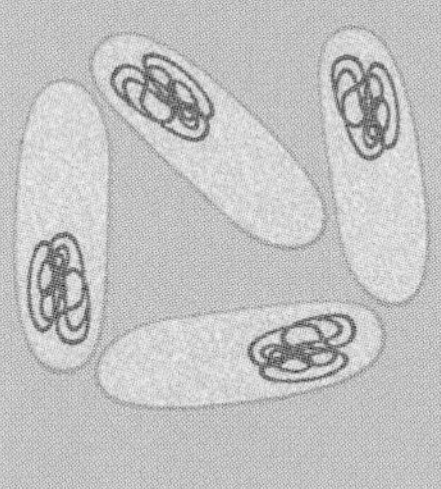	By equipping plants with the ability to produce their own pesticides, scientists hoped to boost crop yields by reducing losses to insects. By the late 1980s scientists working with *Bacillus thuringiensis* (Bt) had pinpointed the genes responsible for producing that bacterium's toxic effects on insects, and had managed to insert the genes into the DNA of crops. The USDA and EPA approved Bt versions of 18 crops for field testing, from apples to broccoli to cranberries. Corn and cotton are the most widely planted Bt crops today. Proponents say Bt crops reduce the need for chemical pesticides. However, critics worry that the continuous presence of Bt in the environment will induce insects to evolve resistance to the toxins and that Bt crops might cause allergic reactions in humans. Another concern is that the crops may harm non-target species. A 1999 study reported that pollen from Bt corn can kill the larvae of monarch butterflies, a non-target species. Monarchs in the wild would be harmed when corn pollen drifts onto milkweed plants monarchs eat, the study's authors maintained. Another study that year showed the Bt toxin could leach from corn roots and poison the soil.

Figure 9.12 The early development of genetically modified foods has been marked by a number of cases in which these products ran into trouble in the marketplace or were opposed by activists. A selection of these cases serves to illustrate some of the issues that proponents and opponents of GM foods have being debating.

breeding by preferentially mating individuals with favored traits so that those traits would be inherited by offspring. Large ears of corn were preferred over small ears, so large-eared individuals were mated and small-eared ones were discarded, leading to a gene pool full of genes for large-eared corn. Early farmers selected plants and animals that grew faster, were more resistant to disease and drought, and produced large amounts of fruit, grain, or meat. In the broad sense, therefore, the genetic modification of organisms by humans is an ancient exercise and one with which we have much experience.

Proponents of GM crops often stress this continuity with our agricultural past and argue that there is little reason to expect that today's GM food will be any less safe than the selectively bred food of past years and centuries. Dan Glickman, head of the USDA from 1995 to 2001, remarked:

Biotechnology's been around almost since the beginning of time. It's cavemen saving seeds of a

Several Notable Examples of Genetically Modified Food Technology	
Food	**Development**
StarLink corn	StarLink corn, a variety of Bt corn, had been approved and used in the United States for animal feed but not for human consumption. In 2000, StarLink corn DNA was discovered in taco shells and other corn products. These products were recalled, amid fears that the corn might cause allergic reactions. No such health effects have been confirmed, but the corn's French manufacturer, Aventis CropScience, chose to voluntarily withdraw the product from the market. Although StarLink corn was grown on only a tiny portion of U.S. farmland, its transgene apparently spread widely to other corn through cross-pollination. This episode cost U.S. taxpayers, because the U.S. government spent $20 million to purchase contaminated corn and remove it from the food supply.
Sunflowers and superweeds	Sunflowers have also been engineered to express the Bt toxin. Research on Bt sunflowers suggests that their transgenes might spread to other plants and turn them into vigorous weeds that compete with the crop. This is most likely to happen with crops like squash, canola, and sunflowers that can breed with their wild relatives. In 2002, Ohio State University researcher Alison Snow and colleagues bred wild sunflowers with Bt sunflowers and found that hybrids with the Bt gene produced more seeds and suffered less herbivory than hybrids without it. They concluded that if Bt sunflowers were planted commercially, the Bt gene would spread into wild sunflowers, potentially turning them into superweeds. Researcher Norman Ellstrand of the University of California-Riverside, meanwhile, had found that transgenes from radishes were transferred to wild relatives 1 km (0.6 mi) away and that hybrids produced more seeds, so the gene could be expected to spread in wild populations. He found the same results with sorghum and its weedy relative, johnsongrass. Such results suggest that transgenic crops can potentially create superweeds that can compete with crops and harm non-target organisms.
Roundup Ready crops	The Monsanto Company manufactures a widely used herbicide called Roundup. Roundup kills weeds, but kills crops too, so farmers must apply it carefully. Thus, Monsanto engineered Roundup Ready crops, including soybeans, corn, cotton, and canola, that are immune to the effects of its herbicide. With these variants, farmers can spray Roundup on their fields without killing their crops, in theory making the farmer's life easier. Of course, this also creates an incentive for farmers to use Monsanto's Roundup herbicide rather than a competing brand. Unfortunately, Roundup is not completely benign; its active ingredient, glyphosate, is the third-leading cause of illness for California farm workers. It also harms nitrogen-fixing bacteria and desirable fungi in soils that are essential for crop production. Biotech proponents have argued that GM crops are good for the environment because they reduce pesticide use. This may often be the case, but some studies have shown that farmers apply more herbicide when they use Roundup Ready crops.
Terminator seeds	In the late 1990s the USDA worked with Delta and Pine Land Company to engineer a line of crop plants that can kill their own seeds. This so-called "terminator" technology would ensure that farmers buy seeds from seed companies every year rather than planting seeds saved from the previous year's harvest. Because GM crops require a great deal of research and development, seed companies reason that they need to charge farmers annually for seeds in order to recoup their investment. Critics worried that pollen from terminator plants might fertilize normal plants, damaging the crops of farmers who save seeds from year to year. Opposition was fierce in the developing world, and some nations, like India and Zimbabwe, banned terminator seeds. These countries saw the efforts of biotech seed companies to sell them terminator seeds as a ploy to make poor farmers dependent on multinational corporations for seeds. In the face of this opposition, in 1999 agrobiotech companies Monsanto and AstraZeneca announced that they would not bring their terminator technologies to market.

high-yielding plant. It's Gregor Mendel, the father of genetics, cross-pollinating his garden peas. It's a diabetic's insulin, and the enzymes in your yogurt. . . . Without exception, the biotech products on our shelves have proven safe.

However, as biotech critics are quick to point out, the techniques geneticists use to create GM organisms do differ from traditional selective breeding in several ways. For one, traditional selective breeding generally mixes genes of individuals of the same species, whereas with recombinant DNA technology, scientists mix genes of different species, even species as different as viruses and crops, or spiders and goats. Even if they had tried, our agricultural ancestors could not have crossed trees with peas, or bats with cats. For another, selective breeding deals *in vivo* with whole organisms living in the field, whereas genetic engineering takes place in the lab, involving *in vitro* experiments dealing with genetic material apart from the organism. And whereas traditional breeding selects from among combinations of genes that come together on their own, genetic engineering creates the novel combinations directly. Traditional breeding thus changes organisms through the process of selection, whereas genetic engineers intervene at the stage of mutation.

Biotechnology is transforming the food and products around us

In just three decades, GM foods have gone from science fiction to big business. As recombinant DNA technology first developed in the 1970s, scientists debated among themselves whether the new methods were safe. They regulated and closely monitored their own research until most scientists were satisfied that reassembling genes into bacteria did not create dangerous superbacteria. Once the scientific community declared itself confident that the technique was safe in the 1980s, industry leaped at the chance to develop hundreds of applications, from improved medicines such as hepatitis B vaccine and insulin for diabetes to designer plants and animals. Traits engineered into crops, such as built-in pest resistance and herbicide resistance, made it easier and cheaper for large-scale commercial farmers to do their jobs, so sales of GM seeds to these farmers in the United States and other countries took off.

Today well over two-thirds of the U.S. harvests of soybeans, corn, and cotton consists of genetically modified strains. Worldwide in 2002, it was estimated that GM crops were planted on 58.7 million ha (145 million acres) of farmland, an area between the sizes of California and Texas. According to an annual report of the International Service for the Acquisition of Agri-biotech Applications, these crops were grown by 5.5–6 million farmers in 16 countries. Four nations (the United States, Argentina, Canada, and China) accounted for 99% of GM crops, with the United States alone accounting for two-thirds of the global total (Figure 9.13). Because these nations are major food exporters, much of the produce on the world market is transgenic for certain crops such as soybeans, corn, cotton, and canola. In addition, other major agricultural nations, such as India and Brazil, are just now beginning to approve and grow GM crops. The global area planted in GM crops has jumped by more than 10% annually every year since 1996. In 2002, for the first time, over half the world's people lived in nations in which GM crops are grown. The market value of these crops grew from $75 million in 1995 to an estimated $4.25 billion in 2002.

As GM crops were adopted, as research proceeded, and as biotech business expanded, however, many scientists, citizens, and politicians became concerned. Some were afraid the new foods might be dangerous for humans to eat. Others were concerned that transgenes might escape, and pollute ecosystems and damage non-target organisms. Still others worried that pests would evolve resistance to the supercrops and become "superpests," or that transgenes would be transferred from crops to other plants and turn them into "superweeds." Some, like Quist and Chapela, worried that transgenes might ruin the integrity of native ancestral races of crops. Because the technology is new and its large-scale introduction into the environment is newer still, there remains

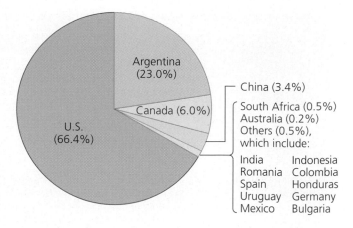

Figure 9.13 At 66.4% of the global total, the United States leads the world in land area dedicated to genetically modified crops. Following, although far behind the United States, is Argentina at 23%. Data from Global area of genetically modified crops in 2002, Biotech, 2003.

a lot scientists don't know about how transgenic crops behave in the field. Certainly, millions of Americans and others eat GM foods every day without outwardly obvious signs of harm, and evidence for ecological effects is limited so far. However, it is still too early to dismiss the concerns discussed above without further scientific research. Therefore, critics argue that we should proceed with caution, adopting the **precautionary principle**, the idea that one should not undertake a new action until the ramifications of that action are well understood.

The best efforts so far to test the ramifications of planting GM crops produced results in 2003. These were three large-scale studies commissioned by the British government as it considered whether to allow the planting of GM crops. The first study, on economics, found that GM crops could produce long-term financial benefits for Britain, although short-term benefits would be small. The second study addressed health risks and found little to no evidence of harm to human health but noted that effects on wildlife and ecosystems should be tested before crops are approved. The third study tested effects of three GM crops on populations of animals such as insects and birds, in field experiments of unprecedented size, involving 19 researchers, 200 sites, and $8 million in funding. Plots of GM maize, beets, and oilseed rape were grown next to plots of the non-GM versions of these crops, and researchers surveyed the plots for weed growth, weed seeds, insects, and other variables. Results showed that fields of GM beets and GM oilseed rape supported less biodiversity than the non-GM versions of these crops, but that GM maize supported more biodiversity than its non-GM version.

These differences were due to herbicide treatments. The GM beets and oilseed rape tolerated herbicides and so received more herbicides than conventional fields, leading to fewer weeds and therefore fewer insects living on the weeds. GM maize, on the other hand, permitted a less-powerful herbicide to be used than with conventional maize, so GM maize fields had more weeds and therefore supported more biodiversity. Although many policymakers were hoping the biodiversity study would end the debate once and for all, the science showed that the impacts of GM crops will be complex, and will likely vary with the conditions under which different crops are grown.

More than science is involved in the debate over GM foods

Much more than science, however, has been involved in the debate over GM foods. Ethical issues have played a large role; for many people, the idea of "tinkering" with the food supply seems dangerous or morally wrong by its very nature. Even though our agricultural produce is the highly artificial product of thousands of years of selective breeding, people tend to think of food as natural. Furthermore, because every person relies on food for survival and cannot choose *not* to eat, the genetic modification of staples such as corn, wheat, and rice has given many people ethical qualms.

The perceived lack of control over one's own food has also driven an equally strong nonscientific component of the debate: concern about domination of the global food supply by a few large businesses. Gigantic agrobiotech companies, among them Monsanto, Syngenta, Bayer CropScience, Dow, DuPont, and BASF, create this technology and stand to profit economically from it. Some activists say these huge multinational corporations threaten the independence and well-being of the small farmer. This perceived loss of democratic local control is a driving force in opposition to GM foods, especially in Europe and the developing world. Biotech critics are also concerned that much of the research into the safety of GM organisms is funded, overseen, or conducted by the corporations that stand to profit if their transgenic crops are approved for human consumption, animal feed, or ingredients in other products.

Finally, public relations has played a role. When the U.S.-based Monsanto Company began developing GM products in the mid-1980s, it foresaw public anxiety and worked hard to inform, reassure, and work with environmental and consumer advocates, who the company feared would otherwise oppose the technology. Monsanto even lobbied the U.S. government to regulate the industry so the public would feel safer about it. These efforts were undermined, however, when the company's first GM product to market, a growth hormone to spur milk production in cows, alarmed consumers concerned about children's health. Then, when the company went through a leadership change, its new head changed tactics and pushed new products aggressively without first reaching out to opponents. Opposition built, and the company lost the public's trust, especially in Europe and in the developing world.

In Canada, Monsanto has been engaged in a high-publicity battle with a third-generation Saskatchewan farmer named Percy Schmeiser. Schmeiser claimed that pollen from Monsanto's Roundup-Ready canola (see Figure 9.12) used by his neighbors blew onto his land and pollinated his non-GM canola. Schmeiser had never purchased Monsanto's patented seed and said he did not want the crossbreeding. Monsanto investigators took seed samples and charged him with violating Canada's law that makes it illegal for farmers to reuse patented seed or grow the seed without a contract with the company. Monsanto sued Schmeiser, and the court sided with Monsanto, ordering the farmer to pay the corporation roughly $238,000. Schmeiser's appeal was initially denied, but in 2003 the Supreme Court of Canada agreed to hear his case in early 2004. Schmeiser has received wide support, a government committee has called for revising the patent law, and the National Farmers Union of Canada has called for a moratorium on GM food. Meanwhile, Monsanto continues to demand that small farmers in Canada and the United States heed the Canadian and U.S. patent laws.

Weighing the Issues:
Early Hurdles for GM Foods

As the vignettes in Figure 9.12 illustrate, a number of GM foods have run into difficulties. Do you think this reveals an underlying problem with the approach, or are these simply examples of unavoidable glitches that occur during the early development of any new technology? Is it right for anti-GM activists to criticize golden rice for not being nutritious enough, if scientists may be able to improve its nutritional quality through further genetic engineering? How might each of the difficulties detailed in Figure 9.12 serve as lessons, so that scientists and biotech companies can avoid such obstacles in the future? Is there reason to expect that debate between proponents and opponents of GM organisms will subside in the future?

Given such developments, the future of GM foods seems likely to hinge on social, economic, legal, and political factors as well as on scientific ones. European consumers have expressed widespread unease about possible risks of GM technologies, while U.S. consumers have largely accepted the GM crops approved by U.S. agencies. Opposition in nations of the European Union resulted in a de facto moratorium on GM foods from 1998 to 2003, blocking the importation of hundreds of millions of dollars in U.S. agricultural products. This prompted the United States to bring a case before the World Trade Organization in 2003, complaining that Europe's resistance was hindering free trade. Europeans now demand that GM foods be labeled and criticize the United States for not joining 100 other nations in signing the Cartagena Protocol on Biosafety, a treaty that lays out guidelines for open information about exported crops.

Transnational spats between Europe and the United States will surely affect the future direction of agriculture, but the world's developing nations could exert the most influence in the end. Recent decisions by the governments of India and Brazil to approve GM crops (following long and divisive debate) could add greatly to the world's transgenic agriculture. The same can be said if China continues aggressively promoting transgenic crops. A counterexample is that of Zambia. This African nation was one of several that refused U.S. food aid meant to relieve starvation during a drought in late 2002. The governments of these nations worried that their farmers would plant some of the GM corn seed meant to be eaten and that GM corn would thereby establish itself in their countries. They viewed this outcome as undesirable because African economies depend on food exports to Europe, which has put severe restrictions on GM food. In the end Zambia's neighbors accepted the grain after it had been milled (so that none could be planted), but Zambia held out. Citing health and environmental risks, uncertain science, and the precautionary principle, the Zambian government declined the aid, despite its 2–3 million people at risk of starvation. Intense debate followed within the country and around the world. Eventually the United Nations delivered non-GM grain, and in April 2003 the Zambian government announced a plan to coordinate a comprehensive long-term policy on GM foods.

The Zambian experience demonstrates some of the ethical, economic, and political dilemmas modern nations face. The corporate manufacturers of GM crops naturally aim to maximize their profits, but they also aim to develop products that can boost yields, increase food security, and reduce hunger. While industry, activists, policymakers, and scientists all agree that hunger and malnutrition are problems and that agriculture should be made environmentally safer, they often disagree about the solutions to these dilemmas and the risks that each proposed solution presents.

Preserving Crop Diversity

As the excitement over Quist and Chapela's controversial findings in Oaxaca demonstrate, one concern many people harbor about transgenic crops is that transgenes might move, by pollination, into local native races of crop plants.

Preserving the integrity of native variants gives us a bulwark against commercial crop failure

As Quist and Chapela argued, the regions where food crops first were domesticated generally remain important repositories of crop biodiversity. Although modern industrial agriculture relies on a small number of plant types, its foundation lies in the diverse varieties that still exist in places like Oaxaca. These varieties contain genes that, through conventional crossbreeding or genetic engineering, might confer resistance to disease, pests, inbreeding, and other pressures that challenge modern agriculture. Monocultures essentially place all our eggs in one basket, such that any single catastrophic cause could potentially wipe out entire crops. Having available wild relatives of crop plants and/or domesticated varieties, or cultivars, of crop plants, gives us the genetic diversity that may include ready-made solutions to unforeseen problems. Because accidental interbreeding such as Quist and Chapela claimed to have found could decrease the diversity of local variants, many scientists argue that we need to protect areas like Oaxaca.

One example of native cultivars potentially coming to the rescue of a major commercial crop involves potatoes. The disease that caused the Irish potato famine of the 1840s has been largely controlled by fungicides since then, but in the 1990s several fungicide-resistant strains spread and cut potato yields by 15%, or $3.25 billion. Scientists in Peru, however, may have just the answer: They have found that some cultivars and wild relatives of potatoes native to the region carry genes that confer resistance to the new disease strains. As another example, when a virus started attacking green revolution rice varieties in Asia in the 1970s, geneticists found resistance to the pathogen in a population of a wild rice species in Uttar Pradesh, India and addressed the problem. They found resistance *only* in that one population, however, and since then that population has vanished.

The scale of the loss of genetic diversity of crop plants can be gauged with a few statistics. The number of wheat varieties in China is estimated to have dropped from 10,000 in 1949 to 1,000 by the 1970s, and Mexico's famed maize varieties now number only 30% of what was extant in the 1930s. The data in Table 9.3 suggest that many fruit and vegetable crops in the United States have decreased in diversity by 90% in less than a century.

A primary cause of this loss of diversity is that market forces have discouraged diversity in the appearance of fruits and vegetables. Commercial food processors prefer items to be similar in size and shape, for convenience. Consumers, for their part, have shown preferences for uniform, standardized food products over the years. This is because items that are easily recognizable are viewed as safe, while consumers are likely to be suspicious of items that look unusual. Now that local

organic agriculture is growing in affluent societies, however, some consumer preferences for diversity are increasing, so future trends remain to be seen.

Seed banks are living museums for seeds

Protecting areas with high crop diversity is one way to preserve genetic assets for our agricultural systems. Another is to collect and store seeds from crop varieties and periodically plant and harvest them to maintain a diversity of cultivars. This is the work of seed banks or gene banks, institutions that preserve seed types as a kind of living museum of genetic diversity. In total these facilities held 6 million seed samples in 1999, keeping them in cold dry conditions to encourage long-term viability. The $300 million in global funding for these facilities is not adequate for proper storage and for the labor of growing out the seed periodically to renew the stocks; therefore, it is questionable how many of these 6 million seeds are actually preserved. Major efforts include large seed banks such as the U.S. National Seed Storage Laboratory at Colorado State University, the Royal Botanic Garden's Millennium Seed Bank in Britain, Seed Savers Exchange in Iowa, and the Wheat and Maize Improvement Center (CIMMYT) in Mexico.

One seed bank, Native Seeds/SEARCH (NS/S) of Tucson, Arizona, shows the many types of biological and cultural conservation efforts that such an organization can pursue. NS/S keeps 2,000 seed collections of 99 species of plants used as traditional foods by Native Americans of the Desert Southwest, a region encompassing Arizona, New Mexico, and northwestern Mexico. Beans, chiles, squashes, gourds, maize, cotton, and lentils are all in its collections, as well as lesser-known plants such as amaranth, lemon basil, and devil's claw (Figure 9.14). Besides promoting protection of wild ancestors of crop plants in nature and encouraging small-scale sustainable agriculture, the organization sells food products to members and donates seeds to Native American farmers. In so doing, it has reconnected many elderly Native Americans with plants they remember from their youth and is helping tribes rediscover and maintain their own cultures. NS/S has also begun a program to record oral histories of uses of native foods before older people and their memories pass from the scene. Reintroducing the region's native people to their traditional foods appears to help them combat the diabetes that is so rampant in their communities that results from the Western diet they have adopted. Traditional foods such as mesquite flour, prickly pear pads, chia seeds, tepary beans, and cholla cactus buds have been medically shown to help fight diabetes.

Table 9.3 Reduction in Diversity of Vegetables in a U.S. Seed Bank, 1903–1983

Vegetable	Number held in 1903*	Number held in 1983*	Percent loss
Asparagus	46	1	98
Bean	578	32	94
Beet	288	17	94
Carrot	287	21	93
Leek	39	5	87
Lettuce	487	36	93
Onion	357	21	94
Parsnip	75	5	93
Pea	408	25	94
Radish	463	27	94
Spinach	109	7	94
Squash	341	40	88
Turnip	237	24	90

*Numbers of varieties are those held at the U.S. National Seed Storage Laboratory, Colorado State University. Data from World Resources Institute; and P. Harrison and F. Pearce, *AAAS Atlas of Population & Environment*, 2000.

(a) Traditional food plants of the Desert Southwest

(b) Pollination by hand

Figure 9.14 Seed banks preserve genetic diversity of traditional crop plants. Native Seeds/SEARCH of Tucson, Arizona preserves seeds of dozens of food plants important to traditional diets in the Desert Southwest (**a**). Care is taken to hand-pollinate varieties (**b**) in order to protect their genetic distinctiveness.

The diet of people of the Desert Southwest shares similarities with that of natives of Oaxaca, where squash, beans, chili peppers, cotton, and maize have long been grown. The Mexican government has recognized that traditional foods have cultural value as well as practical value. For this reason, the Mexican government helped create the Sierra de Manantlan Biosphere Reserve around an area harboring the localized plant thought to be the direct ancestor of maize. For this reason, too, it imposed a moratorium in 1998 on the planting of transgenic corn in the country. The announcement of Quist and Chapela was all the more worrisome because it suggested the alleged transgene transfer had occurred despite the moratorium.

Feedlot Agriculture: Livestock and Poultry

Food from cropland agriculture makes up a huge portion of the human diet, but most people also eat animal products. People don't *need* to eat meat or any other animal products to live full, active, healthy lives, but it is difficult for many people to obtain a balanced diet without incorporating animal products. Most of us do eat animal products, and this choice has significant environmental, social, agricultural, and economic impacts.

Consumption of animal products is growing

As wealth and global commerce increase, so has our consumption of animal products (Figure 9.15). The world population of domesticated animals raised for food increased from 7.3 billion animals to 20.6 billion animals between 1961 and 2000. Most of these animals are chickens. Global meat production has increased fivefold since 1950 and per capita meat consumption more than doubled between 1950 and 2000, with pork being the most-eaten meat per unit weight. Table 9.4 shows other evidence of our growing appetite for animal products. For the many people who enjoy eating meat, milk, eggs, and other animal products, animal husbandry has enriched human lives, bringing us gastronomic pleasure and supplying protein and fat to our diets.

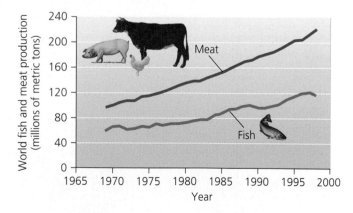

Figure 9.15 The production of meat from farm animals has increased steadily worldwide over the past few decades. The production of fish (marine and freshwater, harvested and farmed) has also risen steadily. Data from *AAAS Atlas of Population & Environment,* University of California Press, 2000.

Table 9.4 Some Statistics on the Consumption of Animal Products

China is home to 450 million pigs.

New Zealand has 12 times as many sheep as people.

U.S. livestock produce 130 times more manure than humans do.

Feedlots produce 43% of the world's meat, up from 33% in 1990.

Livestock account for 16% of total global production of methane, a greenhouse gas 25 times as potent as carbon dioxide.

The average person in the industrialized world eats 77 kg (169 lb) of meat each year.

Heavy consumption of animal products has led to feedlot agriculture

In traditional agriculture, livestock were kept by farming families near their homes or were grazed on open grasslands by nomadic herders or sedentary ranchers. These traditions have survived to the present day, but the advent of industrial agriculture has added a new method. **Feedlots**, also known as *factory farms* or *concentrated animal feeding operations (CAFOs)*, are essentially huge barns or outdoor pens designed to deliver energy-rich food to animals living in extremely high densities (Figure 9.16). The trend has been toward larger feedlots containing more and more animals, and today 43% of the world's meat comes from feedlots.

Feedlot operations allow for greater production of food and are probably necessary for a country with a level of meat consumption like that of the United States. Feedlots have one overarching benefit for environmental quality: taking these animals off the land and concentrating them in feedlots reduces the impact they would otherwise have on large portions of the landscape. In Chapter 8 we evaluated the impacts of grazing on rangelands, the grassland areas devoted to supporting herds of cattle, sheep, goats, and other livestock. We learned that overgrazing can degrade soils and rangeland vegetation and that hundreds of millions of hectares of land are considered overgrazed. Animals that are densely concentrated in feedlots instead will not contribute to overgrazing and soil degradation.

Of course, feedlots are not without impact, and many environmental advocates have attacked them for their contributions to water and air pollution. Waste from feedlots can emit extremely strong odors and can pollute surface water and groundwater, because livestock produce prodigious amounts of feces and urine. One dairy cow can produce about 20,400 kg (44,975 lb) of waste

Figure 9.16 Modern industrial feedlots raise livestock in dense concentrations.

in a single year. Greeley, Colorado, is home to North America's largest meatpacking plant and two adjacent feedlots, all owned by the agrobusiness firm ConAgra. Each feedlot has room for 100,000 cattle that are fed surplus grain and injected with anabolic steroids to stimulate growth. During its stay at the feedlot, a typical steer will eat 1,360 kg (3,000 lb) of grain, gain 180 kg (400 lb) in body weight, and generate 23 kg (50 lb) of manure each day. The amount of manure that 200,000 such animals generate exceeds the amount of waste produced by all the human residents of Atlanta, St. Louis, Boston, and Denver combined. Thus feeding, watering, and raising animals for food on feedlots, as well as slaughtering, butchering, and packaging them in meatpacking plants, can bring substantial environmental and health impacts. Poor waste containment practices at feedlots in North Carolina, Maryland, and other states have been linked to outbreaks of disease, including virulent strains of *Pfisteria,* a microbe that causes life-threatening lesions in fish.

Feedlot impacts can be minimized when properly managed, however. Both the EPA and the states regulate U.S. feedlots, and most farmers comply with regulations. Most manure from feedlots is applied to farm fields as fertilizer, reducing the need for chemical fertilizers. Manure in liquid form can be injected into the ground where plants need it, and farmers can conduct tests to determine amounts that are appropriate to apply.

Weighing the Issues:
Are Feedlots Targets for Terrorism?

In 2001, the United Kingdom experienced a disastrous outbreak of foot-and-mouth disease. Over 4 million cattle were culled, and damages were estimated at $50 billion. Some experts view this event as a blueprint for terrorists wanting to cause economic disruption. Factory farms, feedlots, and monoculture, they claim, make it easy to introduce a disease and cripple an industry. Do you think these threats are realistic? Are they serious enough to recommend alternatives to intensive farming methods, given that this type of production is efficient and cost-effective? What types of steps should be taken to ensure animal and crop safety?

Our food choices are also energy choices

What we choose to eat has ramifications for how we use energy and the land that supports agriculture. Recall our discussion of thermodynamics and trophic levels in Chapter 4 and Chapter 5. Every time energy moves from one trophic level to the next, as much as 90% of the useful energy present in the lower trophic level is lost. For example, if we feed grain to a cow and then eat beef from the cow, we lose a great deal of the grain's energy to the cow's digestion and metabolism. Energy is lost when the cow converts the grain to muscle mass to grow, and further loss occurs as the cow uses its muscle mass on a daily basis to maintain itself. For this simple reason, eating meat is far less energy-efficient than relying on a vegetarian diet. The lower in the food chain our food sources are, the greater the proportion of the sun's energy we put to use as food, and the more people Earth can support.

Today there are more people eating more meat than ever before. To satisfy the increasing demand for meat, we are feeding a larger portion of the world's grain production to animals than ever before. In 1900 we fed about 10% of global grain production to animals. In 1950 this number had reached 20%, and by the beginning of the 21st century we were feeding 45% of global grain production to animals. Global meat production increased by 29% between 1950 and 1995 and will likely double from 1998 levels by 2050. Coupled with the fact that global per capita grain production has been declining since the early 1980s, this increased emphasis on grain-fed animals has further decreased the amount of food energy available to many people. While much of the grain fed to animals is not of a quality suitable for human consumption, the resources required to grow it could have instead been applied toward growing food for people. One partial solution is to feed livestock crop residues, plant matter such as stems and stalks that we would not consume anyway, and this is increasingly being done.

Some animals convert grain feed into milk, eggs, or meat more efficiently than others (Figure 9.17). For this reason, scientists have calculated the relative energy-conversion efficiencies of different types of animals. Such energy efficiencies have ramifications for land use, because land and water is required to raise food for the animals, and some animals require more than others. Figure 9.18 shows the area of land and weight of water required to produce 1 kg (2.2 lb) of food protein for milk, eggs, chicken, pork, and beef. Clearly, producing eggs and chicken meat requires the least space and water, while producing beef requires the most. Such differences make clear that when we choose what to eat, we are also indirectly choosing how to make use of resources such as land and water.

Feed input

Produce output (edible weight)

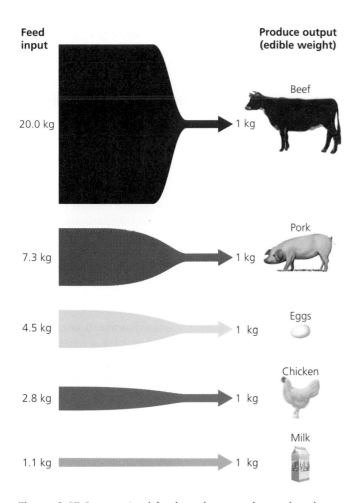

Figure 9.17 Some animal food products can be produced with less input of animal feed than can others. Chickens must be fed 2.8 kg of feed for each 1 kg of resulting chicken meat, for instance, while 20.0 kg of feed must be provided to cattle to produce 1 kg of beef. Go to **GRAPH IT** on the website or CD-ROM. Data from Vaclav Smil, *Feeding the World: A Challenge for the Twenty–First Century,* MIT Press, 2001.

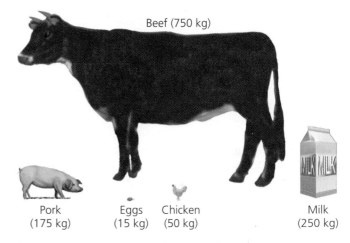

(a) Land required to produce 1 kg of protein

(b) Water required to produce 1 kg of protein

Figure 9.18 Producing different types of animal products requires different amounts of land and water. Raising cattle for beef requires by far the most land and water of all animal products. Go to **GRAPH IT** on the website or CD–ROM. Data from Vaclav Smil, *Feeding the World: A Challenge for the Twenty–First Century,* MIT Press, 2001.

Aquaculture

In addition to plants grown in croplands and animals raised on rangelands and in feedlots, we rely on aquatic organisms as food sources. As with all other types of food, harvesting aquatic food sources requires either their capture in the wild or their propagation in confined facilities. Wild fish populations are plummeting throughout the world's oceans as increased demand and new technologies enable us to catch fish more effectively. The majority of marine fisheries are now overharvested, as we shall see in Chapter 13. This means that raising fish and shellfish on "fish farms" may be the only way to meet the demand for these foods from our growing population.

We call the raising of aquatic organisms for food in controlled environments **aquaculture,** and this practice has increased dramatically in recent years (Figure 9.19). In fact, it is the fastest-growing type of food production, with global output having increased nearly fivefold in 15 years, from 6.9 million tons in 1984 to 33.3 million tons in 1999, when its production was valued at $47.9 billion. Aquaculture today provides a third of the world's fish for human consumption, is most common in Asia, and involves over 220 species. Some, such as carp, are grown for local consumption, whereas others, such as salmon and shrimp, are exported to affluent countries. As aquaculture expands globally, it is bringing

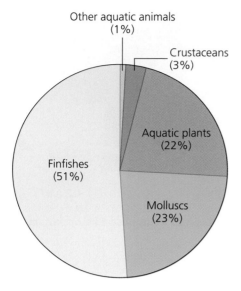

(a) World aquaculture production by major species groups, 2000

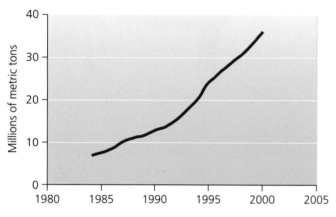

(b) World aquaculture production

Figure 9.19 (a) Aquaculture, or fish-farming, involves a wide diversity of marine and freshwater organisms. (b) Global production of meat from aquaculture has risen steeply in the past two decades. Data from FAO: The State of World Fisheries and Aquaculture, 2002 (a); and Aquaculture Production Statistics, 1984–1993 and Fishery Statistics: Aquaculture Production (b).

benefits of economic and food security to many developing regions of the world, but is also causing a series of environmental problems.

Aquaculture brings a number of benefits

Depending on the species being raised and the location, several types of enclosures may be used in aquaculture. Many aquatic species are grown in open water in large, floating net-pens. Others are raised in land-based ponds or holding tanks. People pursue both freshwater and marine aquaculture.

Aquaculture provides several substantial economic, social, and environmental benefits. When conducted on a small scale by families or villages, as in China and much of the developing world, aquaculture helps ensure people a reliable protein source. This type of small-scale aquaculture can be sustainable, and it is compatible with other activities. For instance, uneaten fish scraps make excellent fertilizers for crops. Aquaculture on larger scales can help improve a region's or nation's food security by increasing overall amounts of fish available. Aquaculture on any scale also has the benefit of reducing fishing pressure on wild stocks, which as we noted are largely already overharvested and declining. Reducing fishing pressure also reduces the by-catch (the unintended catch of non-target organisms) that results from many types of commercial fishing. Furthermore, aquaculture relies far less on fossil fuels than do fishing vessels and provides a safer work environment than commercial fishing. Aquaculture can also be remarkably energy-efficient. Aquaculture ponds may produce as much as 10 times more fish per unit area than can be harvested from oceanic waters on the continental shelf, and up to 1,000 times as much fish as can be harvested from waters of the open ocean.

Aquaculture has negative environmental impacts as well

Despite its many benefits, aquaculture does have several disadvantages. The dense concentrations of farmed animals can increase the incidence of disease, which reduces food security, necessitates antibiotic treatment, and results in additional expense. A virus outbreak wiped out half a billion dollars in shrimp in Ecuador in 1999, for instance. Aquaculture can also produce prodigious amounts of waste. This waste results from the digestion of food by the farmed organisms and, in the case of net-pens, from the large portion of the feed that goes uneaten and decomposes in the water column. Farmed fish in some cases are fed grain and in other cases are fed fish meal. As discussed above, the strategy of growing grain to feed to animals that we then eat reduces the overall energy efficiency of food production and consumption. When fish meal is used, its source is most often wild ocean fish such as herring and anchovies, whose harvest may place additional stress on wild fish populations.

The escape of farmed animals into the environment can have negative consequences. If farmed aquatic organisms escape into ecosystems where they are not native, they can cause unpredictable harm. They may spread disease to native stocks and may outcompete

Organic Farming

Fields of wheat and potatoes, some grown organically and some cultivated with the synthetic chemicals favored by industrialized agriculture, stand side-by-side on an experimental farm in Switzerland. Although conventionally farmed fields receive up to 50% more fertilizer, they produce only 20% more food than organically farmed fields. How are organic fields able to produce decent yields without synthetic agricultural chemicals? The answer, scientists have found, lies in the soil.

Researchers have long documented that organic farming is better for the environment. But in a world with more than six billion people to feed, environmental impact has received less attention than whether our farms produce enough to feed the planet's human population. Organic farming, some have contended, might put fewer synthetic chemicals into the air and water, but it also contributes little to the world's food supply.

To address concerns about crop yields, Swiss researchers at the Research Institute of Organic Agriculture have been comparing organic and conventional fields since 1978, using a series of growing areas that feature four different farming systems. One group of plots mirrors conventional farms, in which large amounts of chemical pesticides, herbicides, and fertilizer are applied to soil and plants. Another set of fields is treated with a mixed approach of conventional and organic practices, including chemical additives, synthetic sprays, and livestock manure as fertilizer. Organic plots use

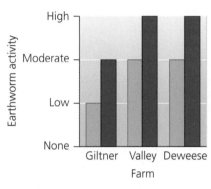

A study by researchers in Nebraska and North Dakota demonstrated that organic farming at three sites increased topsoil depth moderately and activity of earthworms dramatically. Data from M.A. Liebig and J.W. Doran, Impact of organic production practices on soil quality indicators, *Journal of Environmental Quality,* 1999.

only manure, mechanical weeding machines, and plant extracts to control pests. A fourth group of plots follows similar organic practices but also uses extra natural boosts, such as adding herbal extracts to compost. The two organic plots receive about 35–50% less fertilizer than the conventional fields and 97% fewer pesticides. More than 20 years of monitoring has shown that the organic fields yielded 80% of what the conventional fields produced, researchers reported in the journal *Science* in 2002. Organic crops of winter wheat yielded about 90% of the

conventional wheat crop yield. Organic potato crops averaged about 68% of the conventional potato yields. The comparatively low potato yield was due to nutrient deficiency and a fungus-caused potato blight.

Scientists have hypothesized that organic farms keep their yields high, because organic agricultural practices better conserve soil quality, keeping soil fertile over the long-term. In one study that appears to back this hypothesis, U.S. researchers compared five pairs of organic and conventional farms in the mid-1990s in North Dakota and Nebraska. After extracting soil samples from each of the farms, the researchers chemically analyzed the soil for nutrients like carbon and nitrogen, for water-holding ability, and for microbial biomass. Organic farming, they found, produced soils that contained more naturally occurring nutrients, held greater quantities of water, and had higher concentrations of microbial life than conventionally farmed soil. Organic farms also had deeper nutrient-rich topsoil and greater earthworm activity—all signs of soil healthy enough to produce impressive crops without help from synthetic chemicals.

Scientists at the Swiss research project have found similar signs of soil fertility on their organic research plots. Increasingly, researchers are concluding that organically-managed soil supports a more diverse range of microbial and plant life, which translates into increased biodiversity, self-sustaining fields, and strong crop yields. Such findings may be pivotal as large growers increasingly debate whether to practice organic farming.

native organisms for food or habitat. The possibility of competition also arises when the farmed animals have been genetically modified. Like the transgenic corn that Quist and Chapela claimed to have influenced Mexican maize, transgenic fish have become a part of the food production system in recent years. Genetic engineering of Pacific salmon has resulted in the production of transgenic fish that weigh 11 times more than nontransgenic ones, and transgenic Atlantic salmon raised in Scotland have been engineered to grow to 5–50 times the normal size for their species (Figure 9.20). GM fish such as these salmon may outcompete their non-GM wild cousins while also spreading disease to them. They may also interbreed with native and hatchery-raised fish and weaken already troubled stocks. Researchers have concluded that under certain circumstances, escaped transgenic salmon may increase the extinction risk that native populations of their species face, in part because the larger male fish (such as those carrying a gene for rapid and excessive growth) have better odds of mating successfully.

Shrimp farming illustrates some environmental impacts of aquaculture

From salmon to shrimp, many species can thrive in aquaculture. Shrimp aquaculture illustrates some of the unexpected environmental impacts that can result. In the wild, each female shrimp releases upwards of 100,000 eggs, many of which hatch within 24 hours to

Figure 9.20 Efforts to genetically modify important food fish have resulted in the creation of transgenic salmon, which can be considerably larger than wild salmon of the same species.

become plankton-eating larvae. After about 12 days in the larval form in the open ocean, the young shrimp migrate to coastal waters, where they mature in nutrient-rich estuaries. Shrimp aquaculture attempts to mimic this process in coastal enclosures.

In some cases shrimp farmers use farm-raised larvae to begin this process, but in other cases shrimpers capture wild larvae with extremely fine mesh nets. The latter approach also captures many non-target organisms. Thus, ironically, shrimp farming can actually cause by-catch problems more commonly associated with commercial fishing. According to some published reports, for every shrimp captured in the wild to be raised in captivity, up to 100 other non-target organisms may be inadvertently captured and killed.

Because shrimp are raised in land-based enclosures (usually earthen ponds) built in coastal areas, their farming can cause terrestrial as well as aquatic impacts. In many coastal areas, especially Thailand and other parts of Southeast Asia, critical wetland and mangrove habitat has been destroyed to make room for shrimp ponds. Mangroves are salt-tolerant trees that grow along coasts in many tropical and subtropical areas. Mangrove forests protect coastlines from erosion, provide essential habitat for fish and other animals, and prevent nearshore nutrient pollution problems by absorbing large amounts of nutrients dissolved in coastal waters. Like mangroves, wetlands provide water filtration and critical habitat for economically important species. Shrimp aquaculture has displaced both of these critical habitat types.

Having less natural vegetation to filtrate wastes is an even greater problem when shrimp farming itself generates large amounts of waste, including the 30% of food input that goes unconsumed. Shrimp farmers can minimize their adverse environmental impacts by locating containment ponds away from sensitive areas, by ensuring that ponds do not leak, and by ensuring that waste is not discharged directly into shallow nearshore waters, fresh water, or directly onto land.

Sustainable Agriculture

We have seen examples of adverse environmental impacts in many aspects of agriculture, from the degradation of soils in Chapter 8 to problems arising from pesticide use, genetic modification, and intensive feedlot and aquaculture operations. Although many of these developments in intensive commercial agriculture have alleviated certain environmental pressures, they have often exacerbated others. Industrial agriculture in some form seems necessary to feed the planet's 6 billion people, but many feel we will be better off in the long run by practicing less-intensive methods of raising animals and crops.

Indeed, farmers and researchers have made great advances toward sustainable agriculture in recent years. **Sustainable agriculture** is agriculture that does not deplete soils faster than they form. It is farming and ranching that does not reduce the amount of healthy soil, clean water, and genetic diversity essential to long-term crop and livestock production. It is, simply, agriculture that can be practiced in the same way far into the future. For example, the no-till agriculture practiced in southern Brazil that we examined in Chapter 8 appears to fit the notion of sustainable agriculture, as does the traditional Chinese practice of aquaculture of carp in small ponds. The concept of sustainable agriculture is closely related to the concept known as **low-input agriculture.** This term refers to agriculture that utilizes smaller amounts of pesticides, fertilizers, growth hormones, water, and fossil fuel energy than are currently used in industrial agriculture. Food-growing practices that use no synthetic fertilizers, insecticides, fungicides, or herbicides—but instead rely on biological approaches such as composting and biocontrol—are often termed **organic agriculture.**

Organic agriculture is on the increase

Citizens, government officials, farmers, and agricultural industry representatives have debated the meaning of the word *organic* for many years. Experimental organic gardens began to appear in the United States in the 1940s, but it was decades before the U.S. government developed a clear definition of what it meant to be organic. In 1990 the U.S. Congress passed and the president signed the Organic Food Production Act. This law established national standards for organic products and facilitated the sale of organic food. As required by this act, the USDA in 2000 issued criteria by which crops and livestock could be officially certified as organic (Table 9.5). California legislation passed the same year established even more stringent state guidelines for labeling foods organic. Today 16 other U.S states have laws spelling out standards for organic products.

Weighing the Issues:
Do You Want Your Food Labeled?

Recently, the USDA issued labels to certify that produce claiming to be organic met the government's organic standards. Increasingly, critics of GM foods want GM products to be labeled as well. Given that 70% of processed food currently contains GM ingredients, labeling would cause added—and many people think, unnecessary—costs. But the European Union currently labels such foods. Do you want your food to be labeled? Would you choose among foods based on whether they were organic or genetically modified? Do you feel your food choices have environmental impacts, good or bad? Is purchasing power an effective way to make your views heard?

Table 9.5   USDA Criteria for Certifying Crops and Livestock as *Organic*

For crops to be considered *organic* . . .

. . . the land where they are grown must be free of prohibited substances for at least 3 years.

. . . they must not be genetically engineered.

. . . they must not be treated with ionizing radiation (a means of eliminating bacteria in packaged food).

. . . the use of sewage sludge is prohibited.

. . . they must be produced without fertilizer containing synthetic ingredients.

. . . fertility and crop nutrients must be achieved through crop rotations, cover crops, animal and crop wastes, and synthetic materials approved by the USDA.

. . . they must not be produced using most conventional pesticides.

. . . use of organic seeds and other planting stock are preferred.

. . . control of crop pests, weeds, and diseases should be accomplished through physical, mechanical, and biological management practices or synthetic substances approved by the USDA.

For livestock to be considered *organic* . . .

. . . mammals must be raised under organic management from the last third of gestation and poultry no later than the second day of life.

. . . producers must feed livestock 100% organic agricultural feed; however, vitamin and mineral supplements are allowed.

. . . producers of existing dairy herds must provide 80% organically produced feed for 9 months, followed by 3 months of 100% organically produced feed.

. . . use of hormones or antibiotics is prohibited, although vaccines are permitted.

. . . animals must have access to the outdoors.

Data from The National Organic Program, *Organic Production and Handling Standards*, 2002.

Long viewed as a small niche market, the market for organic foods is on the increase. Although it accounts for only 1% of food expenditures in the United States and Canada, sales of organic products increased 20% annually from 1989 to 2001, when global sales of organic products reached $25 billion. In 2001, 3–5% (close to $10 billion) of Europe's food market was organic. Although 3–5% may not seem like much, organic agriculture expanded by a factor of 35 between 1985 and 2001 in Europe, representing an annual growth rate of 30%.

Production is increasing along with demand. Although organic agriculture takes up less than 1% of cultivated land worldwide (11.5 million ha [28.4 million acres] in 2001), this area is rapidly expanding. In the United States and Canada, the amount of land used in organic agriculture has recently increased 15–20% each year, and as of 2001, 550,000 ha (1.36 million acres) of U.S. farmland and 1 million ha (2.47 million acres) of Canadian farmland were in organic production. In Britain, organic agriculture increased from 50,000 to 400,000 ha (123,553–988,422 acres) from 1999 to 2001 alone. Today farmers in more than 130 nations practice organic farming commercially to some extent.

These trends have been fueled by the desire of many consumers to reduce health risks in their diets, because pesticides have been frequently demonstrated to affect health. Consumers also buy organic produce out of a concern for improving environmental quality by reducing chemical pollution. Conversely, many other consumers will not buy organic produce because it often looks less uniform and aesthetically appealing in the supermarket aisle compared to the standard produce of high-input agriculture. Overall, enough consumers are willing to pay more money for organic meat, fruit, and vegetables that many businesses are making organic foods more widely available. In early 2000, one of Britain's largest supermarket brands announced that it would sell only organic food—and that the new organic products would cost their customers no more than had nonorganic products. In addition to food products, many textile makers who use cotton in their products are increasing their use of organic cotton. The Gap, Levi's, and Patagonia clothing manufacturers are among the companies that have started buying organic cotton.

Government initiatives have also spurred the growth of organic farming. For example, several million hectares of land has undergone conversion from conventional to organic farming in Europe since the European Union adopted a policy in 1993 to support farmers financially during the first years of conversion. Such support is important, because conversion often means a temporary loss in income. Increasing numbers of studies, however, are suggesting that because of reduced inputs and higher market prices, in the long run organic farming can be more profitable for the farmer than conventional methods.

Cuba has embraced organic agriculture

Although organic farming is on the increase in Canada, the United States, and much of Europe, perhaps no other nation has implemented organic farming to the extent that Cuba has. Long a close ally of the former Soviet Union, Cuba suffered economic and agricultural upheaval following the Soviet Union's dissolution. In 1989, as the USSR was breaking up, Cuba lost 75% of its total imports, 53% of its oil imports, and 80% of its fertilizer and pesticide imports. Faced with such losses, Cuba's farmers had little choice but to "go organic." Although Cuba's move toward organic agriculture was involuntary, its response to its economic and agricultural crisis illustrates how other nations might, by choice, begin to farm in ways that rely less on enormous inputs of fossil fuels and synthetic chemicals.

Because far less oil was available to fuel Cuba's transportation system, farmers began growing food closer to cities and even within them. By doing so, Cuban farmers returned to a pattern of food production like that before the industrial revolution. By 1998 the Cuban government's Urban Agriculture Department had encouraged the development of more than 8,000 gardens in the capital city of Havana (Figure 9.21). Over 30,000 people worked in these gardens, which covered 30% of the city's available land. Havana's organic urban gardens fall into five categories:

- *Huertos populares* (popular gardens) are worked by city residents.
- *Huertos intensivos* (intensive gardens) use a high compost-to-soil ratio and are operated by the government or by private citizens.
- *Autoconsumos* are owned by workers and produce food served in workplace cafeterias.
- *Campesinos particulares* are small farmers who farm in the greenbelt at the edge of the city.
- *Empresas estatales* are state-run farms that allow farm workers to share in profits the farms generate.

Cuba has taken other steps to compensate for the loss of fossil fuels, fertilizers, and pesticides, including:

- Using oxen instead of tractors

Figure 9.21 Organic gardening takes place within the city limits of Havana, Cuba out of necessity. With little money to pay for the large amounts of fertilizers and pesticides required for industrialized agriculture, Cubans get much of their food from local agriculture without these inputs.

- Using integrated pest management
- Encouraging people to live outside of urban areas and to remain involved in agriculture
- Establishing centers to breed organisms for biological pest control

Cuba's agriculture likely requires more human labor per unit output than do intensive commercial farms of developed nations, and Cuba's economic and agricultural policies are guided by tight top-down control in a rigid state socialist system. Nevertheless, Cuba's low-input farming has produced some positive achievements. The practices have led to the complete control of the sweet potato-borer, a significant pest insect, and in the 1996–1997 growing season the Cuban people produced record yields for ten crops. Cuba's odd experience has shown that by foregoing synthetic fertilizers and pesti-cides, organic farming on a large scale can enable farmers to mitigate some of the problems common in high-intensity agriculture.

Organic and sustainable agriculture will likely need to play a large role in our future

Organic agriculture succeeds in part because it alleviates many problems introduced by high-input agriculture, even while passing up many of the benefits. For instance, although in many cases more insect pests attack organic crops because of the lack of chemical pesticides, biocontrol methods can keep these pests in check, and the lack of synthetic chemicals also maintains soil quality and encourages helpful pollinating insects. In the end, consumer choice will determine the future of organic agriculture. Falling prices and wider availability suggest that organic agriculture will continue to increase. In addition, sustainable agriculture, whether organic or not, will sooner or later need to become the rule rather than the exception.

Conclusion

Many of the agricultural practices we have discussed have substantial negative environmental impacts. At the same time, it is important to realize that many of these developments in intensive commercial agriculture have had positive environmental effects by relieving certain other pressures on land or resources. Whether Earth's natural systems would be under more pressure from 6 billion people practicing traditional agriculture or from 6 billion people living under the industrialized agriculture model is a very complicated question.

What is certain is that if our planet is to support 9 billion people by mid-century without further degradation of the soil, water, pollinators, and other ecosystem services that support our food production, we must somehow find a way to make the shift to sustainable agriculture. Approaches such as biological pest control, organic agriculture, pollinator conservation, preservation of native crop diversity, sustainable aquaculture, and likely some degree of careful and responsible genetic modification of food may all be parts of the game plan we will need to set in motion. What remains to be seen is the extent to which individuals, governments, and corporations will be able to put their own interests and agendas in perspective to work together toward a sustainable future.

REVIEW QUESTIONS

1. What kinds of techniques have humans employed to increase agricultural food production?

2. What determines undernourishment? What is the difference between undernourishment and malnutrition?

3. How did agricultural scientist Norman Borlaug help inaugurate the green revolution?

4. Name several ways in which high-input agriculture can be beneficial for the environment and several ways in which it can be detrimental to the environment.

5. Describe the current trend in per-capita world food supply. Describe the current trend in per-capita global grain supply.

6. What determines what we define as *pests* and *weeds*?

7. What causes pesticide resistance?

8. What factors make for an effective biological control strategy of pest management? What risks are involved in biocontrol?

9. List several components of a system of integrated pest management (IPM).

10. About how many cultivated plants are known to rely on insects for pollination? Why is it important to preserve the biodiversity of native pollinators?

11. What is recombinant DNA? How is it used?

12. What is a transgenic organism?

13. How is genetic engineering different from traditional agricultural breeding? How is it similar?

14. Describe several reasons many people support the development of genetically modified organisms, and name several of their uses so far.

15. Describe the scientific concerns of those opposed to genetically modified crops. Describe some of the non-scientific concerns.

16. Why did Zambia's leaders choose to refuse genetically modified corn from the United States that might have fed its people during a drought in 2002?

17. How are native crop species critical to the preservation of commercial crops? What can seed banks do to help preserve local native varieties?

18. What was the approximate increase in numbers of animals raised for food between 1961 and 2000?

19. Name several positive and negative environmental effects of feedlot operations.

20. Why is beef an inefficient food from the perspective of energy consumption?

21. What are some economic benefits of aquaculture? What are some negative environmental impacts?

22. What are the objectives of sustainable agriculture? How is it different from "low-input" agriculture? How is organic agriculture defined?

23. What factors are causing the practice of organic agriculture to expand?

DISCUSSION QUESTIONS

1. Describe some symptoms of malnutrition. Why do you think malnutrition combined with obesity is becoming a common condition in the United States?

2. From what you have learned in this chapter about the staple crop corn, how would you choose to farm corn, if you had to do so for a living? Would you choose to grow genetically modified corn? How would you manage pests and weeds? Would you grow corn for people, livestock, or both? Think of the various ways corn is grown, purchased, and valued in places such as the United States, Europe, Oaxaca, and Zambia as you formulate your answer.

3. Those who view GM foods as solutions to world hunger and pesticide overuse often want to speed their development and approval. Others adhere to the precautionary principle and want extensive testing for health and environmental safety. How much caution do you think is warranted before a new GM crop is introduced?

4. Imagine it is your job to make the regulatory decision as to whether to allow the planting of a new genetically modified strain of cabbage that produces its own pesticide and has twice the vitamin content of regular cabbage. What questions would you ask of scientists before deciding on whether to approve the new crop? What scientific data would you want to see, and how much would be enough? Would you also consult nonscientists or take ethical, economic, and social factors into consideration?

5. Cuba adopted low-input organic agriculture out of necessity. If the country were to become economically well-off once more, do you think Cubans would maintain this form of agriculture, or do you think they would turn to intensive high-input farming instead? What factors might influence which of the scenarios might happen?

Media Resources *For further review go to the website* **www.envscienceplace.com** *or student CD-ROM, where you will find quizzes, flashcards, a glossary, additional interactive exercises, and links to relevant news and research sources. Also, on the website and CD-ROM is* **GRAPH IT**, *a series of interactive graphing tutorials to help you interpret graphs and plot data.*

10 Toxicology and environmental health

Alligator Hatchling from
Lake Apopka, Florida

This chapter will help you understand:

- The study of toxic substances and their effects

- The types, abundance, distribution, and movement of synthetic and natural toxicants in the environment

- How concerns for wildlife translate into concerns for human health

- Epidemiology, case histories involving toxic agents, animal testing, and dose-response analysis

- Factors affecting toxicity

- Types of environmental health hazards

- Risk assessment and risk management

- Philosophical approaches to risk and environmental health

- Policy and regulation in the United States and internationally

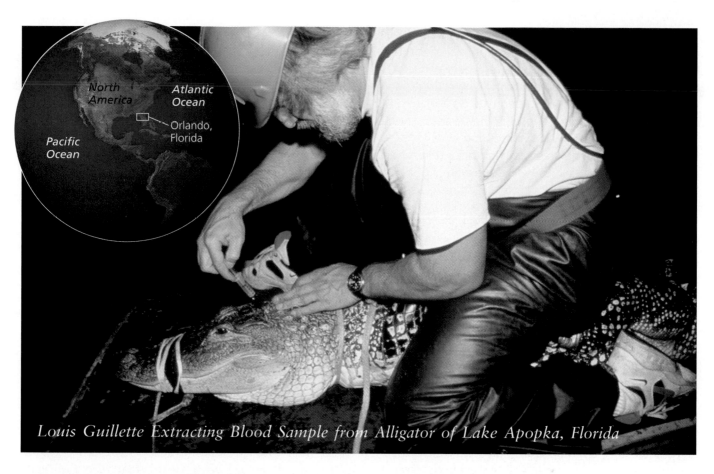

Louis Guillette Extracting Blood Sample from Alligator of Lake Apopka, Florida

Central Case: Alligators and Endocrine Disruptors at Lake Apopka, Florida

"Over the past 50 years we have all been unwitting participants in a vast, uncontrolled, worldwide chemistry experiment involving the oceans, air, soils, plants, animals, and human beings."
—*United Nations Environment Programme, in a guide to the Stockholm Convention on Persistent Organic Pollutants*

"Every man in this room is half the man his grandfather was."
—*Reproductive biologist Louis Guillette to a U.S. Congressional Committee in 1995, regarding declining sperm counts in men worldwide over the previous half-century*

When biologist Louis Guillette first began studying the reproductive biology of alligators in Florida lakes in 1985, he did not foresee where the research would lead. While examining the status of alligator populations in central Florida, he soon discovered that all was not well. Alligators from one lake in particular, Lake Apopka northwest of Orlando, showed a number of bizarre reproductive problems.

Females were having trouble producing viable eggs. Male hatchling alligators had severely depressed levels of the male sex hormone testosterone, and female hatchlings showed greatly elevated levels of the female sex hormone estrogen. The young animals had abnormal gonads, too. Males' testes produced sperm at a premature age, and females' ovaries contained multiple eggs per follicle instead of the expected one egg per follicle.

As Guillette and his co-workers tested and rejected hypotheses for the reproductive abnormalities one by one, they grew to suspect that environmental contaminants might have been somehow responsible. Lake Apopka had suffered a major spill of the pesticides dicofol and DDT in 1980, and yearly surveys thereafter showed a precipitous decline in the number of juvenile alligators in the lake. In addition, the lake received high levels of chemical runoff from agriculture and was experiencing eutrophication (Chapter 6 and Chapter 14) from nutrient input from fertilizers.

The puzzle pieces fell together when a colleague of Guillette's told him that recent research was showing that some environmental contaminants mimic hormones and interfere with the functioning of animal endocrine (hormone) systems. At very low doses, these endocrine disruptors could affect animals, mimicking estrogen and feminizing males. Guillette realized that his alligator observations were similar to results from tests in which rodents were exposed to estrogen during embryonic development. He formulated the hypothesis that certain chemical contaminants in Lake Apopka were disrupting the endocrine systems of alligators during development in the egg.

To test his hypothesis, Guillette and his co-workers went to work comparing alligators from heavily polluted Lake Apopka with those from cleaner lakes nearby. They found that Lake Apopka alligators had abnormally low hatching rates in the years after the pesticide spill and, even as hatching rates recovered in the 1990s, continued to show aberrant hormone levels and bizarre gonad abnormalities. In addition, the penises of Lake Apopka male alligators were 25% smaller than those of male alligators from surrounding lakes.

Similar problems began cropping up in a few other lakes that also experienced runoff of chemical pesticides. In the lab, researchers found that a number of contaminants detected in alligator eggs and young were able to bind to receptors for estrogen and to exist in concentrations great enough to cause sex reversal of male embryos. One chemical in particular, atrazine, the most widely used herbicide in the United States, appeared to disrupt hormones by inducing production of aromatase, an enzyme that converts testosterone to estrogen. In 2003, Guillette reported preliminary findings that nitrate from fertilizer runoff may also act as an endocrine disruptor; when nitrate concentrations in lakes are above the standard for drinking water, juvenile male alligators have smaller penises and 50% lower testosterone levels.

Guillette's results have raised concern not only for alligator health but also for human health. Because alligators and humans share many of the same hormones, many scientists suspect that chemical contaminants could be affecting people, just as they have affected alligators.

The Growth of Environmental Toxicology

The science that examines the effects of poisonous chemicals and other agents on humans and wildlife is **toxicology.** Toxicology has become increasingly important

during the past century as our abilities to produce new chemicals have expanded and as concentrations of chemical contaminants in the environment have increased rapidly. Toxicologists assess and compare substances to determine their **toxicity**, the degree of harm a substance can inflict.

As we review the sometimes poisonous effects of human-made chemicals throughout this chapter, however, it is important to keep in mind that artificially produced chemicals have played a crucial role in giving us the standard of living we enjoy today. Without these chemicals, we would not have the industrial agriculture that produces our food, we would not have many of the medical advances that protect our health and prolong our lives, and we would lack many of the modern materials and conveniences we use every day. It is important to remember these benefits as we examine some of the unfortunate side effects of these advances and search for ever-better alternatives.

Synthetic chemicals are ubiquitous in our environment

Thousands of synthetic (artificial, or human-made) chemicals have been released into our environment (Table 10.1), many of which are **toxicants** that affect the health of wildlife and people. Ubiquitous in the environment, these chemicals can end up far from their origin. For example, seals and whales in the Arctic often contain high levels of synthetic compounds manufactured and applied in urban areas much further south. A 2002 study by the U.S. Geological Survey found that 80% of U.S. streams contain at least trace amounts of 82 wastewater contaminants, including antibiotics, detergents, drugs, steroids, plasticizers, disinfectants, solvents, perfumes, and other

Table 10.1 Estimated Numbers of Chemicals in Commercial Substances during the 1990s

Type of chemical	Estimated number
Chemicals in commerce	100,000
Industrial chemicals	72,000
New chemicals introduced per year	2,000
Pesticides (21,000 products)	600
Food additives	8,700
Cosmetic ingredients (40,000 products)	7,500
Human pharmaceuticals	3,300

Data from Paul Harrison and Fred Pearce, *AAAS Atlas of Population and Environment*, University of California Press, 2000.

Figure 10.1 Synthetic chemicals are everywhere around us in our everyday lives, and some of these compounds may pose environmental or human health risks.

substances. The pesticides we use to kill insects and weeds are some of the most widespread and toxic synthetic chemicals. Although agriculture uses huge amounts of pesticides each year, in the United States more pesticides per unit area are applied by homeowners to their lawns than are applied to farmland by farmers. An even greater concentration of pesticides is applied to golf courses. Every one of us carries traces of numerous industrial chemicals in our bodies. Of the roughly 100,000 synthetic chemicals on the market today, very few have been thoroughly tested for harmful effects (Figure 10.1).

In addition, our environment contains natural chemical substances that may pose health risks. These include everything from oil oozing naturally from the ground, to radioactive radon gas seeping up from bedrock, to toxic chemicals contained in living things, such as the poisons that plants use to ward off herbivores and those that insects use to defend themselves from predators.

The science of toxicology has grown with concern for health and the environment

Toxicologists generally focus on human health, using other organisms as models and test subjects. The growth of the field, together with increased concern about environmental quality, has given rise to the subfield of **environmental toxicology.** Environmental toxicology deals specifically with those toxicants that come from or are discharged into the environment. It includes the study of health effects on humans, other animals, and ecosystems. Animals are studied in environmental toxicology not only out of concern for their welfare, but also because—like canaries in a coal mine—animals can serve as indicators of health threats that may soon come to affect humans.

For example, the endocrine effects on Lake Apopka alligators are not just an isolated instance. The undersized penises of these alligators may be analogous to those of human boys who were known to have been exposed to certain contaminants during their early development. Boys suffering this fate have included sons born to mothers in Taiwan who used cooking oil contaminated with toxicants called polychlorinated biphenyls (PCBs), which are toxic by-products of chemicals used in transformers and other electrical equipment. Furthermore, rates of testicular cancer have risen sharply, and several studies have now shown that human sperm counts are decreasing in many parts of the world—trends many scientists blame on endocrine-disrupting chemicals in the environment.

Among wildlife, Guillette's findings with alligators could be the tip of the iceberg. A flood of scientific studies on endocrine disruptors and wildlife has poured forth in recent years. One set of studies that has made headlines has been that of scientist Tyrone Hayes of the University of California at Berkeley, who has found gonadal abnormalities in frogs similar to those of Guillette's alligators, abnormalities Hayes has attributed to atrazine (Figure 10.2). In lab experiments, male frogs raised in water containing very low doses of the herbicide became feminized and hermaphroditic, developing both testes and ovaries. In field studies, leopard frogs in the wild across North America showed similar problems in areas of atrazine usage. Hayes found these effects at atrazine concentrations well below EPA guidelines for human health and at concentrations lower than commonly exist in U.S. drinking water; thus, he suggests, people may be at risk.

Hayes's work has met with fierce criticism from scientists associated with atrazine's manufacturer, which

Figure 10.2 According to research by University of California, Berkeley, Tyrone Hayes, frogs may suffer reproductive abnormalities from exposure to the best-selling herbicide atrazine. Industry-backed scientists have disputed the findings.

stands to lose millions of dollars if the top-selling herbicide were to be banned in the United States by the EPA. Such battles illustrate dynamics common to environmental toxicology, where the stakes are high, both in terms of economics and human health.

Atrazine and the many other compounds that kill insects and weeds that threaten crops were made possible by advances in chemistry during and following World War II. Mechanization of production and the growth of large corporations enabled the production of huge amounts of synthetic chemicals. As material prosperity grew in Westernized nations in the decades following the war, people began using pesticides not only for agriculture, but also to improve the look of their lawns and golf courses and to fight termites, ants, and other insects inside their homes and offices. It was not until the 1960s that people began to learn about the risks of exposure to many synthetic pesticides. As a result, many people began to protest against indiscriminate pesticide use. In fact, concern over pesticides was a catalyst that helped spur the entire environmental movement in the United

States. The key event was the publication of Rachel Carson's 1962 book *Silent Spring,* which brought the pesticide dichlorodiphenyltrichloroethane (DDT) to the attention of the public.

Silent Spring began the public debate over pesticides

Rachel Carson was a naturalist, author, and government scientist. In *Silent Spring*, she brought together a diverse collection of scientific studies, medical case histories, and other data that no one had previously thought to synthesize and present to the general public. Her message was that DDT in particular and artificial pesticides in general were hazardous to people's health, the health of wildlife, and the well-being of ecosystems. Carson wrote at a time when large amounts of pesticides virtually untested for health effects were indiscriminately sprayed over residential neighborhoods and public areas, on the assumption that the chemicals would kill insects and other pests but do no harm to humans (Figure 10.3). Similarly, most people had no idea that the store-bought chemicals they used in their houses, gardens, and crops might be toxic.

Although challenged vigorously by spokespeople for the chemical industry, who attempted to discredit both the author's science and her personal reputation, Carson's book was a best-seller. Carson suffered from cancer as she finished *Silent Spring*, and lived only briefly after its publication. However, the book eventually helped generate significant social change in views and actions toward the environment. The use of DDT was banned in the United States in 1973 and has been banned in a number of other nations. The United States still manufactures and exports DDT to countries that do use it, predominantly for control of human disease vectors.

Weighing the Issues:
Circle of Poison

It has been called the "circle of poison." Despite the fact that the United States has banned DDT's use, U.S. companies still manufacture and export the compound to many developing nations. Thus, it is possible that pesticide-laden food can be imported back into the United States. How do you feel about this? Is it unethical for one country to sell to others a substance that it has deemed toxic? Are there factors or circumstances that might change the view you take?

Figure 10.3 Before the 1960s, the environmental and health effects of potent pesticides like DDT were not widely studied or publicly known. Public areas such as parks, neighborhoods, and beaches were regularly sprayed for insect control without safeguards against excessive human exposure. Here children on a Long Island, New York, beach are sprayed with DDT from a pesticide spray machine being tested in 1945.

Toxic Agents in the Environment

Determining what types and levels of risk a potential toxicant might pose requires diligent scientific work, both in the laboratory and in the field. Shortly we will look at how scientists study the effects of toxicants in the lab, but first let's quickly survey what kinds of toxic agents exist around us and how they behave and move through the environment.

Toxicants come in many different types

Toxicants can be classified into different types based on their health effects. The best-known are **carcinogens**, chemicals or types of radiation that cause cancer. In cancer, certain malignant cells grow uncontrollably, creating tumors, damaging the body's functioning, and often leading to death. In our society today, the greatest number of cancer cases is thought to result from carcinogens contained in cigarette smoke. Carcinogens can be difficult to identify because there may be a long lag time between exposure to the agent and the detectable onset of cancer. Historically, much toxicological work focused on carcinogens. Now, however, we know that toxicants can produce many different types of effects, so scientists have many more endpoints, or health impacts, to look for.

Mutagens are chemicals that cause mutations in the DNA of organisms (Chapter 4). Although most mutations have little or no effect, some can lead to severe problems, including cancer and many other disorders. Mutations can harm the individual exposed to the mutagen, or if the mutations occur in sperm or egg cells, then the individual's offspring may suffer the effects.

Chemicals that cause harm to the unborn are called **teratogens.** Teratogens that affect the development of human embryos in the womb can cause birth defects. One example involves the drug thalidomide, developed in the 1950s as a sleeping pill and to prevent nausea during pregnancy. Tragically, the drug turned out to be a powerful teratogen, and its use caused birth defects in thousands of babies (Figure 10.4). Even a single dose during pregnancy could result in limb deformities and organ defects. Thalidomide was banned in the early 1960s once the connection with birth defects was recognized. Ironically, today the drug shows some promise in treating a wide range of diseases, including Alzheimer's disease, AIDS, and various types of cancer.

Some toxicants cause harm by affecting the immune system, which protects our bodies from disease. **Allergens** overactivate the immune system, causing an immune response when one is not necessary. One hypothesis for the increase in asthma in recent years is an increase in allergenic synthetic chemicals in our environment. Other toxicants may weaken the immune system, making the body less able to defend itself

Figure 10.4 The drug thalidomide turned out to be a potent teratogen. It was banned in the 1960s, but not before causing thousands of birth defects in babies born to mothers who took the product to relieve nausea during pregnancy. Butch Lumpkin was an exceptional "thalidomide baby" who learned to overcome his short arms and deformed fingers, becoming a professional tennis instructor in Georgia.

against bacteria, viruses, allergy-causing agents, and other attackers.

Still other chemical toxicants, **neurotoxins**, assault the nervous system. Neurotoxins include various heavy metals such as lead, mercury, and cadmium, as well as pesticides and some chemical weapons developed for use in war. A famous case of neurotoxin poisoning occurred in Japan, where a chemical factory dumped mercury waste into Minamata Bay between the 1930s and 1960s. Thousands of people in and around the town on the bay were poisoned by eating fish contaminated with the mercury. First the town's cats began convulsing and dying, and then people began to show odd symptoms including slurred speech, loss of muscle control, sudden fits of laughter, and in some cases death. The company and the government eventually paid out millions of dollars in compensation to affected residents.

Most recently, scientists have recognized the importance of **endocrine disruptors**, toxicants that interfere with the endocrine system. The endocrine system consists of a series of chemical messengers (hormones) that travel though the body. Sent through the bloodstream at extremely low concentrations, these messenger molecules have many vital functions. They stimulate growth, development, and sexual maturity, and they regulate brain function, appetite, sexual drive, and many other aspects of our physiology and behavior. Hormone-disrupting toxicants can affect an animal's endocrine system in various ways, including blocking the action of hormones or accelerating their breakdown (Figure 10.5). Many endocrine disruptors possess molecular structures that happen to be very similar to certain hormones in their structure and chemistry. If a molecule is similar enough, it may mimic a hormone and interact with the receptor molecules for that hormone just as the actual hormone would.

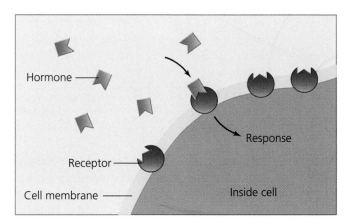

(a) Normal hormone binding

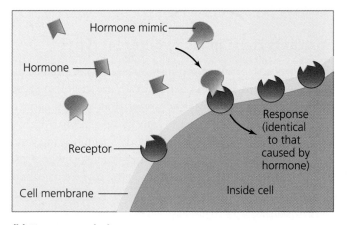

(b) Hormone mimicry

Figure 10.5 Many endocrine-disrupting substances act by mimicking the structure of hormone molecules. Like a key similar enough to fit into another key's lock, the hormone mimic may bind to a cellular receptor for the hormone, causing the cell to react as if it had encountered the hormone.

The potential for endocrine disruption appears to be widespread

Although scientists first noted endocrine-disrupting effects as far back as the 1960s with the pesticide DDT, the idea that synthetic chemicals might be altering the hormones of animals was not widely appreciated until the 1996 publication of the book *Our Stolen Future*, by Theo Colburn, Dianne Dumanoski, and J. P. Myers. Like *Silent Spring*, this volume gathered together a large amount of scientific work from various fields and presented a unified picture that shocked many readers—and brought criticism from some scientists and from the chemical industry. Colburn, a scientist with the World Wildlife Fund, had spent many years preceding the publication of *Our Stolen Future* bringing toxicologists, medical doctors, wildlife biologists, and other scientists together to share their findings. Once she and her co-authors presented the data and stories in *Our Stolen Future* (and on a Web page that continues today, providing updates on new research), the floodgates were opened. Currently a large amount of research is devoted to the hormonal impacts of synthetic chemicals on humans and wildlife.

One common type of endocrine disruption involves the feminization of male animals, as Louis Guillette found with alligators, Tyrone Hayes reported with frogs, and other studies have indicated with fish and other creatures. Feminization may be widespread because a number of chemicals appear to mimic estrogen and bind to estrogen receptors. In the case of the alligators and frogs, however, the herbicide atrazine may instead induce production of the enzyme aromatase, which converts male sex hormones to female ones.

To date, endocrine effects have been most widely found in nonhuman animals, but many scientists attribute the striking drop in sperm counts among men worldwide to endocrine disruptors (Figure 10.6a). In the most celebrated study, Danish researchers reported in 1992 that the number and mobility of sperm in men's semen had declined by 50% since 1938. The research involved a review of 61 studies that included 15,000 men from 20 nations on six continents. Subsequent studies by other researchers—including some who set out to disprove the findings—have largely confirmed the results using other methods and other populations. Although there is tremendous geographic variation that remains unexplained, it seems apparent that human sperm counts have declined notably in many places in the world.

Some researchers have also voiced concerns about rising rates of testicular cancer, undescended testicles, and genital birth defects in men (Figure 10.6b). Meanwhile,

other scientists have proposed that the striking rise in breast cancer rates (one in eight U.S. women today develop breast cancer) may also be due to hormone disruption, because an excess of estrogen appears to feed tumor development in older women.

However, endocrine disruptors can affect more than just the reproductive system. Many concerns center on impairment of the brain and nervous system. North American studies have shown neurological problems

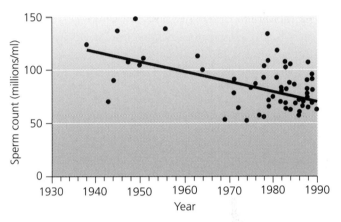

(a) Declining sperm count in humans, based on 61 publications

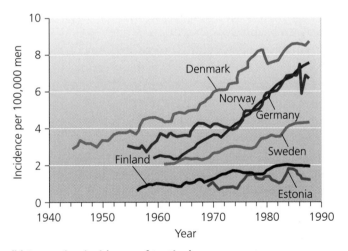

(b) Increasing incidence of testicular cancer

Figure 10.6 A review study in 1992 synthesized the results of 61 studies that reported sperm counts in men from various localities since 1938. The data were highly variable but showed a significant decrease in human sperm counts over time (a). Many scientists have hypothesized that this decrease may result from endocrine-disrupting chemicals in the environment. Some have also hypothesized that endocrine disruptors could be behind the fact that men in a number of countries have shown an increased incidence of testicular cancer in recent decades (b). Data from E. Carlsen et al., *British Medical Journal*, 1992, as adapted by J. Toppari et al., *Environmental Health Perspectives*, 1996. (b) H.O. Adami et al., *International Journal of Cancer*, 1994.

associated with PCB contamination. In one study in Michigan, mothers who ate Great Lakes fish contaminated with PCBs had babies with lower birth weights and smaller heads, compared to mothers who did not eat fish. These babies grew into children who showed weak and jerky reflexes and tested poorly in various intelligence tests.

Endocrine disruption research is ongoing and contentious

Research into hormone disruption is ongoing, and much of it is highly contentious. This contentiousness is due not only to the scientific uncertainty inherent in a developing discipline, but also to the fact that our society has invested so heavily in some of the chemicals that now are suspect. One example is bisphenol-A (see The Science behind the Story). An estrogen mimic, bisphenol-A is used in a great number of diverse plastics products we use daily, from baby bottles to food cans to eating utensils to auto parts, and has even been used in a cavity-preventing coating for children's teeth. The chemical leaches out of many of these products into water and food, however, and recent evidence ties it to birth defects in lab mice. The plastics industry vehemently protests that the chemical is safe, however, and points to other research backing its contention.

The bisphenol-A study results with mice showed effects at extremely low levels of the chemical, as has the research with atrazine and frogs and various other known or purported endocrine disruptors. The likely reason is that the endocrine system is geared to respond to minute concentrations of substances, as it does with hormones in the bloodstream. In other words, because the endocrine system operates with miniscule amounts of chemicals, it may be especially vulnerable to effects from environmental contaminants that are dispersed and diluted through the environment and that reach our bodies in very low concentrations.

Toxicants may become concentrated in surface water or groundwater

Toxicants are not evenly distributed in the environment, and they move about in certain ways (Figure 10.7). For instance, water, in the form of runoff, often carries toxicants from large areas of land and concentrates them in small volumes of surface water. The U.S. Geological Survey findings regarding surface waterways mentioned earlier in this chapter demonstrated this concentrating effect. If chemicals can persist in soil, they can leach down into groundwater and contaminate drinking water supplies.

Many chemicals are soluble in water, and these are often the ones most accessible to organisms, entering organisms' tissues through drinking or absorption. For this reason, aquatic animals such as fish, frogs, and stream invertebrates are especially good indicators of pollution. When aquatic organisms become sick, we can take it as an early warning that something is amiss and conduct research to identify the problem, ideally before it harms humans or the ecological systems that support us. This is why findings showing impacts of low concentrations of atrazine on frogs, fish, and invertebrates have caused such a stir; many scientists see such findings as a warning that humans could be next, because the atrazine that washes into streams and rivers also flows and seeps into our drinking water and drifts through our air.

Airborne toxicants can move widely through the environment

Because many chemical substances can also be transported by air, the possible toxicological effects of chemical usage can be much more widespread than the direct chemical application itself. Wind patterns can carry airborne toxicants long distances. For instance, air pollution from industry in the U.S. Midwest gets blown to northeastern states and southeastern Canada, where it causes acid precipitation (Chapter 11).

Airborne transport of pesticides, or pesticide drift, occurs in the Central Valley of California. This farming region, often called the "fruit basket" of the United States, produces a substantial portion of the nation's crops and is widely considered the most productive agricultural region in the world. But because it is a naturally arid area, food production depends on intensive irrigation, fertilizer use, and pesticide use. Roughly 143 million kg (315 million lbs) of pesticide active ingredients are used in California each year, most of these in the Central Valley. The region's frequent winds often blow the airborne spray—and dust particles containing pesticide residue—for long distances. Families living in towns in the Central Valley have suffered health impacts, and activists for farmworkers maintain that hundreds of thousands of the state's residents are at risk. The human health impacts parallel effects seen in amphibians in the high mountains of the Sierra Nevada, where research has associated pesticide drift from the Central Valley with population declines in four species of frogs.

Most striking, though, is the fact that synthetic chemical contaminants are nearly ubiquitous worldwide.

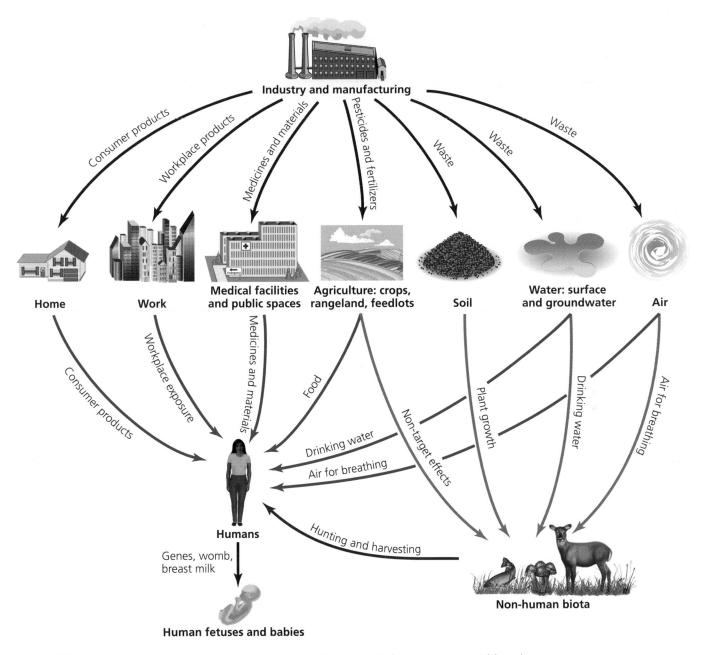

Figure 10.7 Synthetic chemicals take many routes in traveling through the environment. Although humans take in only a tiny proportion of these compounds, and although many compounds are harmless, humans and particularly babies receive small amounts of toxicants from many sources.

Despite being manufactured and applied only in the temperate and tropical zones, contaminants show up in substantial quantities in the tissues of Arctic polar bears, Antarctic penguins, and people living in Greenland. Scientists can travel to the most remote and seemingly pristine alpine lakes in British Columbia and find them contaminated with foreign toxicants, such as PCBs. The surprisingly high concentrations in polar regions result from patterns of global atmospheric circulation that move airborne chemicals systematically toward the poles (Figure 10.8).

Some toxicants persist for a long time

Once toxic agents arrive somewhere, they may degrade quickly and become harmless, or they may remain unaltered and persist for many months, years, or decades. The rate at which chemicals degrade depends on factors such as temperature, moisture, and sun exposure, and how these interact with the chemistry of the toxicant. Those toxicants that persist in the environment have the most potential to harm many organisms over long

Bisphenol A

Can a compound in everyday plastic products damage the most basic biological processes necessary for healthy pregnancies and births? A scientist in Ohio has found evidence that it does—all because one of her lab assistants reached for the wrong soap.

At a laboratory at Case Western Reserve University in August 1998, geneticist Patricia Hunt was making a routine check of her female lab mice, which included the extraction and examination of developing eggs from the ovaries. The results made her wonder what had gone wrong. About 40% of the eggs showed problems with their chromosomes, and 12% had irregular amounts of genetic material, a condition that can cause serious reproductive problems in mice and people alike.

Detective work revealed that a lab assistant had mistakenly washed the lab's plastic mouse cages and water bottles with an especially harsh soap, A-33 manufactured by

Dose and Meiosis Abnormalities in Mice Exposed to Bisphenol A	
BPA Dose*	Chromosomal Problems During Cell Division
Control: 0 mg/g	2/115 (1.7%)
20 ng/g (0.00002 mg)	10/172 (5.8%)
40 ng/g (0.00004 mg)	19/255 (7.5%)
100 ng/g (0.0001 mg)	5/46 (10.9%)

*Oral Reference Dose (or safe intake level) established by the U.S. Environmental Protection Agency = 0.05 mg/kg body weight per day. European Commission's Scientific Committee on Food's Tolerable Daily Intake = 0.01 mg/kg body weight per day (recently revised down from U.S. EPA level of 0.05) Data from Patricia A. Hunt, et al., *Current Biology*, 2003.

Airkem Professional Products. The soap damaged the cages so badly that parts of them seemed to have melted. The cages were made from a type of plastic called polycarbonate, which contains the chemical bisphenol A (BPA). BPA is used to manufacture thousands of plastic products, from baby bottles to food containers to auto parts. The plastics industry uses about 907 million kg (2 billion lb) of BPA every year, according to industry estimates.

The chemical, however, is an estrogen mimic and a long-suspected endocrine disruptor. Some studies link BPA to reproductive abnormalities in mice, such as low sperm counts and early sexual development. Other studies indicate that BPA can leach out of plastic products into water and food. One such study conducted by researchers for the magazine *Consumer Reports* and published in 1999, found that BPA seeped out of the plastic walls of heated baby bottles into infant formula. The plastics industry insisted that BPA was safe and fought back with its own studies. U.S.

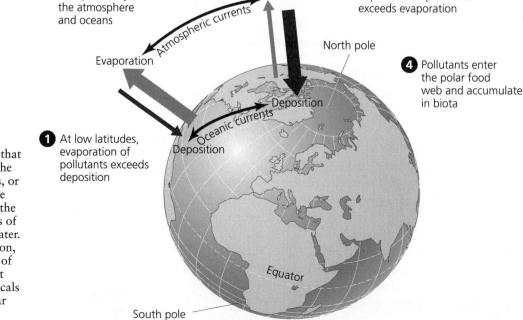

Figure 10.8 In the process of global distillation, pollutants that evaporate and rise high into the atmosphere in lower latitudes, or are deposited in the ocean, are carried preferentially toward the poles by atmospheric currents of air and oceanic currents of water. Polar organisms, for this reason, take in more than their share of toxicants, despite the fact that relatively few synthetic chemicals are manufactured or used near the poles.

2 Pollutants are transported by the atmosphere and oceans

Evaporation

Atmospheric currents

Evaporation

3 At high latitudes, deposition of pollutants exceeds evaporation

North pole

4 Pollutants enter the polar food web and accumulate in biota

Deposition

Oceanic currents

1 At low latitudes, evaporation of pollutants exceeds deposition

Deposition

Equator

South pole

health officials, caught in a scientific stalemate, declined to strengthen regulations on BPA.

Hunt, however, thought the chemical might be adversely affecting her mice. The mice in Hunt's lab were suddenly showing dramatic problems during meiosis, the division of chromosomes during the formation of eggs. The 40% of mouse eggs with meiotic problems represented a huge jump from the 1–2% rate of such defects seen before the cage-washing incident. The 12% with irregular amounts of genetic material, a dangerous cell division error called aneuploidy, would likely lead to nonviable pregnancies or birth defects. In humans, aneuploidy is the main cause of miscarriage and Down's syndrome.

Deciding to recreate the BPA exposure experimentally, Hunt instructed researchers in her lab to intentionally wash polycarbonate cages and water bottles in varying levels of A-33. They then compared mice kept in damaged cages with

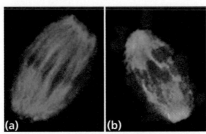

(a) (b)

During normal cell division, chromosomes align properly (a). Exposure to bisphenol A, however, causes abnormal cell division (b), where chromosomes scatter and are distributed improperly and unevenly between daughter cells, a condition called *aneuploidy*.

plastic water bottles to mice kept in undamaged cages with glass water bottles. The developing eggs of mice exposed to BPA through the deliberately damaged plastic showed levels of meiotic problems similar to those in the original A-33 incident, whereas the eggs of mice in the control cages were normal. In an additional round of tests, three sets of female mice were given daily oral doses of BPA over 3, 5, and 7 days, and the

same meiotic abnormalities were observed, although at lower levels. The mice given BPA for 7 days were most severely affected.

Published in 2003 in the journal *Current Biology*, Hunt's findings quickly set off a new wave of concern over the safety of BPA. Hunt wrote in her study, "We have observed meiotic defects in mice at exposure levels close to or even below those considered 'safe' for humans. Clearly, the possibility that BPA exposure increases the likelihood of genetically abnormal offspring is too serious to be dismissed without extensive further study." Although Hunt's research focused on mice, and not humans, the findings were disturbing because sex cells of mice and humans divide and function in similar ways. Regulators in Europe have recently lowered the level of BPA intake they consider safe. In the United States, some scientists are now pushing for more research and renewed federal scrutiny into the safety of BPA.

periods of time. One reason people have been so concerned about DDT is its long persistence time; other highly toxic chemicals, such as PCBs, are also persistent. In contrast, the Bt toxin used in biocontrol and GM crops (Chapter 9) has a very short persistence time. The herbicide atrazine is variable in its persistence, depending on environmental conditions. There are large numbers of persistent synthetic chemicals because many are designed to be persistent. Those used in plastics, for instance, are used precisely because they resist breakdown.

Most toxicants eventually degrade into simpler compounds called *breakdown products*. Often these are less harmful than the original substance, but sometimes they are as toxic as the original chemical. For instance, DDT breaks down into DDE, a highly persistent and toxic compound in its own right. Atrazine produces a large number of breakdown products whose effects have not been fully studied.

Toxicants may accumulate and move up the food chain

Toxicants that organisms absorb, breathe, or consume are subjected to the organism's metabolic processes. Some will be quickly excreted, and others will be degraded into harmless breakdown products. However, some will not be broken down but instead will remain in the body. Toxicants that are fat-soluble or oil-soluble, like DDT and DDE, are absorbed and stored in fatty tissues. Such toxicants, usually organic compounds, may build up in an animal, in a process termed **bioaccumulation.**

Toxicants that bioaccumulate in the tissues of one organism may then be transferred to other organisms as predators consume prey. When one organism consumes another, it takes in any stored fat-soluble toxicants and stores them itself, along with the toxicants it has received from eating many other prey. Thus with each step up the food chain, from producer to primary consumer

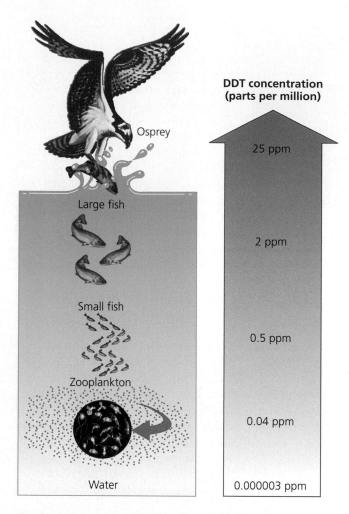

DDT concentration (parts per million)

Osprey

25 ppm

Large fish

2 ppm

Small fish

0.5 ppm

Zooplankton

0.04 ppm

Water

0.000003 ppm

Figure 10.9 Fat-soluble compounds such as DDT bioaccumulate in the tissues of organisms. As organisms lower on the food chain are eaten by animals at higher trophic levels, their load of toxicants is passed up to the consumer. DDT moves from zooplankton through various types of fish, finally becoming highly concentrated in fish-eating birds such as ospreys.

to secondary consumer and so on, the concentrations of toxicants can be greatly magnified. This process, called **biomagnification,** occurred most famously with DDT. Top predators, such as birds of prey, ended up with high concentrations of the pesticide because concentrations were magnified as DDT moved from water to algae to plankton to small fish to bigger fish and finally to fish-eating birds (Figure 10.9).

Because of biomagnification, populations of many North American birds of prey, such as ospreys, peregrine falcons, bald eagles, and brown pelicans, declined precipitously and went locally extinct from the 1950s to the 1970s. The peregrine falcon was almost totally wiped out in the eastern United States, and the bald eagle, the U.S. national bird, was virtually eliminated from the lower 48 states. The brown pelican vanished from its Atlantic Coast range, remaining only in Florida, and the osprey and various other hawks saw substantial population declines. Eventually scientists determined that the main reason for the declines was that DDT was causing these birds' eggshells to grow thinner, so that many eggs were breaking while in the nest.

Such scenarios are by no means a thing of the past. One modern-day story centers on the polar bears of Svalbard Island in Arctic Norway. These bears are at the top of the food chain and feed on seals that have already experienced substantial biomagnification of toxicants. Despite their remote Arctic location, Svalbard Island's polar bears show some of the highest levels of PCB contamination of any wild animals tested, as a result of the process of global distillation shown in Figure 10.8. The contaminants are likely responsible for the immune suppression, hormone disruption, and high cub mortality that the bears seem to be suffering. Furthermore, cubs that do survive receive PCBs in their mothers' milk, so that contamination persists and accumulates between generations.

Not all toxicants are synthetic

Although we have focused on synthetic chemicals thus far, it is vital to recognize that chemical toxicants also exist naturally in the environment around us and in the foods we eat. We have good reason as citizens and consumers to insist on being informed about risks synthetic chemicals may pose, but it is a mistake to assume that all artificial chemicals are unhealthy and that all natural chemicals are healthy. In fact, the plants and animals we eat contain many chemicals that can cause us harm. Recall that plants produce toxins to ward off animals that eat them. In domesticating crop plants, we have selected for strains with reduced toxin content, but we have not eliminated these dangers. Furthermore, when we consume animal meat, we take in toxins the animals have ingested from plants or animals they have eaten.

Scientists have actively debated just how much risk natural toxicants pose. Leading the fray has been biochemist Bruce Ames of the University of California at Berkeley, who maintains that the amounts of synthetic chemicals in our food from pesticide residues are dwarfed by the quantities of natural toxicants. Therefore, Ames reasons, synthetic chemicals are a relatively minor worry. Furthermore, he contends, if fear of pesticide residues or increased cost to remove them causes people to eat fewer fruits and vegetables, then people will actually be at greater risk for disease, because a balanced diet including fruits and vegetables is important for maintaining health.

A respected senior scientist who was formerly a hero of environmentalists for his previous work, today Ames is a target of criticism from many environmental advocates. His critics say that natural toxicants are usually more readily metabolized and excreted by the body than synthetic ones, that synthetic toxicants persist and accumulate in the environment, and that synthetic chemicals expose people (such as farmworkers and factory workers) to risks in ways other than the ingestion of food. What is clear is that more research is required in this area.

Weighing the Issues:
Natural and Synthetic Estrogen Mimics

Skeptics of the idea that synthetic chemicals act as endocrine disruptors point out that humans are exposed to phytoestrogens, or natural estrogens from plants, which are ubiquitous in the environment. However, phytoestrogens are easily broken down by the body and are not stored in tissue. Do you think there is a distinction between synthetic and natural estrogen-mimicking chemicals? Should manufacturers of synthetic chemicals that mimic estrogen be held responsible for health impacts when natural chemicals can also mimic estrogen?

Studying Effects of Toxicants

Determining health effects that particular toxicants produce is a challenging job, particularly because any given person or organism likely has a complex history of exposure to many toxicants throughout life. Toxicologists rely on several different methods, including both correlative surveys and manipulative experiments (Chapter 1).

Wildlife toxicology utilizes careful observations in the field and the lab

When scientists were zeroing in on the impacts of DDT, one key piece of evidence came from museum collections of wild birds' eggs from the decades before synthetic pesticides were manufactured. The eggs from museum collections had measurably thicker shells than the eggs scientists were studying in the field from present-day birds. Scientists have pieced together the puzzle of toxicant effects on alligators by taking measurements

from animals in the wild, then doing controlled experiments in the lab to test hypotheses. With frogs affected by atrazine, scientists first showed toxicological effects in lab experiments, then sought to demonstrate correlations with herbicide usage in the wild.

Often the toxicological study of wildlife advances in the wake of some conspicuous mortality event. Off the California coast in 1998–2001, populations of sea otters fell noticeably, and many dead otters washed ashore. Field biologists documented the population decline, and toxicologists went to work in the lab performing autopsies to determine causes of death. The most common cause of death was found to be infection with the protist parasite Toxoplasma, which killed otters directly and also made them vulnerable to shark attack. Toxoplasma occurs in the feces of cats, so scientists hypothesized that sewage runoff containing waste from litter boxes was entering the ocean from urban areas and infecting the otters.

Human toxicology relies on case histories and on epidemiology

In human toxicology, as in wildlife studies, much knowledge has been gained by studying sickened individuals directly. Medical professionals have long treated victims of poisonings, so the effects of certain common poisons are well known. In addition, autopsies help us understand what constitutes a lethal dose. This process of observation and analysis of individual patients can be labeled a *case history* approach. Case histories have advanced our understanding of human toxicology, but they do not always help us learn the effects of newly manufactured compounds, toxicants people rarely come into contact with, or chemicals existing at low environmental concentrations and exerting minor long-term effects. Case histories also tell us little about probability and risk, such as how many extra deaths we might expect in a population due to a particular cause.

For such situations, which are common in environmental toxicology, epidemiological studies are necessary. **Epidemiological studies** involve large-scale comparisons among groups of people, usually contrasting a group known to have been exposed to some toxicant and a group that has not. Epidemiologists track the fate of all people in the study for a long period of time (generally years or decades) and measure the rate at which deaths, cancers, or other health problems of interest occur in each group. Eventually the epidemiologist analyzes the data, looking for observable differences between the two groups, and statistically tests hypotheses accounting

The Science behind the Story

Pesticides and Child Development in Mexico's Yaqui Valley

With spindly arms and big round eyes, one set of pictures shows the sorts of stick figures drawn by young children everywhere. Next to them is another group of drawings, mostly disconnected squiggles and lines, resembling nothing. Both sets of pictures are intended to depict people. The main difference identified between the two groups of young artists: long-term pesticide exposure.

Children's drawings are not a typical tool of toxicology, but Elizabeth Guillette, an anthropologist, wanted to try new methods. Guillette was interested in the effects of pesticides on children. She devised tests to measure childhood development based on techniques from anthropology and medicine. Searching for a study site, Guillette found the Yaqui Valley region of northwestern Mexico.

The Yaqui Valley is farming country, worked for generations by the indigenous group that gives the region its name. Synthetic pesticides arrived in the area in the 1940s. Some Yaqui embraced the agricultural innovations, spraying their farms in the valley to increase their yields. Yaqui farmers in the surrounding foothills, however, generally chose to bypass the chemicals and to continue following more traditional farming practices. Although differing in farming techniques,

Drawings by children in the foothills

4-year-olds

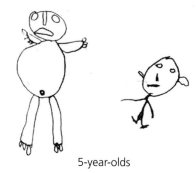

5-year-olds

Drawings by children in the valley

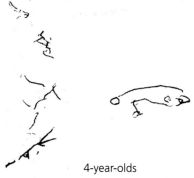

4-year-olds

5-year-olds

Elizabeth Guillette's study in Mexico's Yaqui Valley offers a startling example of apparent neurological effects of pesticide poisoning. Young children from foothills areas where pesticides were not commonly used drew recognizable figures of people. Children the same age from valley areas where pesticides were used heavily in industrialized agriculture could draw only scribbles when asked to draw people. Adapted from Elizabeth A. Guillette, et al., *Environmental Health Perspectives*, 1998.

Yaqui in the valley and foothills continued to share the same culture, diet, education system, income levels, and family structure.

At the time of the study, in 1994, valley farmers planted crops twice a year, applying pesticides up to 45 times from planting to harvest. A previous study conducted in the valley in 1990, focusing on areas with the largest farms, had indicated high levels of multiple pesti-

for differences. When a group exposed to a toxicant shows a significantly greater degree of harm, it suggests that the toxicant may be responsible.

This process is akin to a natural experiment (Chapter 1), in which the experimenter takes advantage of the presence of two groups of subjects made possible by some event that has already occurred. A slightly different type of natural experiment was conducted by anthropol-

ogist Elizabeth Guillette. She conducted qualitative research on how pesticides affected children by comparing children exposed to pesticides in an agricultural valley to children in nearby foothills, who had very little exposure (see The Science behind the Story).

The advantages of epidemiological studies are their realism and their ability to yield relatively accurate predictions about risk. The drawbacks include the need to

cides in the breast milk of mothers and in the umbilical cord blood of newborn babies. In contrast, foothill families avoided chemical pesticides in their gardens and homes.

To understand how pesticide exposure affects childhood development, Guillette and fellow researchers studied 50 preschoolers aged four to five, of whom 33 were from the valley and 17 from the foothills. Each child underwent a half-hour exam, during which researchers showed a red balloon, promising to give the balloon later as a gift, and using the promise to evaluate long-term memory. Each child was then put through a series of physical and mental tests:

- Catching a ball from distances of up to 3 m (10 ft) away, to test overall coordination
- Jumping in place for as long as possible, to assess endurance
- Drawing a picture of a person, as a measure of perception
- Repeating a short string of numbers, to test short-term memory
- Dropping raisins into a bottle cap from a height of about 13 cm (5 in), to gauge fine-motor skills

The researchers also measured each child's height and weight but, because of lack of time and money, stopped short of taking blood or tissue samples to check for pesticides or other toxins. When all tests were completed, each child was asked what he or she had been promised and received a red balloon.

Although the two groups of children were not significantly different in height and weight, they differed markedly in other areas of their development. Valley children were far behind the foothill children developmentally in coordination, physical endurance, long-term memory, and fine-motor skills:

- From a distance of 3 m (10 ft), valley children had great difficulty catching the ball.
- Valley children could jump for an average of 52 seconds, compared to 88 seconds for foothill children.
- Most valley children missed the bottle cap when dropping their raisins, whereas foothill children dropped them into the caps far more often.
- Each group did fairly well repeating numbers, but valley children showed poor long-term memory. At the end of the test, all but one of the foothill children remembered that they had been promised a balloon, and 59% remembered it was red. However, of the valley children only 27% remembered the color of the balloon, only 55% remembered they'd be getting a balloon, and 18% were unable to remember anything about the balloon.

It was the children's drawings, however, that exhibited the most dramatic difference between valley and foothill children. The researchers determined each drawing could earn 5 points, with 1 point each for a recognizable feature: head, body, arms, legs, and facial features. The foothill children drew pictures that looked like people, averaging about 4.5 points per drawing. The valley children, in contrast, averaged 1.6 points per drawing; their scribbles resembled little that looked like a person. By the standards of developmental medicine, the four- and five-year-old valley children drew at the level of a two-year-old.

Some scientists greeted Guillette's study skeptically, pointing out that its sample size was too small to be meaningful. Others said that factors the researchers missed, such as different parenting styles or unknown health problems, could be to blame. Prominent toxicologists argued that without blood or tissue tests on the children, the study results couldn't be tied to agricultural chemicals. Regardless of these criticisms, Guillette maintains that her findings show that nontraditional study methods are a valid way to track the effects of environmental toxins and that pesticides present a complex long-term risk to human growth and health.

---

wait a long time for the results and the fact that epidemiological studies do not address future effects of new products just coming to market. In addition, participants in epidemiological studies encounter many factors that affect their health besides the one under study. Epidemiological studies measure a statistical association between a toxicant and an effect, but they do not confirm that the toxicant is responsible for the effect.

Manipulative experiments are needed to truly nail down causation. However, because subjecting people to massive doses of toxicants in a lab experiment would clearly be unethical, experimenters have traditionally used other animals as subjects to test toxicity. Foremost among these animal models have been laboratory strains of rats and mice.

Animals may serve as models for humans

Although some people feel the use of rats and mice for testing is unethical, animal testing clearly enables scientific advances that would be impossible or far more difficult otherwise. However, new techniques (with human cell cultures, bacteria, or tissue from chicken eggs) increasingly are being devised that may one day replace some live-animal testing.

The logic of using mice and rats is that they are mammals; because of our shared evolutionary history, these creatures possess systems that function similarly to ours. Yet we are also, obviously, different. The extent to which results from animal lab tests can be applied to humans is always somewhat uncertain and can be expected to vary from one study to the next.

Dose-response analysis is a mainstay of toxicology

The standard method of testing with lab animals in toxicology is dose-response analysis. Scientists quantify the toxicity of a given substance by measuring how much effect a toxicant produces at different doses or how many animals are affected by different doses of the toxic agent. The dose is the amount of toxicant the test animal receives, and the response is the degree or type of negative effects the animal exhibits as a result. The response is generally quantified by measuring the proportion of animals exhibiting negative effects. The data are plotted on a graph, with dose on the x axis and quantified response on the y axis (Figure 10.10). The resulting curve is called a **dose-response curve.**

Once they have plotted a dose-response curve, scientists can calculate a convenient shorthand gauge of a substance's toxicity: the amount of toxicant it takes to kill half the population of study animals used. This lethal dose for 50% of individuals is termed the **LD_{50}.** A high LD_{50} indicates a low toxicity, and a low LD_{50} indicates a high toxicity.

Of course, the experimenter may instead be interested in nonlethal health effects. A researcher often will want to document the level of toxicant at which 50% of the population of test animals is affected in whatever way is of interest in the study (for instance, what level of toxicant causes 50% of lab mice to lose their hair?). Such a level is called the **effective dose-50%,** or **ED_{50}.** Chemicals with an especially low LD_{50} and ED_{50} are assumed to be likely poisonous to humans, whereas those with an extremely high LD_{50} and ED_{50} are thought to likely be safe.

Common sense suggests that the greater the dose, the stronger the response will be. However, sometimes re-

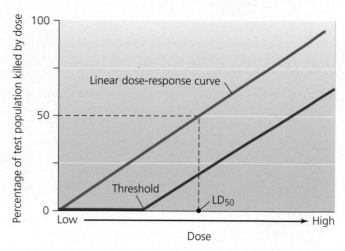

Figure 10.10 A dose-response curve shows how test animals respond to doses of different concentrations of the substance being tested. In a classic linear dose-response curve, the percent of animals killed or otherwise affected by the substance rises with the dose. The point at which 50% of the animals are killed is labeled the lethal-dose-50, or LD50. (shown here pertaining to the blue line). For some toxic agents, a threshold dose exists, below which doses have no measurable effect (shown here with the red line). Go to **GRAPH IT** on the website or CD-ROM.

sponses occur only above a certain dose. Such a **threshold** dose (see Figure 10.10) might be expected if the body's organs can fully metabolize or excrete a toxicant at low doses but become overwhelmed at higher concentrations, or if cells can repair damage to their DNA only up to a certain point.

In addition, sometimes responses decrease with increased dose. Toxicologists are finding that some dose-response curves are U-shaped or J-shaped, or in the shaped of an inverted U. Such counterintuitive curves often appear to apply to endocrine disruptors. As we have seen, these substances interact with the hormone system, which is geared to dealing with extremely low concentrations of hormones. Thus the hormone system appears vulnerable to disruption by toxicants that are at extremely low concentrations themselves. Inverted dose-response curves present a challenge for toxicologists and for policymakers attempting to set safe environmental levels for toxicants. If very many chemicals behave in this way, we may have greatly underestimated the dangers of these compounds in our environment, because many chemicals exist in very low concentrations and are spread over wide areas. Knowing the shape of dose-response curves is important if one is planning to extrapolate from them to predict responses at doses below those that have been tested.

Scientists generally give lab animals much higher relative doses than humans would expect to receive in the environment. This is so that the response is great enough to be measured and so that differences between the effects of small and large doses are evident. A spread of doses thereby helps give shape to the dose-response curve. Once the data from animal tests are plotted, scientists generally extrapolate downward to estimate the effect of still-lower doses on a hypothetically large population of animals. This way they can come up with an estimate of, say, what dose causes cancer in 1 mouse in 1 million. A second extrapolation is then required to estimate the effect on humans, with our higher body weight. Because these two extrapolations go beyond the actual data obtained, they introduce uncertainty into the interpretation of what doses are acceptable for humans. As a result, to be on the safe side, regulatory agencies that set standards for maximum allowable levels of toxicants set the standards well below the minimum toxicity levels estimated from lab studies.

Individuals vary considerably in their responses to toxicants

Different individuals can respond quite differently to identical doses, with some individuals being more sensitive than others. These differences can be genetically based, can result from varying abilities of the body's organs to detoxify substances, or can be due to a person's overall condition. People in poorer health are often more vulnerable to the effects of contaminants.

Sensitivity also can vary with sex, age, and weight. Of greatest concern are fetuses, infants, and young children. Because of their smaller size and rapidly developing organ systems, they tend to be much more sensitive to toxicants than are adults. The degree of this difference was not appreciated for many years. Regulatory agencies, including the U.S. EPA, set standards for adults and extrapolated downward for infants and children, but subsequently found that in many cases their linear extrapolations did not lower standards enough to adequately protect babies. Many critics today contend that despite improvements, regulatory agencies still do not account explicitly enough for risks to fetuses, infants, and children.

The type of exposure can affect toxicity

The toxicity of many substances varies according to whether the exposure occurs in high amounts for short periods of time, known as **acute exposure,** or in lower amounts over long periods of time, known as **chronic exposure.** Incidences of acute exposure are easier to recognize, because they often stem from discrete events, such as an accidental ingestion, an oil spill, a chemical spill, or a nuclear accident. Lab tests generally measure acute toxicity effects, and LD_{50} values express them.

However, chronic exposures are more common—and more difficult to detect and diagnose. Chronic exposure often affects organs gradually, as happens with lung cancer caused by smoking, and liver or kidney damage due to alcohol abuse. Low levels of arsenic in drinking water or pesticide residue on food may also pose chronic risk. Because of the long time periods involved, the relationship between cause and effect may not be readily apparent. In addition, the symptoms of chronic toxicity may appear similar to other diseases.

Mixes may be more than the sum of their parts

It is difficult enough to determine the impact of a single toxicant on an organism, but the task becomes astronomically more difficult when multiple chemicals interact. When mixed, substances may act in concert in ways that cannot be predicted from their effects when alone. Mixed toxicants may sum each other's effects, cancel out each other's effects, or multiply each other's effects. Whole new types of impacts may arise when toxicants are mixed together. Such interactive impacts—those that are more than or different from the simple sum of their constituent effects—are called **synergistic** effects.

Examples of synergistic effects of toxic substances are becoming abundant. With Florida's alligators, several lab experiments have indicated that DDE can either help cause or inhibit sex reversal, depending upon other chemicals with which it interacts. Mice exposed to a mixture of nitrate, atrazine, and aldicarb have been found to show immune, hormone, and nervous-system effects that were not evident from exposure to each of these chemicals alone. Several North American frogs and toads have also shown synergistic effects. Wood frogs are increasingly suffering limb deformities, apparently the result of being parasitized by trematode flatworms. However, being near an agricultural field with pesticide runoff increases the rate of parasitic infection, because, as lab studies have shown, pesticides suppress a frog's immune response, making the amphibian more vulnerable to parasites.

Traditionally, toxicology has tackled only effects of single chemicals one at a time. The complex experimental designs required to test interactions, and the sheer number of chemical combinations, have meant that single-substance tests have received priority. This is changing, but scientists will never be able to test all possible combinations; there are simply too many chemicals in the environment.

Hazards to Environmental Health

Environmental toxicology is one discipline within the broader scope of environmental health. The study and practice of **environmental health** assesses environmental factors that influence human health and quality of life, and seeks to prevent adverse effects on human health and on ecological systems essential to environmental quality and long-term human well-being.

Environmental hazards can be chemical, physical, biological, or socioeconomic

There are many types of environmental health threats, or hazards, besides the synthetic and natural chemical toxicants we have been addressing thus far. Other hazards include the physical or climatic events that we call natural catastrophes. Earthquakes, volcanic eruptions, floods, blizzards, landslides, hurricanes, and droughts are all natural catastrophes—events of nature that we can do little or nothing to predict or prevent.

Another class of hazards consists of biological hazards resulting from ecological interactions between organisms. When we become sick from a virus, bacterial infection, or other pathogen, we are simply the victims of biological elements of our environment that are fulfilling their ecological roles.

Yet another class of environmental hazards can be loosely called cultural or lifestyle hazards, ones that result from the place we live, our socioeconomic status, our occupation, and our behavioral choices. For instance, choosing to smoke cigarettes, or living or working with people who do, can greatly increase our risk of lung cancer. Choosing to smoke is a personal behavioral decision, but secondhand smoke exposure in the home or workplace may not be under one's control and may also be influenced by socioeconomic constraints or demands. Much the same might be said for drug use, diet and nutrition, crime, and mode of transportation. As advocates of environmental justice (Chapter 2) argue, such health factors as living in proximity to toxic waste sites or working unprotected with pesticides are often correlated with socioeconomic deprivation.

Some health threats are part of the natural environment

Some threats to health are unavoidable because they are wholly natural constituents of the natural environment. Electromagnetic radiation is one such example. Getting too much of the sun's rays can cause excessive exposure to ultraviolet light, which can damage DNA in all organisms. Excess UV exposure has been tied to skin cancer, cataracts, and immune suppression in humans, as well as to cancers and cataracts in other animals and reduced productivity in plants.

As another example, the radiation from nuclear power and nuclear weapons that concerns so many of us is not the only radiation risk we face; radiation emanates naturally from minerals such as uranium. Areas with high uranium concentrations in the bedrock and soil give off natural radiation. One such area is the land of the Australian Aborigines we learned of in Chapter 2, where the Mirrar Clan was contesting the Jabiluka and Ranger uranium mines. The Aborigines of the Kakadu area have traditionally called this region "Sickness Country" because they had noticed that people who lived and hunted there tended to become sick. Geologists have since discovered that the volcanic rocks, drinking water, and groundwater of the region are rich not only in uranium but also in radioactive thorium and radon, as well as arsenic, mercury, and fluorine. In addition, the ochre that Aborigines used as pigment in their rock paintings contains high concentrations of lead, arsenic, mercury, and uranium.

On the other side of the globe in some areas of the United States, a similar hazard exists involving natural radiation. Radon is a highly toxic radioactive gas that is colorless and undetectable without specialized kits. Radon seeps up from the ground in areas with certain types of bedrock and can build up inside basements and homes with poor air circulation. The U.S. EPA estimates that slightly less than 1 person in 1,000 may contract lung cancer due to a lifetime of radon exposure at average levels for U.S. homes.

Disease is a major focus of environmental health

Among the biological hazards humans face, disease stands preeminent. Despite all our technological advances, we still find ourselves battling disease, just as all other organisms do. Although humans have little to worry about from other species in the way of predation and competition, we do suffer greatly from parasitism by pathogens, especially the microbes that bring infectious disease.

Infectious disease, also called **communicable** or **transmissible disease**, involves a pathogen that attacks a host. Sometimes the attack is direct, and sometimes it occurs via a **vector**, an organism that transfers the pathogen to the host. Infectious diseases account for

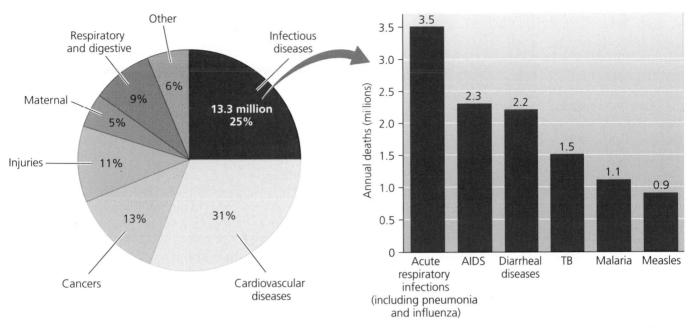

(a) Leading causes of death across the world, 1998

(b) Six diseases account for 90% of deaths by infectious disease

Figure 10.11 Infectious diseases are the second-leading cause of death worldwide, accounting for one-quarter of all deaths per year (**a**). Six diseases—acute respiratory infections, AIDS, diarrhea, tuberculosis (TB), malaria, and measles—account for 90% of all deaths from infectious disease (**b**). Data from World Health Organization, 1999.

25% of the deaths that occur worldwide each year (Figure 10.11)—approximately 13 million people. We have lessened the impact of many infectious diseases and even have eradicated some, but others, such as tuberculosis and AIDS, are increasing. Communicable diseases account for close to half of all deaths in developing countries, but for very few deaths in developed nations. This discrepancy is due to differences in hygiene conditions and access to medicine, which are tightly correlated with relative levels of poverty and wealth.

In order to predict and prevent infectious disease, environmental health experts deal with the often-complicated interrelationships among technology, land use, and ecology. One of the world's leading infectious diseases, malaria (which takes 1.1 million lives each year), provides an example. The microscopic protist that causes malaria depends on mosquitoes as a vector. The protist can sexually reproduce only within a mosquito, and it is the mosquito that injects the protist into a human or other host. Thus the primary mode of malaria control has been to use insecticides such as DDT to kill mosquitoes. Large-scale eradication projects involving insecticide use and draining of wetlands have removed malaria from large areas of the temperate world where it used to occur, such as throughout the southern United States. However, types of land use that create pools of standing water in formerly well-drained

areas can boost mosquito populations and allow malaria to reinvade an area.

Environmental health workers also study nontransmissible diseases, as some of these likewise have strong ties to environmental factors and conditions. For example, malnutrition (Chapter 9) can foster a wide variety of disorders, as can poverty and poor hygiene. Lifestyle choices can affect risks of acquiring certain other noninfectious diseases, such as heart disease. Many noncommunicable diseases have both genetic and environmental components. For instance, whether a person develops asthma is influenced by the genes, but also by the type of environment in which he or she grew up. Children raised on farms suffer less asthma than children raised in cities, studies have shown, apparently because children raised on farms experience more allergens from farm animals at an early age, allowing their bodies to properly adapt.

Environmental health is an issue indoors as well as outdoors

Just as urban and rural lifestyles can pose different health risks, different environmental hazards exist indoors and outdoors (Table 10.2). Outdoor hazards are generally more familiar to us, but we spend most of our lives indoors. Therefore, we must consider the spaces

Table 10.2 Selected Environmental Hazards

Air

Smoking and secondhand smoke
Chemicals from automotive exhaust
Chemicals from industrial pollution
Tropospheric ozone
Pesticide drift
Dust and particulate matter

Water

Pesticide and herbicide runoff
Nitrates and fertilizer runoff
Mercury, arsenic, and other heavy metals in groundwater and surface water

Food

Natural toxins
Pesticide and herbicide residues

Indoors

Asbestos
Radon
Lead in paint and pipes
Toxicants in plastics and consumer products

inside of our homes and workplaces to be part of our environment and, as such, the sources of potential home and occupational hazards. Radon is one such hazard. Asbestos is another.

There are several types of asbestos, each of which is a mineral that forms long thin microscopic fibers. This structure allows asbestos to insulate heat, muffle sound, and resist fire. Because of these qualities, asbestos was used widely as insulation in buildings for many years, as well as in many products. However, its fibrous structure also makes asbestos a great danger when inhaled into the lungs. When lodged in lung tissue, asbestos induces the body to produce acid to combat it. The acid, however, scars the lung tissue while doing little to dislodge or dissolve the asbestos. Consequently, within a few decades the scarred lungs may cease to function, a disorder called *asbestosis*. Asbestos can also cause types of lung cancer. Because of these health hazards, asbestos has been and continues to be removed from many buildings, particularly schools (Figure 10.12).

Another indoor health hazard is lead poisoning. This heavy metal can cause brain damage, learning prob-

lems, behavioral abnormalities, anemia, hearing loss, damage to the liver, kidneys, and stomach, and even death. It has been suggested that the downfall of ancient Roman civilization was caused in part by chronic lead poisoning; the elites of the power structure regularly drank wine sweetened with mixtures prepared in leaden vessels. Certain historical personalities, such as the composer Beethoven, suffered maladies that some historians have attributed to lead poisoning.

Today lead poisoning can result from drinking water passing through the lead pipes common in older houses. Even in newer pipes, lead solder was widely used into the 1980s and is still sold in stores. Lead is perhaps most dangerous, however, to children through its presence in paint. Until 1978, most paints contained lead, and interiors in most houses were painted with lead-based paint. Babies and young children often take an interest in peeling paint from walls and often will ingest some of it or inhale it as dust. Today lead poisoning is thought to affect 3–4 million young children, fully one in six children under age 6.

Many more indoor hazards exist, some of which likely have not yet come to light. One example of a recently recognized hazard is a group of chemicals known as polybrominated diphenyl ethers (PBDEs). These compounds provide fire-retardant properties and are used in a diverse array of consumer products, including computers, televisions, plastics, and furniture. They appear to be released during production and disposal of products and also to evaporate at very slow rates throughout the lifetime of products. These chemicals persist and accumulate in living tissue, and their abundance in the environment and in people in the United States is doubling every few years. Like the chemicals affecting Louis Guillette's alligators, PBDEs appear to be endocrine disruptors. They do not affect reproductive hormones, however; rather, animal testing shows them to affect thyroid hormones. Animal testing also shows limited evidence that PBDEs may affect brain and nervous system development and might possibly cause cancer. Concern about PBDEs rose after a study showed concentrations in the breast milk of Swedish mothers increasing exponentially from 1972 to 1997. The European Union decided in 2003 to ban PBDEs, and industries in Europe had already begun phasing them out. As a result, concentrations in breast milk of European mothers have fallen substantially. In the United States, however, there has so far been little movement to address the issue.

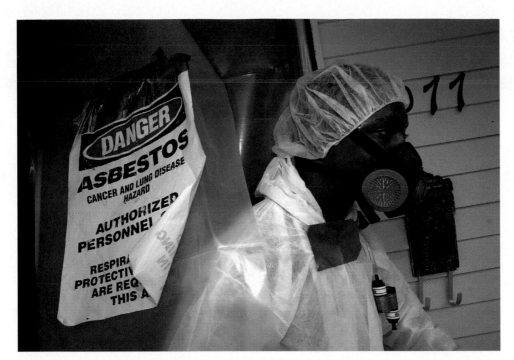

Figure 10.12 Asbestos was widely used in building insulation and other products. A cause of lung cancer and asbestosis, the substance has been removed from many buildings in which it was used.

Risk Assessment and Risk Management

Policy decisions about whether to ban chemicals or restrict their use are not made casually; they generally follow years of rigorous testing and retesting for toxicity. Policy decisions also reach beyond the scientific results on health to incorporate considerations about economics and ethics. The steps between the collection and interpretation of scientific data and the formulation of policy involve assessing and managing risk.

Risk is expressed in terms of probability

Exposure to a toxicant—or any environmental health threat—does not invariably produce some given effect. Rather, it causes some probability of harm, some statistical chance that damage will result. In order to understand the impact of any environmental health threat, a scientist must know not only its strength or toxicity, but also the chances that an organism will encounter it, the frequency with which the organism may encounter it, the amount of substance or degree of threat to which the organism is exposed, and the organism's sensitivity to the threat. These factors help determine the overall risk posed by the substance. Risk can be measured in terms of probability (Chapter 1), an exact quantitative description of how

likely or unlikely a certain outcome is to come about. The mathematical probability that some harmful outcome (for instance, injury, death, environmental damage, or economic loss) will result from a given action, event, or substance expresses the **risk** posed by that phenomenon.

Our perception of risk may not match reality

Every action we take and every decision we make involves some element of risk, some (generally small) probability that things will go wrong. Some actions and decisions are more risky than others, and we try our best in everyday life to behave in order to minimize risk. Interestingly, our perceptions of risk do not always match the statistical reality (Figure 10.13). People will often worry unduly about negligibly small risks but happily engage in other activities that pose high risks. For instance, although most people perceive flying in an airplane as a riskier activity than driving a car, driving a car is statistically far more dangerous.

Psychologists agree that this difference between risk perception and reality stems from the fact that we feel more at risk when we are not controlling a situation and more safe when we are "at the wheel"—regardless of the actual risk involved. When we drive a car, we feel we are in control, even though statistics show we are at greater risk than as a passenger in an airplane. In terms of environmental hazards, this psychology can account

Figure 10.13 Our perceptions of risk do not always match the reality of risk. Listed here are a number of the leading causes of death in the United States, along with a measure of the risk each poses. Risk is measured in days of lost life expectancy, or the number of days of life lost by people suffering the hazard, spread among the entire population—a measure commonly used by insurance companies. One common source of anxiety, airplane accidents, by this measure poses 20 times less risk than home accidents, over 50 times less risk than auto accidents, and over 200 times less risk than being overweight. Data from Bernard Cohen, Catalog of risks extended and updated, *Health Physics,* 1991.

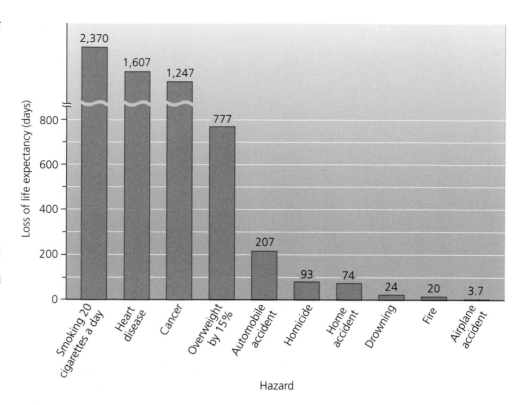

Risk assessment analyzes risk quantitatively

The quantitative measurement of risk and the comparison of risks involved in different activities or substances is termed **risk assessment.** Risk assessment is a way of identifying and outlining problems. In environmental health and toxicology, it is a crucial first step toward ascertaining which substances and activities pose substantial health threats to people or wildlife, and which are largely safe.

Assessing risk for a chemical substance involves several steps. The first steps involve the scientific study of toxicity outlined above—determining whether a given substance has toxic effects and, through dose-response analysis, measuring as precisely as possible how effects on an organism vary with the degree of toxicant exposure. Subsequent steps involve assessing the individual's or population's likely extent of exposure to the substance,

for people's great fear of nuclear power, toxic waste, and pesticide residues on foods—things that are invisible, little known, or little understood, and whose presence in their lives is largely outside their personal control. In contrast, people are more ready to accept the risks of smoking cigarettes, overeating, and not exercising, all voluntary activities statistically shown to pose far greater risks to health.

including the frequency of contact, the concentrations likely encountered, and the length of time the substance is expected to be encountered. Such a multifaceted process is difficult enough when assessing risks for human health, but it is even more complex for environmental health, where even more factors come into play.

Weighing The Issues:
A Nationwide Health Tracking System?

When 16 children were diagnosed with leukemia in the small town of Fallon, Nevada, it raised alarm. The identification of cancer clusters such as that in Fallon has prompted many U.S. scientists and policymakers to back the notion of creating a national environmental health tracking system. Some states already track environmental hazards and illnesses, but comparable statistics are not compiled everywhere. It would take an estimated $1 billion to integrate these disparate sets of information and to maintain records. Do you think the potential benefits are worth it? How might this affect epidemiological studies? How might it affect risk assessments? Could it speed the identification of causes of poisonings or disease outbreaks? How might finding the causes of such events help to mitigate them, or to prevent them in the future?

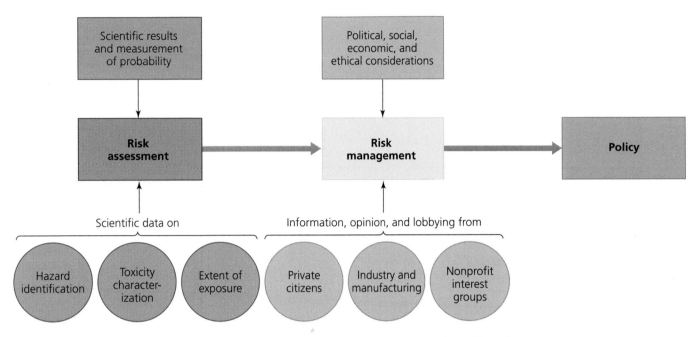

Figure 10.14 Risks must be assessed before policy steps can be taken to minimize them. Thus the first step in addressing the risk of an environmental hazard is risk assessment, a process of quantifying the risk of the hazard and comparing it to other risks. Once science identifies and measures risks, then risk management can proceed. In this process economic, political, social, and ethical issues are considered in light of the scientific data from risk assessment. The consideration of all these types of information is designed to result in policy decisions that minimize the risk of the environmental hazard being addressed.

Risk management combines science and other social factors

Accurate risk assessment is a vital step toward effective **risk management**, which consists of decisions and strategies to minimize risk (Figure 10.14). In most developed nations, risk management is handled largely by federal agencies, such as the EPA and the Food and Drug Administration (FDA) in the United States. In risk management, scientific assessments of risk are considered in light of economic, social, and political needs and values. The costs and benefits of addressing risk in various ways are assessed with regard to both scientific and nonscientific concerns. Decisions whether to reduce or eliminate risk are then made.

In environmental health and toxicology, comparing costs and benefits is often difficult because the benefits are often economic and the costs often revolve around health. Moreover, economic benefits are generally known, easily quantified, and of a definite and stable amount, whereas health risks are hard-to-measure probabilities, often involving a very small percentage of people likely to suffer greatly and a large majority likely to experience little effect. When a government agency bans a pesticide, it generally means measurable economic loss

for the manufacturer and the farmer, whereas the benefits accrue unpredictably in the long term to some percentage of factory workers, farmers, and the general public. Because of the lack of balance inherent in many cost-benefit analyses in risk management, this pursuit frequently tends to stir up disagreements and debate.

Philosophical Approaches to Environmental Health

Because we do not know a substance's toxicity until we measure and test it, and because there are so many untested chemicals and combinations, science will never eliminate the many uncertainties that accompany risk assessment. In such a world of uncertainty, there are two basic philosophical approaches to categorizing substances as safe or dangerous (Figure 10.15).

One approach is to assume that substances are harmless until shown to be harmful. We might nickname this the *innocent-until-proven-guilty approach*. Because thoroughly testing every existing substance

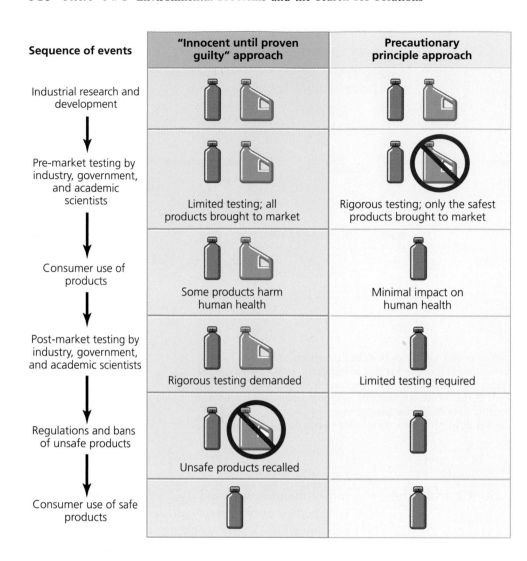

Sequence of events	"Innocent until proven guilty" approach	Precautionary principle approach
Industrial research and development		
Pre-market testing by industry, government, and academic scientists	Limited testing; all products brought to market	Rigorous testing; only the safest products brought to market
Consumer use of products	Some products harm human health	Minimal impact on human health
Post-market testing by industry, government, and academic scientists	Rigorous testing demanded	Limited testing required
Regulations and bans of unsafe products	Unsafe products recalled	
Consumer use of safe products		

Figure 10.15 Testing a new chemical compound or product for toxicity rarely gives a black-or-white answer, and many tests must be run before a substance's properties are well understood. Given such a degree of uncertainty, there are two main approaches that can be taken to introducing new substances on the market. In one approach, substances are innocent until proven guilty; they are brought to market relatively quickly after limited testing. The advantage is that products reach consumers more quickly, but the disadvantage is that some fraction of them may cause harm to some fraction of people. In the other approach, the precautionary principle is adopted, and substances are brought to market cautiously, only after extensive testing. The advantage is that products that reach the market should be safe, while the disadvantage is that many perfectly safe products will be delayed in reaching consumers.

(and combination of substances) for its effects is a hopelessly long, complicated, and expensive pursuit, the innocent-until-proven-guilty approach has the benefit of not slowing down technological innovation and economic advancement. However, it has the disadvantage of allowing certain substances that are dangerous to be put into wide use and perhaps do great harm before their effects are discovered.

The other approach is to assume that substances are harmful until shown to be harmless. This approach follows the precautionary principle discussed in Chapter 9 in regard to genetically modified foods. This more cautious approach should enable us to identify troublesome toxicants before they are released into the environment, but it may also significantly impede the pace of technological and economic advance.

These two approaches are in actuality two ends of a continuum of possible approaches. These two endpoints differ mainly in where they lay the burden of proof—specifically, whether product manufacturers are required to prove safety or whether government, scientists, or citizens are required to prove danger.

Weighing the Issues: The Precautionary Principle

Given the substantial costs of testing chemicals for safety and the increasing concerns about their spread through the environment, should proof of safety be required by government prior to a chemical's release into the environment? Or does this unfairly restrict chemical manufacturers? Should the burden of proof fall to the company that stands to make a profit from a product's release? How do you think adopting the precautionary principle would affect the number of chemicals on the market?

Chemical Product Testing: Industry or Government?

The testing of chemical products for safety can take the so-called "innocent-until-proven-guilty" or "precautionary" approaches. Should manufacturers be held responsible for comprehensive testing of new chemical products before they are introduced to the public? What would be some of the advantages and disadvantages? How extensive should the role of government be in the testing process?

Testing Must Ensure Public Health

Like most things in life, the controversy over product testing arises because both approaches have valid advantages and disadvantages. Allowing industry to follow the innocent-until-proven-guilty approach, with limited testing, reduces development costs for new chemical products and may lead to greater economic activity. If industry were required to comprehensively test chemical product safety before introduction to the public, chemical industry profits could fall and result in job loss and costly new product development. Consumer prices might rise to cover these costs. However, comprehensive testing would lower the number of chemicals that adversely affect biological species.

Our definition of *innocent* is often too narrow. In the innocent-until-proven-guilty approach, *what the consumer actually buys is never tested*, since only ultra pure active ingredients are tested. Surfactants and organic soaps ("other ingredients") are added to improve the active ingredients' lipid or water solubility. Also, production contaminants are not tested and registered, and other ingredients are frequently very active biologically. Therefore, a so-called *innocent* product can cause cancer and reproductive defects. The assumption of a linear dose response is also coming under increased scrutiny, especially at very low physiological doses, where hormonal, immune, and neurological processes respond. At much higher pharmacological doses, where toxicity testing is typically done, the responses of physiological systems to the same chemical can be very different.

Given the inherent inadequacies of the testing process and the uncertainty of the economic impacts, both government and industry should share the responsibility of testing to ensure public safety.

Warren Porter, a toxicologist, evaluates the connection between climate, animal energetics, and behavior using statistical experimental design.

An Industry Perspective

Chemical risk depends on two factors: hazard (toxicity) and exposure. To evaluate a chemical, manufacturers typically start by conducting screening-level toxicological and environmental studies and proceed to more or higher-tier studies as warranted. There is no single comprehensive testing program that is appropriate for all industrial chemicals.

The Toxic Substances Control Act requires almost all new commercial substances to undergo Premanufacture Notification (PMN) review and to describe this preliminary process as an innocent-until-proven-guilty approach is an oversimplification. When the EPA reviews a PMN, it considers the physical and chemical properties of the substance, structural similarity to other compounds of known toxicity, and potential for human exposure and environmental release. If there is no evidence of harm from preliminary testing, longer term or more specialized testing may not be conducted. In some cases, the EPA may require additional testing to determine whether the chemical poses an unreasonable risk to human health or the environment. If the EPA finds risk can be addressed by reducing exposure, it may enter into a binding agreement with the manufacturer to require exposure reduction activities, rather than additional laboratory testing.

Manufacturers often voluntarily conduct new studies to support the continued safe use of their chemicals. The 150 member companies of the American Chemistry Council represent about 90% of U.S. chemical production. These companies are committed to Responsible Care®, under which chemical manufacturers, as good stewards of their products, continue to test as new data and methodologies become available. There is a role for both government and industry in chemical testing, and it is important that the EPA and manufacturers work together in evaluating chemicals to improve health, safety, and the environment.

Marian K. Stanley is Senior Director for the American Chemistry Council, which she joined in 1990. She is responsible for the management of chemical specific issue groups in the Council's self-funded CHEMSTAR Department.

Policy on Toxicants

One's choice of philosophical approach is not simply an academic matter; it has immediate and far-reaching impact on policy decisions, and therefore directly affects what materials are allowed into our environment. In most nations, a blend of these two approaches is followed, but there is marked variation among countries. At the present time, European nations are to a great extent following the precautionary principle regarding the regulation of synthetic chemicals, while the United States is not. Although industry frequently complains that government regulation is cumbersome, environmental and consumer advocates say U.S. governmental policies too often follow the innocent-until-proven-guilty approach and that the burden of proof rarely rests with chemical manufacturers. For instance, although the FDA is required to test chemicals that are added directly to food before these chemicals are brought to market, the many compounds involved in cosmetics require no FDA review or approval before being sold to the public.

In the United States, the tracking and regulation of synthetic chemicals is shared among several federal agencies. The FDA, under the Food, Drug, and Cosmetic Act of 1938 and its subsequent amendments, regulates foods and food additives, cosmetics, and drugs and medical devices. The EPA regulates pesticides under the Federal Insecticide, Fungicide, and Rodenticide Act of 1947 (FIFRA) and its amendments. The Occupational Safety and Health Administration (OSHA) regulates workplace hazards under a 1970 act. Several other agencies regulate the use of other categories of substances. The various remaining synthetic chemicals not covered by other laws are regulated by the EPA under the 1976 Toxic Substances Control Act (TSCA). Examining in more detail the policy process for pesticides and for toxic substances will provide an idea of how government and industry interact.

Pesticides in the United States are registered through the EPA

FIFRA was enacted as the post–World War II U.S. chemical industry was expanding, but well before the age of environmental activism in the 1960s and 1970s that resulted in so many environmental laws. As such, FIFRA was not primarily intended to protect public health or the environment, but to assure consumers that products actually worked as their manufacturers claimed. Subsequent amendments shifted the focus somewhat toward protecting health and safety and charged the EPA with "registering" each new pesticide that manufacturers propose to bring to market.

The registration process involves risk assessment and risk management. In the registration process, the EPA first asks the pesticide manufacturer to provide information, including the results of safety assessments the company has performed according to EPA guidelines. The EPA examines the company's research and all other relevant scientific research. It examines the product's ingredients and how the product will be used and evaluates whether the chemical poses risks to humans, other organisms, or water or air quality. The EPA then approves, denies, or sets limits on the chemical's sale and use and also must approve the language used on the product's label.

Because the registration process takes economic considerations into account, critics say it allows hazardous chemicals to be approved if the economic benefits they offer are great enough. Here the challenges of weighing intangible risks involving human health and environmental quality against the tangible and quantitative numbers of economics become apparent.

The EPA also regulates diverse chemicals under the Toxic Substances Control Act

TSCA directed the EPA to monitor some 75,000 industrial chemicals manufactured in or imported into the United States. These chemicals do not include pesticides, food additives, or drugs, but they do include PCBs and other compounds involved in plastics. The act gave the agency power to regulate these chemicals and ban them if they are found to pose excessive risk. TSCA also was the first law to require screening of these substances before they entered the marketplace.

However, many public health and environmental advocates view TSCA as being far too weak. They note that the screening required of industry is extremely minimal and that in order to mandate more extensive and meaningful testing, the EPA must show proof of the chemical's toxicity. In other words, the agency is caught in a Catch-22: In order to push for studies looking for toxicity, it must have proof of toxicity already.

The result, these advocates say, is that most synthetic chemicals are not thoroughly tested before being put on the market. Of those that fall under TSCA, only 10% have been thoroughly tested for toxicity; only 2% have been screened for carcinogenicity, mutagenicity, or teratogenicity; fewer than 1% are government-regulated; and almost none have been tested for endocrine, nervous, or immune system damage, according to the U.S.

National Academy of Sciences. The EPA reported in 1998 that only 7% of chemicals produced in large amounts have a complete toxicity screening, 50% show incomplete data, and 43% lack all basic data.

Industry's critics say chemical manufacturers should be made to bear the burden of proof for the safety of their products before they hit the market. Industry's supporters say that safety advocates will never be satisfied that industry has done enough. They say that mandating vastly more toxicological research will greatly hamper the introduction of products consumers want and increase the price of products as research costs are passed on to consumers.

Toxicants are regulated internationally

In April 2003, European Union (EU) commissioners proposed legislation that would require chemical manufacturers to test and register 30,000 chemicals already in use and would impose restrictions on 1,500 chemicals already considered hazardous. In announcing the policy, EU environment commissioner Margot Wallström said, "It is high time to place the responsibility where it be-

longs, with industry." Industry was not pleased and estimated that the law would cost it $7–8 billion over a decade. At the same time, however, some people on all sides agreed that the new policy would have the positive effect of spurring research into safer products and creating new markets for them. The proposal was expected to make its way toward becoming law through at least 2005.

Action regarding chemical toxicants has also been taken in the form of international treaties. The Stockholm Convention on Persistent Organic Pollutants (POPs), introduced in 2001, appears on its way to ratification. POPs are toxic chemicals that persist in the environment, bioaccumulate in the food chain, and often can travel long distances. The PCBs and other contaminants found in polar bears are a prime example. Because these contaminants so often cross international boundaries, an international treaty seemed the best way of dealing fairly with such transboundary pollution. The Stockholm Convention aims first to end the use and release of 12 of the POPs shown to be most dangerous, a group nicknamed the "dirty dozen" (Table 10.3). It sets guidelines for phasing out these chemicals and encourages transition to safer alternatives.

Table 10.3 The "Dirty Dozen" Persistent Organic Pollutants (POPs) Targeted by the Stockholm Convention

Toxicant	Type	Description
Aldrin	Pesticide	Kills termites, grasshoppers, corn rootworm, and other soil insects
Chlordane	Pesticide	Kills termites and is a broad-spectrum insecticide on various crops
DDT	Pesticide	Widely used in past to protect against malaria, typhus, and other insect-spread diseases; continues to be applied in several countries to control malaria
Dieldrin	Pesticide	Controls termites and textile pests; also insect-borne diseases and insects in agricultural soil
Dioxins	Unintentional by-product	Produced as a result of incomplete combustion and in chemical manufacturing; released in certain kinds of metal recycling, pulp and paper bleaching, automobile exhaust, tobacco smoke, and wood and coal smoke
Endrin	Pesticide	Kills insects on cotton and grains; also used against rodents
Furans	Unintentional by-product	Result from the same processes that release dioxins; also are found in commercial mixtures of PCBs
Heptachlor	Pesticide	Kills soil insects and termites, cotton insects, grasshoppers, and malaria-carrying mosquitoes
Hexachlorobenzene	Fungicide; unintentional by-product	Kills fungi that affect crops; also released during chemical manufacture and from processes that give rise to dioxins and furans
Mirex	Pesticide	Combats ants and termites; also is a fire retardant in plastics, rubber, and electrical goods
PCBs	Industrial chemical	Used in industry as heat-exchange fluids, in electrical transformers and capacitors, and as additives in paint, carbonless copy paper, sealants, and plastics
Toxaphene	Pesticide	Kills insects on cotton, grains, fruits, nuts, and vegetables; kills ticks and mites on livestock

Data from United Nations Environment Programme (UNEP), 2001.

Conclusion

International agreements such as the Stockholm Convention represent a hopeful sign that governments will act to protect the world's people, wildlife, and ecosystems from harm done by chemicals known to be toxic. At the same time, solutions can often come more easily when they do not arise from government regulation alone. To many minds, consumer choice, exercised through the market, may be the best way to influence industry's decision-making. Consumers of products can make decisions that influence industry when they have full information about the risks involved. Obtaining the information through research can be a long and convoluted process, but this is where the science of toxicology plays a crucial role.

Once scientific results are in, a society's philosophical approach to risk management will then determine what policy decisions are made. Whether the burden of proof is laid at the door of industry or of government, however, it is important to realize that we will never attain complete scientific knowledge of any risk. At some point we must choose whether or not to act on the information available. We must also consider both costs and benefits of any action we take. Synthetic chemicals have brought many good things to our lives. They have brought us innumerable modern conveniences, have resulted in a larger food supply, and have supported medical advances that save and extend human lives. Human society would be very different without them. Yet a safer and happier future, one that safeguards the well-being of both humans and the environment, depends on knowing the risks that some substances pose and on having effective means in place to phase out harmful substances and replace them with safer ones.

REVIEW QUESTIONS

1. Where may toxicants manufactured by humans be found in the environment? List several types of these synthetic chemicals.

2. When did concern over the effects of pesticides start to grow in the United States? What kinds of problems does environmental toxicology study?

3. Describe the argument presented by Rachel Carson in *Silent Spring*. What policy resulted from the book's publication? Is DDT still used?

4. List and describe the six types or general categories of toxicants described in this chapter.

5. When did scientists first notice the effects of endocrine disruption? What endpoints, or health problems, have indicated the prevalence of this problem? What publication helped present a new understanding of the causes of endocrine disruption?

6. What is the current state of research into endocrine disruption, and how has it begun to affect the development of public policy?

7. How do toxicants travel through the environment? Where are they most likely to be found?

8. What are the lifespans of toxic agents? Are their breakdown products also toxic?

9. Describe the processes of bioaccumulation and biomagnification.

10. Name some naturally occurring substances that can act as toxic agents. Explain the arguments of Bruce Ames and those of his critics, regarding the prevalence and effects of natural toxicants.

11. How are studies of wildlife toxicology conducted?

12. What are epidemiological studies, and how are they most often conducted?

13. Why are animals used in laboratory experiments in toxicology?

14. What is the standard method of animal testing in toxicology? Explain the dose-response curve. How is this graph interpreted? Why are high LD_{50}s and ED_{50}s considered safe and low LD_{50}s and ED_{50}s considered unsafe for humans?

15. What factors may affect an individual's response to a toxic substance?

16. Why is chronic exposure to toxic agents often more difficult to measure and diagnose than acute exposure? What are synergistic effects, and why are they difficult to measure and diagnose?

17. What types of health hazards does research in the field of environmental health encompass, aside from toxic chemicals?

18. Why is disease the greatest biological hazard that humans face? What kinds of interrelationships must environmental health experts study in order to learn about how diseases affect human health?

19. Where does most exposure to lead, asbestos, and PBDEs occur? How have each of these exposure problems been addressed?

20. How do scientists present or express risk?

21. How do scientists identify and assess risks from substances or activities that may pose health threats?
22. Why are the decisions and strategies used to minimize risk in environmental health and toxicology likely to create disagreement and debate?
23. Describe the "innocent-until-proven-guilty" and "precautionary principle" approaches to environmental health. Where does each place the burden of proof?
24. Why does the EPA monitor asbestos, radon, lead, and nearly 75,000 other industrial chemicals? Can the agency conduct this monitoring effectively? According to the EPA, what are the approximate rates of toxicity screening for chemicals produced in large amounts?
25. Where has the European Union (EU) decided to place the burden of proof for chemical screening? How is this different from the general approach taken to policy on toxicants in the United States?

DISCUSSION QUESTIONS

1. How is research on endocrine disruption helping to change approaches to the study of environmental toxicants in general? How did Louis Guillette determine that chemical contaminants were responsible for the hormonal changes he observed in alligators at Lake Apopka?
2. Do you feel that laboratory-bred animals should be used in lab experiments in toxicology? Why or why not?
3. Discuss ways that we may cope with the uncertainty of risk assessment for synthetic chemicals in environmental health. Can you think of alternatives to taking one of the two philosophical approaches discussed in the chapter? Should these approaches apply to natural toxicants as well?
4. Examine what you have learned from this chapter regarding the policies of the United States and the European Union toward the study and management of the risks of synthetic chemicals. Which do you believe is more effective, the policies of the United States or the European Union, and why?
5. Describe some environmental toxicants that you think you may be living with indoors. How do you think you may have been affected by indoor or outdoor environmental toxicants in the past? What philosophical approach do you plan to take in dealing with these toxicants in your own life?

Media Resources *For further review go to the website* **www.envscienceplace.com** *or student CD-ROM, where you will find quizzes, flashcards, a glossary, additional interactive exercises, and links to relevant news and research sources. Also, on the website and CD-ROM is* **GRAPH IT**, *a series of interactive graphing tutorials to help you interpret graphs and plot data.*

11 Atmospheric science and air pollution

View of Earth's Atmosphere from Space

This chapter will help you understand:

- The composition, structure, and function of Earth's atmosphere

- Outdoor air pollution

- Indoor air pollution

- Stratospheric ozone depletion

- Acid precipitation

- Solutions to air pollution problems

Police Officer in London's "Killer Smog," 1952

Central Case: The 1952 "Killer Smog" of London

"You had this swirling, like somebody had set a load of car tires on fire."
—*Stan Cribb, eyewitness to the December 1952 "Great Smoke" of London*

"The modern field of environmental health owes much to the tragedy that befell Greater London some 50 years ago."
—*Devra L. Davis, Michelle L. Bell, and Tony Fletcher, in an editorial for the journal Environmental Health Perspectives*

December 5, 1952, was a particularly cold Friday in London, and many residents stoked their coal stoves to keep the chill away. Few people took notice when thick, foul-smelling **smog,** a fog polluted with smoke and chemical fumes, first settled over the city. After all, London had been famous for its "pea souper" fogs for well over a century.

But the smog that December weekend was particularly thick; the city's air quality was ten times worse than usual for that year. Visibility was so poor that pedestrians could not see across the street. Transportation came to a standstill, and roads became clogged with aban-

doned cars. Schools closed, flowers wilted, cattle died of asphyxiation, and an opera was halted because the audience could not see the stage.

A wind finally relieved Londoners of the miserable smog on Tuesday, December 9. By that time, however, authorities estimated that over 4,000 people had died, mostly from lung ailments such as bronchitis that were induced or aggravated by the pollution. A 2002 study by American researchers estimated that the actual death toll, including delayed cases that appeared over the next two months and were considered unrelated at the time, may have been as high as 12,000.

The "killer smog" of 1952 was remarkable, but hardly unique. Similar, although less severe, phenomena had occurred in London as early as 1813 and again in 1873, 1880, 1891, and 1948. Other such events have taken lives in Pennsylvania, New York, Mexico, and Malaysia. London's killer smog, together with other severe smog events, helped change the way the public viewed air pollution. Before the 1950s, most people considered urban smog a necessary burden; today, we recognize the importance of clean air and view air pollution as an environmental challenge that can be overcome.

We have overcome much already; declines in air pollution represent some of the biggest successes of environmental policy. The declines have largely been due to limits on toxic emissions brought about through legislation, such as the British Clean Air Acts of 1956 and 1968 and the U.S. Clean Air Act of 1970 and its amendments in 1990. Today, the air in many American cities is cleaner, and the concentration of airborne particles around London averages one-tenth that of the 1950s.

However, much remains to be done. In 2002, a London governmental body estimated that vehicle emissions contribute to the premature deaths of 380 city residents each year. Furthermore, many cities in developing nations today face conditions similar to those of 1950s London. For example, in 1995 airborne pollution in Delhi, India, was measured at 1.3 times London's average for the year 1952, and air pollution in Lanzhou, China, was measured at 2.7 times London's 1952 level. To understand what caused London's killer smog and how recurrences can be prevented, we must investigate the two factors that contributed to the event: natural atmospheric conditions and human-made pollutants.

Atmospheric Science

Every breath we take reaffirms our connection to the **atmosphere,** the thin layer of gases that surrounds Earth. Humans and other organisms live at the bottom of this layer, drawing from it needed chemicals (such as oxygen for animals and carbon dioxide for plants) and expelling into it the gaseous by-products of our metabolism. The atmosphere supports life in other ways as well: It absorbs potentially dangerous solar radiation, burns up incoming meteors, transports and recycles water and other chemicals, and moderates climate.

The atmosphere's chemical composition has changed over Earth's history (Chapter 4). Oxygen gas (O_2) began to build up about 2.7 billion years ago as a result of the emergence of photosynthetic microbes. Today, nitrogen gas (N_2) constitutes roughly 78% of the atmosphere; most of the rest consists of oxygen gas (Figure 11.1). In addition to these and various natural trace components, the atmosphere has also come to contain small quantities of gases produced by human activities. Furthermore, humans are significantly altering the quantities of atmospheric gases such as carbon dioxide (CO_2), methane (CH_4), and ozone (O_3).

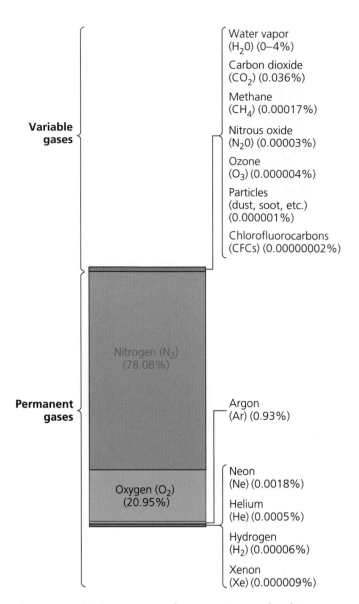

Figure 11.1 Air in our atmosphere consists mostly of nitrogen, secondarily of oxygen, and lastly of a mix of minor gases, some of which are fixed in their concentration and some of which are variable in their concentration, due either to natural causes or human-induced change. Data from Donald C. Ahrens, *Essentials of Meteorology,* second edition, Wadsworth Publishing Company, 1998.

Important atmospheric properties include temperature, pressure, and humidity

Earth's lower atmosphere is highly dynamic, and movement of air within the lower atmosphere results from differences in the physical properties of different air masses. Among the physical properties that influence our dynamic atmosphere are pressure and density, relative humidity, and temperature. Gravity pulls gas molecules toward Earth's surface, causing air to be most dense near the surface and lower as altitude increases.

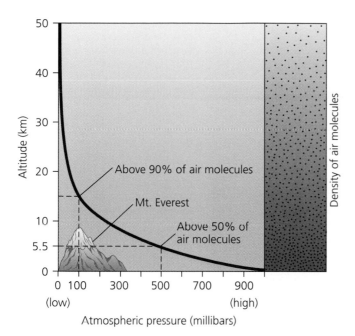

Figure 11.2 As one climbs higher through the atmosphere, gas molecules in the air become less and less densely packed. As density decreases, so does atmospheric pressure. Because the vast majority of air molecules are low in the atmosphere, one needs to be only 5.5 km (3.4 mi) high to be above half the planet's air molecules. Adapted from Donald C. Ahrens, *Essentials of Meteorology*, second edition, Wadsworth, 1998.

Atmospheric pressure, which measures the weight per unit area produced by a column of air, also decreases with altitude, because the higher the altitude, the fewer molecules that are pulled down by gravity (Figure 11.2). At sea level, atmospheric pressure is equal to 14.7 lb/in² or 1,013 millibar (mb). Mountain climbers trekking to Mount Everest can look up and view the world's highest mountain from Kala Patthar, a pass near the base, at roughly 5.5 km (18,000 ft). At this altitude, pressure is 500 mb, and the climber is standing above half the air molecules in the entire atmosphere. A climber who reaches Everest's peak (8.85 km [29,035 ft]), where the "thin air" is just over 300 mb, is standing higher than two-thirds of the atmosphere's air molecules.

Another important property of air is its **relative humidity**, the ratio of water vapor a given volume of air contains to the maximum amount it *could* contain, for a given temperature. Relative humidity of 50%, for example, means the air contains half the water vapor it possibly could at a given temperature. Humans are sensitive to changes in relative humidity because we sweat to cool our bodies. When relative humidity is high, sweat evaporates slowly and the body cannot cool itself efficiently. This is why high humidity makes the weather feel hotter than it really is. Low humidity speeds evaporation and makes it feel cooler than it really

is. Relative humidity varies considerably from place to place and time to time. The average relative humidity in June in Phoenix, Arizona, is only 31%, whereas the average humidity on the island of Guam rarely drops below 88% all year round.

The temperature of air also varies with location and time, and these temperature differences affect air circulation. Temperature varies over Earth's surface because the sun's rays strike some areas more directly than others. Temperature also varies vertically with changes in altitude through the layers of the atmosphere.

The atmosphere consists of several layers

The atmosphere that stretches so high above us and seems so vast is actually just a thin coating about 1/100th of Earth's diameter, like the fuzzy skin of a peach. This coating consists of four layers whose boundaries are not visible to the human eye, but which atmospheric scientists recognize by measuring differences in temperature, density, and composition (Figure 11.3).

The bottommost layer, the **troposphere**, bathes Earth's surface. It is the layer that most directly affects living things: We inhale from the troposphere and exhale into it, and the nutrients from terrestrial and aquatic ecosystems cycle through it (Chapter 6). The movement of air within the troposphere is also largely responsible for the planet's weather. Although it is thin (averaging 11 km [7 mi] high) relative to the atmosphere's other layers, the troposphere contains three-quarters of the atmosphere's total mass, because air is denser near Earth's surface. Tropospheric air temperature declines by about 6°C for each kilometer in altitude (or 3.5°F per 1,000 ft), dropping to a temperature of about −52°C (−62°F) at its highest point. You may have noticed this phenomenon if you've traveled in an airplane that displays the external temperature. At the top of the troposphere, however, temperatures suddenly cease to decline with altitude, marking a boundary called the **tropopause.** The tropopause acts like a cap, largely preventing mixing between the troposphere and the atmospheric layer just above it, the stratosphere.

The **stratosphere** extends from 11 km (7 mi) to 50 km (31 mi) above sea level. Its temperature gradually rises with altitude, reaching a maximum of −3°C (27°F). Similar in composition to the troposphere, the stratosphere is 1,000 times drier and less dense. It is relatively calm with little vertical mixing of air, so that substances entering it tend to remain for a long time. The stratosphere warms with altitude because its ozone absorbs or scatters most of the sun's ultraviolet (UV) radiation. Most of the atmosphere's minute amount of ozone is

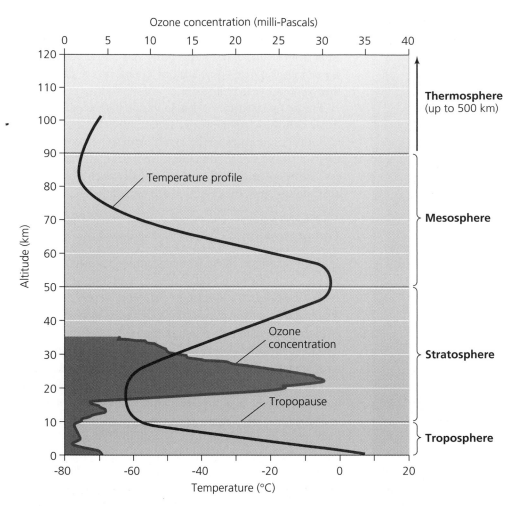

Figure 11.3 Although the atmosphere's molecular composition is largely similar throughout its four layers, certain characteristics do change with altitude. Temperature drops with altitude in the troposphere, rises with altitude in the stratosphere, drops in the mesosphere, then rises again in the thermosphere. Ozone reaches a peak in a portion of the stratosphere, giving rise to the term *ozone layer*. Adapted from Mark Z. Jacobson, *Atmospheric Pollution: History, Science, and Regulation*, Cambridge University Press, 2002; Edward A. Parson, *Protecting the Ozone Layer: Science and Strategy*, Oxford University Press, 2003.

concentrated in a portion of the stratosphere, roughly 17–30 km (10–19 mi) above sea level, a region that has come to be called Earth's **ozone layer.** Because solar UV light can damage living tissues by inducing DNA mutations, the ozone layer's protective effects are vital to maintaining life on Earth.

Above the stratosphere lies the **mesosphere,** which extends 50–90 km (31–56 mi) above sea level. Air pressure is extremely low here, and temperatures fall with altitude, reaching their lowest point (about −90°C [−130°F]) at the top of the mesosphere. From the outer mesosphere, the **thermosphere** extends upward to an altitude of 500 km (300 mi), where solar rays produce temperatures over 1,700°C (3,092 °F). It is largely in the thermosphere that ions (charged particles; Chapter 4) from the sun can react with atmospheric molecules to produce the beautiful haunting displays known as auroras (Figure 11.4). Whereas the chemical composition (see Figure 11.1) of the atmosphere's lower three layers is relatively constant throughout, in the thermosphere the molecules are so few and far between that they collide

Figure 11.4 When charged particles from the sun interact with Earth's magnetic field, a shimmering light show called the *aurora* can be produced in the night sky, generally nearer to the polar regions. The aurora borealis, or northern lights, occurs in the Northern Hemisphere, while the aurora australis, or southern lights, occurs in the Southern Hemisphere.

only rarely. As a result, heavier molecules (such as nitrogen and oxygen) sink, and light ones (such as hydrogen and helium) end up near the top of the thermosphere.

Solar energy heats the atmosphere, helps create seasons, and causes air to circulate

Beyond causing auroras and menacing life with UV rays, radiation from the sun plays a major role in our atmosphere by driving most of its air movement. The sun's energy helps create seasons and drives both weather and climate. An enormous amount of solar energy continuously bombards the upper atmosphere—over 1,000 watts/m², many thousands of times greater than the total output of electricity generated by humans. Of that solar energy, about 70% is absorbed by the atmosphere and planetary surface, while the rest is reflected back (see Figure 12.1, Chapter 12).

The spatial relationship between Earth and the sun determines the amount of solar radiation that strikes each point of Earth's surface. Sunlight is most intense when it shines directly overhead and meets the planet's surface at a perpendicular angle. At this angle, sunlight passes through a minimum of energy-absorbing atmosphere and focuses on a minimum of surface area. Given the Earth's curvature, this means that solar radiation intensity is highest near the equator and weakest near the poles (Figure 11.5).

Because Earth is tilted on its axis (an imaginary line connecting the poles, running perpendicular to the equator) by about 23.5 degrees, the Northern and Southern Hemispheres each tilt toward the sun for half the year. The seasons result from these periodic variations in sun exposure (Figure 11.6). Regions near the equator are largely unaffected by this tilt; they experience about 12 hours each of sunlight and darkness every day throughout the year. Near the poles, however, the effect is strong, and seasonality is pronounced.

Land and surface water absorb solar energy, reradiating some heat and causing some water to evaporate. Air near Earth's surface therefore tends to be warmer and wetter than air at higher altitudes. These differences set into motion the process of air circulation (Figure 11.7). Warm air rises, creating vertical currents. As air rises into regions of lower atmospheric pressure, it expands and cools. Once the air cools, it descends and becomes denser, replacing warm air that is rising. This type of circular current, with warm air rising to be replaced by colder air descending, is called a **convection current**. Convection currents operate not only on columns of air, but also within water bodies such as lakes and oceans (Chapter 13), in columns of magma beneath Earth's surface (Chapter 6), and even in a simmering pot of soup. Convection currents play key roles in guiding both weather and climate.

The atmosphere drives weather and climate

In everyday speech, we often use the terms *weather* and *climate* interchangeably. However, these two words have very distinct meanings to atmospheric scientists. **Weather** consists of the local physical properties of the

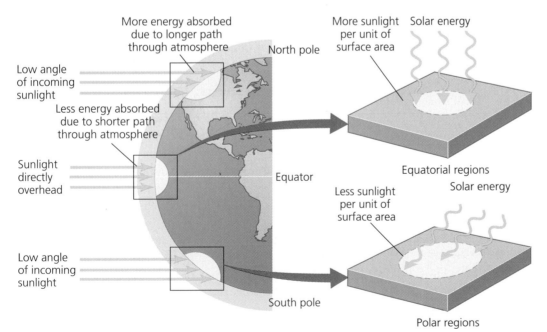

More energy absorbed due to longer path through atmosphere

Low angle of incoming sunlight

Less energy absorbed due to shorter path through atmosphere

Sunlight directly overhead

Low angle of incoming sunlight

North pole

Equator

South pole

More sunlight per unit of surface area

Solar energy

Equatorial regions

Less sunlight per unit of surface area

Solar energy

Polar regions

Figure 11.5 Because of Earth's curvature, on average the polar regions receive less solar energy than equatorial regions. One reason is that sunlight gets spread over a larger area when striking a surface at an angle, a phenomenon that increases with distance from the equator. Another reason is that when sunlight enters the atmosphere at a greater angle near the poles, it must traverse a longer distance before reaching the surface, during which more energy may be absorbed or reflected.

Figure 11.6 The seasons occur because Earth is tilted on its axis by 23.5 degrees. As Earth revolves around the sun, the Northern Hemisphere tilts toward the sun for one half of the year, and the southern half tilts toward the sun for the other half of the year. Summer occurs in each hemisphere during the period in which the hemisphere receives the most solar energy because of its tilt toward the sun.

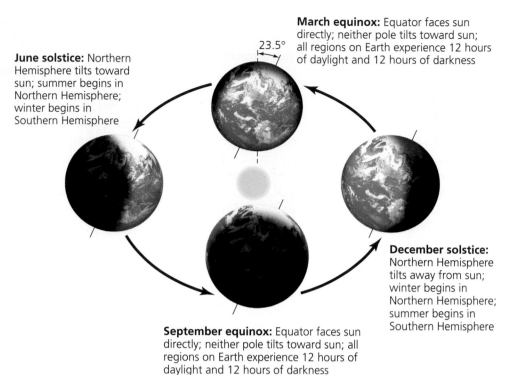

March equinox: Equator faces sun directly; neither pole tilts toward sun; all regions on Earth experience 12 hours of daylight and 12 hours of darkness

23.5°

June solstice: Northern Hemisphere tilts toward sun; summer begins in Northern Hemisphere; winter begins in Southern Hemisphere

December solstice: Northern Hemisphere tilts away from sun; winter begins in Northern Hemisphere; summer begins in Southern Hemisphere

September equinox: Equator faces sun directly; neither pole tilts toward sun; all regions on Earth experience 12 hours of daylight and 12 hours of darkness

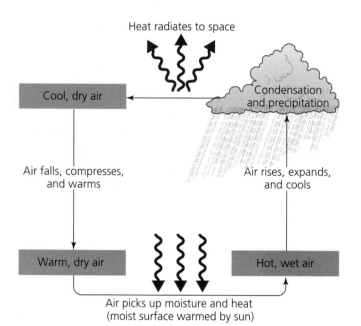

Heat radiates to space

Cool, dry air

Condensation and precipitation

Air falls, compresses, and warms

Air rises, expands, and cools

Warm, dry air

Hot, wet air

Air picks up moisture and heat (moist surface warmed by sun)

Figure 11.7 Weather is driven in part by convection currents, circular patterns of air flow that function on a large scale in the atmosphere. Warm air near Earth's surface picks up moisture as it heats up and begins rising. Once aloft, this air cools and moisture condenses, forming clouds and precipitation. Cool, drying air begins to fall, compressing and warming in the process, and warm dry air near the surface begins the cycle anew.

troposphere, such as temperature, pressure, humidity, cloudiness, and wind. Weather specifies atmospheric conditions over relatively short time periods, typically hours or days, and within relatively small geographic areas. **Climate**, in contrast, describes the pattern of atmospheric conditions found across large geographic regions over long periods of time, typically seasons, years, or millennia. Mark Twain once noted the distinction between climate and weather by saying, "Climate is what we expect, weather is what we get." For example, even very dry climates (such as that of the desert around Phoenix) have occasional wet weather.

Weather is produced by interacting air masses

Most changes in weather occur when air masses with different physical properties meet. The boundary between two air masses that differ in temperature and density is called a **front**. The boundary where a mass of warm air displaces a mass of colder air is termed a **warm front**. Some of the air behind a warm front rises over the cold air mass ahead, and then cools and condenses to form clouds that may produce rain. A **cold front** is the boundary along which a cold air mass displaces a warm air mass. The cold air, being denser, tends to wedge beneath the warm air, pushing the warm air upward, where it cools and expands to form clouds and potentially produce thunderstorms. Once a cold front

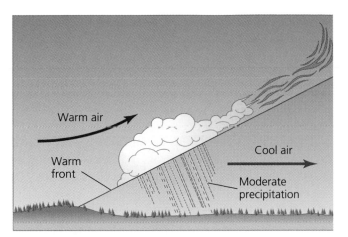

(a) Warm front

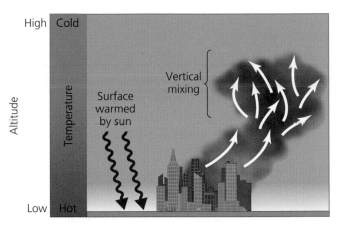

(a) Normal conditions

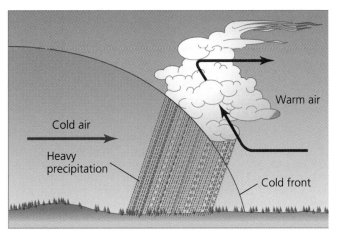

(b) Cold front

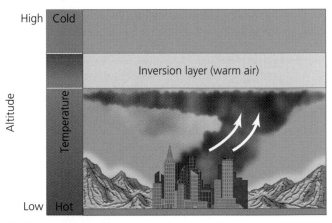

(b) Thermal inversion

Figure 11.8 When a warm front approaches, the warm air rises over cool air, causing moderate precipitation as moisture in the warm air condenses. When a cold front approaches, the cold air pushes under warmer air, forcing warm air upward, causing condensation, and resulting in heavy precipitation.

Figure 11.9 An inversion layer is a natural atmospheric occurrence that can exacerbate air pollution conditions locally. Under normal conditions (a), tropospheric temperature decreases with altitude and air of different altitudes mixes somewhat freely. When an inversion layer forms (b), warm air sits atop cooler air, preventing mixing and trapping the cooler air (and any pollutants within it) nearer the surface.

passes through, however, the sky usually clears and the temperature and humidity drop (Figure 11.8).

Opposing air masses may also differ in atmospheric pressure. An air mass with elevated atmospheric pressure, or a **high-pressure system**, contains air that descends. High-pressure air masses typically bring fair weather. In a **low-pressure system**, air moves toward the low atmospheric pressure at the center of the system and spirals upward. The air expands and cools, and clouds and precipitation often follow.

One type of weather event has implications for environmental health. Under most conditions, the air in the troposphere decreases in temperature as altitude increases, and warm air rises, causing vertical mixing. Occasionally, however, a pocket of relatively cold air

occurs near the ground, with warmer air above it. This departure from the normal temperature distribution is called a **temperature inversion** or **thermal inversion** (Figure 11.9). The cold air, denser than the air above it, resists vertical mixing and creates a pocket of very stable air. Whereas vertical mixing normally allows ground-level air pollution to be diluted upward, thermal inversions trap pollutants near the ground. It was a thermal inversion that sparked London's killer smog. A high-pressure system then settled over the city, acting like a cap on the air pollution and keeping it in place. Although London's killer smog was the worst single such event recorded, inversions regularly cause episodes of smog buildup in many urban areas throughout the world.

Global climate patterns result from the differential heating of Earth's surface

On larger geographic scales, air movements create climatic patterns that are maintained over long periods of time. Sunlight produces global patterns of convection currents (Figure 11.10a). Near the equator is a pair of convection currents called **Hadley cells.** Here, where sunlight is most intense, surface air is warmed, rising and expanding in two giant columns, one heading toward each pole. As each air column rises and expands, it releases moisture, producing the heavy rainfall of the tropical rainforests near the equator. After releasing much of their moisture, these huge columns of air cool and descend back to Earth at about 30° latitude north and south. Because the descending air has a low relative

humidity, it absorbs moisture from the land. The regions around 30° latitude are therefore quite arid, giving rise to deserts. Two pairs of similar but less intense convection currents, called **Ferrel cells** and **polar cells**, force air upward and create precipitation around 60° latitude north and south, and force air downward at around 30° latitude and in the polar regions.

These three pairs of cells account for the latitudinal distribution of moisture across Earth's surface: wet, tropical climates near the equator, arid climates near 30° latitude, somewhat moist regions at 60° latitude, and somewhat dry, arctic conditions near the poles. These patterns, combined with temperature variation, help explain why biomes tend to be arrayed in latitudinal bands (Chapter 6, Figure 6.7).

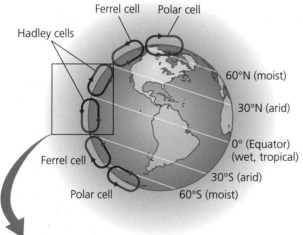

Figure 11.10 A series of large-scale convection currents (**a**) help determine global patterns of humidity and aridity. Warm air near the equator rises, expands, and cools, and moisture condenses, giving rise to a warm, wet climate in tropical regions. Air travels poleward and descends roughly around 30° latitude. This air that lost its moisture in the tropics absorbs moisture from the surface, causing the regions around 30° latitude to be arid. This convection current, a Hadley cell, occurs on both sides of the equator. Between 30° and 60° latitude a convection current operating in reverse occurs, a Ferrel cell, and between 30° and 60° latitude a polar cell convection current occurs. Global wind currents (**b**) show latitudinal patterns as well. Tradewinds near the equator blow westward, while westerlies between 30° and 60° latitude blow eastward.

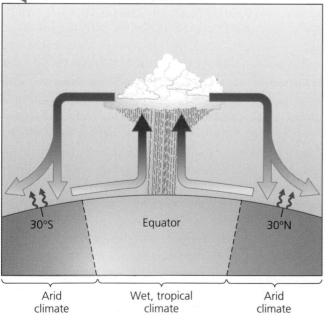

(a) Convection currents

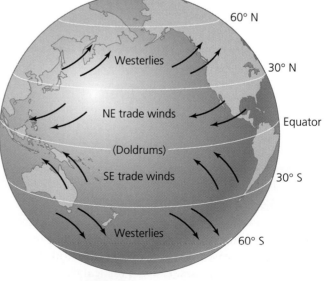

(b) Global wind patterns

Global wind patterns are influenced by Earth's rotation

As cool air moves to replace rising warm air in a convection current, horizontal air currents are produced, which we know as wind. The Hadley, Ferrel, and polar cells produce wind patterns that extend across the planet. The north-south patterns of air circulation within the cells, however, do not translate into north-south surface winds, because of the influence of the **Coriolis effect.** Earth's rotation causes regions near the equator to spin more quickly than regions near the poles, so that north-south winds appear to be deflected from a straight path. To ground-based observers, air currents appear to curve, traveling partly in east-west directions. This apparent deflection is the Coriolis effect.

The interaction of the convection currents and Earth's rotation produces the global wind patterns shown in Figure 11.10b. Rising air near the equator produces a region with few latitudinal winds known as the *doldrums.* Between the equator and 30° latitude lie the west-blowing *trade winds.* From 30° to 60° latitude are the *westerlies,* which originate from the west and blow east. Humans have recognized and used these air circulation patterns for centuries to facilitate ocean-going commerce, particularly in the 15th through 19th centuries, when wind-powered sailing ships were a major form of transportation.

The oceans and their interaction with the atmosphere also affect weather, climate, and the distribution of biomes, as we will see in Chapter 13. In Chapter 12, we will learn how west-blowing trade winds and convection currents in ocean water together maintain normal ocean circulation, which, when interrupted, produces El Niño conditions that have significant impacts on climate, ecological systems, and economies throughout the world. Now, however, we will examine how pollution of the atmosphere, both outdoor and indoor, affects ecological systems, economies, and human health.

Outdoor Air Pollution

Human activities alter virtually every part of the environment, and the atmosphere is no exception. Throughout human history, we have made the atmosphere a dumping ground for our airborne wastes. Whether from primitive wood fires or modern coal-burning power plants, humans have generated significant quantities of **air pollution,** material added to the atmosphere that can affect climate and harm organisms, including ourselves. However, although air pollution remains a problem, government regulation and improved technologies in recent years have greatly diminished it in countries of the developed world.

The majority of outdoor air pollution comes from natural sources

Outdoor air pollution consists of volatile chemicals or particulate matter that readily mix into the troposphere. When we think of outdoor air pollution, we tend to envision huge smokestacks belching black smoke from industrial plants. However, natural processes actually produce the majority of the world's air pollution. For example, the metabolism of plants, the decay of dead plants, and salt from sea spray add chemicals to the air that can harm organisms and affect climate. Besides these ongoing processes, sporadic sources such as dust storms, volcanic eruptions, and forest fires can have major impact. Some of these natural impacts can be exacerbated by human activity and land use.

Winds sweeping over arid terrain can send huge amounts of dust aloft. In 2001 strong westerlies lifted soil from deserts in Mongolia and China. The dust blanketed Chinese towns, spread to Japan and Korea, traveled across the Pacific Ocean to the United States, then crossed the Atlantic and left evidence atop the French Alps. Every year, hundreds of millions of tons of dust are blown in the other direction by trade winds across the Atlantic from northern Africa to the Americas (Figure 11.11a). Fungal and bacterial spores carried along with the dust have been linked to die-offs in Caribbean coral reef systems. Although dust storms are natural, the immense scale of these events results from poor farming and grazing practices that strip vegetation from soil and allow for wind erosion. Continental-scale dust storms took place in the United States in the 1930s, when soil from the drought-stricken Dust Bowl states blew eastward to the Atlantic (Chapter 8). Improved farming practices have since prevented a recurrence of such storms in the United States, although not before much of the nation's best topsoil was lost.

Volcanic eruptions also release large quantities of particulate matter, as well as sulfur dioxide, into the troposphere. Especially large eruptions may blow matter into the stratosphere, where it can remain for months or years. The 1980 eruption of Mount St. Helens in Washington produced 1.1 billion m^3 (1.4 billion yd^3) of dust that circled Earth for 15 days (Figure 11.11b). The

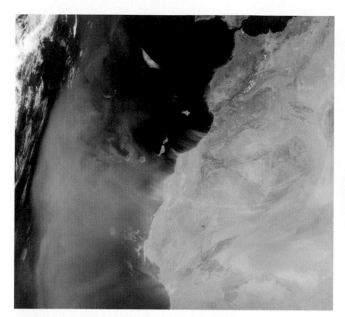

(a) Dust storm from Africa to the Americas

(b) Mount St. Helens eruption 1980

(c) Natural fire in California

Figure 11.11 Massive dust storms, such as this one blowing across the Atlantic Ocean from Africa to the Americas (**a**) are one type of natural air pollution. Volcanoes are another, as shown by Mount St. Helens (**b**), which erupted in Washington State in 1980. A third cause of natural pollution is natural fires in forests and grasslands (**c**).

year (Figure 11.11c). Many fires occur naturally, but most today result from slash-and-burn farming techniques. In 1997, a severe drought brought on by the 20th century's strongest El Niño event (Chapter 12) caused fires in Indonesia to rage out of control. Their smoke sickened 20 million Indonesians and caused collisions between cargo ships and a plane crash in Sumatra. More than 170 million metric tons of carbon monoxide (CO) were released from these fires. These, combined with tens of thousands of other fires in drought-plagued Mexico, Central America, and Africa, released more CO into the atmosphere during 1997–1998 than did the worldwide burning of fossil fuels.

Human activities create various types of outdoor air pollution

Human activity can exacerbate the severity of natural air pollution. Human activity can also introduce new sources of air pollution, and it is these that we have the greatest ability to bring under control. The term **point source** describes a specific spot—such as a factory's smokestacks—where large quantities of pollution are discharged. In contrast, **non-point sources** are more

massive 1883 eruption on the Indonesian island of Krakatau threw enough dust into the atmosphere to produce gorgeous sunsets throughout the world and cause a 1°C drop in global temperature. Still-greater volcanic activity may have disturbed the climate enough to account for some of the mass extinctions (Chapter 5 and Chapter 15) during Earth's history.

The burning of vegetation also pollutes the atmosphere with smoke and soot. Over 60 million ha (150 million acres) of forest and grassland burn in a typical

diffuse, often consisting of many small sources. In 1952 London, coal-burning power plants acted as point sources contributing to the killer smog, while millions of home fireplaces together comprised a potent non-point source. Many sources may emit air pollution while in motion, such as automobiles, aircraft, ships, locomotives, construction equipment, and lawn mowers.

Once a pollutant is in the atmosphere, it may do harm directly or may induce chemical reactions that produce harmful compounds. **Primary pollutants**, such as soot and carbon monoxide, are those emitted into the troposphere in a form that is directly harmful. In contrast, **secondary pollutants** are hazardous substances produced through the reaction of substances added to the atmosphere with chemicals normally found in the atmosphere.

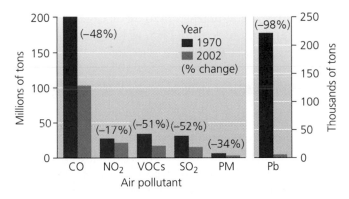

Figure 11.12 Six major types of pollutants are tracked by the U.S. EPA and termed "criteria pollutants." Lead and carbon monoxide have shown substantial declines since the 1970s. Go to **GRAPH IT** on website or CD–ROM. Data from Environmental Protection Agency, 2002 Trends report: Air quality continues to improve.

Six pollutants are closely tracked by the EPA

Of the many substances we discharge into the atmosphere, the U.S. EPA gives special attention to several judged to pose especially great threats to human health and welfare. For these six *criteria pollutants*—carbon monoxide (CO), lead (Pb), nitrogen dioxide (NO_2), ozone (O_3), sulfur dioxide (SO_2), and particulate matter—the EPA has established national air quality standards. The agency monitors levels of these pollutants throughout the country, primarily in urban areas, where they tend to concentrate. In 2003, the EPA reported that at least 133 million Americans (47% of the population) lived in counties where at least one of these six pollutants reached unhealthy levels during 2001. However, the percentage of days citizens were exposed to unhealthy air dropped from 10% in 1988 to 3% in 2001, showing that although air pollution remains a problem, the situation is improving. Figure 11.12 compares 1970 and 2002 emission levels of the six pollutants.

Carbon monoxide Carbon monoxide is a colorless, odorless gas produced primarily by the incomplete combustion of fuels. In the United States in 2001, 121 million tons of CO were released, making the gas by far the most abundant air pollutant by mass. Vehicles account for about 62% of these emissions, but other sources include lawn and garden equipment (10%), forest wildfires (6%), open burning of industrial waste (3%), and residential wood burning (2%). Carbon monoxide is dangerous to humans and other animals, even in small concentrations, because it can bind irreversibly to hemoglobin in red blood cells, preventing the hemoglobin

from binding with oxygen. Since 1982, CO emissions in the United States have decreased by 62%, largely because of cleaner-burning automobile engines.

Sulfur dioxide Like CO, sulfur dioxide is a colorless gas. Of the 15.8 million metric tons of SO_2 released in the United States in 2001, about 70% was created during the combustion of coal for industry and electricity generation. During combustion, elemental sulfur (S), which occurs to varying degrees in coal, reacts with oxygen gas (O_2) to form SO_2. Once in the atmosphere, SO_2 may react to form sulfur trioxide (SO_3) and then sulfuric acid (H_2SO_4), which may then fall back to Earth in the form of acid precipitation, a form of pollution.

Nitrogen dioxide Nitrogen dioxide is a highly reactive foul-smelling reddish gas that contributes to smog and acid precipitation. Nitrogen dioxide, along with nitrogen monoxide (NO), belongs to a family of compounds called nitrogen oxides (NO_x). Nitrogen oxides result when atmospheric nitrogen and oxygen react at the high temperatures created by combustion engines. Of the 22.3 million tons of nitrogen oxides released in the United States in 2001, about 56% resulted from combustion in vehicle engines, with electrical utility and industrial combustion accounting for most of the rest. Although NO_2 emissions have dropped by 14% in the last 20 years, emissions from nitrogen oxides as a group have increased, by 9% from 1982 to 2001. The EPA attributes this increase to engines powering construction and recreation equipment and to diesel engines. Because NO is readily converted to NO_2 in the atmosphere, the entire family of compounds is regulated.

Tropospheric ozone Although ozone in the stratosphere shields organisms from the dangers of UV radiation, O_3 from human activity builds up low in the troposphere and acts as a pollutant. Within the troposphere, this colorless gas results from the interaction of sunlight, heat, nitrogen oxides, and volatile carbon-containing chemicals. Ozone is therefore categorized as a secondary pollutant and often forms over urban areas with a great deal of primary pollution, particularly on hot summer days. A major component of smog, O_3 can bring about health problems as a result of its instability as a molecule; this triplet of oxygen atoms will readily release one of its threesome, leaving a molecule of oxygen gas and a free oxygen atom. The free oxygen atom is then able to participate in reactions that can readily injure living tissues. Although concentrations fell by 11–18% (depending on how they were measured) in the United States from 1982 to 2001, tropospheric O_3 is the pollutant that most frequently exceeds EPA standards in urban areas.

Weighing the Issues:
Ozone as a Rural Pollutant?

A scientific study published in the journal Nature *in 2003 demonstrated that urban production of tropospheric ozone in New York City was responsible for reduced tree growth in nearby rural areas. Regional atmospheric circulation patterns caused surrounding rural areas to be directly affected by the city's pollution. Should this revelation of "downwind effects" change the nature of the debate over air quality? If so, how? If tropospheric ozone can reduce tree growth, can you speculate about other impacts?*

Lead Lead is a metal that enters the atmosphere as a particulate pollutant. The lead-containing compound tetraethyl lead, when added to gasoline, improves engine performance. However, the exhaust from leaded gasoline injects lead into the atmosphere, from which it can settle on land and water. From there, lead can enter the food chain, accumulate within body tissues, and cause central nervous system malfunction, mental retardation among children, and a wide variety of other ailments. Once the dangers of lead were recognized, regulatory action phased out leaded gasoline in the United States and other industrialized nations. Lead emissions in the United States plummeted (Figure 11.13), drop-

ping by 93% from 1982 to 2001. Since 1993, when the U.S. phaseout of leaded gasoline was complete, lead emissions have remained steady and low, demonstrating the substantial effect that federal legislation can have on air quality. Of the 4.2 million tons of lead emitted in the United States in 2000, the industrial smelting of metals accounted for 49%, while only 12% came from fuel combustion. However, many developing nations continue to add lead to gasoline and suffer significantly from lead pollution.

Particulate matter Lead is only one of many types of particulate matter that can enter the atmosphere. Any solid or liquid particles small enough to be carried aloft are considered **particulate matter**, and many can cause damage to respiratory tissues when inhaled. Particulate matter includes primary pollutants, such as dust and soot, as well as secondary pollutants, such as sulfates and nitrates. Of the 31.4 million tons of particulate matter released in the United States in 2001, most was wind-blown dust (60%). Nationally, emissions of large-grained particles decreased by 10% from 1991 to 2000, while small-grained particle emissions increased by 4% during that time period. It was largely the emission of particulate matter from industrial and residential coal-burning sources that produced London's 1952 killer smog and the deaths resulting from that episode (Figure 11.13).

Volatile organic compounds are ingredients in some other pollutants

In addition to the six pollutants discussed above, many governments regulate **volatile organic compounds (VOCs),** a large group of potentially harmful organic (carbon-containing) chemicals used in industrial processes such as dry-cleaning and manufacturing. One group of VOCs is made up of hydrocarbons, molecules containing only hydrogen and carbon (Chapter 4); examples include methane (CH_4 or natural gas), propane (C_3H_8, used as a portable fuel), butane (C_4H_{10}, found in cigarette lighters), and octane (C_8H_{18}, a component of gasoline). Human activity accounts for about half the VOC emissions in the United States, while the remaining VOC emissions come from natural sources. For example, plants produce isoprene (C_5H_8) and terpene ($C_{10}H_{15}$), and animals produce methane. The largest sources of VOC emissions from human activity include industrial use of solvents (28%) and vehicle emissions (27%). VOCs are major contributors to urban smog; once in the atmosphere, VOCs can react to produce secondary pollutants, such as ozone.

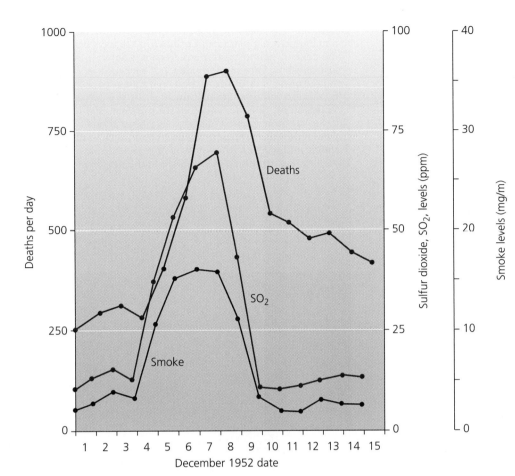

Figure 11.13 London's killer smog of 1952 produced a sharp increase of deaths simultaneously with and just following spikes in smoke and atmospheric sulfur dioxide. Although levels of these pollutants dropped to normal after six days, the death rate remained elevated for many days afterward. Data from Richard A. Anthes et al., *The Atmosphere*, third edition, Charles E. Merrill Publishing Company, 1981.

Various other substances are considered toxic air pollutants

Several other chemicals known to cause serious health or environmental problems are classified as toxic air pollutants. Some toxic air pollutants are produced naturally. For example, hydrogen sulfide gas (H_2S) gives the mud of swamps and bogs the odor of rotten eggs. However, most toxic air pollutants are produced by human activities, such as metal smelting, sewage treatment, and other industrial processes. The U.S. Clean Air Act of 1970 identified 188 different toxic air pollutants, most of which, such as benzene and methylene chloride (found in paint stripper), are VOCs.

Toxic air pollutants gain the most attention when large quantities are released accidentally and force people to be evacuated from an area. The worst such disaster took place in 1984, when 40 tons of methyl isocyanate (used to produce insecticides) leaked from a Union Carbide facility in Bhopal, India. A deadly cloud killed over 3,800 people and injured over 11,000 in less than an hour. In the eyes of many, this disaster underscored a need to carefully regulate facilities that stockpile large quantities of potentially deadly substances. As disastrous as such accidents can be, however, most outdoor air pollution impacts are due to chronic low-level emissions that result in recurrent conditions, such as urban smog.

Industrial smog is produced by burning fossil fuels

In response to the increasing incidence of fogs polluted by the smoke of Britain's industrial revolution, a British scientist coined the term "smog" long before the 1952 event in London. Today the term is used worldwide to describe unhealthy mixtures of air pollutants that often form over urban areas. The smog that enveloped London in 1952 was what we would today call **industrial smog,** or gray-air smog. When coal or oil are burned, some portion is completely combusted, forming CO_2; some is partially combusted, producing CO; and some remains unburned and is released as soot, or particles of carbon. In addition, coal contains varying amounts of contaminants, primarily sulfur, which reacts with oxygen to form sulfur dioxide. Sulfur dioxide can then undergo a series of reactions to form sulfuric acid and ammonium sulfate (Figure 11.14a). These chemicals

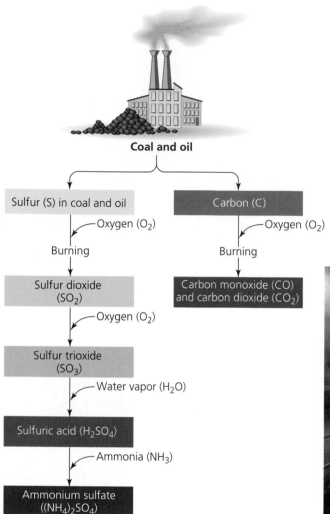

Coal and oil

Sulfur (S) in coal and oil   Carbon (C)

Oxygen (O₂)   Oxygen (O₂)

Burning   Burning

Sulfur dioxide
(SO₂)   Carbon monoxide (CO)
                                    and carbon dioxide (CO₂)

Oxygen (O₂)

Sulfur trioxide
(SO₃)

Water vapor (H₂O)

Sulfuric acid (H₂SO₄)

Ammonia (NH₃)

Ammonium sulfate
((NH₄)₂SO₄)

(a) Burning sulfur-rich oil or coal without adequate pollution control technologies

Figure 11.14 Emissions from the combustion of coal and oil in manufacturing plants and utilities in the days before pollution control technologies often created industrial smog. In industrial smog, carbon monoxide and carbon dioxide were emitted as a result of the carbon component of the fossil fuel. In addition, sulfur contaminants in the fuel when combusted created sulfur dioxide, which in the presence of other chemicals in the atmosphere could produce several other sulfur compounds (a). In certain weather conditions, industrial smog could blanket whole towns or regions, as in the case of Donora, Pennsylvania, shown here in the daytime during its 1948 killer smog event (b).

(b) Donora, Pennsylvania, at midday in the 1948 smog event

and others produced by further reactions, along with soot, are the main components of industrial smog and give the smog its characteristic gray color.

Industrial smog is far less common today than it was 50 to 100 years ago. In the wake of the 1952 London episode and others, the governments of most developed nations began regulating industrial emissions to minimize the external costs (Chapter 2) they impose on citizens. However, in industrializing regions such as China, India, and Eastern Europe, heavy reliance on coal burning (both by industry and by citizens heating and cooking in their homes) combined with lax air pollution controls produce industrial smog that presents health problems in many urban areas.

Although coal burning supplies the chemical ingredients for industrial smog, the weather also plays a role, as it did in London in 1952. A similar event occurred four years earlier in Donora, Pennsylvania: A thermal inver-

sion trapped smog containing particulate matter emissions from a steel and wire factory. Twenty-one people were killed, and over 6,000 people—nearly half the town—became ill (Figure 11.14b). In Donora's killer smog, air near the ground cooled during the night. Normally, morning sunlight warms the land and air, causing the air to rise; however, because Donora is surrounded by mountains, not enough sun reached the valley floor to warm and disperse the cold air. The resulting thermal inversion kept a cloud of smog over the town long enough to impair visibility and cause serious health problems. Mountainous topography like Donora's is a factor in the air pollution of many other cities where surrounding mountains trap air and create inversions. This is true for the Los Angeles basin, which has long symbolized chronic smog problems in American popular culture. Modern-day Los Angeles, however, suffers from a different type of smog, one called photochemical smog.

(a) Photochemical smog over Mexico City

Photochemical smog is produced by a complex series of atmospheric reactions

A photochemical process is one whose activation requires light. **Photochemical smog,** or brown-air smog (Figure 11.15a), is formed through light-driven reactions of primary pollutants and normal atmospheric compounds that produce a mix of over 100 different chemicals, ground-level ozone often being the most abundant among them (Figure 11.15b). Due to high levels of NO_2, photochemical smog forms a brownish haze over cities. Hot, sunny days in urban areas provide perfect conditions for the formation of photochemical smog. Exhaust from morning traffic releases large amounts of NO and VOCs into the troposphere over a

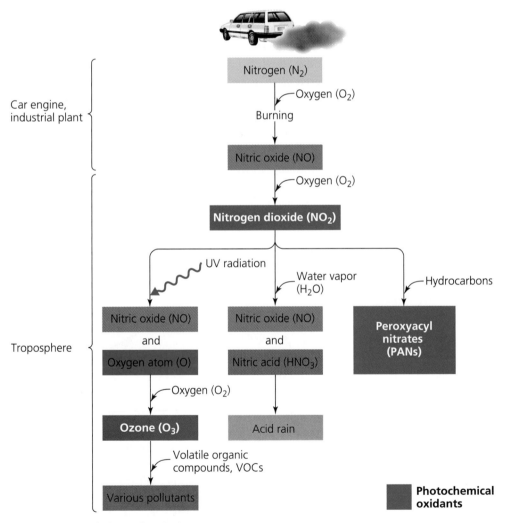

(b) Formation of photochemical smog

Figure 11.15 Photochemical smog is common today over many urban areas, especially where topography or inversion layers encourage it. Cities like Mexico City (**a**) frequently experience photochemical smog. Nitric oxide, a key element of photochemical smog, can start a chemical chain reaction (**b**) that results in the production of other compounds, including nitrogen dioxide, peroxyacyl nitrates, nitric acid, and ozone. Nitric acid can contribute to acid precipitation as well as photochemical smog.

city. Increasing sunlight then promotes the production of ozone and other ingredients of photochemical smog. Levels of photochemical smog in urban areas typically peak around midafternoon, irritating people's eyes, noses, and throats. Air pollutants called *peroxyacyl nitrates,* created by the reaction of NO_2 with hydrocarbons, can spur further reactions that damage building materials and living tissues in animals and plants.

Photochemical smog afflicts many major cities, especially those with topography and weather conditions that promote it. In Athens, Greece, site of the 2004 Olympics, the problem has been bad enough that the city government has provided incentives to replace aging automobiles and has mandated that cars with odd-numbered license plates be driven only on odd-numbered days, and those with even-numbered plates only on even-numbered days. According to Greek officials, pollution levels have been lowered by 30% since 1990 as a result.

Stratospheric ozone depletion is caused by synthetic chemicals

A pollutant at low altitudes, ozone is a highly beneficial gas at altitudes centering around 25 km (15 mi) in the lower stratosphere, where it is concentrated in the so-called ozone layer. Even here, concentrations of ozone are very low, only about 12 parts per million. Despite their rarity, however, ozone molecules are highly effective at absorbing incoming ultraviolet radiation from the sun, thus protecting life on Earth's surface.

Starting in the 1960s, atmospheric scientists began wondering why their measurements of ozone were lower than theoretical models predicted. Researchers wondering whether ozone might be depleted by natural or artificial chemicals finally pinpointed a group of human-made compounds derived from simple hydrocarbons, such as ethane and methane, in which hydrogen atoms are replaced by chlorine, bromine, or fluorine. One class of such compounds, **chlorofluorocarbons (CFCs),** was being mass-produced by industry at a rate of a million metric tons per year in the early 1970s and growing by 20% a year.

Then in 1974, atmospheric scientists Sherwood Rowland and Mario Molina broke the news that CFCs depleted stratospheric ozone by splitting O_3 molecules and creating O_2 molecules from them (Figure 11.16). Three years before Rowland and Molina's study, researcher J. E. McDonald had shown that ozone loss, by allowing more UV radiation to reach the surface, would result in thousands more skin cancer cases each year. These

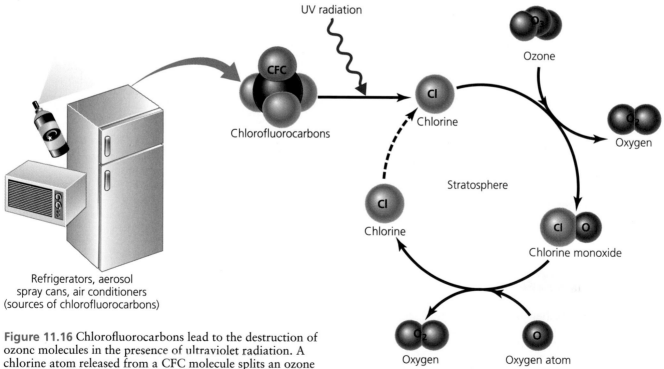

Figure 11.16 Chlorofluorocarbons lead to the destruction of ozone molecules in the presence of ultraviolet radiation. A chlorine atom released from a CFC molecule splits an ozone molecule, forming oxygen gas and a temporary ClO molecule. The oxygen atom in the ClO will then bind with a stray oxygen atom to form oxygen gas, leaving the chlorine atom to start the destructive cycle anew.

results caught the attention of policymakers, environmentalists, and industry alike (see The Science behind the Story).

Then in 1985, scientists from the British Antarctic Survey announced that stratospheric ozone levels over Antarctica had declined 40–60% in the previous decade, leaving a thinned ozone concentration that was soon dubbed the **ozone hole** (Figure 11.17). Research over the next few years confirmed the link between CFCs and ozone loss in the Antarctic and indicated that similar but lesser depletion was occurring in the Arctic and perhaps globally. Already concerned about skin cancer, scientists were also becoming anxious over the possible effects of increased UV radiation on ecosystems. Research was showing a complex multitude of effects, including harm to crops and to the productivity of ocean phytoplankton, the base of the marine food chain.

Ozone depletion was halted by the Montreal Protocol

In light of the science and the ongoing health and ecological concerns, international efforts to restrict CFC production finally bore fruit in 1987 with the Montreal Protocol, which has been signed by over 180 nations. In this convention, nations agreed to cut CFC production in half. Five follow-up agreements strengthened the pact, deepening the cuts, advancing timetables for compliance, and addressing related ozone-depleting chemicals. Today the production and use of ozone-depleting compounds has fallen by 95% since the late 1980s, and scientists can discern the beginnings of long-term recovery of the ozone layer (although much of the 5 billion kg of CFCs emitted into the troposphere has yet to diffuse up into the stratosphere). Industry was able to shift to alternative, environmentally safer chemicals, which have largely turned out to be cheaper and more efficient. For these reasons, the Montreal Protocol and its follow-up amendments are widely considered the most spectacular success story so far in addressing any global environmental problem.

Environmental scientists have attributed this success primarily to two factors:

1. Policymakers engaged industry in helping to solve the problem, and government and industry worked together on developing technological fixes and replacement chemicals. This cooperation greatly reduced the number and severity of the typical battles that erupt between environmentalists and industry.

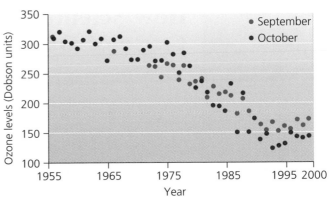

(a) Monthly mean ozone levels at Halley Bay, Antarctica

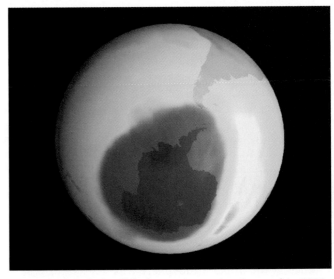

(b) The "ozone hole" (blue) over Antarctica, September 2000

Figure 11.17 The "ozone hole" consists of a region of thinned ozone density in the stratosphere over Antarctica and the southernmost ocean regions that has reappeared seasonally each September in recent decades. Data from Halley Bay, Antarctica **(a)** show a steady decrease in ozone concentrations from the 1960s to 1990, after which ozone-depleting CFCs began to be regulated under the Montreal Protocol. Colorized satellite imagery from September 6, 2000 **(b)**, shows the ozone hole at its maximal recorded extent to date. Data from British Antarctic Survey, 2002.

2. Implementation of the Montreal Protocol after 1987 successfully followed an adaptive management approach, which allows for altering strategies midstream in response to new scientific data, technological advances, or economic figures.

Because of its success in solving the problem of ozone depletion, the Montreal Protocol is widely seen as a model for international cooperation in addressing other pressing global problems, such as persistent organic pollutants (Chapter 10), climate change (Chapter 12), and biodiversity loss (Chapter 15).

How Scientists Identified CFCs as the Main Cause of Ozone Depletion

For half a century after their invention in the 1920s, chlorofluorocarbons (CFCs) were thought to be useful, nontoxic, and environmentally friendly. In the early 1970s, however, scientists became concerned that CFCs could cause long-term damage to the stratospheric ozone layer. By the late 1980s, evidence for such damage had become strong enough to justify a complete ban on CFC production. In their attempts to understand how CFCs influenced the ozone layer, scientists relied on a wide variety of data sources, including historical records, field observations, laboratory experiments, and computer models.

Both stratospheric ozone and CFCs had been the subject of much research before they were linked in the 1970s. Ozone was discovered in 1839, and its presence in the upper atmosphere was first proposed in the 1880s. In 1924, British scientist G.M.B. Dobson built what has become the standard instrument for measuring ozone from the ground. By the 1970s, the Dobson ozone spectrophotometer, as it is known, was being used by a global network of observation stations. Despite this network, scientists were unable to establish a clear picture of global trends in ozone levels because of natural variations in

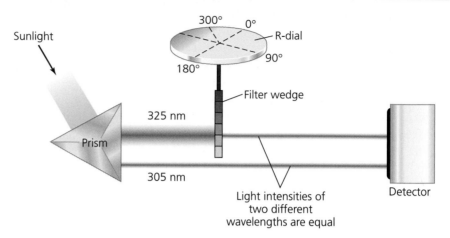

A ground-based instrument, the Dobson spectrophotometer measures atmospheric ozone concentration by manipulating wavelengths of ultraviolet (UV) light in sunlight that has passed through the atmosphere. UV light passes through the prism, which separates wavelengths of 325 nanometers (nm) and 305 nm, and then travels toward the detector. Because ozone absorbs wavelengths of 305 nm but not wavelengths of 325 nm, light that reaches the instrument after passing through the atmosphere contains more light of 325 nm wavelengths. The ratio between the intensities of the two wavelengths of light indicates the amount of ozone in the light's path between the sun and the spectrophotometer. To measure this ratio, the R-dial rotates, causing the filter wedge to block more and more 325-nm light, until the intensities of 325-nm and 305-nm light are equal. At that point the reading on the R-dial is recorded and a conversion is used to calculate the atmospheric ozone concentration. Figure adapted from University of Alaska, Fairbanks, Stratospheric Ozone.

ozone levels and the difficulty of comparing data from different stations.

Research on CFCs also had a long history. First invented in 1928, CFCs were found to be useful as refrigerants, fire extinguishers, and propellants for aerosol spray cans. Starting in the 1960s, CFCs also found wide use as cleaners for electronics and as a part of the process

of manufacturing rigid polystyrene foams. Research on the chemical properties of CFCs showed that they were almost completely inert; that is, they rarely reacted with other chemicals. Therefore, scientists surmised that, at trace levels, CFCs would be harmless to both people and the environment.

However, in June 1974, chemists F. Sherwood Rowland

Weighing the Issues:
Need for International Cooperation to Solve Global Problems

The Montreal Protocol showed how international collaboration, together with technological advances, can drastically and rapidly address a pressing environmental problem. Currently, global problems such as organic pollutants, climate change, and biodiversity loss have not seen the same degree of targeted action. Why do you think this is? Besides the halting of stratospheric ozone depletion, can you name other success stories in addressing major environmental problems? Are any on the horizon?

and Mario Molina published a paper in the journal *Nature,* arguing that the inertness that made CFCs so ideal for industrial purposes could also have disastrous consequences for the ozone layer. More reactive chemicals are broken down to their constituent atoms in the lower atmosphere; CFCs, in contrast, float up to the stratosphere unchanged. Once CFCs reach the stratosphere, intense ultraviolet radiation from the sun breaks them into their constituent chlorine and carbon atoms. Each free chlorine atom, it was calculated, can catalyze the destruction of as many as 100,000 molecules of ozone.

Rowland and Molina were the first to assemble a complete picture of the threat posed by CFCs, but their conclusions would not have been possible without the contributions of other scientists. British researcher James Lovelock (Chapter 6) had developed an instrument to measure extremely small concentrations of atmospheric gases, and American researchers Richard Stolarski and Ralph Cicerone had shown that chlorine atoms can catalyze the destruction of ozone.

Rowland and Molina's finding, which earned them the 1995 Nobel Prize in Chemistry, helped spark a discussion among scientists, policymakers, and industry leaders over limits on CFC production. As a result, the United States and several other nations banned the use of CFCs in aerosol spray cans in 1979. Other uses continued, however, and by the early 1980s total global production of CFCs was increasing. Then, in 1985, a new finding shocked scientists and spurred the international community to take further action.

The unexpected finding came from a study by scientists at a British research station in Antarctica, where ozone levels had been recorded continuously since the 1950s. In May 1985, Joseph Farman and his colleagues reported in *Nature* that Antarctic ozone levels had been declining dramatically since the 1970s. The decline exceeded even the worst-case predictions.

To determine what was causing the ozone hole, expeditions were mounted in 1986 and 1987 to measure trace amounts of atmospheric gases using ground stations and high-altitude balloons and aircraft. Together with other scientists, Dutch scientist Paul Crutzen, who would share the 1995 Nobel Prize with Molina and Rowland, used the data collected on the expeditions to conclude that the ozone hole was the result of a unique combination of Antarctic weather conditions and human-made chemicals. In the frigid Antarctic winter, they found high-altitude clouds, or polar stratospheric clouds, were formed. In the spring, those clouds provided ideal conditions for CFC-derived chlorine and other chemicals to catalyze the destruction of massive amounts of ozone. The problem was exacerbated by the fact that Antarctica's atmosphere was largely isolated from the rest of Earth's atmosphere.

In the following years, scientists used data from ground stations and satellites to show that ozone levels were declining not just over Antarctica but globally. In 1987, those findings helped convince the world's industrialized nations to agree on the Montreal Protocol, which aimed to cut CFC production in half by 1998. Within two years, however, further scientific evidence and computer modeling showed that more drastic measures would be needed if serious damage to the ozone layer was to be avoided. In 1990, the Montreal Protocol was strengthened to include a complete phaseout of CFCs by 2000. By 1998, the amount of chlorine in the atmosphere appeared to have leveled off, suggesting that the agreements had had the desired effect.

Acid precipitation represents another transboundary pollution problem

Just as stratospheric ozone depletion crosses political boundaries, so does another major atmospheric pollution concern—acid precipitation. **Acid precipitation,** which includes acid rain as well as other forms of precipitation such as acid fog and acid snow, forms pollutants (mostly SO_2 and NO_x) that react with water, oxygen, and oxidants. Sulfur dioxide and nitrogen oxides are produced primarily through fossil fuel combustion by automobiles, electric utilities, and industrial facilities. Once emitted into the troposphere, these primary pollutants can react (as described above) to produce acidic compounds, primarily sulfuric acid and nitric acid. Suspended in the troposphere, droplets of these acids may travel for days or weeks, covering hundreds or thousands of kilometers before falling in precipitation (Figure 11.18).

Although acid precipitation does not feel acidic to our touch, its low pH values (Chapter 4) have wide-ranging, cumulative detrimental effects on ecosystems and on our built environment (Table 11.1). Acid precipitation leaches basic minerals such as calcium and magnesium from soil, changing soil chemistry and harming plants and soil organisms. Streams, rivers, and lakes may become significantly acidified from runoff. In fact, thousands of lakes now contain water acidic enough to kill fish. The fish can die when the acidic conditions make aluminum more accessible and toxic to them. Elevated aluminum in the soil hinders water and nutrient uptake by plants. In some regions of the United States, acid fog with a pH of 2.3 (about the same as vinegar) can sit atop forests for extended periods. Certain forests in eastern North America have experienced widespread tree die-offs from these conditions. Besides harming trees, acid precipitation also damages agricultural crops. In addition, it gradually eats away at buildings, cars, and other structures. It is estimated that billions of dollars of damage are being done to ancient cathedrals in Europe, monuments in Washington, D.C., temples in Asia, and stone statues throughout the world, as their features gradually erode away (Figure 11.19).

Because the pollutants leading to acid precipitation may travel long distances, their effects can be felt far from their point sources—a situation that has led to political bickering among states and nations. For instance, the New England states have suffered some of the worst acid precipitation problems because they are downwind from the heavy industry of the upper Midwest. Much of the pollution from factories in Pennsylvania, Ohio, and

---

Table 11.1 Effects of Acid Precipitation on Ecosystems in the Northeastern United States

Acid deposition in northeastern forests has...

accelerated leaching of base cations (ions that counteract acid deposition) from soil

allowed sulfur and nitrogen to accumulate in soil

increased dissolved inorganic aluminum in soil, hindering plant uptake of water and nutrients

leached calcium from needles of red spruce, leading to tree mortality from wintertime freezing

increased mortality of sugar maples due to leaching of base cations from soil and leaves

acidified 41% of Adirondack, New York, lakes and 15% of New England lakes

lowered lakes' capacity to neutralize further acids

elevated aluminum levels in surface waters

reduced species diversity and abundance of aquatic life and negatively affected entire food webs

Source: Adapted from C. T. Driscoll et al., *Acid rain revisited,* Hubbard Brook Research Foundation, 2001.

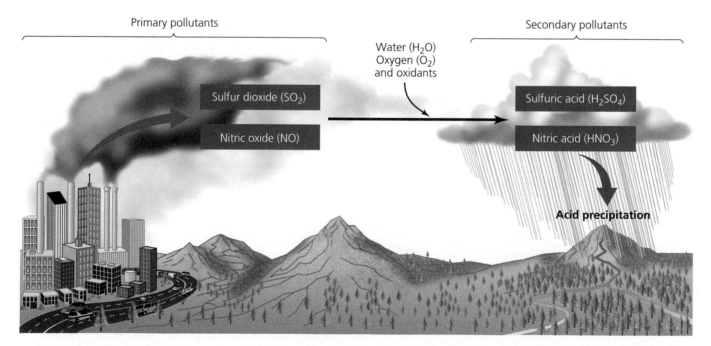

Figure 11.18 Acid precipitation is a phenomenon that can cause problems many miles downwind from its source. Emissions from industries and utilities that include sulfur dioxide and nitric oxide begin the process. These chemicals can be transformed into sulfuric acid and nitric acid through chemical reactions in the atmosphere, and the acidic compounds fall to Earth in precipitation.

Figure 11.19 Acid rain can cause harm to vegetation, can alter soil chemistry, can affect soil- and forest-dwelling animals, and can even eat away at statues and buildings.

Illinois falls out in states to their east, including New York, Vermont, and New Hampshire, as well as in regions to the north, including Ontario, Quebec, and the maritime provinces of Canada. As Figure 11.20 shows, many regions of greatest acidification are downwind of major source areas of pollution.

Acid precipitation has not been reduced as markedly as scientists had hoped

Reducing acid precipitation involves reducing amounts of the primary pollutants that contribute to it. New technology has helped; "scrubbers" that filter pollutants in smokestacks have allowed factories to decrease emissions. As a result of declining emissions of SO_2, average sulfate precipitation in 1996–2000 was 10% lower than in 1990–1994 across the United States, with a 15% reduction in the eastern states. However, average nitrate precipitation increased nationally by 3% between these time periods because of increasing NO_x emissions.

A recent report by scientists at New Hampshire's Hubbard Brook research forest, where acid precipitation's effects were first demonstrated, has disputed the notion that the problem of acid precipitation is being

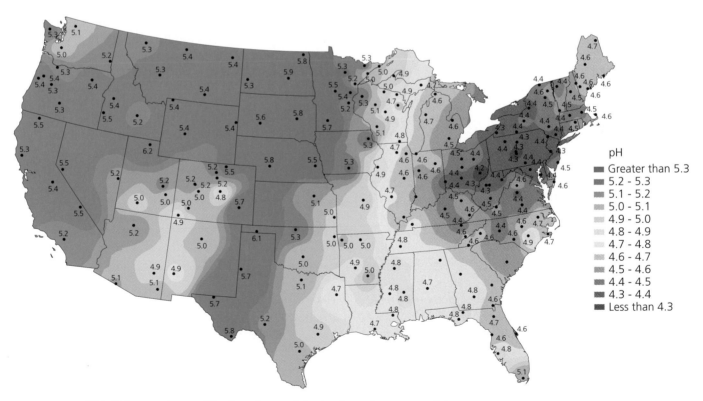

Figure 11.20 This U.S. map shows pH values for precipitation in various areas of the country. Acid precipitation is most serious in parts of the Northeast and Midwest, generally downwind from (roughly east of) areas of heavy industrial development. Data from National Atmospheric Deposition Program, Hydrogen ion concentration as pH from measurements made at Central Analytical Laboratory, 2002.

Acid Rain at Hubbard Brook Research Forest in New Hampshire

The effects of acid precipitation are subtle, involving incremental changes in pH levels that take place over long periods of time and affect species with long life cycles, such as trees. For this reason, no single experiment can give us a complete picture of acid precipitation's effects. Nonetheless, one long-term study conducted in the Hubbard Brook Experimental Forest in New Hampshire's White Mountains has been critically important to our understanding of acid rain in the United States.

Established by the U.S. Forest Service in 1955, Hubbard Brook was initially devoted to research on hydrology, the study of water flow through forests and streams. In 1963, in collaboration with scientists at Dartmouth University, Hubbard Brook researchers broadened their focus to include a long-term study of nutrient cycling in forest ecosystems. Since then, they have collected and analyzed weekly samples of rain, snow, and other forms of precipitation. The measurements make up the longest-running North American record of acid precipitation and have helped shape U.S. policy on sulfur and nitrogen emissions.

At Hubbard Brook, only one form of acid deposition—wet deposition—has been measured

The effects that acid rain has on trees can be seen in this forest on Mt. Mitchell in western North Carolina.

regularly. Wet deposition is the deposition of rain or other precipitation, as well as fog or sea spray, with low pH onto soil, plants, or groundwater. Dry deposition, in contrast, is the deposition of airborne acidic particles. Throughout Hubbard Brook's 3,160 ha (7,800 acres), small plastic collecting funnels, about a foot in diameter at their openings, channel precipitation into clean bottles, which researchers retrieve and replace each week. Hubbard Brook's own laboratory measures acidity and conductivity, which indicates the amount of salts and other electrolytic contaminants dissolved in the water. Concentrations of sulfuric acid, nitrates, ammonia, and

other compounds are measured elsewhere.

By the late 1960s, ecologists Gene Likens, F. Herbert Bormann, and others had found that precipitation at Hubbard Brook was several hundred times more acidic than natural rainwater. By the early 1970s, a number of other studies had corroborated their findings. Together, these studies indicated that precipitation from Pennsylvania to Maine had average pH values around 4 and that individual rainstorms showed values as low as 2.1—almost 10,000 times more acidic than ordinary rainwater.

In 1978, the National Atmospheric Deposition Program was launched to monitor precipitation

solved. Instead, the report said, the effects are worse than first predicted, and the mandates of the 1990 Clean Air Act amendments are not adequate to restore ecosystems in the northeastern United States. At Hubbard Brook, for instance, half the soil's calcium content has leached out, leaving the soil less able to neutralize future acid precipitation. This means that forests affected by acidification may take a long time to recover. An additional 80% reduction in sulfur emissions from

electric utilities would be needed, the report estimates, to allow New Hampshire streams to recover in 20–25 years. The data on acid precipitation show that although there have been many impressive gains in the mitigation of air pollution, more can clearly be done to alleviate outdoor pollution problems. The same can be said for indoor air pollution, a source of human health threats that is less familiar to most of us, but statistically more dangerous.

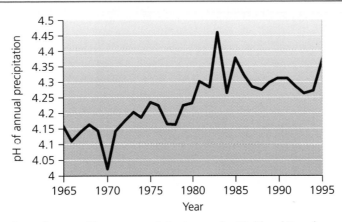

Over the past 40 years, precipitation at the Hubbard Brook Environmental Forest has become slightly less acidic; however, rain at the research forest is still much more acidic than pure rain. Data from C. T. Driscoll et al., Hubbard Brook Research Foundation, 2001.

across the United States. Initially consisting of only 22 sites, including Hubbard Brook, the program now comprises more than 200, each of which gathers weekly acid deposition data.

By the late 1980s, the National Atmospheric Deposition Program had produced a nationwide map of pH values. The most severe problems were found to be in the Northeast, where prevailing west-to-east winds were blowing emissions from fossil-fuel-burning power plants in the Midwest. Scientists hypothesized that when sulfur dioxide, nitrogen oxides, and other polluting gases arrived in the Northeast, they were absorbed by water droplets in clouds, converted to acidic compounds such as sulfuric acid, and deposited on farms, forests, and cities in the form of rain or snow. To some extent, the Clean Air Act of 1970 has helped reduce acid rain in the Northeast. The accompanying figure shows the pH record for an area of Hubbard Brook known as Watershed 6. As you can see, between 1965 and 1995 the average pH increased slightly, from about 4.15 to about 4.35. In 1990, as a consequence of the Hubbard Brook study and the nationwide research that followed, the Clean Air Act of 1970 was amended to further restrict emissions of sulfur dioxide and other acid-forming compounds. Nonetheless, acid deposition continues to be a serious problem in the Northeast.

Some of the long-term consequences of acid rain are now becoming clear as well. In 1996, researchers reported that approximately 50% of the calcium and magnesium in Hubbard Brook's soils had been leached out by acid rain. Meanwhile, acid precipitation had increased the concentration of aluminum in the soil, which can prevent tree roots from absorbing nutrients. The resulting nutrient deficiency slows forest growth and weakens trees, making them more vulnerable to drought and insects. It also reduces the ability of soil and water to neutralize acidity, making the ecosystem increasingly vulnerable to further inputs of acid.

In October 1999, researchers used a helicopter to distribute 50 tons of a calcium-containing mineral called wollastonite over one of Hubbard Brook's watersheds in order to raise the concentration of base cations to estimated historical levels. Over the next 50 years, scientists plan to evaluate the impact of calcium addition on the watershed's soil, water, and life. By providing a comparison to watersheds in which calcium remains depleted, the results should provide new insights into the consequences of acid rain and the possibilities for reversing its negative effects.

Indoor Air Pollution

Indoor spaces generally have higher concentrations of pollutants than outdoor spaces. This is true not only of industrial workplaces and offices, but also of schools and residences. As a result, the health effects from indoor air pollution outweigh those from outdoor air pollution (Figure 11.21). One recent estimate, from Australia's Commonwealth Science Council in 1999, attributed 14 times as many deaths globally to indoor air pollution as to outdoor air pollution, with a total of 2.8 million deaths annually from indoor sources. Another estimate, from the U.N. Development Programme in 1998, attributed 2.2 million deaths to indoor air pollution and 500,000 deaths to outdoor air pollution. Indoor air pollution alone, then, takes 6,000 to 8,000 lives each day.

(a) South African family cooking indoors

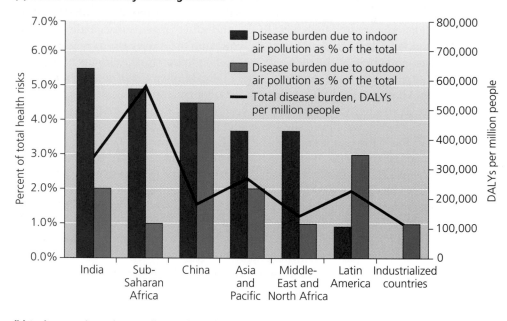

(b) Indoor and outdoor pollution health risk statistics

Figure 11.21 Indoor air pollution is an under-recognized health threat in both developed nations and developing nations. In the developing world, fires for cooking and heating are often built inside homes, as seen here in a South African kitchen (**a**), exposing family members to particulate matter and carbon monoxide. In most regions of the developing world, indoor air pollution is estimated to cause upwards of 3% of total health risks (**b**). In this graph, disability adjusted life years (DALYs) indicate the burden of disease in total number of years of healthy life lost, including premature death and disability over a period of time, due to both indoor and outdoor air pollution. Indians suffer most severely from indoor air pollution, with approximately 650,000 years of life lost per million people, folowed by Sub-Saharan Africans, who suffer approximately 580,000 years loss of life per million people. *Source:* World Bank. Data from United Nations Development Programme, World Bank, Energy sector management assistance programme, 2002.

If the scale of impact from indoor air pollution seems surprising, consider that the average U.S. citizen spends at least 90% of his or her time indoors. Then consider that in the past half-century a dizzying array of consumer products has been manufactured and sold, many of which we keep in our homes and offices and have come to play major roles in our daily lives. Many of these products are made of synthetic materials, and, as we saw in Chapter 10, novel synthetic substances are not comprehensively tested for health effects before being brought to market. Products such as insecticides and cleaning fluids can exude volatile chemicals into the air,

as you might expect, but solid materials such as plastics and chemically treated wood products can, too.

In an ironic twist, some attempts to be environmentally prudent during the "energy crisis" of 1973–1974 worsened indoor air pollution in developed countries. To reduce heat loss and improve energy efficiency, building managers sealed off ventilation in existing buildings, and building designers constructed new buildings with little ventilation and with windows that did not open. These steps may have saved energy, but they also worsened indoor air pollution by trapping stable, unmixed air—and its contaminants—inside.

Indoor air pollution in the developing world arises from fuelwood burning

Indoor air pollution problems have even more impact in the developing world. Data from Australia's Commonwealth Science Council indicates that 9 out of 10 of the 2.8 million annual deaths from indoor air pollution occur in the developing world, most of these in rural areas. Millions of people in developing nations burn wood, charcoal, animal dung, or crop waste in their homes for cooking and heating with little or no ventilation, and in the process inhale dangerous amounts of soot and smoke. In the air of such homes, concentrations of particulate matter are commonly 20 times above the U.S. EPA safe levels, the World Health Organization (WHO) has found. Poverty forces fully half the population and 90% of rural residents of developing countries to heat and cook with indoor fires. Some people will burn almost any available fuel, even discarded plastic, in indoor fires. In doing so, these people are in effect subjecting themselves to a daily dose of London's smog of 1952. Indoor air pollution from fuelwood burning, the WHO estimates, may cause more than 5% of all deaths in some developing nations each year.

Many people who tend indoor fires are simply not aware of the health risks. Many do not have access to the statistics showing that chemicals and soot released by burning coal, plastic, and other materials indoors can increase risks of pneumonia, bronchitis, allergies, sinus infections, cataracts, asthma, emphysema, heart disease, cancer, and premature death. Many also do not know that studies have found young children to be particularly susceptible to such pollution.

Recognizing indoor air pollution as a problem is still fairly novel

Even in the developed world, recognizing indoor air pollution as a problem is still quite novel. The 1970 U.S. Clean Air Act did not even mention indoor air; rather, indoor spaces were assumed to be safe havens from outdoor pollution. Even smoke from indoor fires was long viewed merely as a nuisance, in the absence of data showing it to cause health problems.

Today we know far less about indoor air pollution than we do about outdoor air pollution. In the United States, there is no scheme in place to monitor pollutants comprehensively, as there is for some outdoor substances, such as the EPA's six criteria pollutants. What we know about indoor pollution so far has come mostly from individual independent studies, not comprehensive and coordinated federal or international efforts. Indeed, we are still learning what indoor sources exist. In 1999, EPA scientists reported an unexpected source of indoor air pollution: automatic dishwashers. According to the study, the hot interior of a dishwasher can serve as a reaction chamber for a whole array of pollutants that can then build up inside the home. In the presence of hot water, chlorine-containing detergents react easily with organic food remains on dishes, leading to the formation of VOCs and organochlorine compounds. Other known indoor sources of air pollution in the industrialized world include showers, washing machines, and kitchen faucets (Table 11.2).

Scientists do, however, know enough already to have identified the most deadly indoor threats. Particulate matter and chemicals from wood and charcoal smoke are the primary health risks in the developing world. In developed nations, the top risks appear to be cigarette smoke and radon, a naturally occurring radioactive gas.

Tobacco smoke and radon are the most deadly indoor pollutants in the developed world

The health effects of smoking cigarettes are well known in developed countries, but only recently have scientists quantified the risks of inhaling secondhand smoke. Secondhand smoke, or environmental tobacco smoke, is smoke exhaled by a smoker, which may then be inhaled by a nonsmoker who is nearby or shares the same enclosed airspace. Secondhand smoke has been found to cause many of the same problems as directly inhaled cigarette smoke, ranging from irritation of the eyes, nose, and throat, to exacerbation of asthma and other respiratory ailments, to lung cancer. This hardly seems surprising when one considers that environmental tobacco smoke consists of a brew of over 4,000 chemical compounds, many of which are known or suspected to be toxic or carcinogenic.

Women living with a spouse who smokes have a 24% greater chance of developing lung cancer from secondhand smoke, one study has indicated. Fortunately, the popularity of smoking has declined greatly in the United States and some other nations in recent years. The exposure of young children in the United States has decreased by almost half: A 1998 study found 20% of young children exposed, as compared to 39% in 1986.

After cigarette smoke, radon gas is the second-leading cause of lung cancer in the United States; an authoritative federal study has estimated 15,000–22,000 radon-related lung cancer deaths per year in the United States. As we saw in Chapter 10, radon is a radioactive gas resulting from the natural decay of uranium in soil, rock, or water, which seeps up from the ground and infiltrates buildings. Radon is colorless and odorless, and its occurrence in different locations is variable and hard to predict (Figure 11.22). As a

Table 11.2 Indoor Air Pollutants—At Home

Pollutant	Sources	Possible Health Effects
Asbestos	Deteriorating, damaged, or disturbed insulation, fireproofing, acoustical materials, and floor tiles; exposure to clothing and equipment brought home from job sites with elevated asbestos concentrations	No immediate symptoms; long-term risk of chest and abdominal cancers (lung cancer and mesothelioma) and lung diseases (asbestosis); smokers at higher risk
Biological pollutants	Bacteria; molds; fungi; mildew; viruses; animal dander; cat saliva; house dust; mites; cockroaches; pollen; rats; mice; occur in contaminated central air systems, standing water, and water-damaged materials	Allergic reactions; rhinitis; some types of asthma; infectious illnesses; sneezing; watery eyes; coughing; shortness of breath; dizziness; lethargy; fever; digestive problems
Carbon monoxide (CO)	Unventilated kerosene and gas space heaters; leaking chimneys and furnaces; back-drafting from furnaces, gas water heaters, wood stoves, and fireplaces; gas stoves; generators; gasoline-powered equipment; automobile exhaust from attached garages; tobacco smoke	Impaired vision and coordination; headaches; dizziness; confusion; nausea; fatal at very high concentrations
Combustion by-products	All carbon monoxide sources; nitrogen dioxide; respirable particles (including secondhand smoke); polycyclic aromatic hydrocarbons (PAHs), or gases or particles released from wood-burning heaters; motor vehicle exhaust; cigarette smoke; laying of asphalt roads	Impaired vision and coordination; headaches; dizziness; confusion; nausea; fatal at very high concentrations; many PAHs suspected carcinogens
Endocrine-disrupting chemicals	A range of chemicals, including brominated fire retardants, phthalates, pesticides, and organotin compounds; plastics in computers, televisions, electronic equipment, soft toys; compounds in house dust, carpets, vinyl flooring, polyurethane foam, meat, fish, breast milk	Hormone disruption, including disrupted thyroid function; hindered brain development; altered sex hormone balance; threats to breast-feeding babies; possibility of bioaccumulation

Figure 11.22 The risk from radon depends upon underground geology. This map shows relative levels of risk from radon across the United States, but there is much fine-scale geographic variation from place to place not evident on the map. Testing your home for radon is the surest way to discover whether this colorless odorless radioactive gas could be a problem in your home. Data from U.S. Geological Survey, Generalized geological radon potential of the United States, 1993.

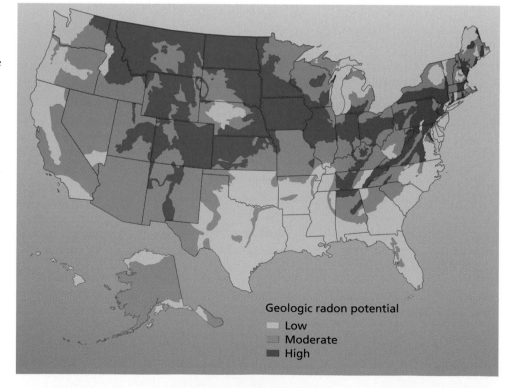

Geologic radon potential

Low
Moderate
High

Table 11.2 *Continued*

Pollutant	Sources	Possible Health Effects
Formaldehyde	Pressed-wood products (wall paneling, particleboard, fiberboard); furniture containing urea-formaldehyde (UF) resins; urea-formaldehyde foam insulation (UFFI); combustion sources; second-hand tobacco smoke; carpets; floor linings; insulation foam; furniture foam; durable press drapes; textiles; glues	Eye, nose, and throat irritation; nausea; dizziness; lethargy; flu-like symptoms; difficulty breathing; asthma attacks; suspected carcinogenicity
Lead	Lead-based paint; contaminated soil, dust, and drinking water	Convulsions; coma; death from high levels of exposure; adverse health effects on the central nervous system, kidney, and blood cells from lower levels of exposure; impaired mental and physical development; fetuses and young children especially susceptible
Nitrogen dioxide	Kerosene heaters; unventilated gas stoves and heaters; tobacco smoke	Eye, nose, and throat irritation; impaired lung function; respiratory infections
Pesticides	Both active and inert ingredients in insecticides, termiticides, and disinfectants	Irritation to eye, nose, and throat; damage to central nervous system and kidney; increased risk of cancer; headaches, dizziness, muscle twitching, weakness, tingling sensations, and nausea from exposure to high levels of cyclodiene pesticides
Radon	Naturally occurring radioactive gas formed by decay of uranium in soil, rock, and water	Lung cancer
Respirable particles	Fireplaces; wood stoves; kerosene heaters; chimneys; secondhand tobacco smoke	Eye, nose, and throat irritation; respiratory infections; bronchitis; pneumonia; ear infections; lung cancer; suspected in Sudden Infant Death Syndrome (SIDS)
Volatile organic compounds (VOCs)	Paints; paint strippers; wood preservatives; aerosol sprays; cleansers; disinfectants; moth repellents; air fresheners; stored fuels; automotive products; hobby supplies; dry-cleaned clothing; cosmetics; soft plastics; glues; markers; carpets	Eye, nose, and throat irritation; headaches; loss of coordination; nausea; damage to liver, kidney, and central nervous system; suspected carcinogenic

Sources: U.S. Environmental Protection Agency, Tools for healthy schools: Toxic hazard management in children's environments, 2002.

result, the only way to determine whether radon is entering a building is to purchase a test kit. Testing in 1991 led the EPA to estimate that 6% of U.S. homes had radon levels exceeding the EPA's safe standard. Since the mid-1980s, 18 million U.S. homes have been tested for radon, and 700,000 have undergone radon mitigation. New homes are now being built with radon-resistant features—over a million such homes since 1990.

A great diversity of volatile organic compounds pollute indoor air

Some indoor pollutants are the same as the pollutants we meet with outdoors. For instance, carbon monoxide commonly occurs indoors. Because CO is so deadly and so hard to detect, many homes are equipped with detectors that sound an alarm if incomplete combustion produces dangerous levels of CO. The most diverse and abundant indoor pollutants are VOCs. These molecules exist in everything from plastics to oils to perfumes to paints to cleaning fluids to adhesives to pesticides. These airborne carbon-containing compounds evaporate or leak from furnishings, building materials, color film, carpets, laser printers, fax machines, and sheets of paper. Some products, such as chemically treated furniture, release large amounts of VOCs when new and progressively less as they age.

Air Pollution

Air pollution leading to photochemical smog remains a major health and environmental concern in and around many major cities. Is more or different government regulation needed to reduce this type of pollution?

A Market Approach to Fighting Smog

Federal and state regulations to control emissions from factories, power plants, and cars have reduced smog. Further progress will require more effective and affordable means to reduce smog "precursors," nitrogen oxides and volatile organic compounds, from old and new sources. Simply tightening the same regulations used in the past and applying them to the same sources is unlikely to significantly reduce smog for three reasons: control costs will be high for many "old" sources already being regulated; emissions from many smaller sources like sport utility vehicles, trucks, lawnmowers, and motorboats will need to be regulated for the first time; and the right mix of reductions in nitrogen oxide and volatile organic compound emissions depends on local circumstances.

For large stationary sources, emissions trading is an economically attractive supplement to current technology-based regulations. It has already proven successful in controlling sulfur dioxide emissions from power plants at far less cost than traditional regulation. Emissions trading works through the following process: First, government regulators set a regional limit on nitrogen oxides. Then, regulators assign or auction to emissions sources only as many allowances (for example, one allowance equals one ton of nitrogen oxide emissions) as the overall limit (or "cap") permits. Lastly, emissions sources could take one of three actions: reduce their own emissions to stay within their allowances; sell or retire any excess allowances; or buy others' excess allowances to meet their own obligations. They would choose according to market prices with the cheapest sources of reductions being made available first.

Small mobile sources that run on diesel fuel and gasoline are also a problem. For now, government still needs to press manufacturers to build cleaner burning vehicles and fuel makers to make cleaner fuels, as the EPA is doing. Meanwhile, more research will help regulators understand how to tailor their efforts to highly variable local conditions.

Debra Knopman is associate director of the science and technology research unit of the RAND Corporation. This statement is based primarily on work she did at the Progressive Policy Institute, where she was director of the Center for Innovation and the Environment from 1995 to 2001.

To Conquer Urban Smog, We Must Force Industry to Do What It Does Not Want to Do

Concise rules are relatively easy for government agencies and private citizens to enforce. Conversely, convoluted rules are relatively easy for polluters and their lawyers to manipulate and evade. Consequently, you will not hear many public health advocates call for an increase in the length or complexity of anti-smog regulations. What we want instead are stricter rules applied more strategically.

The public health community has generally welcomed efforts to reduce air pollution through innovative forms of regulation. Clean air advocates have, for example, supported tax breaks and other incentives designed to encourage people to drive less, because tailpipe emissions are a major contributor to urban smog. Unfortunately, though, these programs have yielded only moderate reductions in smog-forming pollution. And while another innovation, emissions trading, can reduce smokestack emissions on a regional scale, it is not a solution to urban smog. For instance, Houston's air quality will not improve if the refineries surrounding the city pay cleaner facilities elsewhere for the freedom to continue polluting at current levels.

The solution to urban smog is not a greater number of rules, more incentive programs, or further market-based innovations. Rather, it is tougher standards placed where they will produce the largest emissions reductions. Refineries must be forced to curb flaring, power plants must be forced to install pollution controls, and manufacturers must be forced to make cleaner engines. To be sure, tougher requirements will cost industry money, but the benefits to the U.S. economy will swamp the costs. Just ask the Bush administration. The White House Office of Management and Budget recently concluded that over the past decade, strict enforcement of the Clean Air Act cost industry and taxpayers $26 billion while saving the national economy between $120 billion and $193 billion by reducing the number of premature deaths, hospitalizations, and lost workdays.

David McIntosh is a staff attorney at the Natural Resources Defense Council in Washington, DC. He litigates and lobbies to ensure effective implementation of clean air laws and to counter efforts aimed at weakening those laws.

Other items, such as photocopying machines, emit VOCs each time they are used.

We are surrounded by products that emit VOCs, but fortunately they are emitted in very small amounts. EPA and European surveys have both measured levels of total VOCs in buildings and have found them to be nearly always less than 0.1 part per million (ppm). This is, however, a substantially greater concentration than is found outdoors. When the EPA conducted a survey of U.S. office buildings between 1994 and 1998, it found 34 of 48 VOCs it tested for, with at least one type found in 81% of buildings. All of these compounds existed at higher levels indoors than outdoors, suggesting that they originated from indoor sources.

The implications for human health of chronic exposure to VOCs are far from clear. Because they exist in such low concentrations and because individuals regularly are exposed to mixtures of many different types, it is extremely difficult to study the effects of any one pollutant. An exception is formaldehyde, which does have clear and known health impacts. This VOC, one of the most common synthetically produced chemicals, irritates mucous membranes, induces skin allergies, and causes other ailments. Formaldehyde is used in numerous products, but health complaints have mainly resulted from its leakage from pressed wood and insulation. The use of plywood has decreased in the last decade because of health concerns over formaldehyde. Although the United States has no safety standard for formaldehyde outside the workplace, Canada has a non-industrial standard of 0.1 ppm, and the WHO sets its standard at 0.08 ppm.

VOCs also include pesticides, which we examined in Chapter 9 and Chapter 10. Some pesticides are used indoors (three-quarters of U.S. homes use at least one pesticide indoors during an average year), but most are for outdoor use. Thus it may seem surprising that the EPA found in a 1990 study that 90% of people's pesticide exposure came from indoor sources. Households that the agency tested had multiple pesticide volatiles in their air, at levels 10 times above levels measured outside. Some of the pesticides found had apparently been used years previously against termites, then seeped into the houses through floors and walls. DDT, banned 15 years before the study, was found in five of eight homes, probably having been brought in on the soles of occupants' shoes from outdoors.

Living organisms can pollute indoor spaces

Tiny living organisms that we breathe in can also be indoor pollutants. Dust mites and animal dander can increase asthma in children, for instance. Microbial life such as fungi and mold (in particular, their airborne spores) can cause potentially serious health problems, from headaches and allergies to asthma, respiratory diseases, and cardiovascular effects. Certain bacteria that pollute air can also cause allergic reactions or infectious disease. One example is the bacterium that gives rise to legionnaires' disease. Of the estimated 10,000–15,000 annual U.S. cases of legionnaires' disease, 5–15% are fatal. Heating and cooling systems in buildings make ideal breeding grounds for microbes, providing moisture, dust, and foam insulation as substrates and air currents to carry the organisms aloft.

Microbes that induce allergic responses are thought to be one frequent cause of building-related illness, a sickness produced by indoor pollution in which the specific cause is not identifiable. When the cause of such an illness is a mystery, and when symptoms are general and nonspecific, the illness is often called *sick-building syndrome*. VOCs are often thought responsible for sick building syndrome. The U.S. Occupational Safety and Health Administration (OSHA) has estimated that 30–70 million Americans have suffered ailments due to the environment of the building in which they live. We can reduce the prevalence of sick building syndrome by using low-toxicity building materials and ensuring that buildings are adequately ventilated.

Weighing the Issues:
How Safe Is Your Indoor Environment?

Think about the amount of time you spend indoors. Name the potential indoor air quality hazards in your home, work, or school environment. Are these spaces well-ventilated? What could you do to make the indoor spaces you use safer?

We can reduce indoor air pollution

Using low-toxicity material and providing adequate ventilation are key to alleviating indoor air pollution in almost any situation. In the developed world, we can try to limit our use of plastics and treated wood where possible and to limit our exposure to pesticides, cleaning fluids, and other known toxicants by keeping them in a garage or outdoor shed rather than in the house. We can also test our homes and offices for radon. Most important is to keep our indoor spaces as well ventilated as possible to minimize concentrations of chronic contaminants—toxic material we must live with.

Remedies for fuelwood pollution in the developing world include drying wood before burning (which reduces the amount of smoke produced), shifting to less-polluting fuels (such as natural gas), and replacing inefficient fires with cleaner stoves that burn fuel more efficiently. The Chinese government has invested in a program that has placed more-efficient stoves in millions of homes in China, for instance. According to WHO studies, this is a relatively cost-efficient means of reducing the health impacts of indoor biomass combustion. Installing hoods, chimneys, or cooking windows can increase ventilation for little cost, alleviating the majority of indoor smoke pollution.

Conclusion

Indoor air pollution is a potentially serious health threat. However, by keeping informed of the latest scientific findings and taking appropriate actions, we as individuals can significantly minimize the threat to our families and ourselves. Outdoor air pollution has been addressed more effectively by government legislation and regulation. In fact, reductions in outdoor air pollution levels in the United States and other developed nations represent some of the greatest strides in environmental protection to date. Much room for improvement remains, however, particularly in reducing acid precipitation and the photochemical smog resulting from urban congestion. Fortunately, developed nations no longer experience the type of pollution that Londoners suffered in their 1952 killer smog, but avoiding such high pollution levels in the developing world as less-wealthy nations industrialize will continue to be a challenge.

REVIEW QUESTIONS

1. What has been the human impact on the chemical composition of the atmosphere?
2. What is relative humidity?
3. Approximately how thick is Earth's atmosphere? Name one characteristic of each of the four atmospheric layers.
4. Where is the "ozone layer" located? How does ultraviolet (UV) radiation damage living tissues?
5. Describe the general distribution of temperature from warm to cool across the four layers of the atmosphere.
6. How does solar energy warm the air and help to create convection currents?
7. What factors led to the deadly smog in London in 1952? Describe a thermal inversion.
8. How do Hadley, Ferrel, and polar cells help to determine long-term climatic patterns and the location of biomes?
9. Explain how the Coriolis effect creates latitudinal winds.
10. Name three natural sources of outdoor air pollution and three sources caused by human activity.
11. What is the difference between a primary and a secondary pollutant? Of the six criteria pollutants monitored by the EPA, which are primary pollutants, which are secondary pollutants, and which include both?
12. What was the greatest single source of sulfur dioxide pollution in 2001?
13. Why are carbon monoxide and ozone harmful to living tissues?
14. What are the major synthetic sources of volatile organic compounds (VOCs)?
15. What is smog? How is the formation of smog dependent on the weather?
16. What are the primary components of industrial or gray-air smog? How is photochemical or brown-air smog different?
17. How do chlorofluorocarbons (CFCs) deplete stratospheric ozone? Why is this depletion a long-term international problem?
18. According to environmental scientists, what factors have accounted for the success so far of the Montreal Protocol?
19. Why are the effects of acid precipitation often felt in areas far from where the primary pollutants are produced?
20. About how many humans die from indoor air pollution each day? What is the World Health Organization estimate for deaths from fuelwood burning in developing countries?
21. Name some common sources of indoor pollution.

DISCUSSION QUESTIONS

1. Discuss three ways in which solar energy influences Earth's atmospheric conditions.

2. Consider the case of the "killer smog" of 1952 in London. What factors make it particularly difficult to study causes of air pollution and to develop solutions?

3. How may human activity sometimes exacerbate natural forms of air pollution? Discuss two examples.

4. International regulatory action has produced reductions in CFCs, while other transboundary issues such as acid rain and other pollutants have not yet been dealt with as effectively. What type of actions do you feel might be appropriate for pollutants that cross political boundaries in a less conspicuous manner? Consider cases such as formaldehyde and VOCs that may be emitted at very low levels from manufactured products. What do you think are the best ways to lessen the health impacts of indoor pollutants?

Media Resources *For further review, go to the website* **www.envscienceplace.com** *or student CD-ROM, where you will find quizzes, flashcards, a glossary, additional interactive exercises, and links to relevant news and research sources. Also, on the website and CD-ROM is* **GRAPH IT**, *a series of interactive graphing tutorials to help you interpret graphs and plot data.*

12 Global climate change

Muratti Island of the Maldives

This chapter will help you understand:

- Earth's climate and the variety of factors influencing global climate

- Apparent human influences on the atmosphere and global climate

- Effects of climate change

- Potential future impacts of climate change

- The general methods of modern climate researchers

- The scientific, political, and economic debates concerning global climate change

- Potential responses to global climate change

Fishermen of the Maldives

Central Case: Rising Temperatures and Seas May Take the Maldives Under

"The impact of global warming and climate change can effectively kill us off, make us refugees. . . ."
—*Ismail Shafeeu, Minister of Environment, Maldives*

"More people enjoy health and prosperity now than ever, and a warmer environment . . . should sustain life better than the current one does."
—*February 2001, Oil and Gas Journal*

A nation consisting of a chain of low-lying coral islands, or atolls, in the Indian Ocean, the Maldives is known for its spectacular equatorial setting, colorful coral reefs, and sun-drenched beaches. For visiting tourists it is a paradise, and for 320,000 Maldives residents it is home. But residents and tourists alike now feel the Maldives are under threat from global climate change and the rising sea level that global warming is causing. "The impact of global warming and climate change can effectively kill us off, make us refugees," according to Ismail Shafeeu, minister of environment for the island nation.

Nearly 80% of the Maldives' land area of 300 km² (116 mi²) lies less than 1m (39 in) above sea level. In

a nation of 1,190 islands whose highest point is just 2.4 m (8 ft) above sea level, rising seas could be a matter of life or death. The world's oceans have already risen 10–20 cm (4–8 in) during the twentieth century, as water from melting icecaps has flowed into the ocean and as warming temperatures have expanded ocean water. Current projections are that sea level will rise between 9 and 88 cm (20–35 in) more by the year 2100. For obvious reasons, these projections worry citizens of the Maldives.

The island's government has already evacuated residents from four of the lowest-lying islands over the past few years. It is not simply flooding that the Maldivians worry about, but larger more powerful storms, salt water intruding into their drinking water supplies, and damage to the coral reefs that are so crucial to the nation's tourism- and fishing-driven economy.

Maldives islanders are not alone in their worries. Other island nations, from the Galapagos to Fiji to the Seychelles, are also looking anxiously ahead to a changed future, one in which they may be constantly battling the sea. Mainland coastal areas of the world, such as parts of Florida and Louisiana, will likely face the same issues.

Despite the concerns of people from the Maldives to New Orleans, some people say that global warming is a good thing. Less than a year after Shafeeu issued his warning, the editors of *The Oil and Gas Journal* painted a different picture, writing that "More people enjoy health and prosperity now than ever, and a warmer environment . . . should sustain life better than the current one does."

In this chapter we will examine Earth's climate and the disagreements about its future. We will focus on the greenhouse effect, global climate change, and the role that human activities play in these processes. Our discussion will encourage and enable you to draw your own conclusions about the situation in the Maldives and the possible implications—both positive and negative—of global climate change.

Earth's Hospitable Climate

As you learned in Chapter 11, weather describes an area's short-term atmospheric conditions (over hours or days) including temperature, moisture content, wind, precipitation, barometric pressure, insolation (the amount of solar radiation the area receives), and other characteristics. Climate is an area's long-term pattern of atmospheric conditions. **Global climate change** describes changes in Earth's climate, such as temperature, precipitation, and storm intensity.

While most scientists agree that human activities, including fossil fuel combustion, farming, and deforestation, are altering Earth's atmosphere and climate, it would be inaccurate to think of global climate change as something new or as something caused entirely by people. The issue is not whether climate ought to be stable; it never has been. The issue is that our actions could be contributing to climatic changes that could have adverse consequences for ecosystems and for millions of people, including the residents of the Maldives, Florida, Louisiana, and other coastal regions.

The sun and the atmosphere keep Earth warm

Three factors exert more influence on Earth's climate than all others combined. The first is the sun. Without it, Earth would be dark and icy. The second is the atmos-

phere. Without an atmosphere, Earth would be as much as 33°C (59°F) colder, on average, and temperature differences between night and day would be much greater. The third is the oceans, which shape climate by storing and transporting heat and moisture.

The sun is the source of almost all energy that Earth receives. A relatively miniscule amount of energy arrives from other celestial bodies and some arises from within Earth, but these sources have little known effect on global climate. Earth's atmosphere reflects about 30% of incoming solar radiation back into space. The remaining 70% is absorbed by molecules in the atmosphere or reaches Earth's surface, where some is absorbed and the rest is reflected (Figure 12.1).

"Greenhouse gases" warm the lower atmosphere

The shape and composition of some molecules enable them to absorb radiation of specific wavelengths and make them transparent to radiation of other wavelengths. For example, almost all incoming solar radiation with wavelengths shorter than 290 nanometers (nm) (Chapter 4) is absorbed by atmospheric ozone (O_3) and oxygen (O_2). As a result, almost none of this radiation reaches Earth's surface. On the other hand, radiation with wavelengths between 300 and 800 nm is not well absorbed by atmospheric gases, so much of this radiation reaches Earth's surface.

After Earth's surface absorbs radiation, its temperature increases and it emits radiation in the infrared portion of the spectrum. Some gases, including carbon dioxide (CO_2), water vapor, ozone (O_3), nitrous oxide (N_2O), halocarbons (a diverse group of gases that include chlorofluorocarbons (CFCs) and hydrochlorofluorocarbons (HFCs)), and methane (CH_4), absorb infrared radiation from Earth's surface very effectively. Gases that effectively absorb infrared radiation released by Earth's surface and later warm Earth's surface by emitting energy are known as **greenhouse gases.** These gases, with the exception of the anthropogenic halocarbons, have been present in the atmosphere for billions of years, but human activities have increased the concentrations of these gases in the past 250 to 300 years.

When these gases absorb heat, the atmosphere (specifically the troposphere; Chapter 11) warms and then radiates heat back to Earth's surface. This warming of the troposphere and Earth's surface is commonly known as the **greenhouse effect.** Although the greenhouse effect is a natural phenomenon, human activities can and do influence it. When we increase

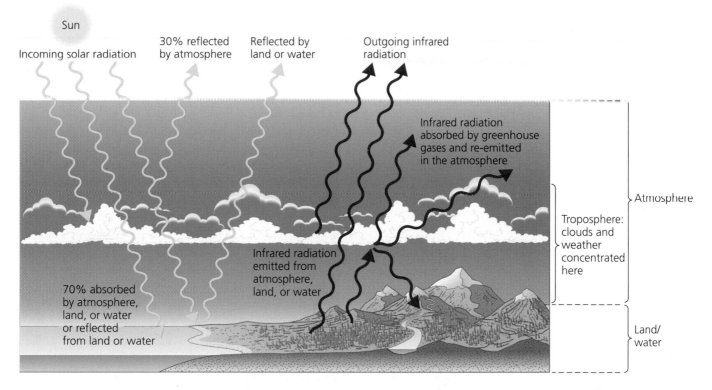

Figure 12.1 Earth's energy balance and average surface temperature depend on the amount of energy Earth receives from the Sun and radiates into space. Solar radiation and atmospheric composition are among the factors that influence this energy balance and global climate.

the concentration of infrared-absorbing gases in the atmosphere, the greenhouse effect is enhanced.

The term *greenhouse effect* is used widely these days, but scientists recognize that it is actually a bit of a misnomer. The analogy of a greenhouse, proposed by French mathematician Jean-Baptiste-Joseph Fourier as early as 1827, does not quite fit the way the atmosphere actually behaves. Greenhouses hold heat in place by preventing warm air from escaping. Greenhouse gas molecules in the atmosphere, in contrast, absorb infrared radiation and then re-emit infrared radiation of slightly different wavelengths; thus air is not trapped, but heat is absorbed, transformed, and radiated. **Global warming**, on the other hand, is an increase in Earth's average surface temperature, whatever the cause.

Table 12.1 shows the relative global warming potential of a number of greenhouse gases. Global warming potential is expressed in relation to carbon dioxide, which is assigned a relative global warming potential of 1. In other words, methane is 23 times as potent a greenhouse gas as carbon dioxide, and nitrous oxide is 296 times as potent as carbon dioxide.

Carbon dioxide is the primary greenhouse gas

Climate change has been a controversial topic, but almost all environmental scientists agree on one point—human activities have contributed to a significant increase in atmospheric concentrations of carbon dioxide. These concentrations have increased by 33%, from around 280 parts per million (ppm) as recently as the late 1700s to 316 ppm in 1959 to 373 ppm in 2002 (Figure 12.2a).

Table 12.1 Global Warming Potentials (CO_2 Equivalents) of Four Greenhouse Gases	
Greenhouse gas	Relative heat-trapping ability
Carbon dioxide	1
Methane	23
Nitrous oxide	296
Hydrochlorofluorocarbons HFC-23	12,000

Data from Intergovernmental Panel on Climate Change, *Climate Change 2001: The Scientific Basis.*

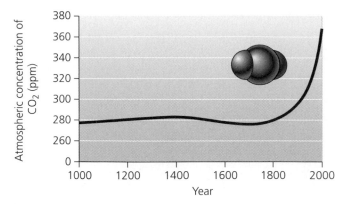

(a) Carbon dioxide

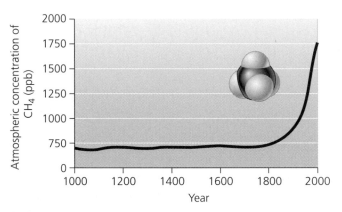

(b) Methane

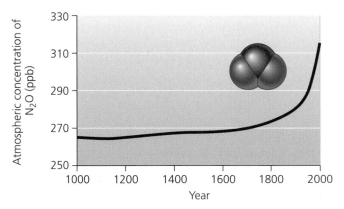

(c) Nitrous oxide

Figure 12.2 Global atmospheric concentrations of **(a)** carbon dioxide (CO_2), **(b)** methane (CH_4), and **(c)** nitrous oxide (N_2O) have increased dramatically since 1800. The increases coincide with the expansion of human industrial activities that produce large quantities of these and other greenhouse gases. Data from the Intergovernmental Panel on Climate Change, 2001.

The 33% increase in atmospheric carbon dioxide concentrations in the last 200 years has brought the concentration of carbon dioxide to its highest level in at least 400,000 years, and likely the highest in the last 20 million years. Moreover, the concentration is increasing faster today than at any time in at least 20,000 years. What has changed since the 1700s to cause the atmospheric concentration of this greenhouse gas to increase so rapidly?

As you may recall from our discussion of the carbon cycle in Chapter 6, and will see further in Chapter 17, a great deal of carbon is stored for long periods in the upper layers of the lithosphere. The deposition, partial decay, and compression of organic matter (mostly plants) that once grew in wetland or marine areas in the Carboniferous Period has led to the formation of the fossil fuels coal, oil, and natural gas in sediments from that time. These processes create carbon reservoirs that would, in the absence of human activity, be practically permanent.

In the past 200 or so years, however, humans have been burning increasing amounts of these fossil fuels in their homes, factories, and automobiles. In the process, we have been transferring large amounts of carbon from one reservoir, the underground deposits that have stored the carbon for millions of years and would have continued to do so, to another, the atmosphere, in the form of carbon dioxide. This movement of carbon from permanent reservoirs into the atmosphere is one reason atmospheric carbon dioxide concentrations have increased so dramatically in such a short time.

At the same time, people have cleared and burned forests to make room for farms, villages, and cities. Forests serve as carbon sinks (Chapter 6), and their removal, especially in areas where they are slow to recover, can reduce the ability of the biosphere to absorb carbon dioxide from the atmosphere. Therefore, deforestation has also contributed to increasing atmospheric carbon dioxide concentrations.

Weighing the Issues:
Greenhouse Gases: Sources and Lifetimes

While concentrations of greenhouse gases derived from fossil fuels continue to increase, so do those of greenhouse gases resulting from agriculture. Agriculture contributes over half of our methane emissions and one-third of our nitrous oxide emissions into the atmosphere. How might increasing emissions of methane and nitrous oxide affect climate in the future? Why are these compounds of great concern? Can you guess the primary agricultural source of nitrous oxide?

Other greenhouse gases add to warming

Carbon dioxide is not the only greenhouse gas increasing in its atmospheric concentration. Methane is also on the rise. We release methane into the atmosphere by tapping into fossil fuel deposits, by raising large herds of cattle that produce methane in their digestive tracts, by disposing of organic matter in landfills, and by growing certain types of crops, especially rice. Atmospheric methane concentrations did not change much between 10,000 years ago and 300 years ago. Since 1750, however, methane concentrations have increased by 151% (Figure 12.2b), and the current concentration is the highest in at least 400,000 years. Human activities cause just more than one-half of all methane emissions today.

Nitrous oxide is another greenhouse gas whose atmospheric concentrations have increased because of human activities. This gas, a by-product of feed lots, chemical manufacturing plants, auto emissions, and modern agricultural practices, has increased by 17% since 1750 (Figure 12.2c). Today, humans are responsible for approximately 33% of total nitrous oxide emissions into the atmosphere (Chapter 6).

Ozone concentrations in the troposphere have increased by 36% since 1750 due to the processes described in Chapter 11. In addition to attacking the respiratory tracts and mucous membranes of humans and other animals and contributing to the formation of photochemical smog, tropospheric ozone is a greenhouse gas.

Another group of greenhouse gases, the **halocarbon gases**, which include the chlorofluorocarbons, or CFCs, add greatly to the atmosphere's heat-absorbing ability (Table 12.1). Their contribution to global warming, however, has begun to slow because of the Montreal Protocol (Chapter 11).

Water vapor is the most abundant greenhouse gas in the atmosphere today. Its concentration in the troposphere varies, but if the concentrations of other greenhouse gases continue to rise, and tropospheric temperatures increase, the oceans and other water bodies may transfer more water vapor into the atmosphere. This potential positive feedback mechanism could amplify the greenhouse effect. Alternatively, increased water vapor concentrations could give rise to increased cloudiness, which could, according to some scientists, slow global warming by reflecting incoming solar radiation back into space.

Aerosols and other elements may exert a cooling effect on the lower atmosphere

While greenhouse gases exert a warming effect, **aerosols,** microscopic droplets and particles, can have either a warming or cooling effect. Generally speaking, soot aerosols, also known as black carbon aerosols, can cause warming, but most tropospheric aerosols lead to climate cooling. Sulfate aerosols produced by fossil fuel combustion may slow global warming, at least in the short term.

When sulfur dioxide enters the atmosphere, it undergoes a number of reactions, and some of these reactions lead to the formation of acid precipitation (Chapter 11). These reactions, along with volcanic eruptions, also contribute to the formation of a sulfur-rich aerosol haze in the upper atmosphere. Such a haze may reduce the amount of sunlight that reaches and heats the surface of Earth. Major volcanic eruptions and the aerosols they release can also exert short-term cooling effects on Earth's climate on the scale of a few years.

Feedback loops complicate climate systems

As we study global climate, it is important to remember the feedback concepts we discussed in Chapter 6. While one particular change may have a direct warming or cooling effect, it may also trigger positive or negative feedback somewhere else in the system.

Positive feedback loops enable relatively small changes in greenhouse gas concentrations to yield large changes in global climate. For example, as global temperature rises, evaporation of seawater and fresh water may increase. As evaporation increases, the moisture content of the atmosphere will increase. Increased moisture content could lead to increased cloud formation, which—depending on whether low- or high-elevation clouds resulted—could either shade and cool Earth (negative feedback) or contribute to warming and accelerate evaporation and further cloud formation (positive feedback). The bottom line is that, because of feedback loops, a relatively minor change in a relatively minor component of the atmosphere can lead to major changes later on.

The atmosphere is not the only factor that influences global climate

So far, we have devoted much attention to the atmosphere's climate-shaping role. It is important to note, however, that the amount of energy released by the sun, as well as changes in Earth's rotation and orbit can also affect climate.

Milankovitch cycles During the 1920s, Serbian mathematician Milutin Milankovitch described three kinds of changes in Earth's rotation and orbit around the sun.

These variations, now known as **Milankovitch cycles**, result in slight changes in the relative amount of solar radiation reaching Earth's surface at different latitudes as shown in Figure 12.3. As these cycles proceed, they change the way solar radiation is distributed over Earth's surface. This, in turn, contributes to changes in atmospheric heating and circulation that have triggered the onset and end of the ice ages and other more subtle climate changes.

Oceanic circulation The oceans also shape climate. For example, the ocean's ability to absorb heat from the atmosphere or from other sources is tremendous. In addition to absorbing heat from the atmosphere, the oceans move energy from one place to another in their currents.

In equatorial regions, such as the area around the Maldives, the oceans receive more heat from the sun and the atmosphere than they emit. At the poles, the oceans emit more heat than they receive. Because cooler water is denser than warmer water, the cooling water at the poles tends to sink, and the warmer surface water from the equator moves over to take its place. This is one of the principles explaining large-scale oceanic circulation that we will examine in Chapter 13.

The best-known interaction between ocean circulation and atmospheric circulation that influences global climate involves the phenomena named El Niño and La Niña (see The Science behind the Story, page 354). These refer to shifts in the temperature of surface waters in the middle latitudes of the Pacific Ocean over periods of 2–7 years. El Niño conditions are triggered when prevailing equatorial winds weaken and allow warm water from the western Pacific to move eastward, eventually preventing cold water from welling up in the eastern Pacific. In La Niña events, cold surface waters extend far westward in the equatorial Pacific. Both these events affect temperature and precipitation patterns around the world in complex ways. Many scientists today are studying whether globally warming air and sea temperatures may be increasing the frequency and strength of El Niño events.

Another possible way in which changing climate could interact with ocean currents involves the northern Atlantic Ocean. Here, warm surface water moves from the equator northward, carrying heat to higher latitudes. The arrival of this heat keeps Europe much warmer than it would otherwise be. As the surface water of the North Atlantic releases heat energy and cools, it becomes more dense and sinks. The deep part of the circulation pattern formed in this way is called the North Atlantic Deep Water (NADW). Figure 12.4

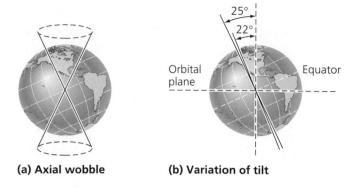

(a) Axial wobble   (b) Variation of tilt

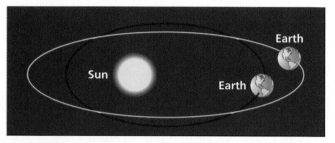

(c) Variation of orbit

Figure 12.3 There are three types of Milankovitch cycles, all of which can influence Earth's climate. (a) The first is an axial wobble that occurs on a 19,000-to-23,000-year cycle. (b) The second is a shift in the tilt of Earth's axis that occurs over a range of 3 degrees on a 41,000-year cycle. (c) The third and longest cycle is a variation in Earth's orbit from almost circular to more elliptical. This cycle repeats itself every 100,000 years.

portrays this circulation pattern as a sort of conveyor belt that moves water and heat from one place to another.

Recently, scientists have realized that an interruption of the NADW formation process could trigger relatively rapid climate change. For example, if global warming caused large portions of the Greenland Ice Sheet to melt, freshwater runoff into the North Atlantic would increase. Surface waters in the North Atlantic could become less dense because of such dilution and warming. (Warm fresh water is much less dense than cold salty water.) This could stop the NADW formation process. Because this process drives a heat pump of sorts that pulls warm, equatorial water northward, a rapid cooling of the North Atlantic region, including much of Europe, might result.

Figure 12.4 The atmosphere and the oceans are closely connected. Warm surface currents carry heat from equatorial waters north toward Europe and Greenland, where they release heat into the atmosphere, then cool and sink into the deep ocean. This is how North Atlantic Deep Water (NADW) is formed. An interruption from melting of the Greenland Ice Sheet could stop the flow of heat from equatorial regions and lead to the rapid cooling of the North Atlantic Ocean and much of Europe.

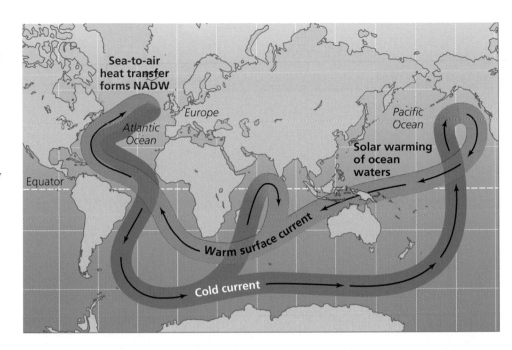

Methods of Studying Climate Change

To understand how climate is changing, scientists must have a good idea of what climatic conditions were like thousands or millions of years ago. Such knowledge of the past is necessary in order to predict climatic changes in the future accurately. Environmental scientists have developed a number of creative methods to decipher clues from the past in order to learn about Earth's climate history.

Geologic records tell us about the past

Earth's ice caps and glaciers hold clues about past climate. Over the ages, these large expanses of snow and ice have accumulated to great depths, preserving within them tiny bubbles of the ancient atmosphere (Figure 12.5). Scientists can extract these trapped air bubbles by drilling into the ice and extracting long columns, or cores. From these **ice cores**, scientists can determine many things, including atmospheric composition, greenhouse gas concentrations, temperature trends, snowfall, solar activity, and even, from layers of trapped soot particles, forest fire frequency.

Besides drilling cores in ice, scientists can also drill cores in beds of sediment beneath bodies of water. These sediments often preserve pollen grains and other remnants from plants that grew in the past, and as we saw with the study of Easter Island in Chapter 1, the analysis

(a) Ice core

(b) Micrograph of ice core

Figure 12.5 (a) At several sites in Greenland and Antarctica, scientists have drilled deep into ancient ice sheets and removed cores of ice like this one, held by Dr. Gerald Holdsworth from the University of Calgary, to extract information about past climates. **(b)** Bubbles (black shapes) trapped in the ice contain small samples of the ancient atmosphere.

Understanding El Niño and La Niña

El Niño was originally the name that Spanish-speaking fishermen gave to an unusually warm surface current that sometimes arrived near the Pacific coast of South America around Christmas time. El Niño literally means "the little boy" or "the Christ child," and the current received this name because of its holiday arrival time. El Niño later became the name reserved for the exceptionally strong warming of the eastern Pacific that occurred every 2 to 7 years and hurt local fish and bird populations by altering the marine food web in the area.

Let's take a look at the science behind El Niño. Under normal conditions, prevailing winds blow from east to west along the equator in the Pacific Ocean. These winds form part of a large-scale convective loop, or atmospheric circulation pattern (a). These winds also push the surface waters westward and cause water to "pile up" in the western Pacific, where some of the warmest ocean water on the planet exists. As a result of these winds, water near Indonesia is, at times, about 50 cm (20 in) higher and approximately 8°C warmer than water along the coast of South America. The prevailing winds that move surface waters westward also make it possible for deep, cold, nutrient-rich water to form an upwelling along the coast of Peru.

When prevailing winds weaken, the warm water that has collected in the western Pacific flows "downhill" and eastward toward South America (b). The presence of warmer surface water and the absence of the normal prevailing winds weaken the upwelling along the coast of Peru. They also change

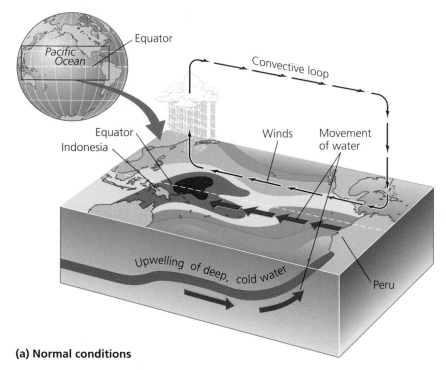

(a) Normal conditions

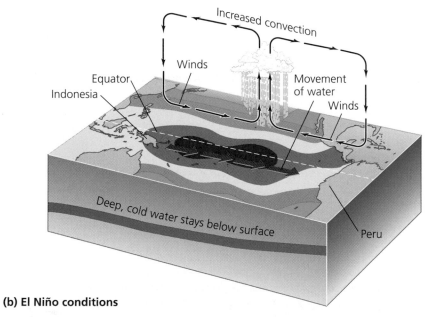

(b) El Niño conditions

In this view, Indonesia is on the left and Peru is on the right. (a) During normal conditions, prevailing winds push warm surface waters toward the western Pacific. In this diagram, red and orange water is warmer than blue and green water. The red and orange waters in the west are also at a higher elevation than those in the east. (b) During El Niño conditions, the prevailing winds have weakened and no longer hold the warm surface waters in the western Pacific. As the warmer water "sloshes" back across the Pacific toward South America, precipitation patterns change.

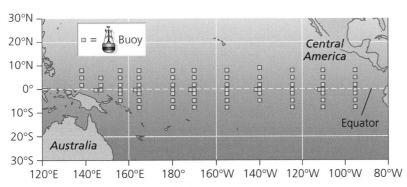

(a) Temperature-sensing buoys: TAO/TRITON array

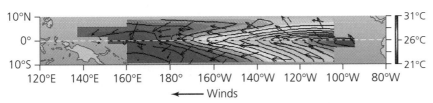

(b) Temperature during normal conditions, December 1993 averages

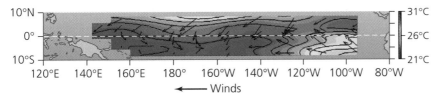

(c) Temperature during El Niño conditions, December 1997 averages

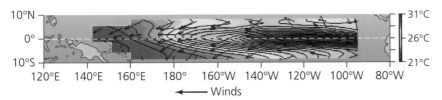

(d) Temperature during La Niña conditions, December 1998 averages

Scientists gather climate data in the Pacific Ocean by using anchored monitoring stations known as the Tropical Atmosphere Ocean Project, or TAO/TRITON, buoys. (a) These buoys are distributed across the Pacific. (b) During normal conditions, cold water reaches the surface of the eastern Pacific via an upwelling, represented by the blue tongue at the right of the graph. (c) During an El Niño event, much warmer water, shown in orange and red, is present in the eastern Pacific and the cold, blue tongue is absent. (d) During a La Niña event, colder-than-normal water is present in the equatorial Pacific Ocean. Data from National Oceanic and Atmospheric Administration, Tropical Atmospheric Ocean Project, 2001.

weather patterns around the world. Because warm surface water can serve as a powerful weather-shaping force by giving rise to heavy precipitation, its movement across the Pacific can create unusually intense rainstorms and floods in areas that are usually quite dry.

Scientists monitor the development of El Niño with an array of wind- and temperature-sensing buoys, known as the Tropical Atmosphere Ocean Project, or TAO/TRITON, that are anchored across the equatorial Pacific (a). Temperature profiles are measured by the buoys during normal conditions (b) and El Niño events (c). Scientists use such data to determine the extent and severity of El Niño events. Awareness of El Niño conditions can enable governments and individuals to better prepare for extreme weather events and changes in ocean conditions.

La Niña, or "the little girl," is the opposite of El Niño. It is characterized by the presence of colder than normal surface water in the equatorial Pacific Ocean (d).

Scientists have made great strides in recent years in understanding El Niño and La Niña. However, a great deal remains to be understood about their causes, their connections to climate change, and their effects on other environmental systems.

of these materials can reveal a great deal about the history of past vegetation. Because the types of plants that grow in an area depend crucially on the area's climate, knowing what plants occurred in a given location in the past can tell us much about what the climate was like at that time (see The Science behind the Story, page 358).

Direct atmospheric sampling tells us about the present

Charles Keeling of the Scripps Institution of Oceanography in La Jolla, California has documented trends in atmospheric carbon dioxide concentrations since 1958. Keeling collects four air samples from five towers every hour from his monitoring station at the Mauna Loa observatory in Hawaii.

Keeling's data show that atmospheric carbon dioxide concentrations have increased from around 315 ppm to 373 ppm since 1958 (Figure 12.6). The seasonal variation in carbon dioxide concentrations that are visible within the overall upward trend are caused by the Northern Hemisphere's vegetation which absorbs much more carbon dioxide during the northern summer when it is more photosynthetically active than it does during the winter.

Coupled general circulation models help us understand climate change

Coupled general circulation models (CGCMs) are computer programs that combine what is known about weather patterns, atmospheric circulation, atmosphere-ocean interactions, and feedback mechanisms to simulate climate change (Figure 12.7). They are called *coupled* models because they are able to couple, or combine, the climate influence of the atmosphere and oceans in a single simulation. This task requires the manipulation of vast amounts of data and complex mathematical equations that were not possible until the advent of the supercomputer.

As of 2001, there were 14 research labs around the world operating CGCMs. The models have been tested to see how closely they simulate known climate patterns when fed the appropriate historical data. In other words, if scientists enter past climate data, and the models produce accurate predictions of current climate, we have reason to believe the models may be realistic and accurate. Some models have produced close matches using this process (Figure 12.8). As computing power increases and our ability to glean data from proxy indicators—pollen, air bubbles from ice cores,

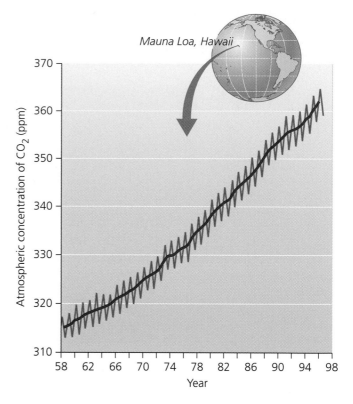

Figure 12.6 Atmospheric concentrations of carbon dioxide have risen steeply since 1958. Go to **GRAPH IT** on the website or CD–ROM. These data were obtained at the Mauna Loa Observatory in Hawaii by the Scripps Institution of Oceanography. Data from National Oceanic and Atmospheric Administration, Climate Prediction Center, 2001.

and other indirect evidence—of past climate improves, the CGCMs are becoming more reliable.

Tests suggest that today's computerized models provide a good approximation of the relative effects of natural and anthropogenic influences on global climate. For example, when CGCMs are used to simulate terrestrial climates since 1850, simulations match the actual meteorological records very closely.

Figure 12.8 shows temperature data from three such simulations. Note that the results in Figure 12.8a are based on natural climate-changing factors alone (such as solar variation and volcanic activity). The results in Figure 12.8b are based on anthropogenic factors only (such as anthropogenic greenhouse gas emissions and sulfate aerosol emissions). The results in Figure 12.8c represent the closest match between predictions and actual climate because they are based on both natural and anthropogenic factors. This is precisely the kind of coupling the CGCMs have been designed to model. Results such as those shown in Figure 12.8c support the reliability and validity of CGCMs and their predictions.

Figure 12.7 Current climate models incorporate many factors, including processes involving the atmosphere, land, oceans, ice, and biosphere, as mathematical equations. Some of these factors are shown here.

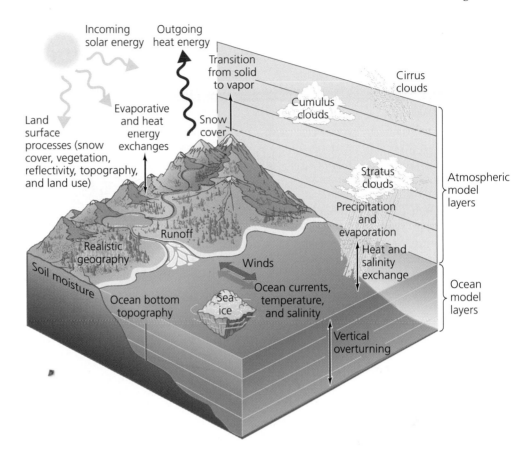

Figure 12.8 Scientists can test their climate models by entering climate data from past years and comparing model predictions (blue areas) with actual observed data (red line). Models that incorporate only natural factors (**a**) or only human-induced factors (**b**) do a mediocre job of predicting real climate trends. Models that incorporate both natural and anthropogenic factors (**c**) are most accurate. Data from the Intergovernmental Panel on Climate Change, 2001.

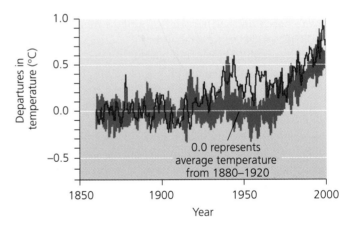

(b) Anthropogenic factors

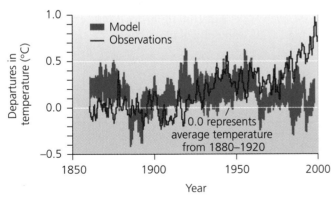

(a) Natural factors

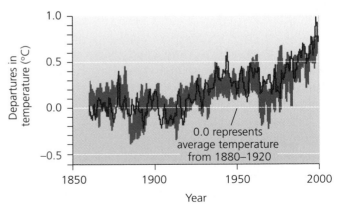

(c) All factors

The Science behind the Story

Scientists Use Pollen to Study Past Climate

Scientists can learn about climate by studying the types of vegetation that grew in particular places in the past. Cores taken from the sediments that collect at the bottoms of lakes and ponds contain pollen and other clues to the plants that lived in those locations at the time each layer was deposited. Tropical plant fossils and pollen tell scientists that warm, wet conditions were present in the past, and fossils of cold-loving plants speak of a cooler climate history.

In 1995, J. S. McLachlan of Harvard University and L. B. Brubaker of the University of Washington published an article about the past climate of Washington state's Olympic Peninsula. The researchers found that as climate and hydrological conditions changed, so did the area's vegetation. Their conclusions were based on the characteristics of pollen, fossils, and charcoal that had accumulated over time on the bottom of a small lake and in a cedar swamp, coupled with regional climate data from other sources. Such sediment deposits serve as a timeline. As scientists dig down through the layers of sediment, they are peering back farther in time.

Sediments deposited on the bottoms of lakes and ponds over thousands of years yield clues about past climates. Pollen and larger plant parts preserved in sediments can tell scientists about past plant communities and, by extension, about past climates. This sediment core sample was taken from Chesapeake Bay in Maryland.

The research team used a long metal cylinder to pull a 1,811-cm (59.4-ft) core of sediment from the middle of Crocker Lake and a 998-cm (32.7-ft) core of sediment from the middle of the cedar swamp. They wrapped the two cores in plastic and aluminum foil and stored them in large freezers at 4°C (39.2°F) to prevent decomposition. Then the team began

removing 1-mm thick slices every 10–30 cm (4–12 in) along the length of the cores and analyzing the pollen and charcoal content of each slice. Pollen is easily transported by the wind and thus can indicate the types of vegetation that occur over a fairly large area. Pollen, therefore, serves as a clue to past regional climates.

The researchers identified between 300 and 1,000 pollen grains to genus and species within each slice. Larger plant fossils, such as those of cones, tree stems, and bark, were also removed and classified. Because these items are not as easily transported as pollen, they are considered to indicate the vegetational and climatological history of smaller areas. Similarly, fine particles of charcoal, which, like pollen, are easily transported through the atmosphere, tell of the history of forest fires on a regional scale, whereas larger pieces suggest fire occurrence in the immediate area.

In combination with other available information, these clues enabled researchers to piece together a record of the general plant and climate history of the Olympic Peninsula.

Climate Change Estimates and Predictions

The most thoroughly reviewed and widely accepted collection of scientific information concerning global climate change is a series of reports issued by the Intergovernmental Panel on Climate Change (IPCC). The IPCC is an international panel of atmospheric scientists, other climate experts, and government officials established by the United Nations Environment Programme (UNEP) and the World Meteorological Organization (WMO) in 1988. Its task is to assess information relevant to ques-

tions of human-induced climate change. In 2001 the IPCC released its Third Assessment Report. This summary of current global trends and probable future trends represents the general consensus of atmospheric scientists around the world.

The IPCC report summarizes evidence of recent changes in global climate

The IPCC report's major conclusions concerning recent climate trends address surface temperature; snow and ice cover; rising sea level and warmer oceans, precipitation patterns and intensity; and effects on wildlife; ecosystems;

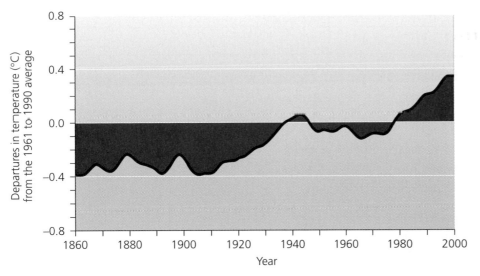

(a) Global temperatures in the past 140 years

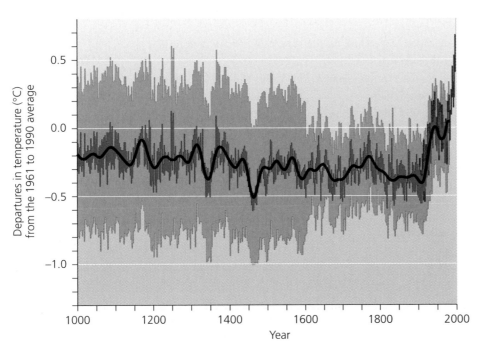

(b) Northern hemisphere temperatures in the past 1,000 years

Figure 12.9 Earth's climate has changed over the past 140 years and over the past 1,000 years. **(a)** Data from thermometers show changes in Earth's average surface temperature since 1860. **(b)** Proxy indicators (blue line) and thermometer data (red line) show average temperature changes in the northern hemisphere over the past 1,000 years. The gray-shaded zone represents the 95% confidence range. Go to **GRAPH IT** on the website or CD–ROM. Data from the Intergovernmental Panel on Climate Change, 2001.

and human societies. The IPCC report is authoritative but, like all science, deals in uncertainties. Its conclusions and predictions are therefore assigned probabilities.

In addition to the data on increases in atmospheric concentrations of greenhouse gases that we discussed earlier, the 2001 IPCC report presented a number of findings on how climate change has already influenced the weather, Earth's physical characteristics and processes, the habits of organisms, and our economies. The report concluded, for instance, that average surface temperatures had increased by 0.6 degrees C (1.0 de-

grees F) during the 20th century (Figure 12.9). It inferred that glaciers, snow cover, and ice caps were melting worldwide. If found that hundreds of species of plants and animals were being forced to shift their geographic ranges and the timing of their life cycles. These findings and many more were based on the assessment of thousands of scientific studies done worldwide over many years, judged by the world's top experts in these fields. Some—but by no means all—of the major findings of the report are shown in Table 12.2.

Table 12.2　Major Findings of the IPCC Third Assessment Report, 2001

Weather indicators

Earth's average surface temperature increased by 0.6 degrees C (1.0 degrees F) during the 20th century (90–99% certainty)

Cold and frost days decreased for nearly all land areas in 20th century (90–99% certainty)

Continental precipitation increased 5–10% over 20th century in the N. hemisphere (90–99% certainty), but decreased in some regions

1990s were the warmest decade of the last 1,000 years (66–90% certainty)

20th-century N. hemisphere temperature increase was the greatest in 1,000 years (66–90% certainty)

Droughts increased in frequency and severity (66–90% certainty)

Nighttime temperatures increased twice as fast as daytime ones (66–90% certainty)

Hot days and heat index increased (66–90% certainty)

Heavy precipitation events increased at northern latitudes (66–90% certainty)

Physical indicators

Average sea level increased 10–20 cm (4–8 in) during 20th century

Rivers and lakes in the N. hemisphere were covered by ice 2 weeks less from beginning to end of century (90–99% certainty)

Arctic sea ice thinned by 10–40% in recent decades, depending on the season (66–90% certainty)

Mountaintop glaciers retreated widely during the century

Global snow cover decreased 10% since satellite observation began in the 1960s (90–99% certainty)

Permafrost thawed, warmed, and degraded in many regions

El Niño events became more frequent, persistent, and intense in the past 40 years in the N. hemisphere

Growing season lengthened 1–4 days per decade in the last 40 years in northern latitudes

Biological indicators

Geographic ranges of many plants, insects, birds, and fish shifted toward the poles and upward in elevation

Plants are flowering earlier, migrating birds are arriving earlier, animals are breeding earlier, and insects are emerging earlier in the N. hemisphere

Coral reefs are experiencing bleaching more frequently

Economic indicators

Global economic losses due to weather events rose roughly ten-fold over the last 40 years, partly due to climate factors

Data from IPCC *Third Assessment Report, 2001.*

Sea-level rise is one of many impacts that climate change will likely bring about

Glaciers are shrinking in many areas around the world, including Glacier Bay National Park and Preserve in Alaska (Figure 12.10). As glaciers shrink, a greater portion of Earth's water enters the liquid phase and becomes available to the oceans and atmosphere. An increased amount of water in the oceans could cause a significant rise in sea level.

Sea level rise will be even greater as the oceans warm along with the rest of Earth. This is due to the fact that water expands as its temperature increases. Most projected sea level rise will result from this thermal expansion, rather than the addition of glacial meltwater to the oceans.

In 1987, unusually high waves struck the Maldives and triggered a campaign to build a large seawall around Male, the nation's capital city. Known as "The Great Wall of Male," the seawall is intended to break up incoming waves and dissipate their energy to prevent the destruction of buildings and roads during storm surges. Storm surges, temporary rises in sea level brought on by the high tides and winds associated with violent storms, would be likely to increase in number and severity as

(a) Glacier Bay National Park and Preserve

(b) Glacier Bay, 1892

Figure 12.10 (a) In Glacier Bay National Park and Preserve in Alaska, ice has retreated more than 105 km (65 mi), (b) since Captain George Vancouver mapped the area in 1794.

global warming progresses. "With a mere one metre rise [in sea level]," Maldives' President Maumoon Abudl Gayoom said in 2001, "a storm surge would be catastrophic, and possibly fatal to the nation."

The Maldives is not the only nation that will suffer from sea-level rise. In fact, among island nations, the Maldives has so far fared better than many others (Table 12.3). The island of Tonga in the South Pacific has been experiencing a long-term sea-level rise of half a centimeter per year. The Galapagos Islands off the Ecuadorian coast saw its sea level rise a reported 5.2 cm (2.2 in) in 2002 alone. The United States will not be immune from loss of its coastline.

The United States has over 153,000 km (95,000 mi) of coastline and more than 8.8 million km² (3.4 million mi²) of marine waters within its territory. Population densities in coastal areas are also relatively high; 53% of the U.S. population lives within the 17% of the nation's land that is in coastal areas.

Figure 12.11a shows the amounts of wetlands and dry lands in the United States that would be inundated by a 51-cm (20-in) sea-level rise. Figure 12.11b shows coastal areas in the United States that are eroding due to sea-level rise and other factors. Figure 12.11c shows the projected effects of sea level rise on one coastal area in particular—the Big Bend area of Florida's Gulf coast.

Table 12.3 Sea-Level Rise in Selected Island Nations

Nation	Long-term sea-level rise (mm)	Average sea-level rise, 2002 (mm/year)
Tonga	4.9	40
Fiji	4.0	2
Japan	3.2	6
French Polynesia	2.5	24
Cook Islands	2.3	12
Galapagos	1.5	52
Tuvalu	0.9	38
Maldives	—	8
Saipan	—	6
Seychelles	—	6
Kiribati	−0.2	35

Data from Worldwatch Institute, *Vital Signs 2003.*

Houston, Texas; New Orleans, Louisiana; and Charleston, South Carolina are among the major cities that are most likely to be affected by rising sea level. Parts of New Orleans protected by seawalls and dikes are already 2.1 m (7 ft) below average sea level. These areas may be as much as 3.3 m (10 ft) below sea level by 2100. Coastal wetlands and mangroves in southern Florida are already being submerged by rising seas, and approximately 2.5 million ha (1 million acres) of Louisiana wetlands have become open water since 1940. Rising sea level has also contributed to increased mortality of trees in coastal areas of Louisiana and southern Florida, where saltwater has intruded into the groundwater on which trees depend. Florida may also lose forests, coastal wetlands, and drinking water sources due to sea-level rise.

The IPCC and other groups project future impacts of climate change

In addition to documenting past changes, IPCC members and others, including a group known as the U.S. Global Change Research Program (USGCRP), have predicted future climate changes and their potential impacts. Congress created the USGCRP in 1990 to coordinate federal climate research.

In 2000, the USGCRP issued a report entitled "Climate Change Impacts on the United States: The Potential Consequences of Climate Variability and Change." This report, based on the work of university researchers, government and industry scientists, and representatives

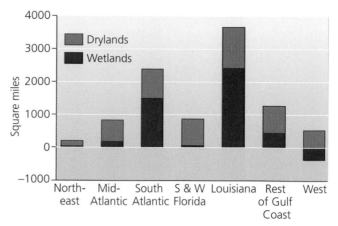

(a) U.S. coastal lands at risk from a 51-cm (20-in) sea-level rise

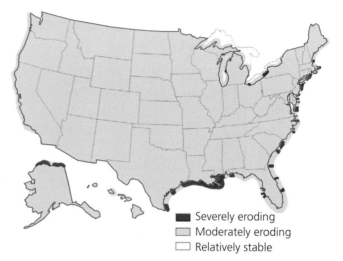

Legend:
- Severely eroding
- Moderately eroding
- Relatively stable

(b) Annual shoreline change

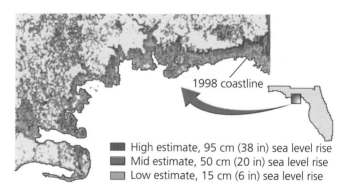

1998 coastline

- High estimate, 95 cm (38 in) sea level rise
- Mid estimate, 50 cm (20 in) sea level rise
- Low estimate, 15 cm (6 in) sea level rise

(c) Changes in Florida's Big Bend

Figure 12.11 The United States will lose coastal areas as a result of a rise in sea level. (**a**) A 51-cm (20-in) sea level rise would inundate wetlands (red) and drylands (purple) on all coasts. (**b**) Coastal areas around the nation are at risk of increased erosion due to sea-level rise. (**c**) In the Big Bend area of Florida, a 15-cm (6-in) sea-level rise would inundate the pale purple areas, a 50-cm (20-in) rise would inundate the lighter and darker purple areas, and a 95-cm (38-in) rise would inundate all three shaded purple zones. Data from U.S. Global Climate Change Research Program, Climate change impacts on the United States: The potential consequences of climate variability and change, 2000.

of non-governmental organizations, highlighted the past and future effects of global climate change on the United States. According to the report, annual average U.S. temperature increased by 0.6°C (1.0°F) during the 20th century. The report also featured a series of predictions for impacts of climate change in the United States (Table 12.4).

Climate change will affect agriculture and forestry

Drought and temperature extremes are among the threats to farms and forests worldwide from climate change. Some croplands already at the limits imposed by heat stress and water availability could be pushed beyond their ability to produce food. If average temperatures increase by more than 3 or 4°C, for example, most tropical and subtropical areas will likely see decreased crop production, and farmlands in mid-latitudes will begin to see significant declines. Agricultural productivity

---

Table 12.4   Predicted Impacts of Climate Change on the United States

- Mountaintop ecosystems will become more endangered.
- Southeastern U.S. forests will break up into savanna/grassland/forest mosaics.
- Northeastern U.S. forests will lose native sugar maples.
- Loss of coastal wetlands and real estate due to sea level rise will continue.
- Snowpack reduction will continue.
- Water shortages will worsen.
- Melting permafrost will undermine Alaskan buildings and roads.
- Skiing and other winter sports will decline.
- Air conditioning use will increase.
- Some crop and forest productivity may increase.
- Average U.S. temperatures will increase 3–5°C (5–9°C) in next 100 years.
- Some ecosystems, including alpine meadows and barrier islands, may disappear altogether while others, such as southeastern U.S. forests, will change dramatically.
- Drought, floods, and snowpack changes will give rise to environmental and social problems.
- Impacts on U.S. food supply are hard to predict. Drought and other factors may present problems while longer growing seasons may benefit some crop producers.
- Forest growth may increase in the short term, but in the long term, drought, pests, and fire may also alter forest ecosystems.
- Greater temperature extremes will cause increased health problems and mortality among human populations. Some tropical diseases will also spread north into temperate latitudes.

---

may increase at higher latitudes because of warmer temperatures, but the overall effect of a warmer climate on total agricultural productivity is difficult to predict. Agricultural productivity might remain somewhat stable globally while increasing in some areas and decreasing in others. This possibility is one reason that some people argue that global warming is a good thing.

Weighing the Issues:
Agriculture in a Warmer World

Some people claim that a warmer, wetter climate in the far North would merely expand potential agricultural lands northward. In fact, the U.S. Corn Belt is already pushing into Canada. Might this lead to greater agricultural production globally? How might such a northward shift of agriculture affect the poorer nations of the lower latitudes that are already suffering from food shortages and agricultural problems? Can you think of any other reasons that a northward shift of agriculture might cause problems?

In addition, the widespread re-growth of forests in eastern North America is one of the primary factors mitigating current U.S. emissions, offsetting an estimated 25% of carbon emissions in the last 40 years. How might a northward shift of agricultural lands affect these carbon sinks in the North and Notheast?

Recent research conducted by the USGCRP indicates that U.S. forests will undergo a transformation due to ongoing climate change. The USGCRP tracks changes in atmospheric composition, ecosystems, the global carbon cycle, and other factors. In its 2000 report, the USGCRP predicted that U.S. forests will become more productive due to a fertilizing effect of additional atmospheric carbon dioxide. As productivity increases, however, forest fire frequency and intensity may increase by 10% or more. Alpine and subalpine forests will become less common as the climate becomes warmer, and these mountaintop forests have nowhere to go. Although some forest types will decline, others, including oak/hickory and oak/pine forests, may expand in the eastern United States (Figure 12.12).

Freshwater ecosystems would also face challenges

In areas where precipitation and stream flow increase, erosion and flooding could threaten the structure and function of aquatic systems. Where agriculture and other human activities have altered the landscape, such flooding could bring increased pollution to freshwater ecosystems.

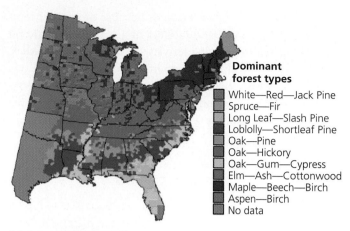

Dominant forest types

- White—Red—Jack Pine
- Spruce—Fir
- Long Leaf—Slash Pine
- Loblolly—Shortleaf Pine
- Oak—Pine
- Oak—Hickory
- Oak—Gum—Cypress
- Elm—Ash—Cottonwood
- Maple—Beech—Birch
- Aspen—Birch
- No data

(a) Current (1960–1990)

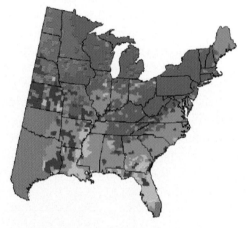

(b) Canadian scenario (2070–2100)

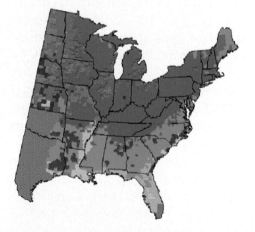

(c) Hadley scenario (2070–2100)

Figure 12.12 While some forest types will decline, others, including oak/hickory and oak/pine forests, may expand in the eastern United States. This figure shows predicted national forest cover changes based on two different modeling scenarios. Data from U.S. Global Climate Change Research Program, Climate change impacts on the United States: The potential consequences of climate variability and change, 2000.

Where precipitation decreases, a different set of problems could arise. The shrinking of lakes, ponds, wetlands, and streams would affect not only the organisms that live, breed, and feed in those habitats but also human health and well-being. In 2001, about 1.7 billion people were living in areas with limited water supplies. By 2025, according to the IPCC, this number will likely increase to 5 billion out of a projected world population of 8.4 billion.

The Maldives is likely to suffer from water-related stresses, because its human population is expanding rapidly at the same time as the rising sea level threatens to bring salt water into the nation's wells. The contamination of groundwater and soils by seawater is called salt water intrusion (Chapter 14). Although this problem occurs in many places, it is particularly threatening to island nations that depend upon small lenses of freshwater that float on saline groundwater.

Marine ecosystems would also be affected

Another possible climate change impact that worries Maldivians is the damage that warming seas, rising sea levels, and storm surges can do to marine ecosystems, including coral reefs. Coral reefs provide habitat for important food fish, reduce wave intensity and protect fragile coastlines from erosion, and provide popular snorkeling and scuba diving sites. Many visitors to the Maldives come expressly to experience the reef environment, and damage to reefs would depress tourism and fishing.

Human health could suffer, or perhaps benefit

Humans could face increased exposure to a wide array of health problems as a result of climate change. These problems include heat stress resulting from the increase in days of high temperatures, humidity, and air pollution; increased global distribution of tropical diseases such as malaria and dengue; diseases and sanitation problems that occur when floods overcome sewage treatment systems; injuries and drowning associated with increased storm frequency and intensity; and hunger-related ailments that become more severe as human population continues to grow and stress on agricultural systems increases.

Figure 12.13a, shows USGCRP projections regarding the 21st century July heat index across the United States. One model predicts that the heat index—a product of temperature and humidity—will be 14°C (25°F) hotter in much of the southeastern United States. Hot weather can lead to high mortality rates in American cities (Figure 12.13b).

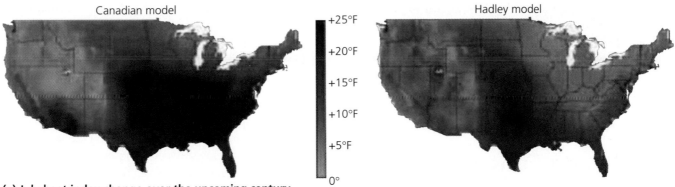

(a) July heat index change over the upcoming century

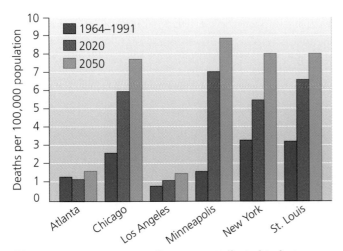

(b) Average summer mortality rates attributed to hot weather episodes

Figure 12.13 (**a**) The Canadian model and the Hadley model are two USGCRP projections of the July heat index across the United States for the upcoming century. The Canadian model predicts that the heat index (a product of temperature and humidity) will be 14°C (25° F) hotter than that of much of the present-day southeastern United States. (**b**) Past and projected future mortality rates attributed to hot weather are shown for various cities across the United States.
Data from National Assessment Synthesis Team, *Climate Change Impacts on the United States: The Potential Consequences of Climate Variability and Change*, U.S. Global Change Research Program, Cambridge University Press, 2000.

Researchers at the Goddard Institute for Space Studies have predicted that potential effects of global climate change on New York City include an increase in instances of heat exhaustion and heat stroke, disease outbreaks, and more widespread irritation of the respiratory system resulting from higher concentrations of air pollution. Figure 12.14 shows the relationship between maximum daily temperature and maximum daily ozone concentrations for New York and Atlanta over a two-year period. Hotter temperatures are generally associated with weaker ground-level air circulation.

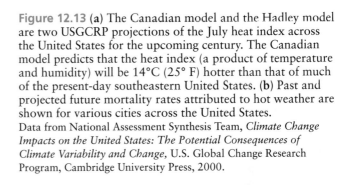

(a) Atlanta, GA

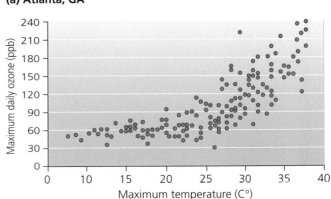

(b) New York, NY

Figure 12.14 This figure shows the relationship between maximum daily temperature and maximum daily ozone concentrations for New York and Atlanta between 1988 and 1990. Hotter temperatures are generally associated with minimal ground-level air circulation and, therefore, higher ground-level concentrations of ground-level ozone, a pollutant that harms human health. Data from National Assessment Synthesis Team, *Climate Change Impacts on the United States: The Potential Consequences of Climate Variability and Change*, U.S. Global Change Research Program, Cambridge University Press, 2000.

At the same time, some scientists, including some IPCC contributors, have argued that a warmer world will present fewer cold-related diseases and injuries. The trade-off between an increase in warm-weather diseases and a decrease in cold-weather health problems is one of the great unknowns of global climate change.

The Debate Concerning Global Climate Change

Nearly all environmental scientists agree that Earth's atmosphere and climate are changing. Despite this consensus, there is a great deal of debate regarding the causes of this change, its extent, and the impact the change will have on human welfare and environmental systems. There is also a great deal of disagreement over how we should respond to these changes.

Scientists agree that climate change is occurring but disagree on some of the details

Almost all scientists have concluded that human activities are altering the atmosphere and that our greenhouse gas emissions are influencing the climate. However, scientists do not agree on the role that clouds, water vapor, particles including soot and sulfate aerosols, and the oceans and NADW formation, play in shaping climate. These uncertainties, coupled with the complex feedback mechanisms on which scientists rely to collect data about climate, make it difficult to predict the future.

Global climate change debates occur in the economic and political arenas as well

Even though scientists agree that the climate is changing and that human activities play a part in this change, policymakers and economists disagree about the appropriate response to these changes. Political and economic disagreements stem from the following questions:

- Would the economic and political costs of reducing greenhouse gas emissions outweigh the costs of unabated emissions and concomitant global climate change?
- Should industrialized nations, developing nations, or both take responsibility for reducing greenhouse gas emissions?

- Should changes to reduce greenhouse gas emissions occur voluntarily or as a result of political and legal pressure or economic sanctions?
- How should we allocate funds and human resources for reducing greenhouse gas emissions and coping with climate change ?

In addition to these questions, some policymakers claim that we need more certainty before we take steps to reduce greenhouse emissions. Others argue that reducing greenhouse gas emissions enough to significantly slow or stop global warming would be more expensive and disruptive than allowing warming to continue and dealing with its impacts. Some citizens and corporations argue against any type of government regulation or guidance that would lead to reductions in greenhouse gas emissions. Still others argue that humans have adapted to significant change in the past and that climate change will simply be another hurdle that we can and will overcome.

Weighing the Issues:
Environmental Refugees

The island citizens of the Maldives need only look southward to find an omen of their future. The Pacific island nation of Tuvalu, which has been losing 9 cm (3.5 in) of elevation per decade to rising seas, may be considered the first casualty of global climate change. Appeals from its 11,000 citizens were heard by New Zealand, which began accepting these environmental refugees in 2003. Do you think this new refugee flow casts doubt on the claims of those who hold that climate change is uncertain? How might the perspective of a Tuvalu resident differ from that of a U.S. oil industry executive? Without political clout, what might individual people of Tuvalu or the Maldives do to save their ways of life?

Technological and Political Methods of Emissions Reduction

Today most scientists are confident that anthropogenic greenhouse gas emissions are changing the composition of Earth's atmosphere. They also conclude that these changes are contributing to global climate change. We don't know the precise extent of our influence on climate, but many people believe that it is time to reduce our emissions. Let us take a look at anthropogenic greenhouse gas emissions in the United States, their sources, and potential means of reducing them.

Since 1990, the generation of electricity, largely through coal combustion, has produced the largest portion (34%, as of 2000) of U.S. emissions (Figure 12.15). Transportation ranks second at 27%; industry is third at 19%; agriculture and residential sources are tied for fourth at around 8% each, and commercial sources account for 5% of total greenhouse emissions in the United States. Let's take a look at some potential reduction methods for electricity generation and transportation, the two largest sources of greenhouse gas emissions. We will evaluate each source from a scientific and technical perspective and from a political and economic perspective as well.

Electricity generation is the largest source of anthropogenic greenhouse gases in the United States

From cooking and heating to the clothes we wear, much of what we own and do depends on electricity. In the United States fossil fuels are the largest source of energy used for electricity generation. In 2000, the United States produced 3.8 trillion kilowatt-hours of electricity. Of this amount, 2.7 trillion kilowatt-hours, or 71%, was generated by fossil fuel combustion. Coal contributed the lion's share of fossil fuel energy (56% of electricity generated by electrical utilities) and most of the greenhouse gas emissions resulting from energy generation.

Obviously, a reduction in the amount of fossil fuels burned to generate electricity could reduce greenhouse gas emissions, as would a decrease in electricity con-

Figure 12.15 Coal-fired electricity-generating power plants, such as this one in Maryland, are the largest contributors to U.S. greenhouse emissions.

sumption. Reduction in electricity consumption could be accomplished through a decline in the number of people using electricity—a population decrease—and restraint in consumption. There are at least two other possible ways to reduce the amount of electricity and, hence fossil fuels, we need.

Electricity conservation and efficiency A solution is to reduce the demand for electricity through increased efficiency and conservation. Efficiency and conservation can arise from new technologies, such as high-efficiency lightbulbs or appliances, or from individual ethical choices to reduce electricity consumption.

In one example of a technological solution, the EPA promotes energy conservation through its Energy Star Program. The Energy Star Program rates household appliances, lights, windows, fans, office equipment, heating and cooling systems, and appliances. It provides this information for homeowners, government agencies, businesses, and churches.

Following are examples of the energy you can save by choosing Energy Star products:

- An Energy Star refrigerator can cut your carbon dioxide emissions by 100 kg (220 lbs) annually.
- An Energy Star washing machine can cut your carbon dioxide emissions by 200 kg (440 lbs) pounds annually.
- Compact fluorescent lights can reduce the energy you use for lighting by 40%.
- Energy Star homes use highly efficient construction, duct work, insulation, heating and cooling systems, and windows to reduce energy use by as much as 30%.

These are only a few of the technological means of increasing energy efficiency.

While technological fixes are popular and profitable for their producers, you can also take a decidedly low-tech approach to energy conservation. In fact, for nearly all of human history, people managed without the electrical appliances that most of us take for granted today. It is entirely possible for each of us to simply choose to use fewer greenhouse gas-producing appliances and technologies or, at the very least, to become familiar with practical steps for using electricity more efficiently.

In the United States and in other industrialized nations, reducing personal consumption of fossil fuels has received much less attention than the development and sale of newer, more efficient technologies. As we discussed in Chapter 3, our societal fixation on gross national product and our neglect of concepts such as

Global Climate Change

A recent issue of the Oil and Gas Journal cited the following regarding global climate change: "More people enjoy health and prosperity now than ever, and a warmer environment richer in carbon dioxide should sustain life better than the current one does." Do you agree? Why or why not?

The Seventh Generation

"In our every deliberation, we must consider the impact of our decisions on the next seven generations." These are the ancient guidelines of the Iroquois Nation. I can think of no finer measure of a society than its legacy. The psychologist Abraham Maslow observed that only after their basic needs are met can humans concentrate on higher values, such as self-actualization, that define their humanity. Global warming is emerging as both a "science" and a "values" issue. Do we want to consume resources with abandon? Do we want to count on technology to gallop to our rescue like a hero in a cowboy melodrama? Conversely, do we want to use our collective intelligence to manage our finite planet and leave to future generations a world as wonderful as the one in which most Americans are fortunate enough to live?

The overwhelming majority of scientists who have looked closely at global warming have concluded that its threat is real, huge, and imminent. Yet it's easy to procrastinate. After all, the biggest threats may be decades in the future, and the earliest and most affected nations are poor and remote. Thus it's no surprise that entrenched industries would rather fight than switch. They're using every trick in the book to obfuscate the issues. They trumpet the alleged pros of warming and downplay the cons.

The next two decades will be a time of decision making. The evidence of global warming and its impact is now strong and will get stronger. We still lack good implementation plans, but these are coming. I believe that by 2020 the United States will be on a pathway to a high-efficiency economy running on renewable fuels and helping the rest of the world to do the same. Through global warming we're learning the practical realities of living aboard "Spaceship Earth."

We're learning that we're an interdependent species capable of destroying our world—as the Easter Islanders destroyed theirs. We're learning that we can only thrive if we work together.

Paul Craig, a physicist by training, is chairman of the Sierra Club's Global Warming and Energy Program.

The Benefits of Global Climate Change

In the rush to judgment that passes for science in the global climate change debate, the potential benefits of global climate change are seldom discussed.

The acknowledgment of the potential benefits of global climate change is critical to the debate, because the most widely prescribed solution to the perceived crisis is a drastic weaning of humanity from an economy based on fossil fuels to one based on non-carbon energy. This transition could wreak worldwide economic devastation the costs of which, in human and environmental terms, could outweigh those of projected global climate change.

Yale University economist William Nordhaus estimates that implementing the Kyoto Protocol on climate change would cost the United States $2.5 trillion over 10 years and the other Kyoto signatories more than $600 billion in that same time frame while making "little progress in slowing global warming. . . ." Surrendering such a massive portion of the developed nations' gross domestic product to a purpose with no significant economic return and little environmental benefit would spawn a global recession for a generation.

What of the benefits of a warmer planet? More of a warmer Earth's surface would be suitable for human-beneficial activities such as farming, aquaculture, and planting forests—aiding the world's ability to feed and house its poor. Although there is mention of the increase in heat-related health effects, what of the correlative decrease in cold-related health effects? Or of the sharply reduced consumption of energy for heating? Weather phenomena are a mixed bag under climate-change scenarios, as loss of some coastal and mountain ecosystems must be weighed against relief in perpetually drought-stricken regions.

There is so much uncertainty over climate change science (solar activity cannot yet be discounted) that it is far better for us first to consider all of the possible effects of climate change—good and bad—before we opt for a possible non-solution that squanders our economic resources and thus our ability to help us adapt to whatever change may come.

Bob Williams is executive editor of the *Oil and Gas Journal* and editor of the *Oil and Gas Journal Latinoamerica*.

the genuine progress indicator, may reflect this bias toward technological solutions rather than lifestyle choices.

Renewable sources of electricity Using electricity more efficiently is one means of reducing greenhouse gas emissions. Another is to develop technologies that generate electricity without use of fossil fuels. Several viable alternatives to fossil fuels exist for the generation of electricity. These include hydroelectric power, which does not produce greenhouse gases but can cause undesirable changes in riverine ecosystems (Chapter 14 and Chapter 18). Hydropower requires the construction of in-stream barriers, which often require concrete, a source of greenhouse gas emissions. Geothermal energy, which uses heat trapped within the earth to turn turbines and produce electricity, is another option. Photovoltaic cells, which turn sunlight into electricity, do not generate greenhouse gases directly, but they do require plastics and metals, which release greenhouse gases during their manufacture. Wind power, which can require large land areas and dependable winds to be economically viable, does not produce greenhouse gases directly, but the metal and other materials required to construct windmills can be a source of greenhouse gases (Figure 12.16). We will examine renewable energy sources in more detail in Chapter 18.

Transportation is the second largest source of greenhouse gases in the United States

Can you imagine life without a car? Most Americans probably can't—a reason why transportation is the second largest source of greenhouse gas emissions in the United States.

In 1970, cars in the United States traveled 1.6 trillion km (1 trillion mi). This number doubled during the 1990s. Approximately one-third of the average American city—including roads, parking lots, garages, and gas stations—is devoted to cars. The average American family makes ten trips by car each day, and governments in the United States spend $200 million per day on road construction and repairs. The current population of registered automobiles in the United States is 220 million, and this number is projected to soon surpass the human population of the country (Figure 12.17).

Obviously, a reduction in the amount of fossil fuels burned to power our vehicles could contribute to a decrease in U.S. greenhouse gas emissions. There are several possible approaches to reduce the amount of fossil fuels we burn in our automobiles.

There are substitutes for our inefficient automobiles

The typical automobile is highly inefficient. Although automobiles promise convenience and time savings, they also waste time and fuel. In fact, close to 85% of the fuel you use does something other than move your car down the road. According to the U.S. Department of Energy, 62% of the fuel energy we put into our cars is spent on overcoming friction within the engine, excess heat, and inefficiencies in the engine. While driving in town, our cars waste 17% of their fuel energy as their engines idle. Each vehicle's basic components, from the water pump to the stereo, consume another 2% of the fuel energy. Drive train friction and inefficiencies claim another 5% and only about 13–14% of the fuel energy actually

Figure 12.16 Low-carbon or carbon-free energy sources include wind power. While alternative energy sources such as wind turbines produce few or no greenhouse gases, they do have other environmental impacts.

Figure 12.17 Traffic such as this is a fact of life for many Americans and for the citizens of many other nations. It is also the source of a great deal of our transportation-related greenhouse gas emissions.

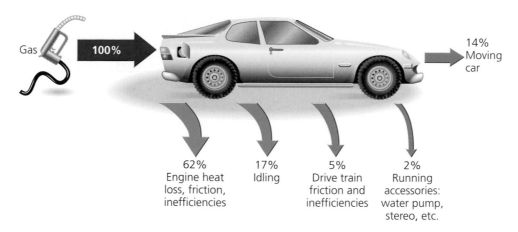

Figure 12.18 Conventional automobiles are extremely inefficient. Almost 85% of useful energy is lost, whereas only 14% actually moves the car down the road.

moves the vehicle and its occupants from point A to point B (Figure 12.18). Although more aerodynamic designs, increased engine efficiency, proper maintenance, and improved tire design may help to reduce these losses, automobiles may always be somewhat inefficient.

Because of these inefficiencies, the high costs of automobile ownership, and concerns regarding traffic and the environmental impacts of automobiles, many people are making lifestyle choices that reduce their reliance on cars. For example, it is possible to choose to live near your place of employment and to make greater use of public transportation or human-powered transportation such as bicycling and walking (Figure 12.19).

Figure 12.19 Sometimes the most effective solutions are the simplest. By choosing human-powered transportation methods, such as bicycles, we can greatly reduce our individual transportation-related greenhouse gas emissions. An increasing number of people are choosing to live closer to their workplaces and to enjoy the dual benefits of exercise and reduced gas emissions by walking or cycling to work or school.

Making automobile-based cities and suburbs more friendly to pedestrian and bicycle traffic, and improving people's access to public transportation, stand as major challenges for progressive city and regional planners throughout the United States and the world (Chapter 16).

Advancing technology is also making possible a number of alternatives to the traditional automobile. These include hybrid vehicles that combine electric motors and gasoline-powered engines for greater efficiency; fully electric vehicles; alternative fuels such as biodiesel and compressed natural gas; and hydrogen fuel cells that use oxygen and hydrogen, producing only water as a waste product. We will examine these and other alternatives in our discussion of renewable energy approaches in Chapter 18.

Public policy and personal transportation choices can have a great impact

While improving the efficiency of automobiles is one way to reduce greenhouse gas emissions, increasing our reliance on mass transit and public transportation is another. In a 2002 study, the American Public Transportation Association concluded that increasing use of public transportation is the single most effective strategy for conserving energy and reducing environmental pollutants. According to the Association, if U.S. residents increased their use of buses, trains, and other forms of public transport to the level these are used by Canadians (7% of daily travel needs) or Europeans (10% of daily travel needs), enormous amounts of energy and greenhouse emissions would be saved (Table 12.5). These savings, the study indicated, could drastically cut down on air pollution, dependence on imported oil, and the

Table 12.5 Energy and Emissions the United States Could Save with Public Transportation

If U.S. residents used public transportation at the same rate as Canadians, the United States could . . .	If U.S. residents used public transportation at the same rate as Europeans, the United States could . . .
Reduce dependence on imported oil by one-half of the amount imported from Saudi Arabia	Reduce dependence on imported oil by >40%, nearly the amount imported from Saudi Arabia
Save nearly the amount of energy used by the U.S. petrochemical industry	Save more energy than the amount used by the U.S. petrochemical industry, and nearly the amount used to produce all U.S. food
Reduce CO_2 emissions by an amount equal to nearly 20% of CO_2 emitted from residential sources or >20% of CO_2 emitted by commercial sources	Reduce CO_2 emissions by >25% of those mandated under the Kyoto Protocol
Reduce CO pollution by twice the combined level emitted by high polluting industries (chemical manufacturing, oil and gas production, metals processing, and industrial coal use)	Reduce CO pollution by three times the combined level emitted by four highest polluting industries (chemical manufacturing, oil and gas production, metals processing, and industrial coal use)
Reduce NOx emissions by 25% of the combined level emitted by the four industries cited above, and cut VOC pollution by almost 60% of the combined level emitted by these industries	Reduce NOx emissions by 35% of the combined level emitted by the four industries cited above, and cut VOC pollution by 84% of the combined level emitted by these industries

Source: American Public Transportation Association, 2002

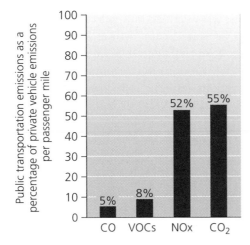

Figure 12.20 Public transportation reduces almost all types of air pollution. Carbon monoxide (CO); Volatile organic compounds (VOCs); Nitrous oxides (NOx); Carbon dioxide (CO_2). Data from Robert J. Shapiro, Kevin A. Hassett, and Frank S. Arnold, Conserving energy and preserving the environment: The role of public transportation, American Public Transit Association, July 2002.

nation's contribution to global climate change (Figure 12.20). Although the American Public Transportation Association findings indicate that there is still a great deal of room for improvement, they also indicate that public transportation is already reducing fossil fuel consumption by 855 million gallons of gas (45 million barrels of oil) each year. Further, mass transit ridership increased by 4.5% between 1998 and 1999 while private vehicle travel increased by just 2%. From 2001 to 2002

bus ridership increased 31%, and in New York City, mass transit participation increased by 7%. At the same time, reliable and convenient public transit is not available in many communities in the United States; this represents a challenge and an opportunity for future solutions to traffic and pollution problems.

In an interesting intersection of transportation policy and economics, it is becoming increasingly common for entities such as universities to fund transit in exchange for unlimited free transit use by students. For example, in 2001 The Pennsylvania State University provided $1 million to the local transit agency in exchange for the elimination of a 40-cent fare on all routes between the university and the city's downtown area. This increased bus ridership by 250% in one year. This type of subsidy is common in mass transit systems and is also the basis of complaints by some mass transit critics. It is common to hear that if mass transit can't pay for itself, it shouldn't receive taxpayer support in the form of subsidies or other funding mechanisms. What such critics frequently forget is that private automobile use and our road networks are heavily subsidized as well.

Beyond technology and personal choice, international climate treaties also play a role

In 1992, the United Nations convened the United Nations Conference on Environment and Development Earth Summit in Rio de Janeiro. Nations represented at the Earth Summit signed five documents including the Framework Convention on Climate Change (FCCC). The FCCC outlined a plan for reducing greenhouse gas emissions to 1990 levels by the year 2000 through a voluntary, nation-by-nation approach.

By the late 1990s, it was apparent that a voluntary approach to slowing greenhouse gas emissions was not likely to succeed. Between 1990 and 2000, for example, U.S. greenhouse gas emissions (in CO_2 equivalents) increased by 12.4%.

However, from 1990 to 1998, Germany demonstrated that economic vitality does not require an ever-increasing rate of greenhouse gas emission. Although Germany has the third most technologically advanced economy in the world and is a leading producer of iron, steel, coal, chemicals, automobiles, machine tools, electronics, textiles, and other goods, it still has managed to reduce its carbon dioxide emissions by 16%. In the same period, the United Kingdom cut its emissions by 8.9%, and the Russian Federation cut its emissions by 58%, although much of that reduction was due to the economic collapse that followed the breakup of the Soviet Union.

After watching the seas rise and observing the refusal of most industrialized nations to cut their emissions, nations of the developing world—the Maldives among them—initiated an effort to create a binding international treaty that, if ratified, would *require* all signatory nations to reduce their greenhouse gas emissions. This effort led to the development of the Kyoto Protocol.

The United States has resisted the Kyoto Protocol

The **Kyoto Protocol** was an outgrowth of the FCCC. Drafted in 1997 in Kyoto, Japan, it was intended to become a binding agreement that would, by 2012, reduce emissions of six greenhouse gases to levels lower than those of 1990 (Table 12.6). However, some industrialized nations, including the United States, have called the treaty unfair because it requires industrialized nations such as the United States to reduce emissions but does not require the same of developing nations, such as China and India. The justification that proponents of the Kyoto Protocol offer for the different requirements is that the developed world created the current problem and therefore should make the sacrifices necessary to solve it. As this book went to press in late 2003, the Kyoto Protocol had been ratified by 111 nations and was awaiting only the ratification of Russia to take effect.

Kyoto Protocol critics and supporters alike acknowledge that even if every nation complied with the limits

Table 12.6 Reductions Required by Kyoto Protocol

Country	Required change 1990–2010	Observed change 1990–1999
United Kingdom	−12.5%	−14.5%
Germany	−21.0%	−18.9%
France	0	−2.2%
Italy*	−6.5%	+5.4%
Canada	−6.0%	+15.0%
Japan	−6.0%	+11.2%
United States	−7.0%	+11.7%
Russian Federation	0	−38.5%
Ukraine[†]	0	−55.0%

*2000 data
[†]1998 data (most recent available)
Data from U.N. Framework Convention on Climate Change Table of National Communications, 1999.

established in this treaty, greenhouse gas emissions would continue to increase—albeit more slowly than they would in the absence of the treaty.

Because resource use and per-capita carbon dioxide emissions are far greater in the industrialized world, governments and businesses in developed nations often feel they have more to lose, economically, from mandatory restrictions on emissions. Ironically, this fear neglects the equally likely probability that developed nations are the ones most likely to gain economically in such a situation, by being best-positioned to invent and develop new technologies that could power the world in a post-fossil fuel era.

The Maldives, which gained its independence from the United Kingdom in 1965, shows characteristics of both "developed" and "developing" nations. On one hand, it has an adult literacy rate of 95.7% and a life expectancy at birth of 64.5 years. It has also put into law significant environmental protections, such as those established for the protection of its coral reefs and fishing resources. On the other hand, the Maldives' annual population growth rate is 3%, making it one of the fastest-growing nations on Earth. The country has no fossil fuel reserves, and uses only about 2,000 barrels of oil per day—less than 0.1% of the world's total energy consumption. Per-capita energy use there is also quite low. In 1999, the average U.S resident used 23.5 times as much energy as the average resident of the Maldives. Thus, like most of the developing world, the Maldives is apt to suffer more than its fair share of consequences from the greenhouse gas emissions of the industrialized world.

Some feel climate change demands the precautionary principle

With regard to global climate change, as with many other important environmental issues, we may never be entirely certain of the precise outcomes of our actions until after they have occurred, and perhaps not even then. With this uncertainty in mind, the drafters of the 1992 Rio Declaration included a passage invoking the precautionary principle (Chapter 9):

> In order to protect the environment, the precautionary approach shall be widely applied by the States according to their capabilities. Where there are threats of serious or irreversible damage, lack of full scientific certainty shall not be used as reason for postponing cost-effective measures to prevent environmental degradation.

That is, advocates of the precautionary principle assert that if a threat is reasonably suspected, we should take precautionary action without waiting for scientific certainty regarding cause and effect. In simplified form, the precautionary principle holds that scientific uncertainty plus suspected harm should yield a precautionary approach.

Weighing the Issues:
The Precautionary Principle

Those who criticize taking a precautionary approach with climate change say that precaution will impede economic growth and innovation. Advocates of the precautionary principle say the stakes are too high to gamble with climate. What do you think? Is the precautionary approach an appropriate guideline given the uncertainty and potential risk? What role should economics play in the discussion?

Conclusion

We have seen that many factors, including human activities, can shape atmospheric composition and global climate. We have also seen that scientists and policymakers are beginning to understand anthropogenic climate change and its environmental impacts more fully. While some people draw optimistic conclusions about the overall benefits and drawbacks of a warmer, more carbon dioxide–rich environment, many scientists and policymakers are deeply concerned. As time passes, fewer and fewer experts are arguing that the changes will be minor. Global climate change and sea level rise will affect far-flung places such as the Maldives, but will also change populated mainland areas such as coastal Florida (Figure 12.21). By becoming familiar with climate-related problems and potential solutions, you will become better equipped to interpret the many messages you will receive regarding global climate change in the coming years.

(a) Maldives

(b) Florida coast

Figure 12.21 Whether you believe that global climate change will ultimately benefit or harm humanity, there is little doubt that it will affect almost everyone in the world either directly or indirectly. Places as far apart as the Maldives and the Big Bend area of Florida will experience changes as humans continue to exert an influence on Earth's climate systems.

REVIEW QUESTIONS

1. What type of solar radiation reaches Earth's surface? How do greenhouse gases warm the lower atmosphere?
2. See Table 12.1. How is global warming potential determined?
3. Why is water vapor considered a greenhouse gas? How can an increase of water vapor create a positive or negative feedback effect, or feedback loop?
4. How could sulfate aerosols slow global warming in the short-term?
5. See Figure 12.3. What are Milankovitch cycles?
6. Why do oceans tend to rise in equatorial regions? How does this explain why Europe is warmer than it could be?
7. How do scientists study the ancient atmosphere?
8. Has simulating climate change with computer programs been effective in helping us predict climate? How do these programs work?
9. How can rising sea levels, caused by global warming, create problems for humans?
10. What may be the effects of a warmer climate on agriculture?
11. How does recent research predict change in forest distribution as a result of warmer climate?
12. How may rising sea levels affect marine ecosystems?
13. What are some of the negative impacts of warmer climate on human health? Are there positive effects?
14. Do all scientists agree that climate is indeed changing? On what counts do they currently disagree?
15. Why is electricity generation the largest source of greenhouse gas emissions in the United States? In what ways can we deal with these emissions?
16. In the United States, transportation is the second largest cause of greenhouse gas emissions. How much fuel energy is actually used to power our automobiles? What are some ways that greenhouse gas emissions from transportation can be reduced?
17. What did the Pennsylvania State University do to increase bus ridership?
18. How have developing nations led the way in developing treaties to regulate climate change and reduce greenhouse gas emissions? Why are these nations more likely to suffer the effects of climate change?

DISCUSSION QUESTIONS

1. As you have seen in many places in this book, people draw dramatically different conclusions about the significance and implications of environmental change. Global climate change and sea level rise are no exception. What are some questions that you might ask of the editors of *The Oil and Gas Journal* and of representatives of the Maldives' government to better understand their positions and their reasons for holding them?
2. Some people argue that it is appropriate, and even helpful, for former fossil fuel company executives to serve in high-level government positions where they can make decisions regarding energy policy, carbon dioxide emissions, and global climate change. Others say this presents a potential conflict of interest and should be discouraged. List some of the arguments justifying each side. What is your opinion?
3. Today, many people argue that "we need more proof," or "better science" before we make large changes in the way we live our lives. How much "science," or certainty do you think we need before we have enough to guide our decisions regarding climate change? How much certainty do you need in your own life before you make a change? Should nations and elected officials follow a different standard? Do you believe that the precautionary principle would be an adequate standard in the case of global climate change?
4. As we have seen in many cases already in this book, it is often impossible to separate environmental issues from political issues. For example, the people of the Maldives, concerned that their home and families are at risk due to rising sea levels, are relying on their political leaders to orchestrate an effective solution. Ismail Shafeeu, the Maldives' Minister

of Environment, believes the solution will require political leadership at the international level on the part of the United States.

We have to strive to point out the problems that are associated with climate change and the responsibilities we feel lie with countries such as the U.S. [because the developed countries burn most of the fossil fuel], particularly the U.S. as a global leader. We would like to see the U.S. take a more constructive approach to these problems than the one they are presently taking. . . . The general impression is that the government of the United States has this sense that it's surrounded by a lot of countries trying to destroy its economy, but I don't think any country is asking of the United States something that they have not offered to do themselves.

Do you agree with Shafeeu's claims? Why or why not?

Media Resources *For further review go to the website* **www.envscienceplace.com** *or student CD-ROM, where you will find quizzes, flashcards, a glossary, additional interactive exercises, and links to relevant news and research sources. Also, on the website and CD-ROM is* **GRAPH IT**, *a series of interactive graphing tutorials to help you interpret graphs and plot data.*

13 The oceans: Natural systems, human use, and marine conservation

Florida Keys

This chapter will help you understand:

- Physical, geological, chemical, and biological aspects of the marine environment

- Ocean currents

- Major types of marine ecosystems

- Historic and current human uses of marine resources

- Human impacts on marine environments

- The current state of ocean fisheries and reasons for their decline

- Marine protected areas and reserves as innovative solutions

- The scientific basis for claims that marine reserves increase fish populations

North
America

Atlantic
Ocean

Florida Keys

Florida Keys National Marine Sanctuary

Central Case: Seeding the Seas with Marine Reserves

"I saw the fisheries spiral down. There's less fish and they're smaller. It's not anything like it was 15 years ago."
— *Don DeMaria, commercial fisherman, Florida Keys*

"We have no water-quality problems. We have no problems with our reef. The sanctuary was shoved down our throats."
—*Bettye Chaplin, Florida Keys realtor and co-founder of the Conch Coalition*

Stretching southwest from the southern tip of Florida for 320 km (200 mi), the string of islands known as the Florida Keys hosts some of North America's richest marine ecosystems. With the world's third-largest barrier coral reef, as well as sea grass meadows, coastal mangrove forests, and estuaries, the Keys draw 3 million tourists each year, many of whom come to enjoy its remarkable marine environment.

As human visitation increased, however, so did human impact. Overfishing, trash and sewage dumping, boat groundings, and careless anchoring were damaging the region's ecosystems while its waters received in-

puts of pesticides, oil, and heavy metals from roads, residential areas, and farms on the islands. Runoff rich in sediments, fertilizer, and nutrients from leaky septic systems were damaging sea grass beds and causing plankton blooms and eutrophication in Florida Bay, between the Keys and the mainland. All these impacts were harming the corals, those living animals whose skeletons make up the structure of coral reefs, which are home to so much biodiversity. From 1966–2000 two-thirds of the area's monitoring sites documented declines in the diversity and area of living corals.

These impacts on water quality, sea grass, and coral reefs in turn depressed fish stocks. Since the 1970s scientists and fishers alike had reported that the Keys' fish and lobster populations were shrinking and that the remaining fish and lobster were smaller in size. Scientists monitoring the Keys' fish populations concluded that reef fish were severely overexploited by commercial and recreational fishing; a report by sanctuary scientists held that 13 of 16 grouper populations, 7 of 13 snapper populations, and 2 of 5 grunt populations had been overfished.

After three ships ran aground on the coral reefs in less than 3 weeks in 1989, Congress established the Florida Keys National Marine Sanctuary (Figure 13.1) to protect the rare and endangered natural and cultural resources that many marine scientists and visitors to the area consider resources of worldwide significance. This sanctuary expands on several previous smaller protected areas and today protects 9,800 km² (3,800 mi²) of marine habitat, including 530 km² (205 mi²) of coral reef and over 6,000 plant, fish, and invertebrate species. Oil exploration, mining, waste-dumping, and large ships are banned. Fishing is allowed throughout most of the sanctuary, but 24 smaller areas recently zoned as reserves protect 65% of the sanctuary's shallow reef habitat and prohibit all harvesting of natural resources, including recreational and commercial fishing.

Although these 24 reserves amount to only 6% of the sanctuary's total area, they have been a magnet for controversy. Two years after the sanctuary was established, a group of Florida Keys residents formed the Conch Coalition to protest the sanctuary and the possibility of no-fishing reserves within it. A conch ("conk") is a marine snail that is a popular food item in the Keys. The Conch Coalition, which eventually claimed 3,000 supporters, publicly criticized the proposal and brought a lawsuit. Conch Coalition supporters hanged and burned in effigy the sanctuary superintendent and another sanctuary proponent during a protest. Today, however, most Keys residents have come to support the sanctuary and its no-fishing reserves as fish populations have begun to increase.

Sanctuaries and reserves are each types of **marine protected areas,** a term used to describe any portion of ocean that is protected from some human activities but may be open to others. Whereas national parks and other types of protected areas (Chapter 15 and Chapter 16) have been established on land for over a century, the oceanic equivalent is quite new. "Most people think it's common sense on land to have areas where we don't hunt," James Bohnsack, a Florida-based National Marine Fisheries Service researcher, explained in 1999.

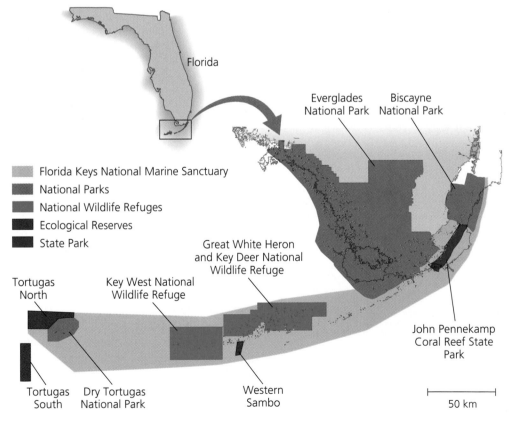

Figure 13.1 The Florida Keys National Marine Sanctuary extends over 9,800 km² (3,800 mi²) of mangrove islands, coral reefs, and shallow waters, from the southeastern tip of Florida westward to the distant keys called *The Tortugas.* The sanctuary is adjacent to or encompasses three national parks, one state park, two national wildlife refuges, and three marine ecological reserves.

"We're now trying to create natural water areas—to see the buffalo roam, so to speak. It's a major change of thinking that protects the ecosystem and biodiversity, but it also protects the fishery."

Oceanography

The fact that people established marine protected areas long after protected areas were widespread on land reflects the fact that humans are terrestrial animals. We—scientists and laypeople alike—have always been more aware of our terrestrial surroundings than of the marine world. Yet the oceans cover the vast majority of our planet's surface, and understanding them is crucial for understanding how our planet's systems work. The oceans influence global climate, teem with biodiversity, facilitate transportation and commerce, and provide us numerous resources. Even landlocked areas far from salt water are affected by the oceans and how we interact with them. The oceans provide fish for people to eat in Iowa, supply oil to power cars and tractors in Saskatchewan, and, by influencing global climate patterns, affect the weather in Nebraska. The study of the physics, chemistry, biology, and geology of the oceans is called **oceanography.** We begin this chapter by examining the structure of the oceans and the way they work, including the marine and coastal ecosystems that are important in the lives of so many people in the Florida

Keys and elsewhere. As we proceed, we will investigate how humans have interacted with the oceans and their resources. We will conclude by examining the science behind efforts in the Florida Keys and elsewhere to establish marine reserves to safeguard the biodiversity and ecosystem functioning of the oceans, as well as the future of fishing economies.

The oceans cover most of Earth's surface

Although we generally speak of the world's oceans (Figure 13.2) in the plural, giving each major basin a name—Pacific, Atlantic, Indian, Arctic, and Antarctic—all these oceans are connected, comprising a single vast body of water. This one "world ocean" covers 71% of our planet's surface and contains 97.2% of its surface water. The oceans largely define the hydrosphere, help shape the atmosphere, interact with the lithosphere, and encompass a large portion of the biosphere, including at least 250,000 species. The world's oceans touch and are touched by every environmental system and every human endeavor.

The oceans contain more than water

Ocean water contains approximately 96.5% H_2O by mass; most of the remaining mass is attributed to dissolved salts (Figure 13.3). Ocean water is salty because the ocean basins are the final repository for water that runs off the land. The water that flows from the continents in rivers carries sediment and dissolved minerals

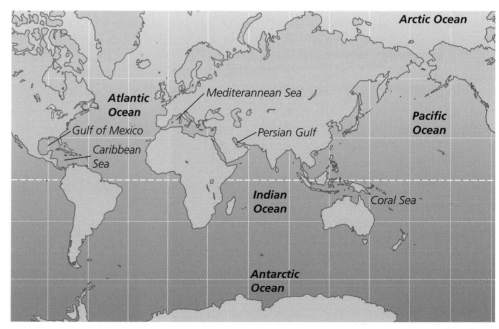

Figure 13.2 The world's oceans are all connected in a single vast body of water but are given different names for convenience. The Pacific Ocean is the largest and, like the Atlantic and Indian Oceans, includes both tropical and temperate waters. The smaller Arctic and Antarctic Oceans include the waters in the north and south polar regions. Many smaller bodies of water are named as seas or gulfs; a selected few are shown here.

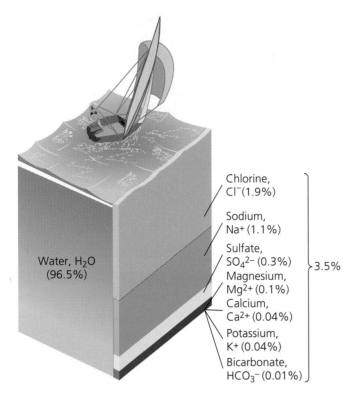

Figure 13.3 Ocean waters consist of 3.5% salts, by mass. Most of this salt is NaCl in solution, making sodium and chloride ions abundant. A number of other ions and trace elements are also present.

Labels in figure: Water, H_2O (96.5%); Chlorine, Cl^- (1.9%); Sodium, Na^+ (1.1%); Sulfate, SO_4^{2-} (0.3%); Magnesium, Mg^{2+} (0.1%); Calcium, Ca^{2+} (0.04%); Potassium, K^+ (0.04%); Bicarbonate, HCO_3^- (0.01%); 3.5%

into the ocean, as do winds. Evaporation from the ocean's surface then removes pure water from the oceans, leaving a high concentration of the salts that had been carried there. If we were hypothetically able to evaporate all the water from the oceans, the world's ocean basins would be completely covered with a layer of dried salt 63 m (207 ft) thick.

The salinity of ocean water varies from place to place because of differences in evaporation, precipitation, and freshwater runoff from land and glaciers, but it generally ranges from 3.3 to 3.7%. Salinity near the equator is lower because of greater precipitation: Rising warm, wet air produces precipitation when it cools (Chapter 11), and because precipitation is relatively salt-free, surface waters in this region are slightly less salty than average. At latitudes roughly 30–35 degrees north and south, as a result of the same process that creates desert climates, evaporation exceeds precipitation, making surface salinity of the oceans slightly higher than average.

Another factor in ocean chemistry is dissolved gas content, particularly the dissolved oxygen on which gill-breathing marine animals depend. Oxygen concentrations are highest in the upper layer of the ocean, reaching

13 ml/L of water. On average, 36% of the gas dissolved in seawater is oxygen, produced by photosynthetic planktonic plants and bacteria and also by diffusion from the atmosphere.

Besides the dissolved minerals shown in Figure 13.3, nitrogen and phosphorus occur in seawater in trace amounts (well under 1 part per million) and play essential roles in the nutrient cycling of marine ecosystems.

Ocean water is vertically structured

Surface waters in tropical regions receive more solar radiation and are therefore warmer; in polar regions, surface water is coldest. In all regions, however, temperature declines with depth (Figure 13.4a). Water density increases as salinity increases and as temperature decreases. These differences give rise to different layers of water; heavier (colder and saltier) water sinks, and lighter (warmer and less salty) water remains nearer the surface (Figure 13.4b). Waters of the surface zone are heated by sunlight each day and are stirred by wind such that they are of similar density throughout, down to a depth of approximately 150 m (490 ft). Below the zone of surface water lies the pycnocline, or thermocline, a region in which density increases and temperature decreases with depth. The pycnocline contains about 18% of ocean water by volume, compared to the surface zone's 2%. The remaining 80% resides in the deep zone beneath the pycnocline. The dense water in this zone is sluggish and not affected by winds and storms, sunlight, and daily temperature fluctuations.

Despite the daily heating and cooling of surface waters, ocean temperatures, even on the surface, are much more stable than temperatures on land. Mid-latitude oceans experience yearly temperature variation of only around 10°C (50°F), and tropical and polar oceans are yet more stable. The reason is that water has a very high heat capacity, a measure of the heat required to increase temperature by a given amount. It takes much more heat energy to increase the temperature of water than it does to increase the temperature of air by the same amount. High heat capacity enables the oceans to absorb a tremendous amount of heat from the atmosphere. In fact, the heat content of the entire atmosphere is equal to that of just the top 2.6 m (8.5 ft) of the oceans. By absorbing heat and later releasing it to the atmosphere, the oceans help shape Earth's climate (Chapter 12). Also shaping climate is the ocean's surface circulation, a system of currents that move in the pycnocline and the surface zone.

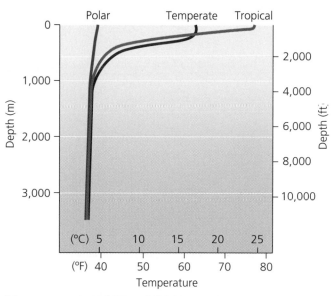

(a) Temperature profiles at polar, temperate, and tropical latitudes

Figure 13.4 Ocean water differs in temperature and density with depth. Water temperatures (**a**) near the surface are warmer due to daily heating by the sun and become rapidly colder with depth over the top 1,000 m (3,300 ft). This temperature differential is greatest in the tropics because of intense solar heating and is least in the polar regions. Deep water at all latitudes is equivalent in temperature. Density decreases unevenly with depth, giving rise to several distinct zones (**b**). Waters of the surface zone are well-mixed and roughly equivalent in density, while density decreases with depth in the pycnocline. Waters of the deep zone resist mixing and are largely unaffected by sunlight, winds, and storms.

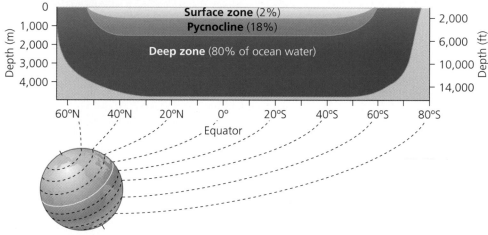

(b) Variation in the average depth of density zones with latitude

Ocean water flows horizontally in currents

Far from being a static pool of water, Earth's ocean is composed of vast river-like flows driven by density differences, sunlight, and wind. Surface **currents** move in the upper 400 m (1,300 ft) of water, horizontally and sometimes for great distances. Wind, heating and cooling, gravity, and the Coriolis effect (Chapter 11) together drive the global system of ocean currents (Figure 13.5). These long-lasting patterns influence global climate and play key roles in El Niño and La Niña events (Chapter 12). They also have been crucial in navigation and human history; seafarers through the ages have lived and died by currents, and they carried Polynesians to Easter Island, Darwin to the Galapagos, and Euro-

peans to the New World. Currents may transport heat, nutrients, or pollution and also may carry the larvae of many marine species, an important consideration in the use of marine reserves to restore depleted populations. Some currents are very slow; others are rapid and powerful. The Gulf Stream is one of the more powerful currents. From the Gulf of Mexico it moves eastward along the southern edge of the Florida Keys, then northward past Miami at a rate of 160 km per day (nearly 2 m/sec, or over 4.1 mi/hr). An average width of 70 km (43 mi) across, the Gulf Stream continues across the North Atlantic to Europe, bringing warm water from the south and moderating Europe's climate, which otherwise would be much colder.

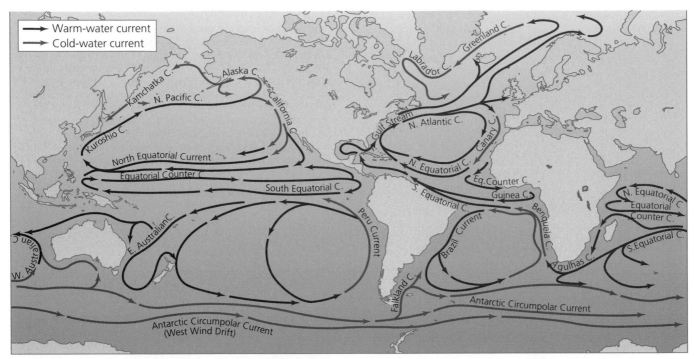

(a) The major surface currents of the world's oceans

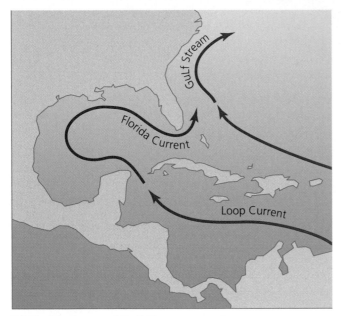

(b) Florida Keys and Gulf Stream currents

Figure 13.5 The upper waters of the oceans move in currents, making for long-lasting and predictable global patterns of water movement (**a**). These warm- and cold-water currents interact with the planet's climate system and have been used in human navigation of the oceans for centuries. The region around the Florida Keys (**b**) experiences the eastward movement of the warm-water Florida Current. This current originates with the Loop Current that loops through the Caribbean Sea and Gulf of Mexico, and it flows into the Gulf Stream, the powerful warm-water current that proceeds up the east coast of North America toward northern Europe. *Source:* Tom S. Garrison, *Oceanography,* third edition, 1999.

Upwellings also occur where strong winds blow away from or parallel coastlines. An example is the northern and central California coast, where north winds and the Coriolis effect move surface waters away from shore, raising nutrient-rich water from below and making for a biologically rich region. The cold water also chills the air along the coast, giving San Francisco its famous fog and cool summers. Conversely, in areas where surface currents converge or come together, surface water is displaced downward, a process called **downwelling.** Downwelling transports warm water rich in dissolved gases, providing an influx of oxygen for deep-water life. Vertical currents also take place in the deep zone, where differences in water density can lead to rising and falling convection currents, such as we have seen with molten rock (Chapter 6) and air in the atmosphere (Chapter 11). North Atlantic deep water (Chapter 12) provides an example of such circulation that could have far-reaching effects on global climate.

Vertical movement of water affects marine ecosystems

Surface winds and heating can also lead to vertical currents in seawater. **Upwelling,** or the flow of cold, deep water toward the surface, occurs in areas where currents diverge or flow away from one another. Because upwelled water is rich in nutrients, upwellings are often sites of high primary productivity and lucrative fisheries (Figure 13.6).

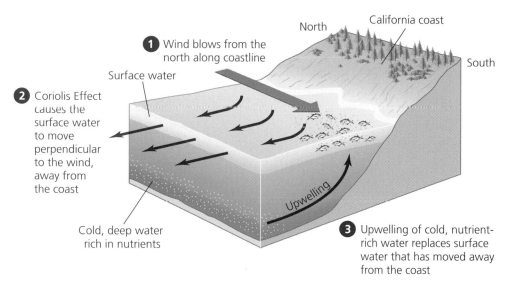

1 Wind blows from the north along coastline

North

California coast

South

Surface water

2 Coriolis Effect causes the surface water to move perpendicular to the wind, away from the coast

Cold, deep water rich in nutrients

Upwelling

3 Upwelling of cold, nutrient-rich water replaces surface water that has moved away from the coast

Figure 13.6 Upwelling is the movement of bottom waters upward, and this type of vertical current often brings nutrients up to the surface, creating rich areas for marine life. One way this can occur is illustrated by the situation off the coast of California. North winds blow along the coastline, while the Coriolis Effect draws water away from the coast. Water is then drawn up from the bottom to replace the water that moves away from shore.

Weighing the Issues:
Why Understand Ocean Currents?

Mapping ocean currents is crucial to understanding where and how the larvae of many fish and marine invertebrates become distributed from place to place, because currents help determine where larvae will settle and mature into adults. Why is this important for people to know? What else can be carried by currents? Describe several other ways in which a greater understanding of ocean currents can be helpful and important to people.

The topography of the seafloor can be rugged and complex

Despite the depiction of oceans as smooth, blue swaths on most maps and globes, the ocean floor is just as rough, rocky, and diverse as the terrestrial portion of the lithosphere. The deepest spot in the oceans, more than 11,000 m (36,000 ft) below sea level in the Mariana Trench in the South Pacific near Guam, is deeper than Mount Everest is high, by over a mile. The planet's longest mountain range is under water—the midocean ridge runs the length of the Atlantic (Figure 13.7). Underwater volcanoes shoot forth enough magma to build islands above sea level, including the Hawaiian Islands, and steep canyons similar in scale to Arizona's Grand Canyon lie just offshore of continents.

We can gain an understanding of the major types of underwater geographic features by examining a stylized map (Figure 13.8) that reflects bathymetry (the study of ocean depths) and topography (physical geography, or the shape and arrangement of landforms). In bathymetric profile, gently sloping **continental shelves** underlie the shallow waters bordering the continents. Continental shelves vary

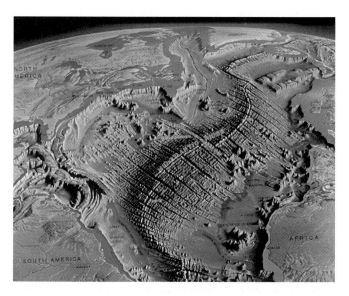

Figure 13.7 The seafloor is every bit as rugged as topography on the continents. The spreading margin between tectonic plates at the mid-Atlantic Ridge gives rise to a vast underwater volcanic mountain chain, cross-hatched by immense perpendicular breaks in the oceanic crust.

Figure 13.8 A stylized bathymetric profile serves to show key geologic features of the submarine environment. Shallow regions of water exist around the edges of continents over the continental shelf, which drops off at the shelf-slope break. The relatively steep dropoff called *the continental slope* gives way to the more gradual continental rise, all of which are underlain by sediments that have run off from the continents. Vast areas of seafloor are flat abyssal plain. Seafloor spreading occurs at oceanic ridges, and oceanic crust is subducted in trenches. Volcanic activity along trenches often gives rise to island chains such as the Aleutian Islands. Features on the left side of this diagram are more characteristic of the Atlantic Ocean, while features on the right side of the diagram are more characteristic of the Pacific Ocean. Adapted from H.V. Thurman, *Essentials of Oceanography,* fourth edition, Macmillan, 1990.

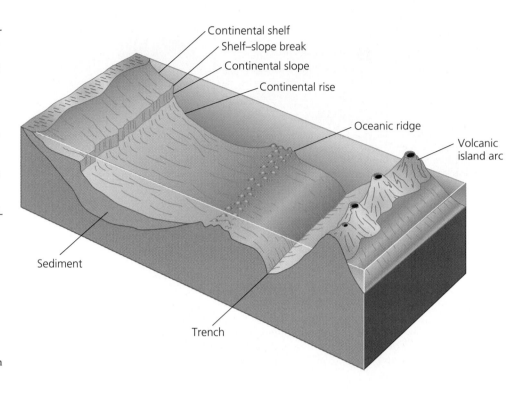

in width from 100 m (330 ft) to 1,300 km (800 mi), averaging 70 km (43 mi) wide, with an average slope of 1.9 m/km (10 ft/mi). These shelves drop off with relative suddenness at the shelf-slope break. The continental slope angles somewhat steeply downward, connecting the continental shelf to the deep ocean basin below.

The deep ocean basins consist of abyssal plains, extremely flat expanses that make up much of the seafloor. Volcanic peaks that rise above the ocean floor provide physical structure for marine animals and are frequently the site of productive fishing grounds. Some island chains, such as the Florida Keys, are formed by reef development and lie atop the continental shelf. Others, such as the Aleutian Islands curving across the North Pacific from Alaska toward Russia, are volcanic in origin, with peaks that rise above sea level. The Aleutians are also the site of a deep trench that, like the Mariana Trench, formed at a convergent tectonic plate boundary, where one slab of crust dives under another in the process of subduction (Chapter 6).

Different regions of the oceans have very different conditions, some of which support life more effectively than others. The uppermost 10 m (33 ft) of ocean water absorbs 80% of the solar energy that reaches its surface. For this reason, nearly all of the oceans' primary productivity occurs in the well-lit top layer of the oceans. Generally, the warm, shallow waters of continental shelves are most biologically productive and support the greatest species diversity. Biological oceanographers, or marine biologists, tend to classify marine habitats and ecosystems into two types. Those occurring between the ocean's surface and the ocean floor are classified as **pelagic**, whereas those that occur on the ocean floor are classified as **benthic.** Each of these major areas contains several vertical zones (Figure 13.9).

Marine Ecosystems

Every region of the oceans is home to life, and the diversity in topography, temperature, salinity, nutrients, and sunlight from place to place means that marine environments boast a wide variety of ecosystems. Like terrestrial ecosystems, most marine ones are powered by solar energy. Sunlight drives photosynthesis among plankton in the euphotic zone, but even the darkest ocean depths provide habitat for some living things. Organisms in benthic habitats dwell in or on the sea floor, pelagic organisms live in open water away from continents and islands, and coastal organisms dwell in shallow waters near the shore.

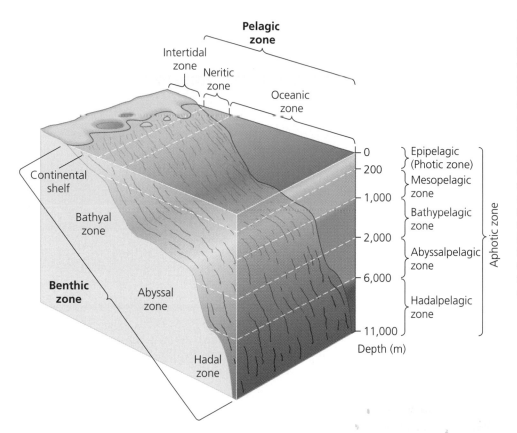

Figure 13.9 Oceanographers classify marine habitats into benthic (seafloor) and pelagic (open-water) categories and subdivide these major categories into zones based upon depth. Pelagic waters extend from the epipelagic zone at and near the surface down to the hadalpelagic zone at depths below 6,000 m (19,700 ft). Near-surface waters that receive adequate light for photosynthesis are also often termed the *photic zone,* and waters above the continental shelves are also said to be in the *neritic zone.* Benthic zones range from the intertidal zone where the ocean meets the land, to the continental shelves, and down to the progressively deeper bathyal, abyssal, and hadal zones.

We will examine coastal ecosystems in most detail, because they are the most productive and diverse and also are often the most vulnerable to human impact.

Open-ocean ecosystems vary in their biological diversity

Biological diversity in pelagic areas of the open ocean is highly variable in its distribution. Plant productivity and animal life near the surface are concentrated in areas of nutrient-rich upwelling. Phytoplankton form the base of the marine food chain in the pelagic zone; these photosynthetic algae, protists, and cyanobacteria feed the zooplankton, tiny floating consumers, which in turn become food for fish, jellyfish, whales, and other free-swimming animals (Figure 13.10). Predators at higher trophic levels include larger fish, sea turtles, and sharks. In addition, many bird species feed on the open ocean, remaining there for most of the year and then returning to islands and coastlines to breed.

In recent years biologists have been learning more about animals of the deep ocean, although tantalizing questions remain and many organisms are not yet discovered. In these deep-water ecosystems, animals are specially adapted to deal with extreme water pressures and

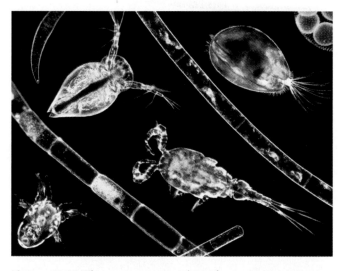

Figure 13.10 The uppermost reaches of ocean water contain billions upon billions of phytoplankton, tiny photosynthetic algae, protists, and bacteria that form the base of the marine food chain, as well as zooplankton, tiny animals and protists that dine on phytoplankton and comprise the next trophic level.

have evolved to live in the dark without food from plants. Some of these often bizarre-looking creatures scavenge detritus (organic particles) that fall from above, some are predators, and others attain food from symbiotic mutualist bacteria. Some species also carry bacteria that produce light chemically by bioluminescence (Figure 13.11).

Finally, as we explored in Chapter 4, some ecosystems cluster around hydrothermal vents, locations where heated water spurts from the seafloor, often carrying minerals that precipitate to form large rocky structures. The tubeworms, shrimp, and other creatures in these recently discovered systems derive their energy ultimately from chemicals in the heated water rather than from sunlight. They manage to thrive within the amazingly narrow zones between water that is scalding-hot and water that is icy cold.

Kelp forests harbor many organisms in temperate waters

Large brown algae or seaweed, or **kelp**, grows from the floor of continental shelves, reaching upward toward the sunlit surface. Some kelp reaches 60 m (200 ft) in height and can grow 45 cm (18 in) in a single day. Dense stands of kelp form underwater forests on the continental shelves in many temperate waters (Figure 13.12). Kelp forests, with their complex structure, supply shelter and food for invertebrates and fish, which in turn provide food for higher-trophic-level predators such as seals and great white sharks. Indeed, kelp forests were the setting for our illustration of a keystone predator in Chapter 5; recall that sea otters served to control sea urchin populations, and when otters disappear, urchins overgraze the kelp, destroying the forests and creating

Figure 13.12 The tall brown algae, called *kelp*, grow from the floor of the continental shelf and gives structure to kelp forests, a key marine ecosystem. Numerous fish and other creatures eat kelp or find refuge among its fronds.

"urchin barrens" in their place. Kelp forests also absorb wave energy and protect shorelines from erosion. People derive alginates from kelp, essential ingredients in a wide range of consumer products, from toothpaste to shampoo to ice cream.

Coral reefs are treasure troves of biodiversity

Moving from temperate waters into subtropical and tropical waters, kelp forests give way to coral reefs. A reef is an outcrop of rock, sand, or other material that rises near the surface of a relatively shallow body of salt water. A **coral reef** is a mass of calcium carbonate composed of the skeletons of tiny colonial marine organisms called *corals* (Figure 13.13a). Coral reefs may occur as an extension of a shoreline, as a barrier island paralleling a shoreline, or as a ring around a sunken island, a formation called an *atoll*.

Corals are invertebrate animals that belong to the phylum *Cnidaria* and are relatives of sea anemones and jellyfish. Some corals have hard external skeletons encompassing a soft hollow body, whereas others have flexible internal skeletons (Figure 13.13b). Corals are sessile (stationary), attached to rock or existing reef and capturing passing food with stinging tentacles. They also derive nourishment from symbiotic (Chapter 5) algae, known as **zooxanthellae**, that inhabit their bodies and produce food through photosynthesis. Most corals are colonial, and the colorful surface of a coral reef is made up of thousands of densely packed individuals. As corals die, their skeletons remain part of the reef while new corals grow atop them, making the reef ever larger.

Figure 13.11 Life is scarce in the dark depths of the deep ocean, but the creatures that do live there often appear bizarre to us. The anglerfish lures prey toward its mouth with a bioluminescent (glowing) organ that protrudes from the front of its head.

(a) Corals

Figure 13.13 Coral reefs are among the richest ecosystems in the world for biodiversity—and among the most threatened. Reefs are built up over many millennia by tiny animals called corals **(a)** that live in colonies, growing atop the skeletons of past corals. Individual coral polyps **(b)** contain stinging cells, called nematocysts, that help the coral capture food. Many also contain zooxanthellae, tiny photosynthetic organisms engaged in a mutualism with the coral.

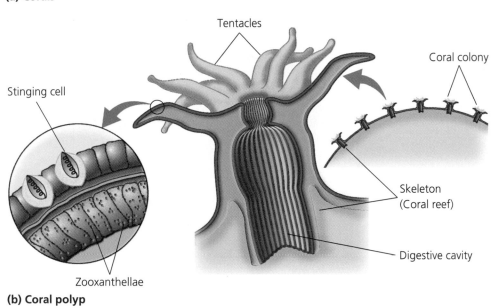

(b) Coral polyp

The reefs of the Florida Keys National Marine Sanctuary host many types of coral (Figure 13.14a and 13.14b). Coral reefs absorb wave energy and protect the shore from damage. They also provide essential habitats for many of the Keys' 6,000 species of marine organisms. Globally, coral reefs host as much biodiversity as any other type of ecosystem. Besides the staggering diversity of anemones, sponges, hydroids, tubeworms, and other sessile invertebrates, innumerable molluscs, flatworms, starfish, and urchins patrol the reefs, and thousands of fish species find food and shelter in reef nooks and crannies. Larger predators, such as sharks and moray eels (Figure 13.14c), feed on the fish.

Coral populations and coral reefs are experiencing worldwide declines, however. Many reefs have undergone "coral bleaching," a process that occurs when zooxanthellae leave the coral, depriving it of nutrition.

Corals lacking zooxanthellae lose color and frequently die, leaving behind ghostly white patches in the reef. Coral bleaching may result from increased sea surface temperatures associated with global climate change or from the influx of artificial pollutants into area waters, although scientists have not ruled out natural causes. Coral reefs also sustain damage when divers stun fish with cyanide in order to capture them for food or for the pet trade, a common practice in waters of Indonesia and the Philippines. It is estimated that for each fish caught in this way, one square meter of reef is destroyed by cyanide poisoning.

A few coral species thrive in waters outside the tropics and build large reefs on the ocean floor at depths of 200–500 m (650–1,650 ft). These little-known reefs, which occur in cold-water areas such as off the coasts of Norway, Spain, and the British Isles, are only now

(a) Elkhorn coral

(b) Sea fan

Figure 13.14 Different species of reef-dwelling corals give rise to a variety of distinct structures that make up reefs in Florida and elsewhere. Elkhorn coral **(a)** is an antler-shaped hard coral that grows in many shallow waters worldwide. Also common in the Florida Keys is a soft coral called the *sea fan* or *gorgonian* **(b)**. Fish take refuge among corals, but reef crevices also hide predators, such as moray eels **(c)**, that may ambush them. Coral reefs face many environmental stresses from natural and human impacts, and many corals have died due to coral bleaching in which corals lose their mutualistic zooxanthellae.

(c) Moray eel among corals

beginning to be studied by scientists, but already many have been badly damaged by trawling, the fishing practice that drags heavy nets along the seafloor. Norway and other countries are now beginning to protect some of these reefs.

Weighing the Issues:
The Coral Crisis

Some scientists are exploring new technologies to enhance coral growth and the abundance of other reef-inhabiting organisms, including developing chemical cues to attract larvae and special wavelengths of light to attract fish. Do you think technology could offer adequate solutions to the declines of coral reefs? Why or why not? What other solutions can you suggest for the coral crisis? What challenges might each of these solutions encounter?

Intertidal zones are dynamic ecosystems that undergo constant change

Where the ocean meets the land, **intertidal** or **littoral** ecosystems lie along shorelines between the farthest reach of the highest tide and the lowest reach of the lowest tide. **Tides** are the periodic rising and falling of the ocean's height at a given location, caused by the gravitational pull of the moon and sun. Intertidal organisms spend part of each day submerged in water, part of each day dry and exposed to the air and sun, and part of each day being lashed and beaten by waves. Subject to tremendous extremes in temperature, moisture, sun exposure, and salinity, these creatures must also deal with onslaughts of marine predators at high tide and terrestrial predators at low tide.

The intertidal environment is a tough place to make a living, but it is home to a remarkable diversity of organisms. Rocky shorelines can be full of life among

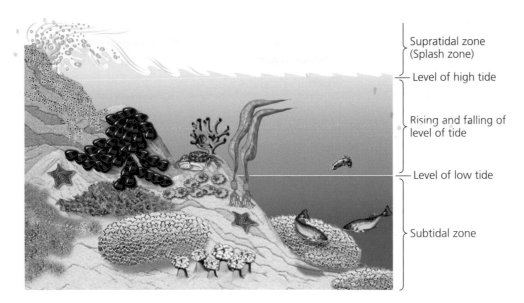

(a) Intertidal zone

(b) Low and high reaches of tides

Figure 13.15 The rocky intertidal zone is the region along a rocky shoreline (a) between the lowest and highest reaches of the tides (b). This is an ecosystem of high biodiversity, typically containing large invertebrates like seastars (starfish), barnacles, crabs, sea anemones, corals, bryozoans, snails, limpets, chitons, mussels, nudibranchs (sea slugs), and sea urchins. Fish swim in the tidal pools, many types of algae may cover the rocks, and insects, birds, and mammals visit in search of food. Areas higher on the shoreline are exposed to the air more frequently and for longer periods, so those organisms that can tolerate exposure best specialize in the upper intertidal zone. The lower intertidal zone is less frequently exposed and for shorter periods, so organisms less able to tolerate exposure thrive here.

the crevices, which provide shelter and pools of water (tide pools) left during low tides. Sessile animals such as anemones, chitons, mussels, and barnacles live attached to rocks, filter-feeding food from the water washing over them. Urchins and sea slugs eat algae growing in the intertidal area. Predatory starfish creep slowly along, preying on the filter-feeders and herbivores at high tide. Crabs clamber around the rocks, scavenging detritus. At low tide, birds come by and pick off animals that have not protected themselves well enough.

The rocky intertidal is so diverse because environmental conditions, including temperature, salinity, and the presence or absence of water, change dramatically from the highest to the lowest reaches (Figure 13.15). This environmental variation gives rise to bands of habitat such that organisms array themselves in zones, according to their own niches and needs. The diversity of organisms is lower in sandy intertidal areas, such as occur in the Florida Keys, yet plenty of organisms burrow into the sand at low tide to await the return of high tide, when they emerge to feed.

Salt marshes cover large areas of coastline in temperate areas

Along many of the world's coastlines, the tides reach inland and spread across large areas, creating **salt marshes**. In high latitudes and temperate regions, salt marshes are common where gently sloping, sandy or silty substrate meets the ocean. Rising and falling tides flow into and out of channels called *tidal creeks* and at highest tide spill over onto elevated marsh flats (Figure

13.16). Marsh flats grow thick with grasses, rushes, shrubs, and other herbaceous plants. Grasses are the dominant vegetation in most salt marshes; the most common species are those of the genera *Spartina* and *Distichlis*. Salt marshes boast very high primary productivity (Chapter 5) and provide critical habitat for shorebirds, waterfowl, and the adults and young of many commercially important fish and shellfish species. In many parts of the world, people have altered salt marshes to make way for coastal shipping, industrial facilities, farming, and other development.

Mangrove forests line coastlines throughout the tropics and subtropics

In tropical and subtropical latitudes, mangrove forests replace salt marshes in gently sloping sandy and silty coastal areas. **Mangroves** are trees with unique types of roots that curve upward like snorkels to attain oxygen lacking in the mud and that serve as stilts to support the tree in changing water levels. Fish, shellfish, crabs, snakes, worms, and other organisms thrive among the root networks, and birds find habitat for feeding and nesting in the dense foliage of these coastal forests. In some parts of the world, mangroves also provide materials that people use for food, medicine, tools, and construction. Mangroves just reach the southern edge of the United States, and the Florida Keys National Marine Sanctuary contains tens of thousands of acres of mangroves, including the red mangrove, black mangrove, and white mangrove—three species adapted to specialize in different tidal zones (Figure 13.17).

In south Florida and elsewhere, however, many mangrove forests have been removed as people have converted coastal areas for residential, recreational, and commercial uses. One of the greatest threats to the survival of mangroves is coastal shrimp farming (Chapter 9), which has deforested 65,000 ha (160,000 acres) in Thailand alone. By 1990, mangrove cover in the Philippines had declined from 448,000 to 110,000 ha (1.1 million to 270,000 acres). It is estimated that half the world's mangrove forests have vanished because of human activity and that mangrove forest area continues to decline by 2–8% per year. As of 2002 only about 1% of the world's remaining mangroves benefited from some sort of protection.

Figure 13.16 Salt marshes occur in temperate intertidal zones where the substrate is muddy, allowing salt-adapted grasses to grow. Tidal waters generally flow into and out of the marshes in channels called *tidal creeks*, amid flat areas called *benches*, sometimes partially submerging the grasses. In this salt marsh in Lewes, Delaware, linear channels have been cut by people for mosquito control.

Figure 13.17 Mangrove forests are important ecosystems along tropical and subtropical coastlines throughout the world. Mangrove trees, such as these off of Lizard Island, Australia, show specialized adaptations for growing in saltwater and provide habitat for many types of fish, birds, crabs, and other animals.

Fresh water meets salt water in estuaries

Many salt marshes and mangrove forests occur in or near **estuaries**, areas where rivers flow into the ocean, mixing fresh water with salt water. Biologically productive ecosystems, estuaries experience significant fluctuations in salinity as tidal currents and freshwater runoff vary daily and seasonally. For shorebirds and for many commercially important shellfish species, estuaries provide critical habitat. In addition, for anadromous fishes (those, like salmon, that spawn in fresh water and mature in salt water), estuaries provide a transitional zone where young fish make the passage from fresh water to salt water.

Estuaries around the world have been affected by urban and coastal development, water pollution, habitat alteration, and overfishing. The estuary of Florida Bay, for instance, where fresh water from the Everglades system mixes with salt water, has suffered pollution and experienced reduced freshwater input from irrigation and fertilizer use by the sugar industry, from leaky septic tanks, and from other human land uses. Coastal ecosystems have borne the brunt of human impact since so many people choose to live along coastlines, and residential and commercial development is especially dense in coastal areas. In fact, roughly two-thirds of Earth's human population lives within 160 km (100 mi) of the ocean.

How Humans Use and Impact Oceans

People have long clustered their settlements near coastlines, and our species has a long history of interacting with the oceans. We have traveled across them, exploited them for their resources, and polluted them with our waste, all the while with fascination for the beauty, power, and vastness of the seas.

The oceans provide transportation routes

The oceans have provided routes for transportation for thousands of years, and continue to provide affordable means of moving people and products over vast distances. The historical impacts of shipping on human culture and commerce are profound, accelerating the global reach of certain cultures and the interaction of long-isolated peoples. Shipping has had substantial impacts on the environment as well. The thousands of ships plying the world's oceans today carry everything from cod to cargo containers to crude oil. Ships transport ballast water as well, which when discharged at ports of destination may transplant species picked up at ports of departure. Some of these species—such as the zebra mussel (Chapter 15)—establish and become invasive in their new homes.

We extract energy and minerals from the oceans

We also use the oceans as sources of commercially valuable energy. By the 1980s about 25% of our production of crude oil and natural gas was coming from exploitation of deposits beneath the seafloor (Figure 13.18). According to recent estimates, offshore areas may contain as much as 2 trillion barrels of oil, roughly half the amount known to exist underground on land. The exploitation of oil and gas deposits in the Gulf of Mexico has triggered some debate in recent years; for example, the risk of an oil spill in the Florida Keys National Marine Sanctuary concerns many sanctuary supporters.

The oceans also hold potential for providing renewable energy sources. Engineers have developed turbines that can generate electricity by utilizing the ebb and flow of the tides (Chapter 18, Figure 18.14, and Figure 18.15). Another possible energy source is ocean thermal energy conversion. According to the U.S. government's National Renewable Energy Laboratory, in a single day the oceans capture as much energy from the sun as is contained in 250 billion barrels of oil. If we could harness just 0.1% of this solar energy, we could use it to generate 20 times the electricity used in the United States each day.

We also extract minerals from the ocean floor. By using large vacuum cleaner–like hydraulic dredges, miners collect sand and gravel from beneath the sea. Also extracted are sulfur from salt deposits in the Gulf of Mexico and phosphorite from many offshore areas including several near the California coast. Other valuable minerals found on or beneath the sea floor include calcium carbonate (used in making cement), silica (used as fire-resistant insulation and in manufacturing glass), and rich deposits of copper, zinc, silver, and gold ore. Many minerals are found concentrated in manganese nodules, small ball-shaped accretions that litter parts of the ocean floor. It is estimated that over 1.5 trillion tons of manganese nodules exist in the Pacific Ocean alone and that their reserves of metal exceed all terrestrial reserves. The logistical difficulty of mining them, however, has kept their extraction uneconomical so far.

Marine pollution threatens resources

Humans have taken from the oceans, and we have also given back—in the form of pollution. Oceans have long been made a sink for human wastes. Even into the

Figure 13.18 Crude oil and natural gas from beneath the seafloor are some of the more economically valuable resources that we have harvested from the oceans. Offshore oil drilling is one of many impacts that humans have on the marine environment. Many maintain that oil platforms such as these off the Santa Barbara, California, coast aesthetically compromise the ocean.

mid-20th century, it was common for coastal cities in the United States to dump trash and pump untreated sewage onto mudflats and into embayments. Fort Bragg, a bustling town on the northern California coast, boasts of its Glass Beach, an area where beachcombers can collect sea glass, the colorful surf-polished glass sometimes found on beaches after storms stir sediments and dislodge debris. Glass Beach is in fact the site of the former town dump, and besides well-polished glass, the perceptive visitor may also spot old batteries, rusting car frames, and all other manner of trash protruding from the bluffs above the beach. Such coastal dumping practices have left a toxic legacy around the world, but marine pollution continues even today. Oil, plastic, industrial chemicals, sewage sludge, excess nutrients, abandoned fishing gear—all eventually makes its way into the oceans.

Oil pollution comes not only from massive spills

Major oil spills such as the one that occurred when the *Exxon Valdez* struck a reef in Prince William Sound, Alaska (Chapter 4), make headlines and cause serious environmental problems on a local or regional scale. Yet it is important to put such accidents into perspective. The majority of oil pollution in the oceans comes not from large spills in a few particular locations, but from the accumulation of innumerable widely spread small sources, including leakage from small boats and runoff from human activities on land. In addition, the amount of petroleum spilled into the oceans each year is equaled by the amount that seeps into the water from naturally occurring seafloor deposits.

Nonetheless, minimizing the amount of oil we release into our coastal waters is important, because petroleum pollution is detrimental to the marine environment and the human economies that draw sustenance from that environment. Petroleum can physically coat and kill intertidal and free-swimming marine organisms, and ingested chemical components in petroleum can poison marine life (Chapter 4). In response to headline-grabbing oil spills, governments around the world have begun to implement more stringent safety standards for tankers, such as requiring industry to pay for tugboat escorts in sensitive and rocky coastal waters, and to develop prevention and response plans for major oil spills. The U.S. Oil Pollution Act of 1990 created a $1 billion prevention and cleanup fund and required that by 2015 all oil tankers in U.S. waters be equipped with double hulls as a precaution against puncture. The oil industry has resis-

ted many such safeguards, and today the ship that oiled Prince William Sound is still plying the world's oceans, renamed the *Sea River Mediterranean*, and still featuring only a single hull. However, over the past three decades, the amount of oil spilled in U.S. waters and world wide has decreased, in part because of an increased emphasis on spill prevention and response (Figure 13.19).

Nets and plastic debris endanger marine life

Plastic bags and bottles, fishing nets, gloves, fishing line, buckets, floats, abandoned cargo, and nearly everything else that humans transport on the sea or dispose into it can present problems for marine organisms. According to the Ocean Conservancy, as much as 89% of the trash in the North Pacific is plastic. Because most plastic is not biodegradable, it can drift for decades before washing up on beaches. Marine animals including seabirds, fish, and endangered sea turtles, can die as a result of ingesting material they cannot digest or expel. Floating plastic debris, for example, may be mistaken for food. Such is the case when clear plastics resemble jellyfish.

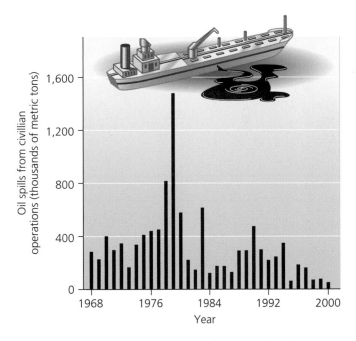

Figure 13.19 Tankers and other vessels account for about half of all the anthropogenic oil pollution of ocean waters; the other half originates in runoff from municipal, industrial, and residential sources on land. The figure shows estimated yearly global amounts from all sources. Less oil is being spilled into ocean waters today, thanks in part to regulations on the oil shipping industry and on terrestrial sources. Data from *Vital Signs,* Worldwatch Institute, 2002.

Fishing nets that are lost during storms, snags, or accidents or are intentionally discarded frequently continue catching fish and other organisms for decades. Ocean Conservancy data indicate that 30,000 northern fur seals die each year from entanglement in nets. Of 115 marine mammal species, 49 sometimes are known to eat or become entangled in marine debris, and 111 of 312 species of seabirds are known to eat plastic. Sea turtles of all five species found dead in the Gulf of Mexico have died consuming or contacting marine debris.

We can all help minimize this type of harm by reducing our use of plastics, by cutting the rings of plastic six-pack holders before throwing them away, and by picking up trash from beaches (Table 13.1). An average of 130 tons of trash is hauled off a single beach near Santa Monica, California, every month. Marine debris hurts people as well as wildlife. A recent survey of fishers off the Oregon coast indicated that more than half had encountered problems from plastic debris, and debris has caused over $50 million dollars in insurance payments.

Table 13.1 Waste Items Picked Up during a Recent National Coastal Cleanup of U.S. Beaches*

Type of item	Number
Cigarette butts	608,759
Miscellaneous plastic pieces	240,820
Plastic foam pieces	206,890
Plastic bags and wrappers	205,762
Plastic lids	179,103
Plastic straws	131,134
Total weight of all trash	*1.3 million kg (2.9 million lb)*

*The cleanup was sponsored by The Ocean Conservancy and involved 151,502 volunteers. Data from The Ocean Conservancy, International coastal cleanup, 2002.

Excess nutrients can cause algal blooms

Pollution from fertilizer runoff or other nutrient inputs can have dire effects on marine ecosystems, as we saw with the Gulf of Mexico's dead zone in Chapter 6. The release of excess nutrients into surface waters can spur unusually high growth rates and population densities of phytoplankton, causing eutrophication, in freshwater or saltwater ecosystems. Such problems have occurred in the Florida Keys, leading the EPA in 2001 to prohibit the release of sewage and other waste into the waters of the Florida Keys National Marine Sanctuary.

Algal blooms may lead to other problems as well, because several species of marine algae produce powerful toxins that attack the nervous systems of vertebrates. When excessive nutrient concentrations give rise to population explosions among these toxic algae, **harmful algal blooms** may occur. These phenomena are also known as **red tides**, because certain algae species produce reddish pigments that discolor surface waters (Figure 13.20). Red tides can cause illness and death among zooplankton, birds, fish, marine mammals, and humans as toxins are passed up the food chain. They can also lead to considerable economic loss for communities dependent on fishing or beach tourism. We can reduce the risk of these outbreaks by reducing nutrient runoff into coastal waters, and we can minimize their health effects by monitoring to prevent human contact with or consumption of affected organisms.

As severe as the impacts of marine pollution can be, however, most marine scientists concur that the more worrisome dilemma is overharvesting. Unfortunately, the old cliché that "there are always more fish in the sea" appears not to be true; the oceans today have been dangerously overfished, and most fishing stocks have collapsed or are on the brink of collapsing.

Emptying the Oceans

The oceans and their biological resources have provided for many of human needs for thousands of years. However, today we are placing unprecedented pressure on marine resources. Half the world's marine fish populations are fully exploited, meaning that we cannot harvest them more intensively without driving them toward extinction, according to a 2001 United Nations Food and Agriculture Organization (FAO) report. An additional 25% of marine fish populations are overexploited and already being driven toward extinction, the FAO reported. Thus only one-quarter of the world's marine fish populations can yield more than they are already yielding without being driven into decline. Total global fisheries catch, after decades of increases, leveled off after about 1988 (Figure 13.21). As our population grows, we will only become more dependent on the oceans' bounty. Existing fishing practices are not sustainable given present consumption rates, many scientists and fisheries managers have concluded. This makes it vital, they say, that we take immediate steps to modify our priorities and improve our use of science in fisheries management.

Dinoflagellate (*Gymodinium*)

Figure 13.20 Red tides are events in which certain types of algae bloom out of control in marine surface waters, producing toxins in the form of tiny round organisms, *Gymodinium,* that harm marine life with pigment that turns the water red.

Red tide, Bountiful Islands, Gulf of Carpentaria, Australia

Overfishing is nothing new

Humans have harvested fish, shellfish, turtles, seals, and other marine animals from the oceans for thousands of years. Archaeological evidence for ancient coastal communities reveals shellfish-rich diets, and many sites around the world are known for their vast middens, or piles, of discarded oyster and clam shells. While much of this harvesting may have been sustainable, historical ecologists are now learning that the depletion of marine animals by humans did not begin with today's industrialized fishing fleets. Rather, overfishing took its toll on a variety of marine species beginning centuries or millennia ago, then accelerated during the colonial period of European expansion before intensifying further in the 20th century.

A recent synthesis of historical evidence by marine biologist Jeremy Jackson and others revealed that ancient overharvesting likely affected ecosystems in astounding ways we only partially understand today. Several large animals, including the Caribbean monk seal, Steller's sea cow, and Atlantic gray whale, were hunted to extinction long enough ago that scientists never studied them or the ecological roles they played. Overharvesting of the vast oyster beds of Chesapeake Bay led to the collapse of the oyster fishery there in the late 19th century. Eutrophication and hypoxia similar to that of the Gulf of Mexico (Chapter 6) have resulted, because there are no longer oysters to filter nutrients from the water.

Florida Bay, according to Jackson and his colleagues, is suffering today from the overhunting of green sea turtles

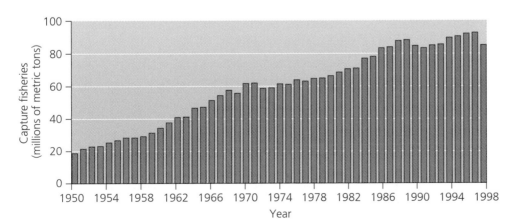

Figure 13.21 The total global fisheries catch has increased over the years, as new technologies enable fishing fleets to find more fish, even as stocks become depleted. In recent years, however, growth in the global catch has stalled, and many fear that further fisheries collapses and a global catch decline are imminent if conservation measures are not taken soon. Data from FAO, World review of fisheries and aquaculture, 2000.

in past centuries. The once-abundant turtles ate sea grass (often called turtle grass) and likely kept it cropped low like a lawn. But with today's turtle population a tiny fraction of what it once was, sea grass grows thickly, dies, and rots in place, giving rise to disease like the sea grass wasting disease that ravaged Florida Bay in the 1980s. A better-known case of historical overharvesting is the near extinction of many species of whales in the world's oceans, which resulted from commercial whaling that began centuries ago and was brought to a halt only in 1986. Although overfishing has a long history, the industrialized methods, new technology, and global reach of today's commercial fleets are making our impact much more rapid and far-reaching than in past centuries.

Modern fishing fleets deplete marine life rapidly

Modern industrialized fishing fleets can deplete marine life rapidly. In a 2003 study (Figure 13.22), fisheries biologists Ransom Myers and Boris Worm analyzed fisheries data from FAO archives, looking for changes in the catch rate of fish in various areas of ocean since they were first exploited by industrialized fishing. For one region after another they found the same pattern: Catch rates dropped precipitously during the first several years of fishing, with 90% of large-bodied fish and sharks eliminated within only a decade. Following that, populations stabilized at 10% of their former levels. This means, Myers and Worm concluded, that the oceans

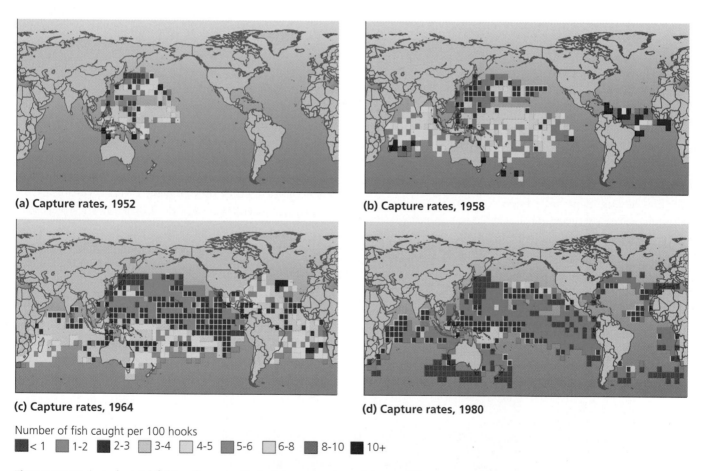

(a) Capture rates, 1952

(b) Capture rates, 1958

(c) Capture rates, 1964

(d) Capture rates, 1980

Number of fish caught per 100 hooks

■ < 1 ■ 1-2 ■ 2-3 ☐ 3-4 ☐ 4-5 ■ 5-6 ☐ 6-8 ■ 8-10 ■ 10+

Figure 13.22 As industrial fishing fleets reached each new region of the world's oceans, capture rates of large predatory fishes were initially high and then within a decade declined markedly. In the figure, reds and oranges signify high capture rates and blues signify low capture rates. High capture rates in the western Pacific in 1952 (**a**) gave way to low ones in later years. Excellent fishing success in the tropical Atlantic and Indian Oceans in 1958 (**b**) had turned mediocre by 1964 (**c**), and poor by 1980 (**d**). High capture rates in the north and south Atlantic in 1964 (**c**) gave way to low capture rates there in 1980 (**d**).

today contain only one-tenth of the large-bodied animals they once did. It also means that declines happened so fast in most areas of the world that scientists never knew the original abundance of these animals.

Many fisheries are collapsing today

The percentage of oceanic fish stocks that are overfished increased tenfold between 1950 and 1994 and threefold from 1974 to 2001 (Figure 13.23). Many fisheries have collapsed, and others are in danger of collapsing. Besides being ecologically devastating, these collapses take a severe economic toll on communities and regions that depend on fishing. In 2001, according to the FAO, 36 million people worked directly in capture fisheries and aquaculture (Chapter 9), while fishing and related industries indirectly employed 400 million people, contributed to food security in many nations, and generated $50 billion annually in international trade.

A prime example of fishery collapse took place in the 1990s affecting groundfish fisheries in the Atlantic off the Canadian and U.S. coasts. The term *groundfish* refers to various species that live in benthic habitats, such as Atlantic cod, haddock, halibut, and flounder. These fish are major food sources that powered fishing economies in Newfoundland, Labrador, the maritime provinces, and the New England states for close to 400 years. Yet

fishing pressure became so intense that most stocks collapsed in recent years, bringing fishing economies down with them. With Canada's cod stocks down 99% and showing no sign of recovery, the Canadian government in 2003 ordered a complete ban on cod fishing in the Grand Banks region off Newfoundland and Labrador. To soften the economic blow, it offered $50 million to affected fishermen of the region.

A ray of hope for such fishermen comes from just south of the border, however. When the groundfish fisheries of Georges Bank in the Gulf of Maine collapsed in the mid-1990s, three areas totaling 17,000 km² (6,600 mi²) were closed to fishing. The closures worked; five years later, haddock, flounder, and yellowtail were recovering, and scallops rebounded strongly, attaining sizes 9–14 times as large as before the closures. Fishermen began having better luck, especially in areas just outside the closed regions, where satellite observations showed that fish were clustering.

Fisheries declines are masked by several factors

Despite the depletion of fish stocks in region after region as industrialized fishing has intensified, the amount of overall global fish production has remained stable for 15 years. You might wonder how this could be. The seeming stability of the total global catch can be

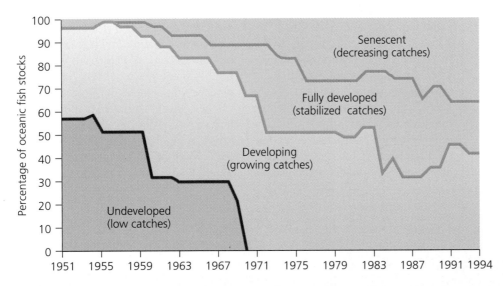

Figure 13.23 Throughout the latter half of the 20th century, marine fisheries have been increasingly exploited. By 1970, undeveloped fisheries, or newly discovered fisheries, which once comprised the largest percentage of all fish stocks, were no longer. Exploration and technology moved nearly 70% of fisheries into the developing stage by 1970, producing steady growth in catches. Further exploitation brought catch rates down and stabilized overall catches, however, leading roughly 40% of fisheries into the fully developed stage by 1986. By 1993, overexploitation pushed nearly 35% of the world's fisheries into the senescent stage, a stage of decreasing catches. Go to **GRAPH IT** on the website or CD-ROM. Data from Food and Agricultural Organization of the United Nations.

The Science behind the Story

China's Fisheries Data

China is responsible for a larger share of the world's marine and inland fisheries catch than any other nation, according to the FAO. Thus China's catch data has a major impact on the FAO's attempts to assess the health of the world's fisheries. In 2001, two fisheries scientists published a paper suggesting that China had exaggerated its catch by as much as 100% during the 1990s. The inflated catch data had led to complacency among the world's fisheries managers and policymakers, they argued, because it led the FAO to overestimate the true amount of fish left in the ocean.

The two authors of the paper, University of British Columbia researchers Reg Watson and Daniel Pauly, were initially suspicious of China's data for several reasons. First, China's coastal fisheries had been overexploited for decades, yet reported catches continued to increase. Second, China's "catch per unit effort" remained unchanged from 1980 to 1995, even though the abundance of fish had decreased. Finally, China's catch statistics suggested that its coastal waters were far more productive than ecologically similar areas fished by other nations.

In the November 29, 2001, issue of the journal *Nature*, Watson and

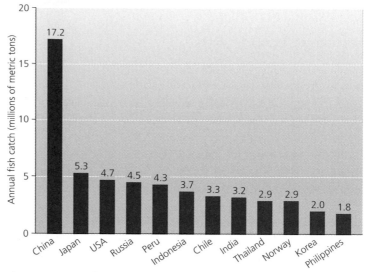

China's reported fish catches in recent years have dwarfed those of other nations, but a recent study asserted that Chinese fisheries managers had been falsely inflating their catches. Data from Food and Agriculture Organization of the United Nations, 2000.

Pauly reported the results of a study testing the hypothesis that China's apparent productivity was unrealistically high. Using a number of different databases, including FAO catch data collected since the 1950s, they built a statistical model to predict global fisheries catch. They divided the world's oceans into approximately 180,000 "cells," each spanning half a degree longitudinally and latitudinally. The cells were then characterized by factors that affect catch size, such as depth, primary productivity, fishing

rights, and species distribution. These factors were then used to predict total annual catch for each cell.

For most cells, the catch predicted by the model was similar to the catch reported by fleets that fished those waters. In large swaths of China's coastal waters, however, predicted catches were over 5 metric tons/km² smaller than reported catches, and in some areas the difference was over 10 metric tons/km². In 1999, the model predicted that China's total catch would be 5.5 million metric tons,

explained by several factors that mask real declines. One is that fishing fleets now venture farther from their home ports in order to find fish. Japanese, European, Canadian, and American boats have been traveling longer distances to reach less-fished portions of the ocean. They have also been spending longer periods of time fishing and setting out greater numbers of nets and lines. In other words, fishing fleets have had to expend more and more effort just to catch the same number of fish.

Improved technology also helps explain high catches despite declining stocks. Today's fishing fleets

can reach almost any spot on the globe with boats that can attain speeds of 80 kph (50 mph). They have access to an array of technologies that militaries have developed for spying and for chasing enemy submarines, including advanced sonar mapping equipment, satellite navigation, and thermal sensing systems. Some fleets rely on aerial spotters to find schools of commercially valuable fish, such as bluefin tuna. Between 1970 and 1990, adult bluefin tuna numbers in the western Atlantic dropped by 90%. Yet another cause of misleading stability in global catch numbers

whereas the figure reported by Chinese fisheries authorities was 10.1 million metric tons.

When Watson and Pauly proceeded to analyze global fisheries trends using the model's predictions rather than China's official numbers, they found that the total annual global catch had decreased by 0.36 million metric tons since 1988, instead of increasing by 0.33 metric tons, as reported by the FAO. Their finding suggested that the total global catch, rather than remaining stable through the 1990s, actually had begun to decline in the 1980s.

Because the FAO is the only organization that collects global catch data, fisheries managers and policy-makers look to it for information about the state of the world's fisheries. By using China's inflated numbers, Watson later wrote, the FAO had encouraged a global "mopping-up operation" in which the last of the world's fish populations were being decimated.

The Chinese government and the FAO were both critical of the study. China's reported catch was "basically correct," said Yang Jian, director-general of the Chinese Bureau of Fisheries. Watson and Pauly's model, he suggested, failed to take into account unique aspects of China's fisheries, such as its large catch of crabs and jellyfish. Yang

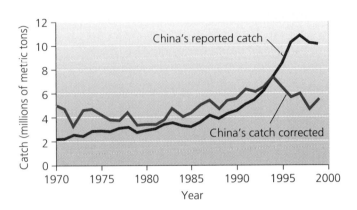

Fisheries biologists used models to estimate the true amount of China's fish harvests and found the estimated amounts to be substantially lower than the reported amounts. Data from Reg Watson and Daniel Pauly, Systematic distortions in world fisheries catch trends, *Nature*, 2001.

also noted that China had already begun to address the problems of overfishing and overreporting that did exist. In 1998, for instance, the Chinese government had promoted a "zero-growth policy" that capped China's total catch at 1998 levels.

For its part, the FAO noted that it had treated China's catch statistics with caution in its analyses. In its recent publications, global fisheries catch had been reported twice, once with and once without China's data. The FAO had also held several meetings with Chinese officials to discuss ways of reducing overreporting. More importantly, the FAO suggested, nothing in its reports had encouraged complacency.

Although it had indeed reported that the global catch remained stable through the 1990s, it also noted that many individual fisheries had declined or collapsed and that others were being aggressively overfished. "The 'news' offered in *Nature*," the FAO concluded, "simply confirms important information already available for many years from FAO."

Assigning blame for the state of the world's fisheries is not the most productive response to Watson and Pauly's findings, says fisheries scientist Andy Rosenberg of the University of New Hampshire. "This is a global problem, not a case of a few bad actors."

is that not all data supplied to international monitoring agencies may be accurate (see The Science behind the Story).

We are "fishing down the food chain"

Overall figures on total global catch do not include information on the species, age, and size of fish harvested—and this is crucial information. Careful analyses of fisheries data have revealed in case after case that the size of fish caught declines with increased fishing. As

mortality rates increase among the fish, the average age and size of individuals available declines. In addition, as particular species become too rare to fish profitably, fleets begin targeting other species that are in greater abundance. Generally this means shifting from large, desirable species to smaller, less-desirable ones. Fleets have time and again depleted popular food fish such as cod and snapper and shifted their emphasis to low-value species. Because this often entails catching species at lower trophic levels, this phenomenon has been termed "fishing down the food chain."

Some fishing practices kill nontarget animals and damage ecosystems

The removal of species at high trophic levels from marine environments, particularly those that act as keystone species (Chapter 5), can have serious ramifications for marine ecosystems. Fishing practices can also harm ecosystems in other ways. Many practices catch more than just the species they target. **By-catch** refers to the capture of animals not meant to be caught, and it accounts for the deaths of many thousands of fish, sharks, marine mammals, and birds each year.

Boats that drag huge driftnets through the water (Figure 13.24a) capture substantial numbers of dolphins, seals, and sea turtles, as well as countless non-target fish. Most of these end up dying from drowning (mammals and turtles need to surface to breathe) or from air exposure on deck (fish breathe through gills in the water). Many nations have banned or restricted driftnetting because of excessive by-catch. The widespread death of dolphins in driftnets also motivated consumer efforts to label tuna as "dolphin-safe" if its capture uses methods designed to avoid dolphin by-catch. Such measures have helped reduce dolphin deaths from an estimated 133,000 per year in 1986 to less than 2,000 per year in 1998.

Similar by-catch problems exist with longline fishing (Figure 13.24b), which involves dragging extremely long lines with baited hooks spaced along their lengths. Besides catching nontarget turtles and sharks, longline fishing kills many albatrosses, magnificent seabirds with wingspans up to 3.6 m (12 ft). It is estimated that 300,000 seabirds of various species die each year from longline fishing. Fortunately, tying colored ribbons on the lines to make them more visible scares bird away from the hooks as they are being reeled in or out of the water.

Other fishing practices can directly damage entire ecosystems. Bottom-trawling (Figure 13.24c) is the practice of dragging weighted nets over the floor of the continental shelf in order to catch such benthic organisms as scallops and groundfish. Trawling crushes many organisms in its path and leaves long swaths of sea bottom damaged, especially those areas with structural complexity, such as reefs, that animals use for shelter. Only in recent years has underwater photography begun to reveal the extent of structural and ecological disturbance done by trawling.

Marine Conservation Biology

Because we bear responsibility and stand to lose a great deal if valuable ecological systems collapse, marine scientists have been working to develop solutions to the

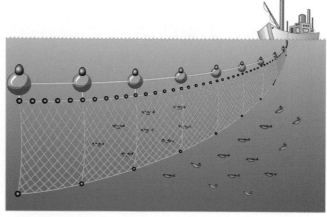

(a) Driftnetting

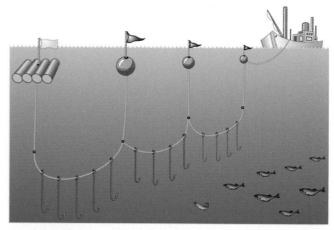

(b) Longlining

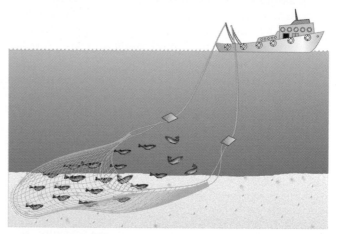

(c) Bottom-trawling

Figure 13.24 Commercial fishing fleets use several methods of capture. In driftnetting (**a**) huge nets are dragged through the open water to capture schools of fish. In longlining (**b**), lines with numerous baited hooks are pulled through the open water. In bottom-trawling (**c**), weighted nets are dragged along the floor of the continental shelf. All methods result in large amounts of bycatch, or capture of non-target animals. Bottom-trawling can also result in severe structural damage to reefs and benthic habitats.

problems that threaten the oceans. Many have begun by taking a hard look at the strategies used traditionally in fisheries management.

Traditional fisheries management is based on maximum sustainable yield

For decades, fisheries management has used scientific assessments to seek to ensure sustainable harvests. Historically, fisheries managers have studied fish population biology and used that knowledge to regulate the timing of harvests, the techniques used to catch fish, and the scale of the harvest. The goal was to allow for maximal harvests of particular populations while keeping fish available for the future, a concept called **maximum sustainable yield.** If data indicated that current yields looked unsustainable, managers might limit the number or total mass of that fish species that could be harvested or might restrict the type of gear fishermen could use.

Despite such efforts, many fish stocks have plummeted. Numerous marine scientists and some managers now feel it is time to rethink fisheries management. One key change these reformers suggest is to shift the focus away from the individual fish species and toward viewing marine resources as elements of larger ecological systems. This means considering the effects of fishing practices on habitat quality, on interspecific interactions, and on other ecological factors that may have indirect or long-term effects on populations. One key aspect of an ecosystem-based approach is to set aside areas of ocean where systems can function without human interference.

We can protect areas in the ocean

Large numbers of **marine protected areas (MPAs)** have now been established, most of them along the coastlines of developed countries. The United States contains by one count nearly 200 MPAs, and a 2000 executive order from the Clinton administration encouraged the establishment of more. However, despite the name, marine protected areas do not necessarily protect their natural inhabitants. In fact, nearly all MPAs allow fishing or other extractive activities. As a recent report from an environmental advocacy group put it, even national marine sanctuaries "are dredged, trawled, mowed for kelp, crisscrossed with oil pipelines and fiber-optic cables, and swept through with fishing nets."

Because of the lack of true refuges from fishing pressure, many scientists have begun agitating for the establishment of areas where no fishing is allowed. Such "no-take" areas have come to be called **marine reserves.** Designed to serve as refuges to preserve entire ecosystems intact without human interference, marine reserves are also intended to improve fisheries. Scientists have argued that marine reserves can act as production factories for fish for surrounding areas, because fish larvae produced inside reserves will disperse outside and stock other parts of the ocean. By serving both purposes, proponents argue, marine reserves are a win-win proposition for both environmentalists and fishermen.

Weighing the Issues:
Preservation on Land and at Sea

Almost 4% of U.S. land area is designated as wilderness, yet far less than 1% of coastal waters are being protected in reserves. Why do you think it has taken so long for the preservation ethic to make the leap to the oceans?

Marine reserves have met forceful opposition

Many fishermen, however, don't agree. Nearly every marine reserve proposed has met with pockets of intense opposition from people and businesses who use the area for fishing or recreation. Opposition comes from industrial fishing fleets that fish commercially as well as from individuals who fish recreationally. They are concerned that marine reserves will not increase fish stocks but will simply put more areas off-limits to fishing.

In the Florida Keys, property rights advocates who had opposed the sanctuary protested the 1998 establishment of reserve zones where fishing was to be prohibited. They rallied citizen opposition in the local newspapers and filed a $27 million lawsuit against the federal government. In some parts of the world, such protests have become violent. Fishermen in the Galapagos Islands have rioted, looted, and destroyed the administration building at Galapagos National Park to protest fishing restrictions. However, as people in the Keys have come to see that the reserves have not threatened their access as much as opponents feared, and as they begin to see improvements in the marine life and ecosystems around them, many former opponents have become supporters. Reserve workers estimated in 2000 that 70% of Keys residents supported the no-take reserves and that more than 50% would support establishing more of them.

Scientists insist that reserves work for both fish and fishermen

Part of the reason fishermen have been reluctant to view no-take marine reserves as a win-win solution that benefits ecosystems, fish populations, and fishing economies

Marine Reserves Work

In November 2001, a team of fisheries scientists published a paper in the journal *Science,* providing some of the first clear evidence that reserves could benefit nearby fisheries. The team, led by York University researcher Callum Roberts, focused on reserves off the coast of Florida and the Caribbean island of St. Lucia.

Following the establishment in 1995 of the Soufrière Marine Management Area (SMMA), a network of reserves intended to help restore St. Lucia's severely depleted coral reef fishery, Roberts and colleague, Julie Hawkins conducted annual visual surveys of fish abundance in the reserves and nearby areas. Within 3 years, they found that the biomass of five commercially important families of fish— surgeonfishes, parrot fishes, groupers, grunts, and snappers— had tripled inside the reserves and doubled outside them. Roberts and Hawkins also interviewed local fishers and found that those with

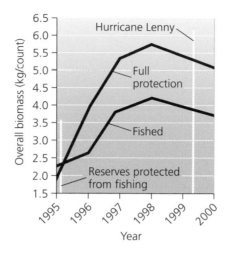

Established in 1995, the Soufrière Marine Management Area (SMMA), along the coast of St. Lucia, had a rapid impact. By 1998, fish biomass within the five reserves tripled and in adjacent, fished areas, it doubled. Data from Callum M. Roberts, Effects of marine reserves on adjacent fisheries, *Science,* 2001.

large traps were catching 46% more fish per trip in 2000–2001 than they had in 1995–1996, and fishers with small traps were catch-

ing 90% more. Roberts and his colleagues concluded that "in 5 years, reserves have led to improvement in the SMMA fishery, despite the 35% decrease in area of fishing grounds."

Roberts and his colleagues also studied the oldest fully protected marine reserve in the United States, the Merritt Island National Wildlife Refuge (MINWR), established in 1962 as a buffer around what is today the Kennedy Space Center on Cape Canaveral, Florida. In a previous study, Darlene Johnson and James Bohnsack of the National Oceanic and Atmospheric Administration and Nicholas Funicelli of the United States Geological Service had found that the reserve contained more and larger fish than nearby unprotected areas and that some of the reserve's fish appeared to be migrating to nearby fishing areas. Bohnsack, Roberts, and their colleagues corroborated the evidence for migration by analyzing trophy records from the International

---

alike is that scientific studies of such protection have only recently begun to yield results. In the past few years, however, large amounts of data have been synthesized from reserves around the world, and these data indicate that marine reserves *do* work as scientists had hoped. The first statement came in 2001, when 161 prominent marine scientists signed a "consensus statement" summarizing the effects of marine reserves. Besides boosting fish biomass, total catch, and recordsized fish, the report stated, marine reserves yield several benefits. Within reserve boundaries, they

- Produce rapid and long-term increases in abundance, diversity, and productivity of marine organisms.
- Decrease mortality and habitat destruction.
- Lessen the likelihood of extirpation of species.

Outside reserve boundaries they

- Can create a "spillover effect" when protected species spread outside reserves.
- Allow larvae of species protected within reserves to "seed the seas" outside reserves.

The consensus statement was backed up by research into reserves in the Caribbean and Florida (see The Science behind the Story) and others worldwide. At Apo Island in the Philippines, fishing improved outside a reserve as biomass of large predators increased eightfold inside the reserve. At two coral reef sites in Kenya, commercially fished and keystone species were up to 10 times more abundant in the protected area as in the fished area. At Leigh Marine Reserve in New Zealand, snapper increased 40-fold, and spiny lobsters were increasing 5–11% yearly. Spillover from this reserve improved fishing and ecotourism, and local residents who were opposed to the reserve now support it. In the

Game Fish Association. They found that the proportion of Florida's record-sized fish caught near Merritt Island increased significantly afte 1962. Nine years after the refuge was established, for instance, the number of spotted sea trout records from the Merritt Island area jumped dramatically. Bohnsack, Roberts, and their colleagues hypothesized that the reserve was providing a protected zone in which fish could grow to trophy size before migrating to nearby areas, where they were caught by recreational fishers.

Not everyone saw the St. Lucia and Merritt Island cases as proof that marine reserves could rescue depleted fisheries. In February 2002, several alternative interpretations were published as letters in *Science*. Mark Tupper, a fisheries scientist at the University of Guam, suggested that the St. Lucia results were relevant only to coral reef fisheries in developing nations, whereas Florida's boost in fish populations was due primarily to

limits on recreational fishing. Karl Wickstrom, editor-in-chief of *Florida Sportsman* magazine, suggested that the increase in trophy fish near MINWR was caused by commercial fishing regulations and changes in how trophies were recorded and promoted. And Ray Hilborn, a fisheries scientist at the University of Washington, challenged the study's scientific methods. In the St. Lucia case, he pointed out, there had been no control condition.

In response, Roberts and his colleagues reaffirmed the validity of their results while acknowledging certain limitations. They agreed with Tupper that marine reserves are not always effective and often need to be complemented by other management tools, such as size limits. "We agree that inadequately protected reserves are useless," they wrote, "but our study shows that well-enforced reserves can be extremely effective and can play a critical role in achieving sustainable fisheries."

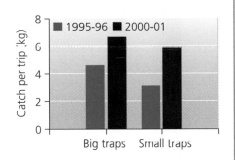

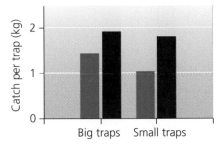

Roberts and his colleagues studied fish biomass of the SMMA over two 5-month periods in 1995–1996 and 2000–2001, by collecting data using large and small fishing traps. For fishers with big traps, catch increased 46%, and for those with small traps, it increased 90%. Per trap, catch increased 36% for big traps and 80% for small traps. Data from Callum M. Roberts, Effects of marine reserves on adjacent fisheries, *Science*, 2001.

Florida Keys, surveys just 3 years after establishment of the no-take zones showed increases in size and number of spiny lobsters and in populations of three of four major reef fish. Keys fishermen are catching more fish, and fishing revenues are up.

The review of data from existing marine reserves as of 2001 revealed that within only 1–2 years after their establishment, marine reserves

- Increased densities of organisms on average by 91%.
- Increased biomass of organisms on average by 192%.
- Increased average size of organisms by 31%.
- Increased species diversity by 23%.

From data like these and considerations of socioeconomic factors, the scientific consensus statement drew a series of conclusions (Table 13.2).

The question is how best to design reserves

If marine reserves work in principle, the question becomes how to best design reserves and arrange them into networks. Scientists today are asking how big the reserves need to be, how many there need to be, and where they need to be placed. Of the 40 studies that have estimated how much area of the ocean should be protected in no-take reserves, estimates range from 10% to 65%, with most falling between 20% and 50%. Other studies are modeling how to optimize the size and spacing of individual reserves so that ecosystems are protected, fisheries are sustained, and people are not overly excluded from marine areas (Figure 13.25). If marine reserves are designed strategically to take advantage of ocean currents, many scientists say, then they may well seed the seas and help lead us toward solutions to one of our most pressing environmental problems.

Table 13.2 Conclusions from the 2001 Scientific Consensus Statement on Marine Reserves

The 161 scientists concluded that . . .

Reserves conserve both fisheries and biodiversity

To meet goals for fisheries and biodiversity conservation, reserves must encompass the diversity of marine habitats

Reserves are the best way to protect resident species and provide heritage protection to important habitats

Reserves must be established and operated in the context of other management tools

Reserves need a dedicated program to monitor and evaluate their impacts both within and outside their boundaries

Reserves provide a critical benchmark for the evaluation of threats to ocean communities

Networks of reserves will be necessary for long-term fishery and conservation benefits

Existing scientific information justifies the immediate application of fully protected marine reserves as a central management tool

Data from The National Center for Ecological Analysis and Synthesis, Scientific consensus statement on marine reserves and marine protected areas, 2001.

	Consequences for conservation	Consequences for small fisheries	Consequences for commercial fisheries	Overall consequences
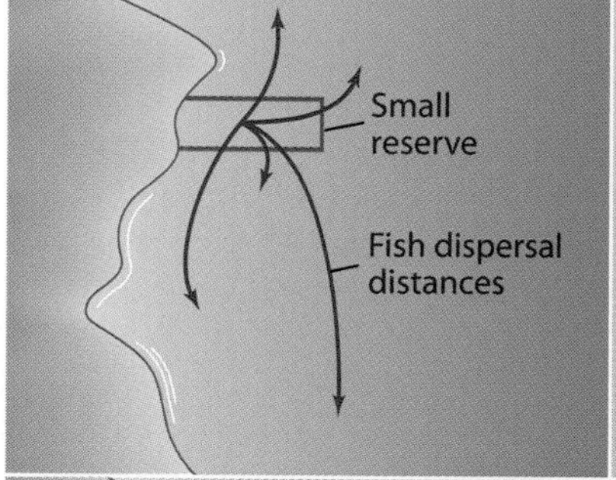 Small reserve / Fish dispersal distances	Reserve is not self-sustaining; most species are lost.	High periphery-to-area ratio makes many fish available to fishers outside reserve. This is unsustainable, however, because small-size reserves offer too little protection for fish populations to sustain themselves.	Provides little or no boost to fish populations outside reserve, but does not appreciably decrease area of fishing grounds.	Too many fish wander out of reserve, yet fisheries experience little gain. Many populations are not sustained. Reserve fails to meet any stakeholder goals.
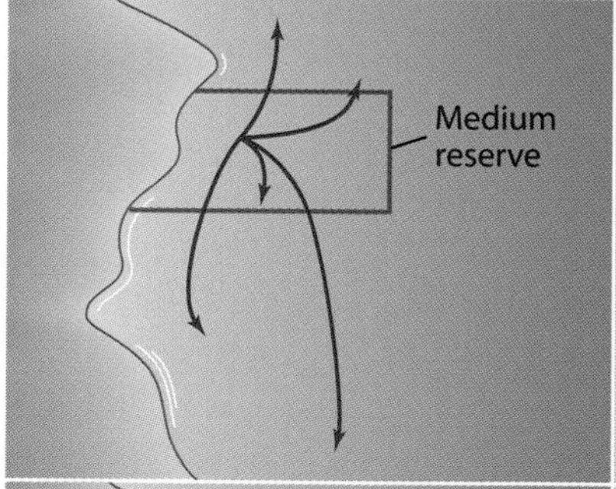 Medium reserve	Reserve is moderately self-sustaining; some species are lost.	Intermediate periphery-to-area ratio makes some fish available to fishers outside reserve, while offering a large-enough area of refuge to protect other fish, thus sustaining many populations.	Provides a significant boost to fish populations outside reserve, with only a moderate decrease in area of fishing ground.	Conservation is effective inside reserve, and fisheries gain fish while not losing too much fishing area. Good balance of benefits for all stakeholders.
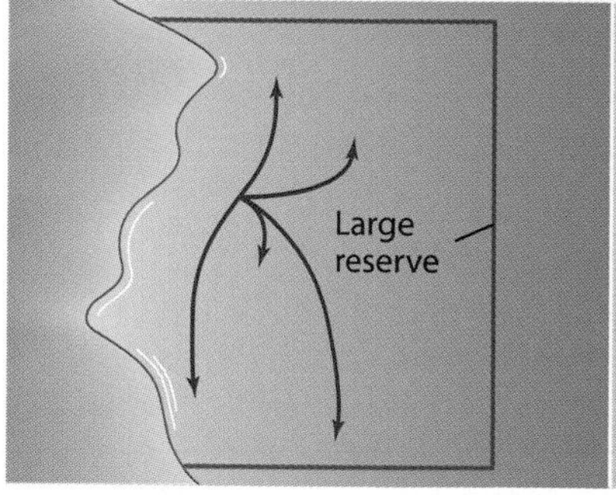 Large reserve	Reserve is completely self-sustaining; all species are retained.	Low periphery-to-area ratio makes relatively few fish available to fishers outside reserve; the "spillover" for fisheries is small in relation to the size of the area protected in the reserve.	Provides relatively little boost to fish populations in regions around reserve, and severely decreases area of fishing grounds.	Preservation is effective inside reserve, but too few fish spill over into fishing grounds that have been much-reduced in area. Outcome not acceptable to fisheries stakeholders.

Figure 13.25 Marine reserves may have different effects on ecological communities and fisheries, depending on the size of the reserves. Young and adult fish and shellfish of different species can disperse different distances, as indicated by the red arrows in the figure. More animals will disperse out of small reserves than out of large reserves. A reserve that is too small may fail to protect animals because too many disperse out of the reserve. A reserve that is too large may protect fish and shellfish very well but will provide relatively less "spillover" into areas that can legally be fished by people. Thus medium-sized reserves may offer the best hope of preserving species and ecological communities while also providing adequate fish to fishermen and human communities. Determining the actual size of such reserves, however, requires a great deal of real data on dispersal distances, water currents, community composition, animal behavior, and human economies. *Source:* B.S. Halpern and R.R. Warner, Matching marine reserve design to reserve objectives, *Proceedings of the Royal Society: Biological Sciences*, 2003.

Marine Reserves

Can marine reserves or other forms of "no-fishing" zones help us solve problems facing the oceans today? Why or why not?

"No-fishing" Zones Do Not Prevent Overfishing

If you close off part of the ocean from all forms of "take," there will obviously be more fish in the protected area. That is just common sense. However, a major problem with our oceans is severe overfishing. Fish are being killed faster than they can replenish themselves. There is a simple way to improve this situation—stop killing so many fish. "No-fishing" zones don't accomplish this. What they do is shift the fishing pressure to another area. Anything short of reducing the number of fish being killed, such as marine reserves, not only won't solve the problem, but it may give a false impression that something is being done, thus delaying action to actually solve the problem.

There are many traditional fishery management tools that can be used to prevent overfishing. These tools include bag limits, size limits, slot limits, closed seasons, protected species, and catch and release only species. For commercial fishermen there are quotas, gear restrictions, trip limits, and limited entry, plus all the management tools listed above for recreational anglers.

A major problem for fishery managers has been politics. Commercial fishing interests have very effective lobbyists who have been instrumental in preventing or delaying needed management restrictions. If lobbying fails, they often challenge regulations through the court system. The result is management regulations that are not restrictive enough. Marine reserves will not solve this problem.

In cases where remote, pristine areas are meant to be preserved, perhaps marine reserves allowing catch and release only could be effective. Examples of this are the remote reefs in northern Hawaii and some snapper spawning areas far west of Key West, Florida. However, "no fishing" zones by themselves will not rebuild fisheries.

Michael Leech joined the International Game Fish Association (IGFA) in 1983 and became President of IGFA in 1992. Under his leadership the IGFA recently constructed their permanent world headquarters in Dania Beach. The IGFA Fishing Hall of Fame and Museum is not only a tourist attraction but serves as the world center for most fishing-related information.

Marine Reserves Restore Ecosytems

Marine reserves are a powerful tool for protecting and restoring marine ecosystems. They are successful because they protect not only species but habitats as well. In the past, there were innumerable naturally recurring marine reserves around the oceans—places that were too far from land, too deep, or too rocky to fish. Modern technology systematically eliminated those reserves, and today we protect far less than 1% of the oceans in established reserves.

Recent scientific studies have demonstrated that reserves provide a clear benefit to conservation of marine organisms: biomass, density, individual size, and diversity all increase inside reserves. It is especially important to protect large marine organisms, which produce disproportionately more offspring than smaller ones. A 37 cm (14.6 in) vermilion rockfish produces 150,000 young, whereas a 60 cm (23.6 in) one produces 1.7 million young! Allowing individuals to get big and fat is very valuable. Modeling results indicate that reserves substantially benefit fisheries, from both spill over (fish spilling from the reserve) and export (larvae produced inside the reserve and transported away by currents). However, not all species will recover immediately when a reserve is established. Species that grow slowly and reproduce late will need longer to colonize and recover.

Networks of marine reserves, connected by the movement of organisms, are likely to provide the best combination of conservation and fishery benefit. A network provides protection for a large total area, and long perimeter over which organisms can escape to reseed adjacent areas. Networks of reserves are not a panacea—they need to be coupled with good fishery management and pollution control—but marine reserves are one of the most promising tools available for solving the problems plaguing ocean environments today.

Jane Lubchenco is a Professor of Marine Biology and Zoology at Oregon State University. She works with policy makers, business leaders, private foundations, religious leaders, other scientists, governmental and non-governmental organizations, and students to help figure out how to make a transition to sustainability. She is President of the International Council for Science and co-founded the Aldo Leopold Leadership Program, PISCO—the Partnership for Interdisciplinary Studies of Coastal Oceans and COMPASS—the Communication Partnership for Science and the Sea.

Conclusion

In the Florida Keys and hundreds of other protected areas around the country, scientists are gradually demonstrating to the public and to wary fishermen that setting aside protected areas of the ocean serves to maintain natural systems and also to enhance fisheries. In some parts of the world, the public and politicians have jumped on board enthusiastically. Off the coast of New Jersey and several other states, government agencies are busy constructing "artificial reefs"—piles of concrete and large, bulky waste items, such as old subway cars (Figure 13.26)—that they hope will serve as homes for marine life and will boost fishing. Although some see such efforts as simply an excuse to dispose of garbage, others hope that they will help create new ecosystems and reinvigorate depleted fisheries. As historical studies reveal more information on how much biodiversity our oceans formerly contained and have now lost, we may increasingly look beyond simply making fisheries stable and instead consider restoring the ecological systems that used to flourish in our waters.

Figure 13.26 The state of New Jersey is dumping old New York City subway cars into the waters off its coast in an attempt to build "artificial reefs" to serve as refuges for marine life and help reinvigorate coastal fisheries.

REVIEW QUESTIONS

1. What is a marine protected area? How does it differ from a marine reserve?

2. What do oceanographers study?

3. About how much of Earth's surface is covered by the "world ocean?" About how much salt does ocean water contain, and why does salinity vary?

4. What produces dissolved oxygen in ocean salt water?

5. How are water density, salinity, and temperature related at each layer of ocean water?

6. How does Earth's atmospheric heat content compare to the heat content of its oceans?

7. What factors drive the system of ocean currents? In what directions do ocean currents move, and how do such movements affect conditions for life in the oceans?

8. What kinds of information are displayed in bathymetric and topographical maps of the oceans?

9. Near what physical structures in the oceans are productive areas of biological activity likely to be found?

10. Describe three kinds of ecosystems found near coastal areas and the kinds of life they support.

11. Why are kelp forests critical ecosystems for both humans and marine life?

12. In what forms may a coral reef occur? What are three reasons many coral reefs are dying?

13. What are the characteristics of a salt marsh?

14. Compare mangrove forests and salt marshes. What is causing the disappearance of these systems?

15. How does the process of shipping contribute to the spread of invasive species worldwide?

16. What kinds of non-renewable energy resources do humans extract from the oceans? What potential renewable sources of energy might the oceans provide?

17. Describe three major forms of pollution in the oceans and the consequences of each.

18. Provide an example of how overfishing can lead to ecological damage and fishery collapse.

19. According to the 2001 U.N. Food and Agriculture Organization (FAO) report, roughly what percentage of the world's fish stocks can experience further harvesting without going into decline? Explain the conclusion of the Myers and Worm study of 2003.

20. Explain how three forms of industrial fishing create bycatch and harm marine life.

21. What do some scientists see as a drawback to the fisheries management approach called "maximum sustainable yield"?

22. What is the only form of marine protected area (MPA) that is completely protected from human activities? How might this kind of region ultimately benefit human fishing interests?

23. Explain each side of the argument for and against marine reserves.

DISCUSSION QUESTIONS

1. What benefits do you derive from the oceans? How does your behavior affect the oceans?

2. We have been able to reduce the amount of oil we spill into the oceans, but petroleum-based products like plastic also continue to litter our oceans and shorelines. Discuss some ways that we can reduce this threat to the marine environment?

3. Why has global fish production remained stable for 15 years?

4. Consider what you know about biological productivity in the oceans, about the data in the chapter on the successes of marine reserves so far, and about the social and political issues surrounding the establishment of marine reserves. What ocean regions do you think it would be particularly appropriate to establish as marine reserves?

5. Why does the 2001 scientific consensus statement argue for networks of reserves to be established? Based on what you know about marine organisms, ecosystems and fisheries, can you think of particular reasons why networks of reserves may be beneficial?

Media Resources *For further review, go to the website* **www.envscienceplace.com** *or student CD-ROM, where you will find quizzes, flashcards, a glossary, additional interactive exercises, and links to relevant news and research sources. Also, on the website and CD-ROM is* **GRAPH IT**, *a series of interactive graphing tutorials to help you interpret graphs and plot data.*

14 Freshwater resources

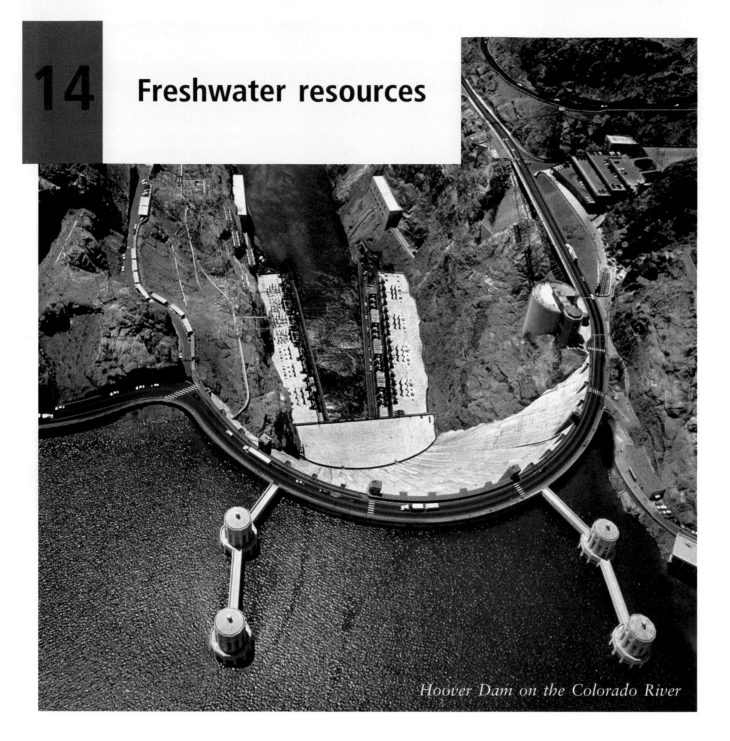

Hoover Dam on the Colorado River

This chapter will help you understand:

- The importance of water to ecosystems, human health, and economic pursuits

- The hydrologic cycle and human interactions with it

- Freshwater distribution on Earth

- Freshwater ecosystems

- How we use water and alter freshwater systems

- Water quantity problems: freshwater depletion

- Water quality problems: water pollution

- Solutions to problems of depletion and pollution

Colorado River Delta

Central Case: Plumbing the Colorado River

"We've gone from being assured that we lived in this magical place where the rules of water didn't apply to [a] wake-up call about the fact that we do live in the California desert. People have lived in this false water utopia."
—*Buford Crites, City Councilor, Palm Desert City, California*

"Water promises to be to the 21st century what oil was to the 20th century: the precious commodity that determines the wealth of nations."
—Fortune *magazine, May 2000*

As the clock struck midnight on New Year's Eve, millions of Californians toasted the arrival of 2003 with champagne. But many people in the state that night had another liquid on their minds: water. Their fears were borne out the next day when the U.S. government followed through on its threat to cut off 15% of California's water supply from the Colorado River—the lifeline which had turned the desert green, which drives one of the world's biggest economies, and without which southern California as we know it could simply not exist. In ordering the cutoff, Interior Secretary

Gale Norton was simply holding up her end of a deal that an irrigation district in California had scuttled. It may seem bizarre that a 3–2 vote of one county irrigation district could block enough water for 1.6 million households in Los Angeles and San Diego, but it is just the latest episode in the colorful history of California water politics and the century-long water wars among the seven states battling for rights to what was once the West's wildest river.

The Colorado River begins in the high peaks of the Rocky Mountains, charges through the Grand Canyon, crosses the border into Mexico, and dumps into the Sea of Cortez, draining 637,000 km² (246,000 mi²) of southwestern North America. Its raging waters have chiseled through thousands of feet of bedrock, creating the Grand Canyon and leaving extraordinary scenery along its 2,330 km (1,450 mi) length. Today, however, the Colorado's waters are pooled up behind the massive dams erected to manage their flow and are siphoned away to irrigate farm fields. Only a small amount of water—and sometimes none at all—reaches the river's mouth at the Sea of Cortez.

The waters of the Colorado River irrigate 7% of U.S. cropland, slake the thirst of over 20 million people, keep hundreds of golf courses green in the desert, and fill the swimming pools and fountains of Las Vegas casinos. The Colorado provides vital water to the rapidly growing desert metropolises of the U.S. Southwest—Phoenix, Tucson, Las Vegas, San Diego, Los Angeles, and many others. Behind the river's five major dams, reservoirs store four times as much water as flows in the river in an entire year. About 80% of this water is stored in Lake Mead behind Hoover Dam (completed in 1936) and in Lake Powell behind Glen Canyon Dam (completed in 1964). The Colorado's dams provide flood control and recreation, produce about 12 billion kilowatt-hours of electricity from hydroelectric power each year, and provide irrigation that makes agriculture possible in this arid region.

For 80 years, the seven states along the Colorado have attempted to divide the river's water among themselves, guided by the Colorado River Compact they signed in 1922, which apportioned water to each state. California has long exceeded its allotted annual amount of 4.4 million acre-feet (an acre-foot is the volume of water that would cover one acre to a depth of one foot). California has been permitted to exceed its allotment because Colorado, Wyoming, Utah, Nevada, New Mexico, and Arizona have not used all of their portions. With the populations of these states booming, however, they will soon need their full shares, so the federal government under Interior Secretary Bruce Babbitt in 2000 pressured California to lower its usage. California agreed to reduce its withdrawals to 4.4 million acre-feet, implemented gradually over 15 years.

California worked hard to get agricultural districts, which controlled most of the water distribution in the state, to agree to sell part of their shares. At the last minute, however, the Imperial Irrigation District backed out, refusing to sell its water to San Diego. The New Year's deadline passed, and 2003 saw the federal cutoff implemented, the irrigation district suing the government, the rest of California angry with the irrigation district, and the Interior Department offering to renew the water deliveries if the state could only get its house in order. At last, 10 months later, one member of the Imperial Irrigation District changed his vote, the deal was patched up, and the federal water cutoff was ended. Southern California's residents were able to continue living—at least for a little while longer—the mirage in the desert that is southern California.

Movement and Distribution of Freshwater

The well-known line from Coleridge's poem *The Rime of the Ancient Mariner*, "Water, water, everywhere, nor any drop to drink," describes the situation on our planet as a whole. Water may seem abundant to us because it comes out of our faucets every time we turn them on, but in the global perspective it is quite rare and limited. About 97.5% of Earth's water resides in the oceans and is too salty for humans to drink or use to water crops. Only 2.5% is considered **freshwater**, water that is relatively pure, holding very few dissolved salts. Most freshwater is tied up in glaciers and icecaps, about a fifth is in underground aquifers, and only 1% resides in surface reservoirs such as lakes, soil, and atmospheric moisture. Only about half of this 1% is in lakes, rivers, and streams, comprising 0.01325% of the planet's total water (Figure 14.1). Thus, just over 1 part in 10,000 of Earth's water is easily accessible to us for drinking and irrigation.

Water moves in the hydrologic cycle

At any given time, each of Earth's water reservoirs contains close to the amounts specified in Figure 14.1, but water is constantly moving among them in what we know as the hydrologic cycle (Chapter 6, Figure 6.23). The hydrologic, or water, cycle interacts with all other biogeochemical cycles, and the hydrosphere is intimately linked with the biosphere, lithosphere, and atmosphere. As water moves, it redistributes heat, erodes mountain ranges, builds river deltas, maintains organisms and ecosystems, shapes civilizations, and gives rise to political conflicts. The general properties of the hydrologic cycle discussed in Chapter 6 can be better understood by examining how the cycle works in the Colorado River watershed in particular.

The Colorado River originates high in the Rocky Mountains of the state of Colorado, from waters that run down the western slope from spring snowmelt and glaciers in the high peaks. The western slope of the Rockies receives precipitation when warm moist winds from the west are forced upward by the rugged topography. As the air rises into regions of lower pressure, it expands, cools, and increases in relative humidity (Chapter 11). Clouds form from the enhanced humidity, and moisture condenses, creating rain or snow. The air mass passes over the crest of the mountains and sinks

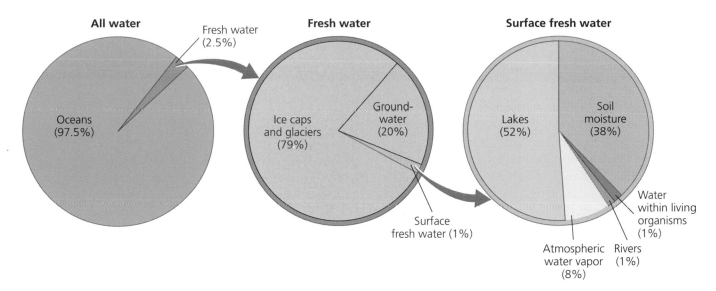

Figure 14.1 Only 2.5% of Earth's water is freshwater, water without dissolved salts. The rest is salty ocean water. Of that 2.5%, most is tied up in glaciers and ice caps, and only about 1% comprises surface water. Of this 1%, the vast majority is in lakes and soil moisture. Data from UNEP and World Resources Institute, as presented by: P. Harrison and F. Pearce, *AAAS Atlas of Population and the Environment*, 2000.

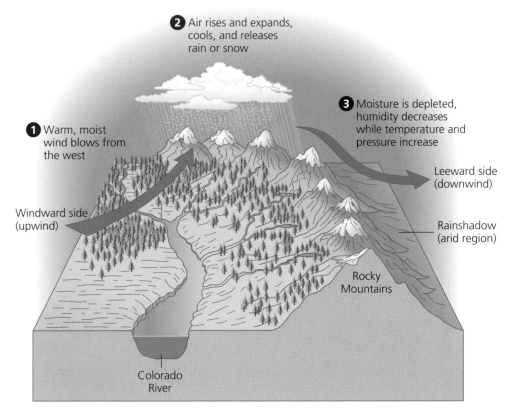

Figure 14.2 The Colorado River originates high in the Colorado Rockies, where precipitation falls predominantly on west-facing slopes. Air in weather systems approaching from the west is pushed up in elevation as it reaches the Rockies. Rain or snow then condenses and falls in mountainous areas, particularly windward slopes. Air depleted of moisture then continues eastward, creating an arid rainshadow on the leeward side of the mountains.

down the eastern slope, depleted of its moisture. Relative humidity decreases, while temperature and pressure increase. As a result, humidity and precipitation are higher on the windward (upwind) side of the mountain range and lower on the leeward (downwind) side, creating on the leeward side an arid area that geographers and atmospheric scientists term a **rainshadow** (Figure 14.2).

From rain, snow, and glacial ice deposited on the western slope of the Rockies, thousands of rivulets and creeks flow downhill into small rivers, which flow into larger rivers, which in turn join the Colorado River. Some water is held in alpine lakes and wet meadows or in marshes at lower elevations. All along the way, surface water infiltrates the soil, some reaching the water table. Other water is taken up by the roots of plants and then is transpired through their leaves, returning to the atmosphere. The water that reaches the Colorado River rages downstream through the continent's most ferocious rapids—or at least it used to before we built the gigantic dams that today make much of the Colorado a series of huge lakes. As the river's water moves slowly from one dam and reservoir to the next, a great deal is removed by pipes and canals for drinking supplies, industrial use, and especially for agricultural irrigation.

In the past, the river's copious supply of water surged into the Sea of Cortez, nourishing one of the continent's greatest estuaries. Today the withdrawals have turned the mighty river into a small stream that barely reaches the sea. What river water does enter the salt water of the Sea of Cortez eventually mixes with water of the open ocean. As ocean water evaporates into the atmosphere, it closes the circle of the hydrologic cycle. Water vapor from the Pacific Ocean is blown inland, some of it eventually being pushed up the west slope of the Rockies, to fall again as precipitation.

Because the Colorado River is located in a region of naturally low precipitation, the movement of its water has a particularly strong influence on the ecology and human communities of its watershed. The area lies in the rainshadow of California's mountain ranges, which moisture-laden air from the Pacific must traverse before reaching the Rockies. In addition, global atmospheric circulation patterns (Chapter 11) make regions around 30° latitude particularly dry. The Colorado River watershed, extending from roughly 32° to 40° N, is at the latitudinal zone of dry sinking air that gives rise to deserts around the world.

Groundwater plays a key role in the hydrologic cycle

It is easy for us to understand the movement of surface water because we witness it all the time, but the role of groundwater in the hydrologic cycle is more difficult to visualize. Any precipitation reaching Earth's land surface that does not evaporate, flow into waterways, or get taken up by organisms infiltrates the surface. Most percolates into the soil layers, moving downward to become **groundwater,** which makes up one-fifth of Earth's freshwater supply and plays a key role in meeting human water needs.

Groundwater is contained within **aquifers:** porous, spongelike layers of rock, sand, or gravel that hold water. An aquifer's upper layer, or zone of aeration, is generally riddled with open spaces that contain both water and air. In the lower layer, or zone of saturation, the spaces contain nearly all water. The boundary between these two zones is the water table. Picture a sponge resting partly submerged in a tray of water; the lower part of the sponge is completely saturated, while the upper portion may be moist but contains plenty of air in its pores. Any geographic area where water infiltrates Earth's surface and reaches an aquifer below is known as an **aquifer recharge zone.**

There are two broad categories of aquifer. A **confined,** or **artesian aquifer,** exists when a water-bearing porous layer of rock, sand, or gravel is trapped between an upper and lower layer of less permeable substrate (often clay). In such a situation, water may be under great pressure because it is trapped between these layers. A layer of less permeable substrate underlies **unconfined aquifers,** but there is no upper layer to confine them. Thus the water they contain is under considerably less pressure than that of confined aquifers (Figure 14.3).

Groundwater is not stationary. Just as surface water becomes groundwater by infiltration and percolation, groundwater becomes surface water through springs and human-drilled wells. Groundwater flows downhill and from areas of high pressure to areas of low pressure. A typical rate of groundwater flow might be about 1 m (3 ft) per day. Because of this slow movement, groundwater may remain in an aquifer for a long time. In fact, groundwater can be ancient. The average age of groundwater has been estimated at 1,400 years, and some is thought to be tens of thousands of years old. Nonetheless, the volumes of groundwater are large enough that each day in the United States alone, aquifers release 1.9 trillion liters (492 billion gallons) of groundwater into bodies of surface water—nearly as much as the daily flow of the Mississippi River. The world's largest known aquifer is the Ogallala Aquifer, which underlies the Great Plains of the United States (Figure 14.4). It spans 453,000 km^2 (176,700 mi^2), is 370 m (1,200 ft) deep at its thickest point, and has a water-holding capacity of 3,700 km^3 (881 mi).

Water is unequally distributed across Earth's surface

Fortunately, the Great Plains and its farmlands have the massive Ogallala Aquifer as a source of freshwater. Many other areas of the continent and the world are not so endowed. Different regions possess vastly different amounts

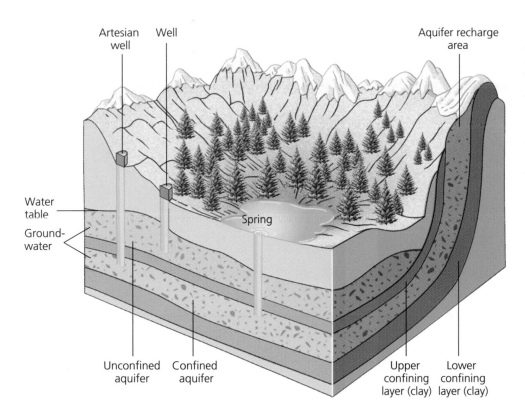

Figure 14.3 Groundwater may occur in unconfined aquifers above impermeable layers or in confined aquifers under pressure between impermeable layers. Water may rise naturally to the surface at springs, and we dig wells to extract groundwater. Artesian wells tap into confined aquifers to mine water under pressure.

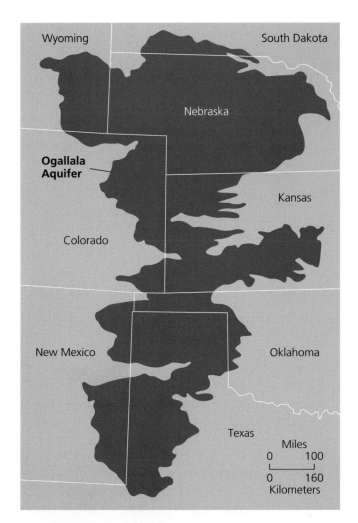

Figure 14.4 The Ogallala Aquifer is the world's largest, holding 3,700 km^3 (881 mi^3) of water before pumping began. This aquifer underlies 453,000 km^2 (175,000 mi^3) of the Great Plains beneath 8 U.S. states from South Dakota to Texas. Overpumping for irrigation is currently reducing the volume and extent of this aquifer.

of groundwater, surface water, and precipitation. Precipitation ranges from about 1,200 cm (470 in) per year at Mount Waialeale on the Hawaiian island of Kauai to virtually zero in the Atacama Desert of Chile. Annual runoff in Europe ranges from 300 cm (118 in) in western Norway to less than 2.5 cm (1 in) in parts of Spain.

People are not distributed across the globe in accordance with water availability. Many areas with high population density are water-poor (Figure 14.5), leading to inequalities in per capita water resources among and within nations. For example, Canada has 20 times more water for each of its citizens than does China. The Amazon River carries 15% of the world's runoff, but its watershed holds less than half a percent of the world's human population. Many densely populated nations, such as Pakistan, Iran, and Egypt, face serious water shortages. Asia possesses the most water of any continent but has the least water available per person, whereas Australia, with the least amount of water, boasts the most water available per person. Because of this mismatched distribution of water and human

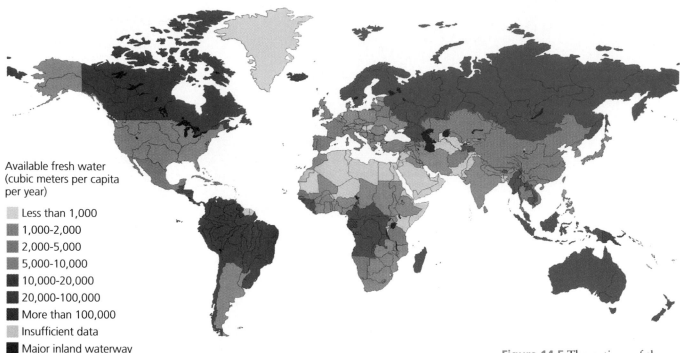

Available fresh water
(cubic meters per capita
per year)

Less than 1,000
1,000-2,000
2,000-5,000
5,000-10,000
10,000-20,000
20,000-100,000
More than 100,000
Insufficient data
Major inland waterway

(a) Water-stressed nations: Available freshwater resources, 2000

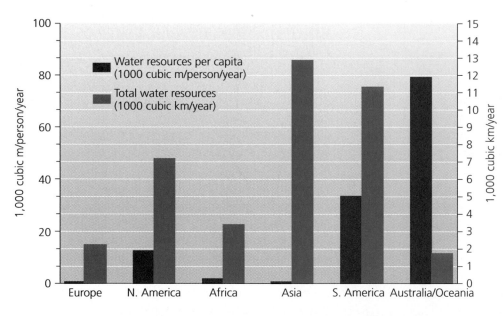

(b) Water-stressed nations: Total water resources and per capita water resources

Figure 14.5 The nations of the world vary tremendously in the amount of freshwater per capita that is available to their citizens. For instance, countries like Iceland, Papua New Guinea, Gabon, and Guyana have more than 100 times as much water as many Middle Eastern and North African countries (**a**). There is just as much variation among continents, which show great imbalances between total water resources and per-capita water resources (**b**). Heavily populated Asia has tremendous total water resources but extremely low amounts per capita, for example, while Australia and the oceanic island nations have little total water but high amounts per capita, due to their low populations. Go to **GRAPH IT** on the website or CD-ROM. Data from (**a**) UNEP and World Resources Institute, as presented by: P. Harrison and F. Pearce, *AAAS Atlas of Population and the Environment*, 2000. (**b**) U.N. Sustainable Development Programme, 2002.

population across the globe, one challenge for human societies has always been to transport freshwater from its source to where it is needed. In nearly every modern country, such transport is important for equalizing access among people in rural and urban settings and in different geographic areas.

The natural distribution of freshwater is uneven across time as well as space; rain does not always fall when humans need it. India's monsoon season brings concentrated storms in which half the country's annual rain may fall during just a few hours. Northwest China receives three-fifths of its annual precipitation during three months when crops do not need it. Unequal distribution of water is one reason humans have erected dams to store water so that it may be distributed when needed. We will examine our strategies of storing, transporting, and using water, but first let's consider how the uneven distribution of freshwater on Earth results in the creation and maintenance of freshwater ecosystems.

Freshwater Ecosystems

Surface freshwater makes up only a miniscule portion of Earth's water, and bodies of freshwater cover only a small percentage of the globe's surface. Yet freshwater ecosystems host a disproportionately large fraction of the planet's biota. Just as Colorado River water is crucial to human economies of the Southwest, freshwater ecosystems support a substantial number of the Earth's living organisms, both in types of species and in numbers of individuals. We will discuss several classes of ecosystems, which you may wish to think of as being like the aquatic equivalent of biomes. These are types of water bodies with similar hydrological and ecological dynamics and that share similar suites of plants and animals (Figure 14.6).

Rivers and streams wind through landscapes

Bodies of actively flowing water comprise one major class of freshwater ecosystems. As water from rain, snowmelt, or springs runs downhill, it tends to come together in small rivulets where the topography dips lowest. Eventually these rivulets join together to form water bodies large enough to be called streams, creeks, or brooks. Aquatic insects such as water striders and water beetles thrive in such habitats, and the larvae of many insects—dragonflies, damselflies, mayflies, and so forth—develop in streams before metamorphosing into adults that live terrestrially or take to the air. The speed at which water flows can influence which animals and plants live in streams. Mosquito larvae prefer placid water, while the larvae of blackflies attach themselves to rocks in fast-moving water. Some vertebrate animals also depend on streams; certain types of salamanders live their entire lives in streams, and others spend their larval stage in streams. Many streams are seasonal, running high in wet seasons and low to not at all in the driest part of the year. In regions moist enough to maintain water flow year-round, fish may thrive in streams. Streams also provide water to many terrestrial animals for drinking and bathing.

As streams flow downhill, they join one another and eventually form larger water channels, or rivers. Small rivers join to form larger rivers, and eventually river water reaches the ocean. A smaller river flowing into a larger one is called a tributary of the larger river. As we learned in Chapter 3, the area of land that includes the land drained by a river and all its tributaries is called that river's watershed. Rivers and streams share some plants and animals. Salmon that migrate from oceans up rivers to spawn often travel up tributaries and streams before mating and laying eggs. Kingfishers, birds that dive into the water to capture fish, patrol the banks of watercourses large and small.

Due to their size and power, rivers often exert significant forces to shape the local landscape. Where a river bends, the force of water rounding the bend gradually eats away at the shore along the outer edge of the bend, eroding soil from the bank. Meanwhile, sediment deposits along the inside of the bend, where water currents are weaker. In this way, over many years, river bends can become more and more exaggerated in shape. Eventually, a bend may become such an extreme loop that water begins to erode a shortcut from one end of the loop to the other. The bend may then be cut off, as the river pursues a direct course, and the bend is left to form an isolated U-shaped water body called an *oxbow lake*. Over many thousands or millions of years, a river may move from one course to another, back and forth over a large area. The region over which it moves becomes flat and laden with silt and is called a **floodplain.** Fertile soil endows floodplains with diverse riparian communities of plants and animals, making such regions productive for agriculture.

Lakes and ponds change with time

Small bodies of water that do not actively flow are often called ponds, and large bodies of still water are termed lakes. Some ponds and lakes are located in topographical depressions from which there is no route of exit; streams or rivers flow into them but do not flow out. Other ponds and lakes have watercourses both entering and exiting them. The depth of ponds and lakes varies greatly, and the type of life within them varies with depth. Shallow waters allow more light to enter, enabling photosynthesis and plant growth. Deep water lacks sunlight and is also colder in temperature. Waters that are clear allow sunlight to penetrate more deeply, whereas turbid waters full of particulate matter do not. The concentration of dissolved oxygen depends on the

(a) Meandering river in Colorado

(b) Torres del Paine National Park, Chile

(c) Dead Sea, Israel

(d) Linyanti Swamp, Botswana

Figure 14.6 Freshwater provides the basis for a number of major ecosystem types that are typically rich in life. Rivers and streams involve water flowing downhill through a landscape; shown in (**a**) is an oxbow of a river. Lakes and ponds (**b**) are large open bodies of water. Some lakes are so large as to be termed inland seas (**c**). Shallow water bodies with ample vegetation are called wetlands (**d**), and include marshes, swamps, and bogs.

amount of oxygen released from photosynthesis and the amount removed by animal and microbial respiration, among other factors.

Both ponds and lakes are potentially subject to the process of eutrophication (Chapter 6, Figure 6.5). If nutrients flow into water bodies at a rate faster than they flow out or are broken down, the water bodies may become increasingly laden with plant material and increasingly lower in dissolved oxygen content. As

lakes or ponds change over time, the species composition of organisms that live within them change as well. Species of fish, plants, and invertebrates adapted to the low-nutrient and high-oxygen conditions of *oligotrophic* lakes and ponds may give way to species adapted to the high-nutrient, low-oxygen conditions of *eutrophic* water bodies. Eventually, water bodies may fill in completely by the process of aquatic succession (Chapter 5, Figure 5.24).

Inland seas are the largest freshwater bodies

Some lakes are so large as to have substantially different characteristics from small lakes, and these are often named seas. North America's Great Lakes are prime examples. These freshwater bodies are so large that we cannot see from shore to shore and it can take many days to traverse their lengths by boat. Because they hold so much water, most of their biota is adapted to open water. Major fish species of the Great Lakes include lake sturgeon, lake whitefish, northern pike, alewife, bass, walleye, and perch. The immense lakes of the Rift Valley of Africa—Lakes Victoria, Malawi, and Tanganyika—are known for being the site of rapid speciation of certain types of cichlid fish, as well as a resource for human beings for millennia, as our species evolved on the African continent. In Asia, Lake Baikal is the world's deepest lake, at 1,637 m (5,370 ft) deep, and the Caspian Sea is the world's largest freshwater body, at 371,000 km^2 (143,000 mi^2) in area.

Wetlands include marshes, swamps, and bogs

Just as ecosystems that straddle the line between ocean and land are some of the most productive and species-rich (Chapter 13), those that combine elements of freshwater and dry land are enormously rich and productive. Often lumped under the term *wetlands,* such areas consist of different types of systems. Freshwater marshes feature water shallow enough to allow plants to grow from the bottom and rise above much of the water surface. North American marshes often fill with marsh vegetation such as cattails. Swamps are similar to marshes, shallow areas rich in vegetation, but swamps often occur in more forested situations. Examples are swamps created when beavers flood areas of forest by building dams with cut tree trunks and cypress swamps of the southeastern United States, where cypress trees are adapted to grow in standing water. Bogs, ponds thoroughly covered with thick floating mats of vegetation, can represent a stage in aquatic succession.

All of these types of wetlands are extremely valuable to wildlife, and all have been extensively drained and filled, largely for agriculture (Chapter 8). It is widely estimated that southern Canada and the United States have lost well over half of their wetland acreage since European colonization. Recent legislation in these and other nations is beginning to slow and reverse this trend. Moves to protect marshes, swamps, and bogs have come as we have gained a better appreciation for the various ways in which we depend on wetland ecosystems and on the water they can provide for our use.

How We Use Water

We use freshwater for drinking, washing, recreation, growing food, powering industry, generating electricity, disposing of waste, and countless other purposes. In our attempts to harness the world's freshwater sources for all these pursuits, we have achieved impressive accomplishments. It is estimated that 60% of the world's largest 227 rivers (and 77% of those in North America and Europe) have been strongly or moderately affected by artificial dams, canals, and diversions.

Water supplies our households, agriculture, and industry

Every one of us uses water in our households for drinking, cooking, and cleaning. As a society we use water for irrigating our crops, watering our livestock, and other agricultural purposes. We also use water in most manufacturing and industrial processes. The proportions of each of these three types of use—residential/municipal, agricultural, and industrial—vary dramatically among nations (Figure 14.7). Nations with arid climates tend to use more freshwater for agriculture, and heavily industrialized nations use a great deal for industry. Globally, we spend about 70% of our annual freshwater allotment on irrigation and other agricultural uses; industry accounts for roughly 20%, and residential and municipal uses for only 10%.

Most freshwater use is **consumptive use**, in which water is removed from a particular aquifer or surface water body and is not returned to it. Examples of consumptive uses can include agricultural irrigation and many industrial and residential uses. **Nonconsumptive use** of water does not remove, or only temporarily removes, water from an aquifer or surface water body. Using water to generate electricity in hydroelectric dams is one example of nonconsumptive use; water is taken in on the upstream side of the dam, passed through the dam machinery to turn turbines, and then released on the downstream side.

Agriculture accounts for 87% of the world's consumptive use of freshwater. The green revolution (Chapter 8 and Chapter 9) required significant increases in irrigation, and 60% more water is withdrawn for irrigation today than in 1960. During this period, the amount of land under irrigation has roughly doubled (Figure 14.8). Expansion of irrigated agriculture has kept pace with population growth; irrigated area per capita has remained stable for at least four decades at around 460 m^2 (4,900 ft^2). About 70% of the world's irrigated area is in Asia. Irrigation, when done properly,

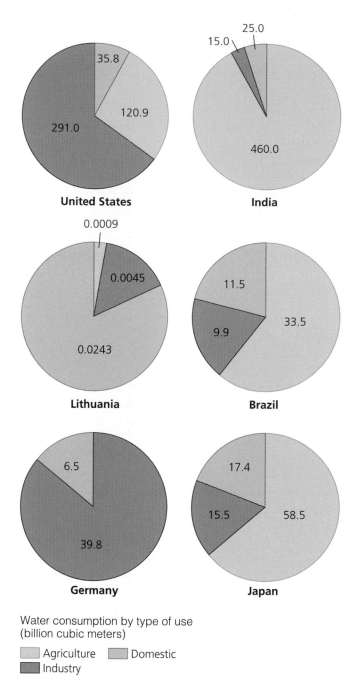

Water consumption by type of use
(billion cubic meters)

- Agriculture
- Industry
- Domestic

Figure 14.7 Different nations apportion their freshwater consumption in different ways. Industry consumes most of the water used in the United States and Germany; agriculture uses the most in India, Brazil, and Japan; and most water in Lithuania goes toward domestic use. Data from World Bank, as presented by: P. Harrison and F. Pearce, *AAAS Atlas of Population and the Environment*, 2000.

can more than double crop yields over those on land fed by rains alone, because irrigation systems give farmers the ability to control the application of water when and where it is needed. The world's 274 million ha (677 million acres) of irrigated cropland make up only 18% of

world farmland but yield fully 40% of world agricultural produce, including 60% of the global grain crop.

However, when water becomes scarce and more valuable, economic considerations favor water use in industry where a given amount of water can produce 70 times more dollar value of output than agriculture (for example, 1,000 tons of water can yield 1 ton of wheat worth $200, while 1,000 tons of water can increase manufacturing output by $14,000). Indeed, over the past 50 years, as agricultural water use doubled, water consumption by industry increased sixfold.

We extract water from aquifers

One-third of Earth's human population—including 99% of the rural population of the United States—relies directly on groundwater for its water needs. To reach groundwater, many people in rural areas of developing countries obtain their water by walking to local wells and hauling or pumping out water. In developed countries, individual homes in rural areas often have their own electrically powered wells. Homes and businesses in urban and suburban areas receive water through complex networks of pipes and pumps generally provided and maintained by their municipal governments.

Today, extraction of water from aquifers is on the rise. In India, the number of wells quadrupled and the area irrigated with groundwater increased sixfold between 1951 and 1997 as the green revolution enabled the nation to boost its agricultural productivity. The number of wells in Pakistan increased by nearly 15 times from 1964 to 1993, and in China they increased 20-fold between 1961 and the mid-1980s.

We divert surface water to suit our needs

In areas near surface water, people have found it far easier to make use of rivers, streams, lakes, and ponds, moving water from these sources to farm fields and houses. People have been moving water from where it is to where it isn't for thousands of years; archaeological evidence shows that many ancient civilizations depended on complex irrigation systems to transport water from surface bodies. The powerful empires of ancient Egypt were built on the Nile River and could not have flourished without intricate networks of canals dug to spread the river's water among the vast agricultural fields that grew food for the society (Figure 14.9).

The Colorado River's water is diverted and utilized every bit as much today (Figure 14.10). Early on in its course, some Colorado River water is piped through a

Figure 14.8 Overall global water consumption rose in tandem with the area of land irrigated for agriculture throughout the 20th century. Data from United Nations, Sustainable development commission on freshwater, 2002.

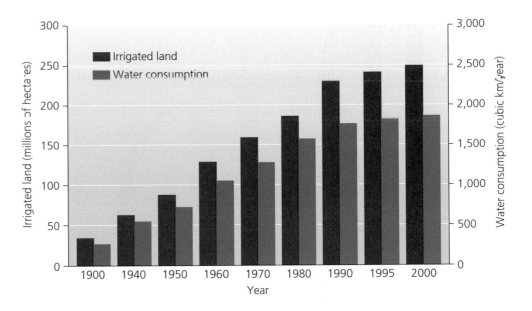

Figure 14.9 In ancient Egypt, shadufs were used to lift water out of the River Nile for crop irrigation, a practice that continued through the 19th century, at the time this photograph was taken.

Figure 14.10 The 2,330-km (1,450-mile) Colorado River drains a watershed of 637,000 km² (246,000 mi²) (a). Thirteen major dams pool water in enormous reservoirs along the lengths of the Colorado River and its tributaries. Several major canals divert water from the lower portion of the river, mostly for irrigating farmland in arid desert regions. The degree to which this once-wild river has been dammed led cartoonist Lester Dore to portray the Colorado and its tributaries as an immense plumbing system (b). *Source:* (b) *High Country Times*, November, 1997.

(a) The Colorado River and its tributaries

mountain tunnel across the Continental Divide and down the Rockies' eastern slope to supply the city of Denver. More is removed for Las Vegas and other cities and for farmland as the water proceeds downriver. When it reaches Parker Dam on the California-Arizona state line, large amounts are diverted into the Colorado River Aqueduct, which brings water to millions in the Los Angeles and San Diego areas in a huge open-air canal. From Parker Dam, Arizona also draws water, moving it in the large canals of the Central Arizona Project. Further south at Imperial Dam, water is diverted into the Coachella and All-American Canals, destined for agriculture, mostly in the Imperial Valley. To make this desert bloom, Imperial Valley farmers soak the earth with subsidized water for which they pay one penny per 795 l (210 gallons).

Today more large-scale diversion projects are under development around the world. One of the biggest plans consists of an aqueduct to bring water from the Yangtze River in southern China to the Yellow River in northern China, where the arid climate and population pressure have led the river to dry up in many places. With such massive projects that transfer water between major watersheds, we are truly "replumbing the planet."

We try to control floods with dikes and levees

While much engineering brainpower and muscle has been applied to transport water from place to place, we have also expended vast efforts to keep water in place. Flood prevention ranks high among reasons we control

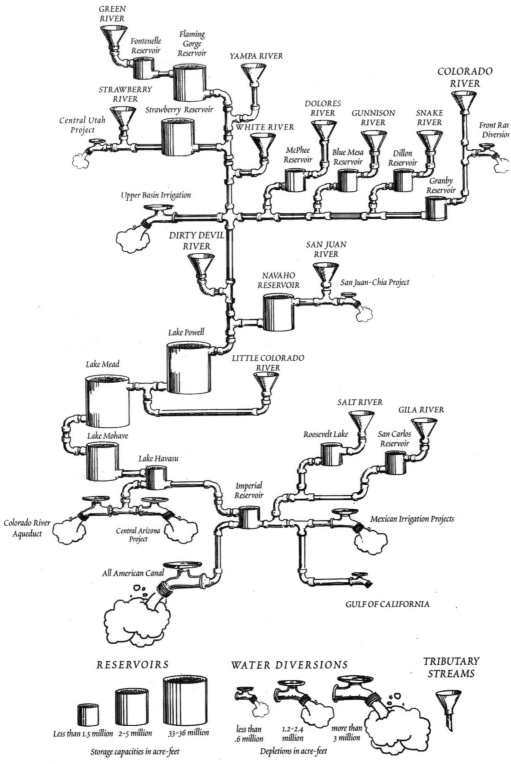

(b) The plumbing of the Colorado River Basin

the movement of freshwater. Towns, cities, and farms have always been attracted to riverbanks for their water supply and for the flat topography and fertile soil of floodplains, the region of land over which a river has historically wandered and over which it periodically floods. Flooding is a normal, natural process during times of high water due to snowmelt or heavy rain. Floodwaters carry nutrient-rich sediments and spread them over large areas of the floodplain. This is why farmers have always been attracted to floodplains. Flooding is of long-term benefit both to natural systems and to human agriculture.

In the short term, however, floods can do tremendous damage to the farms, homes, and property of people who choose to live in floodplains. To protect against floods, people have built dikes and levees (long raised mounds of earth) along the banks of rivers to hold rising water in main channels. Many dikes are small and local, but in the United States thousands of miles of massive levees have been constructed by the Army Corps of Engineers along the banks of major waterways. While these structures prevent flooding at most times and places, they can sometimes exacerbate flooding, because they force water to stay in the channel and accumulate, leading to occasional catastrophic overflow events.

We have erected thousands of dams

A dam is any obstruction placed in a river or stream to block the flow of water so that water can be stored in a reservoir. We build dams to prevent floods, generate electricity, provide drinking water, and ease irrigation (Figure 14.11, Table 14.1). Worldwide, more than 45,000 large (>15 m, or 49 ft, high) dams have been erected across the rivers of over 140 nations. Additionally, tens of thousands of smaller dams have been built. Virtually the only major rivers in the world that remain undammed and free-flowing are in the tundra and taiga of Canada, Alaska, and Russia, and in smaller regions of Latin America and Africa. The largest dams are some of the greatest engineering feats humans have produced. The two behemoths of the Colorado River, Hoover Dam and Glen Canyon Dam, stand 221 m (726 ft) and 216 m (710 ft) high, respectively, stretch for 379 m (1,244 ft) and 476 m (1,560 ft) across, and consist of 6.6 and 7.3 million tons of concrete and steel. Hoover Dam holds back 35.2 km^3 (8.4 mi^3) of water in a reservoir that stretches 177 km (110 mi) long and reaches 152 m (500 ft) deep.

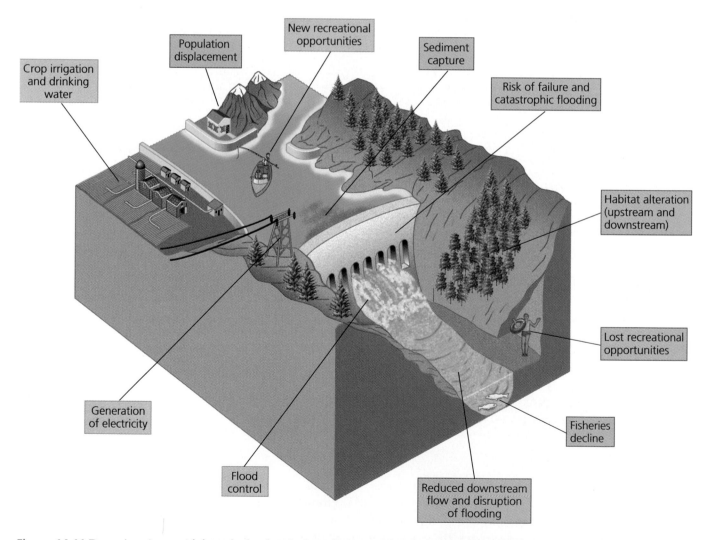

Figure 14.11 Damming rivers with large hydroelectric dams has many diverse consequences for people and the environment. The generation of clean and renewable electricity is one of several major benefits of dams (green boxes); habitat alteration is one of several negative impacts of dams (red boxes).

Table 14.1 Major Benefits and Costs of Dams

Benefits	Costs
Electricity generation. Many dams generate electricity by storing water and then releasing it through turbines that spin to generate electricity. Such hydroelectric dams (Chapter 18) provide considerable amounts of inexpensive electricity to help run our economies and power our modern conveniences. Some nations gain nearly all their energy from hydropower.	*Habitat alteration.* The reservoirs of dams flood riparian habitats, among the most productive habitats for biodiversity, and displace or kill riparian species. Dams also change rivers downstream, making them tranquil and low in dissolved oxygen, so that many fish adapted to fast-flowing rivers cannot survive. Shallow water downstream from a dam warms, but is periodically flushed with cold water from the reservoir, a phenomenon that can stress or kill both cold-water and warm-water species.
Emissions reduction. Hydroelectric power provides electricity without producing emissions that pollute the air. By replacing the combustion of coal or other fossil fuels as an electricity source, hydropower allows us to greatly reduce air pollution and its health and environmental hazards. As such, hydropower not only benefits people but helps keep the natural environment cleaner.	*Fisheries decline.* Fish like salmon and shad that migrate up rivers to spawn encounter dams as a barrier. Although "fish ladders" have been built at many dams to allow their passage, most fish do not make it, and none can pass the largest dams. Population declines of several now-endangered salmon species have had severe impacts on many fishing economies.
Crop irrigation. Water stored by dams can be withdrawn and diverted for irrigating crops. Irrigation is especially vital for agriculture in arid areas; most agriculture in the Colorado River Basin, for instance, could not be sustained without irrigation. By holding large amounts of water, reservoirs can release irrigation water when farmers most need it and can buffer entire regions against the effects of drought.	*Population displacement.* Land flooded by reservoirs typically includes fertile farmland. Dams have submerged many human settlements, even large modern cities, forcing the displacement of people who live and farm there. An estimated 40-80 million people globally have been displaced by dam projects over the past half-century.
Drinking water. Many dams store water in reservoirs for municipal drinking water supplies. Such reservoirs hold plentiful, reliable, and clean drinking water, provided that watershed lands that drain into the reservoir are left in a natural state and not developed or polluted.	*Disruption of flooding.* In the long term, flooding is a valuable ecosystem service. Rivers create much of the world's most productive farmland by depositing rich sediment during floods; without flooding, topsoil is lost and farmland gradually deteriorates in quality.
Flood control. Dams can prevent floods by storing seasonal surges, such as those following snowmelt or monsoon rains. Water can then be released from a dam gradually so as to even out flows downstream. The prevention of floods saves people's lives and prevents property damage.	*Sediment capture.* As fast-flowing rivers give way to slow-flowing reservoirs, sediment falls through the water column and settles, trapped behind dams. Sediment is prevented from nourishing downstream floodplains and estuaries, and reservoirs become clogged with silt.
Shipping. By replacing rocky river beds and rapidly flowing water with deep placid pools, dams make it possible to ship crops and goods by water over longer distances.	*Lost recreational opportunities.* Undammed rivers provide aesthetic, recreational, and economic benefits. When a wild river is dammed, recreational opportunities like tubing, whitewater rafting, and kayaking are lost.
New recreational opportunities. People can pursue activities such as fishing from boats, houseboating, and the use of personal watercraft on dammed reservoirs in regions where such recreation was not possible before.	*Risk of failure.* The risk that a dam may fail catastrophically always exists. While extremely rare, these events can cause massive property damage and loss of life.

Glen Canyon Dam holds back 33.3 km^3 (7.9 mi^3) of water in a reservoir that stretches 300 km (186 mi) long and reaches 171 m (560 ft) deep.

Weighing the Issues:
The Klamath Crisis

In 2001, angry farmers of Klamath County, Oregon disobeyed a federal order to divert irrigation waters downstream to save two endangered species of fish. During that bone-dry year, there simply wasn't enough water to irrigate farmers' fields and keep endangered salmon and suckerfish alive. The government had long ago begun allocating more water than was sustainable in the region. Although the homesteaders enjoyed federal incentives at the turn of the century (the dam was built to encourage settlers to farm the region), their children are now subject to a conflicting federal mandate—the Endangered Species Act. Can you think of solutions to the Klamath crisis? What would you do to reconcile these needs?

China's Three Gorges Dam is the world's largest and most expensive

The complex mix of benefits and costs that dams produce is exemplified by the world's largest dam project (Figure 14.12). The Three Gorges Dam on China's Yangtze River, 186 m (610 ft) high and 2 km (1.3 mi) wide, was completed in 2003. When filled in 2009, the resulting reservoir should be 616 km (385 mi) long, as long as Lake Superior. The reservoir will hold over 38 trillion liters (10 trillion gallons) of water. It will generate hydroelectric power, enable boats and barges to travel farther upstream, and provide flood control. The power generation may be enough to replace dozens of large coal or nuclear plants.

One of the costs of the Three Gorges dam, aside from its $25 billion construction price tag, is that its reservoir will flood 22 cities and the homes of 1.13 million people, requiring the largest resettlement project in China's history. The reservoir behind the dam will also inundate archaeological sites 10,000 years old and submerge productive farmlands and wildlife habitat. In addition, critics hold, the reservoir will slow the flow of the river so that suspended sediment will settle and begin to fill the reservoir as soon as it is completed. Other scientists worry about water quality, saying that the Yangtze's many pollutants will be trapped in the reservoir, making the water undrinkable. Indeed, high levels of bacteria were found in water as it began building up behind the dam. The Chinese government, however, plans to sink $5 billion into building hundreds of sewage treatment and waste disposal facilities.

More and more dams are now being removed

Those who feel that the costs of some dams have outweighed their benefits have been pushing for such dams to be dismantled. By removing dams and letting rivers flow free, these people say, we can restore riparian ecosystems, reestablish economically valuable fisheries, and reintroduce river recreation like fly-fishing and rafting. Increasingly, private dam owners and the Federal Energy Regulatory Commission (FERC), the U.S. government agency charged with renewing licenses for dams, have agreed. Roughly 500 dams have been removed in the United States, nearly 200 of them in the past decade alone. One reason is that many dams have aged and are in need of costly repairs or have outlived their economic usefulness.

The drive to remove dams gathered steam in 1999 with the dismantling of the Edwards Dam on the Kennebec River in Maine, which resulted from the first

(a) Sichuan Province, China

(b) The Three Gorges Dam in Yichang

Figure 14.12 China's Three Gorges Dam, completed in 2003, is the world's largest dam. Well over a million people were displaced and whole cities were leveled for its construction, as was the case in Sichuan Province shown here (**a**). Even as water poured through the floodgates (**b**), the reservoir began filling in 2003, and will continue filling for several years.

FERC determination that the environmental benefits of removing a dam outweighed the economic benefits of relicensing it. Once the 7.3–m (24–ft) high, 279–m (917–ft) long dam was removed, 10 species of migratory fish, including salmon, sturgeon, shad, herring, alewife, and bass, were expected to use the 27–km (17–mi) stretch of river above the dam site. The first of these was spotted doing so soon after the dam's removal.

More small dams will come down in years ahead as over 500 FERC licenses come up for renewal in the next decade. Dam removal proponents have not shied away from urging removal of the biggest dams, either. The Colorado River's Glen Canyon Dam is one such ambitious target, because of the legendary beauty of Glen Canyon, which was submerged.

Depletion of Freshwater: The Challenge of Quantity

Ensuring adequate quantities of freshwater for our basic needs is an endeavor as old as our species. Technological advances will not eliminate this challenge. Human population growth, expansion of irrigated agriculture, and industrial development have caused global annual water use to rise from 2,600 km³ (620 mi³) to nearly 4,000 km³ (1,240 mi³) in the past three decades. It is projected to increase 40% further by 2020, according to one study.

The hydrologic cycle makes freshwater a renewable resource. However, if we extract water from accessible sources faster than it can be replenished, then we effectively lose access to water and diminish the amount that a person or a civilization can use over a lifetime. When our water usage exceeds what a lake, river, stream, or aquifer can provide, we must either reduce our use, find another water source, or be prepared to run out of water. Data indicate that at present our freshwater consumption in much of the world is unsustainable, and we are depleting many sources of surface water and groundwater. This depletion is causing increasing shortages of freshwater. At least 1.7 billion people live in areas of water scarcity, that is, areas where water availability per person is less than 1,000 m³ (35,000 ft³) per year. The number of people facing water scarcity is expected to grow to at least 2.4 billion by 2025.

We are depleting surface water

On most days, what water is left of the Colorado River after all the diversions comprises just a small trickle making its way to the Sea of Cortez over the sediments of the once-rich delta. On some days, the water does not reach the Sea of Cortez at all. The Colorado's plight is not unique. Several hundred miles to the east, the Rio Grande makes a similarly unimpressive entrance into the Gulf of Mexico. A number of times in the last few years, the Rio Grande too has dried up completely before reaching its mouth, the victim of overextraction by both Mexican and U.S. farmers in a time of drought. Across the globe in China, the Yellow River has run dry with increasing frequency in recent years, accounting for the government's planned diversion scheme from Three Gorges Dam. Even the river that has nurtured civilization longer and better than any other, the Nile, now peters out before reaching its mouth.

Nowhere are the effects of surface water depletion so evident, however, than the Aral Sea. Once the fourth-largest lake on Earth, just larger than Lake Huron, it has lost four-fifths of its volume in 40 years and could disappear altogether in the near future (Figure 14.13). This dying inland sea, on the border of present-day Uzbekistan and Kazakhstan, is the victim of farming and irrigation practices. The former Soviet Union instituted large-scale cotton farming in this region by flooding the dry land with water from the two rivers leading into the Aral Sea. For a few decades this action boosted Soviet cotton production, but eventually it led the Aral Sea to shrink, while the soil became waterlogged and salinized. Today what cotton grows on the blighted soil cannot bring the regional economy back, and scientists are struggling with ways to save the Aral Sea, whose ecosystems have been seriously damaged.

We are depleting groundwater

Groundwater is at even greater risk of depletion than surface water because most aquifers recharge either very slowly or not at all. Thus it is easy to withdraw the water, a process called **water mining**, much more rapidly than it can be replenished. If we compare an aquifer to a bank account, we are making more withdrawals than deposits, and the balance is shrinking. Globally, over the past 60 years we have been withdrawing groundwater in amounts that increase 2.5–3% annually, greater than the rate of population growth. Farmers and others today are pulling 160 km³ (5.65 trillion ft³) more water out of the ground each year than is finding its way back into the ground. This degree of overpumping is equal to the quantity of water needed to produce 10% of the world grain supply.

As aquifers are depleted, water tables drop. Groundwater becomes more difficult to extract and eventually can run out. In India, for example, where water is being pumped twice as fast as it is recharged, some water tables are falling 1–3 m (3.3–9.9 ft) per year. China and other Asian nations, as well as Yemen and other Middle Eastern nations, are seeing their water tables drop by more than 1 m (3.3 ft) each year. In Mexico, water tables in 10 well-studied aquifers are falling at rates of 1.8–3.3 m (5.9–10.9 ft) per year. In the United States, by the late 1990s overpumping had drawn the Ogallala Aquifer down by 325 km³ (11.5 trillion ft³). This volume is equal to the yearly flow of 18 Colorado Rivers.

Overpumping of groundwater in coastal areas can cause salt water to intrude into aquifers, making water undrinkable. This has occurred widely in Middle Eastern nations and in localities as varied as Florida, Turkey,

(a) Ships stranded by the Aral Sea's fast-receding waters

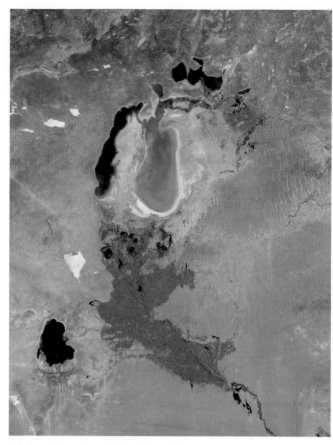

(b) Satellite view of Aral Sea, 2002

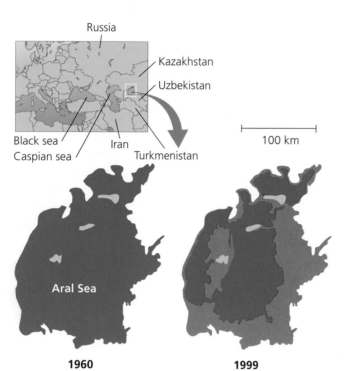

(c) The shrinking Aral Sea, then and now

Russia

Kazakhstan

Uzbekistan

Black sea

Caspian sea Iran Turkmenistan

100 km

North
Aral Sea

Aral Sea

South
Aral
Sea

1960   1999   2002

Figure 14.13 Ships lie stranded in the sand (**a**) because the waters of the Aral Sea have receded so far and so quickly (**b**). Central Asia's Aral Sea was once the world's fourth-largest lake. However, it has been shrinking for the past four decades (**c**) and could disappear completely in the near future. The primary cause has been over-withdrawal of water to irrigate cotton crops.

Bangkok, and Manila. Saltwater intrusion in Manila has contaminated wells 10 km (6.3 mi) inland.

As aquifers lose water, their substrate can become weaker and less capable of supporting overlying strata and any human structures built upon them. In such cases, the surface above may subside. Cities from Venice to Bangkok to Beijing are slowly sinking for this reason. Mexico City offers perhaps the most dramatic example.

The city's downtown has sunk over 10 m (33 ft) since the time of Spanish arrival; streets are buckled, old buildings lean at angles, and underground pipes break so often that 30% of the system's water is lost to leaks. Sometimes subsidence can be sudden and localized, appearing in the form of **sinkholes,** areas where the ground gives way with little warning, occasionally swallowing people's homes (Figure 14.14). Once the ground subsides, soil becomes compacted, losing the porosity that enabled it to hold water in the first place. Recharging a depleted aquifer may therefore become more difficult. It has been estimated that compacted aquifers under California's Central Valley have lost storage capacity equal to that of 40% of the entire state's artificial surface reservoirs.

Falling water tables not only cause problems for farmers and others who depend on wells for their water, but also do vast ecological harm. Permanent wetlands exist where water tables are high enough to reach the surface, so when water tables drop, wetland ecosystems dry up. Wetlands above Spain's Upper Guadiana aquifer, for instance, have been eliminated by groundwater depletion. In Jordan, the Azraq Oasis covered 7,500 ha (18,500 acres) and enabled migratory birds and other terrestrial and aquatic species to find water in the desert. The water table beneath this oasis dropped from 2.5 m (8.2 ft) to 7 m (23 ft) during the 1980s because of increased well use by the city of Amman. As a result, the oasis dried up altogether during the 1990s.

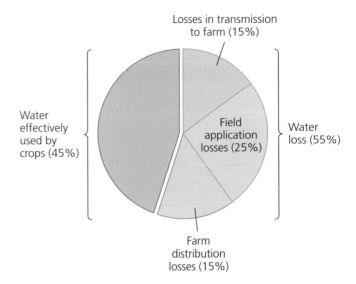

Figure 14.15 Of the water used to irrigate crops globally, only about 45% ends up being effectively utilized by crops. The majority is lost to evaporation and percolation in its application and transport. Data from Food and Agriculture Organization, as presented by: P. Harrison and F. Pearce, *AAAS Atlas of Population and the Environment*, 2000.

Inefficient irrigation wastes water

Worldwide, the amount of land under irrigation has been increasing. Whether from aquifers or surface bodies, the majority of the freshwater we use for irrigation is lost before it ever reaches the crops (Figure 14.15). Inefficient "flood and furrow" irrigation, in which fields are liberally flooded with water that may evaporate from shallow standing pools, accounts for 90% of irrigation worldwide, although new more efficient methods are becoming more popular. Overirrigation leads to waterlogging and salinization (Chapter 8), which affects one-fifth of farmland today and reduces world farming income by $11 billion.

Governments of many nations have subsidized irrigation in order to promote agricultural self-sufficiency. Saudi Arabia used up 13% of its groundwater reserves on wheat farming from 1980 to 1995, and irrigated area has doubled in Syria and Iraq over the past 30 years. Unfortunately, huge amounts of groundwater are being used up for little gain; because of the dry climate and inefficient irrigation methods, only 30–50% of the water actually reaches crops.

Will we see a future of water wars?

Freshwater depletion leads to shortages, and the scarcity of any vital resource can lead to social conflict, including hostility and armed clashes between political states. We have only to look to the Colorado River and the troubled relationships of the states that border it to see evidence of this. In 1933 the governor of Arizona, foreseeing that California's water diversion from the Colorado might endanger Arizona's future allotment, sent the state's National Guard to threaten the construction of Parker Dam. After a long standoff, the U.S. interior secretary halted the project to avoid hostilities while the

Figure 14.14 When groundwater is withdrawn at too great a rate, the land above it may collapse in sinkholes, sometimes bringing buildings down with it.

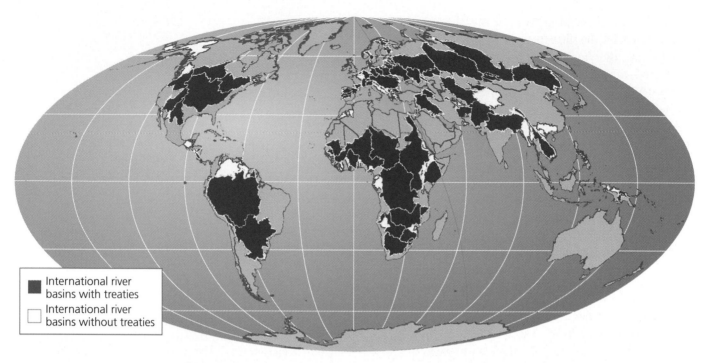

(a) International river basins with treaties

Figure 14.16 Because so many rivers cross national boundaries, many countries have entered into agreements on how to share use of river water. Transboundary river basins covered by international treaties are colored in blue in (a). Most major North American rivers are covered by treaties between Canada, the United States, and Mexico (b). *Source:* Aaron T. Wolf, The transboundary freshwater dispute database project, Water International, 1999.

issue was mediated in court. Arizona won the court case, but California got the dam authorized by the U.S. Congress, and Arizona chose not to tackle the U.S. Army.

Many predict that water's role in regional conflicts will only increase as human population continues to grow in water-poor areas. A total of 261 major rivers whose watersheds cover 45% of the world's land area cross national borders and are shared by at least two nations (Figure 14.16). Transboundary disagreements (such as we saw along the Tijuana River in Chapter 3) are common. The World Bank's vice president for environmental affairs and the World Water Commission's chairman, Ismail Serageldin, has remarked that "the wars of the twenty-first century will be fought over water."

On the positive side, a great many nations have found ways to cooperate with their neighbors to resolve water disputes. For instance, India has struck cooperative agreements over management of transboundary rivers with several of its neighbors, including Pakistan, Bangladesh, Bhutan, and Nepal. In Europe, international conventions have been signed by multiple nations along the Rhine and the Danube Rivers. Such progress gives reason to hope that wars over water in the future will be few and far between.

Weighing the Issues:
How Far Can You Reach for Water?

In 1941, the burgeoning metropolis of Los Angeles needed water, and decided to divert streams feeding into Mono Lake, over 565 km (350 mi) away in northern California. As the lake level fell 14 m (45 ft) in 40 years, the amount of salt doubled and indigenous species suffered. The populations of other desert cities—such as Las Vegas, Phoenix, and Denver—are expected to double in the coming years. Where will they go for the water needed to sustain their communities? What challenges might they face in trying to pipe in water from distant sources? How else could they meet their future water needs?

Legend:
- International river basins with treaties
- International river basins without treaties

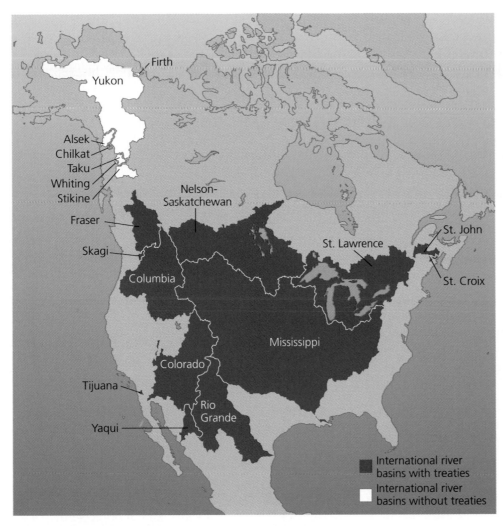

(b) International river basins in North America with treaties

Solutions to Depletion of Freshwater

To address the problem of freshwater depletion, we can either increase supply or reduce demand. Strategies that aim to reduce demand include conservation and efficiency measures. Ultimately, population stabilization or reduction may be required to permanently alleviate freshwater shortages, but in the meantime conservation and efficiency measures can go a long way toward making water use sustainable. Trying to lower demand is often more difficult politically in the short term but will likely be necessary in the long term. In the developing world, international aid agencies are increasingly willing to fund demand-based solutions over supply-based solutions, because demand-based solutions offer better economic returns and cause less collateral ecological and social damage.

Strategies most often used to increase supply in a particular area have involved transporting water through pipes and aqueducts from areas where it is more plentiful or accessible. In some instances this has been done by mutual agreement, but in other instances water-poor regions have forcibly appropriated water from people, communities, or nations too weak to keep the water for themselves. Transporting water from one place to another is not an equitable solution socially or politically if it harms the region that loses the water. Another strategy is to develop new technologies to find or "make" more water.

We can "make" more water by desalination

The best-known technological approach to generate freshwater is **desalination**, or **desalinization**, the removal of salt from seawater. One method is to mimic the hydrologic cycle by hastening evaporation from allotments of ocean water with heat, and then condensing the vapor—that is, distilling freshwater. Another method involves forcing water through membranes to filter out salts, the most common process of which is reverse osmosis. As of 2000, over 7,500 desalination facilities were in operation around the world. Most are in the arid Middle East, with more than 25% in Saudi Arabia alone, and some others are in small island nations that lack groundwater supplies. Desalination works— the largest plant, in Saudi Arabia, produces 485 million liters (128 million gal) of water every day—but it is expensive. In 1992 the world's largest reverse osmosis desalination plant was completed along the Colorado River near Yuma, Arizona. It was intended to reduce the salinity of irrigation runoff from the Wellton-Mohawk Irrigation District, which, when it reentered the Colorado River, significantly increased the river's salinity. Building the plant cost $256 million, but it operated for only 8 months and at only one-third of its capacity. Annual operating costs for the plant, were it to resume operations, would be $25.8 million per year, according to the U.S. Bureau of Reclamation.

Demand by agriculture can be reduced

Although supply-side solutions are worth pursuing, there is a need to reduce demand as well. Because most human water use is for agriculture, it makes sense to look first to agriculture for ways to decrease demand. Farmers can use technology to improve efficiency in a number of ways, including lining irrigation canals to prevent leaks and leveling fields to minimize runoff. Furthermore, some methods of irrigation are more efficient than others. Techniques to increase irrigation efficiency include low-pressure spray irrigation, which sprays water downward toward plants, and drip irrigation systems, which target individual plants and introduce water directly onto the soil (Chapter 8, Figure 8.23). Both methods reduce the amount of water lost to evaporation and surface runoff. Low-pressure precision sprinklers in Texas have efficiencies of 80–95% and have saved 25–37% of water. Drip irrigation, which has efficiencies as high as 90%, could cut water use in half while raising yields by 20–90% and giving developing-world farmers $3 billion in extra annual income, it has been estimated. Such improvements in Jordan more than doubled crop yields from 1973 to 1986.

In addition, choosing crops to match the land and climate in which they are being farmed can save huge amounts of water. Currently, crops that require a great

deal of water, such as cotton, rice, and alfalfa, are often planted in hot and arid areas where irrigation is government-subsidized. As a result, the true cost of water is not part of the costs of growing the crop. Eliminating subsidies and growing crops in climates that provide adequate rainfall could greatly reduce water use in many parts of the world. In addition, the genetic modification of crops (Chapter 9) is resulting in some varieties that require less water.

We can lessen residential, municipal, and industrial freshwater use in many ways

Although we use less freshwater in industry, residences, and municipalities than we do in agriculture, there are many readily available ways in which we can conserve water in these non-agricultural sectors. We can reduce our household water use by installing low-flow faucets, showerheads, washing machines, and toilets. Automatic dishwashers, studies show, use less water than washing dishes by hand. If our homes have lawns, it is best to water them at night when water loss from evaporation is minimal. Better yet, replacing water-intensive lawns with native plants that are adapted to your region's natural precipitation patterns saves the most water. Xeriscaping, landscaping using plants adapted to arid conditions, has become a popular approach in much of the U.S. Southwest.

Industry and municipalities can take water-saving steps as well. Manufacturers have successfully shifted to processes that use less water, and in doing so they have reduced their costs. Something as seemingly obvious as finding and patching leaks in pipes has saved some cities and companies large amounts of water—and money— once they have invested in the search. Another water conservation practice is recycling wastewater for suitable uses. Most water that runs down sinks, showers, and bathtub drains (often called "gray water") can, with little or no treatment, be suitable for irrigation and for some industrial uses. Some cities, among them St. Petersburg, Florida, are recycling all of their wastewater.

In yet another approach, some governments are taking advantage of seasonal differences in precipitation. By capturing excess surface runoff during the rainy season and pumping it into underground aquifers, governments in both Arizona and England are making more efficient use of their available water supply.

Various economic solutions to water conservation are being debated

Many economists have suggested market-based strategies for achieving sustainability in water use, ending government subsidies of inefficient practices, and letting water become a commodity whose price reflects the true costs of its extraction. Many others worry that making water a fully priced commodity would make it less available to the world's poor and thus increase the gap between rich and poor. Because industrial use of water can be 70 times more profitable than agricultural use, for example, market forces alone would likely favor water's use in ways that would benefit wealthy and industrialized people, companies, and nations at the expense of the poor and less industrialized.

Similar concerns surround another potential solution, the privatization of water supplies. During the 1990s a number of public water systems were partially or wholly privatized, their construction, maintenance, management, or ownership transferred to private companies. This was done in hope of increasing the systems' efficiency, but many people worry that firms have little incentive to allow continued equitable access to water for rich and poor alike. Already in some developing countries, rural residents without access to public water supplies who are forced to buy water from private vendors end up paying on average 12 times more than those connected to public supplies. However, other experiences indicate that decentralization of control over water, from the national level to the local level, may help conserve water. In Mexico, the effectiveness of irrigation systems improved dramatically once they were transferred from public ownership to the control of 386 local water user associations.

Regardless of how demand is addressed, the ongoing shift from supply-side to demand-side solutions is beginning to pay dividends. In Europe, the new focus on demand (through both government mandates and public education) has decreased public water consumption, and industries are becoming more efficient in their water use. The United States decreased its total water consumption by 10% from 1980 to 1995, thanks to conservation measures, even while its population grew 16%.

Pollution of Freshwater: The Challenge of Quality

The quantity and distribution of freshwater poses one set of environmental and social challenges. Safeguarding the *quality* of water involves another collection of environmental and human health dilemmas. To be safe for human consumption, water must be relatively free of disease-causing organisms and toxic materials. To be safe for aquatic organisms, water must be freer still of contaminants.

Despite admirable advances in cleaning up water pollution in developed countries, the World Commission on Water in 1999 concluded that over half the world's major rivers are "seriously depleted and polluted, degrading and poisoning the surrounding ecosystems, threatening the health and livelihood of people who depend on them." The largely invisible pollution of groundwater, meanwhile, has been termed a "covert crisis." Major pollutants of ground and surface water include natural sources as well as numerous anthropogenic (human-made) sources, including untreated sewage, agricultural runoff, and petroleum and chemical spills.

Scientists use several indicators of water quality

Scientists measure certain physical, chemical, and biological properties of water to characterize its quality. Biological properties, such as the presence of disease-causing organisms, can indicate how safe water is for human consumption. Among four important chemical properties is pH, a measure of water's acidity or alkalinity (Chapter 4). Normal pH values for naturally occurring water supplies range from 6 to 8, but human activity can raise or lower pH dramatically, as we have seen in the case of acid precipitation (Chapter 11). Another property commonly used to characterize water is **hardness.** Hard water contains higher concentrations of calcium and magnesium ions and prevents soap from lathering. Hard water also leaves hard, chalky deposits behind when heated or boiled. The third chemical characteristic scientists use to classify water is **dissolved oxygen** content. Dissolved oxygen is an indication of aquatic ecosystem health because surface waters low in dissolved oxygen are less capable of supporting aquatic life. Water quality experts also measure taste and odor, because even low concentrations of some chemicals, such as phenolic liquors, can make otherwise high-quality water undesirable for human consumption.

Scientists also classify water according to three physical characteristics. **Turbidity** measures the density of suspended particles in a water sample. Fast-moving rivers that cut through arid or eroded landscapes, such as the Colorado and China's Yellow River, carry a great deal of sediment and are turbid and muddy-looking as a result. Water color is itself an indicator of water quality and can reveal particular substances present in a water sample. Some forest streams run the color of iced tea because of naturally occurring chemicals called tannins present in decomposing leaf litter. Temperature is a third physical characteristic experts use to assess water quality. High temperatures can interfere with certain biological processes, and warmer water holds less dissolved oxygen.

Water pollution comes from point sources and from larger areas

As we saw in Chapter 11, the term *pollution* describes any matter or energy released into the environment, whether from human activity or from natural sources, that causes undesirable impacts on the health and well-being of humans and other organisms. As with air pollution, water pollution can be emitted from point sources—single locations, such as a pipe from a factory—or can consist of nonpoint-source pollution arising from multiple cumulative inputs over larger areas, such as farms, city streets, and residential neighborhoods (Figure 14.17). Point sources are easy to identify but sometimes difficult to regulate, especially if owners or supporters of industrial plants lobby politicians to pressure agencies not to enforce regulations. Nonpoint-source pollution presents a different type of societal challenge, requiring public education and the willingness of citizens to change certain behaviors. It is nonpoint-source pollution that poses the greater threat to water quality in the United States, according to the Environmental Protection Agency (EPA). Many common activities give rise to nonpoint-source

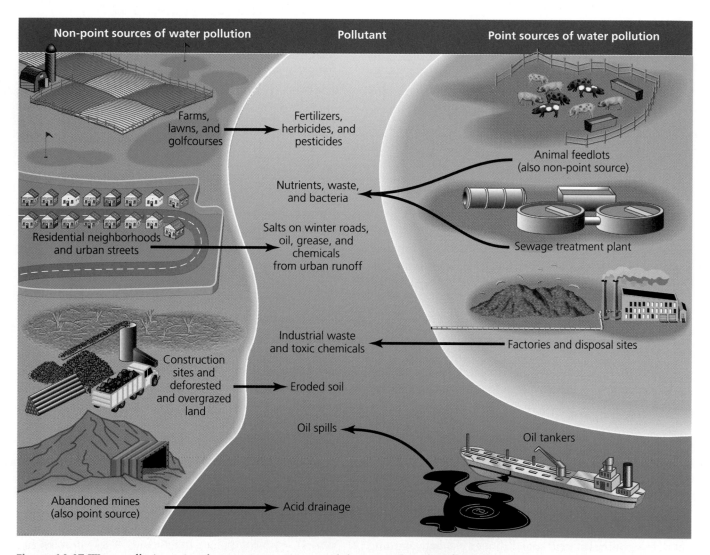

Figure 14.17 Water pollution arises from various sources, and these are often classified as being point sources or nonpoint sources. Point-source pollution comes from specific facilities or locations, usually from single distinct outflow pipes. Nonpoint-source pollution (such as runoff from urban streets, residential neighborhoods, lawns, and farms) originates from numerous sources spread over large areas.

water pollution, including applying fertilizers and pesticides to lawns, applying salt to roads in winter, and changing automobile oil.

Regardless of its source, water pollution comes in many forms and can cause diverse impacts on aquatic ecosystems and human health. We can categorize pollution into several types, ranging from the biological pollution of disease-causing organisms, to natural and artificial chemical pollution, to physical pollution of sediment, to thermal pollution affecting water temperature.

Pathogens and waterborne diseases are a biological form of water pollution

Many disease-causing organisms (including viruses, bacteria, and other pathogens) survive in surface water. Some enter inadequately treated drinking water supplies via human and animal waste. A study of global water supply and sanitation issues by the World Health Organization (WHO) and United Nations Children's Fund (UNICEF) in 2000 showed that despite advances in many parts of the world, major problems still existed. On the positive side, 4.9 billion people (82% of the human population) had access to safe water as a result of some form of improvement in their water supply—an increase from 4.1 billion (79% of the population) in 1990. However, over 1.1 billion people were still without safe water supplies. In addition, 2.4 billion people had no sewer or sanitation facilities. Most of these people were Asians and Africans, and four-fifths of the people without sanitation lived in rural areas. These conditions contribute to 5 million deaths per year, as well as other health impacts. The WHO/UNICEF study report estimated the following:

- Four billion cases of diarrhea each year kill 2.2 million people, most of them children under 5 years old.
- One in 10 people in the developing world is infected by intestinal worms.
- Six million people are blind because of trachoma, a partly waterborne disease caused by microbial eye infection.
- About 200 million people in 74 countries are infected with schistosomiasis, a potentially fatal disease caused by parasitic blood flukes.

As our understanding of the pathogens found in water has advanced, we have developed several strategies for reducing the risks they pose. Collecting sewage and treating it to remove pathogens constitutes one approach. Another strategy is using chemical or other means of disinfecting drinking water. Finally, a variety of hygienic measures can fight waterborne disease.

Among these are public education to encourage attention to personal hygiene, and government enforcement of regulations to ensure the cleanliness of food production, processing, and distribution.

Nutrient pollution can cause eutrophication

In many freshwater systems, biology and chemistry interact to cause pollution of the type that leads to hypoxia in areas such as the Gulf of Mexico's dead zone (Chapter 6). The process of eutrophication, or overnourishment, of surface water and its plants by nutrient pollution occurs readily in freshwater bodies.

The growth of microscopic aquatic algae and of larger aquatic plants on the bottoms of lakes and ponds is limited when certain nutrients are in short supply. In marine systems, nitrogen supply is usually the factor that most limits growth; in freshwater systems, phosphorus supply is generally the limiting factor. When excess levels of nitrogen and phosphorus enter surface waters, they fertilize aquatic plants, increasing their growth rates and boosting their populations. Although plant growth is desirable in aquatic ecosystems, both as a source of oxygen and a source of food for other organisms, too much plant growth can cause problems. Algae that grow too rapidly can cover the water's surface, harming the deeper-water plants that depend on sunlight for survival. As algal blooms approach the carrying capacity of the lake or pond, the algae dies off en masse, providing food for decomposers such as bacteria. Because decomposition of organic matter requires oxygen, the increased bacterial activity drives down dissolved oxygen levels. These levels can often drop too low to support fish and shellfish populations, leading to dramatic changes in aquatic ecosystems. Eutrophication (Figure 14.18 on page 438) is a natural process, but human activities leading to nutrient input can dramatically increase the rate at which it occurs.

Both phosphorus and nitrogen are common in runoff from farms, golf courses, lawns, and sewage. We can reduce the amounts of nitrogen and phosphorus that reach surface waters in several ways. Perhaps the simplest way is to reduce the amounts of fertilizer applied on farms, lawns, and golf courses. We can also plant vegetation to increase the uptake of nutrients in polluted areas. Additionally, we can make sure our sewage and wastewater is treated at wastewater treatment facilities (see The Science behind the Story). Before discharging organic wastes into surface waters, treatment plant operators remove wastes physically and aerate wastewater in tanks to encourage aerobic bacteria to decompose them (Chapter 19).

Removing Phosphorus from Wastewater

In the early 1990s, the Maine Department of Environmental Protection (DEP) launched a program to improve the water quality of the state's lakes and ponds by reducing discharges from wastewater treatment plants. Oakland, Maine, was one community that suffered poor water quality due to wastewater treatment plant discharges. Wastewater dumped into the town's slow-moving Messalonskee Stream contained high levels of phosphorus, which caused algal blooms and eutrophication. Although the Maine DEP initiative formally targeted "lakes and great ponds," the Messalonskee Stream counted as a great pond because it was impounded for part of its length. To reduce phosphorus levels in the Messalonskee, the DEP offered two solutions: (1) ship the town's waste to a regional treatment center or (2) construct an 8 km (5 mi) pipeline to carry treated wastewater to the Kennebec River, where it could safely be released. The Kennebec River is a fast-flow-

Following installation of automatic controls and improved clarifiers

The 1994 installation of automatic controls to regulate ferric chloride release and improvement in clarifiers to redirect currents and solid waste in Maine's Oakland Wastewater Treatment Facility resulted in a significant decrease in phosphate concentration in Messalonskee Stream. Data from James Fitch and Daniel Bolduc, Practical engineering combined with sound operations optimizes phosporus removal, *Water Engineering & Management*, April, 2002.

ing river and thus would dilute waste, thereby eliminating the risk of algal blooms and subsequent eutrophication.

For financial and political reasons, the town rejected both of the DEP's solutions. Instead, a decision was made to reduce phosphorus

Toxic synthetic chemicals pollute our waterways

Besides nutrients and organic wastes derived from living organisms, our waterways have become polluted with toxic substances, including pesticides, petroleum products, and other synthetic chemicals. Many of these can poison animals and plants, can alter aquatic ecosystems, and, when ingested at sufficient concentrations and over long periods, may cause a wide array of human health problems, including cancer (Chapter 10).

Toxic metals, such as arsenic, lead, and mercury, and acids from acid precipitation and acid drainage from mining sites, can also cause a variety of negative impacts on human health and the environment. We can reduce releases of some of these toxic inorganic chemicals by creating and enforcing more stringent regulations of

industries. Better yet, we can modify our industrial processes to rely less on these substances.

Sediment can be a pollutant

Although floodplains can build fertile farmland, sediment that rivers transport can also pollute. Mining, clear-cutting, land clearing for real estate development, and careless cultivation of farm fields all expose soils to wind and water erosion. This can cause elevated levels of sediment to be dumped into nearby bodies of water. Although some water bodies, such as the Colorado River and Yellow River, are typically sediment-rich, many others are not. When a clear-water river undergoes a large influx of sediment, aquatic habitat can change dramatically, and fish that evolved to survive well in clear-water

concentrations by treating waste-water with the chemical ferric chloride. Ferric chloride binds to phosphorus, creating the compound ferric phosphate. Insoluble in water, ferric phosphate precipitates out, removing dissolved phosphorus from water and transforming it into a solid. The solid-phase phosphorus can then be disposed of with other solid wastes.

Although initial tests using ferric chloride indicated that the chemical reduced phosphorus levels significantly, reductions were not significant enough to prevent algal blooms. So the town turned to a consulting company to help determine whether further improvements to the treatment process would allow the plant to meet DEP requirements. The consulting company found that automating the application of ferric chloride prevented algal blooms more effectively than applying it manually. Control devices were installed that automatically adjusted the amount of ferric chloride added to wastewater. When wastewater flow increased through the treatment plant, automatic controls released more ferric chloride. Since the implementation of automatic controls, the Oakland treatment plant has added approximately 205 l (54 gal) of ferric chloride for every 950,000 ls (250,000 gal) of wastewater per day during the summer, when the risk of algal bloom is high. Half that amount is added during the winter.

The consulting company also recommended improvements to the plant's clarifiers, which store water temporarily to allow solid pollutants to settle so they can be removed. Clarifiers comprise one of the final steps in the waste treatment process. By adding dyes to the water stream, the consultants discovered that the clarifiers were short-circuiting: currents along the bottom and over the side of the tanks were preventing solids from settling. Most waste was removed before the water hit the clarifiers, and waste that did reach the clarifiers was in the form of suspended particles, a mix of particulate feces and other wastes as well as remaining ferric phosphate. To reduce currents, the consultants installed angled plates, known as Crosby baffles, around the rims of the clarifiers to redirect currents toward the center, where solids settle. Other improvements reduced the amount of sludge stored in the clarifiers, lowering the risk of discharge of large amounts of solids into the stream during periods of high flow. In upgrading several pieces of equipment, the plant also improved efficiency in filtering residual water from sludge, lowering phosphorus discharges still further. Phosphorus concentrations dropped significantly after 1994, when the first improvements were implemented.

Because of these and other advances carried out between 1994 and 1998, algae blooms have not occurred in Messalonskee Stream since 1993. It's no surprise that the Oakland Wastewater Treatment Facility soon became a model for other small- to moderate-sized towns facing wastewater treatment challenges.

---

environments, such as salmon, may not be able to handle the change. We can reduce sediment pollution problems by more thoughtfully managing farms and forests and by avoiding large-scale disturbances of vegetation.

Weighing the Issues:
Sedimentation Can Affect Water Resources and Habitat

United States Geological Survey (USGS) data show that as the percentage of fine-grained sediment in a streambed increases, salmon eggs become less likely to hatch and successfully develop into fish. Salmon need gravelly stream bottoms that protect their eggs from predation by other fish and prevent them from being swept downstream. Fine sediment particles fill in the voids between stones, creating a smooth bottom that is poor spawning habitat for salmon. Construction projects and the removal of trees and other vegetation increase the amount of sediment entering streams. How might these impacts to salmon be important to humans and other organisms? What other aspects of water quality might sedimentation influence?

Even heat and cold can pollute

Because water's ability to hold dissolved oxygen decreases as temperature rises, some aquatic organisms may not survive temperature increases in surface waters. Many human activities can increase surface water temperatures. For instance, when water is withdrawn from a river or stream and used to cool industrial facilities, it will transfer heat energy from the facility it cools back into the surface water to which it is returned. Humans

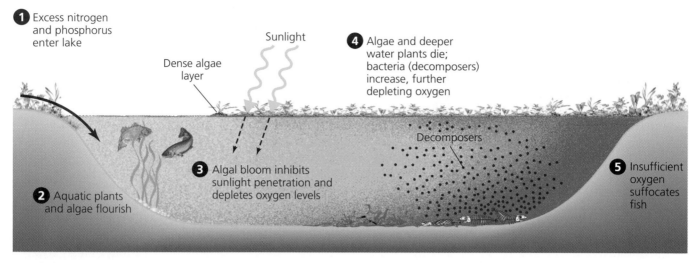

1 Excess nitrogen and phosphorus enter lake

Dense algae layer

Sunlight

4 Algae and deeper water plants die; bacteria (decomposers) increase, further depleting oxygen

Decomposers

3 Algal bloom inhibits sunlight penetration and depletes oxygen levels

2 Aquatic plants and algae flourish

5 Insufficient oxygen suffocates fish

(a) Eutrophication of a freshwater lake

(b) Oligotrophic water body

(c) Eutrophic water body

Figure 14.18 Pollution of freshwater bodies with excess nutrients leads to the process of eutrophication, which we first encountered in the sea water discussion of Chapter 6. Excess nutrients boost algal growth, which blocks sunlight. Material from dead algae later are decomposed by bacteria, and this decomposition sucks up oxygen, lowering dissolved oxygen levels to the point where fish and other aquatic animals perish (**a**). Eutrophication also leads water bodies to fill in with organic material, hastening the process of aquatic succession. An oligotrophic water body (**b**) with clear water and low nutrient content may eventually become a eutrophic water body (**c**) with abundant vegetation and high nutrient content.

also raise surface water temperatures by removing streamside vegetation that shades water.

Too little heat can also cause problems. On the Colorado and many other dammed rivers, water at the bottom of reservoirs is much colder than water at the surface. If dam operators release water from the depths of a reservoir, downstream water temperatures become much lower. In the Colorado River system, these low water temperatures have favored cold-loving invasive trout over an endangered native species of suckerfish.

Groundwater pollution is a serious problem

All the types of pollution we have discussed threaten surface water, and some threaten groundwater as well. Increasingly, groundwater sources once assumed to be ancient and pristine have been contaminated by pollution, much of it from industrial and agricultural practices. Groundwater pollution is a problem largely hidden from view and is extremely difficult to monitor; it can be out-of-sight, out-of-mind for decades until

VIEWPOINTS

Dams
Should we remove dams?

The Case for Dam Removal

The dams currently in existence in the United States were built to provide a variety of services including flood control, water supply, and hydropower, which runs mills and generates electricity. While many dams continue to provide a useful service, a large number are considered obsolete, providing no direct economic, safety, or social function. For example, many mill dams continue to stand across streams and rivers 100 to 200 years after the mill they powered went out of operation or was torn down. These dams should be considered for removal.

Regardless of size, all dams harm riparian environments. Dams block the free flow of water down a stream corridor and create a pool, or impoundment, behind them— an artificial lake in the middle of a stream community. Impounded waters often divide into layers by temperature and depth, with heated waters in the upper layer and oxygen-poor cooler water in the lower layer. The macroinvertebrates that fish depend on for food cannot survive under these lake conditions. Carp and non-native lake fish that can survive in hotter and oxygen-poor waters often displace trout and other cold-water stream species. Dams block the movement of migratory fish and other aquatic species, preventing them from reaching upstream areas to feed, spawn, and successfully reproduce. River sediments that would normally travel downstream and replenish beaches or gravel stream bottoms, where most macroinvertebrates live and fish spawn, are blocked by dams.

Rivers are dynamic systems. They move within floodplains, exchanging nutrients, sediments, and interacting on many levels. When dams interrupt that exchange, river functions are impaired and the fish and wildlife dependent on free-flowing river systems do not thrive as well. Once a dam has outlived its utility, it makes great sense to restore the river back to its original condition.

Sara Nicholas is associate director of dam programs for American Rivers and works out of their Mid-Atlantic office in Harrisburg, Pennsylvania. She has a master of science degree in environmental science from the Yale School of Forestry.

Dams for Today and Tomorrow

Our need for dams today is greater than ever because water is a finite resource. Over the past century, global water use has increased at twice the rate of population growth. Currently, world population is expected to grow by 50%, to 9 billion people in total, by 2050. Dams and reservoirs address the needs of a growing world by efficiently storing and regulating water for multiple uses. Our world relies on the benefits they bring—drinking water, flood control, power generation, irrigation, and recreation.

Ninety percent of the dams in the United States are small, local projects that lack controversy. Consider what dams do every day for millions of people. Along the Mississippi 70% of America's grain exports are barged to the Gulf of Mexico. Dams support 55 million irrigated acres of crop and pasture land (mostly in the arid West). Dams and reservoirs carry water to millions of people via canals and aqueducts. And dams help communities avoid billions of dollars in flood damage.

With dams, however, come environmental concerns such as fish passage, changes to water quality, and altered habitats. The challenge for any community is to balance the economic, environmental, and social considerations in utilizing dams. Today's choices often reflect the values, needs, wealth, and options of different communities and countries. It should be no surprise that in a world where 1.7 billion people are without electricity, hydropower is being developed in 80 countries.

Our challenge is to make decisions that embrace what research, sound science, technological innovation, and engineering prowess offers. We can embrace these things while also staying true to our historic and evolving cultural, environmental, and economic values. We owe it to future generations to make thoughtful, responsible policy choices about dams that not only affect our way of life, but theirs.

Thomas Flint is a fifth-generation farmer, actively farming in Grant County, Washington. He was elected to the Grant County Public Utility District board of commissioners in 2000, is founder and director of the public education effort known as *AgFARMation,* is a grassroots activist, and holds director's positions on the Black Sands Irrigation District and the Columbia Basin Development League.

The Science behind the Story

Arsenic in the Waters of Bangladesh

In the 1970s, UNICEF, with the help of environmental scientists at the British Geological Survey, launched a campaign to improve access to freshwater in Bangladesh. By digging thousands of small artesian wells, the designers of the program hoped to reduce Bangladeshis' dependence on disease-ridden surface waters. In the mid-1990s, however, scientists began to suspect that the very wells that had been dug to improve Bangladeshis' health were contaminated with arsenic, a poison that, if ingested frequently, can cause serious skin disorders and other illnesses, including cancer.

It was a medical doctor, not a hydrologist or a geologist, who sounded the first alarm. In 1983, dermatologist K.C. Saha of the School of Tropical Medicine in Calcutta, India, saw the first of many patients from West Bengal, an area of India just west of Bangladesh, who showed signs of arsenic poisoning. Through a process of elimination, contaminated well water was identified as the likely cause of the poisoning. The hypothesis was confirmed by

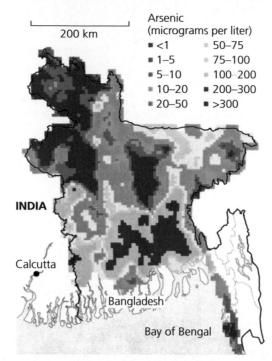

The discovery that thousands of wells dug for drinking water in Bangladesh at the urging of international aid workers turned out to be laced with arsenic is a modern tragedy. This map shows that the highest concentrations of arsenic —from 100 to 300 or more micrograms per liter of water — are found in the southern portion of the country. Source: D.G. Kinniburgh and P.L. Smedley Eds., Arsenic contamination of groundwater in Bangladesh, Department of Public Health Engineering (DPHE) Bangladesh, British Geological Survey Report, 2001.

groundwater testing and by the work of epidemiologists, among them Dipankar Chakraborti of Calcutta's Jadavpur University.

Initially, Chakraborti and other scientists focused their attention on West Bengal, where a number of contaminated wells were discovered. It was not until the late 1990s that large-scale testing of Bangladesh's wells began. In 1998, initial tests were conducted in some of the worst-affected regions and soon provided strong evidence of widespread contamination. By 2001, when the British Geological Survey and the government of Bangladesh published their final report, 3,524 wells had been tested. Of the shallow wells, those less than 150 m (490 ft) deep, 46% exceeded the World Health Organization's maximum recommended level of 10 µg/l of arsenic (1.33×10^{-6} oz/gal, or

widespread drinking water contamination is discovered. People in many countries in both the developing and developed worlds are still spraying tons of persistent pesticides on farm fields with little thought that they could turn up in the groundwater they drink years later.

Groundwater pollution is more serious than surface water pollution because it is longer lasting. Rivers flush their pollutants fairly quickly, but groundwater retains its contaminants until they decompose, which in the case of persistent pollutants can be many years or decades. The long-lived pesticide DDT, for instance, is still found widely in U.S. aquifers even though it was banned 35 years ago. In addition, chemicals are broken down much more slowly in aquifers than in surface water or soils. Groundwater generally contains less dissolved oxygen, microbes, minerals, and organic matter, so decomposition is slower. For instance, concentrations of the herbicide alachlor are reduced to half their original levels after 20 days in soil, but take almost 4 years to do so in groundwater.

There are many sources of groundwater pollution

Some groundwater contamination is natural. A variety of chemicals that are toxic at high concentrations, including aluminum, fluoride, nitrates, and sulfates, can dissolve naturally into groundwater. The poisoning of Bangladesh's wells by arsenic is one high-profile case of

8.35×10^{-8} lb/gal). Extrapolating across all of Bangladesh, the scientists estimated that as many as 2.5 million wells serving 57 million people were contaminated. As the accompanying figure shows, arsenic contamination is most prevalent in southern Bangladesh, but localized hotspots are found even in northern regions of the country.

Scientists have not yet reached a consensus on the complex chemical processes by which Bangladesh's shallow aquifers became contaminated. All agree that the arsenic is of natural origin; what remains unclear is how the low levels of arsenic naturally present in soils were dissolved in the aquifers in elemental and highly toxic form. One initial explanation, suggested by Chakraborti and his colleagues, placed most of the blame on agricultural irrigation. By drawing large amounts of water out of the aquifers during Bangladesh's dry season, they argued, irrigation had permitted oxygen to enter the aquifers and prompted the release of arsenic from pyrite, a common mineral.

Other scientists contend that pyrite oxidation cannot explain the majority of cases of arsenic contamination. In a 1998 paper in the journal *Nature,* Ross Nickson of the British Geological Survey and his colleagues suggested that arsenic was being released from iron oxides carried into Bangladesh by the Ganges River. In support of their argument, they pointed to the results of a hydrochemical survey that measured the chemical composition of aquifers throughout Bangladesh. Contrary to the predictions of the pyrite oxidation hypothesis, the survey found that arsenic concentrations tended to increase with aquifer depth and to be inversely correlated with concentrations of sulfur, a component of pyrite. In light of these findings, Nickson and his colleagues concluded that highly reducing chemical conditions created by buried organic matter, such as peat, had probably leached arsenic from iron oxides over the course of thousands of years.

Recently, Massachusetts Institute of Technology hydrologist Charles Harvey and his colleagues have suggested that irrigation may contribute to the arsenic problem after all, but not because of pyrite oxida-

tion. In a 2002 paper in the journal *Science,* they described an experiment in which more than a dozen wells were dug in the Munshiganj region of Bangladesh, near the capital city of Dhaka. Contrary to the pyrite oxidation hypothesis, and in agreement with Nickson and his colleagues, they found little evidence of a connection between sulfur or oxygen and arsenic.

However, they also found that they could increase arsenic concentrations by injecting organic matter, such as molasses, into their experimental wells. In the process of being metabolized by microbes, the molasses appeared to be freeing arsenic from iron oxides. A similar process might take place naturally, Harvey and his colleagues argued, when runoff from rice paddies, ponds, and rivers recharges aquifers that have been depleted by heavy pumping for irrigation. In support of this hypothesis, they found that much of the carbon in the shallow wells was of recent origin. Other scientists, however, have found arsenic in much older waters. This finding suggests that Bangladesh's arsenic problem may be caused by multiple hydrological and geological factors.

---

natural contamination (see The Science behind the Story). However, groundwater pollution from human activity is widespread. Industrial, agricultural, and urban wastes—from heavy metals to petroleum products to radioactive waste to pesticides—can leach through soil and seep into aquifers. Pathogens and other pollutants can also enter groundwater through improperly designed wells. Some aquifer contamination even results from the intentional pumping of wastes below ground. It has been reported that 60% of liquid hazardous waste in the United States is pumped deep underground (Chapter 19), a total of 34 billion liters (9 billion gallons) per year. In several U.S. states, some of these wastes have apparently found their way into drinking water.

Leakage from underground storage tanks is another major contributor to groundwater pollution. These include septic tanks, tanks of industrial chemicals, and tanks of oil and gas. In the United States, the EPA has embarked on a nationwide cleanup program to unearth and repair these tanks before they do further damage to soil and groundwater quality. After more than a decade of work, the EPA in 2003 had confirmed leaks from 436,000 tanks, had initiated cleanups on 401,000 of them, and had completed cleanups of 297,000. Intercepting pollutants such as chlorinated solvents and gasoline before they reach aquifers is important: Many such chemicals are carcinogenic or otherwise toxic, and once an aquifer is contaminated it is extremely difficult to restore.

Agriculture also contributes to groundwater pollution. Nitrates from fertilizers have leached into groundwater in Canada's prairie provinces and in 49 U.S. states. Pesticides were detected in over half of the shallow aquifer sites tested in the United States in the mid-1990s, although generally below the levels set by the EPA for drinking water. Nitrates in drinking water have been linked to cancers, miscarriages, and the "blue-baby" syndrome that suffocates infants. Agriculture can also contribute pathogens; in 2000, the groundwater supply of Walkerton, Ontario, became contaminated with the bacterium *E. coli,* probably from livestock manure together with well failure, heavy rainfall, and human error. Two thousand people became ill, and seven died.

Manufacturing industries have been heavy polluters through the years. Although legislation has forced them to reduce their discharges, groundwater can bear a toxic legacy for long afterwards. The town of Woburn, Massachusetts, near Boston has a long history of heavy industry, and its dumps received large amounts of industrial waste through the years. In 1979 it was discovered that two municipal drinking water wells were contaminated with trichloroethylene (TCE), chloroform, and other toxicants that had seeped in from the waste. The wells were closed, but an epidemiological study by the state documented a significant clustering of childhood leukemia cases among children whose mothers drank from the well water while pregnant.

Military sites, too, have added to groundwater contamination. During World War II on land near St. Louis, the U.S. Army operated the world's largest facility to produce trinitrotoluene (TNT). Nitroaromatic by-products seeped into the ground and eventually into the drinking water for miles around. The site was high on the list to be cleaned up under the 1980 Superfund legislation (Chapter 19). At one of the best-known Superfund sites, the Hanford Nuclear Reservation in Washington State, vast quantities of radioactive waste have seeped into groundwater, some of it with a half-life of a quarter-million years.

Solutions to Pollution of Freshwater

As numerous as our freshwater pollution problems may seem, it is important to remember that many of them were worse a few decades ago. Citizen activism and government response during the 1960s and 1970s in the United States resulted in legislation such as the Clean Water Act of 1977. The Act, actually a collection of amendments to older water pollution laws, made it illegal to discharge pollution from a point source without a permit, set standards for industrial wastewater, set standards for contaminant levels in surface waters, and funded construction of sewage treatment plants. Thanks to this legislation and others, point-source pollution in the United States has been reduced since the 1970s, and rivers and lakes are cleaner than they have been in decades.

Other developed nations have followed suit with federal laws of their own and have also reduced their pollution levels. In Asia, for example, legislation, regulation, enforcement, and increased investment in sewage treatment have brought about striking improvements in water quality in Japan, Singapore, China, and South Korea. However, nonpoint-source pollution remains a major challenge for developed countries, as do eutrophication and acid precipitation. Furthermore, conditions in developing countries are at risk of deteriorating further rather than improving. Fortunately, international attention is firmly focused on assuring "water security" for the new century (Table 14.2).

The Great Lakes of Canada and the United States represent a success story in fighting water pollution. In the 1970s these lakes, which hold 18% of the world's surface freshwater, were badly polluted with wastewater, fertilizers, and such toxic chemicals as dioxins, dichlorodiphenyl dichlorethene (DDE), and polychlorinated biphenyls (PCBs). Algal blooms occurred along beaches, and Lake Erie was pronounced dead. Now, 30 years later, efforts of the Canadian and U.S. governments have paid off. According to Environment Canada, releases of seven toxic chemicals have been reduced by 71%, municipal phosphorus has been decreased by 80%, and chlorinated pollutants from paper mills are down by 82%. Levels of PCBs and DDE are down 78% and 91%, and bird populations are rebounding. Lake Erie is now home to the world's largest walleye fishery. The Great Lakes' troubles are by no means over—sediment pollution is still high, PCBs still settle on the lakes from the air, and fish are not always safe to eat (Chapter 10)—but the progress so far shows how conditions can improve when citizens push their governments to take action.

In some cases, however, solutions will likely need to address prevention, not simply treatment and cleanup. For instance, a prominent expert on water and soils has said that solving the problem of groundwater pollution will require a complete overhaul of the way we live and

Table 14.2 Goals of World Water Forum 2000

At the Second World Water Forum at The Hague, Netherlands, in March 2000, international ministers declared the following as the main challenges in water management for the new century:

Meeting basic needs of access to safe and sufficient water and sanitation, and empowering people to meet their needs

Securing the food supply by efficient mobilization and use, and equitable distribution, of water

Protecting ecosystems to ensure their integrity, by managing water sustainably

Sharing water resources to promote peaceful cooperation among states or groups sharing transboundary water resources

Managing risks from floods, droughts, pollution, and other hazards

Valuing water to reflect its importance, and pricing water to reflect true costs, while assuring equity and meeting needs of the poor

Governing water wisely so the public and all stakeholders are included in management decisions

Source: Second World Water Forum, 2000.

dispose of waste, not just an "end-of-pipe" approach of piecemeal cleanups. Indeed, preventing pollution in the first place would seem to be the best strategy when one considers the other options for dealing with groundwater contamination:

- We can filter groundwater before distributing it. This can be expensive; facilities in the U.S. Midwest by one estimate spend $400 million annually just to remove the herbicide atrazine from water supplies.
- We can pump water out of the aquifer, treat it, then inject it back in, repeatedly. Most Superfund sites with contaminated groundwater are using this method, but scientists have estimated that it generally takes an impractically long time. So far work has begun on only a little over 1% of all sites with heavily polluted groundwater, and the eventual cleanup bill has been estimated at $1 trillion. Even then, scientists agree, complete cleanup will be impossible.
- We can restrict usage of pollutants on lands above selected aquifers. Restriction would alleviate groundwater contamination in those aquifers; however, such restriction could simply shift the pollution elsewhere.

Although a sea change in our attitudes, behavior, or technology may be necessary to clean up our water completely, there are many things ordinary people can do to help minimize freshwater pollution. One course of action is to exercise the power of consumer choice in the marketplace by purchasing phosphorus-free detergents and other "environmentally friendly" products. Another is to become involved in protecting local waterways. Locally based "riverwatch" groups or watershed associations enlist volunteers to collect data and help state and federal agencies safeguard the health of rivers and other water bodies. Similar programs, such as Australia's Waterwatch and Streamwatch programs, are emerging in many countries throughout the world as citizens and policymakers increasingly demand clean water.

Conclusion

Citizen action, government legislation and regulation, new technologies, economic incentives, and public awareness of the importance of water conservation are all enabling us to confront what will surely be one of the great environmental challenges of the new century: assuring ourselves and the planet's ecosystems adequate quantity and quality of freshwater.

Although accessible freshwater comprises only a minuscule percentage of Earth's total water content, we have grown used to taking it for granted. With our expanding population and still-increasing water usage, however, we are fast beginning to meet scarcity. Water depletion and water pollution are already taking a large toll on the health, economies, and societies of much of the developing world and are beginning to do so in arid areas of the developed world. Continued pressure on our vital water resources may finally make us realize just how dependent we are. As author Marc Reisner wrote in *Cadillac Desert*, the premier account of the water history of the American West, "If the Colorado River suddenly stopped flowing, you would have four years of carryover capacity in the reservoirs before you had to evacuate most of southern California and Arizona and a good portion of Colorado, New Mexico, Utah, and Wyoming." The vision is not unimaginable; in parts of the world where surface water or groundwater has been depleted, people have had to move away, leaving deser- tified landscapes behind them. There is reason to hope that we may yet attain sustainability in our water usage, however. Potential solutions are numerous, and the issue is too important to ignore.

REVIEW QUESTIONS

1. Use the Colorado River watershed to describe water movement through the hydrologic cycle.

2. Define groundwater. What role does groundwater play in the hydrologic cycle?

3. Why are sources of freshwater unreliable for some people and plentiful for others?

4. Describe some consumptive and non-consumptive uses of freshwater.

5. How much of the global population relies on groundwater? What methods do we use to obtain water from aquifers and surface water sources?

6. What are floodplains? In what ways is flooding a good thing? How is flooding prevented or controlled?

7. Name three benefits of damming rivers, and name three costs.

8. How large is China's Three Gorges Dam and its reservoir? What particular environmental, health, and social concerns has it raised?

9. Why do the Colorado, Rio Grande, Nile, and Yellow Rivers now slow to a trickle or even run dry before reaching their deltas?

10. Why are water tables dropping around the world? What are the environmental costs of falling water tables?

11. List several inefficient irrigation methods, and describe the consequences of inefficient irrigation.

12. Why might water cause international tensions to erupt in the future?

13. How might desalination technology help us "make" more water? Describe two methods of desalination. Where is this technology being used?

14. Can agricultural demand for water be decreased? What are some methods that can increase the efficiency of agricultural irrigation?

15. Describe some of the ways that we can reduce household water use. How can industrial and municipal uses of water be reduced?

16. Name the chemical and physical properties of water that scientists use to determine water quality.

17. In what ways can humans help to reduce eutrophication in freshwater systems?

18. Why is sediment a pollutant? How can sediment pollution be prevented?

19. Why do many scientists consider groundwater pollution a bigger problem than surface water pollution?

20. What are some anthropogenic sources of groundwater pollution?

DISCUSSION QUESTIONS

1. Discuss some strategies for equalizing distribution of water throughout the world. Consider supply and transport issues. Have our methods of drawing, distributing, and storing water changed very much throughout history? How is the scale of these methods affecting the availability of water supplies?

2. How can human methods of drawing and using freshwater affect the hydrologic cycle?

3. Why is the industrial use of water more economically profitable than agricultural use? How could this lead to social and political tensions if water scarcity becomes a greater problem in the future?

4. Have the provisions of the Clean Water Act of 1977 been effective? Discuss some of the methods we can use, in addition to "end-of-pipe" solutions, to prevent contamination and ensure "water security."

Media Resources *For further review, go to the website* **www.envscienceplace.com** *or student CD-ROM, where you will find quizzes, flashcards, a glossary, additional interactive exercises, and links to relevant news and research sources. Also, on the website and CD-ROM is* **GRAPH IT**, *a series of interactive graphing tutorials to help you interpret graphs and plot data.*

15 Biodiversity and conservation biology

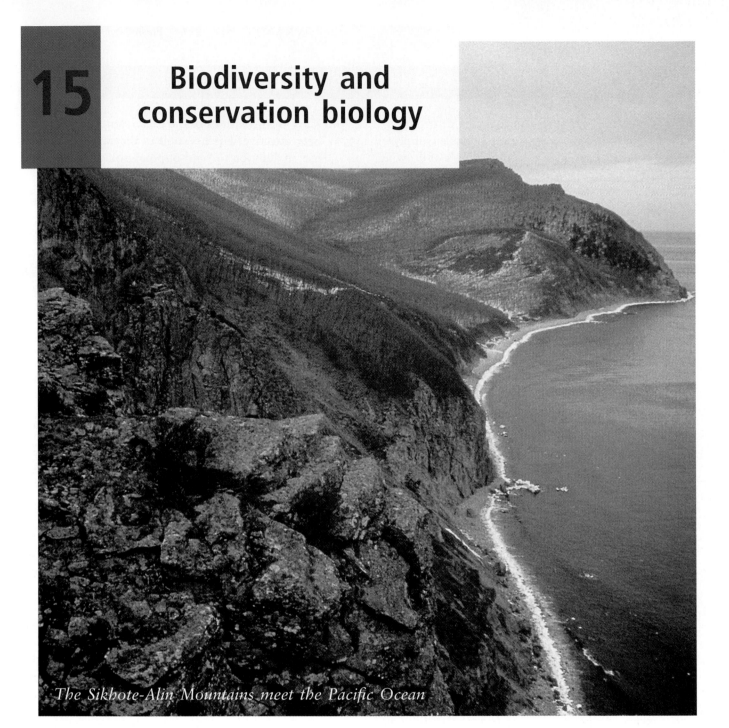

The Sikhote-Alin Mountains meet the Pacific Ocean

This chapter will help you understand:

- The scope of biodiversity on Earth
- Measurements of biodiversity
- Background extinction rates and periods of mass extinction
- Primary causes of biodiversity loss

- Benefits of biodiversity
- Conservation biology and its practice
- Island biogeography theory
- Traditional and more innovative biodiversity conservation efforts

A Siberian tiger in the Sikhote-Alin Mountains

Central Case: Saving the Siberian Tiger

> "Future generations would be truly saddened that this century had so little foresight, so little compassion, such lack of generosity of spirit for the future that it would eliminate one of the most dramatic and beautiful animals this world has ever seen."
> —*George Schaller, wildlife biologist*

> "Except in pockets of ignorance and malice, there is no longer an ideological war between conservationists and developers. Both share the perception that health and prosperity decline in a deteriorating environment. They also understand that useful products cannot be harvested from extinct species."
> —*Edward O. Wilson, Harvard University entomologist and biodiversity expert*

Up until the past 200 years, tigers roamed widely across the Asian continent from Turkey to northeast Russia to Indonesia. Humans have driven the majestic striped cats from most of their historic range, however. As a result, today tigers are exceedingly rare and almost everywhere are creeping toward extinction.

Of the tigers that still survive in small pockets of their former range, those of the subspecies known as the Siberian tiger are the largest cats in the world, with males reaching 363 kg (800 lb) and 3.66 m (12 ft) in length. Also named Amur tigers for the watershed they occupied along the Amur River (which forms part of the present-day boundary between Siberian Russia and Manchurian China), these cats now find their last refuge in the temperate forests and taiga of the remote Sikhote-Alin Mountains of the Russian Far East.

For thousands of years the Siberian tiger coexisted with the Tungus, the native people of what is today the Russian Far East. Tigers held a prominent place in native language and lore. Some natives referred to the tiger as "Old Man" or "Grandfather," and the Manchurian people of the region equated the tiger with royalty and viewed it as a guardian of the mountains and forests. It was uncommon for indigenous people of the region to kill a tiger unless it had preyed upon a person. Even in instances when people singled out a man-eating tiger and killed it, they often conducted ceremonies and expressed respect for the animal and remorse for the hunt.

The Russians who moved into and exerted control over the region in the early to mid 20th century, however, had no such cultural traditions. Communist Party officials and Russian soldiers hunted the tiger for sport and hides, and some Russians reported killing as many as 10 tigers in a single hunt. In addition, poachers began killing tigers in order to sell their body parts for traditional medicine and aphrodisiacs to China and other Asian countries. At the same time, road building, logging, and agriculture began to break up tiger habitat and provide easy access for well-armed hunters. The tiger population dipped to perhaps 20–30 animals.

International conservation groups began to get involved, working with Russian biologists to try to save the dwindling tiger population. One such group was the Hornocker Wildlife Institute, a research and conservation organization then affiliated with the University of Idaho. In 1991 the group helped launch a Russian-American effort, the Siberian Tiger Project, devoted to studying the tiger and its habitat. The team put together a plan to protect the tiger, began educating people regarding the tiger's importance and value, and worked closely with those who live in proximity to the big cats.

Thanks in part to these and other efforts by conservation biologists, today the Siberian tiger population is up to roughly 150–450, and 500 more survive in zoos around the world. The long-term outlook still looks challenging, but many people are involved in the effort to save these endangered creatures. It is one of many efforts around the world today to stem the loss of our planet's biological diversity. In this chapter, we examine this diversity, its value, its loss, and the efforts being made to maintain it.

Our Planet of Life

Although the plight of tigers and other large charismatic mammals has been well publicized, they represent only a small fraction of the number of endangered species. Growing human population and resource consumption are putting ever-greater pressures on the flora and fauna of the planet, from tigers to tiger beetles. Earth's diversity of life makes our planet, as far as we know, unique in the universe, but we have already begun losing the very quality that makes our planet so special.

In Chapter 5 we introduced the concept of biological diversity, or biodiversity, as "the sum total of all organisms in an area, taking into account the diversity of species, their genes, their populations, and their communities." We will now refine this definition and examine current biodiversity trends and their relevance to our lives. We will then explore some science-based solutions to biodiversity loss in the United States and around the world.

What is biodiversity?

Many people who study our planet's life have been inspired by the work of Edward O. Wilson, a professor and curator in entomology at Harvard University (Figure 15.1). A world-renowned expert on ants and a Pulitzer prize–winning writer, Wilson has authored

Figure 15.1 Edward O. Wilson is the world's most recognized authority on biodiversity and its conservation. Pellegrino University Research Professor and Honorary Curator in Entomology at Harvard University, and a world expert on ants, Wilson has written over 20 books and won two Pulitzer Prizes.

many books, including *The Diversity of Life,* which explains the value of biodiversity. In it, he defines biodiversity as

> the variety of organisms considered at all levels, from genetic variants belonging to the same species through arrays of species to arrays of genera, families, and still higher taxonomic levels; [biodiversity] includes the variety of ecosystems, which comprise both the communities of organisms within particular habitats and the physical conditions under which they live.

Definitions of the term *biodiversity* are plentiful, some more technical than Wilson's and some much simpler. The United Nations Environment Programme (UNEP) and several other international organizations have agreed to define it as "the variety of life in all its forms, levels and combinations." As sociologist of science David Takacs explains in his 1996 book, *The Idea of Biodiversity,* different biologists employ different working definitions according to their own aims, interests, and values. No matter which definition you prefer, though, it is clear that biodiversity is not simply a count of species but a concept as multifaceted as life itself.

Biodiversity encompasses several levels of life's organization

Despite the breadth of the term *biodiversity* and the varied ways in which people use it, there is broad agreement that the term applies across several main categories or levels in the organization of life (Figure 15.2). The category that is easiest to visualize and that is most commonly used is species diversity.

Species diversity We can express **species diversity** in terms of the number or variety of species in the world or in a particular region. One component of species diversity is species richness, the number of species, and another is evenness or relative abundance, or the extent to which numbers of individuals of different species are equal or skewed. As you recall from Chapter 5, a species is a distinct type of organism or, more precisely, a population or group of populations whose members uniquely share certain characteristics and can breed with one another and produce fertile offspring.

Taxonomists, the scientists who classify species, use an organism's physical appearance and genetic makeup to determine to which species it belongs. Taxonomists also group species by their similarity into a hierarchy of categories meant to reflect evolutionary relationships. Re-

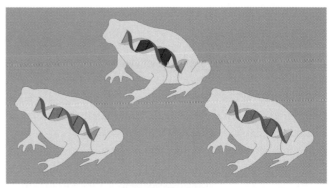

Genetic diversity

Species diversity

Ecosystem diversity

Figure 15.2 The concept of biodiversity encompasses several levels in the hierarchy of life, which we will name genetic diversity, species diversity, and ecosystem diversity. Genetic diversity refers to variation in DNA composition among individuals within a species. Species diversity refers to number or variety of species. Ecosystem diversity refers to variety at levels above the species level, such as the ecosystem, community, habitat, or landscape levels.

lated species are grouped together into *genera* (singular, *genus*), related genera are grouped into families, and so on (Figure 15.3). Every species is given a two-part scientific name in Latin denoting its genus and species. The tiger, *Panthera tigris*, represents one species that is different from the world's other species of large cats such as the jaguar *(Panthera onca)*, the leopard *(Panthera pardus)*, and the African lion *(Panthera leo)*. These four species are closely related in evolutionary terms, as indicated by the genus name they share, *Panthera*. They are more distantly related to cats in other genera such as the cheetah *(Acinonyx jubatus)* and the bobcat *(Felis rufus)*, although all cats are classified together in the family Felidae.

As we saw in Chapter 5, speciation, the generation of new species, adds to species diversity, while extinction, as has apparently befallen the golden toad, decreases species diversity. Although immigration and emigration may increase or decrease species diversity locally, only speciation and extinction change it globally.

Biodiversity exists below the species level in the form of subspecies, populations of a species that occur in different geographic areas and differ from one another in at least some characteristics. Subspecies are formed by the same processes that give rise to speciation, but they represent cases in which divergence has not proceeded

far enough to create separate species. Scientists denote subspecies with a third part of the scientific name. The Siberian tiger, *Panthera tigris altaica*, is one of five subspecies of tiger still surviving. Tiger subspecies differ in color, thickness of coat, stripe patterns, and size. *Panthera tigris altaica* is 5–10 cm (2–4 in) taller at the shoulder than the Bengal tiger *(Panthera tigris tigris)* of India and Nepal, and it also has a thicker coat and larger paws (Figure 15.4).

Genetic diversity Subspecies are designated when scientists recognize substantial genetically based differences among individuals of different populations of a species. However, all species consist of individuals that vary genetically from one another to some degree, and this genetic diversity is an important component of biodiversity. **Genetic diversity** encompasses the differences in DNA composition among individuals within a given species. At the minimum, individuals may be identical or may have genomes that vary by only a single nucleotide, but individuals generally differ far more than this.

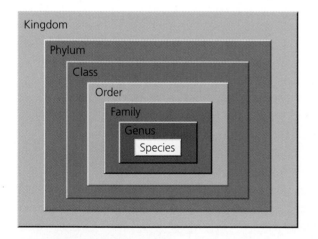

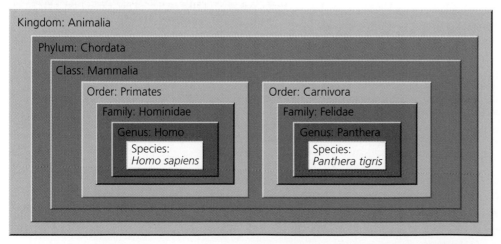

Figure 15.3 Taxonomists classify organisms using a hierarchichal system meant to reflect evolutionary relationships. Species that are similar in their appearance, behavior, and genetics (presumably due to recent common ancestry) are placed in the same genus, and organisms of similar genera are placed in the same family. Families are placed within orders, orders within classes, classes within phyla, and phyla within kingdoms. For instance, humans (*Homo sapiens*, a species in the genus *Homo*) and tigers (*Panthera tigris*, a species in the genus *Panthera*) are both mammals, so are both classified within the Class Mammalia. However, the differences between our two species that have evolved over millions of years have resulted in enough divergence that we are placed in different orders and families. The astounding complexity of Earth's biodiversity has necessitated that taxonomists also create many more subcategories between the major levels shown in the figure.

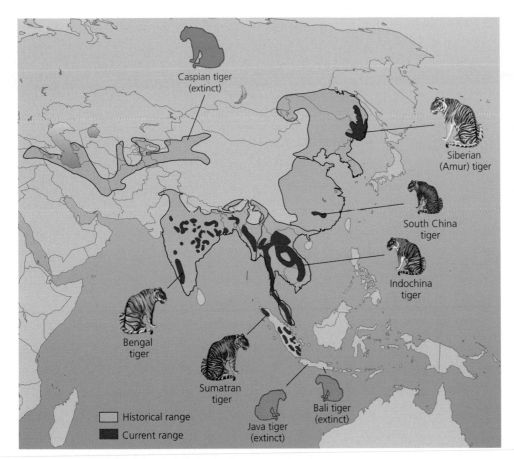

Figure 15.4 Three of the eight subspecies of tiger became extinct during the 20th century. Today only the Siberian (Amur), Bengal, Indochina, South China, and Sumatran tigers persist. The Bali, Javan, and Caspian tigers are extinct. Tigers have been driven from most of the geographic range they historically occupied. This map contrasts the ranges of the eight subspecies in the years 1800 and 2000. Data from The Tiger Information Center, 2003.

Whether the genetic diversity is extremely minor or great enough to warrant subspecies status, such diversity has repercussions for the well-being of a species in at least two major ways. First, as a species becomes adapted to local environmental conditions in the area it lives, its genetic diversity may decrease. Adaptation to local conditions enables the species to do better, as long as those conditions do not change. Whereas Bengal tigers are adapted to warm temperatures, for instance, Siberian tigers are adapted to cold ones; in Siberian tigers, genes for thicker coats of fur have been favored while those for thin coats of fur have been weeded out.

In the long term, however, species with more genetic diversity may have better chances of persisting, because their built-in variation better enables them to cope with environmental change. Species with little genetic diversity are vulnerable to environmental change for which they are not genetically prepared. Species with depressed genetic diversity may also be more vulnerable to disease and may suffer the effects of inbreeding, which occurs when parents that are too genetically similar mate and produce weak or defective offspring. Scientists have sounded warnings over low genetic diversity in species that have dropped to extremely low population sizes in the past, including cheetahs, bison, and elephant seals, but the full consequences of the reduced diversity in these species remain to be seen. Genetic diversity in our crop plants also continues to be a prime concern to humanity, as we saw in Chapter 9.

Ecosystem diversity Biodiversity encompasses more than the species level, including groupings of organisms, their interactions with one another, and even their relationships with their abiotic environment. Scientists have viewed biodiversity on these levels in several ways. **Ecosystem diversity** refers to the number and variety of ecosystems in some specified area, **community diversity** refers to the number and variety of biotic community types, and **habitat diversity** refers to the number and variety of habitats. Each of these concepts can apply to various geographic scales, from small localities to large regions to the global scale. **Landscape diversity** refers to the variety and geographic arrangement of habitats, communities, or ecosystems over a wide area, including the sizes, shapes, and interconnectedness of patches of these entities. Under any of these concepts, a seashore of rocky and sandy beaches, forested cliffs, offshore coral reefs, and ocean waters would hold far more biodiversity than the same area of a monocultural cornfield; similarly, a mountain slope whose vegetation changes from desert to hardwood forest to coniferous forest to alpine meadow would hold more biodiversity than an area the same size consisting of only desert, forest, or meadow.

Measuring biodiversity is not easy

As you might guess, coming up with accurate quantitative measurements to express a region's biodiversity is difficult. This is partly why scientists express biodiversity in terms of its most easily measured component, species diversity, and in particular, species richness. Species richness is indeed a good gauge for overall biodiversity when different areas are being compared, but we still are profoundly ignorant of the number of species that exist worldwide. As of 2002, scientists had identified approximately 1.75 million species of plants, animals, and microorganisms. Estimates for the total number of species that actually exist range from 3 million to 100 million, with our best educated guesses converging somewhere around 14 million.

Species are not evenly distributed among taxonomic groups. In number of species, insects have a staggering predominance over all other forms of life (Figure 15.5). Insects account for more than half of all species in the world. Among insects, about 40% are beetles. Beetles outnumber all noninsect animals and all plants. No wonder the 20th-century British biologist J.B.S. Haldane famously quipped that God must have had "an inordinate fondness for beetles."

Our estimates of species numbers are incomplete for several reasons. One reason is that some areas of Earth remain little explored. We have barely sampled the ocean depths, hydrothermal vents, and the tree canopies and soils of tropical forests. Another reason is that many species are tiny and easily overlooked; these inconspicuous organisms include species of bacteria, archaea, nematodes (roundworms), fungi, protists, and soil-dwelling arthropods. In addition, many organisms are so difficult to identify that species thought to be identical often turn out to be two or more different species once biologists look more closely. This is the case not only with microbes, fungi, and small insects, but also with organisms as large as trees, birds, and whales.

Smithsonian Institution entomologist Terry Erwin carried out one famous attempt to inventory the number of anthropod species in the world. In 1982 Erwin's crews fogged rainforest trees in Central America with clouds of airborne insecticide and then collected insects, spiders, and other arthropods with sheets and funnels as they died and fell from the treetops. The thousands of tiny creatures were then parceled out among specialists for identification. Erwin handled beetles and concluded that there were 163 beetle species limited to just a single tree species, *Luehea seemannii*. If this were typical for a tropical tree, he figured, then the world's 50,000 tropical tree species would hold 8,150,000 beetle species. If beetles represent 40% of all arthropods, then 20 million arthropod species live in tropical canopies, and because canopies hold two-thirds of arthropods, then the total number of arthropod species in tropical forests alone would be 30 million. Many assumptions were involved in this calculation, of course, and several follow-up studies have revised Erwin's estimate, most revising it downward to 5–10 million.

In 2000 an audacious attempt to inventory the world's species was launched. Along with top scientists, the entrepreneur Kevin Kelly, who had co-founded *Wired* magazine, created the All Species Foundation, aiming "to discover and describe all living organisms on Earth, within one human generation, and to make this information available to everyone everywhere." The group figured that it would take 25 years and $1–3 billion to accomplish the feat. The subsequent economic downturn and drop in philanthropy stalled the young effort, but it continues, now backed by 100 collaborating scientists.

You may be able to help measure biodiversity where you live

On a more local scale, a number of efforts are underway to catalog the biodiversity of particular areas. An endeavor called the All Taxa Biodiversity Inventory involves Great Smoky Mountains National Park on the Tennessee–North Carolina border, for instance. Here, scientists and volunteers from the U.S. National Park Service and dozens of universities and organizations are trying to gather and disseminate information on the 100,000 species thought to live in the park.

However, biological diversity is not just something that exists in national parks and tropical rainforests. Biodiversity is everywhere, even in your own backyard. Today, locally based efforts are springing up to measure biodiversity in small, seemingly commonplace areas. One type of effort that has become popular is the bioblitz, an event held in a particular area, such as a state park or a city, in which taxonomists and interested citizens team up and race to survey every species they can find within 24 hours. The first bioblitz, held in Washington D.C.'s Kenilworth Park in 1996, found

(a) Organisms scaled in size to number of species

Figure 15.5 Earth's species are not evenly spread among different groups. (a) A drawing that presents different types of organisms scaled in size to their number of species gives a good visual sense of the disparity in species number among various groups, and makes clear the predominance of insects. In all efforts to count species, it should be remembered that the majority of species are thought not yet to be discovered or described, and that some groups (such as insects, nematodes, protists, fungi, and others) may actually have many more species than scientists have so far described. (b) Three-quarters of known species are animals, and nearly three-quarters of animals are insects, many of them belonging to the single order Coleoptera (beetles). Vertebrates comprise only 3.9% of all animals. Among vertebrates, about half are fishes, and mammals comprise only 9%. Data from B. Groombridge and M.D. Jenkins, Global Biodiversity: Earth's living resources in the 21st century, United Nations Environmental Program-World Conservation Monitoring Centre, 2002.

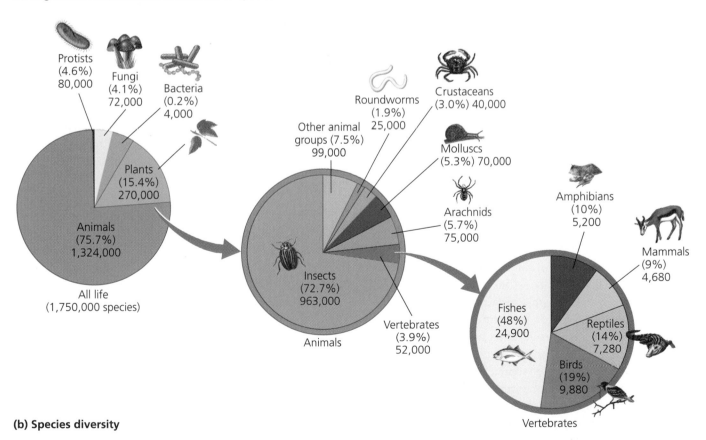

(b) Species diversity

nearly 1,000 species; a Connecticut bioblitz holds the record so far at 2,519 species. As of mid-2003, at least 75 such events had been held, from Kansas to Germany. Bioblitzes are meant to promote public awareness of biodiversity as much as to obtain accurate species counts, yet they have turned up surprising finds. At a 1998 event not far from E. O. Wilson's Harvard office, for example, the ant specialist turned up two ant species that science had not yet named or described.

Global biodiversity is not distributed evenly

Numbers of species tell only part of the story of biodiversity on Earth. Living things are distributed across the planet unevenly, and scientists have long sought to explain the patterns they see in the distribution of biodiversity.

For example, not all groups of organisms contain equal numbers of species. For some groups, only one or a few species exist. Other types of organisms have given rise to many species in a relatively short period of time. One such group is the family Asteraceae, consisting of such flowering plants as daisies, ragweed, and sunflowers. This group, which began with a common ancestor roughly 30 million years ago—not too long ago in geological time—now consists of over 20,000 species. Such an explosion of diversity is called a *radiation*. An **adaptive radiation** occurs when an ancestral species gives rise to many species that fill different niches; each species adapts to its niche by natural selection. Famous examples involve the Galapagos finches first studied by Darwin and the Hawaiian honeycreepers (see Figure 4.23). In both groups of birds, new species evolved various types of bill shapes in adapting to different food sources.

Species diversity also varies according to biome (Chapter 6). Tropical dry forests and rainforests tend to support more species than tundra and deserts, for instance. The variation in diversity by biome is related to one of the planet's most striking patterns of species diversity: the fact that species richness generally increases as one approaches the equator (Figure 15.6). This pattern of variation with latitude, called the **latitudinal gradient,** has been one of the most obvious patterns in ecology, but it also has been one of the most difficult patterns in ecology for scientists to account for. Hypotheses for the cause of this pattern abound, but it seems likely that plant productivity and climate stability play key roles in explaining the phenomenon (Figure 15.7). Greater amounts of solar energy, heat, and humidity at tropical latitudes lead to more plant growth, making areas nearer the equator more productive and able to support larger numbers of animals. In addition, the relatively stable climates of equatorial regions—

their similar temperatures and rainfall from day to day and season to season—help ensure that single species won't dominate ecosystems, but that instead numerous species can coexist. While varying environmental conditions favor generalists—species that can deal with a wide range of circumstances but that do no single thing very well—stable conditions favor organisms with specialized niches that do certain particular things very well. In addition, polar and temperate regions may be relatively depauperate in species because glaciation events repeatedly forced organisms out of these regions and toward more tropical latitudes.

We will discuss further geographic patterns in biodiversity later in this chapter, when we explore solutions to the ongoing loss of global biodiversity that our planet is currently experiencing.

Biodiversity Loss and Species Extinction

Biodiversity at all levels is being lost to human impact, most irretrievably in the extinction of species, which, once vanished, can never return. Extinction (Chapter 5) occurs when the last member of a species dies and the species ceases to exist, as apparently was the case with Monteverde's golden toad. In contrast, the extinction of a particular population from a given area, but not the entire species globally, is called **extirpation.** The tiger has been extirpated from most of its historic range, but it is not yet extinct. Although a species that is extirpated from one place may still exist in others, extirpation is an erosive process that can, over time, lead to extinction.

Extinction is "natural"

Extirpation and extinction are natural processes. If organisms did not naturally go extinct, we would be up to our ears in dinosaurs, trilobites, ammonites, and the millions of other types of organisms that vanished from Earth long before humans appeared. Paleontologists estimate that roughly 99% of all species that have ever lived are already extinct; that is, the wealth of species on our planet today comprises only about 1% of all the species that ever lived. Most extinctions preceding the appearance of humans have occurred one by one, at a rate that paleontologists refer to as the **background rate of extinction.** For example, the fossil record indicates that for both birds and mammals, one species in the world typically became extinct every 500–1,000 years.

Figure 15.6 For many types of organisms, numbers of species per unit area tends to increase as one moves toward the equator. This trend, the latitudinal gradient in species richness, is one of the most readily apparent—yet least understood—patterns in ecology. One example is bird species in North and Central America, which increase from 30-100 recorded in any one spot in arctic Canada and Alaska to over 600 in areas of Costa Rica and Panama. Adapted from R.E. Cook, Variation in species density on North American birds, *Systematic Zoology,* 1969.

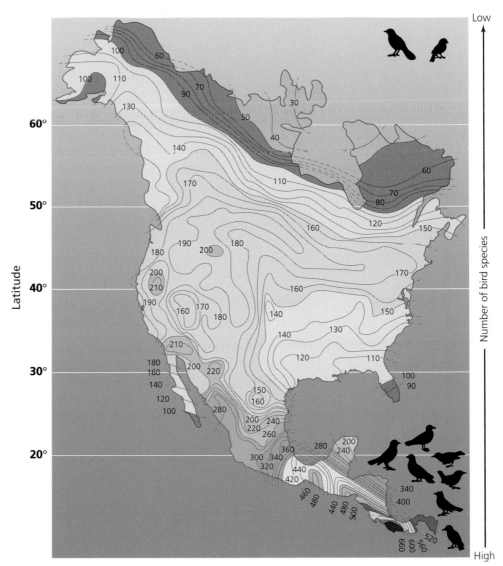

Figure 15.7 Ecologists have offered many hypotheses for the latitudinal gradient in species richness, and one set of ideas is summarized here. The variable climates (across days, seasons, and years) of polar and temperate latitudes favor organisms that can survive a wide range of conditions. Such generalist species have expansive niches; they can do many things well enough to survive, and they spread over large areas. In tropical latitudes, the abundant solar energy, heat, and humidity induce greater plant growth, which supports more organisms. The stable climates of equatorial regions favor specialist species, which have restricted niches but do certain things very well. Together these factors are thought to promote greater species richness in the tropics.

Temperate and polar latitudes
• Variable climate favors fewer widespread generalist species.

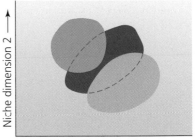

Tropical latitudes
• Greater solar energy, heat, and humidity promote more plant growth to support more organisms. Stable climate favors specialist species. Together these encourage greater diversity of species.

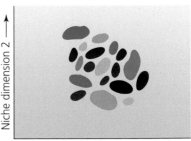

The Earth has experienced five previous mass extinction episodes

Extinction rates have risen far above this background rate during several mass extinction events of Earth's history. In the last 440 million years, there have been five major episodes of mass extinction (Figure 15.8), each of which has taken more than one-fifth of life's families and at least half its species (Table 15.1). The most severe episode occurred at the end of the Permian period, approximately 248 million years ago, when close to 54% of all families, 90% of all species, and 95% of marine species became extinct. The best-known episode occurred at the end of the Cretaceous period, 65 million years ago, when an apparent asteroid impact brought an end to the dinosaurs and many other groups (Chapter 5). In addition, there is evidence for further mass extinctions in the Cambrian period and earlier, more than half a billion years ago.

If current trends continue, the modern era, also known as the Quaternary period, may see the extinction of more than half of all species. Although similar in scale to previous mass extinctions, today's ongoing mass extinction is different in two primary respects. First, humans are causing it. Second, humans will suffer as a result of it.

Humans set the sixth mass extinction in motion years ago

In our written history, we have recorded many instances of human-induced species extinction over the past few hundred years. Sailors documented the extinction of the dodo on the Indian Ocean island of Mauritius in the 17th century, for example, and we still have a few of the dodo's body parts in museums. Among North American birds in the past two centuries, we have lost the passenger pigeon, Carolina parakeet, great auk, Labrador duck, and probably the Bachman's warbler and ivory-billed woodpecker. Several more species, including the whooping crane, California condor, and Kirtland's warbler, teeter on the brink of extinction.

However, species extinctions caused by humans likely precede written history; people may have been hunting species to extinction for thousands of years. Scientists have inferred this from archaeological evidence in areas of the world that humans have colonized relatively recently. In case after case, a wave of extinctions has seemingly followed close on the heels of human arrival (Figure 15.9). After Polynesians reached Hawaii, for instance, many birds went extinct, and evidence of an extinction wave also exists from larger island nations, such as New Zealand and Madagascar. The pattern also

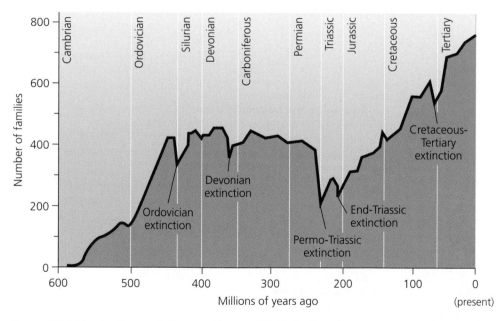

Figure 15.8 The fossil record shows evidence of five episodes of mass extinction during the past half-billion years of Earth history. At the end of the Ordovician, the end of the Devonian, the end of the Permian, the end of the Triassic, and the end of the Cretaceous, 50–95% of the world's species appear to have gone extinct. Each time, biodiversity has later rebounded to equal or higher levels, but the rebound has taken millions of years in each case. Adapted from E.O. Wilson, *The Diversity of Life*, Belknap Press, 1999.

Table 15.1 Mass Extinctions

Event	Date (mya: millions of years ago)	Cause	Types of life most affected	Percent of life depleted
Ordovician	440 mya	unknown	marine organisms; but terrestial record unknown	>20% of families
Devonian	370 mya	unknown	marine organisms; but terrestial record unknown	>20% of families
Permo-Triassic (P-T)	250 mya	possibly volcanism	marine organisms; but terrestial record less known	>50% of families; 80–95% of species
end-Triassic	202 mya	unknown	marine organisms; but terrestial record less known	20% of families; 50% of genera
Cretaceous-Teritary (K-T)	65 mya	asteroid impact	marine and terrestial organisms, including dinosaurs	15% of families; >50% of species
Current	Beginning 0.01 mya	human impact *via* habitat destruction, land and resource use, hunting, and promoting invasive species	large animals, specialized organisms, island organisms, organisms hunted or harvested by humans	ongoing

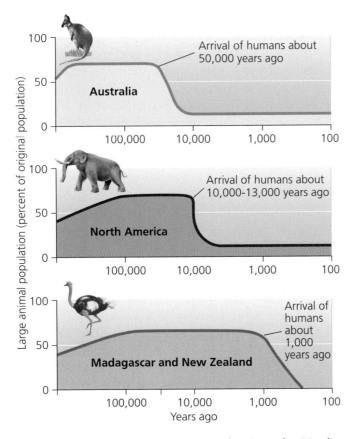

Figure 15.9 Shortly after humans arrived in Australia, North America, and Madagascar and New Zealand a large number of species of large mammals and flightless birds became extinct. It is thought that this did not occur in Africa because humans and other animals evolved together for many millions of years. Adapted from E.O. Wilson, *The Diversity of Life*, Belknap Press, 1999.

holds for at least two continents; large vertebrates died off in Australia after Aborigines arrived roughly 50,000 years ago, and North America lost 35 genera of large mammals after humans arrived in the continent at least 10,000–13,000 years ago. Although some scientists hold that climate variation instead of hunting may be responsible for the North American extinctions, the fact that human arrival co-occurs with extinction waves independently in several regions at different times in history is certainly strongly suggestive.

Current extinction rates are much higher than normal

Humans have raised the rate of extinction above the background rate for centuries now. However, today species loss is proceeding at a still more accelerated pace as our population growth and resource consumption put ever-increasing strain on habitats and wildlife.

In late 1995, 1,500 of the world's leading scientists reported to the United Nations that the current global extinction rate was more than 1,000 times greater than it would have been without human destruction of habitat. The scientists went on to report in their Global Biodiversity Assessment that more than 30,000 plant and animal species face extinction and that in the preceding 400 years, 484 animals and 654 plant species were known to have become extinct. The scientists also reported that mammals—like the tiger—were becoming extinct 40 times faster than ever before. Independent estimates by UNEP in 2002 reported that extinction is occurring 50–100 times faster than the background rate and that 45% of Earth's forests, 50% of its mangrove ecosystems, and 10% of its coral reefs had been destroyed by recent human activity.

To keep track of the current status of endangered species, the World Conservation Union (IUCN, the acronym derived from the name in which the organization was founded—the International Union for Conservation of Nature and Natural Resources (1956)) maintains the **Red List,** an updated list of species facing unusually high risks of extinction. The 2003 Red List reported that 24% (1,130) of mammal species and 12% (1,194) of bird species are threatened with extinction. Among other major groups (for which assessments are not fully complete), estimates of the percentage of species threatened ranged from 39% to 89%. From 1996 to 2003, the total number of vertebrate animals listed as threatened climbed by 6%, from 3,314 to 3,524. Since 1970, at least 58 fish species, 9 bird species, and 1 mammal species have become extinct, and in the United States alone over the past 500 years, 236 animals and 17 plants are confirmed to have gone extinct. For all of these figures, the *actual* numbers of species extinct and threatened, like the actual number of total species in the world, are doubtless greater than the *known* numbers.

Among the 1,130 mammals facing possible extinction on the IUCN's Red List is the tiger, which despite—or perhaps because of—its tremendous size and reputation as a fierce predator, is one of the most endangered large animals on the planet. In 1950 eight tiger subspecies existed. Today three are extinct. The Bali tiger, *Panthera tigris balica,* went extinct in the 1940s; the Caspian tiger, *Panthera tigris virgata,* during the 1970s; and the Javan tiger, *Panthera tigris sondaica,* during the 1980s. In each case, overhunting and habitat alteration were responsible.

The major causes of species loss spell "HIPPO"

The overharvesting that decimated Siberian tiger populations in the early 20th century is only one form of human activity that has reduced the populations of many species. Overexploitation provides the "O" in **HIPPO,** an acronym scientists have coined to denote the five primary causes of species decline and extinction: Habitat alteration, Invasive species, Pollution, Population growth, and Overexploitation. The most prevalent and powerful of these five causes is habitat alteration.

Habitat alteration Nearly every human activity may potentially alter the habitat of the organisms around us. Farming replaces diverse natural communities with simplified ones of only one or a few plant species. Grazing changes the structure and species composition of grasslands. Either type of agriculture—farming or grazing—can lead to desertification. Clearing of forests removes the food, shelter, and other resources that forest-dwelling organisms need to survive. Hydroelectric dams turn rivers into reservoirs upstream and affect water conditions and floodplain communities downstream. Urbanization and suburban sprawl supplant diverse natural communities with simplified human-made ones, driving many species from their homes. Global climate change now threatens to alter habitat on large scales by pushing climatic zones toward higher latitudes and elevations.

All these changes in habitats significantly affect the organisms that depend on them. These effects are generally negative because organisms are already adapted to the habitats in which they live, so any change is likely to render the habitat less suitable for them. Of course, human-induced habitat change may benefit certain species. Animals such as starlings, house sparrows, pigeons, and gray squirrels do very well in urban environments and benefit from our modification of natural habitats. However, these species are relatively few; for every species that benefits, more are harmed. Furthermore, those species that do well in our midst tend to be weedy, cosmopolitan species that are in little danger of disappearing any time soon.

Habitat alteration is by far the greatest cause of species extinction today, biologists agree. It is the primary source of population declines for 83% of threatened mammals and 85% of threatened birds, according to UNEP data. As just one example of thousands, the prairies native to North America's Great Plains have been almost entirely converted to agriculture. The area of prairie habitat has been reduced by more than 99%; as a result, grassland bird species have declined by an estimated 82% to 99%. Many of these species have been extirpated from large areas, and the two species of prairie chickens still persisting on the Great Plains could soon go extinct.

Invasive species The introduction of invasive species (Chapter 5) to new environments has also pushed native species toward extinction. Some introductions have been accidental—for example, the marine organisms that have been transported between continents in the ballast water of ships, the animals that have escaped from the pet trade, and the weed seeds that cling to our socks as we travel from one place to another. Other introductions have been intentional; human immigrants, for example, have introduced many food crops and domesticated animals to new places, and people have transported certain organisms to novel localities for other economic or aesthetic reasons, generally unaware of the ecological consequences that could result.

Most organisms introduced to new areas perish, but the few types that survive may do very well, especially if they find themselves without the predators and parasites that attacked them back home or without the competitors that had limited their access to resources. Once released from the limiting factors of predation, parasitism, and competition, an introduced species can begin increasing rapidly. If it meets little environmental resistance in its new home, it may keep expanding and may displace native species. In countless cases, the introduction of non-native species has led to the extirpation of native species and, in some cases, even to the extinction of native species. In addition, invasive species cause billions of dollars in economic damage each year. Figure 15.10 shows a diverse selection of invasive species.

Modern-day Hawaii showcases the devastation invasive species can cause. One story begins with feral pigs. By rooting through soil, consuming plants, trampling vegetation, and spreading seeds of invasive weeds, pigs have transformed many Hawaiian plant communities. The pigs' hoofprints and wallowing grounds have also created muddy pools of water that are ideal for another invasive species—the mosquito *Culex quinquefasciatus*. This mosquito is now the vector for introduced strains of avian malaria *(Plasmodium relictum)* and avian pox *(Poxvirus avium)*, which are decimating populations of Hawaii's native birds. Many of the same Hawaiian honeycreepers that arose from an impressive adaptive radiation are now going extinct from disease and the complex series of steps that led to it. The native species of the Hawaiian Islands in particular and of islands in general are especially vulnerable to disruption from invasive species because the native species have been in isolation for so long with relatively few parasites, predators, and competitors. As a result, they have not evolved the defenses necessary to resist invaders that are better adapted to these pressures.

Pollution Pollution can also negatively affect organisms in many ways. Air pollution (Chapter 11) can degrade forest ecosystems. Water pollution (Chapter 14) can adversely affect fish and amphibians. Agricultural runoff (including fertilizers, pesticides, and sediments) can harm many terrestrial and aquatic species. Toxic agents such as PCBs and endocrine-disrupting compounds can have substantial toxicological effects (Chapter 10), and the effects of major oil and chemical spills on wildlife are dramatic and well known. However, although pollution is a substantial threat, it tends to be less significant than public perception holds it to be. The damage to wildlife and ecosystems caused by pollution can be severe but tends to be less than the damage caused by habitat alteration, invasive species, or human population growth.

Population growth As we have seen throughout this book, the growth in human population (Chapter 7) has touched just about every environmental issue and exacerbated just about every environmental problem. Our continued population growth poses a threat to other species indirectly through a number of avenues. These include each of the other components of the HIPPO dilemma. More people means more habitat alteration, more pollution, more overexploitation, and more invasive species. Along with the growth in resource consumption by affluent societies, human population growth is the ultimate reason behind many of the proximate threats to biodiversity.

Overexploitation Scientists use the "O" of the HIPPO acronym to refer to two different things, each of which can fall within the meaning of the term *overexploitation*. Some scientists focus on overharvesting of species from the wild, others on overconsumption of resources. Both types of overexploitation have far-reaching effects on biodiversity.

For most species, a high intensity of hunting by humans will not *in itself* pose a threat of extinction, but for some species it can. The Siberian tiger is one such species. Large in size, few in number, long-lived, and raising few young in its lifetime—a classic K-strategist (Chapter 5)—the Siberian tiger is just the type of animal to be vulnerable to population reduction by hunting. Indeed, the advent of Russian hunting nearly drove the animal extinct, but decreased hunting during and after World War II contributed to a gradual population increase. By the mid-1980s the Siberian tiger population was likely up to 250 individuals. The political freedom that came with the Soviet Union's breakup in 1989, however, brought with it a freedom to harvest Siberia's natural resources, the tiger included, without any regulations or rules. This coincided with an economic expansion in many Asian countries whose cultures considered it appropriate to use tiger penises to attempt to

Invasive Species			
Species	Native to...	Invasive in...	Effects
Mosquito fish (*Gambusia affinis*) 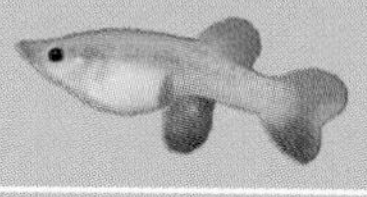	North America	Africa, Asia, Europe, and Australia	Introduced to control mosquito populations, the mosquito fish outcompetes native fish, eats their eggs, and does no better than native species in controlling mosquitoes.
Zebra mussel (*Dreissenna polymorpha*) 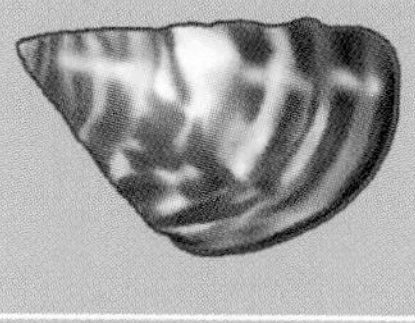	Caspian Sea	Freshwater ecosystems, including the Great Lakes of Canada and the United States	Zebra mussels most likely made their way to North America by traveling in ballast water taken on by cargo ships. They compete with native species and clog water treatment facilities and power plant cooling systems.
Kudzu (*Pueraria montana*) 	Japan	Southeastern United States	A vine that can grow 30 m (100 ft) in a single season, the U.S. Soil Conservation Service introduced kudzu in the 1930s to help control erosion. Adaptable and extraordinarily fast-growing, kudzu has taken over thousands of hectares of forests, fields, and roadsides in the southeastern United States.
Asian long-horned beetle (*Anoplophora glabripennis*) 	Asia	United States	Having first arrived in the United States in the 1990s, these beetles burrow into hardwood trees and interfere with trees' ability to absorb and process water and nutrients. They can destroy the majority of hardwood trees in an area. Several U.S. cities, including Chicago in 1999 and Seattle in 2002, have cleared thousands of trees after detecting these invaders.
Rosy wolfsnail (*Euglandina rosea*) 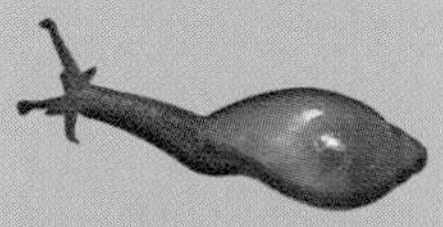	Southeastern United States and Latin America	Hawaii	In the 1950s well-meaning scientists introduced the rosy wolfsnail to Hawaii to prey upon and reduce the population of another invasive species, the giant African land snail (*Achatina fulica*), which had been introduced early in the 20th century as an ornamental garden animal. Within a few decades, however, the carnivorous rosy wolfsnail had instead driven more than half of Hawaii's native species of banded tree snails to extinction.
Cane toad (*Bufo marinus*) 	Southern United States to tropical South America	Northern Australia	Since being introduced 70 years ago to control insects in sugarcane fields, the cane toad has wreaked havoc across northern Australia. Toxins from the poisonous skin of this tropical American toad can kill its predators, and the cane toad outcompetes native amphibians.
Bullfrog (*Rana catesbiana*) 	Eastern North America	Western North America	Although not poisonous, the bullfrog is causing amphibian and reptile declines in western North America. Bullfrog tadpoles grow large and can outcompete and prey on other tadpoles but need to grow a long time in permanent water to do so. Historically most water bodies in the arid West dried up part of the year, making it impossible for bullfrogs to live there, but artificial impoundments—dams, farm pounds, canals—gave the bullfrogs bases from which they could spread.

Figure 15.10 Invasive species are those that thrive in areas where they are introduced, outcompeting, preying on, or otherwise harming native species. Invasive species are one of the primary threats to global biodiversity today. Out of the many thousands of invasive species, this chart shows a few of the best-known.

Invasive Species			
Species	Native to...	Invasive in...	Effects
Gypsy moth (*Lymantria dispar*)	Eurasia	Northeastern United States	In the 1860s a scientist introduced the gypsy moth to Massachusetts in the mistaken belief that it might be bred with others to produce a commercial-quality silk. The gypsy moth failed to start a silk industry and instead spread through the northeastern United States and beyond, where its outbreaks defoliate trees over large regions every few years.
European starling (*Sturnus vulgaris*)	Europe	North America	Starlings were first introduced to New York City in the late 19th century by Shakespeare devotees intent on bringing every bird mentioned in Shakespeare's plays to the new continent. It only took 75 years for the birds to spread to the Pacific coast, Alaska, and Mexico, becoming one of the most abundant birds on the continent.
Indian mongoose (*Herpestes auropunctatus*)	Southeast Asia	Hawaii	Rats that had invaded the Hawaiian islands from ships in the 17th century were damaging sugarcane fields, so in 1883 the Indian mongoose was introduced to control rat populations. Unfortunately, the rats were active at night and the mongooses fed during the day, so the plan didn't work. Instead mongooses began preying on native species like ground-nesting seabirds and the now-endangered Nene or Hawaiian goose (*Branta sandvicensis*).
A green alga (*Caulerpa taxifolia*)	Tropical oceans and seas	Mediterranean Sea	Dubbed the "killer algae," *Caulerpa taxifolia* has spread along the coasts of several Mediterranean countries since it apparently escaped from Monaco's aquarium in 1984. Creeping underwater over the sand and mud like a green shag carpet, it crowds out other plants, is inedible to most animals, and tangles boat propellers. It has been the focus of intense eradication efforts since arriving recently in Australia and California.
Cheatgrass (*Bromus tectorum*)	Eurasia	Western United States	In just 30 years after its introduction to Washington state in the 1890s, cheatgrass has spread across much of the western United States. Its secret: fire. Its thick patches that choke out other plants and use up the soil's nitrogen burn readily. Fire kills many of the native plants, but not cheatgrass, which grows back even stronger amid the lack of competition.
Brown tree snake (*Boiga irregularis*)	Southeast Asia	Guam	Nearly all native forest bird species on the South Pacific island of Guam have disappeared. The culprit is the brown tree snake. The snakes were likely brought to the island inadvertently as stowaways in cargo bays of military planes in World War II. Guam's birds had no evolved defenses against the snake's nighttime predation. The snake also causes numerous power outages each year on Guam and spread to other islands where they repeat their ecological devastation. The arrival of this snake is the greatest fear of conservation biologists in Hawaii.

boost human sexual performance and to use bones, claws, whiskers, and other tiger body parts to treat a wide variety of maladies. Thus, the early 1990s brought an unregulated boom in poaching (poachers killed at least 180 Siberian tigers between 1991 and 1996), as well as a dramatic increase in the logging of the Korean pine forests on which the tigers and their prey depend.

A similar story is playing out in the grasslands of central Asia, where populations of an antelope called the saiga *(Saiga tatarica)* have crashed from one million in 1993 to less than 30,000 only 10 years later. The animals are killed for their horns, which are shipped to China for use in traditional medicine. Ironically, this slaughter began because biologists had promoted using the saiga's horns as an alternative to those of the endangered rhinoceros. Of course, North America has seen its share of overhunting in the past century or two. Bison were hunted to near extinction, mostly for sport. Top predators such as wolves, mountain lions, and grizzly and black bears were extirpated from large portions of their ranges and remain absent from most areas today. Beavers, hunted for their pelts, and coyotes, shot along with wolves, are today recolonizing large areas of the continent from which they had been eliminated.

Over the past century, hunting has led to steep declines in the populations of numerous other K-strategist animals, such as whales. The Atlantic gray whale has gone extinct, and several other whale species remain threatened or endangered. Gorillas and other primates that are killed for their meat may be facing extinction soon. Thousands of sharks are killed each year simply for their fins, which are used in soup. World fish consumption increased by 240% from 1960 to 2001.

Hunting is not the only form of overexploitation. For example, human consumption of paper products tripled from 1970 to 2000, increasing the need for logging and, in turn, accelerating forest loss and thereby causing a number of impacts on many species. In ways such as these our consumption of biotic resources has had impacts on ecosystems and has heightened extinction risks to wildlife.

Sometimes causes of biodiversity loss are difficult to determine

The HIPPO acronym helps us keep track of the primary causes of species decline. However, the reasons for the decline of any given species are often multifaceted and complex and, as such, can be difficult to determine. The current precipitous decline in populations of amphibians throughout the world provides an example. Frogs, toads, and salamanders worldwide are decreasing in number. Several have already gone extinct, and scientists are struggling to explain why. Recent studies have implicated a wide array of factors, including chemical contamination, disease transmission, habitat loss, ozone depletion, and climate change, and most scientists now suspect that such factors may be interacting synergistically. As the declines continue, researchers now are racing against time to pinpoint their causes (see The Science behind the Story).

Benefits of Biodiversity

Scientists all over the world are presenting us with data that confirm what any naturalist who has watched the changes in habitat in his or her hometown already knows: From amphibians to tigers, biodiversity is being lost at all scales, rapidly and visibly within our lifetimes. These observations beg the question, "Does it matter?" There are many ways to answer this question, but we can begin by considering the various ways that biodiversity benefits people. Scientists have offered a number of concrete, tangible reasons for preserving biodiversity, showing how biodiversity directly or indirectly supports the long-term sustainability of human society. In addition, beyond tangible benefits to humans, many people feel that there are ethical and aesthetic dimensions to biodiversity preservation that cannot be ignored.

Biodiversity provides valuable ecosystem services free of charge

Contrary to popular opinion, some things in life can indeed be free, as long as we choose to protect the living systems that provide them. Healthy forests can provide clean air and buffer hydrologic systems against flooding and drought. Native landraces of crops can provide insurance against disease and drought. Abundant wildlife can attract tourists and boost economies of developing nations. Intact ecosystems provide these and other valuable processes, known as *ecosystem services* (Chapter 2), for all of us, free of charge.

Maintaining these ecosystem services is one clear benefit of protecting biodiversity. According to UNEP, biodiversity

- Provides food, fuel and fiber.
- Provides shelter and building materials.
- Purifies air and water.
- Detoxifies and decomposes wastes.
- Stabilizes and moderates Earth's climate.
- Moderates floods, droughts, wind, and temperature extremes.
- Generates and renews soil fertility and cycles nutrients.
- Pollinates plants, including many crops.
- Controls pests and diseases.
- Maintains genetic resources as key inputs to crop varieties, livestock breeds, and medicines.
- Provides cultural and aesthetic benefits.
- Provides us the means to adapt to change.

Biodiversity, in the form of organisms and ecosystems, supports a vast number of valuable processes that humans would have to pay for if nature did not provide them. As we saw in Chapter 2, the annual value of just 17 of these ecosystem services may be in the neighborhood of $16–54 trillion per year.

Clearly, the outright conversion of large swaths of natural habitat could greatly affect a natural system's ability to provide services to our society, but what about the extinction of selected species? Skeptics have rightly asked whether the loss of a few species will really make much difference in an ecosystem's ability to perform its functions.

Top predators play key roles in their ecosystems

The answer to this question appears to depend on which species are removed. Removing a species that can be functionally replaced by others may make little difference to an ecological system. Recall, however, from Chapter 5 our discussion of keystone species. Like the keystone that holds together an arch, a keystone species is one whose removal would result in significant changes in an ecological system. If a keystone species is extirpated or driven extinct, other species may disappear or experience significant population changes as a result.

Often top predators, such as tigers, are considered keystone species. A single top predator may prey on many other carnivores, each of which may prey on many herbivores, each of which may in turn consume many plants. Thus the removal of a single individual at the top of a food chain can have impacts that multiply as they cascade down the food chain. Any resulting change to the plant community may then have effects that work their way back up the food chain. We discussed (Chapter 5)

the case of the sea otter, whose removal allowed sea urchins to thrive and consume so much kelp that kelp forests were destroyed, depriving many other species of crucial habitat. This effect was first seen when humans hunting for otter for their pelts depleted otter populations. Recent research by marine biologist James Estes of the University of California Santa Cruz and his colleagues reveals yet another part of the story: Orcas (killer whales) in some areas have turned to eating sea otters because their preferred prey, seals, became rare. Because this new top predator influences the population levels of otters, a single orca appears able to have a substantial impact on kelp forest communities.

Although top predators may be relatively few in number, they are often among the species most vulnerable to human impact. Large animals are frequently hunted because they are a source of meat, or are thought to pose a danger to humans, or threaten livestock. Large predators such as tigers, wolves, and grizzly bears also need large areas of habitat through which to wander and find food, making them especially vulnerable to habitat loss and habitat fragmentation. Top predators are also particularly vulnerable to the buildup of toxic pollutants in their tissues through the process of biomagnification (Chapter 10). All these pressures can take an especially heavy toll because top predators are often K-strategists, living long lives and producing few offspring; thus, the removal of even a few individuals can make a big difference.

Top predators are not the only species that exert far-reaching influence over their ecosystems: the influence of other species can be equally significant. Environmental systems are complex, and it can often be difficult to predict in advance which particular species may be important to an ecosystem's functioning. Thus, many people prefer to apply the precautionary principle in the spirit of Aldo Leopold (Chapter 2), who advised, "To keep every cog and wheel is the first precaution of intelligent tinkering."

Biodiversity gives us natural classrooms

Our understanding of such ecological phenomena as keystone species has been possible only because ecologists have had natural areas and diverse communities available to study. Although the disturbance of natural communities may be necessary to show the effects of human impact, ecologists need relatively pristine areas to serve as reference states in order to understand such impact. More generally, the parks and reserves established to protect biodiversity can serve as natural classrooms for all citizens. These areas and their flora and

Amphibian Diversity and Amphibian Declines

Amphibians illustrate the two most salient aspects of Earth's biodiversity today: more and more species are being discovered by scientists, while more and more populations and species are vanishing before our eyes.

While new species of most classes of vertebrates are discovered at a rate of only one or a few per year, the number of known amphibian species (including frogs, salamanders, and others)—about 5,500 as of 2003—has jumped nearly 35% just since 1985. At the same time, however, at least 200 species are in steep decline, suggesting to researchers that they may be naming some species just before they go extinct and losing others before they are even discovered. More than 30 species extant just years or decades ago, including the golden toad (Chapter 5), are now altogether gone. These losses are especially worrying because amphibians, such as frogs and salamanders, are widely regarded as "biological indicators" that can indicate whether an ecosystem is in good shape or is becoming degraded. Relying on both aquatic and terrestrial environments, often breathing and absorb-

The odd-looking purplish frog *Nasikabatrachus sahyadrensis*, recently found in India, is one of many new amphibian species being discovered.

ing water through their skins, they are sensitive to pollution and other environmental stresses. The link between amphibians and environmental quality suggests that studying their biodiversity patterns and the reasons for their declines can tell us much of general interest about the state of our environment.

In countries such as Sri Lanka, intensive scientific scrutiny and improved technology have revealed amphibian "hot spots." In the 1990s, an international team of scientists decided to determine

whether Sri Lanka, a large tropical island off the coast of India, held more than the 40 frog species that were already known. Researcher Madhava Meegaskumbura and his team combed through trees, rivers, ponds, and leaf litter for eight years, collecting more than 1,400 frogs at 300 study sites. The scientists analyzed the physical appearance, habitat use, and vocalizations of the frogs they found to known species. They also examined the frogs' genetic make-up, by obtaining sequences of nucleotides in several regions of their DNA. They then compared these genetic, physical, and behavioral characteristics to those of known species of frogs.

The research team found that the DNA from many of their frogs didn't match that of known species. And they found that many of their frogs looked different, sounded different, or behaved differently than known species. Clearly, they had discovered new species of frogs unknown to science. Some of these novel species live on rocks, with leg fringes and markings that help them disguise themselves as clumps of moss. Other species are tree frogs that lay their eggs in baskets they construct. In all, more than 100

---

fauna provide a wide array of tangible educational and social benefits, including the study of biology, natural history, ecology, and chemistry, as well as painting and photography.

Biodiversity enhances food security

Biodiversity benefits our agriculture as well. As our discussion of native landraces of corn in Oaxaca, Mexico, in Chapter 9 showed, genetic diversity within crop species and their ancestors is enormously valuable both to agriculture and to an area's economy. In 1995, Turkey's wheat crops received at least $50 bil-

lion worth of disease resistance from wild wheat strains. California's barley crops annually receive $160 million in disease resistance benefits from Ethiopian strains of barley. During the 1970s a researcher discovered a maize species in Mexico known as *Zea diploperennis*. This maize is highly resistant to disease, and it is a perennial, meaning it will grow back year after year without being replanted. At the time of its discovery, its entire range was limited to a 10-ha (25-acre) plot of land in the mountains of the Mexican state of Jalisco.

Other potentially important food crops await utilization (Figure 15.11). The babassu palm *(Orbignya*

new species of amphibians were discovered—all on an island only slightly larger than the state of West Virginia. When reported in the journal *Science* in 2002, the study entitled "Sri Lanka: An amphibian hot spot," caught the attention of conservation biologists worldwide.

Such promising discoveries, however, come against a backdrop of distressing declines. Observed numbers of amphibians are down around the globe. Scientists are racing to pin down the causes, zeroing in on several likely factors as culprits: habitat destruction, pollution from contaminants such as pesticides, disease, aggressive invasive species, and climate change. Most worrisome to scientists are species that decline even when no direct damage, such as habitat loss, is apparent. In some such cases, researchers surmise that a combination of factors may be at work. In one study, researchers Rick Relyea and Nathan Mills presented young frogs with two common dangers—pesticides and predators—to see how the mix affected their survival.

The team collected 10 pairs of tree frogs from a wildlife area in Missouri and placed their eggs in clean water. When tadpoles emerged from the eggs, groups of 10 were each put in different tubs of water. Some tubs were left as is, others contained varying levels of carbaryl, a common pesticide, and still others contained the harmless solvent acetone as a control. To some of each of these three types of tubs, the researchers added a hungry, caged predator—a young salamander. The salamander couldn't reach the tadpoles, but the tadpoles were aware of its presence. In a series of experiments, some lasting as long as 16 days, researchers watched to see how many tadpoles survived the different combinations of stress factors.

Their results, published in the *Proceedings of the National Academy of Sciences* in 2001, revealed that tadpoles that withstood one such stress might not survive two. As expected, all the tadpoles in clean water with no predators survived, and all the tadpoles exposed to high concentrations of carbaryl died within several days, regardless of predator presence. But when carbaryl levels were lower, the presence of the salamander made a noticeable difference. In one trial, about 75% of tadpoles survived the pesticide if no predator was present, but in the presence of the salamander, survival rates dropped to about 25%. Thus, when both stresses were present (a condition likely to arise in the tadpoles' natural habitat), death rates increased by two to four times.

One year later, a study published in the same journal by herpetologist Joseph Kiesecker found similar results with pathogens and pesticides. His field and lab experiments revealed that wood frogs were more vulnerable to parasitic infections that cause limb deformities when they were exposed to water containing pesticides.

As scientists learn more about how such factors combine to threaten amphibians, they are gaining a clearer picture of how the fate of these creatures may foreshadow the future for other organisms. "Amphibians have been around for 300 million years. They're tough, and yet they're checking out all around us," says David Wake, a biologist at the University of California at Berkeley, who was among the first to note the creatures' decline. "We really do see amphibians as biodiversity bellwethers."

phalerata) of the Amazon produces more vegetable oil than any other plant. The serendipity berry *(Dioscoreophyllum cumminsii)* produces a sweetener 3,000 times sweeter than sucrose (table sugar). Several species of salt-tolerant grasses and trees are so hardy that farmers can irrigate them with saltwater. These same plants also produce animal feed, a substitute for conventional vegetable oil, and other economically important products. Such species could be of immeasurable benefit to areas undergoing soil salinization due to excessive and poorly managed irrigation (Chapter 8). Clearly, biodiversity benefits our agricultural systems—but only if we conserve it.

Biodiversity provides traditional medicines and high-tech pharmaceutical products

Wild species yield new products, including pharmaceuticals, fibers, crops, and petroleum substitutes (Figure 15.12). People have made medicines from plants for centuries, and many of today's widely used drugs were discovered by studying chemical compounds present in wild plants, animals, and microorganisms. It can be argued that every species that goes extinct represents one lost opportunity to find a cure for cancer or AIDS.

In 1997, 10 of the 25 best-selling pharmaceuticals owed their origin to wild species, and each year such

Food Security and Biodiversity: Potential new food sources		
Species	Native to...	Potential uses and benefits
Amaranths (3 species of *Amaranthus*) 	Tropical and Andean America	Grain and leafy vegetable; livestock feed; rapid growth; drought-resistant
Buriti palm (*Mauritia flexuosa*) 	Amazon lowlands	"Tree of life" to Amerindians; vitamin-rich fruit; pith as source for bread; palm heart from shoots
Maca (*Lepidium meyenii*) 	Andes Mountains	Cool-resistant root vegetable resembling radish, with distinctive flavor; near extinction
Tree tomato (*Cyphomandra betacea*) 	South America	Elongated fruit with sweet taste
Babirusa (*Babyrousa babyrussa*) 	Indonesia: Moluccas and Sulawesi	A deep-forest pig; thrives on vegetation high in cellulose and hence less dependent on grain
Capybara (*Hydrochoeris hydrochoeris*) 	South America	World's largest rodent; meat esteemed; easily ranched in open habitats near water
Vicuna (*Lama vicugna*) 	Central Andes	Threatened species related to llama; valuable source of meat, fur, and hides; can be profitably ranched
Chachalacas (*Ortalis*, many species) 	South and Central America	Potential tropical equivalent of chickens; thrive in dense populations; adaptable to human habitations; fast-growing
Sand grouse (*Pterocles*, many species) 	Deserts of Africa and Asia	Pigeon-like birds; adapted to harshest deserts; domestication a possibility

Figure 15.11 By protecting biodiversity we can enhance food security. The wild species shown here are a tiny fraction of those plants and animals that may someday supplement our food supply. Adapted from E.O. Wilson, *The Diversity of Life*, Belknap Press, 1999.

Medicines and Biodiversity: Natural sources of pharmaceuticals		
Plant	Drug	Medical application
Pineapple (*Ananas comosus*) 	Bromelain	Controls tissue inflammation
Autumn crocus (*Colchicum autumnale*) 	Colchicine	Anticancer agent
Yellow cinchona (*Cinchona ledgeriana*) 	Quinine	Antimalarial
Common thyme (*Thymus vulgaris*) 	Thymol	Cures fungal infection
Pacific yew (*Taxus brevifolia*) 	Taxol	Anticancer (esp. ovarian cancer)
Velvet bean (*Mucuna deeringiana*) 	L-Dopa	Parkinson's disease suppressant
Common foxglove (*Digitalis purpurea*) 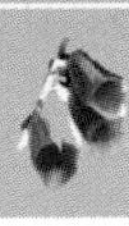	Digitoxin	Cardiac stimulant

Figure 15.12 By protecting biodiversity, we can enhance our ability to treat diseases and to provide economic growth in the pharmaceutical sector. Shown here are just a few of the plants so far found to provide chemical compounds of medical benefit. Adapted from E.O. Wilson, *The Diversity of Life,* Belknap Press, 1999.

products generate $75–150 billion in sales. Even lowly aspirin came originally from a plant called meadowsweet (*Filipendula ulmaria*). Leeches were the original source of hirudin, an anticoagulant that treats a number of circulatory disorders. Besides the thousands of products and billions of dollars wild organisms have provided to multinational pharmaceutical companies and their customers, 75% of the people on our planet use biological resources directly in traditional medicine.

In Australia, where the government has made research into the products of rare and endangered species a high priority, a rare species of cork, *Duboisia leichhardtii,* now provides medical researchers with hyoscine, a compound that physicians use to treat cancer, stomach disorders, and motion sickness. Another Australian plant, *Tylophora,* provides a drug that effectively treats lymphoid leukemia; and *Solanum aviculare* and *S. laciniatum,* known as kangaroo apples, produce the compound salsodine, a precursor to a useful steroid. Researchers are also testing the secretions of bulldog

ants *(Myrmecia)* for use as industrial disinfectants. Other researchers are exploring the potential of the compound prostaglandin E2 in treating gastric ulcers. This compound was first discovered in two frog species unique to the rainforest of Queensland, Australia. Scientists believe that both species are now extinct.

The rosy periwinkle *(Catharanthus roseus)* is a small plant with pink flowers. It produces compounds that are highly effective in treating Hodgkin's disease and a particularly deadly form of leukemia. Had this native of Madagascar become extinct prior to its discovery by medical researchers, two deadly diseases would have claimed far more victims than they have to date.

Weighing the Issues:
Bioprospecting in Costa Rica

Bioprospectors search for organisms that can provide new drugs, foods, or other valuable products. Scientists working for pharmaceutical companies, for instance, scour biodiversity-rich countries for potential drugs and medicines. Many have been criticized for harvesting indigenous species to create commercial products that do not benefit the country of origin. To protect its biodiversity, Costa Rica proactively reached an agreement with the Merck pharmaceutical company in 1991. The nonprofit National Biodiversity Institute of Costa Rica (INBio) allowed Merck to evaluate a limited number of Costa Rica's species for their commercial potential in return for $1 million, plus equipment and training for Costa Rican scientists. The agreement is largely thought to be a success. What do you think? Do both sides win? What if Merck discovers a compound that could be turned into a billion-dollar drug? Does this provide a good model for other countries? For other companies?

Biodiversity provides economic benefits through tourism and recreation

In addition to providing for our food and health, biodiversity left undisturbed can represent a direct source of income through tourism, particularly for developing countries in the tropics that have impressive species diversity and beautiful landscapes. As we saw in Chapter 5 with Costa Rica, many wealthy people like to travel to experience protected natural areas, and in so doing they create economic opportunity for residents living near those natural areas. Visitors spend money at local businesses, hire local people as guides, and support the parks that employ local residents. Ecotourism thus can bring jobs and income to areas that otherwise might be poverty-stricken.

Ecotourism has become a vital source of income for nations such as Costa Rica, with its rainforests; Australia, with its Great Barrier Reef; Belize, with its reefs, caves, and rainforests; and Kenya and Tanzania, with their savanna wildlife. The United States, too, benefits from ecotourism; its national parks draw millions of visitors domestically and from around the world. Because of the economic benefits it brings, ecotourism serves as a powerful incentive for nations, states, and local communities to preserve natural areas and reduce impacts on the landscape and on native species.

As ecotourism increases in popularity, however, a number of critics have warned that too many visitors to natural areas can degrade the outdoor experience and disturb wildlife. Anyone who has been to Yosemite or the Grand Canyon on a crowded summer weekend can attest to this. Ecotourism's effects on species living in parks and reserves are much debated and likely vary enormously from one case to the next. As ecotourism continues to increase, so will debate over its costs and benefits for local communities and for biodiversity.

People value and seek out connections with nature

Not all of the benefits of biodiversity to humans can be expressed in the hard numbers of economics or in the day-to-day practicalities of food and medicine. Some scientists and philosophers argue that there is a deeper importance to biodiversity. E. O. Wilson has described a phenomenon he calls **biophilia**, "the connections that human beings subconsciously seek with the rest of life." Wilson and others have cited as evidence of biophilia our affinity for parks and wildlife, our keeping of pets, the high value of real estate with a view of natural landscapes, and our interest—despite being far removed from a hunter-gatherer lifestyle—in hiking, birdwatching, fishing, hunting, backpacking, and similar outdoor pursuits.

Weighing the Issues:
Biophilia

What do you think of Wilson's biophilia concept? Have you ever felt a connection to other living things that you couldn't explain in scientific or economic terms? Do you think that an affinity for other living things is innately human? How could you determine whether or not most people in your community feel this way?

Do we have an ethical responsibility to prevent species extinction?

If Wilson and others are right, then biophilia not only may be the root of such phenomena as ecotourism and real estate prices, but also may influence our ethics. When Maurice Hornocker and his associates first established the Siberian Tiger Project, he explained his reasons for working to reverse the tiger's decline: "Saving the most magnificent of all the cat species and one of the most endangered should be a global responsibility. . . . If they aren't worthy of saving, then what are we all about? What is worth saving?"

As our society's sphere of ethical consideration has widened over time, and as more of us take up biocentric or ecocentric worldviews (Chapter 2), more people have come to feel that other organisms have an inherent right to exist. On one hand, we humans are part of nature, and like any other animal we need to use resources and consume other organisms to survive. In that sense, there is nothing evil or immoral about our doing so. On the other hand, we have conscious reasoning ability and are able to control our actions and consciously choose from among options. Our ethical sense has developed from this intelligence and ability to choose. The ethical justifications for conserving biodiversity have evolved fairly quickly and recently in Western culture. Aldo Leopold's land ethic is one example of an ethical argument for protecting other species. As we noted in Chapter 2, Leopold argued that humans and "the land" (his term in an era prior to the existence of the term *biodiversity*) were members of the same community and that, therefore, we have an ethical responsibility to them.

Weighing the Issues:
Fragmentation and Biodiversity

Suppose a critic of conservation tells you that human development increases biodiversity, because when a forest is fragmented, new habitats like grassy lots and gardens may be introduced to an area and allow additional species to live there. How would you respond?

Despite our ethical convictions, however, and despite biodiversity's many benefits—from the pragmatic and economic to the philosophical and spiritual—the future of biodiversity is far from secure. In fact, even our protected areas and national parks are not big enough or protected well enough to ensure that biodiversity is fully safeguarded within their borders. The search for solutions to today's biodiversity crisis is an exciting and active one, and scientists are playing a leading role in developing innovative approaches to maintaining the diversity of life on Earth.

Conservation Biology: The Search for Solutions

Although efforts to protect the Siberian tiger and many other species from extinction are relatively new, people have been thinking and writing about conserving nature at least since the time of ancient Greece. Aristotle worried about the rapid destruction of forests he witnessed in and around Greece. Later in Europe, game reserves and royal hunting areas were used for hundreds of years to protect wildlife for exclusive recreational use by the nobility. In the United States, conservationist and preservationist ethics (Chapter 2) inspired the establishment of some of the first parks and protected areas in response to the rapid expansion of human population and industry.

At the present time, more scientists and citizens perceive a need to do something to stem the loss of biodiversity. In his 1994 autobiography, *Naturalist*, E. O. Wilson wrote:

> When the [20th] century began, people still thought of the planet as infinite in its bounty. The highest mountains were still unclimbed, the ocean depths never visited, and vast wildernesses stretched across the equatorial continents. . . . In one lifetime exploding human populations have reduced wildernesses to threatened nature reserves. Ecosystems and species are vanishing at the fastest rate in 65 million years. Troubled by what we have wrought, we have begun to turn in our role from local conqueror to global steward.

Conservation biology arose in response to increasing extinction rates

The urge to act as responsible stewards of natural systems and to use science as a tool in that endeavor, helped spark the rise of conservation biology. **Conservation biology** is a scientific discipline devoted to understanding the factors, forces, and processes that influence the loss, protection, and restoration of biological diversity within and among ecosystems. It arose over the past few

decades as biodiversity losses accelerated and as many scientists became increasingly alarmed at the degradation of the natural systems they had spent their lives studying.

Conservation biologists choose their questions and pursue their research with the aim of developing solutions to problems such as habitat degradation and species loss. Thus, certain values and ethical standards are implicit in this discipline. No conservation biologist studies extinction without hoping to prevent extinction.

Conservation biology is thus clearly an applied and goal-oriented science. Indeed, many have viewed it as an alliance of science and activism. Largely because of this perceived element of activism, conservation biology in its early years drew disdain from many scientists who considered it lacking in objectivity. However, as scientists have come to recognize the human impacts on the planet, more of them have directed their own work to address environmental problems, and the reputation of conservation biology as a discipline has risen. Today it is a thriving pursuit that is central to environmental science and to achieving the goals of preserving biodiversity and making society sustainable.

Conservation biology began to take shape as a discipline in 1959, when Raymond Dasmann's book *Environmental Conservation* was published. The next step in its development was the 1970 publication of David Ehrenfeld's *Biological Conservation*, and during the 1980s conservation biology became a widely recognized discipline. Conservation biologists attempt to integrate an understanding of evolution and extinction with ecology and the dynamic nature of environmental systems. They use field data, lab data, theory, and experiments to study the impacts of humans on other organisms. They also attempt to design, test, and implement ways to mitigate human impact (Figure 15.13).

Island biogeography theory is a key component of conservation biology

Much of modern conservation biology is based on the **equilibrium theory of island biogeography**. This theory, first developed by E. O. Wilson and ecologist Robert MacArthur in 1963, was initially applied to oceanic islands to explain how species come to be distributed among them.

Since its development, researchers have increasingly applied the theory's tenets to other types of islands, including islands of habitat—patches of one type of habitat isolated within vast "seas" of others. The Sikhote-Alin Mountains, last refuge of the Siberian tiger, are an

Figure 15.13 Conservation biologists conduct biological science in the field and the lab with the aim of understanding how to maintain and restore species and ecosystems. Here, a biologist weighs and tags salmon smolts and enters data into a streamside computer station on the East Fork of Idaho's Salmon River.

example of a habitat island. This mountain range is isolated from others by topography and the logging of forests, bounded by ocean to the east and otherwise surrounded by lowlands populated with people.

Island biogeography theory predicts the number of species on an island based on the island's size and its distance from the nearest mainland. The number of species on an island results from a balance between the number being added by immigration and the number being lost by extinction (or more precisely in most cases, extirpation from the particular island). Because immigration and extinction are ongoing dynamic processes, the balance between them represents an equilibrium state.

Several patterns are apparent from the theory of island biogeography. We can understand these patterns by referring to the graphs in Figure 15.14 and working through them one at a time. Figure 15.14a shows that immigration rates are greater for islands with few species (because each new immigrant species represents a high proportion of the island's total species). Secondly, it shows that extinction rates are greater for islands with many species (because interspecific competition may make resources limited, keeping population sizes of any one species small and more vulnerable to extinction). These two trends, illustrated by the figure's immigration and extinction curves, provide the baseline expectations

for examining the effects of island size and distance from the mainland.

Figure 15.14b shows that immigration rates are higher for large islands than for small islands. This is because large islands present fatter targets for organisms to encounter when they disperse or wander, so that more organisms will tend to discover large islands. The figure also shows that large islands have lower rates of extinction than small islands. This is because more space allows for larger populations, and larger populations are less vulnerable to dropping to zero by chance. Together, these trends mean that large islands tend to host more species at equilibrium than small islands. Very roughly, the number of species on an island is expected to double as island size increases by 10 times. This **area effect** can be illustrated with graphs of species-area curves (Figure 15.15). Another reason large islands generally contain more species than small ones is that large islands tend to possess more habitats than smaller islands, thus providing suitable environments for a wider variety of arriving species.

The distance between an island and the nearest continent also affects the number of species on the island. Because of the **distance effect,** the farther an island is from a continent, or a source of immigrants, the fewer species live on the island. The reason is that remote is-

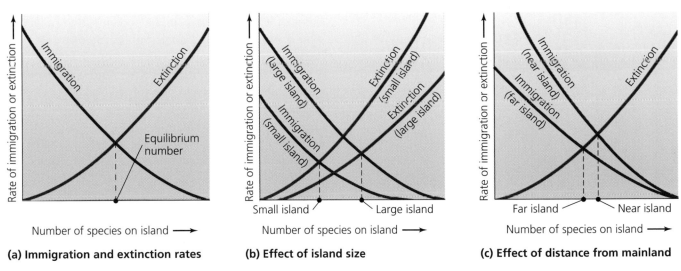

(a) Immigration and extinction rates **(b) Effect of island size** **(c) Effect of distance from mainland**

Figure 15.14 The equilibrium theory of island biogeography explains species richness on islands as a function of immigration rates and extinction rates interacting with island size and distance from the mainland. The graph in (**a**) shows how immigration rates are highest for islands with few species and how extinction rates are greatest on islands with many species. These trends set the stage for examining the effects of island size and distance. In (**b**), immigration rates are higher on large islands because large islands present bigger targets for arriving organisms, while extinction rates are lower on large islands, because their size allows populations to grow larger. Together these trends cause large islands to have a higher number of species at equilibrium, a pattern called the "area effect." In (**c**), immigration rates are higher for islands near the mainland, while extinction rates are not affected. This "distance effect" means that islands closer to mainlands tend to hold more species.

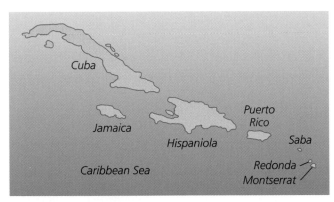

(a) Caribbean islands

Figure 15.15 The prediction of an area effect from island biogeography theory is borne out by observational data from islands around the world. Consider the Caribbean islands (**a**). Such data can be plotted on species-area curves (**b**). By plotting the number of species on various Caribbean islands as a function of the areas of these islands, for instance, it can be seen that species richness increases with area. The increase is not linear, but logarithmic; note the logarithmic scales of the axes. Go to **GRAPH IT** on the website or CD-ROM. Data from E.O. Wilson, *The Diversity of Life*, Belknap Press, 1999.

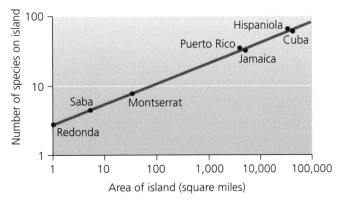

(b) Species area curve

lands are more difficult for individuals to reach. This is shown in Figure 15.14c; the graph's curves show that immigration rates are greater for islands near mainlands, but that proximity to mainlands has no effect on extinction rates.

All these patterns from theory have now been widely supported by empirical data from the real-life study of species on islands (see The Science behind the Story). These patterns hold up for terrestrial habitat islands, such as forests fragmented by logging and road building (Figure 15.16). Small islands of forest lose their diversity fastest, starting with those large species that were few in number to begin with. In a landscape of fragmented

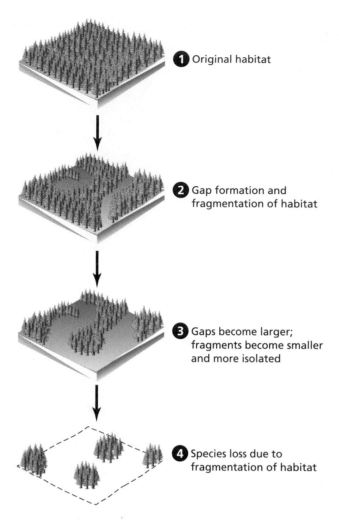

1 Original habitat

2 Gap formation and fragmentation of habitat

3 Gaps become larger; fragments become smaller and more isolated

4 Species loss due to fragmentation of habitat

Figure 15.16 Forest-clearing, farming, road-building, and other types of development and human land use can fragment natural habitats. Habitat fragmentation usually begins with the creation of gaps within a natural habitat. As development proceeds and these gaps expand, the gaps join together, and eventually dominate the landscape, stranding islands of habitat in their midst. As habitat becomes fragmented, many species have a difficult time persisting. The dynamics of species loss from terrestrial habitat islands can be studied using island biogeography theory.

habitat, species requiring the habitat will gradually disappear from the landscape, winking out from one island after another over time.

Some species act as "umbrellas"

Large species that roam great distances, such as the Siberian tiger, require large areas of habitat, and can thus be excellent tools for conservation. Many of them are ecologically important as keystone species. Moreover, meeting the habitat needs of these so-called "um-

brella species" automatically helps meet those of thousands of less charismatic animals, plants, and fungi that would never elicit as much public interest. This approach to conservation is evident in the long-time symbol of the World Wildlife Fund, the panda. The panda is a large, endangered animal requiring sizeable stands of undisturbed bamboo forest. Furthermore, the fact that it appears cute and lovable has made it a favorite with the public—and an effective tool for soliciting funding for conservation efforts that protect much more than just the panda.

Should endangered species be the focus of conservation efforts?

Using large and charismatic mammals as the spearhead for biodiversity conservation efforts has often been an effective strategy. Still, some people feel that the single-species approach has sometimes been carried too far. Currently the primary legislation for protecting biodiversity in the United States is the **Endangered Species Act (ESA)**. Passed in 1973, the ESA forbids the government and private citizens from taking actions (such as developing land) that would destroy endangered species or their habitats. The ESA also forbids trade in products made from endangered species. The aim is to prevent extinctions, stabilize declining populations, and, when possible, to enable populations to recover to the point they no longer need protection. As of 2003 there were 1,263 species in the United States listed as "endangered" or as "threatened," the status considered one notch less severe than endangered.

The ESA has had a number of notable successes. Following the banning of DDT and years of management programs, the peregrine falcon, brown pelican, bald eagle, and other birds have recovered and been taken off the endangered list. Intensive management efforts with other species, such as the red-cockaded woodpecker, have held formerly declining populations steady in the face of continued habitat degradation. In fact, roughly 40% of declining populations have been held stable, despite the fact that the U.S. Fish and Wildlife Service and the National Marine Fisheries Service, the agencies responsible for upholding the Act, have been perennially underfunded for the job. These agencies have faced repeated budgetary shortfalls for endangered species protection, and efforts to reauthorize the ESA ran up against stiff opposition in the U.S. Congress in the late 1990s and early 21st century.

Although most Americans support endangered species protection, some have vocally opposed provisions of the ESA. Some of the resentment results from the perception that the ESA is focused only on single species and values the life of an endangered species over the life or livelihood of a person. Opposing the ESA's protection of the spotted owl is easier if one believes that only the owl is at stake. It becomes harder if one accepts that what's really at stake is the entire old-growth rainforest community of which this umbrella species is a part.

Most popular resentment toward the ESA, however, has stemmed from worries of landowners that federal officials will restrict the use of private land if threatened or endangered species are found on it. This has led in many cases to a practice described as "Shoot, shovel, and shut up," among landowners who want to conceal the presence of such species on their land.

Many ESA supporters argue that such fears are overblown, since the ESA has stopped only a miniscule percentage of development projects proposed. Moreover, a number of provisions of the ESA and its amendments attempt to soften the blow for landowners. Habitat conservation plans and safe harbor agreements are two ways in which the government reaches agreement with landowners, allowing them to harm species in some ways if they voluntarily improve habitat for the species in others.

The controversies the U.S. law has engendered have affected Canada as well. When Canada enacted its long-awaited endangered species law in 2002, the **Species at Risk Act (SARA),** the Canadian government was careful to stress cooperation with landowners and provincial governments, rather than presenting the law as a decree from the national government. Canada's environment minister, David Anderson, wanted none of the hostility the U.S. act had engendered. Environmentalists and many scientists, however, protested that the act was too weak and failed to protect habitat adequately.

Can captive breeding, reintroduction efforts, or cloning help save endangered species?

In the effort to bring back endangered species, biologists have gone to impressive lengths. Zoos and botanical gardens have become centers for the captive breeding of endangered species, so that large numbers of individuals can be raised and then reintroduced into the wild (Figure 15.17). One example in the United States is the extensive program to save the California condor, North America's largest bird. Condors were persecuted in the early 20th century, have collided with telephone and electrical wires, and are highly susceptible to lead poisoning because they scavenge on carcasses of animals,

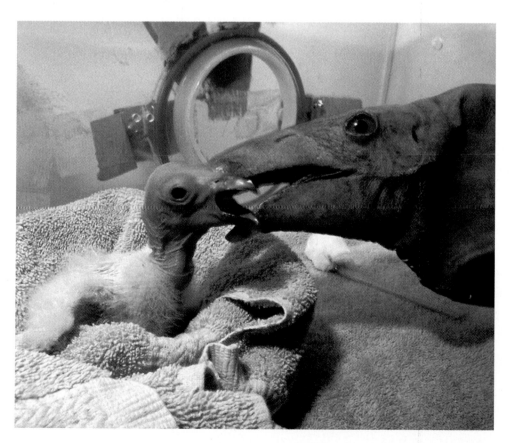

Figure 15.17 Efforts to save the California condor from extinction have involved raising dozens of chicks in captivity with the help of hand puppets designed to look and feel like the heads of adult condors. Biologists feed the growing chicks with these puppets in an enclosure and shield the chicks from all contact with humans, so that when the chick is grown it does not feel an attachment to people.

Testing and Applying Island Biogeography Theory

The researchers who first tested the equilibrium theory of island biogeography experimentally and applied it to conservation biology leaned heavily on their own resourcefulness, along with some plastic, some pesticides, and a small station wagon.

Robert MacArthur and Edward O. Wilson had originally developed island biogeography theory using observational data from oceanic islands, correlating numbers of species found on islands with island size and distance between landmasses. Yet as compelling as island biogeography theory seemed to most biologists in the 1960s, no one had tested its precepts in the field with a manipulative experiment. Wilson decided to remedy that.

Wilson began looking for a group of islands in the United States where he could run an experimental test: to remove all animal life from islands and then observe and measure their recolonization. To be suitable, islands would need to be small, be relatively easy to study, contain few forms of life, and be situated close to the mainland to ensure an influx of immigrating species. Wilson found his research sites off the coast of southern Florida: six small mangrove islands 11–18 m (36–59 ft) in diameter, located in shallow water, and home only to trees and a few dozen species of insects, spiders, centipedes, and other arthropods.

Wilson and Daniel Simberloff, his graduate student at the time, concocted a way to survey, fumigate, and monitor the island ecosystems. Beginning in 1966, Simberloff painstakingly counted each island's arthropods, breaking up bark and poking under branches to find every mite, midge, and millipede. Then with the help of professional exterminators, they wrapped each island in a plastic tarpaulin and gassed the interior with the pesticide methyl bromide. After removing the tarpaulins, Simberloff checked to make sure no creatures were left alive. Wilson and Simberloff then waited to see how life on the islands would return.

The wait lasted two years, with Simberloff repeatedly scrambling up trees and turning over leaves to look for any new creature from the mainland that might have colonized the islands. What he found, documented by elaborate notes and photography, provided the first evidence from a manipulative experiment for the predictions of island biogeography theory. Life on most of the islands recovered within a year, regaining about the same number of species and total number of arthropods the islands had sheltered originally. Larger islands once again became home to a greater number of species than smaller islands. Outlying islands recovered more slowly and reached lower species diversities than did islands near the mainland. These results supported the theoretical predictions with solid quantitative data.

After its publication in the journal *Ecology* in 1969 and 1970, Simberloff and Wilson's research gave new empirical rigor to a set of ideas that was increasingly help-

Daniel Simberloff and E.O. Wilson found mangrove islands in the Florida Keys were perfect to test the tenets of island biogeography theory.

ing scientists understand geographic patterns of biodiversity.

Their research also fueled new questions that were of pressing concern for conservation biology: How widely could the equilibrium theory of island biogeography be applied? Would the theory also hold true for "islands" of habitat on continental mainlands? Could it apply to isolated, fragmented habitats anywhere? At the University of Michigan, biology graduate student William Newmark set out to address these questions.

Newmark had discovered that many North American national parks kept wildlife records that documented sightings of animals over the course of the parks' histories. By examining these historical records, he surmised, he would be able to infer which animal species had vanished from parks, which were new arrivals, and roughly when these changes occurred. Assuming the parks, increasingly surrounded by human development, were like terrestrial islands isolated by farms, roads, and towns and cities, Newmark hypothesized that island biogeography theory would apply to national parks.

In 1983, Newmark drove his Toyota station wagon to 24 parks in the western United States and Canada. At each park, he studied the wildlife sighting records, focusing on larger mammals such as bear, otter, and lynx (but not those species, such as wolves, that had been deliberately eradicated from regions by hunting). Newmark found a few species missing from many parks, and they added up to a troubling total. Forty-two species had disappeared in all, and not as a result of direct human action.

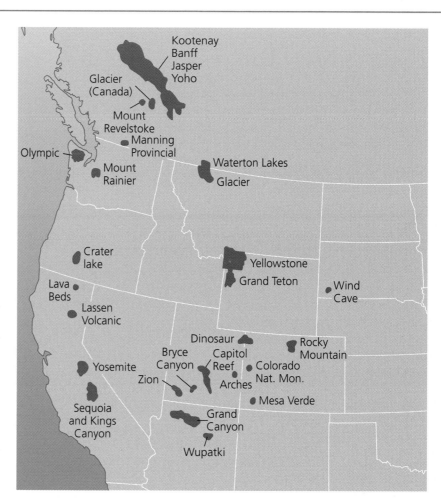

William Newmark visualized the national parks of the western United States and Canada as islands in a sea of development. His data showed that mammal species were disappearing from these recently created terrestrial "islands," in accordance with island biogeography theory. Adapted from David Quammen, *The Song of the Dodo,* 1997.

The red fox and river otter had vanished from Sequoia and Kings Canyon National Parks in California, for example, and the white-tailed jackrabbit and spotted skunk no longer lived in Bryce Canyon National Park in Utah. As theory predicted, the smallest parks showed the greatest number of losses, and the largest parks retained a greater number of species.

The creatures disappeared because the parks, Newmark concluded, were too small to sustain their populations in the long term. Moreover, the parks had become island habitats too isolated to be repopulated by new arrivals. Newmark's findings, published in the journal *Nature* in 1987, put the equilibrium theory of island biogeography squarely on the mainland. Today, this theory informs national park policy, species preservation plans, and the direction of wildlife conservation around the world.

including those killed with lead shot. By 1982 only 22 condors remained, and biologists decided to take all birds into captivity in the hope of boosting numbers and then releasing them. The ongoing program shows signs of success; in recent years, roughly 80 of the 200 birds raised in captivity have been released into the wild at sites in California and Arizona. In 2003 one nesting pair raised the first condor chick in the wild since 1984. Other reintroduction programs have been more controversial. The program reintroducing wolves to Yellowstone National Park has proven popular with the public, but the program reintroducing them to sites in Arizona and New Mexico has met stiff resistance from ranchers. Many ranchers fear the wolves will attack their livestock, and a number of the wolves have been shot.

The newest idea for a technological fix for species extinction is to clone endangered animals. DNA of species is being saved at some facilities in the hope that future genetic technology can enable the viable production of individuals cloned from the DNA. Scientists are even talking of trying to recreate species that have gone extinct in the past, by recovering DNA from preserved body parts. Regardless of whether such efforts can succeed from a technical standpoint, however, most biologists agree that DNA cloning efforts are not an adequate response to biodiversity loss. Without ample habitat and protection in the wild, having a cloned animal in a zoo does little good. Many biologists even worry that speaking of cloning as a viable solution to species loss threatens to distract us from taking measures to prevent species extinction in the first place.

International conservation efforts include widely signed treaties

On the international level, endangered species protection has come in the form of several treaties facilitated by the United Nations since the early 1970s. The 1973 **Convention on International Trade in Endangered Species of Wild Fauna and Flora (CITES)** protects endangered species by banning the international transport of their body parts. CITES can protect the tiger and other rare species when nations choose to enforce it adequately. The 1979 Convention on the Conservation of Migratory Species of Wild Animals is intended to protect migratory animals as they move from one nation to another.

In 1992 the leaders of many nations agreed to the **Convention on Biological Diversity,** a treaty outlining the importance of conserving biodiversity and committing signatory nations to conserving this diversity.

The Convention embodies three goals:

1. To conserve biodiversity
2. To use biodiversity in a sustainable manner
3. To ensure the fair distribution of biodiversity's benefits

According to UNEP, the Convention on Biological Diversity addresses a number of topics, including the following:

- Providing incentives for biodiversity conservation
- Managing access to and use of genetic resources
- Transferring technology, including biotechnology
- Promoting scientific cooperation
- Assessing the effects of human actions on biodiversity
- Promoting biodiversity education and awareness
- Providing funding for critical activities
- Encouraging every nation to report regularly on their biodiversity conservation efforts

In the time since the treaty was first proposed, UNEP has also identified a number of successful biodiversity conservation efforts, including the following:

- A Ugandan program that ensured native people would share in the economic benefits of wildlife-rich protected areas
- A Costa Rican law that rewards forest managers for keeping constant or increasing forested area
- Increasing global markets for "shade-grown" coffee and other crops that are grown without completely removing native forests
- The replacement of pesticide-intensive farming practices with sustainable, ecologically oriented practices in rice-producing Asian nations

As of September 2003, 188 nations had become parties to the Convention on Biological Diversity. Those choosing *not* to become parties to the Convention included Iraq, Somalia, the Vatican, and the United States. For a variety of reasons, the U.S. government is no longer widely regarded as a leader in biodiversity conservation efforts.

Nongovernmental organizations also play a role

Despite, or perhaps because of, the reluctance of the U.S. government to join the 188 nations as parties to the Convention on Biological Diversity, several U.S.–based conservation organizations have taken leadership roles in biodiversity conservation within U.S. borders and abroad. These nongovernmental organizations (NGOs) include such groups as the Wildlife Conservation Society, World Wide Fund for Nature,

Conservation International, World Wildlife Fund-U.S., and The Nature Conservancy. These organizations use principles of conservation biology, including the involvement of local people in conservation planning and implementation, to protect biodiversity around the world. Many of these NGOs have annual budgets of $50–100 million. In addition to purchasing habitat outright, these groups have been active in providing training for park managers around the world, promoting environmental education, and developing innovative approaches to conservation.

In addition to efforts by such organizations and by the United Nations, various countries have taken steps to protect biodiversity within their borders. The Russian government, for instance, issued Decree 795 in 1995, creating a Siberian tiger conservation program and declaring the tiger one of the nation's most important natural and national treasures. Time will tell how effective this decree will be.

Several current efforts in biodiversity conservation are now moving beyond the single-species approach. The Nature Conservancy, for instance, has in recent years moved toward focusing on whole communities and landscapes. In North America, the Wildlands Project is an ambitious effort to restore huge amounts of the continent's land to its presettlement state. We will touch on some of these larger-scale approaches in Chapter 16.

Biodiversity hotspots pinpoint areas of high diversity

One influential international biodiversity conservation approach that is oriented around geographic regions, rather than single species, has been the effort to map **biodiversity hotspots.** The concept of biodiversity hotspots was introduced in 1988 by British ecologist Norman Myers as a way of prioritizing areas that are most important globally for biodiversity conservation. A hotspot is an area that supports an especially great diversity of species, particularly species that are **endemic** to the area, that is, found nowhere else in the world (Figure 15.18). To qualify as a hotspot, a location must harbor at least 1,500 endemic plant species, or 0.5% of the world total. In addition, a hotspot must have suffered extensive habitat alteration or other human impact already and be in danger of suffering still more. To qualify, a location must also have already lost 70% of its original habitat. Although plants are used to determine hotspots, plant diversity is generally correlated with animal diversity, and biodiversity hotspots do indeed host large numbers of endemic vertebrates.

Figure 15.18 The golden lion tamarin (*Leontopithecus rosalia*) of Brazil is one of the world's most endangered primates. Highly successful captive breeding programs have led to roughly 500 individuals in zoos, but the tamarin's Atlantic rainforest habitat is fast disappearing.

The nonprofit group Conservation International maintains a list of 25 biodiversity hotspots (Figure 15.19). This handful of areas, covering only 1.4% of the planet's land surface, is collectively home to 44% of all plant species and 35% of all terrestrial vertebrate species. The hotspot concept serves as an incentive for conservation efforts to focus on these areas of endemism, where the greatest number of species can be protected with the least amount of effort. Indeed, a recent estimate by experts has suggested that to preserve these hotspots completely would be less expensive than first imagined. To preserve their 1.2 million km^2 (386,102 mi^2) would require a one-time investment of $24 billion. This is, of course, a great deal of money, but it represents a rather large bang for the buck, because the total is less than 0.1% of the gross world product and less than 0.1% of the estimated value of the areas' ecosystem services.

Community-based conservation is a popular approach today

Taking a global perspective and prioritizing optimal locations to set aside as parks and reserves makes good sense. However, when it comes to actually setting aside

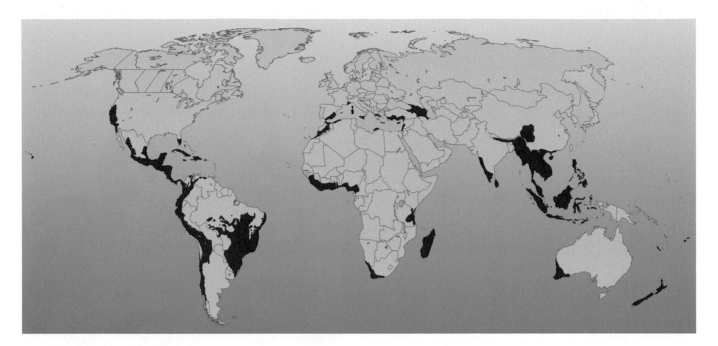

Figure 15.19 Certain areas of the world possess exceptionally high numbers of endemic species, species found nowhere else. As a strategy for saving the most species from extinction, some conservation biologists have suggested prioritizing the preservation of habitat in such areas, which have been dubbed "biodiversity hotspots." This map shows the 25 biodiversity hotspots mapped by the World Conservation Union. Data from Center for Applied Biodiversity Science at Conservation International, 2003.

land for preservation, this affects the lives of people that live in and near the areas under consideration. Indeed, in past decades many conservationists from developed nations, in their zeal to preserve regions in other nations, too often neglected the needs of people in the areas they wanted to protect. Many developing nations came to view this international environmentalism as a kind of neo-colonialism.

Today, however, this has changed, and many conservation biologists actively engage local people in efforts to protect the land and wildlife in their own backyards, in an approach called **community-based conservation.** Although setting aside land for preservation can rob local people of access to natural resources, it can also guarantee that these resources will not be used up or sold to foreign corporations and can instead be sustainably managed. In addition, parks and reserves draw ecotourism, which can support local economies. Community-based conservation sometimes involves collaboration between a local community and an international conservation organization, and sometimes is controlled entirely by the local community.

In the small Central American country of Belize, conservation biologist Robert Horwich and his Wisconsin-based group Community Conservation Inc. have helped start a number of community-based conservation projects. Of the 11 ongoing projects, the Community Baboon Sanctuary has met with particular success. The sanctuary consists of tracts of riparian forest that area farmers have agreed to leave intact, to serve as homes and traveling corridors for the black howler monkey, a centerpiece of ecotourism. The fact that the reserve uses the local nickname for the monkey seems itself a sign of respect for residents, and today a local women's cooperative is running the project. A museum was built, and local residents receive income for guiding and housing visiting biological researchers and ecotourists. Some other projects have not turned out as well, however. Nearby on the Belizean coast, efforts to create a locally run reserve for the manatee foundered due to shortfalls in funding. Community-based conservation has been tremendously successful in some cases and unsuccessful in others.

Community-based conservation is not just happening in developing countries, but also in the United States and Canada. In Horwich's home state of Wisconsin, for instance, several efforts are underway. In one, citizens are restoring the land around a decommissioned Army ammunition plant into a prairie reserve. In another, conservationists teach and work with interested landowners

VIEWPOINTS

Biodiversity

Is establishing parks and protected areas the best strategy to protect biodiversity? Are these adequate to prevent the widespread loss of species?

Parks: The Best Way to Protect Biodiversity?

Even though parks and reserves cover 10% of Earth's land surface, there is no doubt we need many more of them. For instance, consider the 25 "biodiversity hotspots." They contain the last habitats of 44% of Earth's plant species and 35% of vertebrate species (except fish) in just 1.4% of Earth's land surface. If we could safeguard these relatively small areas (no more than twice the size of Alaska), we could reduce the number of extinct species by at least one-third. Yet only a little over one-third of these hotspots are protected. The cost of protecting the rest would be roughly $1 billion per year for five years, which is just one-tenth of all conservation funding worldwide. What a massive need and massive opportunity for protected areas!

However, in the tropics, where most biodiversity is found and where it is most threatened, no island is an island, so to speak. For example, the Kruger Park in South Africa is drying out because rivers arising outside the park are losing their waters to ranches and other development works. The park is also being overtaken by acid rain from South Africa's main industrial region upwind.

Moreover, global warming will shift temperature bands and, consequently, vegetation communities, away from the equator and toward the poles. Plants and animals in Hawaii will have nowhere to go but into the sea, as will those in the southern tip of Africa, northern Philippines, and dozens of other unfortunate localities. Even if they were to be turned into giant parks and perfectly protected on the ground, they would still be vulnerable to global warming   half of which is caused by carbon dioxide emissions from fossil fuels.

Biodiversity enthusiasts, here's a key question for you: which is the country with less than one-twentieth of the world's population and causing one-quarter of the world's carbon dioxide emissions?

Norman Myers is a Fellow of Oxford University and a member of the U.S. National Academy of Sciences. He works as an independent scientist, advising the United Nations, the World Bank, the World Conservation Union, the World Wildlife Fund, and dozens of other conservation bodies around the world.

Setting Aside Nature: Why Nature Parks Will Never Be Enough

Parks and nature reserves are the jewels of conservation, but our scientific discussion often dwells on park size, location, and their corridors rather than questioning the efficacy of protected areas.

Parks are without doubt a valuable strategy for conservation, but they are not the most important strategy. Even the most ambitious conservation plans aspire to having no more than 20-30% of the world's lands protected as nature reserves, and the reality is that closer to 5-10% of land is currently under protection. If we put a fence around our parks to keep humans out, our nature preserves will still end up deteriorating because of how we treat the remaining 70-90% of Earth. Nature cannot be sequestered in some small portion of the globe, while the bulk of land and water are left at the mercy of humans. For example, the effects of global warming, groundwater depletion, and invasive weeds can still alter preserved areas. Conservation will never be attainable unless humans learn to live in, work in, exploit, and harvest nature in a manner that does not ravage biodiversity.

An additional 3 billion people are expected in the next fifty years, bringing the total population to 9 billion. The surging population may have to find food and fiber inside park boundaries. Conservation was founded by population geneticists and taxonomists with a love of rare species and the noble goal of making sure the last remnants of biodiversity did not vanish. Modern conservation still attends to these remnant species but also draws on economics and social science as a means for factoring humans into the solution. Conservation cannot just be about setting aside nature in parks but rather must address working landscapes and human use of biological resources.

Peter Kareiva is Lead Scientist for The Nature Conservancy and an Adjunct Professor at University of California at Santa Barbara and Santa Clara University. Dr. Kareiva has also served as Director of the Division for Conservation Biology at NOAA Fisheries, in the Northwest Fisheries Science Center. His research has been concerned with organisms as diverse as whales, owls, Antarctic seabirds, ladybug beetles, butterflies, wildebeest, and genetically engineered microbes and crops.

along the Wisconsin River to help them better manage their land and restore habitat. In another, environmental advocates have joined farmers who had opposed the establishment of a national wildlife refuge on their lands, working with them to conserve wetlands on private lands in a cooperative manner. In a world of increasing human population, many conservation biologists now feel that parks and preserves will not be adequate for biodiversity and that locally based management that meets people's needs sustainably will be essential.

Further innovative strategies are being employed

As conservation moves from the single-species approach to the hotspot approach to the approach of maximizing the involvement of local people, some new strategies are being attempted. An innovative strategy begun a number of years ago is the **debt-for-nature swap.** These swaps occur when an NGO raises money and then offers to pay off a portion of a developing country's international debt, in exchange for a promise by the recipient country to set aside reserves, fund environmental education, and better manage protected areas.

A newer strategy that Conservation International has pioneered is the **conservation concession.** Nations often sell concessions to corporations, allowing them to extract resources from the nation's land. A nation can, for instance, make money for its treasury by selling to an international logging company the right to log its forests. Conservation International has stepped in and paid nations for concessions for conservation rather than resource extraction. The nation gets the money *and*

keeps its natural resources intact. The money is invested in a foundation that supplies the country with income each year, and the government is provided with funds to pay for expert advice on nonextractive economic activities such as ecotourism and sustainable agriculture. The South American country of Surinam, which still has extensive areas of pristine rainforest, entered into such an agreement, and it has virtually halted logging while pulling in $15 million. It remains to be seen how large a role such innovative strategies will play in the future protection of biodiversity.

Conclusion

The erosion of biological diversity on our planet threatens to result in a mass extinction event equivalent to the mass extinctions of the geological past, most scientists agree. Human-induced habitat alteration, invasive species, pollution, population growth, and overexploitation of biotic resources are the primary causes of biodiversity loss. This loss matters, scientists agree, because biodiversity's pragmatic benefits are such that human society could not function without them. As a result, many conservation biologists and others are rising to the challenge of conducting science aimed at saving endangered species, preserving their habitats, restoring populations, and keeping natural ecosystems intact. The innovative strategies of these scientists hold promise to slow the erosion of biodiversity that threatens life on Earth.

REVIEW QUESTIONS

1. What is biodiversity, and what benefits does it provide? Give as many examples of potential benefits of biodiversity as you can.
2. What is the difference between a species and a subspecies?
3. List and describe the three levels of biodiversity.
4. What is the difference between extinction and extirpation? How are they related?
5. Some people argue that we shouldn't worry about endangered species or extinction today because extinction has always occurred. How would you respond to this view?
6. How did the Tungus people of Siberia view the tiger in their traditional culture? How did this change with the arrival of the Russians early in the 20th century?
7. What is "HIPPO," and how does it relate to biodiversity conservation?
8. List and describe five invasive species and the adverse effects they have had.
9. Define the term *ecosystem services*. Give five examples of ecosystem services that humans would have a hard time replacing if their natural sources were eliminated.

10. What is the relationship between biodiversity and food security? Give at least three examples of potential food security benefits of conserving biodiversity.

11. What is the relationship between biodiversity and medicines and pharmaceuticals? Give at least three examples of medical benefits of conserving biodiversity.

12. Describe four reasons why people suggest biodiversity conservation is important.

13. What is the difference between an umbrella species and a keystone species? Could one species be both an umbrella species and a keystone species?

14. Do you agree with E. O. Wilson's biophilia hypothesis? How could you use science to test his hypothesis?

15. Why do the Viewpoints authors feel that many traditional parks and protected areas fail to conserve biodiversity adequately?

16. Explain the theory of island biogeography. Use the example of the Siberian tiger to support your explanation.

17. Name two successful accomplishments of the U.S. Endangered Species Act and two reasons why many people have criticized it.

18. What is a biodiversity hotspot? What is community-based conservation?

19. What are debt-for-nature swaps, and how do they work?

DISCUSSION QUESTIONS

1. In one of the quotes that open this chapter, biologist and biodiversity advocate E. O. Wilson argues that "except in pockets of ignorance and malice, there is no longer an ideological war between conservationists and developers. Both share the perception that health and prosperity decline in a deteriorating environment." Do you agree or disagree? How do people in your community view biodiversity?

2. Many arguments have been advanced for why preserving biodiversity and preventing extinction are important. Which argument do you think is most compelling, and why? Which argument do you think is least compelling, and why?

3. Imagine that you are an influential legislator in a country that has no endangered species act and that you want to introduce legislation to protect your country's vanishing biodiversity. Consider the U.S. Endangered Species Act and the Canadian Species At Risk Act, as well as international efforts such as the Convention on International Trade in Endangered Species of Wild Fauna and Flora (CITES) and the Convention on Biological Diversity. What strategies would you write into your legislation? How would your law be similar to and different from the U.S., Canadian, and international efforts?

4. Environmental advocates from developed nations who want to preserve biodiversity globally have long argued for setting aside land in biodiversity-rich regions of developing nations. Many leaders of developing nations have responded by accusing the advocates of neocolonialism. "Your nations attained their prosperity and power by overexploiting their environments decades or centuries ago," these leaders asked, "so why should we now sacrifice our own development by setting aside our land and resources?" What would you say to these leaders of developing countries? What would you say to the environmental advocates? Do you see ways that both preservation and development goals might be reached?

5. Compare the biodiversity hotspot approach to the approach of community-based conservation. What are the advantages and disadvantages of each? Can we—and should we—follow both approaches?

Media Resources *For further review, go to the website* **www.envscienceplace.com** *or student CD-ROM, where you will find quizzes, flashcards, a glossary, additional interactive exercises, and links to relevant news and research sources. Also, on the website and CD-ROM is* **GRAPH IT**, *a series of interactive graphing tutorials to help you interpret graphs and plot data.*

Land use, forest management, and creating livable cities

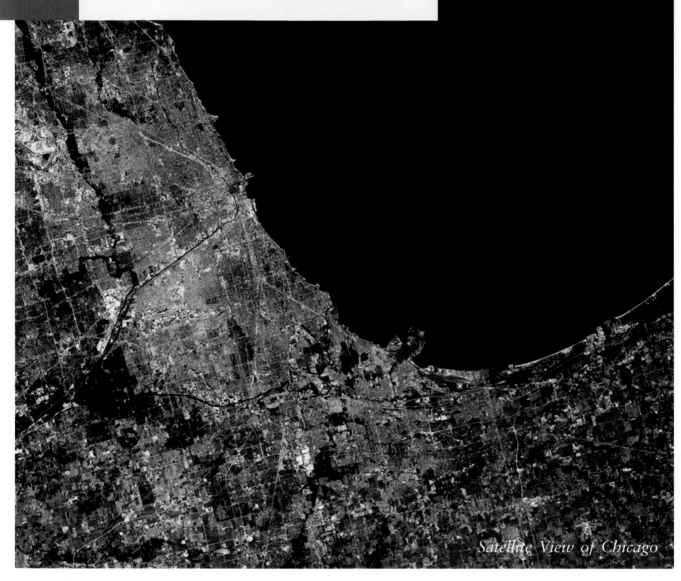

Satellite View of Chicago

This chapter will help you understand:

- Land use decisions
- Urbanization and urban sprawl
- Forestry and forest management

- Agricultural land use
- Parks and reserves
- Planning for livable cities

Children in a Chicago-Area Forest Preserve

Central Case: The Chicago-Area Forest Preserve System

"The woodlands should be brought within easy reach of all people, especially the wage earners."
—*Chicago architect Daniel Burnham, 1909*

"We must pay close attention to our treatment of the land [and] on how we transform it for our use. Right now, our footprint is too big. Going barefoot is not the answer but the time has come to trade in our jackboots for the grace and elegance of ballet slippers."
—*Ecologist Michael L. Rosenzweig, 2003*

For many people, the suburbs of Chicago, Illinois, present an endless expanse of roads, shopping malls, and cookie-cutter houses devouring old farm fields as they spread across the landscape. But winding through the homogeneous sprawl of lawns and pavement is a necklace of jewels. A network of forest preserves surrounds the city in a great crescent, meandering through its suburbs. Collectively these forest preserves hold 40,000 ha (100,000 acres) of woodlands, fields, rivers, marshes, and prairie wildflowers, comprising the largest holding of locally owned public conservation land in the United States.

Because of these lands, Chicago-area residents have places nearby to hike, fish, bird-watch, study nature, or just relax and escape the noise and commotion of the urban environment. Because of these lands, Chicago-area children can grow up with some knowledge of and personal experience with the natural environment of their region. Amid miles of otherwise unbroken development, residents, thanks to the forest preserves, can sense some connection to nature and enjoy a healthier environment.

Not every city is so fortunate. Chicago owes its forest preserve system to two men of uncommon foresight and persistence: landscape architects Dwight Perkins and Jens Jensen. Perkins and Jensen saw clearly at the end of the 19th century what the fast-growing city by the lake would become in the 20th century. They also knew and loved the local landscape, with its mix of prairies and forests, and feared that future generations of Chicagoans might not experience it if some system of reserves were not set aside and protected.

Civic improvement was garnering interest and support at the turn of the 20th century in urban America, as politicians and citizens alike yearned for ways to make their crowded and dirty cities more livable. Chicago's civic leaders welcomed the idea of an outer ring of forest preserves to complement the parks of the inner city that were then being developed. Jensen and Perkins drew up a blueprint that was incorporated into a historic city-planning document produced by architect Daniel Burnham, the 1909 *Plan of Chicago*. With broad public support and a solid financial base, the forest preserve effort purchased over 8,000 ha (20,000 acres) in the several years after 1916, and millions of visitors came to the new preserves for recreation.

Despite many challenges over the years from periods of neglect, invasive species, fire suppression, and deer overpopulation, the forest preserves are being revitalized. With recent land acquisitions and new programs in ecological restoration and environmental education, the Chicago-area forest preserve system remains one of the region's most valuable assets.

Our Urbanizing World

Chicago's forest preserves illustrate one of the many types of parks and reserves we have set aside in order to protect natural lands from development and to preserve open space (Figure 16.1). In the wake of the urbanization and sprawl that have resulted from industrialization and human population growth, people of every industrialized society in the world today have, to some degree, chosen to set aside such land.

Urban environments need natural land

An urbanized society, in which people cluster together in settlements and municipalities, needs stretches of uninhabited and undeveloped land. There are several reasons:

- Large stretches of undeveloped land outside cities provide natural resources needed to support urban populations. Cities and towns cover less than 2% of Earth's surface, and urban dwellers cannot live off this small amount of land. Thus, cities are sinks for resources, drawing resources in from huge areas. For example, London requires a land area estimated at 58 times its size in order to supply its citizens with timber and food alone, not to mention other resources.

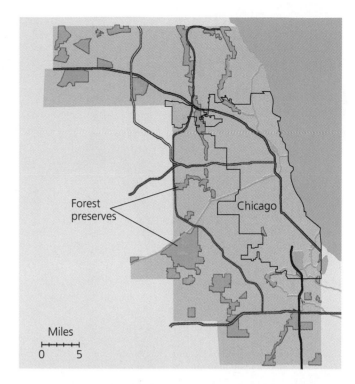

Figure 16.1 The forest preserves of Cook County, DuPage County, Lake County, and Will County wind through the suburbs surrounding the city of Chicago.

- Cities need areas of natural land to provide ecosystem services, including purification of water and air, nutrient cycling, and waste treatment. Major cities such as New York, Boston, San Francisco, and Los Angeles are utterly dependent on the water they pump in from far-away watersheds they have commandeered. When New York City was confronted in 1989 with an Environmental Protection Agency (EPA) order to build a $6 billion filtration plant to protect its citizens against water-borne disease, the city opted instead to purchase and better protect watershed land, for a fraction of the cost (Figure 16.2).
- Natural lands provide escape from the stresses of urban life. They provide open space, greenery, scenic beauty, and places for recreation. They also provide habitat for wildlife, which serves to satisfy biophilia (Chapter 15), our natural affinity for contact with other organisms.
- In addition to the utilitarian reasons cited above, many people feel an ethical obligation to preserve wild landscapes and biodiversity for the sake of non-human organisms and/or the integrity of ecological systems.

Just as the pragmatic aspects of protecting natural lands become more important as human societies become more urbanized, so do the philosophical aspects as many people come to feel increasingly isolated and disconnected from nature.

Land use and resource management are intertwined

Urbanized societies have engaged in decisions that involve setting aside undeveloped land. These societies have also struggled with other **land use** issues, those issues that deal with the ways we apportion areas of land for different functions. Land use and management of natural resources are often intertwined. **Resource management** encompasses strategic decision-making about who should extract resources and in what ways, so that resources are used wisely and not wasted. In this chapter we will examine urbanization, the lands that supply our society's resources, and efforts to provide for human needs in urban centers.

In the United States and other developed nations today, most of us live in cities and towns. But had we been born a century earlier, chances are we'd be living on a farm. Arguably the single greatest societal change over the past century has been the shift from rural to urban living brought about by industrialization.

Cities are not new, but the scale and speed of our urbanization are new

It is important to remember that cities are nothing novel. Many ancient civilizations included vast urban centers where the population—and political power—was concentrated. The ruins of ancient civilizations of Mediterranean cultures, the great Chinese dynasties, and the Mayan and Incan empires all testify that urban centers have been part of human culture for several thousand years. These cities apparently supported themselves, as cities do today, by drawing in resources from outlying rural areas through trade, persuasion, or conquest. In turn, cities have historically exerted influence on the ways people use land in areas surrounding the cities (Figure 16.3).

Although cities themselves are not new, what is new is the current scale of urbanization and the speed at which change is occurring. Population growth has placed more people in towns and cities than ever before. In 1850, the U.S. Census Bureau classified only 15% of U.S. citizens as urban dwellers. That percentage passed

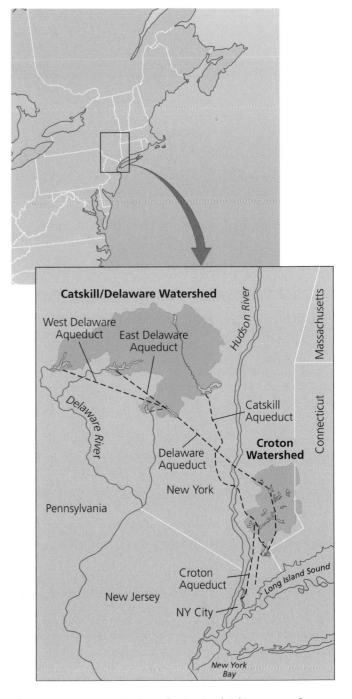

Figure 16.2 New York City obtains its drinking water from upstate reservoirs in the Catskill/Delaware Watershed and the Croton Watershed. This water is piped in via aqueducts to the city. The city has gone to lengths to acquire, protect, and manage watershed land so as to minimize pollution of these water sources.

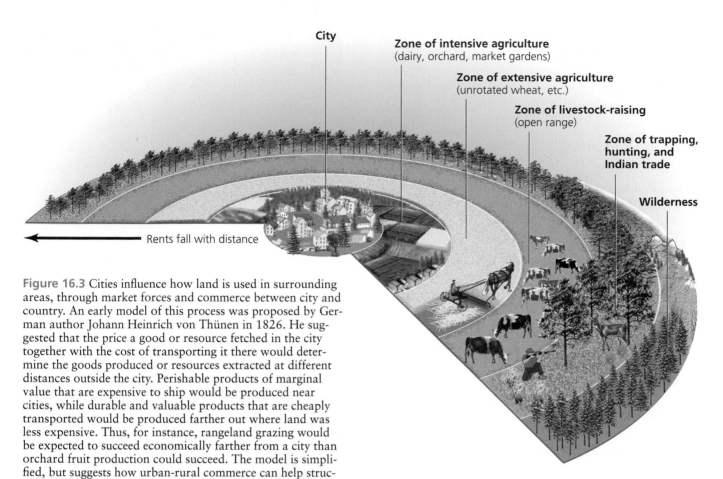

City

Zone of intensive agriculture
(dairy, orchard, market gardens)

Zone of extensive agriculture
(unrotated wheat, etc.)

Zone of livestock-raising
(open range)

Zone of trapping, hunting, and Indian trade

Wilderness

← Rents fall with distance

Figure 16.3 Cities influence how land is used in surrounding areas, through market forces and commerce between city and country. An early model of this process was proposed by German author Johann Heinrich von Thünen in 1826. He suggested that the price a good or resource fetched in the city together with the cost of transporting it there would determine the goods produced or resources extracted at different distances outside the city. Perishable products of marginal value that are expensive to ship would be produced near cities, while durable and valuable products that are cheaply transported would be produced farther out where land was less expensive. Thus, for instance, rangeland grazing would be expected to succeed economically farther from a city than orchard fruit production could succeed. The model is simplified, but suggests how urban-rural commerce can help structure land use far beyond a city's borders. *Source:* von Thünen, as adapted by W. Cronon, *Nature's Metropolis: Chicago and the Great West*, W.W. Norton, New York, 1991.

50% shortly before 1920 and now stands at 75%. Worldwide, the proportion of population that is urban rose from 30% half a century ago to 47% today. Roughly three-quarters of people in Canada and Mexico, Europe, and Latin America are urban. Among developing countries of Africa and Asia, only 4 of every 10 people live in cities and towns, but this figure is rising (Figure 16.4). In addition, the world's cities are interconnected to an unprecedented degree. Although many ancient civilizations traded with one another, today's forces of globalization—commerce, jet travel, diplomacy, television, radio, phone, and Internet communications—have connected societies more than ever before. As wealth has increased, so too has the consumption of resources, the inexorable pull of raw materials from the countryside to the cities.

The world's most populous city, Tokyo, Japan, grew from only 690,000 people in 1800 to 1.5 million a century later. By the early 21st century, its population had

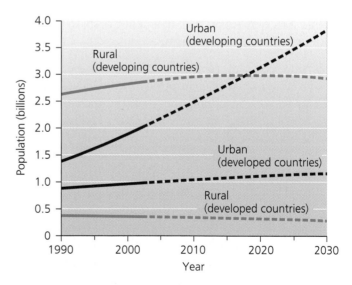

Figure 16.4 In developing countries, urban populations are growing quickly, while rural populations are growing more slowly, largely due to migration from farms to cities. In developed countries, urban populations are growing slowly while rural populations are falling. Solid lines in the graph indicate past data, and dashed lines indicate projections of future trends. Data from United Nations Population Division, as presented in *AAAS Atlas of Population and Environment*, University of California Press, 2000.

reached 28 million. Chicago, although its population is smaller, showed an even faster rise. Incorporated in 1833 with a population of 350, the windy frontier town grew by leaps and bounds as railroads funneled through it the resources of the vast lands to the west on their way to the cities and consumers to the east. Chicago became a center for grain processing, livestock slaughtering, meatpacking, and much else. Employment opportunities soared, as did the city's population, which reached 500,000 in 1880, 1.1 million in 1890, 1.7 million in 1900, and 3.5 million during World War II (Figure 16.5).

Today's urban areas sprawl

The increased numbers of people in the world's cities is only part of the story of urban growth. Urban areas have grown spatially as well, and in many cases this geographic growth has been more dramatic than the growth in numbers. This is especially true of American cities, which have seen many of their residents move into expanding rings of suburbs. Chicago's population has declined to 80% of its peak because so many residents moved from the city to its suburbs. This movement into suburbs, which has taken place throughout the United States and other developed countries, has been driven by

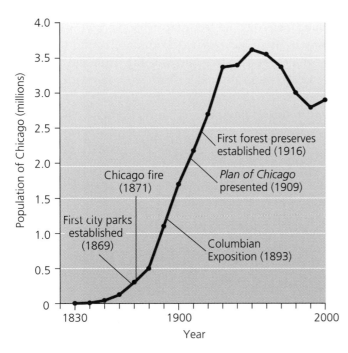

Figure 16.5 Chicago grew at one of the fastest rates of any major city in the world in the late 19th and early 20th centuries. In the second half of the 20th century, the city's population dropped somewhat as many wealthier residents moved into the fast-growing suburbs ringing the city. Data from the Chicago Municipal Reference Library, 2003.

people's desire to live in cleaner, less crowded conditions and in more park-like areas. It has been made possible by the increase in the population's affluence as these nations proceeded through the demographic transition (Chapter 7). Suburban growth has spread human impact more broadly across the landscape. The suburban areas surrounding cities soon grew larger than the cities themselves, and towns ran into one another.

Chicago's metropolitan area now consists of more than 9 million people spread over 23,000 km² (9,000 mi²)—an area 40 times the size of the city proper. Each person in the suburban region takes up an average of 11 times as much space as do residents of the city itself. Today, many cities in the U.S. Southwest are going through growth spurts like Chicago did. Phoenix, Arizona, grew from 105,000 residents spread over 44 km² (17 mi²) in 1950 to 1.3 million residents over 1,200 km² (470 mi²) in 2002. Since 1970, the metropolitan area of Las Vegas, Nevada, has more than tripled in size, while its population has increased by more than five times (Figure 16.6). Between 1950 and 1990, the population of 58 major U.S. cities rose by 80%, but the land area they covered rose 305%. Even in 11 urban areas where population declined between 1970 and 1990, the amount of land covered increased. Houses and roads supplant over 1 million ha (2.5 million acres) of U.S. farmland each year. The phenomenon of urban and suburban spread across the landscape has been given a name: **sprawl.**

Economists, politicians, and city boosters have almost universally encouraged this unbridled growth of cities and suburbs. The conventional assumption has been that all growth is good and that attracting business, industry, and residents will unfailingly increase a community's economic well-being and its political power and influence.

Today this assumption is increasingly being challenged. As the negative effects of sprawl on citizens' lifestyles accumulate, growing numbers of people have begun to question the mantra that all growth is good. Efforts to control growth have sprung up throughout the United States, including local referenda put up for popular vote in California and other states. Some towns have revolted against the "growth mentality." In the late 1990s, the 3,000 residents of the Sonoran Desert community of Tortolita, on the outskirts of rapidly expanding Tucson, fought off efforts of two other suburbs to annex their land. The Tortolitans did not want their scenic and peaceful community to be invaded by the strip malls and parking lots they associated with sprawl in the surrounding towns. Tortolita's residents won protracted court battles and legally incorporated their

(a) Las Vegas, Nevada, 1972

(b) Las Vegas, Nevada, 2002

Figure 16.6 Satellite images show the type of rapid urban and suburban expansion that many people have dubbed "sprawl." Las Vegas, Nevada, is currently one of the fastest-growing cities in North America. Between 1972 (**a**) and 2002 (**b**), the population has increased more than fivefold and the developed area has risen more than threefold.

town, seeking to keep it largely undeveloped and at low population density.

Weighing the Issues:
Sprawl and Transportation

Consider the average shopping center—one-third of the area is the actual mall, whereas two-thirds is dedicated to parking. Urban planners recognize that building our environment around the car has led to nationwide traffic problems and is one of the primary contributors to urban sprawl. How can sprawl be alleviated? How can today's urban planners efficiently meet city-dwellers' transportation needs? Are more roads the answer? Will carpools become mandatory? What is the role of mass transit? How will urban planning need to change in order to connect people and places? Think of your favorite cities—what makes them livable?

Although suburbs tend to have more vegetation and open space than cities, suburban expansion has also put more distance between urban and suburban dwellers and the natural lands outside metropolitan areas. As a

result, many people feel that urban sprawl has caused many of us to feel isolated from nature, and to be less aware of where our resources come from and how our lifestyle choices affect the larger environment on which we ultimately depend. We will return to urban issues later in this chapter, but let's now examine where some of the resources we need for modern urban life come from, and how land use policy helps determine how, where, and by whom these resources are obtained. In this chapter, our discussion of resource management will focus on forest management.

Forestry, Land Use, and Resource Management

The wood that goes toward building homes, producing paper, and many other uses in the United States comes from large tracts of land, primarily in the western and southern portions of the country. Timber harvesting, or logging, is also widespread in other countries of the world. Forest products have helped our society achieve the standard of living we enjoy today, but **deforestation,**

Suburban Sprawl

Urban and suburban sprawl has affected the landscape and the lifestyles of millions of people in recent decades. Is sprawl a problem, or might it actually be good? What strategies should we follow in our city and regional planning and land use?

How We Grow Big

A small town mayor in Minnesota once said to me: "It's not how big you grow, but how you grow big." His point was that most communities will experience growth, but few communities understand how to manage it.

Many elected officials see a building go up and assume that's progress. However, a recent study in Massachusetts showed that a 47,400-m² (510,000-ft²) retail megastore would create no net gain in jobs or property taxes because of the offsetting losses to other businesses. Sprawl masquerades as a form of economic development, but it is really a form of economic displacement.

The low-density, automobile-dependent sprawl that now chokes our suburbs comes at a high price. Not only is there the cost of new water, sewer, and road infrastructure, there is also the loss of character and sense of place. You can't buy small-town quality of life at Wal-Mart. Once it's gone, they can't sell it back to you, at any price. Thousands of acres of open land have been paved over to build megastores that less than ten years later are abandoned. Wal-Mart has nearly 400 "dark stores" on the market today.

For several decades, real estate and development interests have generally dominated zoning. More recently, communities have responded to the saturation of sprawl with the "smart growth" movement. Today we have better tools to manage growth, ranging from urban growth boundaries to placing caps on the size of stores. Citizens' groups have enacted moratoriums, written major development review ordinances, mandated economic impact requirements, and promoted other zoning techniques to slow the spread of sprawl.

We can have growth *and* live-and-let-live communities. Development does not have to be synonymous with sprawl. Land use is one of the few arenas left in which citizens can retain some measure of local control. We can determine how we will grow big.

Al Norman is an activist who has helped citizens groups in 43 states and 3 foreign countries challenge sprawl. Norman is the author of the 1999 book *Slam-Dunking Wal-Mart*. He has been called the "guru of the anti-Wal-Mart movement" by *60 Minutes*.

To Sprawl or Not to Sprawl: Is that *Really* the Question?

Is sprawl good or bad? Conventional wisdom says it's bad. But how reliable is conventional wisdom? For example, after five decades of rapid urban decentralization, air quality has improved in virtually all metropolitan areas; less than 6% of the nation's land mass is developed, and only about 4% is urbanized; housing continues to be affordable in most areas of the United States, in large part because outward expansion has kept land costs in check; agricultural productivity is marching upward; and families continue to move in droves to suburbs despite widely publicized problems such as traffic congestion.

Nevertheless, suburbanization may be one of the most contested public policy issues of the day, particularly on the local level. Why? The answer may have more to do with politics than evidence.

The definition of the term *sprawl* changes depending on where one sits (or lives). Planners typically define sprawl as low-density residential and commercial development that relies primarily on the automobile for transportation. Others, particularly those involved in policy debates, apply different definitions. An editorial board of a major "anti-sprawl" newspaper, for example, defined sprawl as the outward development of strip malls and other low-density commercial uses (not residential development). Farming organizations define sprawl as any development that might infringe on farming operations. Some environmental groups define sprawl as any development outside existing town boundaries.

Without a clear definition of sprawl, public policy becomes problematic: Is the goal to save farmland, stop outward housing development, increase densities, or reduce automobile use? If the answer is "yes" to each of these, then public policy may well be ignoring the housing preferences of the vast majority of Americans. Any given land-use pattern, whether sprawl or high-density, is not necessarily good or bad; the key is to ensure that policy reflects the values and choices of millions of Americans.

Samuel R. Staley is president of The Buckeye Institute for Public Policy Solutions and a senior fellow in Urban Policy at Reason Public Policy Institute in Los Angeles. His applied policy research spans three decades and his most recent book, co-edited with Randall G. Holcombe, is *Smarter Growth: Market-Based Strategies for Land-Use Planning in the 21ˢᵗ Century*, Greenwood Press, 2001.

the clearing and loss of forests, has altered the landscape and ecosystems of much of the planet. Continued deforestation threatens to lead to dire ecological and economic consequences.

Forests are ecologically valuable, but timber is economically valued

Most of the world's forests occur as taiga, the biome (Chapter 6) that stretches across much of Canada, Russia, and Scandinavia, and as tropical rainforest, the biome that occurs in South and Central America, equatorial Africa, and Indonesia and southeast Asia. Most logging today takes place in Canada, Russia, and other nations that contain large expanses of boreal forest, and in tropical countries with large amounts of rainforest, such as Brazil. Temperate forests cover less area, in part because so many have already been cleared by humans over past centuries. In total, depending on one's definition, from one-fifth to one-third of Earth's surface is currently covered with forest (see The Science behind the Story).

Because of their structural complexity and their ability to provide many niches for organisms, forests comprise some of the richest ecosystems for biodiversity. Past forest clearing is known to have caused many species population declines and extinctions. People clear forest for many reasons, but primarily do so to make way for agriculture. Other major causes of deforestation are commercial logging (generally in areas where large stands of old trees remain) for the sale of timber products, and the cutting of fuelwood for subsistence (mainly in arid or desertified areas of the developing world, where few trees grow).

Deforestation has occurred on all continents and, as we saw in Chapter 1 with Easter Island and various Mediterranean cultures, has in some cases helped bring whole civilizations to ruin. Today forests are being felled at the fastest rates in the tropical rainforests of Latin America and Africa (Figure 16.7). Developing countries in these regions are striving to expand settlement areas for their burgeoning populations and to expand their economies and pay off debts by extracting natural resources and selling them abroad. In many cases the resources are being extracted by foreign multinational corporations, which have paid fees to the developing nation's government for a concession or right to extract the resource. At the same time, areas of Europe and eastern North America are actually gaining forest cover slowly as they recover from severe deforestation during past decades and centuries.

The growth of the United States was fed by the deforestation of its land

Timber harvesting propelled the growth of the United States throughout its phenomenal expansion across the continent in the 19th century and into the 20th century. Chicago was built with timber felled in the vast pine and hardwood forests of Wisconsin and Michigan. Those forests were virtually stripped of their trees in the 19th century. This deforestation followed the clear-cutting of the forests farther east, which had been largely replaced by small farms. Logging operations then moved south to the Ozarks of Missouri and Arkansas to harvest more wood for the growing nation. The pine woodlands of the South were logged and many of them converted to

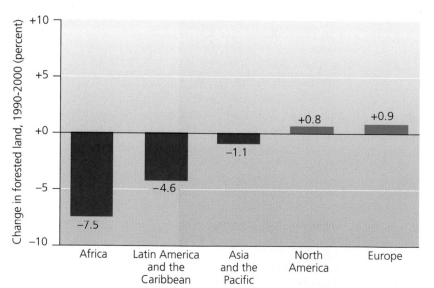

Figure 16.7 Nations in Africa and in Latin America are experiencing rapid deforestation as they attempt to develop, extract resources, and provide new agricultural land for their growing populations. In parts of Europe and North America, meanwhile, forested area is slowly increasing as some former farmed areas are abandoned and allowed to grow back into forest. Go to **GRAPH IT** on the website of CD-ROM. Data from United Nations Environmental Programme, *Global Environmental Outlook 3*, Earthscan Publications, 2002.

Figure 16.8 A mule team drags logs of Douglas fir from a clearcut in Mason County, Washington, in 1901. Early forestry practices in North America caused significant environmental impacts and removed nearly all large virgin timber from one region after another.

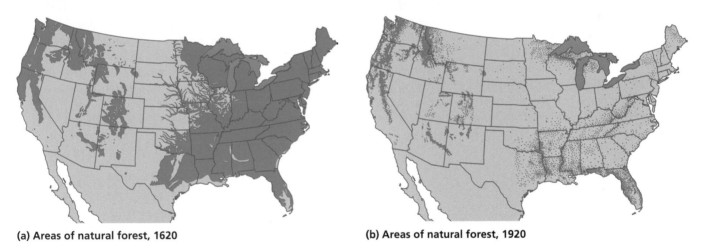

(a) Areas of natural forest, 1620

(b) Areas of natural forest, 1920

Figure 16.9 When Europeans first were colonizing North America, the entire eastern half of the continent and substantial portions of the western half were covered in forest (**a**). By the early 20th century, the vast majority of this forest had been cut, replaced mostly by agriculture and other human land uses (**b**). Even within the remaining forested areas shown on the map, very few large trees remained, and most vegetation was second growth. Source: M. Williams, *Americans and Their Forests*, Cambridge University Press, 1989, as adapted by A. Goudie, *The Human Impact*, MIT Press, 2000.

pine plantations. When the largest, most valuable trees had been removed from these areas, timber companies moved west, cutting the continent's biggest trees in the Rocky Mountains, the Sierra Nevada, the Cascade Mountains, and the Pacific Coast ranges (Figure 16.8). By the mid-20th century almost no virgin timber was left in the lower 48 states (Figure 16.9). The largest oaks and maples found in the East today, and even most red-

woods of the California coast, are merely second-growth trees: trees that have sprouted and grown to partial maturity in the wake of the cutting of virgin timber. The size of the enormous trees they replaced can be seen in the stumps that are left in the more recently logged areas of California and the Pacific Northwest.

The fortunes of timber companies have risen and fallen with the availability of big trees. As each region

Using Geographic Information Systems to Survey Earth's Forests

One map is washed in splashes of green, showing broad stretches of largely undisturbed forests. Another outlines political boundaries. A third map marks human population centers, denoting their density with different colors. A fourth lays out parks, reserves, and other land protected from development. On its own, each map shows only one physical characteristic of Earth. But layered together through the use of a powerful technology, these maps reveal something entirely different: the condition of the world's last intact forests.

Used in a groundbreaking study published by UNEP in 2001, these maps, in conjunction with a geographic information system (GIS), have become an essential tool for environmental planners and resource managers around the world. A GIS is software that builds layered maps, using different sets of data to compile various views of landscapes.

A wide range of data can be used in GIS maps. Environmental researchers most often use information on the status and uses of natural resources. In the case of the UNEP study, researchers wanted a comprehensive look at global forests—where they are, which countries govern them, which forests face the most pressure from encroaching human settlements, and which are most protected. The researchers espe-

cially wanted to know the status of "closed" forests, those with a healthy tree canopy, cover of more than 40% of the forest. Conducting such a survey on the ground, going from forest to forest and country to country, would have been neither practical nor accurate, and synthesizing the findings from such research into a comprehensive paper map would have been almost impossible. With a GIS, however, researchers pull data such as satellite imagery and computerized maps into an all-inclusive view of the area they are studying. Simple GIS maps are generated within a day, and more complex ones with many layers are compiled in a matter of months.

Some of the researchers' most critical data are gathered via a U.S. sensor, an advanced very high resolution radiometer (AVHRR), orbiting the Earth. An AVHRR consists of a rotating mirror, a telescope, and various internal electronics. As it orbits Earth, the AVHRR picks up and measures energy wavelengths rising from the surface. Different types of substrates, from water to vegetation to rocks, release different energy wavelengths, which the AVHRR captures and translates into color images. Different colors presented in the satellite imagery denote whether an area is covered by water, rocks, houses, trees, or vegeta-

tion. For more than a decade, scientists have taken advantage of AVHHR to produce a global plant cataloging project, the Normalized Difference Vegetation Index (NDVI). NDVI researchers survey forests and plants and gauge changes to different vegetation.

Using the NDVI's catalog of worldwide vegetation images, UNEP researchers loaded forest data into their GIS. They then added three other sets of electronic data into the GIS: the distribution of the world's population, global political boundaries, and land protected against development.

In computer imaging systems such as a GIS, data sets are stored as separate data layers, which can be reviewed individually or merged with other layers to build a multifaceted map. GIS software then converts data layers into images on maps, drawing out political boundaries, for example, and then adding images of vegetation for the same area.

In the UNEP study, researchers first analyzed their data layers individually to verify the accuracy of datasets. They scrutinized the satellite vegetation data, for example, to try to make sure it showed forests and not other types of similar-looking ground cover, such as farms or commercial tree plantations.

When compiled, the data layers offered the UNEP researchers a

was denuded of its timber, the industry declined in that region and moved on, while local loggers lost their jobs. If the remaining ancient trees of North America—most in British Columbia and Alaska—are cut, U.S. and Canadian loggers will likely be out of jobs once again, while their employers move on to the developing world, as many already have. Today the North American timber industry focuses more on max-

imizing production from the southern pine plantations, where even-aged monocultural stands of fast-growing tree species are cut in rotation after certain numbers of years and seedlings replanted. Many ecologists view these plantations more as crop agriculture than as ecologically functional forests, because they provide little support for species diversity, and because there is little variation in the age of the trees. For these reasons, the

first-of-its kind look at forests, which accurately guided conservation efforts. About 20% of Earth's surface remains covered by closed forest, the study indicated. Moreover, as previous research had uncovered, forests in such densely populated countries as India and Indonesia are under pressure from expanding human settlement. Such countries require stronger and immediate conservation efforts, researchers determined. The team also found that more than 80% of the world's intact forests were concentrated in just 15 countries and that 88% of that forest was sparsely inhabited by people. These findings could prove critical in starting conservation efforts in regions that are at risk, offering hope that further forest degradation can be prevented.

GIS is not a foolproof technology, as the UNEP researchers noted. GIS maps are only as good as the data that go into them. Some of the UNEP data—those on population size and protected areas—may not be reliable, researchers warn, because the countries providing the information may not always keep accurate records. Some satellite images used in the UNEP study were ambiguous, as was the case with the maps indicating forest cover; the team couldn't be sure every tree farm or plantation had been removed from that set of data before loading into the GIS.

However, the UNEP scientists also stressed that without GIS such a major forest assessment would never have been possible. "It is critical to assess the extent and distribution of such areas using the latest scientific information," UNEP executive director Klaus Toepfer wrote in the introduction to the forest report. Because of the GIS-driven findings, forest planners can focus conservation priorities on areas with the best prospects for continued existence.

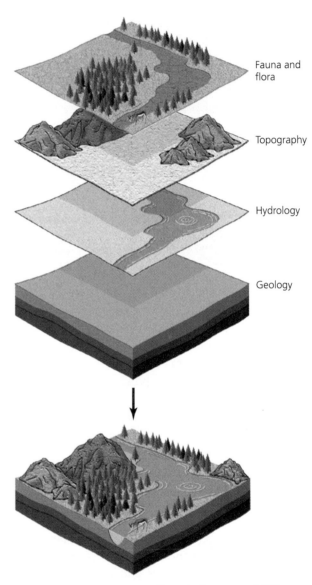

Geographic information systems (GIS) allow the layering of different types of data on natural landscape features and human land uses so as to produce maps integrating this information. GIS can then be used to explore informative correlations between these data sets.

tree plantations do not offer many forest organisms the habitat they need.

Timber harvesting takes place on public lands

Most timber harvesting in the United States takes place on private land, including land owned by timber companies, but some takes place on public lands, notably the national forests. The **national forest** system consists of 77 million ha (191 million acres) in many tracts spread across all but a few states in the country (Figure 16.10). The system, which is managed by the U.S. Forest Service, covers over 8% of the nation's land area.

The depletion of the eastern forests and the fear of a "timber famine" spurred the formation of the Forest Service in 1905 under the leadership of Gifford Pinchot

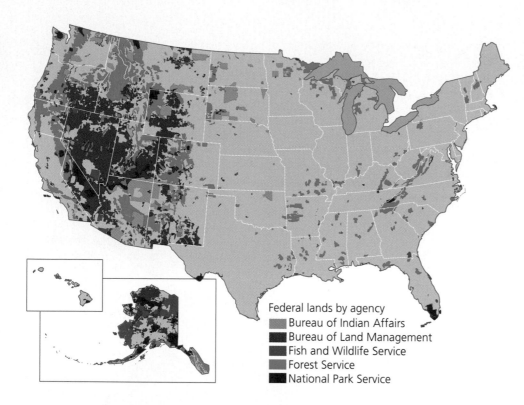

Figure 16.10 Federal agencies own and manage well over 250 million ha (600 million acres) of land in the United States, particularly in the western states. These include national forests, national parks, national wildlife refuges, Native American reservations, and Bureau of Land Management lands. Data from United States Geological Survey, 2002.

Federal lands by agency
- Bureau of Indian Affairs
- Bureau of Land Management
- Fish and Wildlife Service
- Forest Service
- National Park Service

(Chapter 2). Pinchot and others developed the concept of resource management and conservation during the Progressive Era in American politics, a time of social reform when people had confidence that the application of science and education to public policy could greatly improve society. In line with Pinchot's conservation ethic, the Forest Service aimed to manage the forests for "the greatest good of the greatest number in the long run." Pinchot believed the nation should extract and use resources from its public lands, so timber harvesting was, from the start, the goal behind the establishment of the national forests. But conservation meant planting trees as well as harvesting them, and the Forest Service intended to seek restrained use and wise management of timber resources, which are not renewable if not given time to grow back.

When timber companies harvest trees, any of several methods may be used (Figure 16.11). In **clear-cutting**, all trees in an area are cut, leaving only stumps (Figure 16.11a). Although it is the most cost-efficient method in the short term, clear-cutting is also the most environmentally damaging. Other methods cut most trees in an area, but leave a few standing. In seed-tree cutting (Figure 16.11b), mature seed-producing trees are left standing so that they can reseed the area. In shelterwood cutting (Figure 16.11c), mature trees are harvested in two or three cuttings over many years, leaving strips of trees

that can both reseed an area and protect against soil erosion. In selective cutting (Figure 16.11d), only a few trees in a forest are cut, leaving the canopy mostly intact. Although the selective logging technique may appear to do little damage, it involves cutting down the largest, healthiest trees, and the trucks and machinery needed to access individual trees disturb much of the forest floor.

In 1996, the most recent year for which good data are available, timber companies extracted 23.5 million m^3 (828 million ft^3) of live timber from national forests. Although this is a large amount, it is considerably less than the amount cut from private lands and other public forests (Figure 16.12). Timber harvesting declined on national forests during the 1980s and 1990s (Figure 16.13), and in 1996 tree regrowth outpaced tree removal on these lands by five to one. Whereas overall timber production in the United States and other developed countries has remained more or less stable for the past 40 years, it has more than doubled in developing countries.

Although Pinchot's conservation goals would fall within most scientists' definitions of sustainability today, and although regrowth data are encouraging, many environmental advocates and scientists from the 1960s on have debated whether the Forest Service has in fact managed the forests sustainably. Scientists who have

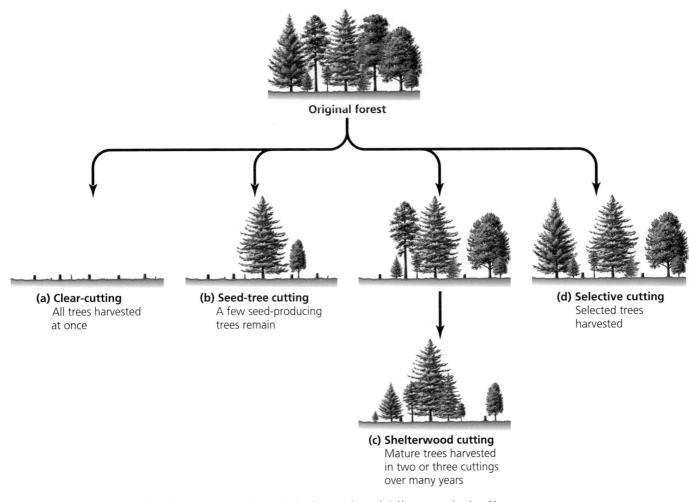

Original forest

(a) Clear-cutting
All trees harvested
at once

(b) Seed-tree cutting
A few seed-producing
trees remain

(c) Shelterwood cutting
Mature trees harvested
in two or three cuttings
over many years

(d) Selective cutting
Selected trees
harvested

Figure 16.11 Foresters and timber companies have devised a number of different methods of harvesting timber from forests. In clear-cutting (**a**), all trees are cut from an area, extracting a great deal of timber inexpensively but leaving a vastly altered landscape. In seed-tree cutting (**b**), small numbers of large trees are left in clearcuts to help re-seed the area. Shelterwood cutting (**c**) involves cutting mature trees gradually, removing them in several cuttings over a number of years. In selective cutting (**d**), selected large trees are removed from forests, while most others are left standing. These latter methods involve less environmental impact than clear-cutting, but all methods can cause significant changes to forest structure and the natural communities of forests.

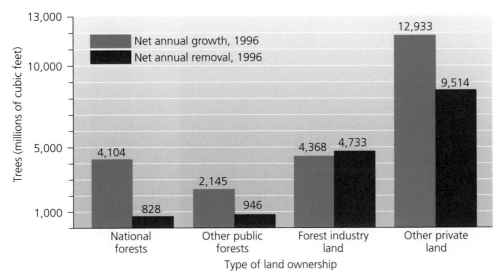

Figure 16.12 Forest Service data indicate that in the United States, trees (measured in cubic feet of wood biomass) are growing at a faster rate than they are being removed. The exception is on land privately owned by the timber industry, where extraction is narrowly outpacing growth. Data are for 1996, the most recent year for which such data are available. Go to **GRAPH IT** on the website or CD-ROM. Data from United States Forest Service, 1996.

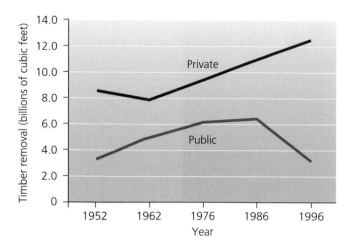

Figure 16.13 Logging has increased on private land in the United States over the past half-century. On public land, it peaked in the 1980s and decreased in the 1990s. Data from U.S. Department of Agriculture Forest Service, U.S. forest facts and historical trends, 2001.

analyzed government subsidies to timber companies conclude that the Forest Service loses at least $100 million of taxpayers' money per year by selling timber well below its costs for marketing and administering the harvest and for building access roads.

National forests are beginning to be managed for public recreation and ecosystem integrity

All methods of logging result in some habitat disturbance. All methods change forest structure and composition. Most methods increase soil erosion, leading to the siltation of waterways, which can degrade habitat and affect drinking water quality. Most methods also speed runoff, sometimes causing flooding. In extreme cases, such as when steep hillsides are clearcut, landslides can result. Significant habitat disturbance almost invariably affects the plants and animals inhabiting an area. In recent decades, increased awareness of these problems has prompted many citizens to protest the way that national forests are managed. These critics have urged that the Forest Service manage forests for recreation, wildlife, and ecosystem integrity, rather than timber. They want forests managed as ecologically functional entities, not as cropland for trees.

The Forest Service has responded to some extent, developing new programs to manage wildlife and endangered species, including nongame species. Forest Service scientists in some districts have become involved in extensive programs of **ecosystem management**, attempt-

ing to protect, recover, or restore whole plant and animal communities that had been lost or degraded. Some of these efforts, ironically, run counter to the Forest Service's best known symbol, Smokey the Bear. The cartoon bear wearing a ranger's hat who advises us to fight forest fires is widely recognized, but unfortunately Smokey's message is badly outdated, and many scientists assert that it has done great harm to American forests.

For over a century, the Forest Service and other land management agencies have suppressed fire whenever and wherever it has broken out. Yet ecological research now clearly shows that many ecosystems depend on fire to maintain themselves. Certain plants have seeds that germinate only in response to fire, and researchers studying tree rings have documented that many ecosystems historically experienced fire with some frequency. Burn marks in a tree's rings reveal past fires, giving scientists an accurate history of fire events extending back hundreds or even thousands of years. Researchers have found that North America's open pine woodlands burned regularly. Grasslands burned most frequently, sometimes every few years. Ecosystems dependent on fire are adversely affected by its suppression; open pine woodlands become cluttered with hardwood understory that ordinarily would be cleared away by fire, for instance, and animal diversity and abundance decline in such cluttered habitats.

In addition, fire suppression in the long term can lead to catastrophic fires that truly do damage forests, destroy human property, and threaten human lives. Fire suppression allows limbs, logs, sticks, and leaf litter to build up on the forest floor over the years—excellent kindling for a catastrophic fire. Such fuel buildup helped cause the 1988 fires in Yellowstone National Park (Figure 16.14), the 2003 fires in southern California, and thousands of other fires across the continent, and it has made catastrophic fires significantly greater problems than they were in the past.

To reduce this fuel load and improve the health and safety of forests, the Forest Service and other agencies have in recent years been burning areas of forest under carefully controlled conditions. These prescribed burning programs have worked very effectively where they have been applied, but they require so much time and effort that they have been implemented on only a relatively small amount of land. In addition, these efforts have been impeded by public misunderstanding and by interference from politicians who have not taken time to understand the science behind the approach.

In the wake of the 2003 California fires, the U.S. Congress passed the Healthy Forests Restoration Act. While this legislation encourages some prescribed burning, it

Figure 16.14 Forest fires are a natural phenomenon to which many plants are well-adapted and that maintain many ecosystems. The suppression of fires by humans over the past century has led to a buildup of leaf litter and young trees, which serve as fuel to increase the severity of fires when they do occur. As a result, catastrophic forest fires (such as this one in Yellowstone National Park in 1988) have become more common in recent years. These unnaturally severe fires can do great damage to ecosystems and human communities. The best solution, most fire ecologists agree, is to forego suppressing natural fires as much as possible and to institute controlled burning to reduce fuel loads and restore forest ecosystems.

relies primarily on the physical removal of small trees and underbrush by timber companies to make forests less fire-prone. Many scientists and environmental advocates have criticized the law, saying it will increase commercial logging in national forests, decrease oversight and public participation in enforcing environmental regulations, and do little to reduce catastrophic fires near populated areas.

Deforestation is proceeding rapidly in many countries

Many current U.S. trends are being paralleled internationally, but some developing nations, such as Brazil, are in the position the United States faced a century or two ago: having a vast frontier that they can develop for human use. A key difference is that today newer technology has allowed deforestation to proceed in these countries more quickly. Furthermore, much of the cutting is being done not by the citizens of those countries but by multinational corporations that export the products elsewhere.

In Sarawak, the Malaysian portion of the island of Borneo, foreign corporations that were granted logging concessions have deforested several million hectares of tropical rainforest since 1963. The clearing of this forest—one of the world's richest, hosting such organisms as orangutans and the world's largest flower, *Rafflesia arnoldii*—has had direct impacts on the 22 tribes of people who live as hunter-gatherers in Sarawak's rainforest. The Malaysian government did not consult the tribes about the logging, which decreased the wild game these people depended on. Oil palm agriculture established afterward caused pesticide and fertilizer runoff that killed fish in local streams. The tribes protested peacefully and finally began blockading logging roads. The government, which at first jailed them, now is negotiating, but it insists on converting the tribes to a farming way of life. For Sarawak and other places in the world, such conflicts over resources, land, and lifestyle may potentially be minimized by the introduction of sustainable forestry practices (Chapter 20). In the case of Sarawak's tribes, agroforestry involving both forest and farms could ease the shift to agriculture and be more lucrative than conventional monocultures for the tribespeople.

Agricultural Land Use

Having replaced many forests, agriculture now covers more of the planet's surface than does forest. Thirty-eight percent of Earth's terrestrial surface is devoted to agriculture—more than the area of North America and Africa combined. Of this land, 26% supports pasture, and 12% consists of crops and arable land. Agriculture is the most widespread type of human land use and causes the greatest human impact on land and ecosystems (Figure 16.15). Although such agricultural methods as organic farming and no-till farming (Chapter 8 and Chapter 9) can be sustainable, the majority of the world's cropland hosts intensive traditional agriculture and monocultural industrial agriculture, involving heavy use of fertilizers, pesticides, and irrigation, and too often producing soil erosion, salinization, and desertification.

The marketplace should, in theory, discourage people from farming with intensive methods on degraded land if such practices are not profitable. But agriculture in many countries is supported by massive subsidies. Governments of 30 developed nations, for instance, handed out $311 billion in farm subsidies in 2001, averaging $12,000 per farmer. Roughly one-fifth of the income of the average farmer from Canada or the United States comes from subsidies. Agricultural lobbyists and other proponents of such subsidies justify them by stressing that the vagaries of weather make profits and losses from farming unpredictable from year to year, such that a system without some way of compensating farmers for bad years might not survive in the long term. Opponents of subsidies, however, argue that subsidizing environmentally destructive agriculture will also be unsustainable.

Vast areas of wetlands have been drained to accommodate farming

Few farmers have had the good fortune to plow the kind of rich and fertile virgin prairie soil that settlers found west of Chicago in the 19th century. Instead, much of today's cropland exists on plots farmed for decades or centuries, or on marginal lands converted from pasture, forest, or wetland. Much of today's farmland exists on the site of former wetlands (Chapter 14)—swamps, marshes, bogs, and river floodplains—that were drained and filled in (Figure 16.16). Throughout recent history governments have encouraged such laborious efforts to drain wetlands. To promote settlement and farming, and to eradicate malaria (because mosquito vectors breed in wetlands), the United States passed a series of laws known as the Swamp Lands Acts in 1849, 1850,

Figure 16.15 Cropland agriculture exerts dramatic effects over a large portion of Earth's landscape. The huge green circles visible from any airplane trip over the Great Plains are created by center-pivot irrigation. Immense sprinklers pivot around a central point, watering a circular area. Shown are fields in eastern Oregon.

Figure 16.16 Most of North America's wetlands have been drained, filled, and the land converted to agricultural use. The northern Great Plains region of Canada and the United States was pockmarked with thousands of "prairie potholes," water-filled depressions that served as nesting sites for most of the continent's waterfowl. Today many of these have been lost; shown are farmlands encroaching on prairie potholes in North Dakota.

and 1860. These laws encouraged draining wetlands and converting them to agriculture. The government transferred over 24 million ha (60 million acres) of wetlands to state ownership (and eventually to private hands) to stimulate drainage, conversion, and flood control. In the Mississippi River valley, the Midwest, and a handful of states from Florida to Oregon, these transfers resulted in the near-eradication of malaria and the creation of over 25 million acres of new farmland. In the 20th century, surveys by the U.S. Department of Agriculture (USDA) and other agencies recognized tens of millions more ha of wetlands that could be converted to agriculture. A USDA program in 1940 provided monetary aid and technical assistance to farmers draining wetlands on their land, resulting in the conversion of almost 23 million ha (57 million acres) of wetlands. Today, less than half the original wetlands in the lower 48 states and southern Canada remain. However, today many people have a new view of wetlands. Rather than viewing them as worthless swamps, science has made clear that they are valuable ecosystems. This scientific knowledge, combined with preservation ethics, has induced policymakers to develop regulations to safeguard remaining wetlands. Because of loopholes and continued debate over the worth of wetlands relative to agriculture and development, however, many wetlands are still being lost.

Weighing the Issues:
Subsidies, Soil, and Wetlands

Started in 1985, the Conservation Reserve Program was a landmark U.S. agricultural initiative designed to protect highly erodible lands. Years of tilling soils had increased soil erosion rates dramatically—increasing sedimentation into rivers and degrading soil fertility. The program provided farmers with a different kind of subsidy—it paid them to take erodible lands out of production and encouraged them to reintroduce wildlife to the areas. Now, under the Wetland Reserve Program, the U.S. government is offering voluntary 30-year subsidies to landowners who do not develop wetland areas. In what ways do you think this program might affect farmers? How might it affect rural communities? How might it affect wildlife and ecosystems in rural areas?

Livestock graze one-fourth of Earth's land

As severe as its ecological impacts have proven to be, cropland agriculture uses less than half the land taken up by livestock grazing, which covers a quarter of the world's land surface. Human use of rangeland, however, does not necessarily exclude its use by wildlife or its continued functioning as a grassland ecosystem. Grazing can be sustainable if done carefully and at low intensity. In the American West, ranching proponents claim that cattle are merely taking the place of the vast herds of bison that once roamed the plains. Indeed, most of the world's grasslands have historically been home to large herds of grass-eating mammals, and grasses have adapted to herbivory. Nonetheless, poorly managed grazing, as we saw in Chapter 8, can have adverse impacts on soil and grassland ecosystems.

Most U.S. rangelands are federally owned and managed by the Bureau of Land Management (BLM). The BLM is, in fact, the nation's single largest landowner; its 106 million ha (261 million acres) are spread across 12 western states (see Figure 16.10). Ranchers are allowed to graze cattle on BLM lands for inexpensive fees, a practice that many public lands advocates see as an inducement to overgrazing. Thus ranchers and environmentalists have traditionally been at loggerheads. In the past several years, however, ranchers and environmentalists have been finding common ground, as they team up to preserve ranchland against what each of them views in common as a threat—the encroaching housing developments of suburban sprawl. Although developers will often pay high prices for ranchland, many ranchers do not want to see the loss of the wide open spaces and the ranching lifestyle that they cherish.

Land use in the American West might have been better managed

Grazing, farming, timber harvesting, and other land uses need not have strongly adverse impacts on the landscape. It is not these activities in themselves that cause environmental problems, but the overexploitation of resources beyond what ecosystems can handle. In the American West, a great deal of damage was done to the land by poor farming practices, overgrazing, and attempts to farm arid lands that were more suitable for grazing or preservation. All of these phenomena have at some time been aided by U.S. government policy.

Most land to the west of the 100th meridian, the longitudinal line splitting the Great Plains in half from Manitoba south through Texas, receives less than 50 cm (20 in) of rain per year, making it too arid for nonirrigated agriculture. One man in U.S. history recognized this fact and attempted to reorient policy so that the West could be settled in a way that allowed farmers to succeed. John Wesley Powell, an extraordinary individual who explored the raging Colorado River by boat despite having lost an arm in the Civil War, undertook vast surveys of the Western lands for the U.S. government in the mid- to late 19th century. A pioneer in calling for government agencies to base their policies on science, Powell maintained that lands beyond the 100th meridian were too dry to support farming on the 65-ha (160-acre) plots the government parceled out through the Homestead Act (Chapter 3). Plots in the West, he said, would have to be 16 times as large and would require irrigation. Moreover, in order to prevent individuals or corporations from monopolizing scarce water sources, he urged the government to organize farmers into cooperative irrigation districts, with each district encompassing a watershed.

Powell's ideas were too revolutionary for the entrenched political interests and the prevailing misconceptions of his time, which held that the West was a utopia for frontier settlement. The ideas in Powell's 1878 *Report on the Lands of the Arid Region of the United States* were for the most part never implemented.

For agriculture and forestry, debates continue today over how best to utilize land and manage resources. The extraction of resources from public lands in the United States has helped propel the economy of the country, and many other nations have done and are doing the same. But as resources dwindle, as forests and soils are degraded, and as the landscape fills with more people, the arguments for conservation of resources—for their sustainable use—have grown stronger. Also growing stronger is the argument for preservation of land—setting aside tracts of relatively undisturbed land intended to remain undeveloped.

Parks, Reserves, and Wildlands

Preservation has been part of the American psyche ever since John Muir rallied support for saving scenic lands in the Sierras (Chapter 2). For ethical reasons as well as pragmatic ecological and economic ones, Americans and many others worldwide have chosen to set aside tracts of land in perpetuity to be preserved and protected from development.

Why have we created parks and reserves?

What specifically in this mix of ethics, ecology, and economics has driven so many cultures to refrain voluntarily from exploiting land for material resources? The historian Alfred Runte has cited four traditional reasons that parks and protected areas have been established:

1. Many people believe that enormous, beautiful, or unusual features such as the Grand Canyon, Mount Rainier, or Yosemite Valley should be protected—an impulse sometimes termed *monumentalism* (Figure 16.17).
2. Parks have been created as a way to make use of sites lacking economically valuable material resources; land that holds little monetary value is easy to set aside because no one wants to buy it.
3. Protected areas offer utilitarian benefits. For example, a watershed protected from development can provide cities with clean drinking water and a buffer against floods.
4. Protected areas offer recreational value to tourists, hikers, fishermen, hunters, and others.

To these four traditional reasons, a fifth has been added and given higher priority in recent years: the preservation of biodiversity. As we saw in Chapter 15, human impact alters habitats in myriad ways and has led to countless population declines and species extinctions. A park or reserve is widely viewed as a kind of Noah's Ark, an island of habitat that can, scientists hope, maintain species that might otherwise disappear.

Figure 16.17 The awe-inspiring beauty of certain regions of the western United States was one reason for the establishment of national parks. Images of scenic vistas such as this one of Bridal Veil Falls in Yosemite National Park, portrayed by the landscape painter Albert Bierstadt, have induced millions of people from North America and abroad to visit these parks.

Federal parks and reserves began in the United States

The striking scenery of the American West impelled the U.S. government to create the world's first **national parks,** largely for reasons of monumentalism and recreation. In 1872 Yellowstone National Park was established as "a public park or pleasuring-ground for the benefit and enjoyment of the people." Yosemite, General Grant, Sequoia, and Mount Rainier National Parks followed after 1890. The Antiquities Act of 1906 gave the president authority to declare selected public lands as national monuments, which can be an interim step to national park status. Presidents from Theodore Roosevelt to Bill Clinton have since used this authority to expand the nation's system of protected lands.

The National Park Service (NPS) was created in 1916 to administer the growing system of parks and monuments, which today numbers 388 sites totaling 32 million ha (79 million acres) and includes national historic sites, national recreation areas, national wild and scenic rivers, and other types of areas (see Figure 16.10). This most widely used park system in the world received 277 million reported recreation visits in 2002—about as many visits as there are U.S. residents. While the high visitation rates signal the success of the park system, they also create overcrowded conditions at some parks. Many observers have therefore suggested that the parks' popularity indicates a pressing need to expand the system.

Other types of protected areas include the system of **national wildlife refuges,** begun in 1903 by President Theodore Roosevelt. Now totaling 37 million ha (91 million acres) spread over 541 sites (see Figure 16.10), the system is administered by the U.S. Fish and Wildlife Service (FWS). In its guiding principles for managing the system, the FWS calls its employees "land stewards, guided by Aldo Leopold's teachings that land is a community of life and that love and respect for the land is an extension of ethics." It also claims its management ranges "from preservation to active manipulation of habitats and populations." Indeed, these refuges not only serve as havens for wildlife, but also in many cases encourage hunting, fishing, wildlife observation, photography, environmental education, and other public uses. Many wildlife advocates find it ironic that hunting is allowed at many refuges, but hunters have traditionally supplied a great deal of the funding that has gone toward land acquisition and habitat management for the refuge system. Like other federal land management agencies, the FWS has increasingly moved toward managing for nongame species as well as game species, and toward managing on the habitat and ecosystem levels.

For example, the agency has engaged in ecological restoration and reintroducing species that have undergone population declines.

Wilderness areas have been established on various federal lands

In response to the public's desire for undeveloped areas of land, in 1964 the U.S. Congress passed the Wilderness Act, which allowed areas of existing federal lands to be designated as "wilderness areas." These areas are off-limits to development of any kind, but they are open to public recreation such as hiking, nature study, and other activities that minimize impact on the land (Figure 16.18). Some pre-existing extractive land uses such as grazing and mining were "grandfathered in," or allowed to continue, within wilderness areas as a political compromise to get the act passed. Congress declared that wilderness areas were necessary "in order to assure that an increasing population, accompanied by expanding settlement and growing mechanization, does not occupy and modify all areas within the United States and its possessions, leaving no lands designated for preservation and protection in their natural condition." Wilderness areas have been established within portions of national forests, national parks, national wildlife refuges, and BLM land, and are overseen by the agencies that administer these areas.

Weighing the Issues:
Managing Wilderness?

Within wilderness areas in the western United States, it is common practice for state fish-and-game agencies to encourage recreation by stocking high-elevation lakes with fish — an expensive practice that often requires helicopters to drop loads of fish from the air. The practice is increasingly controversial; non-native fish are generally used, and 95% of the 16,000 lakes historically lacked fish of any kind, leading ecologists to worry about the effects of potentially invasive species. Fish-stocking is just one example of efforts to manage wildlife in wilderness areas. The Wilderness Act of 1964 mandated both the human use and enjoyment of wilderness and its protection and preservation — to many people, two seemingly opposing goals. Which goal do you feel is more important? Are there ways in which they can be made compatible? Do you feel that management practices such as fish-stocking should be allowed in wilderness areas? Are there other management practices that you would feel differently about?

Figure 16.18 Wilderness areas were designated on various federally managed lands in the United States following the 1964 Wilderness Act. These include areas relatively little disturbed by human activities, such as the Selway-Bitterroot Wilderness in Idaho.

Not everyone supports land set-asides

The restriction of activities in wilderness areas has helped generate opposition to the land protection policies of the U.S. government. Sources of such opposition include some state governments of western states, where large portions of land are federally owned. State governments would naturally like decision-making power over all lands in their states, yet when those states came into existence the federal government was already holding much of the acreage inside their borders. Idaho, Oregon, and Utah own less than 50% of the land within their borders, and in Nevada 80% of the land is federally owned. Western state governments have traditionally sought to obtain land from the federal government and to encourage resource extraction and development on it.

The drive to extract more resources, obtain greater local control of lands, and obtain greater recreational access to public lands is epitomized by the **wise use movement,** a loose confederation of individuals and groups that coalesced in the 1980s and 1990s in response to the increasing success of environmental advocacy. Wise use advocates include those dedicated to protecting private property rights; those opposing government regulation; those wanting federal lands transferred to state, local, or private hands; and those desiring more motorized recreation on public lands. They include farmers, ranchers, trappers, and mineral prospectors at the grassroots level who live off the land, as well as groups representing the large corporations of industries that extract resources. Debate between mainstream environmental groups and wise use spokespeople has been vitriolic, with each side claiming to represent the will of the people and painting the other as the oppressive establishment. Wise use advocates opposed to wilderness and wildlife protection have played key roles in ongoing debates on the use of national parks, such as whether recreational activities that disturb wildlife, such as snowmobiles and jet-skis, should be allowed.

Land is also protected by various non-federal entities

Efforts to set aside land, and the debates over such decisions, at the federal level are paralleled at the state and local levels. Each U.S. state has agencies that manage land and resources on state lands, as do many counties and municipalities. The Chicago area's forest preserve system is one such example. Even before that system was established, New York State had created Adirondack State Park, a mountainous area whose streams converge to form the headwaters of the Hudson River, which flows south past Albany to New York City. Seeing the need for river water to power industries, keep canals filled, and provide drinking water, the state set the land aside, a farsighted decision that has paid dividends through the years.

Private nonprofit groups, such as land trusts, also engage in land conservation. These are local or regional organizations that accept donations and aim to preserve lands valued by members of the region where the trust is based. In most cases land trusts purchase land outright with the aim of preserving it in its natural condition. The Nature Conservancy may be considered the world's largest land trust, but smaller ones are springing up throughout North America. By one estimate, there are 900 local and regional land trusts in the United States that together own 177,000 ha (437,000 acres) and have helped preserve an additional 930,000 ha (2.3 million acres), including well-known scenic areas such as Big Sur on the California coast, Jackson Hole in Wyoming, and Maine's Mount Desert Island.

Parks and reserves are increasing internationally

Land preservation is clearly an international pursuit. Many nations have established national park systems and are benefiting from ecotourism as a result—from Costa Rica (Chapter 5) to Ecuador to Thailand to Tanzania. Efforts are underway to gain further protection for many areas designated as biodiversity hotspots (Chapter 15). Regions vary in the percentage of land that is protected, from Europe's 5% to the Middle East's 23%. The total worldwide area in protected parks and reserves has increased from 2.78 million km^2 (1.07 million mi^2) in 1970 to 12.18 million km^2 (4.70 million mi^2) in 2000. The world's 38,536 protected areas in 2003 covered 1.3 billion ha (3.2 billion acres), or 9.61% of the globe's land area. Unfortunately, parks in developing countries do not always receive the funding they need to manage resources, provide for recreation, and protect wildlife from poaching and timber from logging. Thus many of the world's protected areas are "paper parks"—protected on paper but not in reality.

Some types of protected areas fall under national sovereignty but are designated by, or their protection assisted internationally by, the United Nations. World heritage sites are an example; currently over 560 across 125 countries are listed for their cultural value and nearly 150 for their natural value. One such site is Australia's Kakadu National Park, which we discussed in Chapter 2.

Biosphere reserves are tracts of land with exceptional biodiversity that couple preservation with sustainable development to benefit local people. Each reserve consists of a core area that preserves biodiversity, a buffer zone that allows local activities and limited development that do not hinder the core area's function, and an outer transition zone in which agriculture, human settlement, and other land uses can be pursued in a sustainable way (Figure 16.19).

In addition, many nations have cooperated to establish parks and reserves along the boundaries between them. Transboundary parks, many of them quite large, account for 10% of protected areas worldwide and involve 113 countries. Examples are Waterton-Glacier National Parks on the Canada–U.S. border, and the mountain gorilla reserve shared by three African countries. Some of these lands function as "peace parks," helping ease tensions by acting as buffers between nations that have quarreled over boundary disputes. This is the case with Peru and Ecuador as well as Costa Rica and Panama, and many people hope that peace parks can help resolve conflict between Israel and her neighbors.

The design of parks and reserves has important consequences for biodiversity

Often it is not outright destruction of habitat that threatens species, but rather fragmentation of habitat (Chapter 15). Expanding agriculture, spreading cities, highways, logging, and many other impacts have chopped up large contiguous expanses of habitat into small disconnected ones (Figure 16.20). When this happens, many species

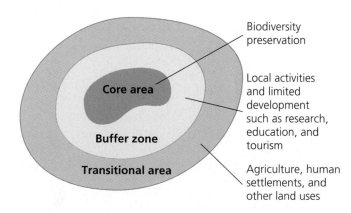

Figure 16.19 Biosphere reserves are international efforts that couple preservation with sustainable development to benefit local residents. Each reserve includes a core area that preserves biodiversity, a buffer zone that allows limited development, and a transition zone that permits sustainable agriculture, settlement, and other human land uses.

suffer. Bears, mountain lions, and other animals that need large ranges in which to roam disappear when their habitat is broken into small fragments. Bird species that depend on being in the interior of a forest may fail to reproduce when forced near the edge of a fragment; their nests may be attacked by predators and parasites that favor open habitats surrounding the fragment or that travel along habitat edges.

Because habitat fragmentation is such an important issue for biodiversity conservation, and because there are limits on the amount of land that can be set aside, conservation biologists have argued heatedly about whether it is better to make reserves large in size and few in number or many in number but small in size. This debate has been nicknamed the **SLOSS** dilemma, for "Single Large Or Several Small." The debate is ongoing and

(a) Mount Hood National Forest, Oregon

Figure 16.20 As human populations have grown and human impacts have increased, most large expanses of natural habitat have become fragmented into smaller disconnected areas. Forest fragmentation—such as that caused by clear-cutting in Mount Hood National Forest, Oregon (**a**)—can have significant impacts on forest-dwelling wildlife. Forest fragmentation has been extreme in the eastern and midwestern United States; shown in (**b**) are historical changes in forested area in Cadiz Township, Wisconsin, between 1831 and 1950.
(b) *Source:* John T. Curtis, "The modification of mid-latitude grasslands and forests by man." In W.L. Thomas, Jr., ed., *Man's role in changing the face of the Earth*, 1956.

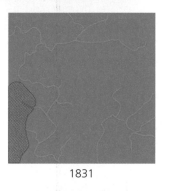

1831

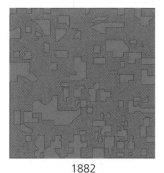

1882

1902

1950

(b) Fragmentation of wooded area (green) in Cadiz Township, Green County, Wisconsin

complex, but it seems clear that large species that roam great distances, such as the Siberian tiger (Chapter 15), benefit more from the "single large" approach to reserve design. Insects that live as larvae in small areas, in contrast, may do just fine in a number of small isolated reserves, if they can disperse as adults by flying from one reserve to another.

A related issue is whether **corridors** of protected land are important for allowing animals to travel between islands of protected habitat. For many species, greater connectivity of fragments potentially means access to more habitat and more assurance of gene flow to maintain populations in the long term. Just how much organisms actually use corridors, however, is an active area of study. Many land management agencies and environmental groups try, when possible, to join new reserves to existing reserves or to connect state reserves to federal reserves.

William Newmark's research (see Chapter 15, The Science behind the Story) suggested that North America's national parks were too isolated from one another to sustain populations of large mammals in the long run. Because of such concerns, one group has recently emerged with a revolutionary plan to ensure the long-term sustainability of North America's wildlife and ecosystems. The Wildlands Project proposes setting aside vast swaths of western North America for biodiversity preservation, with each area connected to others in a network. The gigantic scale of the project and its emphasis on preservation make it unlikely to gain wide support, at least in the near future. Yet, many scientists think such drastic measures may be required if we are to save a great deal of the natural heritage of the continent.

Creating Livable Cities

Some of the same concerns that led to the establishment of national parks and other wildland reserves throughout the United States during the Progressive Era also led to the establishment of parks on smaller scales in the cities of the growing nation.

City parks were widely established at the turn of the last century

In the late 1800s, public parks and gardens began to be established in eastern U.S. cities, using aesthetic ideals borrowed from European parks, gardens, and royal hunting grounds. The lawns, shaded groves, curved pathways, and pastoral vistas we see today in many American city parks and cemeteries originated with the nation's leading landscape architect, Frederick Law Olmsted, who designed New York's Central Park in 1853 and a host of park systems afterwards (Figure 16.21). Olmsted and his followers sought to create landscapes that provided tranquility, greenery, and an escape from the pollution and stress of the urban setting.

Two sometimes-conflicting goals motivated the establishment and design of the early city parks. On one hand, the parks were meant to be "pleasure grounds" for the wealthy, who helped support their establishment financially and who would ride the parks' winding roadways in carriages. On the other hand, the parks were meant to alleviate congestion and allow some escape for the many poverty-stricken immigrants who lived in America's cities at the time. These park users were more interested in active recreation, such as ballgames, than in carriage rides. At times the aesthetic interests of the elite and the recreational interests of the laboring class came into conflict—a friction that survives today in debates over recreation in city and national parks.

In Chicago, civic boosters striving to overcome the city's reputation as a rough-and-ready frontier town of slaughterhouses and meatpacking plants staged the World's Columbian Exposition in 1893. This grand spectacle of a world's fair was meant to show the world the vigor and enterprise of the fast-growing city. To construct the fairgrounds, landscape architect Daniel Burnham turned swampy Jackson Park into a magnificent faux-city of immaculate buildings rising from the lakefront in classic European style. The success of the Exposition, attended by millions of people from around the globe, set Chicago on the road to redesigning itself.

Burnham's 1909 *Plan of Chicago* represented the first thorough city planning program for an American city and was largely implemented over the following years and decades. This plan expanded city parks and playgrounds, cleared industry and railroads from the lakeshore in order to give the public access to Lake Michigan, improved neighborhood living conditions, and streamlined traffic systems. It also included the program for a greenbelt of forest preserves circling the city's outskirts. The greenbelt idea was not brand new; architects Perkins and Jensen based it in part on Boston's system after Perkins's wife had visited that city and found it "so arranged that parks are accessible from all parts of the city, and it is difficult to think of any Boston child as shut away from the beauties of nature."

Figure 16.21 City parks were developed in many fast-growing urban areas in the late 19th century to provide citizens aesthetic pleasure, recreation, and relief from the stresses of the city. Manhattan's Central Park, shown here in a 19th-century engraving, was one of the first and one of the largest.

City and regional planning are tools for creating livable urban areas

The Burnham plan and the forest preserves set early standards for **city planning**, the professional pursuit that attempts to design cities in such a way as to maximize their efficiency, functionality, and beauty (Figure 16.22). City planning grew in importance throughout the 20th century as urban populations expanded, inner cities decayed, and wealthier residents fled to the suburbs. In today's world of sprawling metropolitan areas, regional planning has become every bit as important. Although many cities and regions did not plan as well as did Chicago, planners often have opportunities to correct past mistakes or make improvements. One tool planners use is zoning, the practice of classifying areas for different types of development and land use. Industrial plants may be kept out of districts zoned for residential use in order to preserve the cleanliness and tranquility of residential neighborhoods, for instance.

Zoning boards and planners face many constraints, and one of the biggest is the road-based transportation system that encourages individuals to drive their own automobiles. Once a road system has been developed, and countless businesses and homes built alongside roads, it can be difficult to replace or complement the road system with a more efficient and environmentally friendly mass transit system. Compared to mass transit rail systems, road networks take up more space and emit more pollution. The fuel and productivity lost to traffic jams are thought to cost the U.S. economy $74 billion each year.

Cities with severe traffic problems such as Bangkok, Thailand, and Athens, Greece, have recently opened new rail systems that carry hundreds of thousands of commuters a day. Light rail use is increasing in Europe, and U.S. ridership is now rising faster than is the rate of new car drivers. Installing mass transit systems, however, is not the only way to make cities run more efficiently. Governments can also raise fuel taxes, tax inefficient modes of transport, reward carpoolers with carpool lanes, encourage bicycle use and bus ridership, charge trucks for road damage, minimize investment in infrastructure that encourages sprawl, and stimulate investment in renewed urban centers.

As discussed above, large city parks are a key component of a healthy urban environment. Even small spaces can make a big difference, however. Urban community

Figure 16.22 Daniel Burnham's 1909 *Plan of Chicago* included parks and greenways, efficient transportation routes, and increased access to the lakefront for city residents. The plan has served as a model for city planners ever since.

gardens allow people to grow their own vegetables and flowers in a friendly neighborhood setting. Another type of green space is the corridor or strip of land that connects parks or neighborhoods, often called a *greenway*. The Rails-to-Trails movement has spearheaded the conversion of abandoned railroad right-of-ways into trails for walking, jogging, and biking. To date, nearly 24,000 km (15,000 mi) of 1,200 rail lines have been converted across the United States. In addition to new rail-trails, communities in the Chicago area enjoy miles of trails alongside old canals. Such greenways can also serve as corridors for the movement of wildlife. Accomplishments like these have made American cities much more livable than they once were, and we should be encouraged about such progress. While air and water pollution was substantially cleaned up thanks to environmental legislation, on the local level city parks and city planning have improved urban life, often as a result of the vision and hard work of particular individuals. Further vision and hard work will be required to deal with the challenges posed by urban and suburban sprawl.

Several success stories stand out

One shining model for creating livable cities is Curitiba, Brazil. This city of 2.5 million people has an outstanding bus system that is used each day by three-quarters of the population. The well-planned bus routes accompany measures to encourage bicycles and pedestrians, all of which has resulted in a steep drop in car usage despite a rapidly growing population. The city provides services ranging from recycling to environmental education to job training for the poor to free health care, and surveys show that its citizens are unusually happy and better off economically than elsewhere in the country.

In the United States, Portland, Oregon, stands out for the success of its planning. In the 1970s, residents of Portland and other areas in Oregon foresaw the coming of suburban sprawl, and they acted. A statewide zoning system was established, and cities were required to set urban growth boundaries to limit sprawl, forcing developers to invest in city neighborhoods. A regional government entity called *Metro* was set up in the Portland area to oversee land use, setting aside land that can be developed, and purchasing land to be set aside for preservation. Besides the zoning laws, an innovative light rail system was developed, and Metro policy encouraged the development of self-sufficient neighborhood communities along the rail lines (Figure 16.23).

While some cities are creating new types of urban spaces, others are working hard to increase the "naturalness" of their park environments. At Chicago-area forest preserves, scientists and volunteers now practice ecological restoration by restoring prairies with prescribed

Figure 16.23 Portland, Oregon's light rail system is one component of an urban planning strategy that has helped make it one of North America's most livable cities.

burns. Elsewhere, people are creating new park land by adopting the techniques of ecological restoration. In San Francisco, an old abandoned airstrip has recently been cleared away and a degraded tidal lagoon restored at a site called *Crissy Field*. Birds now flock to the revitalized lagoon, as do thousands of people each day, to walk, jog, and lie on the beach, while taking in a stunning view of San Francisco Bay and the Golden Gate Bridge.

Conclusion

As half the human population has shifted from rural to urban lifestyles, the nature of our impact on the environment has changed. As urban-dwellers, our impacts are less direct but are often more far-reaching. Resources must be not only extracted but also delivered to us over long distances, requiring the use of still more resources. Managing those resources, many of which are extracted from public lands, must be made sustainable if we wish our society to thrive in the future.

Furthermore, if we are to remain an urbanized society, cities must be made increasingly livable. One key component of this effort involves establishing parks or other green spaces within our urban centers to keep us from becoming wholly isolated from nature. In a 2003 book, University of Arizona ecologist Michael Rosenzweig coined the term *reconciliation ecology* to refer to the science of establishing and maintaining habitats for wild species in the midst of the areas where humans work and live. His approach accepts that humans have had extraordinary impact on the planet and argues that the best hope for biodiversity may lie less in setting aside reserves than in adapting our land use practices to be more amenable to wild species. If we have too often eliminated or altered natural areas, then at least we can try to make our cities and suburbs more natural. In so doing, we will be benefiting not only wildlife and ecosystems but also ourselves.

REVIEW QUESTIONS

1. Why do urban areas require large amounts of natural (or undeveloped) land?

2. Why are our cities and suburbs growing in geographic scale so much more rapidly than before? How has urban and suburban growth, or sprawl, begun to affect human perceptions of land use and resource management?

3. Name some major causes of deforestation. Describe the ecological and economic results of deforestation in the United States.

4. Where does logging take place in the United States? Describe the various methods of logging.

5. What are some ecological effects of logging? What has been the U.S. Forest Service's response to public concern over the ecological effects of logging?

6. Are forest fires a good thing? Explain your answer.

7. What factors are causing increased deforestation in developing nations?

8. Approximately what percentage of Earth's land is used for agriculture?

9. What is a wetland? What kinds of policies have caused the conversion of wetlands to agriculture in the United States?

10. What is the BLM? What does it do?

11. Describe John Wesley Powell's solution to the issue of agricultural land use in the western United States.

12. Name five reasons for which parks and reserves have been created.

13. What was the first national park, and when was it established? When was the system of national wildlife refuges established, and which agency manages this system?

14. Why did Congress determine in 1964 that wilderness areas were necessary? How are these areas different from national parks or national wildlife refuges?

15. Why is there opposition to federal land protection in the United States?

16. How do groups other than the federal, state, and local governments also protect land in the United States?

17. Roughly what percentage of Earth's land is protected? What types of protected areas have been organized in countries outside the United States?

18. Explain the SLOSS dilemma and why habitat fragmentation is a critical issue. What is the corridor approach?

19. Describe Daniel Burnham's 1909 *Plan of Chicago*.

20. How are city parks thought to make urban areas more livable? What conflicts have sometimes arisen between users of city parks?

21. What is zoning, and what roles does it play in city and regional planning? What have Curitiba, Brazil, and Portland, Oregon, done to improve the quality of life for their citizens?

DISCUSSION QUESTIONS

1. Compare and contrast the experiences of Chicago, Tortolita, and Sarawak. What forms of reconciliation have occurred in each conflict over land use and resource management?

2. Can you think of a land use conflict that has occurred in your region? How was it resolved, or is it unresolved?

3. What are some ecological effects of farm subsidies? Propose arguments for and against subsidies from an ecological point of view.

4. Discuss some recent forms of urban or suburban green spaces, apart from city parks, that have become popular in planning cities and regional areas. How is transportation a major factor governing the development of these areas?

5. Is Michael Rosenzweig's concept of reconciliation ecology an effective way of approaching conflicts over land use? Should we set aside more reserves or more frequently adapt areas of human habitation for wild species?

Media Resources *For further review, go to the website* **www.envscienceplace.com** *or student CD-ROM, where you will find quizzes, flashcards, a glossary, additional interactive exercises, and links to relevant news and research sources. Also, on the website and CD-ROM is* **GRAPH IT**, *a series of interactive graphing tutorials to help you interpret graphs and plot data.*

17 Nonrenewable energy sources and their environmental impacts

Oil Production Facility at Prudhoe Bay, Alaska

This chapter will help you understand:

- The nonrenewable energy sources that fuel modern civilization

- The history of energy use

- Patterns of energy production and consumption today

- Crude oil, its origins, and consequences of its use

- Coal, its origins, and consequences of its use

- Natural gas, its origins, and consequences of its use

- Environmental impacts of fossil fuel use

- Political, social, and economic impacts of fossil fuel use

- Nuclear energy, its origins, and its history

Caribou Crossing Haul Road Near Trans-Alaska Pipeline

Asia
Arctic Ocean
Alaska's North Slope
Alaska
North America
Pacific Ocean

Central Case: Oil or Wilderness on Alaska's North Slope?

"The roar alone of road building, drilling, trucks, and generators would pollute the wild music of the Arctic and be as out of place there as it would be in the heart of Yellowstone or the Grand Canyon."
—*Former President Jimmy Carter, 2000*

"There is absolutely no indication that environmentally responsible exploration will harm the 129,000-member porcupine caribou herd."
—*Alaska Senator Frank Murkowski, 2002*

Above the Arctic Circle, at the top of the North American continent, the land drops steeply down from the jagged mountains of Alaska's spectacular Brooks Range and stretches north in a vast flat expanse of tundra until it meets the icy waters of the Arctic Ocean (Figure 17.1). Few Americans have been to this remote region, yet it has come to symbolize a struggle between two values in our modern life.

For some U.S. citizens, Alaska's North Slope is one of the last great expanses of wilderness in their sprawling industrialized country, one of the last places humans have left untouched. For these millions of Americans, simply knowing that this wilderness still exists is of

tremendous value. For millions of others, this land represents something else entirely—a source of petroleum, the natural resource that, more than any other, fuels our society and shapes our way of life. To these people, it seems wrong to leave such an important resource sitting unused in the ground. Those who advocate drilling for oil here accuse wilderness preservationists of neglecting the country's economic interests, while advocates for wilderness argue that drilling will sacrifice the nation's natural heritage for little gain.

Ever since oil was found seeping from the ground in this area a century ago, these two visions for Alaska's North Slope have competed and now exist side by side across three regions of this vast swath of land. The westernmost portion of the North Slope was set aside in 1923 by the U.S. government as an emergency reserve for petroleum. This parcel of land, the size of Indiana, is today called the National Petroleum Reserve-Alaska and was intended to remain untapped for oil unless the nation faced an emergency. Government petroleum geologists have explored much of the land to estimate where productive deposits of oil, coal, and natural gas might lie. Commercial oil and gas development began in portions of the reserve in the 1980s, but so far much of the region's 9.5 million ha (23.5 million acres) remains undeveloped.

Adjacent to the National Petroleum Reserve are state lands in the central portion of Alaska's North Slope that experienced widespread oil development and extraction after oil was discovered at Prudhoe Bay in 1968. Since drilling began in 1977, over 12.8 billion barrels (1 barrel = 159 l or 42 gal) of crude oil have been extracted from 19 oil fields spread over 160,000 ha (395,000 acres) of this region. The oil is transported across the state of Alaska by the 1,300-km (800-mi) trans-Alaska pipeline south to the port of Valdez, where it is loaded onto tankers for shipment to the lower 48 states.

East of the Prudhoe Bay region lies the Arctic National Wildlife Refuge (ANWR), an area the size of South Carolina consisting of federal lands set aside in 1960 and

1980 mainly to protect wildlife and preserve pristine eco-systems of tundra, mountains, and seacoast. This scenic region is home to 160 nesting bird species, numerous fish and marine mammals, grizzly bears, polar bears, Arctic foxes, timber wolves, musk oxen, and other animals. In most years, thousands of caribou arrive from the south to spend the summer in the Prudhoe Bay region, giving birth to and raising their calves. Because of the vast caribou herd and the other large mammals, ANWR has been called "the Serengeti of North America."

ANWR has been the focus of debate for decades. Advocates for oil drilling have tried to open its lands for development, while proponents of wilderness preserva-tion have fought for its protection. Scientists, oil indus-try experts, politicians, environmental groups, citizens, and Alaska residents have all been part of the debate. So have the two native Alaskan groups in the area, the

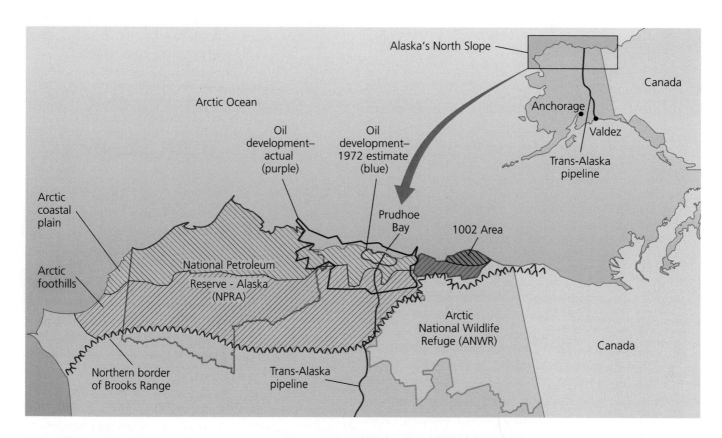

Figure 17.1 Alaska's North Slope is the site of both arctic wilderness and oil exploration. In the western portion of this region, the National Petroleum Reserve-Alaska was established by the fed-eral government as an area in which to drill for oil if the nation needs it in an emergency. It is well-explored but not widely developed. To the east of this area, the Prudhoe Bay region is the site of widespread oil extraction and since 1977 has produced over 12.8 billion barrels of oil. Develop-ment for oil has expanded far beyond the area experts estimated in 1972. Further east lies the Arc-tic National Wildlife Refuge, home to untrammeled Arctic wilderness and the focus of debate for years. Proponents of oil extraction and proponents of wilderness preservation have been battling over whether the 1002 Area of the coastal plain north of the Brooks Range should be opened to oil development.

Gwich'in and the Inupiat, who disagree over whether the refuge should be opened to oil development. The Gwich'in depend on hunting caribou and fear that oil industry activity will reduce caribou herds, whereas the Inupiat see oil extraction as one of the few opportunities for economic development in the area.

In a compromise in 1980, the U.S. Congress passed legislation that put most of the refuge off limits to oil but reserved for future decision-making a 600,000-ha (1.5 million-acre) area of coastal plain. This region, called the 1002 Area (after Section 1002 of the bill that established it), remains undeveloped for oil but can be opened for development by a simple vote of both houses of Congress. Its unsettled status has made it the center of the oil-versus-wilderness debate, and Congress has been caught between passionate feelings on both sides for a quarter of a century. In 2003, the stalemate continued. The House of Representatives voted to open the 1002 Area for drilling, while the Senate voted against it.

Behind the noisy policy debate over ANWR, scientists have attempted to inform the dialogue through research. Geologists have tried to ascertain how much oil lies underneath the refuge, and biologists have tried to predict the potential impacts of oil drilling on Arctic ecosystems. In addition, scientists and nonscientists alike are debating the relevance of the oil beneath the refuge for the security and prosperity of the nation. We will examine these questions by revisiting Alaska's North Slope as we survey the energy we use to heat and light our homes, power our machinery, and provide the comforts, conveniences, and mobility to which technology has accustomed us.

Sources of Energy

The debate over drilling for oil in the Arctic National Wildlife Refuge is a thoroughly modern debate, pitting the culturally new concept of wilderness preservation against the desire to exploit a resource that has come to guide the world's economy only in the past 150 years. However, humans have used—and fought over—energy in one way or another for all of our history.

Humans have long exploited energy sources

Ever since our ancestors first discovered fire, humans have extracted energy from natural resources to cook food and to light and heat dwellings (Figure 17.2). Since then, wood and wood products have been our primary sources of energy for heating and cooking, while we have harnessed animals, wind, and water as energy sources for mechanical work in fields, mills, and granaries. The development of the steam engine in the 18th century allowed the use of high-quality energy sources such as coal for a variety of mechanical purposes as economies industrialized. Meanwhile, deforestation in many areas of the world led to dwindling wood supplies, giving many societies incentive to shift to energy sources other than wood.

In the 20th century, fossil fuels—highly combustible substances formed from the remains of animals and plants from past geological ages—became the dominant source of power in industrialized countries, and then in developing nations. Of the three main fossil fuels (coal, oil, and natural gas), oil predominates today largely because it is most efficient to burn, ship, and store. Throughout the world, these fuels are used to generate electricity, a form of energy that is easier to transfer over long distances and to apply to a variety of uses. Increasingly, renewable energy sources are being developed as alternatives to nonrenewable energy sources, as we will see in Chapter 18.

A variety of renewable and nonrenewable energy sources are available today

Energy on Earth comes from a number of sources (Table 17.1). Some energy rises from Earth's core as heat, enabling us to harness geothermal power. A much smaller amount occurs in the form of the gravitational pull of the moon and sun, and efforts are starting to be made to harness tidal power. An immense amount resides within the bonds among protons and neutrons in atoms, and this energy provides us with nuclear power. The remainder of our energy sources come ultimately from the sun. Besides solar power that we can harness directly, sunlight makes possible the growth of plants, from which we take wood as a fuel source, and which after their death may impart their stored chemical energy to form oil, coal, and natural gas. Solar radiation also helps drive wind patterns and the hydrologic cycle, enabling other forms of energy such as wind power and hydroelectric power.

As we first noted in Chapter 1, energy sources such as sunlight, geothermal energy, and tidal energy are considered renewable because their supplies will not be depleted by our use of them. Other sources such as timber are renewable if we do not harvest them at too great a

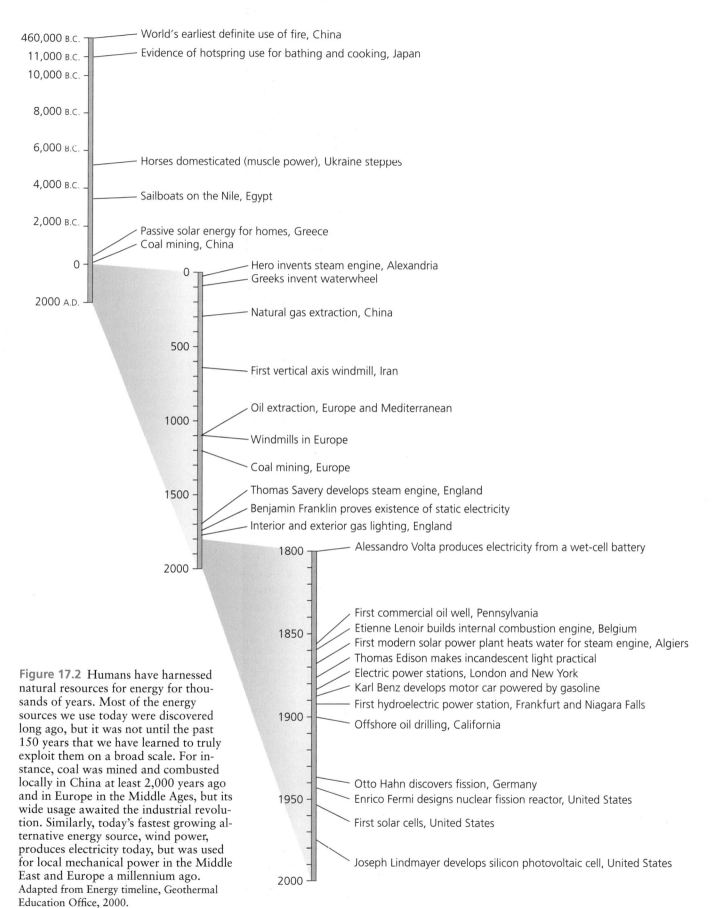

460,000 B.C. — World's earliest definite use of fire, China
11,000 B.C. — Evidence of hotspring use for bathing and cooking, Japan
10,000 B.C.

8,000 B.C.

6,000 B.C. — Horses domesticated (muscle power), Ukraine steppes

4,000 B.C. — Sailboats on the Nile, Egypt

2,000 B.C.
— Passive solar energy for homes, Greece
— Coal mining, China
0
— Hero invents steam engine, Alexandria
— Greeks invent waterwheel
2000 A.D.

0

— Natural gas extraction, China

500

— First vertical axis windmill, Iran

— Oil extraction, Europe and Mediterranean
1000
— Windmills in Europe

— Coal mining, Europe

1500 — Thomas Savery develops steam engine, England
— Benjamin Franklin proves existence of static electricity
— Interior and exterior gas lighting, England
— Alessandro Volta produces electricity from a wet-cell battery
2000

1800

— First commercial oil well, Pennsylvania
1850 — Etienne Lenoir builds internal combustion engine, Belgium
— First modern solar power plant heats water for steam engine, Algiers
— Thomas Edison makes incandescent light practical
— Electric power stations, London and New York
— Karl Benz develops motor car powered by gasoline
— First hydroelectric power station, Frankfurt and Niagara Falls
1900 — Offshore oil drilling, California

— Otto Hahn discovers fission, Germany
1950 — Enrico Fermi designs nuclear fission reactor, United States
— First solar cells, United States

— Joseph Lindmayer develops silicon photovoltaic cell, United States
2000

Figure 17.2 Humans have harnessed natural resources for energy for thousands of years. Most of the energy sources we use today were discovered long ago, but it was not until the past 150 years that we have learned to truly exploit them on a broad scale. For instance, coal was mined and combusted locally in China at least 2,000 years ago and in Europe in the Middle Ages, but its wide usage awaited the industrial revolution. Similarly, today's fastest growing alternative energy source, wind power, produces electricity today, but was used for local mechanical power in the Middle East and Europe a millennium ago. Adapted from Energy timeline, Geothermal Education Office, 2000.

Table 17.1 Energy Sources We Use Today

Energy source	Description	Type of energy
Crude oil	Fossil fuel extracted from ground	Nonrenewable
Natural gas	Fossil fuel extracted from ground	Nonrenewable
Coal	Fossil fuel extracted from ground	Nonrenewable
Nuclear energy	Energy from atomic nuclei of processed uranium mined from ground	Nonrenewable
Hydropower	Energy from running water	Renewable
Solar energy	Energy from sunlight directly	Renewable
Wind energy	Energy from the power of wind	Renewable
Geothermal energy	Earth's internal heat rising from core	Renewable
Biomass energy	Chemical energy stored in plant matter from photosynthesis	Renewable
Tidal and wave energy	Energy from tidal forces and ocean waves	Renewable
Electricity	Energy generated from use of primary energy sources	Secondary
Hydrogen	Stored energy from primary energy sources can be stored in fuel cells	Secondary

rate. In contrast, energy sources such as oil, coal, and natural gas are considered nonrenewable, because at our current rates of consumption we will use up Earth's accessible store of them in a matter of decades to centuries. Nuclear power as currently harnessed through fission of uranium can be considered nonrenewable to the extent that uranium ore is in limited supply. Although these nonrenewable sources result from ongoing natural processes, the timescales on which they are created are so long that, once depleted, they could not be replaced in any time span useful to our civilization. It takes a thousand years for the biosphere to generate the amount of organic matter that must be buried in order to produce a single day's worth of fossil fuels for the world. To replenish the amount of fossil fuels currently remaining in the world would take many millions of years.

Developed nations consume more energy than developing nations

Although people everywhere use energy to cook their food, light their homes, and keep themselves warm or cool, developed nations generally consume far more energy than do developing countries, and their major uses for energy are characteristically different. The most-developed nations use up to 100 times as much electricity per capita as do the least-developed nations, and North America consumes more than five times the world average in energy use per capita (Figure 17.3). Developing nations devote a greater proportion of their energy use to subsistence activities such as food preparation, home heating, and food-growing, whereas industrialized countries use a greater proportion for transportation

and industry (Figure 17.4). In addition, developing countries often use manual or animal energy sources instead of automated and technological ones. For instance, rice farmers in Bali plant rice by hand, but industrial rice growers in California use airplanes. Because industrialized nations rely more on equipment and technology, they use more fossil fuels. For example, fossil fuels supply 87% of the United States' energy needs. At 19.7 million barrels per day, oil is the most heavily used fossil fuel, constituting 40% of U.S. energy use, compared with 24% for natural gas and 23% for coal.

Fossil Fuels

Worldwide, consumption of the three main fossil fuels has been rising for years. Oil use has been rising steadily, coal's growth has only recently begun to level off, and natural gas is increasing fastest and has nearly surpassed coal consumption (Figure 17.5).

Fossil fuels are indeed fuels created from "fossils"

The fossil fuels we burn today in our cars and electrical power plants were formed from the tissues of organisms that lived 100–500 million years ago. The energy these fuels contain came originally from the sun and was converted to chemical-bond energy as a result of plants' photosynthesis (Chapter 4). The chemical energy in these organisms' tissues was then concentrated as these

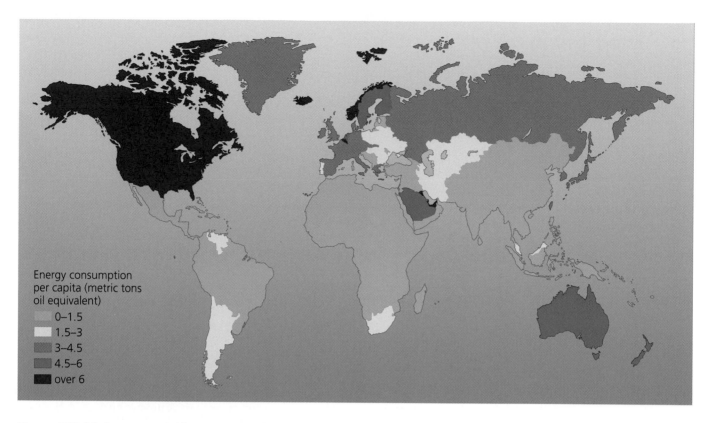

Figure 17.3 Nations vary greatly in their consumption of energy per person. Developed nations, led by the United States, consume the most. The map combines all types of energy, standardized to metric tons of "oil equivalent," that is, the amount of fuel needed to produce the energy gained from combusting one metric ton of crude oil. Data from British Petroleum, *Statistical review of world energy*, June 2003.

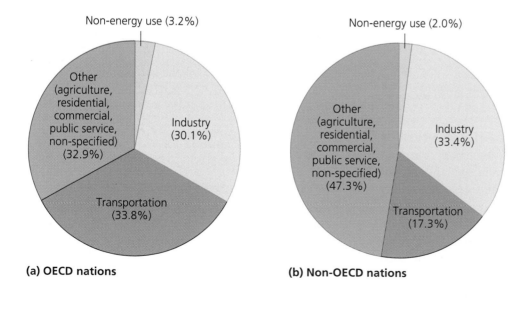

(a) OECD nations

(b) Non-OECD nations

Figure 17.4 Developing nations and developed nations show somewhat different profiles of energy use. These data from 2000 show that developed nations of the Organization for Economic Co-operation and Development (OECD) (a) use nearly twice as much energy for transportation as do non-OECD nations (b), while non-OECD nations use much more energy for agricultural, residential, and other uses. The OECD includes 30 of the most industrialized nations, from Europe and North America plus Japan, Korea, Turkey, Australia, and New Zealand. Non-OECD nations include some large highly industrialized countries such as Russia, China, Brazil, and Argentina, but mostly many smaller and developing nations. Data from International Energy Agency, *Key world energy statistics*, 2002.

CHAPTER SEVENTEEN Nonrenewable energy sources and their environmental impacts **527**

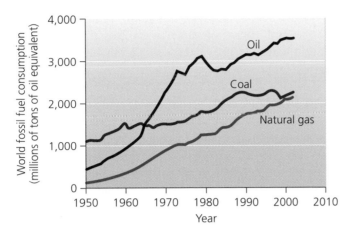

Figure 17.5 Global consumption of fossil fuels has risen greatly over the past half-century. Usage of coal, the leading energy source 50 years ago, has leveled off over the past decade, while usage of natural gas has nearly caught up to it. Oil use rose steeply during the 1960s to overtake coal, and today remains our leading energy source. Data from Worldwatch Institute, *Vital Signs*, 2003.

tissues decomposed and their hydrocarbon compounds were altered and compressed. Most organisms that die do not end up as part of a coal, gas, or oil deposit. A tree that falls in the forest and decays as a rotting log undergoes **aerobic** decomposition; in the presence of air, bacteria and other organisms that use oxygen will break plant and animal remains down into simpler carbon molecules that are recycled through the ecosystem. Fossil fuels are produced only when organic material is broken down in an environment that has little or no oxygen, a condition described as **anaerobic.** Such environments include the bottoms of deep lakes, swamps, and shallow seas. Over millions of years, organic matter that accumulated at the bottoms of such water bodies has been converted into crude oil, natural gas, and coal. Which fossil fuel was formed depends on the chemical composition of the starting material, the temperatures and pressures to which the material was eventually subjected, the presence or absence of anaerobic decomposers, and the passage of time (Figure 17.6).

When little decomposition takes place because the material cannot be digested or appropriate decomposers are not present, then **coal** generally results. Coal is organic matter that was compressed under very high pressure to form dense solid carbon structures. In some pieces of coal it is even possible to see the fossil outlines of plants that died and were buried.

Natural gas is primarily methane, CH_4, which is produced as a by-product when bacteria decompose organic

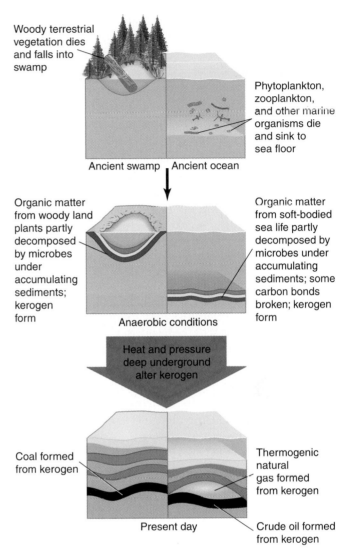

Figure 17.6 The fossil fuels we use for energy today consist of the remains of organic material from plants (and to a much lesser extent, animals) that died millions of years ago. The process begins when organisms die and end up in oxygen-poor conditions, such as when trees fall into lakes and are buried by sediment, or when phytoplankton and zooplankton drift to the seafloor and are thereafter buried. Organic matter that undergoes slow anaerobic decomposition and becomes buried under sediments forms kerogen, the organic source material for the sludge-like substance we know as crude oil. Geothermal heating acts on kerogen to create the oil, and this same thermal process produces natural gas. Natural gas can also be produced nearer the surface by direct bacterial anaerobic decomposition of organic matter. Oil and gas come to reside in porous rock layers beneath dense impervious layers. Coal is formed by plant matter that is compacted so tightly that there is little decomposition.

material under anaerobic conditions. Some natural gas is produced near the surface by bacterial action. Other natural gas results from compression and heat deep below ground, rather than from bacterial action. As organic matter becomes buried more and more deeply under sediments, the pressure exerted by the overlying sediments grows, and temperatures increase. Carbon bonds in the organic matter begin breaking, and the organic matter turns to a substance called kerogen, which acts as a source material for both natural gas and crude oil. Further heat and pressure act on the kerogen to degrade complex organic molecules into various simpler hydrocarbon molecules. At very deep levels (below about 3 km (1.9 mi)) the high temperatures and pressures tend to form natural gas.

The sludgelike liquid we know as **crude oil**, or **petroleum**, tends to form within a window of temperature and pressure conditions often found 1.5–3 km (1–2 mi) below the surface. Crude oil is a mixture of hundreds of different types of hydrocarbon molecules characterized by carbon chains of different lengths (Chapter 4). A chain's length affects its chemical properties, which has consequences for human use, such as whether a given fuel burns cleanly in a car engine. Oil refineries sort the various hydrocarbons of crude oil, separating those intended for use in gasoline engines from those, such as tar and asphalt, used for other purposes.

The crude oil of Alaska's North Slope was formed when dead plant material (and very small amounts of animal material) drifted down through coastal marine waters millions of years ago and was buried in sediments on the ocean floor. These organic remains were then transformed by time, heat, and pressure into the crude oil of today. The shelves of sedimentary rock that now lie beneath Alaska's coastal plain are believed to hold the largest remaining onshore petroleum resource in North America.

Petroleum geologists infer the location and size of fossil fuel deposits

Because fossil fuels form only under certain conditions, they occur concentrated in isolated deposits. Geologists searching for oil, gas, or coal use a number of techniques to map underground rock formations, understand geological history, and predict where fossil fuel deposits might lie. They look not only for old sedimentary rocks, but also for certain arrangements of geologic structures that can trap fossil fuels in particular locations. Geologists use ground surveys, air surveys, drilling of rock cores, or seismic surveys followed by exploratory well drilling to confirm the presence of suspected deposits.

Seismic surveying involves sending sound waves into the ground (by exploding dynamite, thumping the ground with a large weight, or using an electric vibrating machine) and measuring their return to the surface at receiving stations (Figure 17.7). Sound waves reflect off denser layers below ground, but also refract, traveling along denser layers before returning upward. By analyzing how sound waves behave underground, geologists can infer the nature of subterranean structures and draw these structures in cross-sectional seismic profiles. By using such techniques, geologists from the U.S. Geological Survey (USGS) in 1998 estimated, with 95% certainty, the total amount of oil underneath ANWR's 1002 Area to be between 11.6 and 31.5 billion barrels. The geologists' average estimate of 20.7 billion represents their best guess as to the number of barrels of oil the 1002 Area holds.

Some portion of the total amount of oil, whatever it may be, will be impossible to extract using current technology and may have to wait for future technological advances in extraction equipment or methods. Thus, estimates are generally made of "technically recoverable"

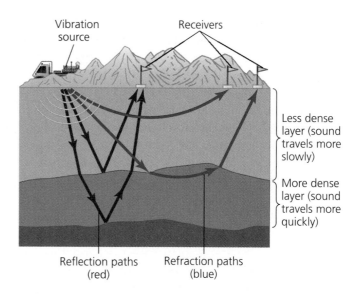

Figure 17.7 Petroleum geologists use seismic surveying to locate promising fossil fuel deposits. One method is to create powerful vibrations at the surface in one location and measure how long it takes the seismic waves produced to reach receivers at other surface locations. Seismic waves travel more quickly through denser layers, and waves may reflect off layers or refract or bend due to density differences in the substrate. Scientists and engineers interpret the patterns of wave reception in order to infer the densities, thicknesses, and location of underlying geological layers — which in turn provide clues about the location and size of fuel deposits.

amounts of fuels. In its 1998 estimates, the USGS calculated technically recoverable amounts of oil under the 1002 Area to be between 4.3 and 11.8 billion barrels, with an average estimate of 7.7 billion barrels. Because certain Native lands and state-owned offshore areas cannot be developed unless the 1002 Area is developed, the USGS also surveyed these areas, and estimated that 1.4-4.2 billion additional barrels likely lay beneath them.

However, oil companies will likely not be willing to extract these entire amounts. Some oil would be so difficult to extract that the expense of doing so would exceed the income the company would receive from the oil's sale. In addition, costs for transporting North Slope oil and natural gas are quite high. Thus the amount a company chooses to drill for will be determined by the expense of extraction and transportation, together with the current price of oil on the world market. Because the price of oil fluctuates, the amount of oil that is "economically recoverable" to extract from a given deposit fluctuates as well. USGS scientists calculated economically recoverable amounts of oil beneath the 1002 Area. At a price of $30 per barrel, 3.0-10.4 billion barrels would be economically worthwhile to recover, they estimated. At prices less than $13 per barrel, none of the oil would be worth extracting.

Thus, technology sets a limit on the maximum that *can* be extracted, whereas economics determines how much *will* be extracted. The amount of a given fossil fuel in a deposit that is technologically and economically feasible to remove under current conditions is termed the **proven recoverable reserve** of that fuel.

Fossil fuel reserves are unevenly distributed

Because fossil fuel deposits are localized and unevenly distributed over Earth's surface, some nations have substantial, proven reserves of fossil fuels while others have very few (Figure 17.8). In addition, the presence of oil, coal, and natural gas is not always correlated, so many nations that are rich in one are poor in the others. How long each nation's fossil fuel reserves will last depends

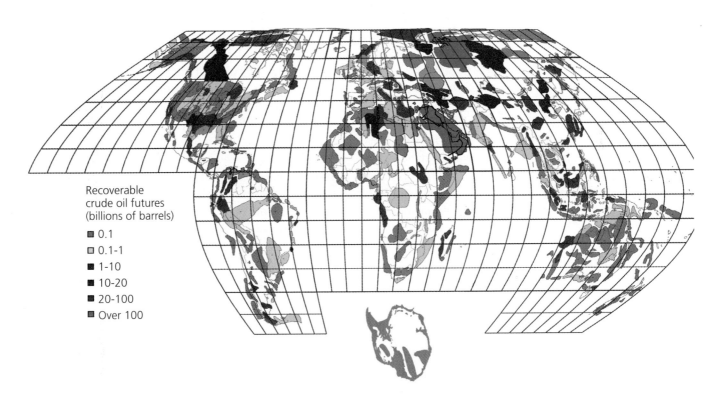

Recoverable
crude oil futures
(billions of barrels)

- 0.1
- 0.1-1
- 1-10
- 10-20
- 20-100
- Over 100

Figure 17.8 Fossil fuel deposits are not spread evenly across Earth but are clustered according to the geological history of each region. Oil exploration so far has enabled the world's petroleum experts to put together a map of estimated reserves of recoverable crude oil across the globe. *Source:* C.D. Masters, D.H. Root, and R.M. Turner, World conventional crude oil and natural gas: Identified reserves, undiscovered resources and futures, United States Geological Survey, 1998.

Table 17.2 Nations with Largest Proven Reserves of Fossil Fuels

Oil (% world reserves)	Natural Gas (% world reserves)	Coal (% world reserves)
Saudi Arabia, 25.0	Russian Federation, 30.5	United States, 25.4
Iraq, 10.7	Iran, 14.8	Russian Federation, 15.9
United Arab Emirates, 9.3	Qatar, 9.2	China, 11.6
Kuwait, 9.2	Saudi Arabia, 4.1	India, 8.6
Iran, 8.6	United Arab Emirates, 3.9	Australia, 8.3
Venezuela, 7.4	United States, 3.3	Germany, 6.7
Russian Federation, 5.7	Algeria, 2.9	South Africa, 5.0
United States, 2.9	Venezuela, 2.7	Ukraine, 3.5
Libya, 2.8	Nigeria, 2.3	Kazakhstan, 3.5
Nigeria, 2.3	Iraq, 2.0	Poland, 2.3

Data from British Petroleum, Statistical review of world energy, 2003.

on how much the nation extracts and produces, how much it consumes, and how much it imports from and exports to other nations. Nearly two-thirds of the world's proven reserves of crude oil are in the Middle East. The Middle East is also rich in natural gas, but Russia contains more than twice as much natural gas as any other country. Russia is also rich in coal, as is China, but the United States possesses more coal than any other nation (Table 17.2).

Oil

Oil has been the world's most-used fuel since the 1960s, when it eclipsed coal. It now accounts for two-fifths of the world's commercial energy consumption. Its use worldwide over the past decade has grown over 1.1% per year—slightly slower than natural gas but much faster than coal.

The age of oil began in the mid-19th century

Historical evidence suggests that people used solid forms of oil (such as tar and asphalt) as long ago as 4,000 B.C., taking them from deposits that were easily accessible at Earth's surface. However, the modern extraction and use of petroleum for energy began in the 1850s. In Pennsylvania, miners drilling for salt occasionally encountered oily rocks instead. At first, entrepreneurs bottled the crude oil from these deposits and sold it as a healing aid, unaware that crude oil is carcinogenic when applied to the skin and poisonous when ingested. Soon, however, a Dartmouth College scholar named George Bissell realized that this so-called rock oil could be used to light lamps and lubricate machinery, and in 1854 Bissell started the Pennsylvania Rock Oil Company. By the time the firm developed its drilling technology and struck oil, Bissell was no longer at the company. Instead, Edwin Drake is credited with drilling the world's first oil well, in Titusville, Pennsylvania, in 1859 (Figure 17.9 and see Figure 17.2).

We drill to extract oil

As geological knowledge has advanced, so have our abilities to locate oil reserves hidden from sight. Geologists' educated reading of the structure and history of subterranean rock to estimate likely repositories of oil is extremely important for petroleum companies, which invest millions of dollars to drill at the particular sites geologists recommend. Oil generally forms from organic source material buried under kilometers of sedimentary rock. Geothermal heating at this depth can separate hydrocarbons from the source material. While too much heating will destroy these hydrocarbons, the right amount will produce the mixture known as crude oil. This liquid oil migrates upward through rock pores, sometimes assisted by seismic faulting. It tends to collect in porous layers beneath dense impermeable layers. Once geologists have identified a location likely to contain a profitable oil field, the oil company will typically conduct exploratory drilling. Holes drilled during this phase are usually small in circumference and often descend to great depths. If oil is encountered in adequate quantities, extraction begins.

Although we may picture oil deposits as large black underground lakes, most oil deposits actually consist of small droplets adhering to the surfaces of holes in porous rock, like a sponge whose pores are full of oil. Just as you would squeeze a sponge to remove its liquid, pressure is required in order to extract oil from porous

VIEWPOINTS

Drilling for Oil in ANWR
Should we drill for oil in the Arctic National Wildlife Refuge?

Preserving the Refuge More Valuable Than Short-Term Oil Fix

It depends on what you value more: a 6-month supply of oil or one of the last pristine wilderness areas on earth. According to the USGS, the lower estimate for economically recoverable oil (95% chance of success at $30 per barrel) is about 3.2 billion barrels, amounting to about 6 months of U.S. consumption. We could easily match this amount of oil with improvements in automotive fuel efficiency and using energy resources more conservatively. However, once the wilderness is gone, we can't get it back.

Given the increase of greenhouse gas emissions into the atmosphere, we should concentrate on developing alternative energy, not extracting more oil. The only sure winner from opening ANWR to drilling would be the multinational oil corporations that benefit by maintaining America's dependence on oil.

The ANWR is teeming with large herds of migrating ungulates, predator-prey interaction, and millions of birds migrating to their nesting grounds in tundra flats. Oil development (such as that found on neighboring Prudhoe Bay) would include not just the "small footprint" offered by proponents but include a vast myriad of pipelines, roads, airfields, and drilling complexes spread across the coastal plain. The Gwich'in Indians refer to the coastal plain as the heart of the Refuge because it provides seasonally critical habitat to nesting birds and to calving and post-calving caribou, denning habitat for maternal polar bears, and year-round habitat to a recovering population of muskox.

Disruption of the native flora and fauna by oil development would be inevitable. We must not yield to this "quick fix" scheme for meeting our energy needs. Instead, we must look at the big picture and apply our ingenuity, technical prowess, and the power of the market to harness new and less destructive alternatives for meeting our energy needs. Doing so will allow us to have our energy and the Arctic too.

Mark L. Shaffer is senior vice president of programs for Defenders of Wildlife. He supervises habitat and species conservation activities for a growing regional staff in eight states.

More Domestic Oil Means Greater American Energy Independence

Environmentalists will find a reason to oppose drilling anywhere. I once asked a Sierra Club spokesman where he would support drilling in lieu of drilling in ANWR, and he couldn't name a place.

So beware when environmental advocates invoke ANWR's caribou as a reason not to drill on what they call "America's Serengeti"—a phrase that fits ANWR primarily because most Americans are as unlikely to visit ANWR as they are to visit the African plains. Under a House proposal, no more than 2,000 acres of ANWR's 19 million acres would be developed. The rest would remain pristine.

Critics point to a March 2002 U.S. Geological Survey that supposedly undercuts President Bush's case for drilling in ANWR. However, the report wasn't as damaging as some reported it to be. As *The Washington Post* noted, the report "argued that the effects of drilling could be minimal if conducted in a sensitive manner over a limited area. Two of its five scenarios for caribou showed negligible effects."

Indeed, there are nine times as many caribou at Prudhoe Bay, where they drill for oil, than there were in 1974, according to George Ahmaogak, the Inupiat Eskimo mayor of the North Slope Borough.

In the end, the environmental argument rests on an insistence that Americans use less energy. Consumers should conserve more, it's true. But even with conservation, in the real world, Americans will still drive to work and businesses will rely on cheap transportation. Petroleum is the fuel that keeps America's economy humming. So the oil has to come from somewhere. Those who truly care about the planet ought to prefer highly regulated drilling in Alaska to under-regulated drilling in foreign lands.

Debra J. Saunders is a syndicated columnist for the *San Francisco Chronicle.* She has written for the *Wall Street Journal,* the *National Review, The Weekly Standard, Reader's Digest,* and *Reason* magazine. She has also appeared on "Politically Incorrect," CNN, BBC radio, and "The News Hour" on PBS. Her book, *The World According to Gore,* was published by Encounter Books in 2000.

Figure 17.9 The world's first commercial oil well was drilled at Titusville, Pennsylvania in 1859. Over the next 40 years, Pennsylvania's oil fields produced half the world's oil supply and helped establish a fossil-fuel-based economy that would hold sway for decades to come.

rock. The oil in deposits is typically already under pressure from one of several sources. It may be pressured from above by the weight of rock, from below by groundwater, from above by trapped gas, or internally from natural gas dissolved in the oil. All these forces are held in place by the surrounding rock until drilling punctures the rock. Therefore, when a drill reaches an oil deposit and relieves the pressure, oil will often rise to the surface of its own accord.

Once pressure is relieved, oil becomes more difficult to extract, and may need to be pumped out. Because oil occurs in small droplets, much of it remains stuck to rock surfaces even after pumping. As much as two-thirds of a deposit may remain in the ground after **primary extraction,** the initial drilling and pumping of available oil (Figure 17.10a). Companies may then begin **secondary extraction,** in which solvents are used or underground rocks are flushed with water or steam to remove additional oil (Figure 17.10b). At Prudhoe Bay, for example, large quantities of seawater are piped in from the coast and pumped into wells to flush out remaining oil. Even after secondary extraction, however, quite a bit of oil can remain; we lack the technology to remove every last drop of oil. Secondary extraction is more expensive than primary extraction. Thus, many U.S. deposits did not undergo secondary extraction when they were first drilled be-

cause the price of oil was too low to make the procedure economical. When oil prices rose in the 1970s, however, many deposits were reopened for secondary extraction.

Offshore drilling produces much of our oil

Drilling for oil and natural gas takes place not just on land but also under many meters of ocean water, in the seafloor on the continental shelves. This is because petroleum deposits are buried under sedimentary rock that may today be located either on land or under the sea. Offshore drilling has required the development of technology that can withstand the considerable forces of wind, waves, and ocean currents. Some drilling platforms are fixed standing platforms built with unusual strength. Others are floating platforms that are resilient and are anchored in place above the drilling site. Offshore oil drilling provides a substantial portion of the world's petroleum. Over 25% of the crude oil and natural gas extracted in the United States comes from offshore drilling sites, primarily in the Gulf of Mexico and off the southern California coast. Offshore sites are estimated to contain about 15% of the nation's proven reserves of oil and natural gas and over half of the remaining undiscovered reserves of these fossil fuels.

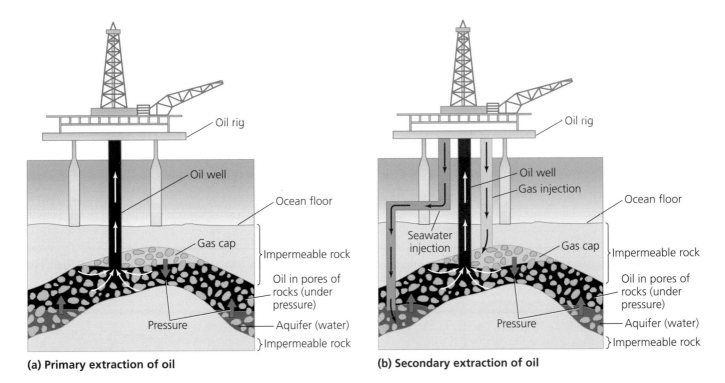

(a) Primary extraction of oil

(b) Secondary extraction of oil

Figure 17.10 The more oil is extracted from a deposit, the more difficult it becomes to extract oil further. Extraction therefore often takes place in two stages. In primary extraction (a), oil is drawn up through the well by keeping pressure at the top lower than pressure at the level of the oil deposit. Once the pressure in the deposit drops, however, new material must be injected into the deposit in order to increase the pressure. Thus, secondary extraction (b) involves injecting seawater just beneath the oil, and/or gases just above the oil, in order to force the remaining oil up out of the deposit.

Oil and petroleum products have many uses

We don't directly use the raw crude oil extracted from the ground. Rather, it must be put through refining processes first (see The Science behind the Story). Because crude oil is a complex mix of hydrocarbons, it has been possible to create many types of petroleum products by separating different components. Initially the primary oil product was kerosene, used for lighting, heating, and cooking. With the development of the internal combustion engine in the late 1800s, the market for petroleum products began to grow. Since the 1920s, refining techniques and chemical manufacturing have greatly expanded uses of petroleum to include a wide array of products (Figure 17.11). We use certain components of crude oil, such as gasoline and diesel oil, to fuel cars, trucks, buses, and ships. Jet fuel powers our airplanes. We use other components to make lubricants, plastics, pharmaceuticals, and fertilizers. Various grades of petroleum are processed into a wide array of consumer products, from computers to polyester shirts to

Figure 17.11 A bicyclist may not use gasoline to power her bicycle but may use a number of other products derived in part from refined components of crude oil, including her water bottle, her clothing, her sunglasses, her helmet, and grease for the chain and gears.

How Crude Oil Is Refined

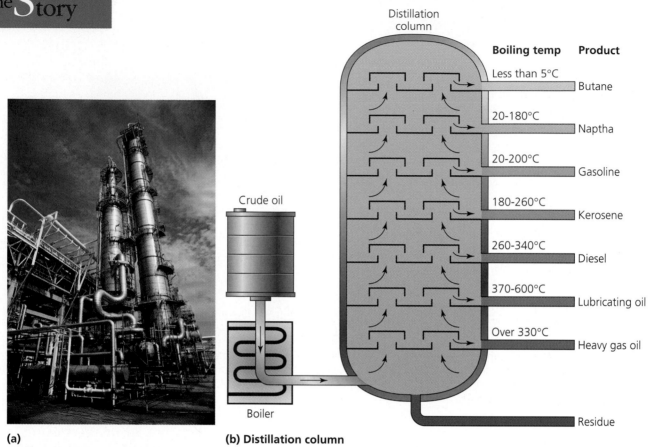

Boiling temp	Product
Less than 5°C	Butane
20-180°C	Naptha
20-200°C	Gasoline
180-260°C	Kerosene
260-340°C	Diesel
370-600°C	Lubricating oil
Over 330°C	Heavy gas oil
	Residue

Distillation column

Crude oil

Boiler

(a)

(b) Distillation column

Crude oil is shipped to petroleum refineries (**a**), where it is refined into a number of different types of fuel. Crude oil is boiled, causing its many hydrocarbon constituents to volatilize and proceed upward through a distillation column (**b**). Those constituents that boil only at the highest temperatures and condense readily once the temperature drops will condense at low levels in the column. Those constituents that volatilize readily at lower temperatures will continue rising through the column and condense at higher levels where temperatures are lower. In this way heavy oils (generally consisting of long hydrocarbon molecules) are separated from lighter oils (generally those with short hydrocarbon molecules).

Crude oil is a complex mixture of thousands of different kinds of hydrocarbon molecules. Through the process of refining, these hydrocarbon molecules are separated into different size classes and chemically transformed to create specialized fuels for heating, cooking, transportation, and energy produc-

tion, as well as lubricating oils, asphalts, and the precursors of plastics and other petrochemicals. To maximize the production of marketable products while minimizing negative environmental impacts, petroleum engineers have developed a variety of refining techniques.

The first step in processing crude oil is distillation, or fractionation. This process depends on the fact that different components of crude oil boil at different temperatures. In refineries, the distillation process takes place in tall columns filled with perforated horizontal trays. When heated crude oil is introduced into the column, lighter components rise as vapor to the upper trays, condensing into liquid as they cool, while heavier components drop to the lower trays. Light gases, such as butane, boil at less than 32° C (90° F), and heavier oils, such as industrial fuel oil, boil only at temperatures above 343° C (650° F).

Since the early 20th century, light gasoline, which is used in automobiles, has been in much higher demand than heavier hydrocarbons and most other derivatives of crude oil. The demand for high-performance, clean-burning gasoline has also risen. To meet these demands, refiners have developed several techniques to convert heavy hydrocarbons into gasoline.

The general name for processes that convert heavy oil into lighter oil is cracking. One of the simplest methods is thermal cracking, in which long-chained molecules are broken into smaller chains by heating in the absence of oxygen. (The oil would ignite if oxygen were present.) Catalytic cracking, a related method, uses catalysts—substances that promote chemical reactions without being consumed by them—to control the cracking process. The result is an increase in the amount of a desired product from a given amount of heavy oil. Today, the most widely used form of cracking is fluidized catalytic cracking, in which a finely powdered catalyst that behaves like a fluid is fed continuously into a reaction chamber with heavy oils. In the reaction chamber, the catalyst facilitates chemical reactions that transform heavy oils into lighter

oils and gases. The products of cracking are then fed into a distillation column.

Refiners can also change the chemical composition of oil through a process called catalytic reforming. Like catalytic cracking, catalytic reforming uses catalysts to promote a chemical reaction that produces a given by-product of heavy oil. In the case of catalytic reforming, the goal is to chemically transform certain hydrocarbons that are slightly heavier than gasoline so that they can be blended with gasoline to obtain higher octane ratings. The octane rating reflects the amount of compression gasoline can undergo before it spontaneously ignites. An octane rating of 92 indicates that a gasoline blend is equivalent to a mixture of 92% octane, which spontaneously ignites only at very high compression levels, and 8% heptane, which ignites more easily. Other things being equal, high-compression engines are more powerful than low-compression engines, so gasolines that can withstand high levels of compression—that have high octane ratings—are preferred. As of 2001, an estimated 30–40% of gasoline used in the United States was produced through catalytic reforming.

High-octane gasoline can also be produced by combining smaller hydrocarbons to synthesize larger molecules through the process of alkylation. In this process, molecules such as isobutylene and isobutene, which each have four carbon atoms, are combined in the presence of catalysts to form

molecules with eight carbon atoms. The resulting high-octane alkylate can then be blended with lower-octane gasoline.

In addition to distilling crude oil and altering the chemical structure of some of its components, refineries also remove contaminants. Sulfur and nitrogen compounds, which can be harmful when released into the atmosphere, are the two most common contaminants in crude oil. Government regulations stemming from legislation such as the Clean Air Act have forced refiners to develop methods of removing such contaminants, particularly sulfur.

One of these methods is hydrotreating, or hydrodesulfurization, in which oil and hydrogen gas are combined at moderate temperatures (260–430°C, or 500–800°F) in the presence of a catalyst. During hydrotreating, sulfur and hydrogen combine to form a toxic gas, hydrogen sulfide (H_2S), some of the nitrogen is converted to ammonia (H_3N), and metals in the oil are deposited onto the catalyst. A method known as the Claus process is then used to convert the toxic gas into water and metallic sulfur, which can be recycled for use in various industrial processes. Together with a secondary cleaning unit, the Claus process can remove more than 98% of the incoming sulfur. Any remaining gas is burned and vented to the atmosphere through stacks. Regulations require these stacks to be tall enough to minimize environmental impact in the local area.

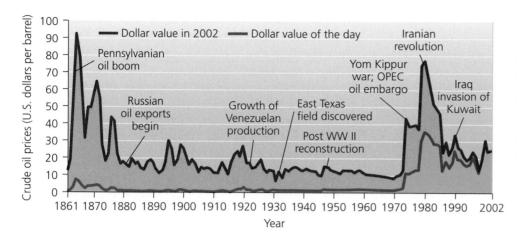

Figure 17.12 World oil prices have gyrated greatly over the decades, with political and economic events in oil-producing countries exerting the most influence. The greatest price hikes in recent times have resulted from wars and unrest in the oil-rich Middle East. Data from British Petroleum, Statistical review of world energy, 2003.

nylon jackets to dishwashing liquid. Oil is also used as a fuel in some power plants and to heat 10% of U.S. homes. Of the various ways in which oil is used, transportation represents the largest share (58% globally in 2000). Because of its low perceived cost and ease of use, oil is nearly our sole fuel source for transportation.

Oil supply and prices affect the economies of nations

Oil is the substance that lubricates the world's economy. So many of our modern technologies and services depend on oil that nations, corporations, and institutions that control the trade in oil exercise extraordinary power. The "energy crisis" of 1973–1974 in the United States demonstrated how the price of oil can affect U.S. government policies and the energy-using habits of the nation.

By 1973, domestic U.S. sources of oil were peaking, and the nation was importing more and more of its oil, depending on a constant flow from abroad to keep cars on the road and machines running. In addition, at that time a greater percentage of homes and electrical plants were run on petroleum than today. Then, in 1973, the predominantly Arab nations of the Organization of Petroleum Exporting Countries (OPEC) resolved to stop selling oil to the United States. The move was prompted by OPEC's desire to raise prices by restricting supply, and its opposition to U.S. support of Israel in the Arab-Israeli Yom Kippur War. The embargo created panic in the West and caused oil prices to skyrocket (Figure 17.12). Short-term oil shortages drove American consumers to wait in long lines at gas pumps.

In response to the embargo, the U.S. government enacted a series of policies designed to reduce reliance on foreign oil. These included developing additional domestic sources (such as those on Alaska's North Slope), resuming extraction at sites shut down after primary ex-

traction had ceased being cost-effective, capping the price that domestic producers could charge for oil, and beginning to import oil from a greater diversity of nations. The government also established a stockpile of oil as a short-term buffer against future shortages. Stored underground in salt caverns in Louisiana, this is called the *Strategic Petroleum Reserve*. Currently the Reserve contains over 600 million barrels of oil; at present rates of U.S. consumption (19.7 million barrels/day), this is equivalent to one month's supply.

Conservation in the United States has been a function of economic need

The policies enacted in response to events in 1973 also included conservation measures, such as a mandated increase in the mile-per-gallon (mpg) fuel efficiency of automobiles, a reduction in the national speed limit to 55 miles per hour (at that time, the most efficient speed to drive a car), and funding of research into non-oil-based energy sources, such as solar power.

Thirty years later, many of the conservation policies developed after the 1973 oil crisis have been abandoned. Without high market prices and an immediate threat of shortages, people lack motivation to conserve. Government funding for research into alternative energy sources has decreased, speed limits have increased, and a bill to raise the mandated average fuel efficiency of vehicles to 35 mpg recently failed in Congress. The average fuel efficiency of new vehicles has fallen from a high of 22.1 mpg in 1988 to 20.8 mpg in 2003 (Figure 17.13). This decrease is due to increased sales of light trucks (averaging 17.7 mpg), including sports utility vehicles, relative to cars (averaging 24.8 mpg). The failure to improve fuel economy of vehicles over the past 20 years, despite the existence of technology to do so, has added greatly to U.S. oil consumption. Transportation

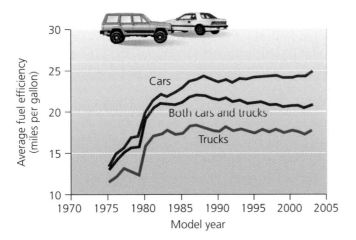

Figure 17.13 Fuel-efficiency for automobiles in the United States rose dramatically in the late 1970s as a result of legislative mandates but has declined slightly since 1988. The decline is due to a lack of further legislation for improved fuel economy and to the increased popularity in recent years of sports-utility vehicles. Data from Light-duty automotive technology and fuel economy trends, Environmental Protection Agency, April 2003.

accounts for 67% of U.S. oil use, and passenger vehicles consume more than half of this energy.

Weighing the Issues:
More Miles, Less Gas

If you drive an automobile, what gas mileage does it get? How does it compare to the vehicle averages in Figure 17.13? If your vehicle's fuel efficiency were 10 mpg greater, how many gallons of gasoline would you no longer need to purchase each year? How much money would you save? If all U.S. vehicles were mandated to increase fuel efficiency by 10 mpg, how much gasoline do you think the over 200 million Americans who drive could conserve? What other strategies can you think of to conserve fossil fuels, and how might they compare in effectiveness to a rise in fuel efficiency standards?

Many critics of oil drilling in the ANWR point out the vast amounts of oil wasted by fuel-inefficient automobiles and argue that a small amount of conservation would save the nation far more oil than it would ever obtain from ANWR. Indeed, the average estimate for recoverable oil in the 1002 Area, 7.7 billion barrels, represents just over one year's supply for the United States at current consumption rates. Spread over a period of extraction of many years, the proportion of U.S. oil needs that ANWR would fulfill appears strikingly small (Figure 17.14).

Oil consumption in the United States shows little sign of abating. The United States consumes over one-fourth of the world's oil, despite the fact that the nation's proven reserves are only one-ninth of Saudi Arabia's and 1/35th of global reserves, and although it produces only one-tenth of the world's oil supply. U.S. consumption has increased by 15.7% over the past decade, slightly outpacing the rest of the world. Table 17.3 shows the top oil-producing and oil-consuming nations.

We may have already depleted half our oil reserves

Levels of consumption are such that many experts calculate that we are already roughly midway through our exploitation of the world's oil reserves. In 1956, geologist

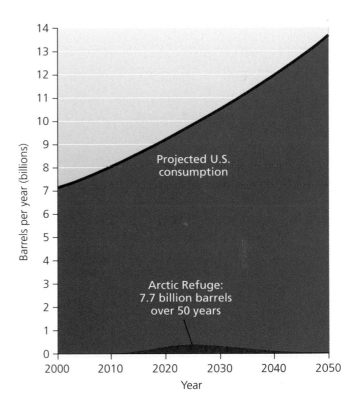

Figure 17.14 Opponents of oil drilling in the Arctic National Wildlife Refuge contend that the amount of oil estimated to be recoverable would make only a small contribution toward the overall U.S. oil demand. In this graph, the best USGS estimate of oil from ANWR's 1002 Area is shown in red, in the context of total U.S. oil consumption if current consumption trends are extrapolated into the future. The actual ANWR contribution, if it comes to pass, would depend greatly on the amount of oil actually present under ANWR, the time it would take to extract it, and future trends in consumption. Adapted from Oil and the Arctic National Wildlife Refuge, Natural Resources Defense Council, April 2002; Arctic national wildlife refuge, 1002 area, petroleum assessment, 1998, including economic analysis, United States Geological Survey, April 2001.

Table 17.3 Top Producers and Consumers of Oil

Production (% world production)	Consumption (% world consumption)
Saudi Arabia, 11.8	United States, 25.4
Russian Federation, 10.7	China, 7.0
United States, 9.9	Japan, 6.9
Mexico, 5.0	Germany, 3.6
China, 4.8	Russian Federation, 3.5
Iran, 4.7	South Korea, 3.0
Norway, 4.4	India, 2.8
Venezuela, 4.3	France, 2.6
Canada, 3.8	Italy, 2.6
Great Britain, 3.3	Canada, 2.5

Data from British Petroleum, Statistical review of world energy, 2003.

M. King Hubbert calculated that U.S. oil production would peak about 1970. His prediction was ridiculed at the time, but it proved to be accurate; U.S. production did peak, and it continues to fall. In 1974 Hubbert used the same type of analysis of technology, economics, and geology to predict that global oil production would peak in 1995. We may just now be witnessing the global peak; after years of increases, global oil production fell slightly in 2001 and again in 2002 (Figure 17.15).

Based on actual production data, experts estimate that at current levels of exploitation, world supplies of recoverable crude oil will last about another 40 years. Supplies of natural gas are projected to last a bit longer, roughly 60 years. Supplies of coal, meanwhile, could last 200–250 years.

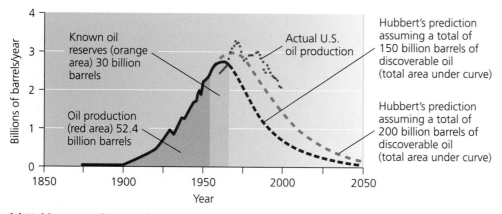

(a) Hubbert's prediction of peak in U.S. oil production and actual data

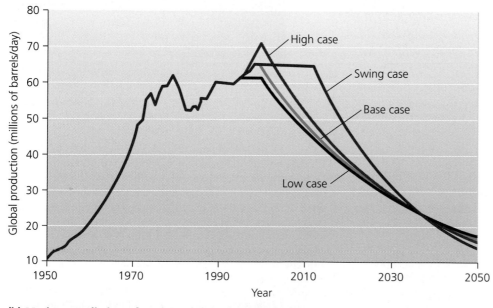

(b) Modern prediction of peak in global oil production

Figure 17.15 Because fossil fuels are a nonrenewable resource, oil supplies at some point pass the midway point of their depletion, and annual production begins to decline. The peak in U.S. production was predicted by geologist M. King Hubbert decades ago and is referred to as "Hubbert's peak" (a). The actual U.S. peak came in the early 1970s only a few years after Hubbert's predicted date. Today many analysts believe global production is about to peak. Shown (b) is a projection from 1996 forecasting a global peak around the turn of the century. Four different scenarios are modeled, each of which predicts an end to growth around 2000 and falling production after that. Go to **GRAPH IT** on the website or CD-ROM. Data from (a) Kenneth S. Deffeyes, *Hubbert's Peak: The Impending World Oil Shortage*, 2001. (b) Colin Campbell, *The Twenty-first Century: The World's Endowment of Conventional Oil and Its Depletion*, 1996.

Coal

Coal is the world's most abundant fossil fuel. The proliferation 300–400 million years ago of swampy environments where organic material could be deposited has resulted in substantial coal deposits throughout the world. Fully one-quarter of the world's coal is located in the United States, and coal provides for one-quarter of the world's commercial energy consumption.

Coal use has a long history

Coal has been used longer than any other fossil fuel. In places where coal veins are exposed at Earth's surface, people have been burning coal for thousands of years. In Arizona, today's Hopi Indians, like their ancestors, use coal to fire pottery, cook food, and heat their homes. The Romans used coal as a heat source in the second and third centuries in Britain, as have people in parts of China for 2,000–3,000 years. Once commercial mining began in the 1700s, coal began to be used widely as a heating source. Coal soon found an expanded market after the invention of the steam engine, as it was used to boil water to produce steam. Coal-fired steam engines performed work previously done by people or horses, including manufacturing, harvesting, and powering transportation such as trains and ships. The birth of the steel industry in 1875 increased demand even further because coal fueled the furnaces used to make steel.

Beginning in the 1880s, people began to put coal to use in generating electricity. In coal-fired power plants, coal combustion converts water to steam, which turns a turbine to create electricity (see The Science behind the Story). Coal provided over half the electrical generating capacity in the United States in 2000 and supplied 43% of electrical power plants. Coal continues to be popular in the United States because the country has such large reserves of it, and it is relatively cheap. The market price of coal, however, does not reflect the full costs of extracting and using coal because it does not take into account external costs (Chapter 2), such as impacts on human health and the environment. Today China and the United States are the biggest producers and consumers of coal (Table 17.4).

Coal varies in its qualities

Coal varies from deposit to deposit in many ways, including the amount of potential energy it contains, its water content, and the amount of sulfur contaminants it contains (Figure 17.16). Scientists classify coal into

Table 17.4 Top Producers and Consumers of Coal	
Production (% world production)	Consumption (% world consumption)
China, 29.5	China, 27.7
United States, 24.0	United States, 23.1
Australia, 7.7	India, 7.5
India, 7.1	Japan, 4.4
South Africa, 5.3	Russian Federation, 4.1
Russian Federation, 4.8	Germany, 3.5
Poland, 3.0	Poland, 2.4
Indonesia, 2.7	Australia, 2.1
Germany, 2.3	South Korea, 2.0
Ukraine, 1.8	Ukraine, 1.6

Data from British Petroleum, Statistical review of world energy, 2003.

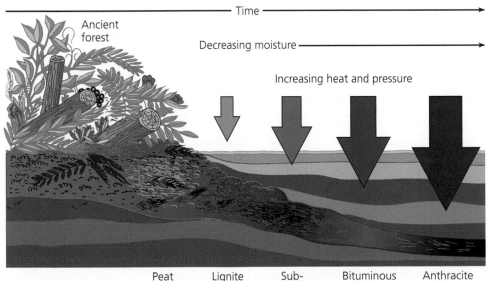

Time

Decreasing moisture

Increasing heat and pressure

Ancient forest

Peat   Lignite   Sub-bituminous   Bituminous   Anthracite

Figure 17.16 Coal is formed as ancient plant matter is compacted below ground. Scientists categorize coal into several different types, depending on the amount of heat, pressure, and moisture involved in its formation. Anthracite coal is formed under greatest pressure, where temperatures are high and moisture content is low. Lignite coal is formed under conditions of much less pressure and heat, but more moisture. The peat that makes up bogs in parts of the British Isles is also part of this continuum, representing plant matter that is minimally compacted.

How Electricity is Generated

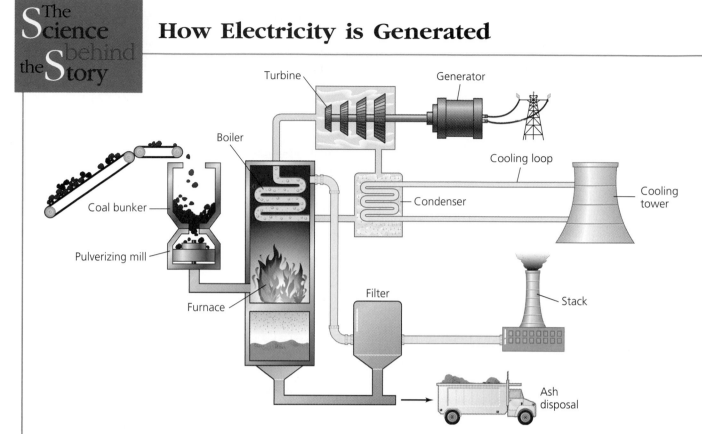

Coal is the primary fuel source used to generate electricity in the United States. Pieces of coal are pulverized and blown into a high-temperature furnace. Heat from the combustion boils water, and the resulting steam turns a turbine, generating electricity by passing magnets past copper coils. The steam is then cooled and condensed in a cooling loop and returned to the furnace. "Clean coal" technologies help filter out pollutants from the combustion process, and toxic ash residue is disposed of in hazardous waste disposal sites.

As the United States tries to balance its growing demand for electricity with rising concerns about the environmental and health impacts of coal combustion, power plants are continuing to rely heavily on coal while scientists work to limit the pollution that use of this fuel creates.

Every new U.S. housing subdivision, cell phone, and DVD player means a greater draw on the country's power grid. Between 1970 and 1998, electricity consumption in the United States jumped by 133%, according to federal statistics, and was expected to increase another 34% from 1998 to 2020. To generate this much electricity using domestic resources, coal has emerged as a central factor in recent national energy policy.

Coal is used to generate electricity in a process that dates back more than a century. Once mined, coal is hauled to power plants, where it is pulverized to ready it for combustion. The crushed coal is blown into a boiler furnace on a super-heated stream of air and burned in a blaze of intense heat—typical furnace temperatures often flare at 815°C (1500° F). Water, circulating around the boiler, absorbs the heat and is converted to high-pressure steam. The steam is injected into a turbine, a type of rotary device that converts the kinetic energy of a moving substance such as steam into mechanical energy. The turbine's key components are a drive shaft and a series of fan-like blades attached to the drive shaft. As steam from the boilers exerts pressure on the blades of the turbine, they spin, turning the drive shaft.

The drive shaft, in turn, is connected to a generator, which also features two important components—a rotor that rotates and a stator that remains stationary. Different types of generators have different designs, but they all make use of the same principle: Moving magnets and coils of copper wire alongside one another will cause electrons in the copper wires to move and thus generate alternating electric current. In some generators the rotor consists of magnets and spins within a stator of coiled copper wire. As the turbine's drive shaft rotates, it causes the rotor to revolve, creating a magnetic field and causing electrons in the stator's copper wires to move, creating an electrical current. This current flows into electrical transmission lines that travel from the power plant out to the customers who use the plant's electricity.

Because coal is the country's largest domestic fossil fuel source but also a major source of pollution, U.S. policymakers have arrived at a dual approach: to continue relying on coal while trying to make its use less toxic. A national energy policy moving toward legislative passage in 2003 set aside $2 billion for research into further development of "clean coal" technologies. These efforts largely focus on approaches to rid the coal-to-electricity process of toxic chemicals, either before or after the coal is burned. Some such technologies focus on "scrubbers," or materials based on minerals such as calcium or sodium that absorb and remove sulfur dioxide from smokestack emissions. Others use chemical reactions to strip away NO_X, breaking it down into elemental nitrogen and water. Still more innovations use multi-layered filtering devices to catch tiny particles of ash before they escape.

Research has driven growing interest in such innovations. One study on removing sulfur dioxide and NO_X from emissions at the Niles Station power plant in Niles, Ohio, found a way to remove about 95% of sulfur dioxide and 90% of nitrogen oxides from the plant's emissions. In the study, conducted from 1988 to 1996, researchers began by installing smokestacks with more efficient filter bags, leading to the capture of significant amounts of ash and gases before emission. The filters also proved effective at removing dangerous substances such as mercury and selenium. The captured ash and gases were reheated, and small amounts of ammonia were added to chemically convert the NO_X into harmless nitrogen gas and water vapor. Further addition of oxygen to the sulfur dioxide in the smokestack gas converted the sulfur compounds into sulfuric acid, which the plant bottled and sold for use in fertilizer and manufacturing processes. Not only did the gas-scrubbing process work well enough for the plant to begin using permanently, but several power plants in Europe have adopted the technology.

Some energy analysts and environmental advocates question such a policy emphasis on clean coal, however. Coal, they maintain, is an inherently dirty way of generating power and should be replaced outright with cleaner energy sources. Others doubt that the George W. Bush Administration's push for clean coal technologies represents a true commitment to cleaner air from coal, since the administration also eased Clean Air Act requirements that would have forced power companies to upgrade their pollution control technology. Without mandates, many power generators have been slow to adopt clean coal technology on their own. However, the research will press ahead, as lawmakers seemed set in late 2003 to approve funding for more than 35 clean coal studies, slated to be conducted over a ten-year period.

four types; from least to most energy-rich, these are lignite, sub-bituminous, bituminous, and anthracite. Organic material that is broken down anaerobically but remains wet, near the surface, and not well compressed may be called **peat.** A kind of precursor stage to coal, peat has been widely used as a fuel in Britain and other locations in the world. As peat further decomposes, as it becomes buried more deeply under sediments, as pressure and heat increase, and as time passes, water is squeezed out of the material, and carbon compounds are packed more tightly together, forming coal. Lignite is the least-compressed type of coal, and anthracite is the most-compressed type. The greater the compression, the greater is the energy content per unit volume.

In each of these coal types, sulfur content varies, depending in part on whether the coal was formed in freshwater or saltwater sediments. Coal in what is today the eastern United States is high in sulfur because it was formed in marine sediments, where sulfur from seawater was present. When high-sulfur coal is burned, it produces sulfate air pollution, leading to episodes such as the industrial smog of Donora, Pennsylvania, and to chronic acid precipitation problems downwind of coal-fired power plants (Chapter 11). Scientists are seeking ways to clean coal of sulfur content prior to burning so that it can continue to be used as an energy source while minimizing impact on the environment (see The Science behind the Story).

Coal is mined from the surface and from below ground

Different methods of extracting coal may be used (Figure 17.17). To reach subsurface deposits, shafts are dug deep into the ground, and networks of tunnels are dug or blasted out to follow coal seams. The coal is removed systematically and shipped to the surface. When coal deposits are at or near the surface, strip-mining methods are used. In **strip-mining,** heavy machinery removes huge amounts of earth to expose the coal, which is mined out directly. The holes are then refilled with the soil that had been removed. Strip-mining operations can occur on gigantic scales; in some cases entire mountaintops are lopped off. This environmentally destructive process, called mountaintop removal, has become more common recently in parts of the Appalachian Mountains.

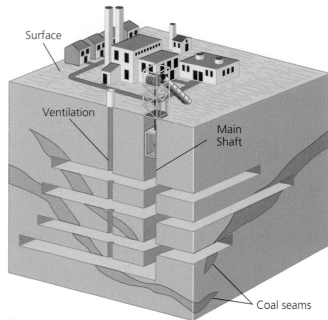

(a) Subsurface mining

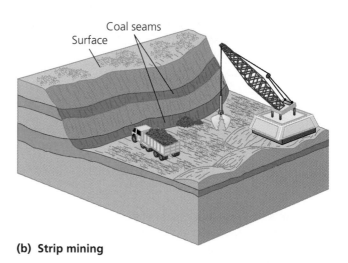

(b) Strip mining

Figure 17.17 Coal is mined in two major ways. In subsurface mining **(a),** miners work below ground in shafts and tunnels blasted through the rock, passageways that provide access to underground seams of coal. This type of mining poses dangers and long-term health risks to miners. In strip-mining **(b),** soil is removed from the surface, exposing the coal seams from which coal is then mined. This type of mining can cause substantial environmental impact to surrounding areas.

Natural Gas

Coal production has leveled off in recent years. Natural gas, in contrast, is the fastest-growing fossil fuel in use today. It now provides, like coal, for one-quarter of global commercial energy consumption.

Natural gas was long known but has only recently been widely used in homes

Naturally occurring seeps of natural gas had occasionally been ignited by lightning and could be seen burning in parts of what is now Iraq, inspiring the Greek essayist Plutarch around A.D. 100 to describe their "eternal fires." Such natural seepages of gas from underground deposits through cracks and passages were used to light nearby streets and buildings in parts of the United States in the 1800s.

The first commercial extraction of natural gas took place in 1821, but during much of the 19th century its use was localized because technology did not exist to pipe gas safely over long distances. Natural gas was used to fuel lamps along city streets, but when electric lights replaced most gas lamps in the 1890s, gas companies began marketing gas for heating and cooking. The first long-distance gas pipeline was built in 1891 to carry gas from deposits in Indiana to homes 195 km (120 mi) away in Chicago, Illinois. However, natural gas did not come to replace coal and oil in homes until after World War II, when wartime improvements in welding and pipe-building made the transport of gas safer and more economical. During the 1950s and 1960s, thousands of miles of underground pipelines were laid throughout the United States. If laid end to end, our current network of natural gas pipelines would extend to the moon and back twice.

Natural gas is formed in two ways

Natural gas can arise from either of two processes. **Biogenic** gas is created at shallow depths by the anaerobic decomposition of organic matter by bacteria. An example is the "swamp gas" you can sometimes smell when stepping into the muck of a swamp. In contrast, **thermogenic** gas is formed at deep depths as geothermal heating separates hydrocarbons from organic material. Thermogenic gas may be formed directly, along with crude oil, or may be formed from crude oil that is altered by heating. While biogenic gas is nearly pure methane, thermogenic gas contains small amounts of a number of other hydrocarbon gases as well as methane. Most gas extracted commercially is thermogenic and is most often found above deposits of crude oil or seams of coal, so its extraction often accompanies the extraction of those fossil fuels. Natural gas deposits are greatest in Russia and the United States, and these two nations lead the world in both gas production and gas consumption (Table 17.5).

Natural gas extraction becomes more challenging with time

In order to access some natural gas deposits, prospectors need only drill an opening to allow gas to flow to the surface, because pressure drives the gas upward naturally. The first gas fields to be tapped were of this type. Most fields remaining today, however, require that gas be pumped to Earth's surface. In Texas, Kansas, California, and other areas of the United States, it is common to see a pumping device called a horse-head pump (Figure 17.18). This pump moves a rod in and out of a shaft, creating pressure to pull both oil and natural gas to the surface. As with oil and coal, many of the most easily accessible natural gas reserves have already been exhausted. Thus, much natural gas extraction today makes use of sophisticated techniques to break into gas-containing rock formations and pump gas to the surface. One such "fracturing technique" is to pump salt water under high pressure into the rocks to crack them. Sand or small glass beads are inserted to hold the cracks open once the water is withdrawn.

Until the 1970s, when natural gas became a saleable commodity because of its expanded use in heating, cooking, and electricity, much of the gas that rose to the surface during oil drilling was simply burned off. Capturing the gas and transporting it to markets where it could be sold as a fuel was too expensive. In Alaska, where natural gas transport remains prohibitively expensive, captured gas is reinjected into the ground for later extraction.

Table 17.5 Top Producers and Consumers of Natural Gas

Production (% world production)	Consumption (% world consumption)
Russian Federation, 22.0	United States, 26.3
United States, 21.7	Russian Federation, 15.3
Canada, 7.3	Great Britain, 3.7
Great Britain, 4.1	Germany, 3.3
Algeria, 3.2	Canada, 3.2
Indonesia, 2.8	Japan, 3.1
Norway, 2.6	Ukraine, 2.8
Iran, 2.6	Iran, 2.7
Netherlands, 2.4	Italy, 2.5
Saudi Arabia, 2.2	Saudi Arabia, 2.2

Data from British Petroleum, Statistical review of world energy, 2003.

Figure 17.18 Horsehead pumps are used to extract natural gas as well as oil and are a commonly seen feature of the landscape in areas such as west Texas. The pumping motion of the machinery draws gas and oil upward from below ground.

Biogenic natural gas is also produced as part of the decay process in landfills, and certain landfill owners are now capturing this gas to sell as fuel (Chapter 19). Besides decreasing energy waste and being profitable, capturing landfill gas helps alleviate the release of methane into the atmosphere. This may help mitigate climate change, because methane is a greenhouse gas that, molecule for molecule, has a greater impact in increasing global warming than does carbon dioxide (Chapter 12).

Environmental Impacts of Fossil Fuel Use

Climate change is one of several major environmental impacts that most scientists attribute to fossil fuel use. Our society's love affair with fossil fuels and the many petrochemical products we have developed from them has boosted our material standard of living beyond what our ancestors could have dreamed, has eased constraints on travel, and has helped lengthen our lifespans. It has had downsides as well, however, including harm to the environment and human health. Concern over these impacts is the prime reason many scientists, environmental advocates, businesspeople, and policymakers are increasingly looking toward renewable sources of energy that exert less impact on natural systems.

Fossil fuel emissions cause air pollution, drive climate change, and affect the carbon cycle

Carbon dioxide released from the burning of fossil fuels has been inferred to affect global climate (Chapter 12). Because global climate change may potentially have many widespread effects, carbon dioxide pollution is fast becoming recognized as the greatest unintended environmental impact of fossil fuel use. When we combust fossil fuels, we alter the balance of Earth's carbon cycle (Chapter 6). We take carbon that has been effectively retired into a long-term reservoir underground and release it into the air, where it can change atmospheric composition, alter climate, and affect organisms in ways scientists are only beginning to understand. Carbon from fossil fuels does return to the atmosphere naturally when sediments are uplifted and eroded, but we are accelerating the rate at which this happens, without fully understanding what the eventual results may be. Any meaningful future reduction of anthropogenic carbon dioxide emissions will need to involve major changes in energy policies and energy technologies among the world's nations.

Fossil fuels release more than carbon dioxide when they burn. Air pollution from fossil fuel combustion can have serious consequences for human health as well as the natural environment. The burning of oil and coal releases sulfur dioxide and nitrogen oxides, which contribute to smog and acid rain. The combustion of gasoline in vehicles can release ozone, which in the troposphere is an eye and lung irritant, a carcinogen, and a component of photochemical smog (Chapter 11). Gasoline combustion in automobiles also releases pollutants that irritate the nose, throat, and lungs.

Water pollution results from fossil fuel use

Fossil fuel use can also pollute rivers, lakes, and oceans. In addition to the effects of acid precipitation on freshwater ecosystems, oil enters bodies of fresh water and the ocean in a number of ways. Surprisingly little is from large headline-grabbing oil spills like that of the *Exxon Valdez* in 1989 (Chapter 4). Rather, small spills and chronic non-point-source pollution account for most oil that reaches the oceans. Oil from industries, homes, automobiles, gas stations, and businesses runs off roadways and enters rivers and sewage treatment facilities to be eventually discharged into the ocean. The U.S. National Academy of Sciences has estimated that nonpoint sources in the United States account for fully 85% of oil that enters the oceans, while spills from tankers and pipelines account for only 8%. A total of 700,000 barrels of oil from human activities reaches the ocean each year in the United States, while roughly 5 million barrels reach the ocean worldwide. However, natural seeps from the seafloor also contribute roughly equivalent amounts of oil to the marine environment; the Academy

has estimated that 1.1 million barrels in the United States and 4.3 million barrels globally enter ocean waters naturally from seeps.

Large, catastrophic oil spills can, however, have significant impacts on the marine environment. Major spills discharge such large amounts of oil locally that they have immediate and major impacts that can last for years or decades. Crude oil's toxicity to most plants and animals frequently leads to high mortality among organisms exposed.

Coal mining affects the environment

The mining of coal can also have substantial impacts on natural systems. Surface strip-mining can destroy large swaths of habitat and cause massive soil erosion and chemical runoff into waterways. Regulations in the United States now require mining companies to restore strip-mined land following mining, but impacts are severe and long-lasting just the same, and most other nations exercise less oversight. Mountaintop removal (Figure 17.19) can have even greater impacts than conventional strip-mining. When tons of soil are removed from the top of a mountain, it is difficult to keep rock and soil from sliding downhill, where large areas of habitat can be degraded or destroyed and creek beds polluted and clogged. Loosening of U.S. government restrictions in 2002 enabled mining companies to legally

Figure 17.19 Strip-mining in some areas is taking place on massive scales, such that entire mountain peaks are leveled, as at this site in West Virginia. Such "mountaintop removal" can cause enormous amounts of erosion into streams that flow from near the mine into surrounding valleys, affecting ecosystems over large areas.

dump mountaintop rock and soil into valleys and rivers below, regardless of the consequences for ecosystems, wildlife, and local residents.

Coal and oil each can pose threats to human health

Coal mining can threaten human health as well as environmental quality, although it is subsurface mining that raises the greatest health concerns. Underground coal mining is one of the world's most dangerous occupations. Besides risking injury or death from collapsing shafts and tunnels and from dynamite blasts, miners constantly inhale coal dust in the enclosed spaces of mines, which can lead to respiratory diseases, including black lung disease, that can shorten miners' lives.

Some chemical components of oil can also affect human health (and the health of other animals). As we saw in Chapter 4, certain hydrocarbons, such as benzene and toluene, appear to be carcinogens, because repeated skin exposure has been shown to cause cancer in laboratory animals. In addition, gases such as hydrogen sulfide can evaporate from crude oil, irritate the eyes and throat, and cause asphyxiation. Crude oil also often contains trace amounts of known poisons such as lead and arsenic. As a result, people working at drilling operations, gasoline refineries, and other jobs that entail frequent exposure to oil or its products can develop serious health problems, including cancer. In addition, leaks from oil operations have been known to contaminate soil and water supplies, posing dangers for the public. As we saw in Chapter 14, thousands of underground storage tanks of petroleum products and other substances have leaked, threatening groundwater supplies. The costs of alleviating such health and environmental impacts are high, and the public eventually pays them in an inefficient manner, because costs are generally not internalized (Chapter 2) in the relatively cheap prices of fossil fuels.

Oil extraction can harm the environment

Oil field development has also been shown to have negative environmental impacts. Drilling activities themselves have fairly minimal impact, but much more than drilling is involved in the development of an oil field. (Figure 17.20). Road networks must be constructed, and many sites may be explored in the course of prospecting. These activities can fragment habitats and can be noisy and disruptive enough to affect wildlife. The extensive infrastructure that must be erected to support a full-scale drilling operation typically includes housing for workers, access roads, transport pipelines, and waste piles for removed soil. In addition, ponds may be constructed for collecting sludge, the toxic leftovers that remain after the useful components of oil have been removed.

Many onshore North American oil reserves are located in arctic or semi-arid areas. With their low rainfall and harsh conditions, plants grow slowly in tundra and

Figure 17.20 Much more than drilling takes place at an oil field. Road networks, pipelines, waste piles, and housing for workers are all required as part of the infrastructure necessary to extract oil. Shown here is one small part of the development at Prudhoe Bay.

semi-desert biomes. As a result, ecosystems in these areas are fragile in the sense that even minor changes can have long-lasting repercussions. For example, tundra vegetation at Prudhoe Bay still has not fully recovered from temporary roads last used 25 years ago during the exploratory phase of development.

Whether Prudhoe Bay's oil operations have had a negative impact on the region's caribou is widely debated. Surveys show that the region's summer population of caribou has increased in the 25 years since Prudhoe Bay was developed. Many supporters of oil drilling have used this trend to argue that oil development does no harm to caribou. Other studies, however, show that female caribou and their calves avoid all parts of the Prudhoe Bay oil complex, including the roads laid down to support it, sometimes detouring miles to do so. These studies also show that the reproductive rate of female caribou in the Prudhoe Bay region is lower than for those in undeveloped areas in Alaska. As a result, although the herd near Prudhoe Bay has increased over the past 25 years, it has not increased as much as herds in some other parts of Alaska. Because there is no way of knowing how the particular herd of the Prudhoe Bay region would have performed in the absence of development (that is, there is no control, as there would be for a manipulated experiment), it is difficult to draw conclusions about the actual impacts, if any, of oil development on caribou. (The Porcupine Herd that visits ANWR's undeveloped 1002 Area rose from 100,000 animals in the 1970s to 178,000 in 1989 and has since declined to 123,000, all for unknown reasons.)

Many scientists anticipate negative environmental impacts of drilling in ANWR

To predict the possible ecological effects of drilling in ANWR's 1002 Area, scientists have examined the effects of development on arctic vegetation, air quality, water quality, and wildlife, including caribou, grizzly bears, and a wide variety of bird species, in Prudhoe Bay and other Alaska locales where the environment is similar to that of ANWR's coastal plain (Figure 17.21). Scientists have compared different areas, and have compared single areas or populations before and after drilling, as described above with caribou. In addition, scientists have run small-scale manipulative experiments when possible, for example, to study the effects of ice roads and secondary extraction methods such as seawater flushing. In one way or another they have examined the effects of road building, oil pad construction, worker presence, oil spills, accidental fires, trash build up, permafrost melts, off-road vehicle trails, and dust from roads.

Based on these studies, if drilling takes place in ANWR, many scientists anticipate damage to vegetation and wildlife. Vegetation can be killed when saltwater pumped in for flushing deposits is spilled or when plants are buried under gravel pits or roads. Air and water quality can be degraded by fumes from equipment and drilling operations, burning of natural gas associated with oil extraction, sludge ponds, waste pits, and oil spills. Other scientists, however, contend that drilling operations in ANWR would have little environmental impact. The roads will be built of ice that will melt in the

Figure 17.21 Alaska's North Slope is home to a variety of large mammals, including grizzly bears, polar bears, wolves, arctic foxes, and large herds of caribou. Whether and how these animals may be negatively affected by oil development is highly controversial, and scientific studies are ongoing. The caribou herd near Prudhoe Bay has increased since oil extraction began there, but has not increased as much as herds in other parts of Alaska. Grizzly bears like these shown here can sometimes be found near, or even walking atop, the trans-Alaska pipeline.

summer, they point out, and most drilling activity will be confined to the winter, when caribou are elsewhere. In addition, drilling proponents maintain, Prudhoe Bay is not an appropriate model for ANWR because Prudhoe Bay's development is larger than that projected for ANWR. In addition, they say, much of the technology used at Prudhoe Bay is now outdated, and ANWR would be developed with more environmentally sensitive technology and approaches.

Weighing the Issues: The Science of Oil's Arctic Impacts

Much evidence from correlative and experimental studies indicates that certain arctic plants and animals can be affected in negative ways by oil development. On the other hand, some scientists say that ice roads, winter drilling, and new technologies will minimize or eliminate such impacts. Given the evidence and arguments spelled out in the text, suggest several further scientific studies that you believe could be undertaken to help resolve some of the questions being debated.

Political, Social, and Economic Impacts of Fossil Fuel Use

The political, social, and economic consequences of fossil fuel use are numerous, varied, and far-reaching. Our discussion here focuses on several of the negative consequences of fossil fuel use and dependence, but it is important to bear in mind that their use has enabled much of the world's population to achieve a higher material standard of living than ever before. It is also important to ask in each case whether or not switching to more renewable sources of energy would solve existing problems.

Nations can become overly dependent on foreign oil

Putting all of one's eggs in one basket is always a risky strategy. The fact that so many nations' economies are utterly tied to fossil fuels means that those economies are tremendously vulnerable to supplies becoming suddenly unavailable or extremely costly. Such reliance means that seller nations can potentially control the price of

oil, forcing buyer nations to pay more and more as supplies dwindle. With the formation of OPEC in 1960, oil-producing nations tried to exercise even greater power over oil prices by regulating production (although OPEC so far has had only marginal success in this).

In the United States, concern over reliance on foreign oil sources has repeatedly driven the proposal to open ANWR to drilling, despite critics' charges that such drilling would likely do little to decrease the nation's dependence. The United States currently imports over 60% of its crude oil. With the majority of world oil reserves lying in the politically unstable Middle East, crises such as the 1973 oil embargo are a constant concern for U.S. policymakers. The United States has cultivated a close relationship with Saudi Arabia, the owner of 25% of world oil reserves, despite the fact that that country's political system allows for little of the democracy the United States claims to cherish and promote. The world's second-largest holder of oil reserves, at 11%, is Iraq.

To counter dependence on a few major suppliers of oil, the United States has diversified its sources of oil, and receives much of it from non-Middle Eastern nations, including Venezuela, Canada, Mexico, and Nigeria. Major trade relations among nations and regions of the world are depicted in Figure 17.22.

As much as U.S. policymakers worry about dependence on other nations for fossil fuels, several other nations are far more vulnerable in their energy dependence. Germany, France, South Korea, and Japan stand out as nations that consume far more energy than they produce, and thus rely almost entirely on imports for their continued economic well-being. Figure 17.23 contrasts consumption and production for nations and regions worldwide.

Figure 17.23 *(opposite page)* Some nations consume far more energy than they produce, while others produce more than they consume and are able to export fuel to high-consumption countries. This diagrammatic map contrasts each nation's or region's production and consumption of commercially traded fuel (in millions of metric tons of oil equivalent). Shaded portions show the amount of fuel produced by the nation or region, while outlines show the amount of consumption. The Middle East produces more than it consumes, while Japan consumes far more than it produces. Nations such as China, India, Australia, and the United Kingdom are fairly well-balanced. Data from British Petroleum, as presented by P. Harrison and F. Pearce, *AAAS Atlas of Population and the Environment,* 2000.

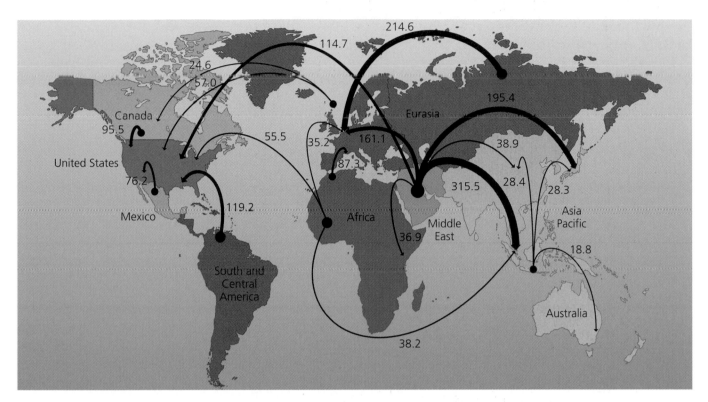

Figure 17.22 The global trade in oil is lopsided, with relatively few nations accounting for most exports, and some nations being highly dependent on others for energy. The United States gets most of its oil from Venezuela and Saudi Arabia, followed by Canada, Mexico, Nigeria, and the North Sea. Canada imports some North Sea oil while exporting more to the United States. Numbers in the figure represent millions of metric tons. Data from British Petroleum, *Statistical review of world energy, 2003.*

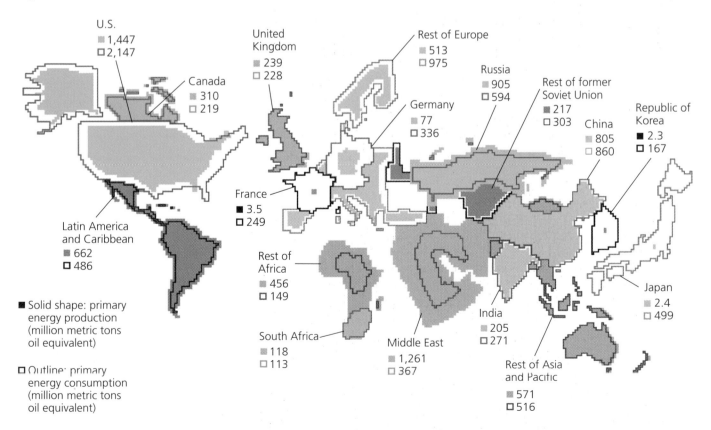

People in regions with oil reserves may or may not benefit from them

Oil production can be extremely lucrative; many of the world's wealthiest corporations deal in oil or oil-related industries, such as the automobile industry. These industries collectively provide jobs to millions of employees and provide dividends to millions of investors. For people who live in oil-bearing areas, development can potentially yield significant economic benefits as well. Since the construction of the trans-Alaska pipeline in the 1970s, the state of Alaska has received $55 billion in oil revenues. Alaska's state constitution requires that one-quarter of state oil revenues be placed in the Permanent Fund, which pays yearly dividends to all Alaska residents. So far the Permanent Fund has earned over $24 billion, and yearly per-capita dividend payouts have ranged from several hundred dollars to $1,000 dollars. Development of ANWR would both add to this fund and create jobs; some estimates anticipate creation of many thousands of jobs and millions of dollars of income. For this reason, many Inupiat who live on the North Slope support oil drilling. The income could be used to pay for health care, police and fire protection, and other services that are currently scarce in this remote and impoverished region.

Alaska's distribution of oil revenue among its citizenry is a rare example. In most parts of the world where oil has been extracted, local residents have not seen great benefits, but instead have frequently suffered. All too often, when multinational corporations have extracted oil in developing countries, paying those countries' governments for access, the money has not trickled down to residents of the regions where oil is extracted. Furthermore, oil-rich developing countries, such as Ecuador and Nigeria, tend to have few environmental regulations, and existing regulations may go unenforced if a government does not want to risk losing the large sums of money associated with oil development.

In Nigeria, oil was discovered in 1958 in the territory of the Ogoni, one of Nigeria's native people, and the Shell Oil company moved in to develop oil fields. Although Shell extracted $30 billion of oil from Ogoni land over the years, the Ogoni still live in poverty, with no running water or electricity. The profits from oil extraction on Ogoni land have gone to Shell and to the military dictatorships of Nigeria. Moreover, the development resulted in oil spills, noise, and constantly burning gas flares, all of which caused illness among people living nearby. From 1962 until his death in 1995, Ogoni activist and leader Ken Saro-Wiwa worked for fair compensation to the Ogoni for oil extraction and environ-

mental degradation on their land. After years of persecution by the Nigerian government, Saro-Wiwa was arrested in 1994, given a trial widely regarded in the international human rights community as a sham, and put to death by military tribunal.

Conversion to renewable energy lies in the future

Knowing that fossil fuel supplies are limited and that their use has health, environmental, political, and socioeconomic consequences, the nations of the world have several policy options for guiding future energy use. One option is to commit to using fossil fuels until they are no longer economically practical, and to develop other energy sources only when supplies have dwindled. A second option is to fund development of alternative energy sources now and attempt to reduce our reliance on fossil fuels gradually. Third, as some environmentalists advocate, we could try to end our usage of fossil fuels as soon as possible. Alternatives to fossil fuels include renewable sources such as wind, sun, moving water, biomass, and Earth's internal heat. We could also institute efficiency and conservation measures. We will explore these options in Chapter 18. First, we will examine a nonrenewable alternative to fossil fuels, nuclear power.

Weighing the Issues:
Will Capitalism Drive a Shift to Renewables?

Many experts say that governments need not take steps to hasten the readiness of renewable energy sources to replace fossil fuels as they decline, because forces of supply and demand in market capitalism will automatically take care of this shift. As fossil fuel supplies dwindle and prices rise, the argument goes, industries and consumers will switch to renewable sources as they become more economical. Do you agree with this outlook? Do you see any possible problems with it? Do you think governments should take any steps to encourage the development of renewable energy sources? Why or why not?

Nuclear Power

While the vast majority of the nonrenewable energy we use comes from fossil fuels, nuclear power contributes significantly to the energy we consume and the electricity we produce. Nuclear power accounted for 6.8% of the world's total primary energy supply in 2000, and for

16.9% of the world's electricity production. Nuclear power has expanded 15-fold since 1970, experiencing most of its growth during the 1980s. The United States leads the world in producing electricity from nuclear power, with nearly a third of the world's production, and is followed by France and Japan. However, only 20% of the United States' electricity comes from nuclear sources. Nations such as Ukraine, Korea, Germany, and Japan rely more heavily on nuclear power, and France receives fully 77% of its electricity from nuclear sources.

Nuclear power comes from uranium via a long process

We generate electricity from nuclear power in reactors, but this is only one step in a longer process sometimes called the nuclear fuel cycle (Figure 17.24). This process begins when the naturally occurring element uranium is mined from underground deposits, as we saw with the mines on Australian Aboriginal land in Chapter 2. Uranium in nature occurs in different isotopes, with over 99% occurring as uranium-238. The isotope uranium-235 (with three fewer neutrons; see Chapter 4) makes up less than 1% of the total, but is the source of energy for nuclear power. Thus uranium ore must be processed in order to enrich the concentration of ^{235}U, to about 3%. The enriched uranium is incorporated into fuel rods for use in nuclear reactors as described below. After several years in the reactor, the fuel loses its ability to generate adequate energy, and must be replaced with new fuel. Some of the spent fuel is reprocessed, in order to recover what usable uranium and plutonium may be left. Other spent fuel is disposed of as radioactive waste. Radioactive waste disposal is discussed in Chapter 19.

Fission releases energy we can harness

Nuclear energy, strictly defined, is the energy that holds together protons and neutrons within the nuclei of atoms. In order to harness this energy, we convert it to thermal energy, which can be used to drive the generation of electricity (or, alternatively, to fuel the terrible destructive power of nuclear weapons). Two processes can convert the energy within an atom's nucleus into usable thermal energy (Figure 17.25). One is **fusion**, in which the small nuclei of lightweight elements are forced together. Because overcoming the mutually repulsive forces of protons is difficult, this process has not yet been developed for power generation. Instead, the key chemical reaction that drives the release of nuclear energy in our power plants is **fission**, the splitting apart of atomic nuclei.

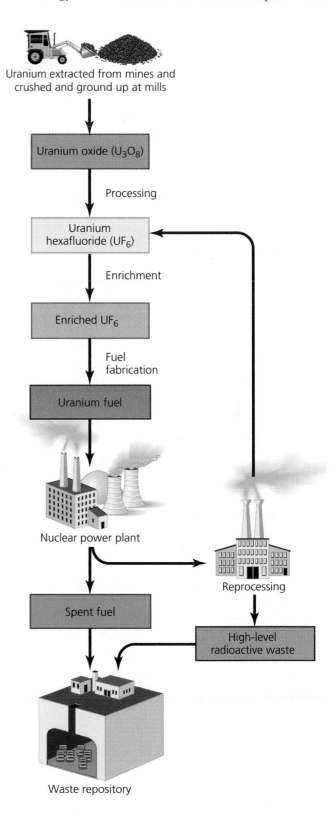

Figure 17.24 Harnessing nuclear energy involves a long, partially cyclical, process sometimes called the nuclear fuel cycle. After uranium ore is mined from the earth, it is processed and enriched, undergoing chemical changes at each step. In its final form as uranium oxide fuel, it consists of roughly 3% ^{235}U and 97% ^{238}U. After fuel rods are depleted during their use in a nuclear power plant, they are either disposed of as radioactive waste or are reprocessed to recover leftover uranium for further use.

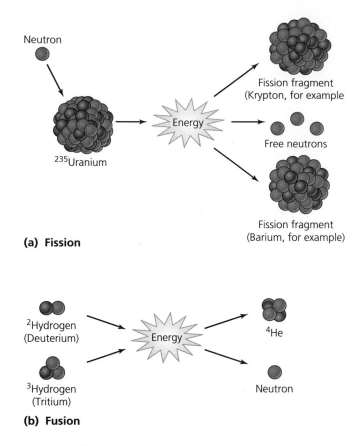

Neutron

Energy

²³⁵Uranium

Fission fragment
(Krypton, for example

Free neutrons

Fission fragment
(Barium, for example)

(a) Fission

²Hydrogen
(Deuterium)

Energy

⁴He

³Hydrogen
(Tritium)

Neutron

(b) Fusion

Figure 17.25 Fission and fusion are two approaches for releasing nuclear energy. In fission (**a**), atoms of uranium-235 are bombarded with neutrons. Each collision results in the uranium atom splitting into smaller atoms and releasing two or three neutrons, along with energy and radiation. Because the neutrons can split further uranium atoms and set in motion a runaway chain reaction, engineers at nuclear plants must absorb excess neutrons in order to regulate the rate of the reaction. In fusion (**b**), two small atoms such as the hydrogen isotopes deuterium and tritium are fused together, causing the loss of a neutron and the release of energy. So far, however, scientists have not been able to fuse atoms without supplying far more energy than the reaction produces, so this process is not used commercially.

In fission, the nuclei of large, heavy atoms such as uranium and plutonium are bombarded with neutrons. Ordinarily neutrons move at too fast a speed to split these nuclei when they collide with them, but if the neutrons are slowed down they can fission, or break apart, the nuclei. Each nucleus that breaks apart into smaller atoms emits multiple neutrons, together with substantial heat and radiation. The multiple neutrons (two or three in the case of fissile ^{235}U) can in turn bombard other ^{235}U atoms in the vicinity. Thus, if not controlled, this chain reaction becomes a runaway process of positive feedback that releases enormous amounts of energy. Inside a nuclear reactor, however, fission is controlled so that the chain reaction maintains a

constant output of energy rather than an explosively increasing output.

In order for fission to begin, the neutrons must be slowed down. In reactors, this is accomplished with a substance called a moderator, which is most often water or graphite. With a moderator allowing fission to proceed, it becomes necessary to soak up the excess neutrons produced when uranium nuclei split, so that on average only a single uranium atom from each split nucleus goes on to split another nucleus. For this purpose, rods of a metallic alloy that absorbs neutrons are placed into the reactor among the water-bathed fuel rods. Engineers can move these control rods so as to regulate and maintain the reaction at the desired rate.

All this takes place within the reactor core, and is the first step in the electricity-generating process of a nuclear reactor (Figure 17.26). In the most common type of reactor, the pressurized light water reactor shown in the figure, energy from the fission reaction heats the moderator water, which is carried through a pressurized circuit called a primary loop. The heat from this loop heats water in a secondary loop, creating steam. The steam is used to turn a turbine, generating electricity. The steam is then cooled by a third loop in a cooling tower, condenses, and is returned to the containment building to be heated again by the primary loop.

Nuclear power grew fast, but ran into problems

As nuclear power was developed in the second half of the 20th century, it showed great promise for being able to generate electricity cleanly, without the air pollutants emitted by fossil fuel combustion. Uranium supplies also promised to last longer than fossil fuel supplies. In the 1970s and 1980s, nuclear power plants were constructed in many developed countries, generally with the help of enormous government subsidies. Over the years nuclear power proved its ability to produce electricity more cleanly than coal and other fossil fuels.

Since then, the industry's growth has stalled. One primary reason is cost overruns. Building, maintaining, operating, and assuring the safety of nuclear power plants all turned out to be more expensive than expected, even with the subsidies. Nuclear power today remains more expensive than coal and other electricity sources, and investors lost interest. Concerns over safety provided the second primary reason for the slowing of nuclear power's growth. Although environmental scientists calculate that chronic health impacts from nuclear plants are probably much less than those of coal combustion,

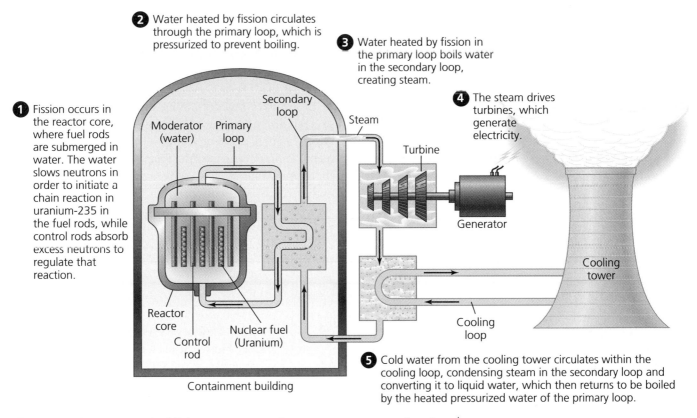

2 Water heated by fission circulates through the primary loop, which is pressurized to prevent boiling.

3 Water heated by fission in the primary loop boils water in the secondary loop, creating steam.

1 Fission occurs in the reactor core, where fuel rods are submerged in water. The water slows neutrons in order to initiate a chain reaction in uranium-235 in the fuel rods, while control rods absorb excess neutrons to regulate that reaction.

4 The steam drives turbines, which generate electricity.

Secondary loop

Moderator (water)   Primary loop   Steam

Turbine

Generator

Reactor core
Control rod
Nuclear fuel (Uranium)

Cooling tower

Cooling loop

Containment building

5 Cold water from the cooling tower circulates within the cooling loop, condensing steam in the secondary loop and converting it to liquid water, which then returns to be boiled by the heated pressurized water of the primary loop.

Figure 17.26 In a pressurized light water reactor, the most common type of nuclear reactor, uranium fuel rods are placed in water, which slows neutrons so as to enable fission. Control rods that can be moved in and out of the reactor core absorb excess neutrons to regulate the chain reaction. Water heated by fission circulates through the primary loop and warms water in the secondary loop, which turns to steam. Steam drives turbines, which generate electricity. The steam is then cooled by water from the cooling tower, and returns to the containment building to be heated again by heat from the primary loop.

the possibility of catastrophic accidents has created a great deal of public anxiety about nuclear power.

In 1979 a partial meltdown at the Three Mile Island nuclear plant in Pennsylvania released radiation into the environment, necessitating evacuation of area residents and causing public alarm. Although the accident was brought under control, and although no significant health effects to residents have been proven, the event put safety concerns squarely on the map for U.S. citizens and policymakers. Then in 1986 an explosion at the Chernobyl plant in Ukraine (part of the Soviet Union at the time) caused the most severe nuclear accident yet recorded. Many thousands of area residents died or became ill from radiation poisoning, several hundred thousand people were evacuated from their homes, and radioactive contamination spread over Ukraine, Russia, eastern and northern Europe, and beyond. The region around Chernobyl remains contaminated today.

Growth in nuclear generating capacity slowed to a crawl during the 1990s, and today, more than 100 of the world's nuclear plants have been decommissioned, leaving nearly 450 operational. The nuclear industry stopped building plants in the United States following Three Mile Island, in western Europe not a single reactor is under construction today, and some nations have moved to phase out nuclear power altogether. The International Energy Agency forecasts that nuclear production will peak at about 7% of world energy use and 17% of electricity production in 2010, then decline to around 5% of world energy use and 9% of electricity production by 2030.

Conclusion

Harnessing energy sources has played a major role in the development of human cultures around the world. Over the past 200 years, fossil fuels have helped us to build complex industrialized societies capable of exploring (and exploiting) all parts of the world, and even venturing beyond our planet. Today, however, we are approaching a turning point in history as our supplies of fossil fuels begin to decline and our awareness of their negative impacts increases. We can respond to this new challenge in creative ways and begin developing alternative energy sources, or we can continue our current dependence on fossil fuels and wait until they near depletion to develop new technologies and ways of life. The path we choose will have far-reaching consequences for human health and well-being, for Earth's climate, and for our natural environment.

The ongoing debate about the future of the Arctic National Wildlife Refuge is a microcosm of this debate over our energy future. Should we drill for oil in the refuge when we are not even sure how much is there? Can oil extraction be completed safely, supplying jobs and income to Alaskans and reducing U.S. reliance on foreign oil? Or are conservation measures more effective in ensuring national security? And is now the time to shift our focus to other sources of energy and keep one of the world's few remaining pristine wilderness areas protected? Fortunately, it is not simply a tradeoff between benefits of energy for us and harm to the environment, climate, and health. Instead, as evidence builds that renewable energy sources are becoming increasingly feasible and economical, it becomes easier to envision giving up our reliance on fossil fuels and charting a win-win future for humanity and the environment.

REVIEW QUESTIONS

1. What natural resource have humans relied on most for energy throughout history? Why are fossil fuels our most prevalent source of energy today?

2. Why are fossil fuels considered nonrenewable sources of energy? What makes renewable energy sources renewable?

3. How do developed and developing countries differ in their overall rates of energy consumption and in the ways they use energy?

4. How are fossil fuels formed? How do environmental conditions determine what type of fossil fuel is formed in a given location?

5. How do geologists locate fossil fuel deposits? Describe the process of seismic surveying.

6. Why are fossil fuels concentrated in localized deposits? In what regions are the majority of these deposits located?

7. When and why was the first oil company founded and what was it called? When was the first oil well drilled? What is the relationship between oil and rocks?

8. How are petroleum products created? Name some of these products.

9. What is the approximate current rate of oil consumption in the United States? How much of this consumption is allocated to transportation? Compare these levels of consumption with U.S. oil reserves and production levels.

10. How long has coal been used by humans as an energy source? How is it used to create electricity?

11. How do scientists classify coal? What determines coal's potential energy?

12. How did natural gas companies maintain their share of the energy market in the face of increasing electricity use in the nineteenth and twentieth centuries?

13. Why is natural gas often extracted simultaneously with other fossil fuels? What constraints on its extraction does it share with oil?

14. What are some of the effects of fossil fuel emissions in the atmosphere?

15. In what ways can crude oil pollute water?

16. Compare the effects of coal mining and oil extraction on the environment.

17. What are some of the health risks of coal mining and oil extraction?

18. Compare some of the contrasting views of scientists regarding the environmental impacts of drilling for oil in the Arctic National Wildlife Refuge (ANWR).

19. About how much of the oil used in the United States is imported? Why could this be a problem?

20. How has the experience of the Ogoni people of Nigeria been similar to or different than that of other people in countries where oil is produced?

DISCUSSION QUESTIONS

1. How did geologists from the U.S. Geological Survey (USGS) estimate the total amount of oil underneath the Arctic National Wildlife Refuge (ANWR) 1002 Area? How is this amount different from the "technically recoverable" and "economically recoverable" amounts of oil? How long would this oil last? Why might an oil company be interested in drilling for this amount of oil, and under what circumstances might it choose not to?

2. About how much oil is left in the world, and how much longer can we expect to use it? How effective were the conservation methods adopted by the United States in response to the "energy crisis" of 1973-1974? Do you think the future shortage of oil and other fossil fuels also represents a crisis? Why or why not?

3. Compare the effects of coal and oil consumption on the environment. Which process do you think has ultimately been more detrimental to the environment, oil extraction or coal mining, and why?

4. Imagine we were living in the 1950s and the United States was facing an oil crisis. Do you think the nation would be debating over whether to drill in ANWR? Why are so many people today concerned about wildlife like caribou and about wilderness preservation?

5. If ANWR were located in a developing country, do you think the citizens and policymakers of that country would choose to drill there?

6. If the United States and other developed countries relinquished dependency on foreign oil and on fossil fuels in general, do you think that their economies would benefit or suffer? Might your answer be different for the short term and the long term? What factors come into play in trying to make such a judgement?

Media Resources *For further review, go to the website* **www.envscienceplace.com** *or student CD-ROM, where you will find quizzes, flashcards, a glossary, additional interactive exercises, and links to relevant news and research sources. Also, on the website and CD-ROM is* **GRAPH IT**, *a series of interactive graphing tutorials to help you interpret graphs and plot data.*

18 Renewable energy sources

Iceland's Blue Lagoon and Svartsengi Geothermal Power Plant

This chapter will help you understand:

- Sources of renewable energy
- Biomass energy
- Hydroelectric energy
- Solar energy
- Wind energy

- Geothermal energy
- Ocean energy sources
- Hydrogen fuel cells and new transportation options
- Principles of energy conservation

Hydrogen Fueling Station, Hamburg, Germany

Central Case: Iceland Moves toward a Hydrogen Economy

"I believe that water will one day be employed as fuel, that hydrogen and oxygen which constitute it, used singly or together, will furnish an inexhaustible source of heat and light. . . . Water will be the coal of the future."
—Jules Verne, in The Mysterious Island, 1874

"Our long-term vision is of a hydrogen economy."
—Robert Purcell Jr., Executive Director, General Motors National Petrochemical and Refiners Association Group

The Viking explorers who centuries ago first set foot on the remote island of Iceland were trailblazers. Today the citizens of the nation of Iceland are blazing a bold new path, one they believe the rest of the world may follow. Iceland aims to become the first nation to leave fossil fuels behind and convert to an economy based completely on renewable energy.

Iceland is essentially a hunk of lava the size of Kentucky that has risen out of the North Atlantic from the rift between tectonic plates known as the mid-Atlantic Ridge. Most of Iceland's 290,000 people live in the capital city of Reykjavik, leaving much of the island an unpeopled and starkly beautiful landscape of volcanoes, hot springs, and glaciers.

The magma that gave birth to the island heats its groundwater in many places, giving Iceland some of the world's best sources of geothermal energy. Iceland has also dammed some of its many rivers for hydropower. Together these two renewable energy sources account for 73% of the country's energy supply and virtually all its electricity generation. Yet the nation, like most others, is also dependent on fossil fuels. It uses oil for its automobiles, some of its factories, and its economically vital fishing fleet—and together these have given Iceland one of the highest per capita rates of greenhouse gas emissions in the world. Because it possesses no native fossil fuel reserves, all must be imported, making for a weak link in an otherwise robust economy that has given its citizens one of the highest per capita incomes in the world. When Iceland signed on to the Kyoto accord on climate change, it received a special exemption from emissions reductions, but still the country will be hard pressed to meet its obligations to minimize greenhouse gas emissions.

Enter Bragi Árnason, a University of Iceland professor popularly known as "Dr. Hydrogen." Árnason began arguing in the 1970s that Iceland could achieve independence from fossil fuel imports, boost its economy, and serve as a model to the world by converting from fossil fuels to a renewable energy economy based on hydrogen. He suggested zapping water with Iceland's cheap and renewable electricity in a chemical reaction called electrolysis, splitting the hydrogen from the oxygen and then using the hydrogen to power fuel cells that would produce and store electricity. The process is clean; nothing is combusted, and the only waste product is water.

In the late 1990s, the interest of international energy companies in his ideas, and the nation's "Kyoto dilemma" finally forced Arnason's countrymen to pay attention. The nation's leaders decided to embark on a grand experiment to test the efficacy of switching to a "hydrogen economy." By setting an example for the rest of the world to follow, and by getting a headstart at producing and exporting hydrogen fuel, these leaders hoped Iceland could become a "Kuwait of the North," getting rich by exporting hydrogen to the rest of an energy-hungry world, as Kuwait has done by exporting oil.

Those planning the shift to a hydrogen economy sketched a stepwise transition, with fossil fuels being phased out over 30–50 years. The process was to begin by introducing three hydrogen-run buses in a demonstration project called *ECTOS (Ecological City Transportation System)*. If all went well, this would be followed by the conversion of the entire Reykjavik bus fleet. After that, Iceland's 180,000 private cars would be powered by fuel cells, and then the fishing fleet would be converted to hydrogen. The final stage would be the export of hydrogen fuel to mainland Europe.

To make this happen, Icelanders in 1999 teamed up with corporate partners, who were looking to develop technology for the future. Auto company Daimler-Chrysler is producing hydrogen buses, oil company Royal Dutch Shell is building and running the hydrogen filling stations to fuel the buses, and Norsk Hydro is providing the electrolysis technology. Costs are split among Iceland and the three corporate partners, and the European Union has also provided funding to get the project rolling.

So far, the plan is nearly on schedule. In April 2003, the world's first commercial hydrogen filling station opened in Reykjavik, and in October three hydrogen-fueled buses went into operation. The public-private consortium, named *Icelandic New Energy (INE)*, will monitor the technology's effectiveness and the costs of developing infrastructure, as the ECTOS project—and hopefully future steps in the plan—proceed. Iceland's citizens are behind the effort; a recent poll showed 93% support among Icelanders.

Iceland may already be serving as a model for the world. Hydrogen-fueled buses are being developed by a number of companies, and prototypes have operated, are operating, or will soon start running in Munich, Berlin, and other German cities; in Torino, Italy; in Copenhagen, Lisbon, and Madrid; in Tokyo; in Perth, Australia; in Chicago and Vancouver; in San Jose, California; and in Winnipeg, Manitoba. Nine European cities are slated to receive Daimler-Chrysler buses powered by compressed hydrogen, in part thanks to Iceland's venture. Hydrogen refueling stations are being demonstrated in Japan, Singapore, and the United States, and fuel-cell passenger automobiles are being tested in Japan and California. A global hydrogen economy may be closer than we suspect.

Renewable Energy Sources

Iceland's bold drive toward a hydrogen economy is only one facet of the global move toward renewable energy sources, those that will not be depleted by use. Although fossil fuels have helped drive the industrial revolution and the increase in material prosperity for the past few centuries, they will not last forever. As we saw in Chapter 17, oil production is thought to be peaking now, and supplies of oil and natural gas may not last half a century more. This is why most energy experts and many businesspeople today accept that the world's economies and technologies will need to shift from fossil fuels to energy sources that cannot be depleted. Although a transition to renewable energy sources appears inevitable, there are open questions, of which two seem most important: First, which renewable sources will we use, and how? Second, will we be able to shift to renewable energy sources soon enough and smoothly enough so as to avoid widespread war, social unrest, and further damage to the natural environment? The answers to these questions will determine much in regard to quality of life in the coming decades.

Renewable energy currently provides only a small portion of our power

Today's economies are powered by fossil fuels, with 80% of all primary energy coming from oil, coal, and natural gas. Nuclear energy provides 6.8%, leaving renewable energy sources to account for 13.8%, most of which is provided by fuelwood and other biomass (Figure 18.1).

Developing countries account for most use of combustible renewables, or biomass, such as fuelwood. Developed nations account for most use of renewable energy sources that require higher technology, such as those used to harness solar energy, wind energy, geothermal energy, and the power of running water. For example, renewables in the year 2000 furnished 16.8% of Canada's energy

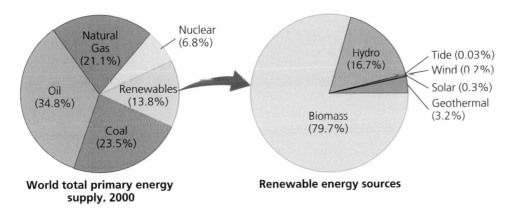

World total primary energy supply, 2000

Renewable energy sources

Figure 18.1 Fossil fuels account for the vast majority of the world's total supply of primary energy (left chart), while, as of 2000, only 13.8% of energy supply came from renewable energy sources. Of these renewable sources (right chart), roughly 80% comes from biomass, a resource that is renewable only to the extent that we do not deplete it. Hydropower accounts for most of the remainder, while the so called "new renewables" of solar, wind, tide, and geothermal are still very minor components of our energy spectrum. Data from International Energy Agency Statistics, 2002.

use, mostly provided by water powering electricity at dams, some by biomass, and small amounts by solar, wind, and other methods. In contrast, developing countries in West Africa, including Benin, Togo, and Senegal, obtain well over half their energy from fuelwood and other biomass but virtually none from technologically advanced methods of renewable energy production. The United States has been a leader in developing technology for harnessing renewable energy, but its entrepreneurs have done so largely in the absence of government encouragement. Only 1.6% of the United States's primary energy comes from high-technology renewable sources.

Renewable energy is used for heating buildings and for generating electricity, with very little going to transportation. Electricity generation is a vital component of the

way we power our economies. Nearly 64% of this total comes from burning fossil fuels, 17% comes from nuclear energy, and 19% comes from renewable energy. Of the renewable sources, solar, wind, geothermal, and biomass energy are all used, but it is hydropower that comprises 90% of all renewable electricity generation (Figure 18.2).

Renewable energy has multiple benefits, and its use is growing.

Renewable energy use is growing fast as a result of diminishing fossil fuel supplies, advances in technology and increasing concern over environmental impacts of fossil fuel combustion. Unlike fossil fuels, renewable sources are inexhaustible on time scales relevant to human societies

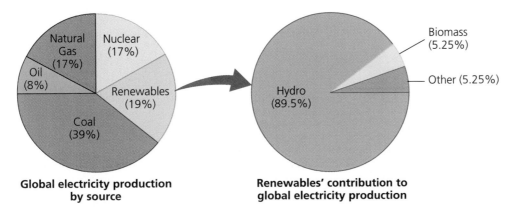

Global electricity production by source

Renewables' contribution to global electricity production

Figure 18.2 While most electricity is produced by combusting fossil fuels, nuclear and renewable energy sources also contribute substantially to our electricity generation (left chart). Hydropower generates nine-tenths of the electricity produced from renewable sources (right chart). Data from International Energy Agency Statistics, 2002.

and can help alleviate air pollution and the greenhouse gas emissions that are driving global warming. Renewable energy can also bring economic benefits. Developing renewables can help diversify an economy's mix of energy, lowering price volatility and protecting against shocks like the 1973 oil embargo (Chapter 17) by lessening dependence on foreign fuel. New energy sources also can create new employment opportunities and sources of income and property tax for local communities, especially rural areas passed over by other types of economic development. In many rural areas and developing countries, locally based renewable sources may also prove cheaper to use than extending the electricity grid infrastructure out from cities. For all these reasons, renewable sources, including solar, wind, and geothermal energy, have grown far faster in recent years than has the overall primary energy supply, with renewables rising 9.4% each year over the past three decades. Among the renewable energy sources, the leader in growth is wind power, with 52% annual growth (Figure 18.3).

Rapid growth in renewable energy sectors seems likely to continue, as population and consumption grow, global energy demand expands, fossil fuel supplies decline, and citizens demand cleaner environments. More governments, utilities, corporations, and consumers are now promoting and using renewable energy sources, and, as a result, the costs of renewables are falling, just as the prices of nonrenewable sources may begin to rise.

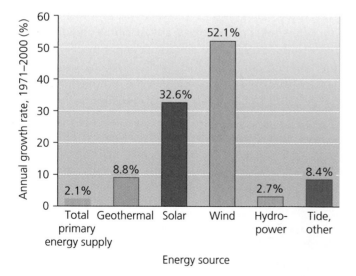

Figure 18.3 The "new renewable" energy sources are growing substantially faster than the total primary energy supply. Solar power has grown over 32% each year since 1971, and wind power has grown over 52% each year. Because these sources began from such low starting levels, however, their overall contribution to our energy supply is not yet large. Go to **GRAPH IT** on the website or CD-ROM. Data from International Energy Agency Statistics, 2002.

The transition cannot be immediate, but it must be soon

Some environmentalists advocate switching completely to renewable energy sources overnight, but there are both technological and economic barriers. Many renewables are expensive, lack adequate technological development, or lack infrastructure to transfer energy on the required scale. However, dramatic improvements in technology and infrastructure in recent years suggest that most barriers to expanding our use of renewables are political. Renewable energy sources have received far less in subsidies, tax breaks, and other incentives from governments than have conventional sources. By one estimate, of the $150 billion in subsidies the U.S. government provided to nuclear, solar, and wind power between 1947 and 1999, the nuclear industry received 96%, whereas the solar industry received 3% and the wind industry less than 1%. Traditionally, research and development of renewable sources have suffered from the continuing availability of oil made inexpensive in part by government policy. Some corporations in the fossil fuel and automobile industries have understood for some time that they will eventually need to switch to renewable sources and that when the time comes, they will need to act fast to stay ahead of their competitors. However, in light of continuing short-term profits, as well as unclear or conflicting policy signals, companies have not been eager to transition until necessary.

Under these circumstances, the best hope may be for a gradual transition from fossil fuels to renewable energy sources. However, if the transition proceeds too slowly, falling fossil fuel supplies may outpace our ability to develop new sources, and we may find our economies greatly disrupted. In the following sections we examine the major renewable sources of energy, beginning with somewhat renewable biomass energy and hydroelectric power, and then several newer forms of renewable energy that hold promise for bringing us a vigorous and sustainable energy economy without the environmental impacts of fossil fuels.

Biomass

Biomass consists of those organic substances produced by recent photosynthesis—unlike fossil fuels, which received their photosynthetic sustenance millions of years ago. Humans harness biomass energy from many types of living plant and animal matter, including wood, charcoal, and animal waste products (Table 18.1). As Figure 18.1 shows, biomass constitutes nearly 80% of all

Table 18.1   **Major Types of Biomass**

Freshly cut fuelwood

Charcoal

Wood from brush-thinning operations in fire-prone forests

Wastes from logging and timber processing

Urban yard waste

Construction waste

Crop residues, such as corn stalks

Trees or other plants grown as crops specifically for biomass energy

Liquid fuels, such as ethanol and biodiesel

Manure from domestic animals

Methane from bacterial decomposition in landfills

---

world's biomass energy consists of wood or charcoal to heat homes and cook food. Over 1 billion people in developing countries still use wood as their principal power source. In addition to this traditional use, a number of innovative uses have recently been developed (Table 18.2).

In developed nations, biomass use often involves some of these innovative approaches and constitutes only a small percentage of overall energy use. In developing nations, however, biomass accounts for fully 35% of energy use, much of which, however, involves the cutting of forests that may not be replenished. In the poorest nations, up to 90% of energy comes from this type of biomass. As these nations become more developed, fossil fuels are slowly replacing these traditional energy sources. As a result, biomass use is growing more slowly worldwide than overall energy use.

primary renewable energy used worldwide; it is far and away the preeminent renewable energy source. Considering what we have learned about the loss of forests (Chapter 16), however, it is fair to ask whether biomass should really be considered a renewable resource. Indeed, biomass is renewable only if it is not used too quickly. At rates of moderate usage, living matter can be replenished over months to decades, rather than the millions of years fossil fuels require. However, sometimes biomass is not replenished, as when forests are cut, soil erodes, and forests fail to grow back. The potential for deforestation makes use of biomass energy not as sustainable a practice as switching to other renewable sources, particularly as human population continues to increase.

Biomass energy brings a mix of benefits and drawbacks

Biomass energy can be efficient in terms of both energy use and cost; biomass tends to be the least expensive type of fuel for combustion in power plants, and improved energy efficiency can lead to lower prices for consumers. Where waste residues from crops or timber products are used, biomass energy can be virtually free, because these by-products would otherwise be discarded. Where waste costs money to dispose of, using waste for fuel can save money. Liquid fuels made from biomass can be similar in market price to petroleum gasoline, given government subsidies and mandates for their consumer use. Biodiesel from soybeans runs roughly $2 per gallon, and ethanol has ranged between $0.90 and $1.90 per gallon over the past decade. Alcohol can fuel cars, stoves, and heaters. Gasohol, gasoline containing 10% ethanol, burns more cleanly than conventional gasoline. Biodiesel does not need to be newly

Biomass can be used for energy in diverse ways

Biomass sources can provide heat energy as fuel, or they can be burned to create steam for turning turbines in a magnetic field, thus creating electricity. Much of the

Table 18.2   **New Uses of Biomass Energy**

Energy may be obtained from biomass by:

- Converting solid biomass into liquid fuels to power automobile and other engines. Ethanol is produced by fermenting starches and sugars, typically from wheat, sugar cane, corn, or grapes. Biodiesel is produced by mixing ethanol and other alcohols with oil from soybeans, rapeseed, animal fat, or other biomass.
- Collecting methane that anaerobic bacteria produce within landfills and using it as natural gas for fuel cells or power plants. Some Chinese farmers harvest such biogas from animal manure and use it for cooking and for heating their homes.
- Burning biomass in power plants built specifically for biomass, generating steam to power turbines.
- Burning small portions of biomass along with coal in conventional power plants (a process called co-firing).
- Converting solid biomass into gaseous form to improve electricity generation.
- Finding ways to use all the heat, steam, and power that comes from combusting biomass in a power plant, such as using these to heat nearby homes and businesses (a process often called *combined heat-and-power*).

synthesized from plants, however. It is possible to use waste substances such as used vegetable oil to fuel vehicles. Increasing numbers of environmentally conscious individuals in North America and Europe are fueling their cars with biodiesel from waste oils, and some buses and recycling trucks are also now running on biodiesel.

In terms of environmental impacts, biomass energy has a mixed record. Unsustainable wood harvesting can lead to deforestation, soil erosion, and desertification, damaging landscapes for thousands of years, eliminating biodiversity, and impoverishing human societies dependent on an area's resources. The combustion of biomass can also lead to health hazards from indoor air pollution, as we saw in Chapter 11. Among the newer uses of biomass, however, capturing landfill gas yields one positive environmental effect: reducing emissions of methane, a powerful greenhouse gas that contributes to climate change.

A larger issue is that the combustion of biomass does not reduce emissions to the extent that other renewable energy sources do. Because carbon is burned, carbon dioxide and other gases are released as they are in the combustion of fossil fuels. However, if the biomass being burned is fully "replaced" by the growth of new plants, then, because plants remove carbon dioxide from the air during photosynthesis, the net effect on emissions can be close to zero.

Growing crops for energy establishes monoculture agriculture, with all its impacts (Chapter 8 and Chapter 9), on precious land that could otherwise be left fallow or used to grow food. Producing ethanol for energy provides farmers with additional markets for their crops, but yields only small amounts of ethanol per acre of crop grown (for instance, 1 bushel of corn creates only 2.5 gallons of ethanol). It is not efficient to grow high-input, high-energy crops merely to reduce them to a few gallons of fuel; such a process is less efficient than directly burning biomass. Another problem is that many farmers apply petroleum-based pesticides and fertilizers to their crops to increase their yields, meaning that a shift to ethanol as a fuel would not eliminate our reliance on fossil fuels. On the positive side, crops grown for energy typically receive lower inputs of pesticides and fertilizers than those grown for food.

Weighing the Issues:
Ethanol

Some observers of U.S. politics argue that subsidies for ethanol and mandates for its addition to gasoline in the United States result in part from promises made to farmers in Iowa every four years during that state's influential caucuses that lead off the presidential primaries. Do you think producing and using ethanol made from corn and other crops is a good idea? Can you suggest ways of using liquid fuels or other types of biomass that would minimize environmental impacts?

Hydroelectric Power

If you have ever driven across Hoover Dam on the Colorado River (Chapter 14), or have seen any other large dam, you have witnessed an impressive example of hydroelectric power in action. In hydropower, the kinetic energy of moving water is used to turn turbines and create electricity. Most hydropower plants have dams that collect a supply of water and release it to run through turbines when power is needed (Figure 18.4). The amount of power generated depends on the distance the water falls and the volume of water released.

Hydroelectric power is widely used

Next to biomass, hydropower is the second largest source of renewable energy, accounting for 2.3% of the world's primary energy supply. Hydroelectric power also accounts for fully 17% of the world's electricity production. For nations with large amounts of water and the economic resources to build dams, hydroelectric power has been a keystone of their development and wealth. Brazil, Norway, Sweden, Austria, Switzerland, Canada, and many other nations today obtain large amounts of their energy from hydropower. Iceland developed hydropower very early on and is planning to use hydroelectric power as a primary source of electricity for powering its production of hydrogen. The great age of dam-building for hydroelectric power (as well as flood control and irrigation) in the United States began in the 1930s, when dams were built as public projects, partly to employ people and help end the economic depression of the time. Dam-building peaked in 1960, when 3,123 dams were completed in a single year.

Hydropower brings benefits but has negative environmental impacts

For producing electricity, hydropower has two clear advantages over fossil fuels. It is renewable; as long as rain still falls from the sky and fills rivers and reservoirs, we can use water to turn turbines. When the hydrologic cycle has been influenced by human activity, or when large

(a) Ice Harbor Dam, Washington

(b) Turbine generator inside MacNary Dam

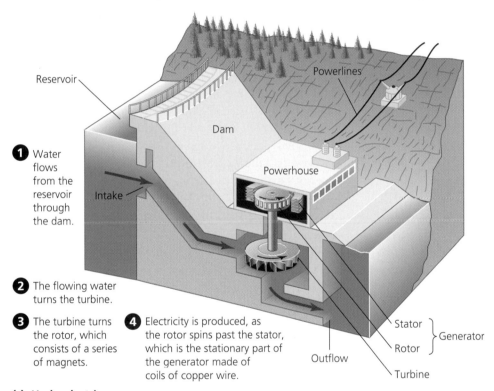

Reservoir

Powerlines

Dam

1 Water flows from the reservoir through the dam.

Intake

Powerhouse

2 The flowing water turns the turbine.

3 The turbine turns the rotor, which consists of a series of magnets.

4 Electricity is produced, as the rotor spins past the stator, which is the stationary part of the generator made of coils of copper wire.

Outflow

Stator ⎫
    ⎬ **Generator**
Rotor ⎭

Turbine

(c) Hydroelectric power

Figure 18.4 Large dams such as the Ice Harbor Dam on the Snake River near Pasco, Washington, (a) generate substantial amounts of hydroelectric power for people in many nations. Inside these dams, water is used to turn turbines (b) to generate electricity. Water is funneled from the reservoir through a portion of the dam (c) and used to rotate turbines, which turn rotors containing magnets. The spinning rotors generate electricity as their magnets pass coils of copper wire. Electrical current is transmitted away through power lines, and the river's water flows out through the base of the dam.

amounts of water are withdrawn from rivers upstream from dams, however, reservoir levels can drop low enough to interfere with power production. Thus when pressure from human population or consumption is intense enough, water supplies may be used faster than they are replenished and hydropower may not be sustainable. However, in most cases hydropower continues today to provide the renewable energy dams were built to provide.

The second advantage of hydropower over fossil fuels is its cleanliness. Because no carbon compounds are being burned in the production of hydropower, no carbon dioxide or other pollutants are being emitted into the atmosphere (although recent evidence indicates that some large reservoirs may release the greenhouse gas methane due to anaerobic decay in deep water). Because it is cleaner than fossil fuel combustion, hydropower has won praise for protecting air quality, climate, and human health. However, hydropower is far from innocuous. Although it produces little air pollution, it does cause thermal pollution, because water exiting a dam is cooler than water entering its reservoir. Dams and the reservoirs they hold back have had far-reaching disruptive effects on the local ecology and economy of riverside areas.

Hydropower may not expand much more

Use of hydropower is growing slightly more quickly than overall energy use, and some gargantuan projects are being planned and carried out, one of which is China's Three Gorges Dam. However, unlike other renewable energy sources, hydropower is not likely to expand very much more. This is because the majority of the world's large rivers that offer excellent opportunities for hydropower have already been dammed. Hydropower will likely continue to increase in developing nations that still possess undammed rivers, but in developed nations growth in hydropower will likely slow or stop. Moreover, in developed nations awareness of the environmental impacts of dams is causing people to propose removing many of the dams that exist (Chapter 14). Iceland will need considerably more electricity than it currently generates in order to power its hydrogen economy, but many Icelanders are opposed to building new dams and flooding scenic areas. While plans for new dams are mostly going ahead, some Icelanders have instead urged obtaining additional electricity from new geothermal and wind sources.

In the United States, 98% of rivers appropriate for dam construction already are dammed, and many of the remaining 2% are protected from dam-building under the Wild and Scenic Rivers Act. In the future the number of dams in the United States is likely to decrease as hydropower licenses expire and are not renewed. The challenge and the opportunity at that time will be to find ways of restoring river habitats after dams have been removed, while harnessing energy from renewable sources such as wind, geothermal, and solar.

Solar Energy

The sun provides energy for almost all biological activity on Earth (Chapter 4) by converting hydrogen to helium in nuclear fusion (Chapter 17). An immense amount of energy travels outward from the sun in all directions as electromagnetic radiation, but only an infinitesimal fraction reaches Earth. That relatively minuscule amount, however, is all that is needed to power life on our planet. About 1 kilowatt/m^2 of solar energy reaches Earth's surface, 17 times the energy of a light bulb in each square meter. The average house has enough roof space to generate all its power needs using solar panels. The sun's raw energy is so strong that if only 0.1% of Earth's surface—roughly the combined area of New Mexico and South Dakota—were covered with solar panels, we would have enough solar energy to power all the world's electrical plants. The amount of energy Earth receives from the sun each day, if hypothetically collected in full for our use, would be enough to power human consumption for 27 years.

Clearly, the potential for using sunlight to meet our energy needs is tremendous. As you will see, however, all this "free" energy from the sun cannot be harnessed just yet; we are still in the process of developing appropriate solar technologies and learning the most effective ways to put the sun's energy to use. A number of approaches, however, have been successfully developed.

We tend to think of solar power as a novel technology, but solar energy was first harnessed for power using active technology in 1767, when Swiss scientist Horace de Saussure built a thermal solar collector to heat water and cook food. In 1891 U.S. inventor Clarence Kemp claimed the first commercial patent for a solar-powered water heater. Two California entrepreneurs soon bought the patent rights and outfitted one-third of the homes of Pasadena, California, with solar water heaters by 1897.

The most commonly used approach to solar energy collection is **passive solar energy collection,** in which buildings are designed and building materials chosen to maximize their direct absorption of sunlight in winter, even as they keep the interior cool in the heat of summer. This approach can be contrasted with **active solar energy collection,** which makes use of technological devices to focus, move, or store solar energy.

Passive solar heating is simple and effective

Passive solar design techniques include using heat-absorbing construction materials and installing low, south-facing windows to maximize sunlight capture in the northern-hemisphere winter. In addition, thermal masses made of straw, brick, or concrete may be strategically located so that in cold weather the mass receives a great amount of sunlight and can radiate this heat in the interior of the building. In warm weather, the mass can be located away from windows so that it absorbs warmed air in the interior to cool it. Passive solar design can also involve use of certain types of vegetation planted in certain locations around a building. By heating buildings in cold weather and cooling them in warm weather, passive solar methods help conserve energy and reduce energy costs. Passive solar design principles can also be used to heat water.

Active solar energy collection can be used to heat air and water in buildings

Among active methods of harnessing solar energy, solar panels or flat-plate solar collectors generally consist of dark-colored, heat-absorbing metal plates mounted in large flat boxes covered with glass panes and are most often installed on rooftops. Water, other liquids, or air is run through tubes that pass through the collectors, transferring heat to places inside a building. Heated water can be pumped to tanks designed to store the heat for later use and through pipes designed to release the heat into the building. Such systems have proven especially effective for heating water for residences. By the early 1990s, more than 4.5 million Japanese buildings and 83% of Israeli households were using them.

Active solar energy is being used for heating, cooling, and water purification in Gaviotas, a small remote town in the high plains of Colombia far from any electrical grid. Engineers have developed inexpensive solar panels that harvest enough solar energy, even under cloudy skies, to boil drinking water for a family of four. They also have designed, built, and installed a unique solar refrigerator in a rural hospital, along with a large-scale solar collector to boil and sterilize water. Their innovations show that solar power does not have to be expensive or confined to regions that are always sunny (Figure 18.5). Many of the solar panels produced in Gaviotas have since been installed on roofs in Colombia's capital city, Bogotá.

Concentrating solar rays can magnify the energy received

The strength of solar energy can be magnified by gathering sunlight from a wide area and focusing it on a single point. This is the principle behind solar cookers, simple

Figure 18.5 Engineers in Gaviotas, a remote highland town in Colombia, have developed inexpensive solar panels that power numerous residences and businesses, providing active solar energy for heating, cooling, and water purification.

portable ovens that use reflectors to focus sunlight onto food and cook it (Figure 18.6). Such cookers are becoming extremely useful and widespread in parts of the developing world.

The principle of concentrating the sun's rays has also been put to work by utilities in large-scale, high-tech approaches to producing electricity from solar energy. Large reflectors that look like satellite dishes can focus sunlight on a receiver, where fluid is heated and expanded to drive pistons or turbines in an engine to produce electricity. Another approach is the use of solar-trough collection systems, which consist of mirrors that gather sunlight and focus it on oil in troughs. The super-heated oil is then used to create steam that drives turbines, as in conventional power plants. In yet another approach, mirrors concentrate sunlight sharply onto a receiver atop a tall "power-tower" (Figure 18.7). From the receiver, heat is transported through fluids (often molten salts) piped to a steam-driven generator to create electricity. These solar power plants can harness light from large mirrors spread across many hectares of land. The world's largest so far, a collaboration among government, industry, and utility companies in the California desert, produces enough power for 10,000 households.

Figure 18.6 Portable solar cookers, such as this one boiling water in a kettle in Nepal, are becoming popular in some parts of the developing world. These inexpensive ovens work by greatly concentrating the sun's rays onto a small area.

Figure 18.7 The principle of portable solar cookers—focusing the sun's rays on a small area —is also used on a large scale in several high-tech approaches. One involves concentrating sunlight on a receiver atop a "power-tower," often from mirrors spread across large expanses of land. Heat is then transported through fluid-filled pipes to a steam-driven generator that produces electricity. Shown is the Solar One facility in the desert of southern California. This landmark power tower project, operated by Southern California Edison from 1982 to 1988, marked the first time a public utility was able to use solar energy to generate a substantial portion of its power. The facility was superceded by Solar Two, which could more effectively store heat for use at night and during cloudy periods.

Photovoltaic cells produce electricity directly from sunlight

A more direct approach to producing electricity from sunlight involves photovoltaic (PV) systems. **Photovoltaic (PV) cells** collect sunlight and convert it to electrical energy directly by making use of the photovoltaic, or **photoelectric effect**, which was first proposed in 1839 by French physicist Edmund Becquerel. This effect is produced when light strikes one of a pair of negatively charged metal plates in a PV cell, causing the release of electrons. The electrons are attracted by electrostatic forces to opposing plates. The repeated flow of electrons from one plate to the other forms an electrical current, which can be converted into alternating current (AC) and used for residential and commercial electrical power (Figure 18.8).

The plates of a PV cell are made primarily of silicon, enriched on one side with phosphorus and on the other with boron. Silicon is a semiconductor, conducting and controlling the flow of electricity. The chemical properties of boron and phosphorus result in the boron-enriched side carrying a positive electrical charge and the phosphorus-enriched side carrying a negative charge. When sunlight strikes the cell surface, it transfers energy to electrons within the PV cell, causing the electrons to move. Because of the difference in charges, when wires connect the two sides, electricity will flow. Photovoltaic cells can be connected to batteries that are able to store the accumulated charge until it is needed.

Multiple PV cells are arranged in modules, which can comprise panels, which can be gathered together in flat arrays. Many houses in the United States and elsewhere now have photovoltaic arrays on their roofs. There are also a few large-scale photovoltaic installations run by electrical utilities. In some remote areas, such as Xcalak, Mexico, PV systems are being used in combination with wind turbines and a diesel generator to power entire villages.

One drawback of photovoltaic systems is that they rely on batteries, and once their batteries are fully charged, no additional energy can be stored. For this reason, most PV systems today are connected to the regional electric grid. Such a connection also enables two-way metering, which allows owners of houses with PV systems to sell their excess solar energy to their local

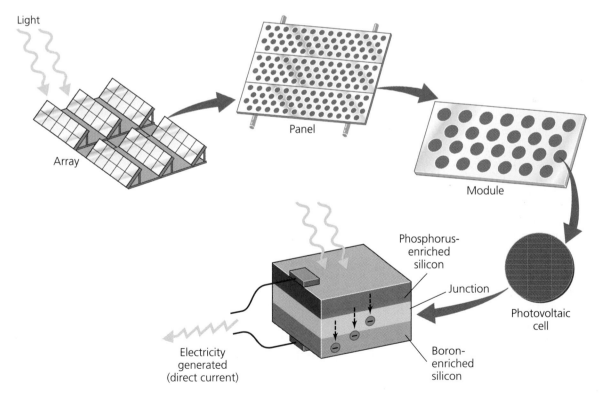

Figure 18.8 A photovoltaic cell converts sunlight to electrical energy. When sunlight hits a layer of silicon infused with phosphorus, electrons are released and travel toward the positively charged layer of silicon laced with boron. This movement of electrons induces an electric current, producing usable electricity. Photovoltaic cells are grouped in modules, which can comprise panels, which can be erected in arrays.

power utility. The value of power sold to the utility by the consumer is subtracted from the consumer's monthly utility bill.

Solar power is little used but fast growing

Although solar technology dates from the 19th century, it was pushed to the sidelines as fossil fuels gained a stronger foothold in our energy economy. Even as solar technology was being refined for applications ranging from hand-held calculators to spacecraft, it was not being developed for the roles that fossil fuels conventionally played. In recent U.S. history, funding for research and development of solar technology has been irregular. After the 1973 oil embargo, the U.S. Department of Energy funded installation and testing of over 3,000 PV cell systems, providing a boost to companies in the solar industry. As oil prices declined, however, so did government support for solar power, and funding for solar has remained far below that for fossil fuels.

Largely because of the lack of investment, solar power presently contributes only a minuscule portion of our energy production. In 2000, solar accounted for only 0.039%—less than 4 parts in 10,000—of the global primary energy supply. For electricity generation in the United States, solar made up only 0.24% of power generation from renewables in that year. However, solar energy's use has been expanding rapidly. Worldwide, it has grown by nearly one-third annually since 1971, a growth rate second only to that of wind power. Solar is proving especially attractive in developing countries, many of which are rich in sun but poor in power infrastructure and where hundreds of millions of people are still without electricity. Some multinational corporations that built themselves on fossil fuels are now helping to spread alternative energy. BP Solar, British Petroleum's solar energy wing, recently completed $30 million projects in the Philippines and Indonesia and—in what may be the largest solar power project ever—is working on a $48 million project to supply 400,000 people in 150 villages with electricity.

Sales of PV cells are also growing fast—by 25% per year in the United States, for instance, and 63% annually in Japan, where they are now sometimes built into roofing tiles. U.S. government energy experts project that 25% annual growth in the PV industry can be sustained through at least 2020. Use of solar technology is widely expected to continue increasing, as prices fall, technologies improve, economic incentives are enacted, and people increasingly desire renewable energy to replace fossil fuels. However, despite all the growth, the very small amount of energy currently produced by solar means that its market share will likely remain small for years or decades to come—unless governments, businesses, and consumers become more motivated by the benefits that solar can provide.

Solar power offers many benefits

The fact that the sun will continue burning for another 4–5 billion years makes it practically inexhaustible as an energy source for human civilization. Additionally, the amount of solar energy reaching Earth's surface should be enough to power our civilization once solar technology is adequately developed. Although these over-arching benefits of using solar energy are clear, there are also many benefits involving the technologies themselves. PV cells and other solar technologies use no fuel, are quiet and safe, contain no moving parts, require little maintenance, and do not even require a turbine or generator to create electricity. An average unit can produce energy for 20–30 years.

Another advantage of solar systems is that they allow for local, decentralized control over power. Individual homes and businesses and isolated communities can use solar to produce their own electricity and may not need to be near a power plant or connected to the grid of a large city. This self-sufficiency can reduce an area's economic dependence on other areas and also reduce the need for extensive networks of transmission lines for electricity. In the developed world, homes with two-way metering that collect solar energy for PV cell use and also are connected to the local grid can save money by selling electricity to their utility. In developing countries, solar cookers enable families to cook food without having to gather fuelwood, lessening people's labor obligations and helping reduce deforestation. In locations such as refugee camps, solar cookers are greatly relieving social and environmental stress. The low cost of solar cookers—many can be built locally for $2–$10 each—has made them available for purchase or donation in many impoverished areas.

Finally, a major benefit of solar power over fossil fuels is that it does not pollute the air with emissions and thus greatly reduces levels of greenhouse gases and pollutants released into the atmosphere relative to fossil fuels. The manufacture of photovoltaic cells *does* require fossil fuel use at this time, but once up and running, a PV system produces no emissions. It is possible to calculate the amount of emissions prevented by various types of solar technologies. The website of BP Solar provides a service with which consumers can calculate the economic and environmental impacts of installing

PV solar cells in any location in the United States, by entering their location and the details of the system. For example, as of August 2003, installing a 5-kilowatt PV system in a home in Dallas, Texas, could provide 51% of annual power needs, save $391 per year on energy bills, and prevent 9,023 lbs of carbon dioxide from being added to the air. Even in rainy Seattle, Washington, a 5-kilowatt system producing 32% of energy needs and saving $336 per year could prevent the emission of 6,419 lbs of carbon dioxide from burning of fossil fuels.

Location and cost can be the disadvantages of solar power

Solar power currently has two major disadvantages. One is that not all regions are sunny enough to provide adequate power with current technology. While Earth as a whole receives vast amounts of sunlight, not every location on Earth does. Cities such as Seattle, Washington, might find it difficult to harness enough sunlight most of the year to depend on solar power. Other areas, such as the southwestern United States, the Middle East, and many areas in Australia, Africa, Asia, and Latin America, receive plenty of sunshine for solar power. Daily or seasonal variation in sunlight can also pose problems for stand-alone solar systems if storage capacity in batteries or fuel cells is not adequate or if backup power is not available from a municipal electricity grid.

The primary disadvantage, however, to current solar technology—as with other renewable sources—is the up-front cost of investing in the equipment. The investment cost for solar remains higher than that for fossil fuels, and indeed, solar is still the most expensive way to produce electricity. Proponents of solar power argue that its noncompetitiveness with fossil fuels and nuclear power results from decades of government promotion of these nonrenewable sources. Fossil fuels and nuclear energy have received many types of financial breaks that solar has not. Because the external costs of nonrenewable energy have largely not been included in market prices, these energy sources have remained relatively cheap, so that governments, businesses, and consumers have had little economic incentive to switch to solar and other renewables.

However, decreases in price and improvements in energy efficiency of solar technologies so far are encouraging, even in the absence of significant financial commitment from government and industry. At their advent in 1954, solar technologies had efficiencies of around 6%, while costing $600 per watt. Recent single-crystal silicon PV cells, however, are showing 24% efficiency in the lab and 15% efficiency commercially, suggesting that future solar technologies may be more efficient than any energy technologies we have today. Solar systems have become much cheaper over the years and now can often pay for themselves in 10–15 years. After that time, they provide energy virtually for free as long as the equipment lasts. With future technological advances, the time to recoup investment could fall to 1–3 years, some experts believe.

Wind Energy

Wind energy can be thought of as another form of solar energy, because it is the sun's differential heating of air masses on Earth that causes wind to blow (Chapter 11). We can harness power from wind by using devices called **wind turbines**, mechanical assemblies that convert wind's kinetic energy, or energy of motion, into electrical energy.

Wind has long been used for energy

Today's wind turbines have their historical roots in Europe, where wooden windmills have been used for 800 years. The Netherlands in particular is famous for its widespread use of windmills, which have generally been used to pump water for irrigating crops or draining wetlands and to grind grain into flour. In each application, wind causes a windmill's blades to turn in a circle, driving a shaft connected to several cogs that turn wheels that either grind grain or pull buckets from a well. In the United States, countless ranches in arid areas of the West and Great Plains feature windmills that draw water up from the ground to supply thirsty cattle.

The first wind turbine built for the generation of electricity was constructed in the late 1800s by inventor Charles Brush, who designed a turbine 17 m (50 ft) tall with 144 rotor blades made of cedar wood. Technology advanced slowly through the 20th century, but it was not until after the 1973 oil embargo that wind energy was funded by governments in North America and Europe. This moderate infusion of funding for research and development boosted technological progress and helped to enable the cost of wind power to be cut in half in less than 10 years. Today wind power at favorable locations is nearly competitive with conventional electricity generation, and modern wind turbines appear more reminiscent of airplane propellers or sleek new helicopters than of romantic old Dutch paintings.

Modern wind turbines convert kinetic energy to electrical energy

Wind turbines use the motion of wind passing through their blades to create electricity. Wind blowing into a turbine turns the blades of the rotor, which rotate the machinery inside a compartment called a *nacelle,* which sits atop a tall tower (Figure 18.9). Inside the nacelle are a gearbox and a generator, as well as equipment to monitor and control the turbine's activity. The towers are built high to maximize wind capture; most today range from 40 to 100 m (131 to 328 ft) tall. Higher is generally better, to minimize turbulence and maximize wind speed. Most rotors consist of three blades and measure 42–80 m (138–262 ft) across. Turbines are designed to yaw, or rotate back and forth in response to changes in wind direction, ensuring that the motor faces into the wind at all times. Turbines can be erected singly, but are most often erected in groups called wind parks, or **wind farms** (Figure 18.10). The world's largest wind farm, at Altamont Pass in California, contains several thousand turbines spread across a landscape of rolling, grassy hills.

Applying principles of modern aerodynamics, engineers have designed turbines to begin turning at specific wind speeds to harvest wind power as efficiently as possible. Some turbines create low levels of electricity by turning in light breezes. Others are programmed to rotate only in strong winds, operating less frequently but

Figure 18.10 Wind turbines are often erected in large groups (called wind farms), situated in areas of the landscape that provide reliably high wind velocities.

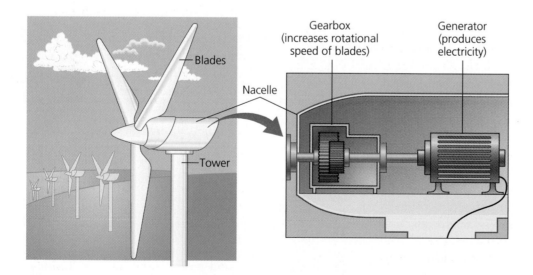

Figure 18.9 A wind turbine converts energy of motion from wind into electrical energy. Wind causes the blades of a wind turbine to spin, turning a shaft that extends into the nacelle that is perched atop the tower. Inside the nacelle, a gearbox converts the rotational speed of the blades (which can be up to 20 revolutions per minute (RPM) or more) into much higher rotational speeds (over 1,500 RPM). These high speeds provide adequate motion for a generator inside the nacelle to produce electricity.

generating large amounts of electricity in short time periods. Slight differences in wind speed can yield substantial differences in power output, for two reasons. First, the energy content of a given amount of wind increases as the square of its velocity; thus if wind velocity doubles, energy quadruples. Second, an increase in wind speed causes more air molecules to pass through the wind turbine per unit time, making power output equal to wind velocity cubed. Thus a doubled wind velocity actually results in an eight-fold increase in power output.

Wind power is the fastest-growing energy sector

Like solar energy, wind provides only a minuscule proportion of the world's power needs, but wind power is growing fast. As of 2000, wind power accounted for only 0.026% of global primary energy supply—less than 3 parts in 10,000. Globally, at the end of 2002, wind supplied electricity equivalent to supply 6 million American homes. Wind provided over 1.5% of U.S. renewable electricity generation in 2000—a small amount but six times more than solar. So far wind energy production is geographically concentrated; only five nations account for 84% of the world's wind energy output (Figure 18.11). Within the United States, California and Texas account for two-thirds of the nation's wind power. Denmark is a leader in wind power; there, a series of land-based and offshore wind farms supply 10% of the nation's power needs.

Wind is the fastest-growing source of energy in the world today, having expanded at an astonishing annual rate of 52% over the past 30 years. Wind is also the world's fastest-growing source of electricity; generating capacity grew at an annual rate of 25% from 1990 to 2000. Experts agree that the growth will continue, because only a very small portion of this resource is currently being tapped. Meteorological evidence suggests that wind power could be expanded in the United States to meet the electrical needs of the entire country (see The Science behind the Story).

Wind power has many benefits

Like solar power, wind produces no emissions once wind power equipment is manufactured and installed. As a replacement for fossil fuel combustion in the average U.S. utility generator, running a 1-megawatt wind turbine for one year can prevent the release of more than 1,500 tons of carbon dioxide, 6.5 tons of sulfur dioxide, 3.2 tons of nitrogen oxides, and 60 lbs of mercury, the U.S. EPA has calculated. The 10 million megawatt-

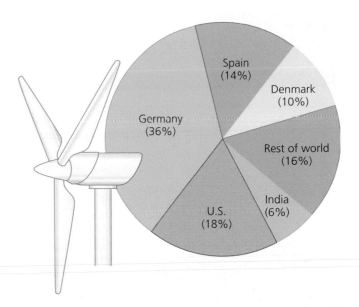

Figure 18.11 Most of the world's fast-growing wind power generating capacity is concentrated in a handful of countries. While tiny Denmark obtains the highest percentage of its energy needs from wind, the larger nations of Germany, the United States, and Spain have so far developed more total wind capacity. Data from American Wind Energy Association (AWEA), Global wind energy market report, 2002.

hours of electricity produced each year by wind farms in the United States prevents emissions of 6.7 million tons of carbon dioxide, 37,500 tons of sulfur dioxide, and 17,750 tons of nitrogen oxides. The amount of carbon pollution prevented from entering the atmosphere is equivalent to a freight train of 50 cars, with each car holding 100 tons of solid carbon, each and every day.

Wind power appears considerably more energy-efficient than conventional power sources. One recent study comparing the amount of energy produced by a technology to the amount consumed by the technology found that wind turbines produced 23 times as much as they consumed. For nuclear energy, the ratio was 16:1; for coal it was 11:1; and for natural gas it was 5:1. Wind farms also use less water than do traditional generating plants.

Wind turbine technology can be used on many scales, from a single tower for local use to fields of thousands for utilities to supply large regions. Small-scale turbine development can help make local areas more self-sufficient, just as solar energy can. For instance, the Rosebud Sioux Tribe of Native Americans in 2003 set up a single turbine on their reservation in South Dakota. The turbine will produce electricity for 220 homes and bring the tribe an estimated $15,000 per year in revenue. Wind resources are rich in this region, and the tribe hopes to erect more turbines in coming years.

The Science behind the Story

Idaho's Wind Prospectors

Where does wind translate into energy? In Idaho, where resource planners decided the state's power future lies in generating electricity from the wind. Now scores of Idahoans, from small farmers to Native American tribes, have joined in the search for gusts with energy potential.

By handing out wind measuring devices to interested landowners, Idaho has turned its citizens into "wind prospectors," who pinpoint potential areas for wind farms. Idaho first launched the public wind prospecting program in 2001, after joining several other northwestern states in a regional research effort. People who join the program must collect data on wind speed and direction, share that data with the state, and agree to make it public.

Finding a promising wind farm site in Idaho rests on the same essentials used by wind prospectors anywhere. Wind farms require some human infrastructure, such as roads for erecting wind turbines and transmission lines for sending out power the turbines generate. But the single most important factor operates solely at the behest of the Earth's climate—the speed and frequency of the wind. Effective commercial wind farms must have a steady flow of wind just above ground level, with regular gusts of at least 20.9 km (13 mi) per hour at a height of about 50 m (164 ft) above the surface.

Determining whether a possible site merits further study starts with analysis of existing data. In many parts of the developed world, decades worth of weather information have been compiled into computerized wind maps that indicate general wind conditions. In Idaho, energy planners provide starter maps that divide the state into seven

To determine where to build wind farms, Idaho's wind prospectors use anemometers, relatively simple devices that collect and relay wind data. Cup wheels rotate to indicate wind speed, and a vane turns on a vertical axis to reveal wind direction.

wind "classes" and reveal which general areas might have enough wind to make a wind farm worthwhile. Areas listed as "Class 3" or higher, with wind speeds of about 23 km (14.3 mi) per hour at 50 m (164 ft) above ground, offer the best possibilities for wind prospectors.

Such maps, however, may not provide enough detail about a specific location. A piece of property, for example, may sit in a Class 3 area but be sheltered by a small hill that blocks the wind. Knowing that kind of detail requires site-specific research. To make site-specific research possible, Idaho landowners in areas listed as Class 3 or higher can borrow devices from the state that measure wind speed and direction. The tools, called *anemometers,* allow novice wind prospectors to set up their own wind research labs.

Anemometers are centuries-old devices that have morphed into contemporary technology. The

Idaho program uses a common cup anemometer, with an array of three or four hollow cups set to catch the wind and rotate around a vertical rod. The force of the wind on the cups causes them to rotate at a speed proportional to the wind speed; the greater the wind, the faster the cups rotate. Wind direction is measured by a vane that turns on a vertical axis pointing directly into the wind. The cup wheel and wind vane are connected electrically to speed and direction dials, which relay wind data.

In Idaho, more than 80 landowners borrowed anemometers from the state in the wind prospecting program's first year, sending collected data from their studies to the state every 60 days for review by energy planners and for subsequent posting online. The studies have generated new funding and wind farm plans in the state. In the fall of 2003, one farm near Idaho Falls won a $500,000 federal grant to help build a 1.5-megawatt wind farm that could supply power for approximately 500 homes.

In eastern Idaho, five anemometers set up on Shoshone-Bannock tribal lands have revealed good prospects for a commercial wind farm on two Native American reservations. The research effort has shown that the lands are "world-class sites" for wind power, according to a state energy official. With average wind speeds in the 29 km (18 mi) per hour range, further study of the tribal lands revealed possible sites for large-scale commercial wind farms, which could mean jobs and revenue for the reservations. Similar wind prospecting programs are now underway on other reservations, as well as in other states, including Utah, Oregon, Virginia, and Missouri.

Another societal benefit of wind power is that land-owners can lease their land for wind development, which provides them extra revenue while also increasing property tax income for rural communities. A single large turbine can bring in $2,000–$4,500 in annual royalties while occupying just a quarter-acre of land. Because each turbine takes up only a small area, most of the land can still be used for farming, ranching, or other uses.

Economically, wind energy involves up-front costs for the erection of turbines and the expansion of infrastructure to allow electricity distribution, but over the lifetime of the project it requires only maintenance costs. Unlike fossil fuel power plants, the turbines incur no ongoing fuel costs. Currently, start up costs of wind farms generally are higher than fossil-fuel-driven plants, but wind farms incur fewer expenses once the plants are up and running. Advancing technology is driving down the costs of wind farm construction; as large wind farms become more efficient, the cost of each unit of electricity produced is dropping.

Wind energy has some downsides

Unlike power sources that can be turned off and on at will, humans have no control over when wind will occur; it is an intermittent resource. This stands as a disadvantage of relying solely on wind, but poses little problem if wind is one of several sources contributing to a utility's power generation. In addition, technologies are available that could effectively address the problem of reliance on intermittant wind resources; for example, batteries or hydrogen fuel can store energy generated by wind and release it later when needed.

Just as wind varies from time to time, it varies from place to place; some areas are simply windier than others. Global wind patterns combine with local topography—mountains, hills, water bodies, forests, cities—to create local wind patterns, and companies study these patterns closely before investing in a wind farm. Extensive meteorological research has given us excellent information with which to judge prime areas for locating wind farms. A map of average wind speeds across the United States (Figure 18.12a) shows that mountainous regions and areas of the Great Plains are best. Based on such information, the young wind power industry has located much of its generating capacity in states with high wind speeds, and is looking to further expand in the Great Plains and mountain states (Figure 18.12b). Provided that wind farms were strategically erected in optimal locations, an estimated 15% of U.S. energy demand could be met by placement on only 43,000 km² (16,600 mi²) of land (with less than 5% of this land area actually occupied by turbines, equipment, and access roads).

Good wind resources, however, are not always near population centers that need the energy. Thus transmission networks will need to be greatly expanded. Moreover, when wind farms *are* proposed near population centers, local residents have often opposed them. In many cases, people have objected to wind farms for aesthetic reasons, feeling that the structures clutter the landscape, particularly because turbines are often located on exposed, conspicuous sites. Although polls show wide public approval of wind projects in regions where wind power has already been introduced, the so-called not-in-my-backyard (NIMBY) syndrome seems to kick in frequently with new wind projects. A proposal for North America's first offshore wind farm, in Nantucket Sound between Cape Cod and the islands of Nantucket and Martha's Vineyard, was facing opposition from area residents in 2003. As a writer for the *New York Times Magazine* perceptively portrayed, wealthy area residents who like to think of themselves as environmentalists opposed an environmentally beneficial development simply because it was located close to where they lived.

Wind turbines are also known to pose a threat to flying birds, which can be killed by the rotating blades, a problem that arose at California's Altamont Pass wind farm. This region had one of the densest populations of golden eagles in the country, and many eagles and other raptors were killed by wind turbines during the 1990s. Studies since then at other sites have suggested that bird deaths are not a widespread problem. It has been estimated that only one to two bird deaths occur per turbine per year—far less than the millions already being killed each year by television, radio, and cell phone towers, tall buildings, cars, domestic cats, pesticides, and other human causes. Thus the key seems to be selecting sites for wind farms that are not in the midst of prime habitat for certain species that are likely to fly into the blades. Protecting birds from wind turbines remains an area of ongoing research.

Offshore sites can be promising

Wind speeds over water are often greater than those over land; notice how speeds increase over the Great Lakes in Figure 18.12. Wind speeds on average are roughly 20% greater over water, producing 50% more power through a turbine. Turbulence near the surface is also reduced, so towers can afford to be lower. Although costs to erect and maintain turbines in water are higher, the stronger, less turbulent winds produce more power and make offshore wind potentially more profitable. For these reasons, offshore wind turbines are becoming popular. Currently offshore wind farms are limited to shallow water,

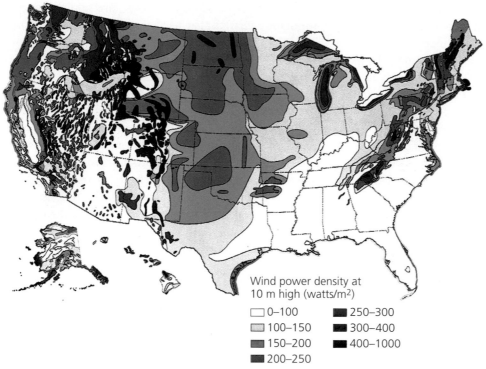

Wind power density at
10 m high (watts/m²)

☐ 0–100	■ 250–300
☐ 100–150	▨ 300–400
▨ 150–200	■ 400–1000
▨ 200–250	

(a) Annual average wind power

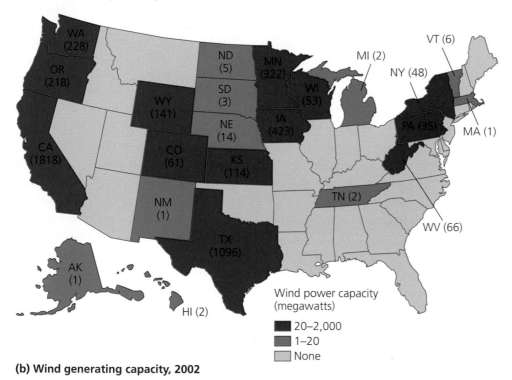

Wind power capacity
(megawatts)

■ 20–2,000	
▨ 1–20	
▨ None	

(b) Wind generating capacity, 2002

Figure 18.12 Wind speeds vary in different geographic areas, so meteorologists have used wind speed measurements to calculate the potential generating capacity from wind in different areas. The map in (a) shows average wind power in watts per square meter at a height of 10 m (33 ft) above ground across the United States. Such maps are used to help guide placement of wind farms. The development of U.S. wind power so far is summarized in (b), which shows the megawatts of generating capacity developed in each state as of 2002. *Sources:* (a) D.L. Elliott et al., *Wind Energy Resource Atlas of the United States,* Solar Energy Research Institute, Golden, Colorado, 1987; (b) National Renewable Energy Laboratory, U.S. Department of Energy, 2002.

where towers are drilled into sediments singly or with a tripod configuration to stabilize them. However, in the future towers may be placed on floating pads anchored to the seafloor in deep water. At great distances from land, it may be best to store the electricity generated as hydrogen and then ship or pipe this to land, rather than try to build submarine cables to carry electricity to shore, but further research is needed on this option.

Denmark erected the first offshore wind farm in 1991. Over the next decade, nine more came into operation across northern Europe, where the North and Baltic Seas offer strong winds. The power output of these farms increased 43% annually, as larger turbines were erected. Several northern European nations are projecting and encouraging continued rapid growth in the near future. Wind advocates in Iceland are considering developing 240 offshore wind turbines in the nation's waters to meet future electricity demand for its hydrogen economy by supplementing power from hydroelectric and geothermal sources.

Geothermal Energy

Geothermal energy is one form of renewable energy that does not originate from the sun. Instead, it is generated from deep within Earth. The radioactive decay of elements amid the extremely high pressures and temperatures deep in Earth generate heat that rises to the surface through magma (molten rock, Chapter 6), and through fissures and cracks. Where this energy heats groundwater, natural spurts of heated water and steam are sent up from below. Terrestrial geysers and submarine hydrothermal vents (Chapter 4) are the surface manifestations of these processes. Iceland exists because of the mid-Atlantic Ridge (Chapter 6), the area of volcanic activity at the spreading boundary of two tectonic plates. The island is in fact the one stretch of the mid-Atlantic Ridge that has built itself up above the surface of the ocean. Because of the geothermal heat in this region, Iceland is filled with volcanoes and geysers. In fact, the word "geyser" originated from the Icelandic *Geysir*, the name for the island's largest geyser, which recently resumed its periodic eruptions after many years in dormancy.

Geothermal power plants use the energy of naturally heated water to generate power. Rising underground water and steam are harnessed to turn turbines and create electricity. Although geothermal energy is renewable in principle (its use does not affect the amount of heat produced in Earth's interior), the power plants we build

to use this energy may not all be capable of operating indefinitely. If a geothermal plant uses heated water at a rate faster than the rate at which groundwater is recharged, the plant will eventually run out of water. This is occurring at The Geysers, in Napa Valley, California, where the first generator was built in 1960. In response, operators have begun injecting municipal wastewater into the ground to replenish the supply. This is being carried out in geothermal power plants in many parts of the world now. After the water is used, it is injected back into aquifers to help maintain pressure and thereby sustain the resource. A second reason geothermal energy may not always be renewable is that patterns of geothermal activity in Earth's crust shift naturally over time, so an area that produces hot groundwater now may not always do so.

Geothermal energy is harnessed for heating and electricity

Geothermal energy can be taken directly from geysers at the surface, but most often tapping the energy of geothermal sources requires that wells be drilled down hundreds or thousands of meters toward the heated groundwater. Generally, water at temperatures of 300–700°F or more is brought to the surface and converted to steam by lowering the pressure in specialized compartments. The steam is then employed in turning turbines to generate electricity (Figure 18.13).

Hot groundwater can also be used directly for heating homes, offices, and greenhouses; for driving industrial processes; and for drying crops. Iceland heats most of its homes through direct heating with piped hot water. Iceland began putting geothermal energy to use in the 1940s, and today 30 municipal district heating systems and 200 small private rural networks supply heat to 86% of the nation's residences. Other locales are benefiting in similar ways; the Oregon Institute of Technology heats its buildings with geothermal energy for 12–14% of the cost it would take to heat them with natural gas. Such direct use of naturally heated water is cheap and efficient, but it is feasible only in areas such as Iceland or parts of Oregon, where geothermal energy sources are abundant and near to where the heat must be transported.

Thermal energy from either water or solid earth can also be used to drive a heat pump to provide energy. Geothermal ground source heat pumps (GSHPs) use thermal energy from near-surface sources of earth and water rather than the deep geothermal heat for which utilities drill. Roughly half a million GSHPs are already used to heat U.S. residences. Compared to conventional electric heating and cooling systems, GSHPs heat spaces 50–70%

Figure 18.13 Geothermal energy facilities convert heat energy from underground into electrical energy **(a)**. Magma heats groundwater deep in the earth, some of which is let off naturally through surface vents such as geysers. Geothermal facilities tap into heated water below ground and channel steam through turbines in buildings above ground to generate electricity. After being used, the steam is often condensed and pumped back into the aquifer so as to maintain pressure. **(b)** At Nesjavellir geothermal power station in Iceland, steam is piped from the boreholes of four wells to a condenser at the plant where cold water pumped from lakeshore wells 6 km (3.7 mi) away is heated. The water, heated to 83°C (181°F), is sent through an insulated 270-km (170-mi) pipeline to Reykjavik and environs, where residents use it for washing and space heating.

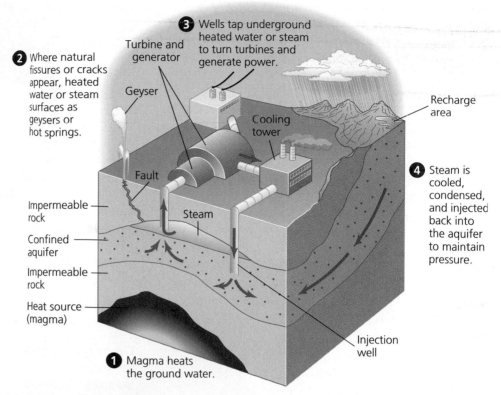

(a) Geothermal energy

(b) Nesjavellir geothermal power station, Iceland

more efficiently, cool them 20–40% more efficiently, can reduce electricity use by 25%–60%, and can reduce emissions by up to 72%. The principle by which these pumps work is the fact that soil does not vary in temperature from season to season as much as air does. The pumps heat buildings in the winter by transferring heat from the ground into buildings; they cool buildings in the summer by transferring heat from buildings into the ground. Both types of heat transfer are accomplished by a single network of underground plastic pipes with water circulating through them. Because heat is simply moved from place to place rather than being produced using outside energy inputs, heat pumps can be highly energy-efficient.

Use of geothermal power is growing

Geothermal energy provides less than half of 1% of total primary energy used worldwide. It provides nearly seven times the output of solar and wind combined, but only a small fraction of the energy from hydropower and biomass. Geothermal energy in the United States in 1999 provided enough power to supply electricity to over 1.4 million U.S. homes. At the world's largest geothermal power plants, The Geysers in northern California, generating capacity has declined by more than 50% since 1989 as steam pressure has declined, but The Geysers still provide enough electricity to supply a million residents. From 1985 to 1999, geothermal energy provided 7.3% of California's electricity. Worldwide, geothermal energy has grown by 8.8% annually over the past 30 years. Currently Japan, China, and the United States lead the world in use of geothermal power.

Geothermal power has benefits and limitations

Like other renewable sources, geothermal greatly reduces emissions relative to fossil fuel combustion. Geothermal sources can release variable amounts of gases dissolved in their water, including carbon dioxide, methane, ammonia, and hydrogen sulfide. However, these gases are generally in very small quantities, and it has been estimated that geothermal facilities on average release only a sixth of the carbon dioxide produced by plants fueled by natural gas. Geothermal facilities using the latest filtering technologies produce even fewer emissions. By one estimate, each megawatt of geothermal power prevents the emission of 7.8 million lb of carbon dioxide emissions and 1,900 lb of other pollutant emissions from gas-fired plants each year.

On the negative side of the ledger, geothermal sources, as we have seen, may not always be truly sustainable. In addition, the water of many hot springs is laced with salts and other minerals that corrode equipment and can pollute the air. These factors may shorten the lifetime of plants, increase maintenance costs, and add to pollution.

In addition, use of geothermal energy is limited to the particular areas where the energy can be tapped. Unless technology is developed to penetrate far more deeply into the ground than we can at present, geothermal energy use will remain more localized than solar, wind, biomass, or hydropower. Although there are places, like Iceland, that are especially rich in geothermal energy sources, most of the world is not. In the United States, geysers exist in some areas, such as Yellowstone National Park, and hot water and steam exist underground in various locations in the western part of the country. Nonetheless, many more hydrothermal resources remain unexploited around the world, awaiting improved technology and economic and policy encouragement by governments.

Ocean Energy Sources

Several very underexploited energy sources reside in the oceans. These sources all utilize continuous natural processes that might, if adequately developed, potentially provide sustainable energy for our needs. Of the three approaches most-developed so far, one involves temperature and two involve motion. Just as dams on rivers use flowing fresh water to generate hydroelectric power, some scientists, engineers, companies, and governments are developing ways to use the motion of ocean water to generate electrical power. Two types of kinetic energy show the most promise so far: the energy of wave motion and the energy of tidal motion.

We can harness energy from the tides

The rising and falling of ocean tides twice each day at coastal sites throughout the world can move large amounts of water past any given coastal point. Differences in height between low and high tides are especially great in long narrow bays like Alaska's Cook Inlet or the Bay of Fundy between New Brunswick and Nova Scotia. Such locations are best for harnessing tidal energy, accomplished by erecting dams across the outlets of tidal basins. The incoming tide is allowed through sluices past the dam, and the outgoing tide is passed through the dam, turning turbines to generate electricity as is done in dams for conventional hydroelectric power (Figure 18.14a). Some newer designs allow for electricity generation from water moving in both directions. The world's largest tidal generating facility is the La Rance facility in France (Figure 18.14b), which has operated for over

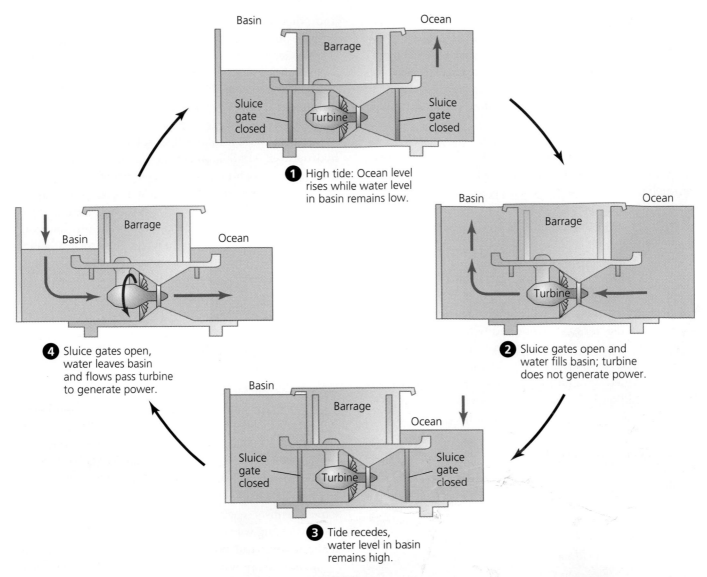

1 High tide: Ocean level rises while water level in basin remains low.

2 Sluice gates open and water fills basin; turbine does not generate power.

3 Tide recedes, water level in basin remains high.

4 Sluice gates open, water leaves basin and flows pass turbine to generate power.

(a) Tidal energy: Ebb-generating power plant with a bulb turbine

Figure 18.14 Energy can be extracted from the movement of the tides at coastal sites where tidal flux is great enough. One way of doing so is with bulb turbines utilizing the outgoing tide (**a**). At high tide ocean water is let through the sluice gates, filling an interior basin. At low tide the basin water is let out into the ocean, spinning turbines to generate electricity. The La Rance tidal power station in France (**b**) is one of the world's largest and is able to spin turbines with both incoming and outgoing tides.

(b) La Rance tidal power station, France

30 years. Smaller facilities now operate in China, Russia, and Canada. Tidal stations release few or no pollutant emissions, but they can have substantial impacts on the ecology of estuaries and tidal basins.

We can harness energy from waves

Wave energy could be developed at a greater variety of sites than could tidal energy. The principle is to harness the motion of wind-driven waves at the ocean's surface and convert this mechanical energy into electricity. Many designs for machinery to harness wave energy have been invented, but few have been adequately tested. Some designs are for offshore facilities, involving, for instance, floating devices that move up and down with the waves. Wave energy is greater at deep-ocean sites, but transmitting the electricity produced to shore would be expensive. Other designs are for coastal onshore facilities. Some of these designs funnel waves from large areas into narrow channels and into elevated reservoirs, from which water is then allowed to flow out, generating electricity as hydroelectric dams do. Other coastal designs use rising and falling waves to push air in and out of chambers, turning turbines to generate electricity (Figure 18.15). No commercial wave energy facilities are operating yet, but a number have been deployed as demonstration projects in several western European nations. Potential negative impacts include harm to coastal and marine ecosystems, hazards to boats, and aesthetic impacts on scenic beaches.

We can use thermal energy from the ocean

Besides the motion of tides and waves, other oceanic energy sources have been envisioned, including the motion of ocean currents, chemical gradients in salinity, and the immense thermal energy contained in the oceans. The concept of ocean thermal energy conversion (OTEC) has been most-developed. Each day the tropical oceans absorb an amount of solar radiation equivalent to the heat content of 250 billion barrels of oil—enough to provide 20,000 times the amount of electricity used daily in the United States. The ocean's sun-warmed surface is higher in temperature than its deep water, and OTEC approaches are based upon this gradient in temperature. In the *closed cycle* approach, warm surface water is piped into a facility to evaporate chemicals, such as ammonia, that boil at low temperatures. These evaporated gases spin turbines to generate electricity. Cold water piped in from ocean depths then condenses the gases so they can be reused. In the *open cycle* approach, the warm surface water is evaporated in a vacuum, and its steam turns the turbines and then is condensed by the cold water. Because the ocean water loses its salts upon evaporating, water can be recovered, condensed, and sold as desalinized fresh water for drinking or agriculture. Research on

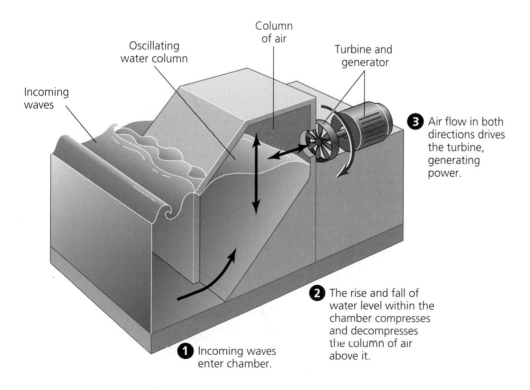

Figure 18.15 Coastal facilities can make use of energy from the continuous motion of ocean waves. As waves are let into and out of a tightly sealed chamber, the air inside is compressed and decompressed, creating air flow that rotates turbines to generate electricity.

Column of air

Oscillating water column

Turbine and generator

Incoming waves

3 Air flow in both directions drives the turbine, generating power.

2 The rise and fall of water level within the chamber compresses and decompresses the column of air above it.

1 Incoming waves enter chamber.

OTEC systems has been conducted in Hawaii and other locations, but costs remain too high, and as of yet no facility is commercially operational.

Weighing the Issues:
Your Island's Energy?

Imagine you have just been elected the president of an island country the size of Iceland, and your nation's Congress is calling on you to propose a national energy policy. Unlike Iceland, your country is located in equatorial waters. Your geologists do not yet know if there are fossil fuel deposits or geothermal resources under your land, but your country gets a lot of sunlight, a fair amount of wind, and has broad shallow shelf regions surrounding its coasts. Your island's population is moderately wealthy but is growing fast, and importing fossil fuels from mainland nations is becoming more and more expensive. What approaches would you propose in your energy policy? What specific steps would you urge your Congress to fund immediately? What trade relationships would you seek to establish with other countries? What questions would you ask of your economic advisors? What questions would you fund your country's scientists to research?

Hydrogen

All the renewable energy sources we have discussed can be used to generate electricity—and to generate it more cleanly than can fossil fuels. As phenomenally useful as electricity is to us, however, it cannot be stored easily in large quantities for use when and where needed. This is why vehicles rely on fossil fuels for their power. However, the development of fuel cells and hydrogen fuel show promise to store energy conveniently and in considerable quantities and to produce electricity at least as cleanly and efficiently as renewable energy sources.

In the "hydrogen economy" that Iceland's leaders and many energy experts worldwide now envision, hydrogen fuel, together with electricity, will serve as the basis for a clean, safe, and efficient energy system. This system will use as a fuel the universe's simplest and most abundant element, a constituent of water and of virtually all organic molecules. In this system, electricity generated from renewable sources that are intermittent,

such as wind or solar, can be used to produce hydrogen. Fuel cells can then employ hydrogen to produce electrical energy as needed to power vehicles, computers, cell phones, home heating, and countless other applications. Fuel cell technology has been used since the 1960s in NASA's spaceflight programs, including today's space shuttle. Basing an energy system on hydrogen could alleviate nations' dependence on foreign fuels and could help fight climate change by eliminating emissions from fossil fuel combustion. For these reasons, the U.S., Canadian, European, and Asian governments are all funding research into hydrogen and fuel cell technology, and every major automobile company is investing in research and development to produce vehicles that run on hydrogen. The commercial development of practical fuel cells still faces many hurdles, but even without fuel cells, hydrogen fuel could still serve us; some experts envision hydrogen coming to be used in the internal combustion engines of the future.

Hydrogen fuel may be produced from water or from other matter

Hydrogen gas (H_2) does not tend to exist freely on Earth; rather, hydrogen atoms tend to bind to other molecules, becoming incorporated in everything from water to organic molecules. Therefore, to obtain the hydrogen gas we want to use as fuel, we must force these other substances to give off their hydrogen atoms, and this requires an input of energy. A number of potential ways of producing hydrogen exist and are being studied (see The Science behind the Story). In **electrolysis**, the process being pursued by Iceland, electricity is input to split hydrogen atoms from the oxygen atoms of water molecules:

$$2H_2O \rightarrow 2H_2 + O_2$$

Electrolysis produces pure hydrogen, and does so without emitting the carbon- or nitrogen-based pollutants of fossil fuel combustion. However, whether this strategy for hydrogen production will cause pollution over its entire life cycle depends on the source of the electricity used for the electrolysis. If coal is burned in order to create the electricity needed to produce hydrogen by electrolysis, then the entire process will not reduce emissions compared with reliance on fossil fuels. If, however, the electricity is produced by wind power or some similarly less-polluting renewable source then hydrogen production by electrolysis would create much less pollution and greenhouse warming than reliance on fossil fuels.

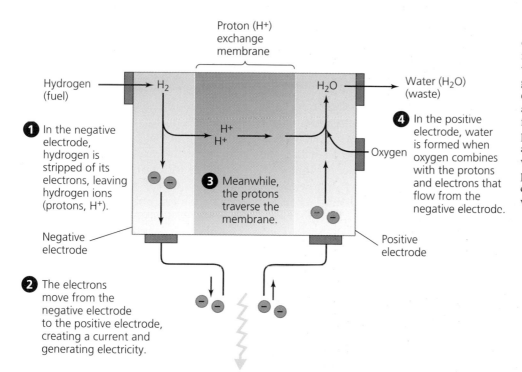

Figure 18.16 Hydrogen fuel drives electricity generation in a fuel cell, creating water as a waste product. Atoms of hydrogen gas are first stripped of their electrons, and then pass through a membrane. The electrons move from a negative electrode to a positive one, creating a current and generating electricity. Meanwhile, the hydrogen ions, after passing through the membrane, combine with oxygen to form water molecules.

The "cleanliness" of a future hydrogen economy in Iceland or anywhere else could, therefore, depend largely on the source of electricity used in electrolysis.

The environmental impact of hydrogen production will also depend on the source material for the hydrogen. Besides water, hydrogen can also be obtained from organic molecules, including biomass and fossil fuels. These processes generally require less energy input, but they result in emissions of carbon-based pollutants. For instance, extracting hydrogen from the methane (CH_4) in natural gas entails producing one molecule of the greenhouse gas carbon dioxide for every four molecules of hydrogen gas:

$$CH_4 + 2H_2O \rightarrow 4H_2 + CO_2$$

Thus, whether a hydrogen-based energy system is environmentally cleaner than a fossil fuel system depends on how the hydrogen is extracted. In addition, some new research suggests that leakage of hydrogen from the production, transport, and use of the gas at Earth's surface could potentially have consequences in the atmosphere, including increasing the depletion of stratospheric ozone and lengthening the atmospheric lifetime of the greenhouse gas methane. Research into these questions is ongoing, because scientists do not want society to switch from fossil fuels to hydrogen without first knowing the possible risks from hydrogen.

Fuel cells produce electricity by joining hydrogen and oxygen

Once hydrogen gas has been isolated, it can be used as a fuel to produce electricity within fuel cells. The overall chemical reaction involved in a fuel cell is simply the reverse of that shown for electrolysis; an oxygen molecule and two hydrogen molecules each split so that their atoms can bind to one another and form two water molecules:

$$2H_2 + O_2 \rightarrow 2H_2O$$

The way this occurs within one common type of fuel cell is shown in Figure 18.16. Hydrogen gas (usually compressed and stored in an attached fuel tank) is allowed into one side of the cell, whose middle consists of two electrodes that sandwich a membrane that only protons (hydrogen ions) can move across. One electrode, helped by a chemical catalyst, strips the hydrogen gas of its electrons, creating two hydrogen ions that begin moving across the membrane. Meanwhile, on the other side of the cell, oxygen molecules from the open air are split into their component atoms along the other electrode. These oxygen ions soon bind to pairs of hydrogen ions coming across the membrane, forming molecules of water that are let out as waste, along with heat. While this is occurring, the electrons from the hydrogen atoms have traveled to a device that completes an electric current between the

Algae as a Hydrogen Fuel Source

As scientists search for new ways to generate energy, some are looking past wind farms and solar panels to an unlikely but promising power source—pond scum. In California and Colorado, algae is being studied as an innovative way to generate the large amounts of hydrogen needed to move society towards a more sustainable energy source.

Hydrogen could potentially provide a major source of renewable energy that is far cleaner than current fuel sources. These benefits, however, hinge on how hydrogen fuel is produced. Some current methods release substantial amounts of carbon dioxide, while other non-polluting processes can be costly. These drawbacks have kept scientists searching for new hydrogen sources.

At the University of California at Berkeley, plant biologist Anastasios Melis thought one possible hydrogen source might be a single-celled aquatic plant known to be a capable, if sporadic, hydrogen producer. The alga *Chlamydomonas reinhardtii* was known to emit small amounts of hydrogen for brief periods of time when deprived of light.

Melis hypothesized that the alga could be encouraged to produce hydrogen in large amounts. He set up an experiment with energy experts at the National Renewable Energy Laboratory in Colorado, aiming to develop ways to tweak the alga's basic biological functions so that the plant produced greater quantities of hydrogen.

Green algae, like terrestrial green plants, photosynthesize, drawing in carbon dioxide and water, absorbing energy from light that converts those nutrients into food, and then expelling oxygen as a waste product. Additional nutrients from the soil or water, and catalysts called *enzymes* within the plant, keep this process running effectively. Plants vary in how much of a particular nutrient they need and respond in specific ways to certain enzymes. In the case of *Chlamydomonas reinhardtii,* the element sulfur is a nutrient needed for effective photosynthesis. The algae also contains an enzyme called *hydrogenase,* which can trigger the algae to stop producing oxygen as a metabolic by-product and start releasing hydrogen instead.

Hydrogenase is normally active only after *Chlamydomonas reinhardtii* has been deprived of light. When deprived of light, the light-dependent reactions of photosynthesis ebb, little oxygen is produced, and hydrogenase is activated. When light returns and the algae begins producing oxygen again, hydrogenase is promptly deactivated and its associated hydrogen release stops.

Melis and his fellow researchers wanted to activate hydrogenase in the algae, so that more hydrogen would be produced. But simply keeping the algae in the dark would not escalate hydrogen production, as the algae's metabolic functions slowed without light, resulting in small amounts of released hydrogen. They decided to try limiting the algae's oxygen output another way, by putting it on a sulfur-free, bright-light regimen. The lack of sulfur would hinder photosynthesis, limiting oxygen output enough to activate hydrogenase and trigger hydrogen production. The presence of light would keep the algae metabolically active and releasing large amounts of by-products.

At labs in California and Colorado, the researchers cultured large quantities of the algae in bottles. Then they deprived the cultures of sulfur, but kept the plants exposed to light for long periods of time—in some cases up to 150 continuous hours. After the sustained light exposure, gas and liquids were extracted from the culture bottles and analyzed.

That analysis confirmed the team's hypothesis. Without sulfur or photosynthesis, the algae were not producing oxygen. This low-oxygen, or anaerobic, environment had induced hydrogenase, which spurred the algae to begin splitting water molecules and releasing gas. The plants had released amounts of hydrogen that were substantial relative to the size of the algae cultures. Hydrogen also dominated the algae's emissions—in gas collection analysis, approximately 87% of the gas was hydrogen, 1% was carbon dioxide, and the remaining 12% was nitrogen with traces of oxygen. The research teams published their findings in the journal *Plant Physiology* in 2000.

Many questions remain about algae-derived hydrogen, especially regarding how much fuel can be harvested continuously using this photobiological production process. Nevertheless, the research results so far are helping give the vision of a hydrogen economy its growing momentum. Within 30 years, some federal energy experts predict photobiological methods for generating hydrogen could be commonplace, meaning cars on future freeways might be powered by pond scum.

two electrodes. The movement of the hydrogen's electrons from one electrode to the other creates the output of electricity.

Hydrogen and fuel cells have many benefits

Hydrogen has a number of benefits as a fuel on which to base a future energy system. We will never run out of hydrogen; it is the most abundant element in the universe. It can be clean and nontoxic to use, and depending on the source of the hydrogen and the source of electricity for its extraction, may produce few greenhouse gases and other pollutants. Instead, pure water and heat may be the only waste products from a hydrogen fuel cell, along with negligible traces of other compounds. Hydrogen is also safe to transport and store, both for people and for the environment. Hydrogen can catch fire, but kept under pressure, it is probably no more dangerous than gasoline in tanks. Fuel cells are silent and nonpolluting, and they allow energy to be stored in the form of hydrogen. Unlike batteries (which also produce electricity through chemical reactions), fuel cells have hydrogen as an external fuel source and will generate electricity whenever fuel is supplied, without ever needing recharging.

Finally, hydrogen and its use in fuel cells is energy-efficient. Depending on the type of fuel cell, anywhere from 35% to 70% of the energy released in the reaction can be utilized. If the system is designed to capture heat as well as electricity, then the energy efficiency of fuel cells can go up to 90%. These rates are comparable or superior to most nonrenewable alternatives. All told, many feel the benefits of hydrogen justify taking the initial steps toward pursuing Iceland's dream, while continuing active research on developing technologies and on predicting environmental impact.

Weighing the Issues:
Precaution Over Hydrogen?

The early steps that Iceland and other nations are taking toward a hydrogen economy have delighted environmental advocates critical of fossil fuels. However, some environmental scientists have recently warned that we do not yet know enough about the environmental consequences of replacing fossil fuels with hydrogen fuel. An increase in tropospheric hydrogen gas would deplete hydroxyl (OH) radicals, they hypothesize, possibly leading to stratospheric ozone depletion and global warming from increased concentrations of methane. Some scientists say such effects will be small; others say there could be further unknown effects we haven't thought of. Do

you think we should apply the precautionary principle to the development of hydrogen fuel and fuel cells? Or should we embark on pursuing a hydrogen economy before knowing all the scientific answers? What factors inform your view?

Energy Conservation

Until our society reaches a point at which we are using solely renewable energy sources, we will be facing the future depletion of nonrenewable sources. In the face of dwindling resources of fossil fuels, we will need to find ways to minimize the extent to which we expend energy. **Energy conservation** is the practice of reducing energy use as a way of extending the lifetime of our fossil fuel supplies, of being less wasteful, and of reducing our environmental impact.

In the summer of 2001, a combination of circumstances, including energy deregulation, a drought, and the shutdown of power plants for repairs, resulted in an energy shortage in California. Television advertisements encouraged people to "flex their power" and conserve energy; rolling blackouts were imposed (that is, power was deliberately shut off in certain areas for short periods of time); and Governor Gray Davis deliberated ways of avoiding similar crises in the future. While the state focused on measures promoting energy conservation as a solution to the energy crisis, the federal government, in the person of Vice President Dick Cheney, stated that although conservation might be virtuous, it would not solve the problem.

Personal choice and increased efficiency are two routes to effective energy conservation

Energy conservation can be accomplished in two primary ways. As individuals, we can make conscious choices to adjust our behavior by taking steps to reduce energy consumption, for example, by turning off lights and cutting back on using machines and appliances. Reducing personal use of power can be cost-effective for an individual or for a business and can also help conserve resources for society.

We can also conserve energy as a society by making our energy-consuming devices and processes more efficient. Many of our current uses are not as efficient as they could be. In the case of automobiles, which use 31% of fossil fuels consumed in the United States, we

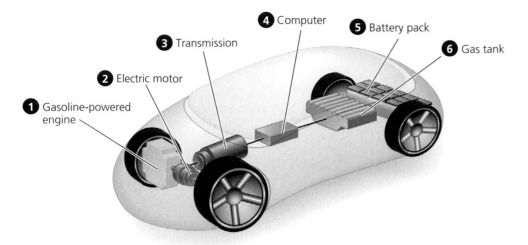

Figure 18.17 A hybrid vehicle uses (1) a gasoline-powered engine and (2) an electric motor to power (3) its transmission and turn the vehicle's wheels. (4) An on-board computer system determines the best balance and timing of gas and electric power. Typically, the electric motor provides the power for low-speed around-town driving and adds extra power on hills. Hybrids are gas-powered and never need to be plugged in. They use their own motion to charge (5) an on-board battery pack that drives the electric motor. All of their energy comes from gasoline that is carried in (6) a typical gas tank.

already possess the technology to increase fuel efficiency far above the current average of 22 mpg. We could accomplish this improvement with cars that have more efficient gasoline engines, that are electric/gasoline hybrids (Figure 18.17), or that run on hydrogen fuel cells. The efficiency of utility plants can be improved through the use of **cogeneration,** a practice in which the extra heat generated in the production of electricity is captured and put to use heating workplaces and homes, as well as producing other kinds of power. Cogeneration can almost double the efficiency of a power plant. Among consumer products, a great number of appliances, from refrigerators to light bulbs, have already been reengineered through the years to increase energy efficiency, but there remains room for improvement. Energy-efficient lighting, for example, can reduce energy use by 80%, and new federal standards for energy-efficient appliances have already reduced per-person home electricity use below what it was in the 1970s. In homes and public buildings, a significant amount of heat is lost in winter and gained in summer because of inadequate insulation (Figure 18.18). The energy required for heating and cooling homes and offices can be reduced by improvements in building design that include the building's location, the color of its roof (light colors keep buildings cooler by reflecting the sun's rays), and its insulation.

While manufacturers can improve the energy efficiency of appliances, consumers also need to "vote with their wallets" by purchasing these energy-efficient appliances. Decisions by consumers to purchase energy-efficient products are crucial in keeping those products commercially available. The U.S. Environmental Protection Agency (EPA) estimates that if all U.S. households purchased energy-efficient appliances, the national annual energy expenditure would be reduced by $200 billion. On an individual consumer's level, studies show that the slightly higher cost of buying energy-efficient washing machines is rapidly offset by savings in water and electricity bills. On the national level, France, Great Britain, and other developed countries have standards of living equal to that of the United States, but use much less energy per capita. This disparity indicates that U.S. citizens could significantly reduce their energy consumption without affecting their quality of life.

Both conservation and renewable energy are needed

It is often said that reducing our energy use is equivalent to finding a new oil reserve. Indeed, effective energy conservation in the United States alone could save 6 million barrels of oil a day. Such a savings would likely far more than offset the energy produced by any oil under the Arctic National Wildlife Refuge (Chapter 17), while also reducing the negative side-effects of fossil fuel exploitation. Conserving energy is even better than finding a new reserve, because it lengthens our access to fossil fuels and also lessens impacts on the environment.

If reducing use and increasing efficiency can cut fuel consumption by such large amounts, why did Vice Pres-

VIEWPOINTS

Hydrogen and Renewable Energy

Is establishing a "hydrogen economy," as Iceland is trying to do, the best way to reduce the use of fossil fuels?

The Role of Renewable Energy for the Hydrogen Economy

Abundant, reliable, and affordable energy is an essential component to a healthy economy. Because hydrogen can be produced from a wide variety of domestically available resources and can be used in heat, power, and fuel applications, it is uniquely positioned to contribute to our growing energy demands, particularly in the environmentally- and traditionally resource-constrained scenarios facing many communities. However, if we are to realize the true benefits of a hydrogen economy, other renewables must play a substantial role in the efficient and affordable production of the hydrogen.

Renewable options that could make a substantial impact in the production of hydrogen are electrolysis powered by wind, photovoltaic, solar-thermal electric, hydropower, and geothermal energy; direct water splitting using microorganisms and semiconductors; and the thermal and biological conversion of biomass and wastes. Researchers around the globe are working on improving these renewable technologies. As a result, costs continue to drop. Renewable hydrogen production technologies, coupled with the advances in hydrogen production equipment, such as electrolyzers, can supply cost-competitive hydrogen and will, ultimately, play a substantial role in our energy supply.

In addition to the potential supply of affordable hydrogen, these technologies also offer a wide variety of opportunities for the development of new centers of economic growth. Most renewable energy investments are spent on materials and workmanship to build and maintain the facilities, rather than on costly energy imports. Therefore, renewable energy investments are usually spent regionally and even locally, leading to new jobs and investments in local economies. Because of this synergistic relationship, realizing the goal of a hydrogen economy will naturally facilitate the advancement of renewable energy. By diversifying our energy supply, we will not only reduce our dependence on imported fuels but will also benefit from cleaner technologies and investment in our communities.

Susan Hock directs the Electric and Hydrogen Technologies and Systems Center of the National Renewable Energy Laboratory. The Center conducts research activities in four areas: distributed power systems integration, hydrogen technologies and systems, geographic information system analysis, and solar measurements and instrumentation.

Is Hydrogen the Answer?

Eventually, fossil and nuclear fuels will become too expensive to extract, or politics will again make one or more of them unavailable (as happened in the 70s), leaving us to contemplate what we will do for energy. Of course, what we should do immediately is employ all practical energy conservation strategies, because Mother Nature is out there making more fossil fuels as we speak, but we don't have time to wait the few million years that will take. Let's consider the short list of renewable energy alternatives: solar, wind, hydro, biomass, geothermal, waves, tides, and ocean thermal energy conversion. These are all relatively benign and abundant.

However, what about hydrogen? As a resource, it's effectively infinite. Ocean water is full of hydrogen. Hydrogen can either be burned or electrochemically used in a fuel cell to provide useful energy. The by-product or "exhaust" is water. You start with water, produce some energy, and end up with water, making it renewable. Another form of hydrogen energy is fusion, hydrogen atoms fusing to form helium plus a lot of energy for no extra charge. That's the way the sun creates energy. What's the catch? It takes about as much energy to extract hydrogen gas from water molecules (by electrolysis) as you get back from your energy conversion device. Until it becomes cheaper to extract hydrogen (both in the economic and net energy senses), the fossil fuels will continue to rule the energy world. The vital breakthrough for fusion may involve using our renewable energy resources mentioned above to separate hydrogen from other molecules.

In my opinion, to reduce our dependence on fossil fuels the alternative energy priority list for this country should be (from highest to lowest): energy conservation, wind, passive solar, biomass, active solar, hydrogen (chemical), hydroelectricity, and, finally, hydrogen (fusion).

Daryl Prigmore has studied energy and the environment since the late 1960's. After receiving bachelor and master of science degrees in mechanical engineering from Colorado State University, he spent 10 years in the energy industry with a company developing solar, geothermal, and low pollution automotive power systems. He has taught energy science classes for the past 20 years, 17 of those years at the University of Colorado at Colorado Springs.

Figure 18.18 Many of our homes and offices could be made more energy-efficient. One way to determine how much heat a building is losing is to take a photograph of the building that records energy in the infrared portion of the electromagnetic spectrum (Chapter 4). In such a photograph, or *thermogram,* shown here, white, yellow, and red signify hot and warm temperatures at the surface of the house, while blue and green shades signify cold and cool temperatures. The white, yellow, and red colors indicate areas where heat is escaping.

ident Cheney say that conservation was not a solution to California's energy shortage? Strictly speaking, energy conservation does not add to the supply of available fuel; thus in the long term it cannot be the sole solution to our energy problems. Regardless of how much we conserve, we will still need energy, and it will have to come from somewhere. This is where renewable energy sources come in. The only sustainable way of guaranteeing ourselves a reliable long-term supply of energy is to speed development of renewable energy sources, so that we can shift away from the use of dwindling and environmentally hazardous fossil fuels.

Conclusion

The coming decline of fossil fuel supplies and the increasing concern over air pollution and global climate change driven by fossil fuel combustion have convinced many people that we will need to shift to renewable energy sources that will not run out and will not pollute. Sources such as biomass and hydropower have already been playing important roles in our energy use and electricity pro-

duction, although these sources are not always strictly renewable. Renewable sources with more promise for sustaining our civilization far into the future, without greatly degrading our environment, include solar energy, wind energy, ocean energy sources, and geothermal energy. Furthermore, by using electricity from renewable sources to produce hydrogen fuel, we may be able to produce electricity when and where it is needed and help convert our transportation sector to a non-polluting renewable basis. Most renewable energy sources have been held back for a variety of reasons, including a lower level of funding for research and development relative to nonrenewable sources, and artificially cheap market prices for nonrenewable resources that do not include external costs. Despite this, our technologies for renewable sources have progressed far enough as to offer hope that we can make the transition from fossil fuels to renewable energy with a minimum of economic and social disruption. Whether we can also limit environmental impact will depend on how soon and how quickly we make the transition, and how much we put efficiency and conservation measures into place. Iceland's pursuit of a hydrogen economy may help show the way for the rest of our global society.

REVIEW QUESTIONS

1. About how much of the world's energy now comes from renewable sources? What is the most prevalent form of renewable energy used in the world? What form of renewable energy is used to generate electricity most often?

2. What is causing renewable energy sectors to expand? What renewable source is experiencing the most rapid growth?

3. What factors are preventing an expedient transition from fossil fuel use to renewable sources of energy?

4. Define biomass and explain why it is such a prevalent and useful form of energy.

5. Describe the positive and negative environmental impacts of biomass.

6. How does hydroelectric power work? About how much of the world's electricity does hydroelectric power generate?

7. What are the advantages and disadvantages of hydropower? Why may we be limited in our ability to use this resource?

8. How does passive solar heating work?

9. How does active solar heating work? Is this a new technology?

10. How can solar rays be captured to maximize the amount of energy produced?

11. Describe the photoelectric effect. How are photovoltaic (PV) cells used?

12. What are the environmental and economic advantages and disadvantages of solar power?

13. How long has wind energy been used? When did wind energy first begin to receive government funding for research and development?

14. How do modern wind turbines generate electricity? How does wind speed affect the process?

15. What are the environmental and economic benefits of wind power?

16. Explain the disadvantages of wind energy.

17. Define geothermal energy and explain how it is obtained and used.

18. Are geothermal power sources truly sustainable? Why or why not?

19. How is hydrogen fuel produced? Is this a clean process? What factors determine the amount of emissions of pollutants hydrogen production will create?

20. What benefits can energy conservation bring? How can it be accomplished?

DISCUSSION QUESTIONS

1. Why might a hydrogen economy be closer than we think? Why might it instead not come to pass? Do you think water could be the coal of the future? Why or why not?

2. How is energy-efficiency in an electricity-generating system determined? For each source of renewable energy discussed in this chapter, how do you think efficiency could be improved?

3. Do you think development and implementation of renewable energy resources to replace fossil fuels can be moved forward without great social, economic, and environmental disruption? What steps would need to be taken to accomplish this? What obstacles stand in the way? Will market forces alone suffice to bring about this transition?

4. Explain how amounts of fossil fuel emissions can be reduced by the use of renewable energy. Visit the BP Solar website; can you determine the potential reductions in emissions that could come about by using solar in your particular area?

5. Name five ways that you yourself can conserve energy in your everyday life. Name three ways that manufacturers of products can increase energy efficiency. Do you think energy conservation is based on personal choices? Does it involve personal sacrifice? Can it bring personal benefit?

Media Resources *For further review, go to the website* **www.envscienceplace.com** *or student CD-ROM, where you will find quizzes, flashcards, a glossary, additional interactive exercises, and links to relevant news and research sources. Also, on the website and CD-ROM is* **GRAPH IT**, *a series of interactive graphing tutorials to help you interpret graphs and plot data.*

Waste management

Recyclable Containers

This chapter will help you understand:

- The types of waste we generate

- The scale of the waste problem

- Municipal solid waste, industrial waste, and hazardous waste

- Wastewater and sewage treatment

- Conventional waste disposal methods: landfilling and incineration

- Waste reduction solutions

- Composting approaches

- Recycling approaches

People Scavenging at the Payatas Dump

Central Case: Dump-Dwellers and Manila's Garbage Crisis

"An extraterrestrial observer might conclude that conversion of raw materials to wastes is the real purpose of human economic activity."
—*Gary Gardner and Payal Sampat, Worldwatch Institute*

"Recycling is one of the best environmental success stories of the late 20th century."
—*U.S. Environmental Protection Agency*

It was a tragedy waiting to happen, and on July 10, 2000, a week of typhoon rains finally triggered it. Loosened by the downpour, an avalanche of wet garbage slid from the 15-m (50-ft) slopes of the Payatas Dump and smashed into a shantytown of squatters living below. At least 200 people were smothered, although the full death toll will never be known.

Located in Quezon City, 20 km (12.4 mi) from downtown Manila in the Philippines, the gigantic trash heap at Payatas was home to roughly 80,000–90,000 people. Many lived in the shantytown at its base called *Lupang Pangako* ("Promised Land"). Each day these people and their children ascended the dump and roamed about in search of salvageable items. Collecting discarded goods

and pieces of plastic, metal, and glass, they would sell these to local junk dealers, eking out a living on 100–200 pesos (U.S. $2.00–4.00) per day. Children generally needed to scavenge all day to help the family make ends meet and thus could not go to school. People suffered from disease and malnutrition, infant mortality was shockingly high, and many children were born with disabilities.

After the deadly avalanche, the government closed the dump. Without the 3,000 tons of new trash brought in daily to Payatas, the scavengers were deprived of their livelihood. Dump-dwellers staged a demonstration and marched on the capitol.

The fact that so many people will live at such a site testifies to the severity of poverty in the Philippines, not to mention other nations in the world. Moreover, the fact that so many people *can make a living* in such places demonstrates the enormous magnitude of waste generated by wealthier segments of the population, and the sheer amount of reusable goods that are needlessly discarded.

As material consumption has risen among Filipino citizens and as the population has continued to grow, so has the generation of waste. Each day, a typical Filipino of moderate wealth generates 0.7 kg (1.5 lbs) of trash. Filipino households generate 10 million tons of solid waste a year, with metro Manila accounting for one-quarter of that total. As the population of the Philippines grows and consumes more, waste generation is projected to rise 40% within a decade.

When the government shut down Payatas in the wake of the garbage avalanche, it turned metro Manila's garbage problem into a full-blown crisis. The country's first two properly designed sanitary landfills had been closed. Their location and management had been poor, and local residents finally barricaded the entrance to one of the sites in protest. With these two landfills and the Payatas dump (a legal but uncontrolled open dump) closed, only one legal open dump remained, and Manila was running out of places to put its garbage. Trash began accumulating on roadsides, illegal dumping became rife, and people openly burned trash in order to reduce its volume, all of which resulted in growing sanitation, health, and environmental hazards.

Because of landfill closures and deterioration of garbage collection services, today only 70% of municipal solid waste is successfully collected and disposed of in Philippine cities, and a mere 40% in rural areas. The nation's one properly designed sanitary landfill and 17 "controlled" dumps with environmental safeguards can handle just 2% of the nation's solid waste.

Air pollution laws prohibit the incineration of waste, but there is promise that recycling and composting will reduce waste. In 2001 metro Manila was recycling only 13% of its waste. However, this rate was double what it was just a few years earlier, thanks to efforts by leaders to boost recycling as a solution to the waste problem. The poverty-stricken people dwelling at Payatas and other dumps, rather than being a burden on the state, actually contribute to solving the solid waste problem. Of the 327 tons per day of waste that was recycled in metro Manila in 1997, 127 tons (39%) was recycled by scavengers.

Today, Filipinos are doing their best to meet the ambitious goals of the national waste control legislation passed in 2000 and signed into law in 2001. The Ecological Solid Waste Management Act requires municipalities to divert 25% of solid waste out of the waste stream by means of recycling, composting, reuse, or reduction within 5 years, and it includes economic incentives for doing so. It prohibits opening new unsafe dumps and requires transformation of existing dumps into landfills that meet health and environmental guidelines. Using a carrot-and-stick approach—tax incentives plus fines and fees—the legislation aims to clean up the Manila area

and set it on a path to deal sustainably with the waste generated by the city's economic advancement.

Types of Waste

The rising material consumption in Philippine society parallels that in nations the world over. In both the developing and developed worlds, people have been producing, selling, and buying increasing amounts of goods every year. This increasing consumption has corresponded with a rise in the standard of living for millions of people. As societies consume more resources and manufacture more products, however, more waste is generated.

Waste refers to any unwanted item or substance that results from a human activity or process. Some waste results from manufacturing procedures and represents resources that do not end up in the final product. Other waste consists of manufactured goods or materials that people discard. Still other waste is made up of materials used for packaging products for manufacture, transport, purchase, and use. Some waste is sewage. For management purposes, waste is divided into several main categories: municipal solid waste, industrial solid waste, hazardous waste, and wastewater.

Wastewater refers to any water that is used in our households, businesses, industries, or public facilities and is drained or flushed down our pipes, as well as to the polluted runoff from our streets and storm drains. **Hazardous waste** refers to waste that is toxic, chemically reactive, flammable, or corrosive. It can include everything from paint and household cleaners to hospital medical waste to industrial solvents. Solid waste, nonliquid waste that is not especially hazardous but that may occur in large quantities, makes up the remainder. **Municipal solid waste** comes from homes, institutions, and small businesses, whereas **industrial solid waste** includes waste from production of consumer goods, mining, petroleum extraction and refining, and even agriculture.

There are several reasons for managing waste

The generation of waste can degrade water quality, soil quality, and air quality, and can thereby affect human health while causing ecological harm. Waste is also a symptom of inefficiency, so reducing waste will in many cases be economically advantageous. Industry, municipalities, and consumers can potentially save money by reducing waste. In addition, waste is generally unpleasant aesthetically. For these and other reasons, waste management has become an important pursuit for cities, industries, and individuals.

There are several ways to manage waste

The goals of waste management are to dispose of waste safely and effectively and to reduce the amount of waste generated. Several approaches to meeting these goals have been developed over the years, but the single most effective approach is to reduce the amount of waste material generated in the first place (Figure 19.1). Manufacturers can reduce waste directly by increasing efficiency so that fewer by-products are produced. Citizens can reduce waste by making the decision to consume less—to

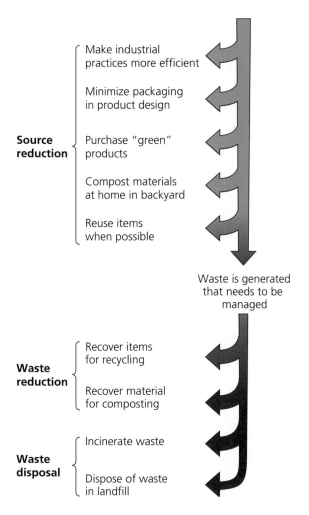

Figure 19.1 The most effective way to reduce waste is to minimize the amount of material that goes into products. There are many approaches that industries and consumers can take to achieve "source reduction." The next-best solution is to reduce the amount of waste that needs to be discarded, through recycling or composting. For all remaining waste, waste managers attempt to find disposal methods that minimize impact to human health and environmental quality.

purchase fewer goods, to purchase goods involving less packaging, and to use those goods longer. Reusing goods—for instance, purchasing used items—also helps reduce the amount of material entering the waste stream.

Recycling is widely viewed as the next-best approach to managing waste. Recycling involves sending used goods to facilities that extract and reprocess raw materials to manufacture new goods. Newspapers, white paper, cardboard, glass, metal cans, and plastic containers have all become increasingly recyclable as recycling technologies are developed and as markets for recycled materials grow. Organic forms of waste can also be reduced or recycled by composting, that is, allowing them to biodegrade. Sewage treatment also involves encouraging efficient biodegradation. Broadly construed, the idea of recycling is nothing humans have invented—recycling is a key component of natural systems.

Regardless of how effectively we can encourage reduction, reuse, composting, and recycling, there will always be some amount of waste we will need to discard. Thus, there will always be a place for what waste managers consider the least-preferred option, conventional disposal methods. These include burying waste in landfills and burning waste in incinerators. In this chapter we will examine these options for managing municipal solid waste and then address other categories of waste: industrial solid waste, hazardous waste, and wastewater.

Municipal Solid Waste

Municipal solid waste is waste produced by consumers, public facilities, and small businesses. It is what we commonly refer to as "trash" or "garbage." Everything from paper to food scraps to roadside litter to bulky items like old appliances and furniture is considered municipal solid waste.

Patterns in the municipal solid waste stream vary from place to place

In North America, paper, yard debris, food scraps, and plastics are the principal components of municipal solid waste, together accounting for over 70% of the waste stream (Figure 19.2). Even after recycling, paper is the largest component of municipal solid waste in the United States, comprising 37% of all waste produced.

The majority of municipal solid waste comes from packaging and from nondurable goods (products meant to be discarded after a short period of use). In addition,

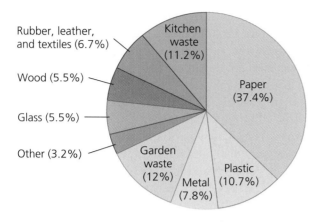

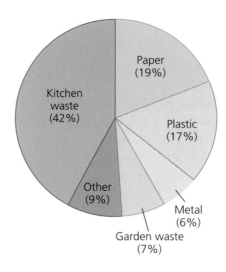

Figure 19.2 Paper products comprise the largest component of the municipal solid waste stream in the United States, followed by yard waste, kitchen waste, and plastic. In total, each U.S. citizen generates close to one ton of solid waste each year. Data from U.S. EPA, 2000.

Figure 19.3 Kitchen waste comprises the largest component of the municipal solid waste stream in the metro Manila region in the Philippines, followed by paper products, and plastic. In total, each citizen of metro Manila generates roughly 219 kg (353 lb, or just over one-sixth of a ton) of solid waste each year. Data from Philippines Environment Monitor 2000, World Bank, 2001.

old durable goods and outdated equipment are thrown away as consumers purchase new products. Peoples' drive for more goods, as well as for the newest technologies, has resulted in a great deal of waste generation. In 2000, a total of 232 million tons of municipal solid waste was produced in the United States, almost one ton per person per year. The average American produces over 2.0 kg (4.5 lb) of trash per day.

Following the United States in per capita solid waste generation are Canada, with 1.7 kg (3.75 lb) per day, and the Netherlands, with roughly 1.4 kg (3.0 lb) per day. Of developed nations, Germany and Sweden produce the least per capita waste, each country generating just under 0.9 kg (2.0 lb) per day. The relative wastefulness of the U.S. lifestyle, with its excess packaging and reliance on nondurable goods, has caused critics to label the United States "the throwaway society."

Consumption and waste patterns are somewhat different in developing countries. In the Philippines, kitchen waste is the primary contributor to solid waste, making up 42% of the municipal waste stream in metro Manila. Paper, in contrast, comprises only 19% of the waste generated (Figure 19.3). The average Manila resident generates 0.60 kg (1.3 lb) of trash daily, similar in magnitude to that generated by citizens of Rome, Jakarta, and many other major urban centers of the world that contain both fabulously rich citizens who consume a great deal, and desperately poor ones who consume very little and waste less.

In both developed and developing nations, consumption and generation of waste have been rising. Waste generation has increased in the United States (Figure 19.4)

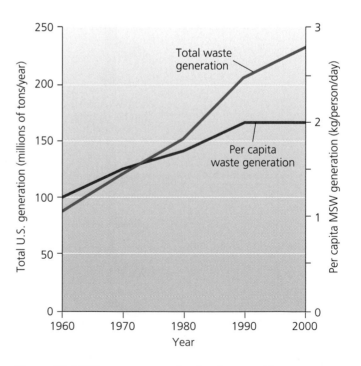

Figure 19.4 U.S. waste generation has increased by more than 2.6 times since 1960, and U.S. per capita waste generation has risen by 67%. Per capita waste generation has flattened out in recent years largely due to recycling and source reduction efforts. Data from U.S. EPA.

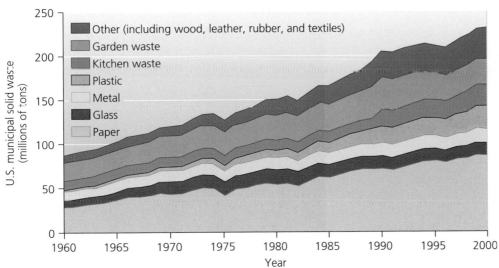

Figure 19.5 All types of waste have been growing in the United States over the past 40 years, but paper waste is the only type that has been taking up a substantially greater share of the waste stream through time. Data from Municipal solid waste in the United States: 2000, Facts and Figures, Environmental Protection Agency, June 2002.

by more than 2.6 times since 1960, and per capita waste generation among U.S. residents has risen by 67% during the same time period. Plastics, which came into wide consumer use only after 1970, have accounted for the greatest relative increase in the waste stream during the last several decades (Figure 19.5).

Although waste has been increasing for decades and is currently rapidly increasing in developing countries, per capita generation rates have slowed and in some cases decreased in many developed nations in recent years. For instance, note in Figure 19.4 that per capita waste production for the United States flattened out during the 1990s. This was due largely to the increased availability of recycling options. Nondisposal solutions, such as recycling, reduction, reuse, and composting, today are taking care of an increasingly larger portion of waste. Before we investigate nondisposal approaches to waste management, let's examine how we dispose of waste.

Open dumping of the past has given way to improved disposal methods

Historically, people have dumped their garbage wherever it suited them. As population densities increased, municipalities took on the task of consolidating trash into concentrated open dumps in order to keep other areas clean. To decrease the volume of trash, these dumps would be burned from time to time. (Open dumping and burning still occur throughout much of the world, and such practices were standard at Payatas.) The rise in population and consumption increased waste, and

dumps grew larger. At the same time, expanding cities and suburbs forced more people into the vicinity of operating dumps and exposed them to the noxious smoke of dump burning. In reaction to opposition from residents living near dumps, and to the rising awareness of health and environmental threats of open dumping and burning, many nations have improved methods for disposing of their waste. Primarily, these methods include burying waste in sanitary landfills and burning waste in incineration plants.

In the 1980s in the United States, waste generation increased while incineration was restricted, and recycling was neither economically feasible, nor widely popular. As a result, available landfill space was limited, and there was much talk of a solid waste "crisis." This scenario is quite similar to what the Philippines and other developing nations are experiencing today, as resource consumption outpaces recycling efforts. Since the late 1980s, however, recovery of materials for recycling has greatly expanded in the United States, decreasing the pressure on landfills (Figure 19.6a). As of 2000, U.S. waste managers were landfilling 55% of municipal solid waste, incinerating 15% of it, and recovering 30% for composting and recycling (Figure 19.6b).

Sanitary landfills are regulated by health and environmental guidelines

In modern **sanitary landfills**, waste is buried in the ground or piled up in large mounds. In contrast to open dumps, however, every effort is made to prevent waste

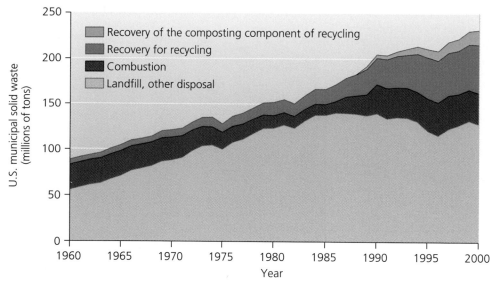

(a) U.S. recycling, 1960–2000

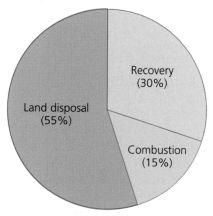

(b) U.S. municipal waste disposal, recovery, and combustion

Figure 19.6 Since the late 1980s, recycling has grown notably in the United States, allowing a smaller proportion of waste to go to landfills (**a**). As of 2000, 55% of U.S. municipal solid waste was going to landfills and 15% to incinerators, while 30% was being recovered for composting and recycling (**b**) Go to **GRAPH IT** on the website or CD-ROM. Data (a,b) from Municipal solid waste in the United States: 2000, Facts and Figures, Environmental Protection Agency, June 2002.

from contaminating the environment (Figure 19.7). First, the site of a landfill is chosen with waste containment in mind. U.S. federal regulations require that landfills be located away from wetlands and earthquake-prone faults and that they be at least 6 m (20 ft) above the water table to avoid groundwater contamination. The primary safeguard against groundwater contamination is to line the landfill with plastic and with 60-120 cm (2-4 ft) of impermeable clay. Regulations require that the bottom and sides of all sanitary landfills be lined in this way in order to prevent wet contaminants from seeping out of the trash and into aquifers. Regular monitoring of area groundwater supplies is also required. In addition, sanitary landfills have systems to collect **leachate**, liquid that results when substances from the trash dissolve in water as rainwater percolates downward through the landfill. Typically these systems consist of pipes running from the bottom of the landfill to leachate collection ponds. Landfill managers are required to maintain leachate collection systems for 30 years after a landfill has closed.

Guidelines also specify how waste is added to the landfill; it cannot simply be heaped on. Each layer of waste is covered with several inches of soil, which helps keep the material dry and thus reduces leachate. Soil also speeds decomposition of the waste, reducing odor and infestation by rats and other pests. After a landfill is closed, it must be capped with an engineered cover that must be maintained. In the United States, most municipal landfills are regulated locally or by the states, but they also must meet national standards set by the EPA. This arrangement was set forth by the federal Resource Conservation and Recovery Act, enacted in 1976 and amended in 1984.

Following regulations to ensure public health and safety is not cheap, and in developing countries, landfill

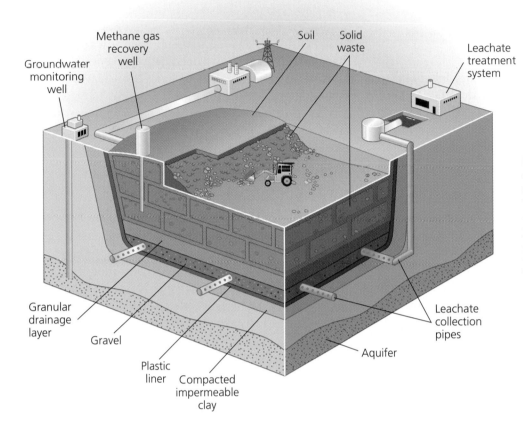

Methane gas recovery well

Soil

Solid waste

Leachate treatment system

Groundwater monitoring well

Granular drainage layer

Gravel

Plastic liner

Compacted impermeable clay

Leachate collection pipes

Aquifer

Figure 19.7 Sanitary landfills are engineered to prevent waste from contaminating soil and ground-water in the area. Waste is laid in a large depression lined with plastic and impervious clay de-signed to prevent liquids from leaching out. Pipes of a leachate collection system serve to draw out these liquids from the bottom of the landfill. The waste is lay-ered along with soil until the de-pression is filled and then continues to be built up until it is capped. Landfill gas produced by anaerobic bacteria may be recov-ered from the landfill, and waste managers must monitor ground-water for contamination.

operators and governments often lack the technology or the funds to safely dispose of waste. The Philippines' first two sanitary landfills, Carmona and San Mateo, were widely viewed as unsafe for the environment and for nearby residents. Carmona, which handled 2.1 mil-lion tons of waste in its most heavily used year, was shut down in 1998, and San Mateo, which took 3.3 million tons in its final year, was closed in December 2000, half a year after the Payatas avalanche. After their closure, inadequate safeguards and malfunctioning treatment systems enabled leachate to seep out of these landfills and threaten groundwater supplies. Today there is only one sanitary landfill in the Philippines. Most garbage goes to "controlled dumps" with limited safeguards or to open dumps with none. The country's Solid Waste Management Act of 2000 mandates that all existing dumps be upgraded to sanitary landfill status in 5 years and that no new open dumps be established.

While the Philippines and many other nations need more sanitary landfills, the United States has greatly re-duced its number. U.S. landfill sites are becoming con-solidated as smaller landfills are closed and the trash stream shifts to a smaller number of much larger land-

fills. In 1988 the country hosted 8,000 landfills, and in 2000 it had fewer than 2,000. Some old landfill sites are today being developed for other uses (Figure 19.8). One in Cambridge, Massachusetts, has been turned into a city park with ball fields and running paths. One tall

Figure 19.8 Old landfills, once properly capped, can serve other purposes. A number of them have been developed into areas for human recreation. Shown is an observation board-walk and platform at the Stevens Creek tidal marsh at the site of an old dump on San Francisco Bay.

landfill near Detroit, Michigan—a site of considerable value in an area without much hilly terrain—was converted into a ski slope.

Landfills have drawbacks

For all their advantages, however, sanitary landfills also have many drawbacks. For example, despite improvements in liner technology and landfill siting, many experts believe that leachate will very likely escape from even well-lined landfills. Liners can get punctured, and leachate collection systems eventually cease to be maintained. While landfills are kept dry to reduce leachate, the bacterial decomposers that break down material thrive in wet conditions. Dry landfills, therefore, slow waste decomposition. In fact, it is surprising how slowly some materials biodegrade when tightly compressed in a landfill. Innovative archaeological research by University of Arizona's William Rathje has revealed that it is possible to find 40-year-old newspapers that are still readable in landfills (see The Science behind the Story).

Another problem is that finding areas to locate new landfills has become increasingly difficult. Most communities do not want landfills in their midst for aesthetic reasons and for fear of health risks. This not-in-my-backyard (NIMBY) reaction (Chapter 18) is a major reason the Philippines has had such trouble keeping landfills open and establishing new ones. It was largely local opposition from residents that forced the closure of Carmona and San Mateo. Moreover, those landfills were in fact unsound, a finding that only strengthened citizens' suspicions and fears about future landfills. In North America, as a result of the NIMBY syndrome, landfills are rarely sited in wealthy, educated neighborhoods with the political power to keep them out, but instead are disproportionately sited in poor and minority communities. Besides the obvious injustice of this pattern, an uneven distribution of landfills also decreases efficiency by requiring that waste be shipped long distances from where it is produced to where it is ultimately disposed.

One famed case of long-distance waste transport illustrates the unwillingness of many communities to accept garbage. In the late 1980s, landfills across the United States were beginning to fill up. In Islip, New York, the town's landfills were full, prompting town administrators to ship waste to a methane production plant in North Carolina. On March 22, 1987, a barge full of garbage left Islip en route to North Carolina. Prior to the barge's arrival, however, it became known that the shipment was contaminated with 16 bags of medical waste, including syringes, hospital gowns, and diapers. Because of the medical waste, the methane plant rejected the entire load. The barge sat in a North Carolina harbor for 11 days before heading for Louisiana. Louisiana, however, would not let the barge dock, either. The barge traveled down toward Mexico, but the Mexican Navy came out to prevent it from entering Mexican waters. Overall the barge traveled 9,700 km (6,000 mi) before eventually returning to New York, where, after several court battles, the waste was finally incinerated at a facility in Queens.

Incinerating trash reduces pressure on landfills

Just as sanitary landfills are an improvement over open dumping, incineration in specially constructed facilities is an improvement over open-air burning of trash. **Incineration**, or combustion, is a controlled process of burning in which mixed garbage is combusted at very high temperatures (Figure 19.9). At incineration facilities, waste is first sorted and metals are removed. Remaining, metal-free waste is chopped into small pieces to aid combustion and then burned in a furnace. Mixes rich in paper content incinerate well because of paper's flammability, while mixes rich in food scraps do not. Incinerating waste reduces its weight by up to 75% and its volume by up to 90%. The remaining material, in the form of ash, is generally disposed of in a landfill.

However, simply reducing the volume and weight does not get rid of components of trash that are toxic. In addition, combustion can create new chemical compounds that can be health hazards. When trash is burned, many hazardous chemicals can be released in the smoke that is generated, from dioxins to PCBs (Chapter 10) to heavy metals. Early incinerators did not include mechanisms for mitigating air pollution and thus released toxic gases and particulate matter into the atmosphere. Such releases caused a backlash against incineration from citizens fearful of the health hazards. Most developed nations now regulate incinerator emissions. The Philippines is the only nation to have banned the incineration of municipal solid waste completely, as part of its clean air legislation. Ironically, the ban, which so many environmental advocates support, is contributing to the waste problem by exacerbating the volume of garbage without a proper disposal site.

As a result of real and perceived health threats from incinerator emissions and community opposition to these plants, several technologies have been developed to mitigate emissions. Filters remove particulate matter, and scrubbers spray plants with liquid formulated to

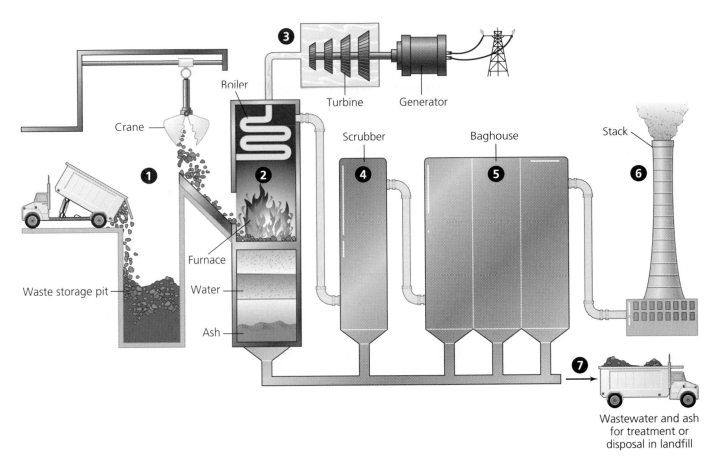

Figure 19.9 Incinerators greatly reduce the volume of solid waste by burning it (although they also may emit toxic compounds into the air through smokestack emissions). Many incinerators are waste-to-energy (WTE) facilities that use the heat of combustion to generate electricity. In a WTE facility, solid waste is burned at extremely high temperatures, heating water, which turns to steam. The steam turns a turbine to create electricity. The toxic gases and particulate matter produced are mitigated chemically and physically before being emitted from the stack. Ash remaining from the combustion process is disposed of in a landfill.

neutralize certain acidic gases. In addition, burning garbage at especially high temperatures can destroy certain pollutants. Even with all these measures, however, not all toxic emissions are eliminated. Just as landfills are required to monitor their leachate, incinerators are required to test their emissions and to make sure that the ash discarded into landfills meets safety standards.

Many incinerators burn waste to create energy

Incineration was initially practiced simply to reduce the volume of waste, but today it very often serves to generate electricity as well. Most North American incinerators today use the heat generated by waste combustion to create energy. So-called **waste-to-energy (WTE)** facilities are incinerators that use heat from the furnace to boil water to create steam that drives electricity genera-

tion or that fuels heating systems. When burned, waste generates approximately 35% of the energy of burning coal. In 2000 there were 102 operational WTE facilities across the United States with a total capacity to process 96,000 tons of waste per day, according to the EPA. In that year, one-seventh of all municipal solid waste was combusted in incinerators.

Although burning waste is an effective means of reducing its volume, the considerable financial cost of incineration is not offset by power generation, and it can take many years for a WTE facility to become profitable. Because of this, many companies that build and operate these facilities require that communities contracting with the facility agree to provide it a certain amount of garbage reliably over many years. In a number of cases, these long-term commitments have interfered with communities' later efforts to reduce their waste through recycling and other waste-reduction strategies.

Digging Garbage: The Archaeology of Solid Waste

Garbage and *knowledge* are two words rarely put together. But when scientist William Rathje dons trash-flecked clothes and burrows into a city dump, he gleans valuable information about how modern Americans live.

By pulling tons of trash out of disposal sites over the course of decades, Rathje has turned dumpster-diving into a noteworthy field of scientific inquiry that he calls *garbology*. An archaeologist by training who has been called "the Indiana Jones of solid waste," Rathje has brought exacting archaeological techniques to the contents of trash cans.

Garbology got its start in Rathje's classroom. As a professor at the University of Arizona in the early 1970s, Rathje wanted his students to learn a technique common among archaeologists — sorting through ancient trash mounds to understand the lives of past cultures. With few ancient civilizations or their trash close at hand, however, he arranged for his students to dig through their neighbors' garbage. In 1973, he gave that effort a name, "The Garbage Project," and began a methodical study of the contents of modern trash.

"Garbologist" William Rathje of the University of Arizona has pioneered the study of our own culture through the waste we generate.

Rathje asked nearby communities to divert some of their garbage trucks and bring trash straight to his study teams. With rakes and notebooks, the researchers sorted, weighed, itemized, and analyzed the refuse. They then visited the homes of the people who had generated the trash and asked residents about their shopping and consumption habits.

Then, in 1987, amid growing debates about how quickly U.S. landfills were filling up, Rathje decided to see what was taking up space in the dumps. With heavy-duty digging tools, The Garbage Project headed to landfills and went below ground using a truck-mounted bucket auger—a large drill commonly employed by geologists and construction crews to handle everything from excavating

Landfills can also produce energy when gas is collected

Combustion in WTE plants is not the only way to gain energy from waste. Deep inside landfills, bacteria help decompose waste in an oxygen-deficient environment. This anaerobic decomposition produces landfill gas, which is slowly released into the atmosphere. Landfill gas consists of roughly half methane, a potent greenhouse gas that contributes to global warming (Chapter 12). Methane can contribute to smog formation and can even cause explosions. But landfill gas can be collected, processed, and used as natural gas, one of our prime sources of fossil-fuel energy (Chapter 17). Today more than 330 operational projects collect landfill gas in the United States. Certain other countries have taken advantage of this resource as well. In Chile, four facilities in Valparaiso and Santiago supply 40% of the region's demand for natural gas. Some analysts have recommended that the Philippines try to harvest landfill gas from its dumps. One source estimated that Payatas and the defunct Carmona and San Mateo sanitary landfills alone could contribute enough gas to generate power for 8,000 Manila-area households.

soil samples to creating new water wells. Rathje and his researchers dug into landfills around North America, boring as far as 30 m (100 ft) down, often drilling approximately 15 to 20 garbage "wells" at each site, with each well yielding up to 25 tons of trash.

Once excavated, landfill contents were sorted, weighed, and identified. Rathje's teams sometimes froze the trash before they worked with it to make the garbage easier to separate and to limit odor and flies. Large pieces were set aside for identification. Smaller bits of trash were put through sieves, and sometimes washed with water to make them easier to label. Rathje has excavated at least 21 dumps, including sites in Tucson, San Francisco, Chicago, Phoenix, New York City, and Philadelphia, uncovering a host of interesting facts in the process:

- **Not much rot.** Trash doesn't decay much in closed landfills, Rathje has found. In the low-oxygen conditions inside most closed dumps, trash turns into a sort of time capsule. Rathje's teams have found whole hot dogs in most digs, intact pastries that are decades old, and grass clippings that are often still green. Decades-old newspapers are legible and can be used to date layers of trash.

- **Paper rules.** Paper products and construction debris take up the most landfill space. Paper-based products make up more than 40% of most landfill content, and construction debris makes up about 20%. Newspapers are often a high-volume item, averaging about 14% of landfill space.

- **Plastic packaging no problem.** Rathje says plastic packaging is not the landfill problem many believe it to be. Plastic packaging makes up only about 4.5% of landfill content, and that figure has not increased substantially since the 1970s, Rathje reported in 1997. Fast-food packaging, polystyrene foam, and disposable diapers also aren't a major problem, making up only about 3% of landfill content. If all plastic packaging were to be replaced by containers made of glass, paper, steel, or similar materials, Rathje claims, the packaging discarded by U.S. households would more than double.

- **Poison in small bottles.** Toxic waste comes in all sizes, Rathje has discovered. If nail polish were sold in 55-gallon drums, its chemical composition would make it illegal to throw out in a regular dump. Nail polish, however, is tossed in small bottles. In 1991, after studying Tucson dumps, Rathje calculated that about 350,000 bottles were getting thrown away each year. Luckily, however, the potentially toxic ingredients in nail polish don't always spread far, he found. Paper, diapers, and other nontoxic garbage often absorb toxic materials in landfills and keep the poisons from leaching out.

Through garbology, Rathje has gleaned unique insights into how we can change our often-wasteful habits. Now a visiting scholar at Stanford University's Archaeology Center, Rathje has emerged as a leading expert on how to reduce waste. In Phoenix, for example, city leaders stepped up recycling programs after a Rathje study found that at least half of the city's trash could be recycled.

Reducing waste is a better option than disposal

Because both landfills and incinerators have drawbacks, these are not ideal strategies for dealing with solid waste. A better alternative is to reduce the amount of waste that is generated in the first place. Reducing the amount of material that enters the waste stream is the best solution to the solid waste problem because it avoids the costs of disposal and recycling, helps conserve resources, minimizes pollution, and can often save consumers and businesses money. Preventing waste generation in this way is known as **source reduction**.

There are a number of strategies for source reduction, some of which involve the operating procedures of manufacturers and businesses, and some of which involve the behavior of consumers. Because much of our waste stream consists of the materials used to package goods, reducing this packaging can have a major effect on the volume of the waste stream. The trend in most developed countries has been for manufacturers to produce goods with greater amounts of packaging. However, consumers can exercise power by choosing minimally packaged goods, by buying fruit and vegetables that are not wrapped, and by buying food in bulk. In addition, manufacturers can reduce the size or weight of goods and materials, as they already have with many items, such as lighter-weight tin, aluminum cans, plastic soft drink bottles, and smaller personal computers. Increasing the longevity of goods can also help reduce waste. Consumers will generally choose to purchase goods that last longer, all else being equal. To maximize sales, however, companies often produce short-lived goods that need to be replaced frequently. Thus, increasing the longevity of goods is largely up to the consumer; if demand is great enough, manufacturers will respond.

Reuse is one main strategy for source reduction

Using already-used goods is one major way to reduce waste, and there are many ways to reuse items, including saving them to use again or substituting disposable goods with durable ones. Habits as simple as bringing your own coffee cup to coffeeshops or bringing sturdy reusable cloth bags to the grocery store can over the years have a substantial impact. Other ways of encouraging reuse include donating unwanted items to people and institutions that can use them and shopping for used items at resale centers yourself. Over 6,000 reuse centers exist in the United States, including stores run by organizations that resell donated items, such as Goodwill and the Salvation Army. In addition to doing good for the environment, source reduction is often economically advantageous. Used items are quite often every bit as functional as new ones and are much cheaper. Studies from U.S. communities estimate that 2–5% of the waste stream is potentially reusable. In light of the fact that so many people can make a living by selling items scavenged from dumps like Payatas, however, it would not be surprising if such percentages were underestimates.

Although reuse and source reduction comprise the most-preferred option for addressing the solid waste problem, it can be difficult for those of us in developed nations to battle our own consumerist tendencies. As noted, the United States is often described as a "throw-away society" because so many of its goods are made to be used once and then discarded. Most of us are so used to this model of one-time use that we fail to realize just how unusual and unprecedented it is. For instance, in the United States people are accustomed to drinking from a soft drink bottle once and then discarding or recycling it. In most of the world, however, soft drink bottles are returned to the retailer once the buyer drinks from them, and those bottles are cleaned by the manufacturer and used over and over again. In general,

developing nations reuse goods far more efficiently than do developed nations.

Financial incentives can tackle overconsumption in a throwaway society

Some states in the United States have developed ways of being efficient with beverage containers. Ten states have "bottle bills," laws that allow consumers to return bottles and cans to stores after use and receive a refund—generally 5 cents per bottle or can. The first bottle bills, in the 1970s, were designed to cut down on litter by providing people a financial incentive not to discard cans and bottles on roadsides. As waste problems mounted in the 1980s, bottle bills also served to cut down on the waste stream. In states where they have been enacted, these laws have proved profoundly effective and resoundingly popular; they are recognized as among the most successful pieces of legislation enacted by states in recent years (Figure 19.10).

It is a testament to the lobbying power of the beverage industries, which have traditionally opposed passage of bottle bills, that more states do not have such legislation. States that do have such laws now face at least two challenges: to amend the laws to include new kinds of containers and to adjust the refunds for inflation. In the three decades since Oregon passed the nation's first bottle bill, for instance, the value of a nickel has dropped such that today, the refund would need to be 22 cents to reflect the refund's original intended value.

Another approach that uses a financial incentive to influence consumer behavior is the "pay-as-you-throw" approach to garbage collection. In these programs,

Figure 19.10 New York is one of the 10 U.S. states with a bottle bill. Here in New York City residents can receive a nickel per can or bottle that they redeem.

municipalities charge residents for home trash pickup according to the amount of trash they put out. The less waste the household generates, the less the resident has to pay. These programs have been effective, and over 4,000 already exist in the United States. Besides reducing waste, such programs promote equity and fairness; people pay for disposal in accordance with how much they use the service.

Composting is a strategy for reducing and recycling organic waste

Organic waste, such as yard trimmings and food scraps, can be reduced and recycled by composting. **Composting** is the conversion of organic waste into mulch or humus by encouraging, in a controlled manner, the natural biological processes of decomposition. The mulch or humus can then be used to enrich the soil. Householders can bury waste in a compost pile, in underground compost pits, or in a specially constructed composting container. As wastes are added heat builds in the interior from microbial action, and decomposition proceeds. Banana peels, coffee grounds, grass clippings, and countless other organic items (Table 19.1) all can be converted into rich, high-quality soil, given enough time, through the actions of earthworms, bacteria, soil mites, sow bugs, and other detritivores and decomposers. Home compost-

ing not only reduces a household's waste stream, but also is a form of recycling. Indeed, it is a prime example of how natural systems operate in recycling loops and how we can live more sustainably by mimicking natural cycles or incorporating them into our daily lives.

At the municipal level, many communities are reducing waste through community composting programs—3,800 across the United States at last count. In these programs, food and green wastes are removed from the waste stream and diverted to central composting facilities, where they decompose into mulch that community residents can use for gardens and landscaping. Nearly half of U.S. states now ban yard waste from the municipal waste stream, helping accelerate the drive toward composting. It is estimated that approximately 20% of the U.S. waste stream is made up of materials that can be easily composted—and fully two-thirds if one includes all paper products. Nations like the Philippines, whose municipal waste tends to be moister and include more organic matter, can do especially well with composting. Expanding composting is one element of the current effort to address Manila's waste problem. Wherever it is done, composting has many benefits: It reduces landfill waste, enriches the soil, improves resistance to erosion, encourages soil biodiversity, makes for healthier plants and more pleasing gardens, and reduces the need for chemical fertilizers and pesticides.

Recycling consists of three steps

Recycling, too, offers many benefits. **Recycling** consists of collecting materials that can be broken down and reprocessed in order to manufacture new items. Together with composting, recycling diverted 64 million tons of materials away from incinerators and landfills in the United States in 1999, the EPA estimates, and today almost one in three waste items is recycled.

There are three basic steps in the recycling loop, which have given rise to the three elements of the commonly seen recycling symbol (Figure 19.11). The first step is collection and processing of recyclable goods and materials after they have been used. Many communities designate locations where residents can drop off recyclables or receive money for them. Many of these drop-off and buy-back centers have been replaced by the more convenient option of curbside recycling, however. In curbside recycling, trucks pick up recyclable items in front of houses, usually in conjunction with municipal trash pickup programs. Curbside recycling has grown rapidly, and its convenience has also helped boost recycling rates. About half of all Americans are now

Table 19.1 Constituents of a Good Compost Mix	
Some materials to include	Some materials not to include, because they can attract pests or lower compost quality
Fruit and vegetable scraps	Meats
Eggshells	Dairy foods
Coffee grounds with filters	Fats
Tea bags	Oils (including peanut butter and mayonnaise)
Fireplace ash	Grease
Leaves	Pet excrement
Grass	Fish scraps
Yard clippings	Diseased plants
Vacuum cleaner lint	Bones
Wool and cotton rags	
Sawdust	
Nonrecyclable paper	

Source: U.S. Environmental Protection Agency, 2003.

Collection and processing
of recyclable materials
by municipalities
and businesses

Consumer
purchase of products
made from recycled
materials

Use of recyclables
by industry to
manufacture
new products

Figure 19.11 The familiar recycling symbol consists of three arrows to represent the three components of a successful and sustainable recycling strategy: collection and processing of recyclable materials, use of the materials in making new products, and consumer purchase of these products.

served by more than 9,000 curbside recycling programs across all 50 U.S. states.

Items collected through these methods are taken to materials recovery facilities (MRFs), where workers sort items into different types. Although most sorting is still done manually, automated processes are used, including magnetic pulleys, optical sensors, water currents, and air classifiers that separate items by weight and size (Figure 19.12). The facilities then clean the materials, shred them, and prepare them for reprocessing into new items.

Once readied, these materials are used in manufacturing new goods. Newspapers and many other paper products utilize recycled paper, many glass and metal containers are now made from recycled materials, and an increasing number of plastic containers are of recycled origin. Some large objects, such as benches and bridges in city parks, are now made from recycled plastics, and glass is sometimes mixed with asphalt (creating "glassphalt") for paving roads and paths. The pages in this textbook are made from up to 20% post-consumer waste, and the ink used upon them is soy-based.

If the recycling loop is to work and become economically sustainable, consumers and businesses must then purchase the products made from recycled materials. Buying recycled goods provides economic incentive for industries to use recycled products and for new recycling facilities to start or existing ones to expand their operations. In this arena, individual consumers have great power to encourage environmentally friendly options through the market system. Buying goods made with recycled materials has caught on, and many businesses now advertise their use of recycled materials, a widespread instance of "green labeling." As markets for products made with recycled materials expand, prices continue to drop.

Recycling has grown rapidly

Today's thousands of curbside recycling programs have sprung up only in the last 20 years. The 480 MRFs in operation in the United States by 1999 were able to process 62,000 tons of materials per day. Recycling in the United States has risen at a greater-than-exponential rate, from 6.4% of the waste stream in 1960 to 30.1% in 2000, according to EPA data (Figure 19.13). The EPA calls the growth of recycling "one of the best environmental success stories of the late 20th century."

Recycling rates vary greatly from one product or material type to another and from one location to another. Rates for different types of materials and products range from nearly zero to almost 100% (Table 19.2). Although the technology and the infrastructure will probably never exist to achieve 100% recycling for most items, for many items there is room for recycling to expand considerably. Recycling rates among U.S. states (Figure 19.14) also vary greatly, from less than 10% to greater than 40%. Some of this variation is due to infrastructure constraints; note in the map that states with higher population densities and more urbanized areas tend to have higher recycling rates. But this correlation is surprisingly weak; rather, the map suggests that other social, cultural,

Figure 19.12 Recyclable materials are sorted manually and with automated machinery at most materials recovery facilities. Here, plastic bottles are sorted as they travel along a conveyor belt.

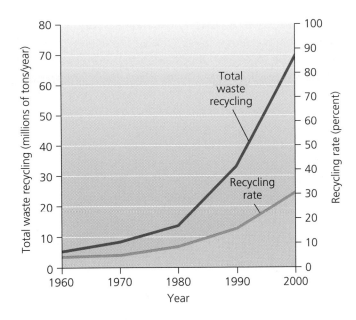

Figure 19.13 Recycling has risen sharply in the United States over the past 40 years. Today more than 70 million tons of material is recycled, comprising roughly 30% of what could be recycled. Data from U.S. EPA, 2000.

Table 19.2 Recycling Rates for Various Materials in the United States, 2001

Auto batteries	93.5
Newspapers	60.2
Steel cans	58.1
Yard trimmings	56.5
Aluminum beer and soft drink cans	49.0
Paper and paperboard	45.4
Tires	38.6
Plastic soft drink containers	35.6
Magazines	31.9
Glass containers	21.2

Data from Recycling rates of selected materials 2001, U.S. Environmental Protection Agency, November 2003.

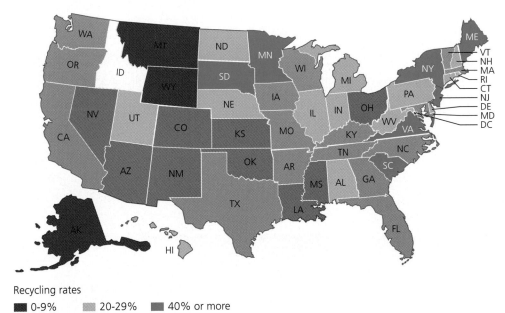

Figure 19.14 U.S. states vary greatly in the rates at which their citizens recycle. Densely populated states such as New Jersey and New York tend to have infrastructure that makes recycling programs easier, but other states such as Minnesota and South Dakota also show strong recycling rates. Data from *BioCycle Magazine*, April 1999.

Recycling rates

- 0-9%
- 10-19%
- 20-29%
- 30-39%
- 40% or more
- Unavailable

and economic factors may be at least as important.

Between 1998 and 2000 alone, total waste recovery (discarded material that was repurchased) in metro Manila rose from 69,400 tons to 101,850 tons. Dump scavengers—who have been recycling all along—provided the foundation for those numbers, but the increase was due to conditions similar to those responsible in developed countries: a mix of entrepreneurial opportunities, governmental incentives for business, and a growing environmental concern among citizens that has influenced individual behavior and consumer choice. As part of current waste management efforts in the Philippines, recycling is being encouraged and seems likely to continue growing.

The economics of recycling is in transition

The growth of recycling has been propelled in part by economic reasons, as established businesses see opportunities to save money and as entrepreneurs see opportunities to start new businesses. However, much recycling has been driven by municipalities' need to reduce waste and by the satisfaction people feel in taking the responsibility to recycle. These two forces have driven recycling's rise in spite of the fact that it has often not been financially profitable. In fact, many of the increasingly popular municipal recycling programs are run at an economic loss. The expense required to collect, sort, and process recycled goods is often more than recyclables are worth in the current market. Furthermore, the more people recycle, the more glass, paper, and plastic are available to manufacturers for purchase, further driving down prices.

However, as more manufacturers use recycled products and as more technologies and methods are developed to use recycled materials in new ways, the market will likely continue to expand, and new business opportunities may arise. We are still at an early stage in the shift from an economy that moves linearly from raw materials to products to waste, to an economy that moves circularly, using waste products as raw materials for new manufacturing processes. The steps we have taken in recycling so far are central to this transition, which many analysts view as key to building a sustainable economy (Chapter 20).

Weighing the Issues:
Recycling Pays

In 2000, recycling resulted in an annual energy savings of at least 660 trillion British thermal units (BTUs)—enough to power 6 million households a year. Recycling aluminum cans alone saves 95% of the energy required to make the same amount of aluminum from bauxite, its source material. Where do you think these energy savings come from? In addition to the energy savings, how else can reducing production from virgin materials have consequences for the environment?

One Canadian city showcases the shift from disposal to reduction and recycling

Waste management strategies can be shifted toward the more sustainable model in a hurry when there is the political will and the funding dedicated to do so. Edmonton, Alberta, has created one of the world's most advanced programs for waste management. As recently as 1998, fully 85% of the city's waste was going to its landfill, and space was running out. Today, only 35% goes to the new sanitary landfill, while 15% is recycled, and an impressive 50% is composted. Edmonton's citizens are proud of the program, and the city boasts an 81% participation rate in its curbside recycling program.

When Edmonton's residents put out their trash, city trucks take it to their new co-composting plant—at the size of eight football fields, the largest in North America. The waste is dumped on the floor of the facility, and large items, such as furniture, are removed and placed in the landfill. The bulk of the waste is mixed with dried sewage sludge in a particular ratio for 1–2 days in five large rotating drums, each the length of six buses. The resulting mix travels on a conveyor to a screen that removes nonbiodegradable items. It is aerated for several weeks in the largest stainless steel building in North America (Figure 19.15). The mix is

(a) Composting facility, Edmonton, Alberta

Figure 19.15 Edmonton, Alberta, boasts one of North America's most successful waste management programs. Edmonton's gigantic composting facility (a) is the size of eight football fields. Inside the aeration building (b), which is the size of 14 professional hockey rinks, mixtures of compostable solid waste and sewage sludge are exposed to oxygen and composted for 14–21 days.

(b) Aeration building, Edmonton composting facility

then passed through a finer screen to remove more nonbiodegradable items, and finally is left outside for 4–6 months. The resulting compost is made available to area farmers and residents. The facility even filters the air it emits with a 1 m (3.3 ft) layer of compost, bark, and wood chips, which eliminates the release of unpleasant odors into the community.

Besides the co-composting facility and the sanitary landfill, Edmonton's waste program also includes a state-of-the-art MRF that handles 30,000–40,000 tons of waste annually and is owned and managed in a public-private partnership, a leachate treatment plant, a research center, public education programs, and a wetland and landfill revegetation program. In addition, 100 pipes collect enough landfill gas for energy for 4,000 homes, bringing in thousands of dollars to the city and helping power the new waste management center. Five area businesses reprocess the city's recycled items. Newsprint and magazines are turned into new newsprint and cellulose insulation. Cardboard and paper are converted into building paper and shingles. Household metal is made into rebar and blades for tractors and graders. And recycled glass is used for reflective paint and signs.

Industrial Solid Waste

The solid waste generated by industry is another major contributor to the waste stream. Each year, U.S. industrial facilities generate about 7.6 billion tons of waste, according to the EPA, about 97% of which is wastewater. Thus, very roughly, 228 million or so tons of solid waste are generated by 60,000 facilities each year—an amount about equal to that of municipal solid waste. In the United States, industrial solid waste is that solid waste that is considered neither municipal solid waste nor hazardous waste under the federal Resource Conservation and Recovery Act.

Waste managers in the United States differentiate between municipal and industrial solid waste largely because these categories are regulated differently. Whereas the U.S. federal government regulates municipal solid waste, only state or local governments regulate industrial solid waste (although sometimes with federal guidance). Industrial waste includes more than just waste from factories. Broadly construed, it includes waste from everything from manufactured consumer goods to mining activities to petroleum extraction and agricultural waste. Waste is generated at several points along the process from raw materials extraction to manufacturing to sale and distribution (Figure 19.16).

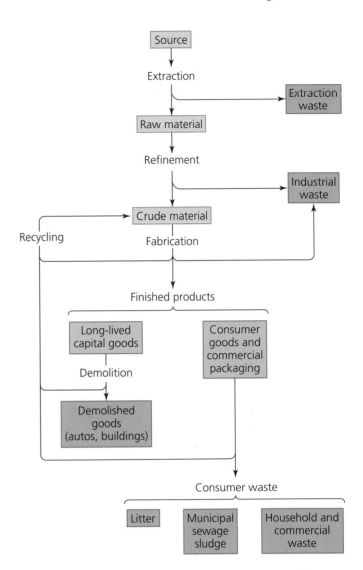

Figure 19.16 Industrial and municipal waste is generated at a number of stages throughout the life cycles of products. Waste is first generated when raw materials needed for production are extracted. Further industrial waste is produced as the raw materials are processed and as products are manufactured. Waste results from the demolition or disposal of products once they are used by businesses and individuals. At each stage there are often opportunities for efficiency improvements, source reduction, or recycling.

Regulation and economics both influence industrial waste generation

Many of the methods and strategies of waste disposal, reduction, and recycling by industry are similar or identical to those for municipal solid waste. For instance, businesses that take care of their own waste management on site most often dispose of their waste in landfills, and companies must design and manage their landfills in ways that meet state, local, or tribal guidelines for groundwater protection and other issues. Other businesses pay to have their waste disposed of at municipal disposal sites. Regulation varies greatly from state to state and area to area, but in most cases, state and local regulation of industrial solid waste is less strict than federal regulation of municipal solid waste. This is one reason that industry continues to suffer public criticism for its waste and pollution impacts. Industries in many areas are not required to have permits, are not required to install landfill liners or leachate collection systems, and do not need to monitor groundwater for contamination.

More than government regulation influences industrial waste generation, however. The amount of waste generated by a manufacturing process is one measure of its efficiency, since waste represents resources that are being lost in an industrial process. The less waste produced per unit or volume of product, the more efficient that process is, from a physical standpoint. However, physical efficiency is not always equivalent to economic efficiency. Often it is cheaper for industry to manufacture its products or perform its services quickly but messily. That is, it can be cheaper to generate waste than to avoid generating waste. In such cases economic efficiency is maximized, but physical efficiency is not. The frequent mismatch between these two types of efficiency is a major reason that the output of industrial waste is so great. While this is largely because market prices do not reflect external societal costs, it is market prices that drive the decisions of industry.

As waste disposal costs rise, however, this increases the financial incentive to decrease waste and increase physical efficiency. Once either the market or government make the physically efficient use of raw materials also economically efficient then businesses have financial incentives to reduce their own waste.

Industrial ecology seeks to reduce waste and make industry more sustainable

In an effort to reduce waste from industrial processes, growing numbers of people today are engaged in industrial ecology. **Industrial ecology** involves modifying techniques of processing and manufacturing and finding new uses for materials previously considered waste. A holistic approach to industry that integrates principles from engineering, chemistry, ecology, economics, and other disciplines, industrial ecology seeks to redesign industrial systems in order to reduce resource inputs and minimize physical inefficiency while maximizing economic efficiency. Industrial ecologists would reshape industry so that almost everything produced in a manufacturing process is used, either within that industry or in a different manufacturing process. The larger idea behind industrial ecology is that human industries should function more like ecological systems, in which almost everything produced is used by some organism, with very little being wasted. Applied to human industry, this makes recycling, or "closing the loop," the preeminent guiding principle. It also brings industry closer to the ideal of ecological economists, in which human economies attain sustainability by functioning in a circular fashion rather than a linear one (Chapter 2 and Chapter 20).

Industrial ecologists pursue their goals in several ways. For one, they examine the entire life cycle of a given product—from its origins in raw materials, through its manufacturing, to its use and finally its disposal—and look for ways to make the process more ecologically efficient. This strategy is called *life-cycle analysis*. A current U.N. Environment Programme initiative encourages companies to conduct and publicize life-cycle data for their products so that consumers can make informed decisions when they purchase products.

Industrial ecologists also examine industries broadly and try to identify points at which waste products from one manufacturing process can be used as raw materials for a different process. For instance, used plastic beverage containers cannot be refilled because of the potential for contamination, but they can be shredded and reprocessed for use in making other plastic items, such as benches, tables, and decks. Some industrial ecologists examine industrial processes with the goal of eliminating products considered harmful to long-term environmental quality. American Airlines replaced ozone-damaging CFC-based aerosols and penetrants (Chapter 11) with nonaerosol, water-based paints and penetrants. Finally, industrial ecologists study the flow of materials through industrial systems to look for ways to increase efficiency by altering manufacturing processes to create products that are more recyclable or reusable. Goods that are currently thrown away when they become obsolete, such as computers, automobiles, and certain appliances, could be designed to be more easily broken down, and their component parts recycled.

These approaches signal a radical departure from previous attitudes toward waste generation. Early in the history of industry, waste was practically ignored, discarded into the environment under the assumption that it would be diluted, decay, or otherwise disappear. As our air, water, and soil became polluted and human health was affected, however, waste managers began to collect and dispose of waste, and policymakers enacted laws to regulate disposal and fine companies and individuals for improper disposal. This end-of-pipe approach has helped reduce pollution and waste, but many people consider financial incentives and market-based solutions to be better instruments for positive change (Chapter 3). Attentive businesses are taking advantage of the insights of industrial ecology to reduce waste and save money while lessening their impact on the environment and human health. As one example, American Airlines switched from hazardous to nonhazardous materials in its Chicago facility, decreasing its need to secure permits from the EPA. American Airlines also employed over 50,000 reusable plastic containers to ship goods, reducing packaging waste by 90%. Its Dallas-Fort Worth headquarters recycled enough aluminum cans and white paper in 5 years to save $205,000. And roughly 3,000 broken baggage containers were recycled into lawn furniture. Furthermore, a program to gather suggestions from employees resulted in over 700 suggestions to reduce waste. Fifteen of these suggestions saved the company over $8 million in their first year of implementation.

Hazardous Waste

Solid waste from industrial and municipal sources is a problem largely because of the volumes in which it can accumulate. Some wastes, however, are a problem because of their chemical nature. **Hazardous waste** is waste that poses a danger or potential danger to human health. Public awareness of hazardous waste has greatly increased in recent decades, driven by numerous highly publicized instances of toxic contamination at abandoned industrial sites across the United States and other nations. The EPA defines hazardous materials based on four criteria:

1. *Ignitability*. Substances that easily catch fire (for example, natural gas or alcohol)
2. *Corrosivity*. Substances that corrode metals in storage tanks or equipment
3. *Reactivity*. Substances that are chemically unstable and readily react with other compounds, often explosively or producing noxious fumes
4. *Toxicity*. Substances that are harmful to human health when they are inhaled, are ingested, or contact human skin

Hazardous wastes are diverse in their chemical composition, and include liquids, solids, and gases. Although many hazardous chemicals become less hazardous over time as they break down, two classes of chemicals are particularly hazardous because their toxicity persists over time. These are heavy metals and organic compounds.

Heavy metals and organic compounds are primary types of hazardous waste

Heavy metals, such as lead, chromium, mercury, arsenic, cadmium, tin, and copper, are used widely in industry for wiring, electronics, metal plating, and metal fabrication. Many brightly colored heavy metal compounds are used in paints, pigments, and dyes. Heavy metals enter the environment when paints, electronic devices, batteries, and other materials are disposed of improperly. Lead from hunters' lead shot and lead fishing weights has accumulated in many rivers, lakes, and forests. Heavy metals are particularly prone to bioaccumulate (Chapter 10) because they do not break down over time. Small amounts of heavy metals consumed by bacteria and invertebrates accumulate in increasingly larger quantities up the food chain, poisoning organisms at high trophic levels. In California's Coast Range, high amounts of mercury occur in rivers and streams that originate near abandoned mercury mines left over from the 19th-century Gold Rush. Mercury ore (cinnabar) was mined from the Coast Range and shipped to the Sierras to be used in extracting gold from gold ore. Once the hundreds of small mercury mines were abandoned, the remaining ore corroded and washed downhill, accumulating in low-elevation lakes and rivers, where fish are now unsafe to eat.

Today in our day-to-day lives, we rely on the capacity of synthetic organic compounds and petroleum-derived compounds to resist bacterial, fungal, and insect activity. Items such as plastic containers, rubber tires, pesticides, solvents, and wood preservatives all are useful to us because they resist decomposition. We use these substances to protect our buildings from decay, kill pests that attack crops, and keep stored goods intact. However, their resistance to decay is a two-edged sword, for it also makes them persistent pollutants. Many synthetic organic compounds are toxic because they can be readily absorbed through human skin and can act as mutagens, carcinogens, or teratogens (Chapter 10), and can interfere with the functioning of the excretory, reproductive, and neurological systems.

These and other types of toxic substances have many negative impacts on human health and environmental quality. Flammable and explosive materials can cause ecological damage and atmospheric pollution. For instance, large-scale fires in tire dumps in California's Central Valley have exacerbated air pollution and caused highway closures. Hazardous wastes in lakes and rivers have caused large-scale fish die-offs and the closure of important domestic fisheries, such as those in Chesapeake Bay.

Hazardous wastes have diverse sources

The many sources of hazardous wastes include industries, households, small businesses, utilities, and building demolition. Industries produce the largest amounts of hazardous waste, but their hazardous waste generation and disposal is now regulated in most developed nations. These regulations have reduced the amount of hazardous waste entering the environment from industrial activities. Currently the largest source of unregulated hazardous waste is household hazardous waste.

Household hazardous waste results from materials commonly used in and around the home that nonetheless contain toxic, reactive, corrosive, or ignitable ingredients. These include a wide range of items, such as paints, batteries, oils, solvents, cleaning agents, lubricants, and pesticides. In the United States, citizens generate 1.6 million tons of household hazardous waste annually, and the average home contains close to 45 kg (100 lb) of it in sheds, basements, closets, and garages.

Several steps precede the disposal of hazardous waste

As with nonhazardous waste, for many years we generated hazardous waste and discharged it carelessly into the water or air or soil without special treatment. In many cases (for example, that of asbestos), people did not know that hazardous substances were harmful to human health. In other cases, their danger was known or suspected, but it was assumed that they would disappear or be sufficiently diluted in the environment so as not to cause problems. Incidents such as the resurfacing of toxic chemicals after years of being buried at Love Canal in New York State demonstrated to the North American public that hazardous waste needs special treatment.

Today, a number of disposal methods for hazardous waste have been developed. Although all of them are improvements on the old habit of open dumping, none of them is completely satisfactory. Since the 1980s, many communities have designated sites or special collection days for disposing of household hazardous waste, or even facilities for exchange and reuse of substances (Figure 19.17). Edmonton, Alberta, established year-round drop-off stations for household hazardous waste, and in 2001 these stations logged over 67,000 visits from residents. Once consolidated in such sites, the waste is transported to an ultimate disposal site.

U.S. law mandates that hazardous materials be tracked "from cradle to grave." When hazardous waste is generated, the producer must log with the EPA the amount of material generated and the location and way in which it is being stored. Once the waste is transferred to a carrier for transportation to a disposal site, the carrier must report the amount and destination of the waste. Finally, the disposal facility records and reports the amount and origin of waste received. This is intended to prevent industries from illegally dumping hazardous waste and encourages them to use reputable waste carriers.

Figure 19.17 Many communities designate collection sites or collection days for household hazardous waste. These boxes of household hazardous waste were turned in during such a collection day.

Because current U.S. laws make disposing of hazardous waste quite costly, irresponsible companies have sometimes been guilty of illegally and anonymously dumping waste on abandoned property, creating health risks for residents and financial headaches for local governments forced to deal with the mess (Figure 19.18). However, high costs have also encouraged responsible businesses and industries to work to transform hazardous materials into nonhazardous substances prior to discarding them. Many biologically hazardous materials can be broken down by treatment in a cement kiln, where high temperatures reduce the volume of waste and can destroy or reduce its potential hazards. Some hazardous materials can be treated by exposure to bacteria able to break down the harmful components and synthesize them into new compounds. Thus, government regulation, in tandem with economic forces, has encouraged source reduction as a preferred method of hazardous waste management, thereby minimizing the costs of disposal or recycling.

Figure 19.18 Unscrupulous individuals or businesses occasionally dump hazardous waste illegally to avoid disposal costs. Pittsfield, Massachusetts, Mayor Gerald Doyle inspects 400 barrels of chemical waste abandoned at the city's former dump in 1998. Cleanup of the site will cost hundreds of thousands of dollars.

There are three main disposal methods for hazardous waste

There are three primary means of hazardous waste disposal in the United States and most other developed countries: landfills, surface impoundments, and injection wells. Secure landfills are frequently used to store hazardous waste. Standards for landfills that receive hazardous waste are higher than those for ordinary sanitary landfills. Landfills receiving hazardous waste must have several impervious liners, have leachate removal systems, and be located far from aquifers. Dumping of hazardous waste in ordinary landfills is a problem everywhere, particularly in developing nations that barely have the resources to maintain sanitary landfills for nonhazardous waste. At the Payatas dump and other Philippine dumps and landfills, hazardous waste was mixed with nonhazardous waste, posing a threat not only to groundwater supplies but also to the thousands of scavengers who came in contact with hazardous waste day after day. Secure landfills for hazardous waste in the developed world do nothing to lessen the hazards of the materials, but they do help to keep these materials isolated from people, wildlife, and natural systems.

A method for storing liquid hazardous waste or waste in dissolved form, is in ponds or **surface impoundments.** To create a surface impoundment, a shallow depression is dug and lined with plastic and an impervious material, such as clay. Water containing small amounts of hazardous waste is placed in the pond and allowed to evaporate, leaving a residue of solid hazardous waste on the bottom (Figure 19.19). This process is repeated indefinitely. Impoundments are not an ideal method of disposal for several reasons. The underlying layer can crack and readily leak waste into the ground. Additionally, some material may evaporate into the atmosphere, or some may be blown into surrounding areas after water evaporates. Heavy rainstorms can also cause waste to overflow an impoundment and contaminate nearby areas. For these reasons, surface impoundments are currently used only for temporary waste storage.

The third disposal method is intended to be a long-term disposal method that takes waste far away from contact with the surface environment. In **deep-well injection,** a well is drilled deep beneath an area's water table. The well must reach below an impervious soil layer and must enter into porous rock. Once the well has been properly drilled, wastes are injected into it. The intention is that the wastes will be absorbed into the porous rock and remain deep underground, isolated from groundwater and human contact (Figure 19.20). This idea seems attractive in principle, but in practice wells can become corroded and can leak

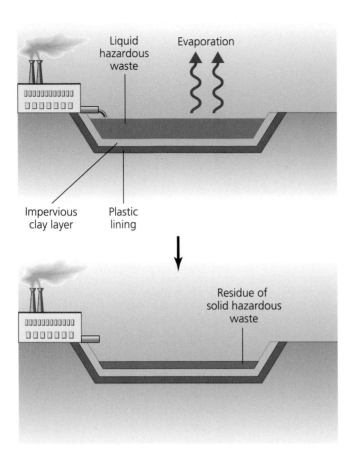

Liquid hazardous waste

Evaporation

Impervious clay layer

Plastic lining

Residue of solid hazardous waste

Figure 19.19 Surface impoundments are one strategy for disposing of liquid hazardous waste. The waste, mixed with water, is poured into a shallow depression lined with plastic and clay to prevent leaking. When the water evaporates, leaving a crust of the hazardous substance, new liquid can be poured in and the process repeated. This method alone is not satisfactory since some waste could possibly leak, overflow, evaporate, or blow away.

wastes into soil along upper levels of the well, allowing waste to enter groundwater reservoirs. Currently the amount of waste disposed of in injection wells is declining, but approximately 34 billion l (9 billion gal) of hazardous waste continue to be placed in U.S. injection wells each year.

Radioactive waste is a special type of hazardous waste

Radioactive waste, a special type of hazardous waste , is particularly dangerous to human health and persistent in the environment. For these reasons, the dilemma of safe disposal has dogged the nuclear energy industry

and the U.S. military for years. Because NIMBY opposition prevents the location of nuclear waste disposal sites just about everywhere in the United States, the U.S. government has been trying for years to establish a single large site for radioactive waste disposal. Having a single site would also make it easier to secure against sabotage. Thus the U.S. government years ago proposed siting a repository at Yucca Mountain, Nevada, a remote uninhabited location in the desert northwest of Las Vegas. Even this plan, however, has run into strong opposition from Nevada residents.

Currently, a site in the Chihuahuan Desert in southeastern New Mexico serves as a permanent disposal site for radioactive waste. The Waste Isolation Pilot Plant (WIPP) is the world's first underground repository for transuranic waste left over from nuclear weapons development. Underground in this case means *deep* underground; the mined caverns holding the waste are located 655 m (2,150 ft) below ground in a huge salt formation thought to be geologically stable. Twenty years in the planning, WIPP became operational in 1999 and will receive thousands of shipments of waste from 23 other locations over the next three decades.

During the Cold War, radioactive material was disposed of unsafely in both the United States and the Soviet Union, particularly during the early years of research into radioactivity. At the University of California at Davis, for instance, radioactive waste from a research program was placed in the campus dump with normal waste. The radioactivity began to contaminate surrounding soil, vegetation, and waterways. The area is now a Superfund site, and all material from the landfill is being excavated, placed in lead-lined containers, and shipped to a hazardous waste disposal site.

One example from the former Soviet Union involves the Mayak Nuclear Complex in Chelyabinsk, Russia, where radioactive wastes were long dumped into local rivers and lakes used as drinking supplies by 24 villages. Doctors estimate that people living in the region and consuming contaminated water have four times the radiation exposure of victims of the better-known Chernobyl nuclear accident in 1986. In 1957 one of the facility's waste containment units malfunctioned and exploded, exposing over a quarter of a million people to dangerous levels of radiation. Fish caught since then in nearby Lake Karachay were examined with a Geiger counter and found to be off the charts in radioactivity. Despite the fact that releases of radioactive water into the lake stopped in 1957, the radioactivity remains in

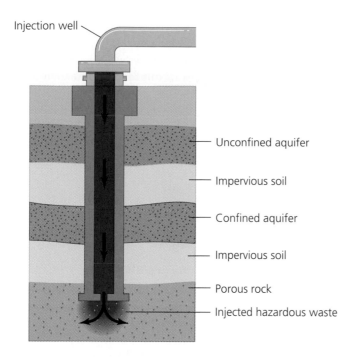

Figure 19.20 A seemingly more satisfactory way of disposing of liquid hazardous waste is to pump it deep underground, in deep-well injection. The well must be drilled below any aquifers, into porous rock separated by impervious clay. The technique is expensive, however, and leakage into groundwater from the well shaft may occur.

Labels in figure:
- Injection well
- Unconfined aquifer
- Impervious soil
- Confined aquifer
- Impervious soil
- Porous rock
- Injected hazardous waste

the environment and continues to accumulate in both fish and people.

Contaminated sites are being cleaned up, slowly

Many thousands of former military and industrial sites lie contaminated today with hazardous waste in Russia, in eastern Europe, in the Philippines, in the United States, and in virtually every other nation on Earth. For most nations, dealing with these messes is simply too difficult, time-consuming, and expensive. The United States, however, has been making an attempt to systematically clean up such sites. In response to deteriorating environmental quality, in the 1970s the U.S. Congress passed a series of laws regulating waste in general and hazardous waste in particular. The 1970 Resource Recovery Act gave the federal government jurisdiction, through the EPA, over waste management, and emphasized development of recycling programs. In 1976 Congress passed the Resource Conservation and Recovery Act (RCRA), giving the federal government jurisdiction over hazardous waste. In 1980 the Comprehensive Environmental Response Compensation and Liability Act (CERCLA) was enacted. This legislation held polluters responsible for their waste and established a federal program to clean up sites polluted with hazardous waste from past activities.

This program, called the **Superfund**, is administered by the EPA. Under EPA auspices, experts identify sites polluted with hazardous chemicals, protect groundwater near these sites, and clean up the pollution. Identified sites are called Superfund sites and include abandoned factory locations, old landfills, and abandoned utilities (Figure 19.21). The Superfund program uses money from the federal budget as well as a trust established from a tax on chemical raw materials. Although the objective of the act is to charge responsible parties for cleanup of sites, in the case of many polluted sites, the responsible parties cannot be found or held liable. Federal money is used to identify sites, maintain data on polluted sites, and pay for cleanup of areas for which no responsible party is found. So far, approximately 25% of Superfund activities have been covered by federal (taxpayers') funds. The funding for this program has grown from $300 million in the early 1980s to $1.4 billion in 2000.

Once a Superfund site has been identified, the EPA determines whether it is a threat to human health. EPA scientists evaluate how close the site is to human habitation, whether wastes are currently moving or are likely to move, and whether the site threatens groundwater or surface water supplies. Sites that appear to be harmful are placed on a prioritized list, with sites that pose the greatest threat to human health highest on the list. As money becomes available, cleanup begins on a site-by-site basis. Since the program's inception in 1980, over 700 Superfund sites have been cleaned up, but roughly 12,000 sites remain on the national priority list. The average cleanup has cost $20 million and has taken 12 years. Clearly, cleanup of hazardous waste sites is extremely expensive and time-consuming. In addition, there are many hazardous chemicals with which we have no effective way of dealing. In such cases, cleanups simply involve trying to isolate waste from human contact, either by building trenches and clay or concrete barriers around a site or by excavating contaminated material, placing it in industrial-strength containers, and shipping it to a hazardous waste dump. Therefore, the current emphasis in the United States and elsewhere is on preventing hazardous waste site formation in the first place.

(a) Berkeley Pit Superfund site, Montana

(b) Superfund cleanup site, Orange County, California

Figure 19.21 Hazardous waste at roughly 12,000 Superfund sites across the United States awaits cleaning up. (**a**) The Berkeley Pit is a Superfund site near Butte, Montana, that is gradually filling with contaminated water and has begun to attract migrating waterfowl. In October 2002, Montana Resources President Steve Walsh (left) and purchasing agent Ron Benton boarded a boat to survey the waterbody for ducks and geese. (**b**) At a Superfund cleanup site in a residential area of Orange County, California, an EPA worker in protective gear sprays down an area of excavated soil.

Wastewater and Sewage Treatment

Wastewater refers to water that has been used by humans in some way, and includes human sewage; water from showers, sinks, washing machines, and dishwashers; water used in manufacturing or cleaning processes by businesses and industries; and stormwater runoff. Stormwater runoff that flows from streets into sewer systems after rains can carry with it a great deal of oil, chemicals, and trash from roadways, parking lots, lawns, and rooftops. Runoff from especially heavy rains can sometimes overwhelm a sewer system's ability to handle wastewater. Such events were the cause of some of the sewage problems we saw with the Tijuana River in Chapter 3.

Until recent decades, wastewater was not treated before being disposed of, even in wealthy countries like the United States. Sewage and other forms of wastewater were simply piped into the nearest river or the ocean. This is still the case in some areas of the United States, and in most developing countries. While moderate amounts of wastewater can be processed by natural systems, the large amounts generated by our densely populated areas can harm ecosystems and pose threats to human health. Thus, attempts are now widely made to treat wastewater before releasing it into the environment.

Wastewater treatment involves several steps

The various pollutants in wastewater can be removed by physical means, chemical means, and biological means. Municipal wastewater treatment often consists of a combination of all of these strategies. Upon its arrival at a municipal treatment plant, wastewater first passes through bar screens that block the passage of large pieces of debris. The debris is removed and is generally landfilled. The wastewater then undergoes **primary treatment**, which involves the physical removal of contaminants. Wastewater flows into a series of tanks in which matter such as sewage solids, grit, and particulate matter are allowed to settle to the bottom. Greases and oils, meanwhile, float to the surface and can be skimmed off. Primary treatment generally succeeds in removing roughly 60% of suspended solids from wastewater.

Wastewater from the primary treatment process then proceeds to **secondary treatment**, which involves using biological means to further remove pollutants. Wastewater is aerated in the presence of bacteria by being stirred up, maximizing the water's exposure to the air. Aerobic bacteria degrade organic pollutants in the water. The wastewater from this process then passes to another settling tank, where remaining solids drift to the bottom (roughly 90% of suspended solids are generally removed after secondary treatment).

Recycling

The EPA has called recycling "one of the best environmental success stories of the late 20th century." Does the recycling movement deserve such acclaim?

Recycling's Percentage Paradox

In 1994 the U.S. recycled 35% of Municipal Solid Waste (MSW) paper—more than double the 1970 paper recycling rate of 15%. More specifically, we recycled 28 million tons of waste paper in 1994, compared to 6-plus million tons of waste paper recycled in 1970. Whew! That's five times more paper recycled by weight. That increase took tremendous effort and dedication by consumers, waste haulers, recycling centers, recycling middlemen, and the paper industry. We must be winning the battle against our garbage!

However, in the twenty-four years between 1970 and 1994, the quantity of paper and paperboard thrown into MSW nearly doubled, burgeoning from 44 million to 81 million tons. This is recycling's percentage paradox—a high recycling rate but more wastes thrown away. Does that add up to reducing waste? The absolute increase in the paper wastes recycled (22 million tons of paper) was not as great as the absolute increase in the paper wastes not recycled (37 million tons)! The paper wastes that went into the environment through landfilling or incineration represented a hefty increase—15 million tons—over the paper wastes that went into the environment in 1970.

So, which of the two years was less demanding on the environment in terms of the disposal of paper wastes? I would say 1970, because even though it had a far lower recycling rate than 1994, far less paper waste actually ended up in the environment in 1970. I strongly applaud both the high level of public participation and the increase in the percent and quantity of recycling in the 1990s. However, I believe that the public should be presented the facts about recycling more clearly in terms of the absolute quantities of wastes not recycled. We are winning the war on waste in terms of increases in recycling quantities, but we are not reducing overall wastes.

William Rathje is founding director of the Garbage Project at the University of Arizona and co-author with Robert Lilienfeld of "Use Less Stuff" (Fawcett Books).

Reduce and Reuse—Absolutely—and Then Recycle What's Left

Recycling yields enormous environmental benefits compared to other waste management options. These benefits relate not just to avoiding landfills but to the conservation of energy and natural resources and the prevention of pollution that result from the use of recycled rather than virgin raw materials in manufacturing. Recovered materials have already been refined and processed once, so manufacturing the second time around is usually much less energy-intensive and much cleaner than the first. Moreover, recycling-based manufacturing reduces the need for energy-intensive and polluting activities like oil and gas extraction, mining, and intensive forest management used to acquire virgin raw materials.

Numerous analyses have compared recycling to other waste management options, examining every major material type—steel and aluminum cans, plastic bottles, glass containers, newsprint, office paper, and boxboard—recycled in the U.S. today. These studies include all component activities of recycling: collection, processing, transport of processed materials back to manufacturers, and remanufacturing. Impacts of these recycling-based systems are compared to the impacts from landfilling or incinerating the same materials and replacing them with new materials made from virgin resources. These studies consistently show that recycling-based manufacturing holds strong environmental advantages over "one-way" systems based on virgin material manufacturing and waste disposal. And these benefits far outweigh the environmental burdens resulting from the collection, processing, and transport of materials recovered in recycling programs.

A typical benefit of recycling is conservation of energy. Manufacturing each of the major material types using recycled instead of virgin materials saves energy. The net reduction in energy use due to recycling averages about 17 million BTUs for every ton recycled. At current recycling levels, the United States is saving an amount of energy equivalent to that needed to supply all the electricity consumed by more than 10 million homes.

Richard Denison is a senior scientist at Environmental Defense specializing in waste reduction, recycling, and other waste management technologies; lifecycle assessment and lifecycle impacts of pulp and paper production use, recycling, and disposal; and environmental considerations in consumer product design. He holds a Ph.D. in Molecular Biophysics and Biochemistry from Yale University.

Using Nature to Treat Our Wastewater

On a stretch of northern California's scenic Redwood Coast, a beautiful marsh hugs the waterfront lining the town of Arcata. Bulrushes, cattails, and wildflowers wave in the breeze. Locals come to jog or walk their dogs. Thousands of birds forage for food in the shallow waters. Few visitors would guess the whole thing is built on sewage.

About 450 km (280 mi) up the coast from San Francisco, the Arcata Marsh and Wildlife Sanctuary is a 121-ha (300-acre) site where science and nature merge to clean up pollution. The marsh is a human-engineered "treatment" wetland created to filter wastewater. In such engineered marshes, the ecosystems of destroyed natural wetlands are recreated and used for bioremediation (Chapter 4), in which aquatic plants and microbes perform secondary wastewater treatment, cleaning partially treated sewage water.

Arcata, a town of 16,000 people that sits on Humboldt Bay, built its treatment wetland in an effort to reverse decades of environmental damage. Original saltwater marshes in the area were first diked in the 1870s by settlers seeking land for farming and grazing. As logging increased, lumber mills and a railway were built near the shoreline, only to be abandoned when logging became less profitable. An early sewage treatment plant, built in 1949, discharged Arcata's inadequately treated wastewater straight into Humboldt Bay.

Then, in 1972, federal lawmakers passed the Clean Water Act, and California policymakers decided that statewide sewage discharge practices needed to be improved for California to comply with the new law. The new requirements that took effect in 1974 forced Arcata to confront the run-down state of its waterfront and sewage treatment plant. The area's first plan to meet the new requirements called for building a larger treatment plant that would cost more than $50 million and pipe wastewater far out into the ocean. Local residents opposed the plan, pushing for less expensive options more in step with the local environment. Two environmental scientists from nearby Humboldt State University, Robert Gearheart and George Allen, came forward, saying wetlands might provide the water treatment Arcata needed.

Gearheart's and Allen's Marsh Pilot Project began in 1979 with 10 small constructed wetlands measuring 6.1 m by 61 m (20 ft by 200 ft) each. The project handled 10% of Arcata's wastewater. After two years, the pilot study revealed that wetlands could clean wastewater and also restore coastal wetland habitat. Arcata's civic leaders voted to use marshland to solve their sewage treatment problem. Wetland construction started in 1981, and the Arcata marsh system opened in 1986.

Treatment begins with heavy filtering at a conventional wastewater plant that can handle up to 18.9 million L (5 million gal) of wastewater per day. The water then gets a final cleaning in Arcata's chain of constructed marshes, which are punctuated with tall stands of water plants and small artificial islands. Some of the system's hardest workers are the microorganisms that grow around the marsh's aquatic plants. These break down nitrogen compounds, pathogens, and suspended solids that initial sewage treatment fails to catch. Nitrogen compounds can be hazardous to human health in drinking water and can kill life in waterways (Chapter 6). Pathogens in sewage, such as coliform bacteria, can sicken people and sometimes harm wildlife.

To remove these contaminants, Gearheart and Allen engineered a system of plants and microbes in the 20 ha (49 acres) of aeration ponds, where oxygen-breathing bacteria break down sewage, and in the six treatment and enhancement marshes, which receive the water for further filtering. The marshes have been lined with a mix of aquatic plants, including duckweed, pennywort, and a native species, hardstem bulrush.

The Arcata Marsh and Wildlife Sanctuary is the site of an artificially constructed wetland that helps treat this northern Californian city's wastewater.

The roots and stems of these plants form a dense network of underwater vegetation, hosting a wide variety of single-celled microbes and fungi. An especially productive zone lies in the plants' rhizome network, where tiny root hairs form a thick web that shelters numerous types of algae, bacteria, and other microscopic sewage-treaters.

Some of the roots and microbes in this network trap suspended solids, holding them until the plants can use them as fertilizer. Other microbes break down nitrogen compounds such as nitrate through denitrification (Chapter 6). In this process, nitrate is chemically broken down into oxygen and nitrogen by the respiratory action of a series of microbes, and the nitrogen emerges as a gas. Other microbes consume coliform bacteria, which may also die when exposed to sunlight in the marsh's shallow waters. Scientists who monitor the marshes know the cleaning system is working well when the roots and stems of marsh plants are slimy and slippery: evidence that cleansing microbes are growing and feeding and that the rhizome network is trapping suspended solids. After approximately two months of treatment, the wastewater is released into Humboldt Bay.

The marshes are carefully monitored to avoid overwhelming the wetland's cleaning abilities. Too much nitrogen can lead to unhealthy algae blooms. If initial sewage treatment does not catch contaminants such as lead, these toxins can accumulate at dangerously high levels and harm microbes or wildlife. Factors such as dissolved oxygen, temperature, and nitrogen content are measured regularly, and marsh plants are thinned every few years to keep them from choking out other forms of life in the wetland.

The system's low maintenance needs and simplicity have won awards and saved money — the cost of creating the marshes was approximately $7 million, far less than that of the conventional sewage plant that was first proposed. The marsh treatment system has also brought the Arcata waterfront back to life. More than 100,000 people visit the marsh each year, and more than 250 species of birds have been observed there. Other cities have used the Arcata Marsh and Wildlife Sanctuary as a model for their own wastewater treatment wetlands, and the marsh is a site for ongoing research on wetlands and solutions to water pollution problems.

Finally, the clarified water from secondary treatment is treated with chlorine and sometimes ultraviolet light in order to kill bacteria. Further techniques can be used to remove residual pollutants (such as filtering wastewater through carbon to remove organic pollutants), but these methods are expensive. Most often, the treated water, or effluent, is piped into water bodies such as rivers or the ocean following primary and secondary treatment. Sometimes, however, "reclaimed" water is used to water lawns of public spaces and golf courses, for irrigation, or for industrial purposes such as cooling water in power plants.

The solid material removed as water is gradually purified through the treatment process is termed sludge. The sludge is sent to digesting vats, where microorganisms decompose much of the matter, producing methane, water, and other simple compounds. The result, a wet solution of "biosolids," is then dried. It is eventually disposed of in a landfill, incinerated, or used as fertilizer on cropland. Each year about 6 million dry tons of sludge is generated in the United States.

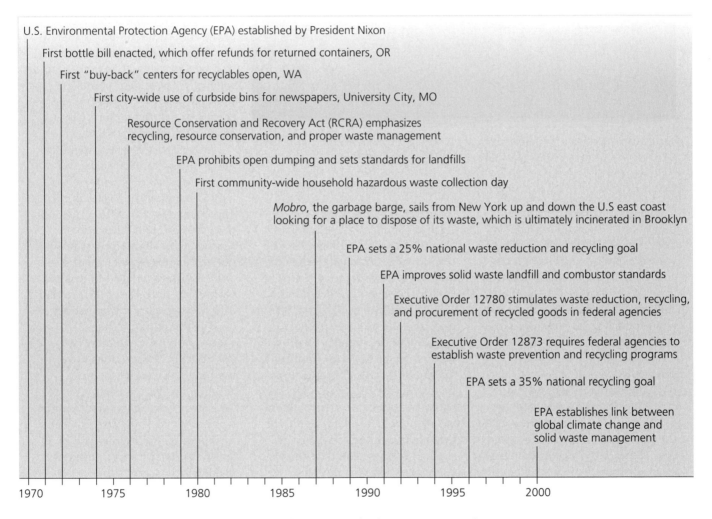

U.S. Environmental Protection Agency (EPA) established by President Nixon

First bottle bill enacted, which offer refunds for returned containers, OR

First "buy-back" centers for recyclables open, WA

First city-wide use of curbside bins for newspapers, University City, MO

Resource Conservation and Recovery Act (RCRA) emphasizes recycling, resource conservation, and proper waste management

EPA prohibits open dumping and sets standards for landfills

First community-wide household hazardous waste collection day

Mobro, the garbage barge, sails from New York up and down the U.S east coast looking for a place to dispose of its waste, which is ultimately incinerated in Brooklyn

EPA sets a 25% national waste reduction and recycling goal

EPA improves solid waste landfill and combustor standards

Executive Order 12780 stimulates waste reduction, recycling, and procurement of recycled goods in federal agencies

Executive Order 12873 requires federal agencies to establish waste prevention and recycling programs

EPA sets a 35% national recycling goal

EPA establishes link between global climate change and solid waste management

1970   1975   1980   1985   1990   1995   2000

Figure 19.22 A number of key laws and regulatory decisions marks the past 30 years of waste management in the United States, where a solid waste crisis has been mitigated by the growth of recycling and moderate steps toward source reduction. Source: Milestones in garbage: A historical timeline of municipal solid waste management, U.S. Environmental Protection Agency, 2002.

The application of sewage sludge, or biosolids, onto land has been a controversial practice. It is estimated that anywhere from 38% to over half of the sludge produced each year is used as fertilizer on agricultural lands. This practice makes productive use of the sludge, increases crop output, and conserves landfill space, but many people have voiced concern over accumulation of toxic metals, proliferation of dangerous pathogens, and odors. Do you feel this practice represents an efficient use of resources or an unnecessary risk? What further information would you want to know to inform your decision?

Artificial wetlands can aid the treatment process

Increasingly, scientists and engineers looking for better solutions to treating ever-growing amounts of wastewater have come to let nature do the work, by using wetlands as filtering devices. Wetlands in nature already perform the ecosystem service of water purification, and wastewater treatment engineers are now manipulating wetlands and even constructing wetlands *de novo* so as to employ them as tools in the quest to cleanse wastewater. At the same time, these wetland areas serve as havens for wildlife and areas for human recreation. A project in Arcata, California, was one of the first such attempts (see The Science behind the Story), but today over 500 artificially constructed or restored wetlands in the United States are performing this type of phytoremediation (Chapter 4).

Conclusion

Our societies have made great strides in addressing many of our waste problems. Our modern methods of waste disposal are far safer for humans and gentler on the natural environment than our past practices of open dumping and open burning of solid and hazardous waste. Increasingly, the techniques and strategies used in developed nations are being implemented in developing nations such as the Philippines, which are today dealing with waste crises similar to the one that afflicted the United States in the 1980s.

In many countries, recycling and composting efforts are making rapid strides. The United States has gone in a few decades from a country that virtually did not recycle to a nation in which 30% of all solid waste is diverted away from disposal (Figure 19.22). This percentage, moreover, seems likely to grow considerably higher. The rapid and continuing growth of recycling, driven by market forces, government policy, and consumer behavior, shows potential to further alleviate our waste problems.

Despite these advances, our prodigious consumption habits have created a greater abundance of waste than ever before. Moreover, our waste management efforts are marked by a number of difficult dilemmas, such as the cleanup of Superfund sites, the safe disposal of hazardous and radioactive waste, and local opposition to disposal sites. These challenges make it even more apparent that the best solution to our waste problems is to find ways to reduce our generation of waste. Some source reduction has already occurred, but the increases in consumption that have so far accompanied increases in wealth in all countries still threaten to deplete many of our valuable raw materials and convert them to waste. The best hope may lie in adopting new ways of producing and using goods, changing our linear economic model to a circular one. This shift is the key, many environmental scientists maintain, to building a truly sustainable human society on our planet.

REVIEW QUESTIONS

1. Why do we need to practice waste management? Define waste management.
2. Describe five major methods of managing waste.
3. Why have some labeled the United States "the throwaway society?" How much solid waste do Americans generate, and how does this compare to other countries?
4. How did the United States sidestep a solid waste "crisis" in the 1980s like that currently being experienced in the Philippines and some other developing countries where open dumping is a problem?
5. By what standards are sanitary landfills regulated?
6. Name at least five problems with landfills.
7. Describe the process of incineration or combustion. How does incineration reduce the weight and volume of waste? How is ash disposed of? What is one drawback of incineration?
8. What does a WTE facility do? Is this an effective strategy for managing waste? Why or why not?
9. How can source reduction benefit consumers?
10. Describe some financial incentives that help consumers use less and create less waste.
11. What is composting? How is it useful in reducing waste?
12. What are the three elements of a sustainable process of recycling?
13. What forces are responsible for promoting recycling amongst businesses and municipalities, in spite of the frequent lack of profitability?
14. How is industrial solid waste defined under the federal Resource Conservation and Recovery Act, and how is it regulated? What are the goals of industrial ecology in managing industrial solid waste?
15. What are the four criteria the EPA uses to define hazardous waste? Why are heavy metals, synthetic organic compounds, and petroleum-derived compounds particularly hazardous?
16. What is the largest source of unregulated hazardous waste? Of regulated hazardous waste?
17. Describe three ways to dispose of hazardous waste. How is radioactive waste from weapons production disposed of in the United States?
18. What is the Superfund program?

DISCUSSION QUESTIONS

1. How much waste are you generating? Look into your waste bin at the end of the day and categorize and measure the waste there. Consider other waste you may have generated in other places throughout the day. How much waste could you have prevented? How much could be reused or recycled? Did you dispose of any hazardous household waste?

2. What kind of solid waste is best for incineration, and why is the United States a better place to practice incineration than the Philippines? What well-intentioned regulation in the Philippines has exacerbated the waste problem? Can you think of any solutions in addition to those measures currently being taken by the Philippine government to deal with the problem?

3. Describe some ways that waste management can help us produce energy or reduce energy use.

4. Compare the success of recycling in the United States and in Manila. How have the incentives for recycling been similar and different?

5. Can manufacturers and businesses benefit from source reduction if consumers stop buying as many products? How? Given what you know about industrial ecology, what do you think the future of sustainable manufacturing may look like?

Media Resources *For further review, go to the website* **www.envscienceplace.com** *or student CD-ROM, where you will find quizzes, flashcards, a glossary, additional interactive exercises, and links to relevant news and research sources. Also, on the website and CD-ROM is* **GRAPH IT**, *a series of interactive graphing tutorials to help you interpret graphs and plot data.*

20 Sustainable solutions

Downtown Johannesburg, South Africa

This chapter will help you understand:

- The concept of sustainable development

- How humans are part of the environment

- How promoting economic welfare and protecting the environment are compatible

- How growth is not equivalent to progress

- The roles that consumption, population, and technology play in human impact on the environment

- Several key approaches to designing sustainable solutions

- That time is limited, but the human potential to solve problems is tremendous

Traditional African Dancers before the World Summit

Central Case: The World Summit in Johannesburg

"From the African continent, the cradle of humankind, we solemnly pledge to the peoples of the world and the generations that will surely inherit this Earth that we are determined to ensure that our collective hope for sustainable development is realized."
—*Declaration signed by 193 nations represented at the World Summit, 2002*

"We hope—surely we must believe—that our species will emerge from the environmental bottleneck in better condition than when we entered . . . taking as much of the rest of life with us as possible."
—*Edward O. Wilson, Consilience, 1998*

It was the largest-ever international gathering of its kind, held to address perhaps the world's most pressing issue. For 10 days in August and September 2002, the World Summit on Sustainable Development drew thousands of people to Johannesburg, South Africa. More than 10,000 delegates from nearly 200 countries attended, along with 8,000 representatives of nongovernmental organizations. More than 100 world leaders addressed the gathering. The world's attention was drawn as well; for every nine participants there were two journalists to relate news of the proceedings across the globe.

The attendees gathered to hammer out agreements for moving their nations and our global society toward sustainable development.

Johannesburg, a city of 3.2 million people, seemed an appropriate location for this United Nations-sponsored conference. It is located on the continent whose people suffer a lower standard of living than any other and might benefit most from sustainable development. The economic deprivation of the African populace is in many ways tied to environmental problems, including desertification, deforestation, and biodiversity loss, rapid population growth, and environmental health hazards. Economically, socially, and politically, Johannesburg is a symbol of hope and progress, a thriving city in a country that so successfully and peacefully threw off the injustice of apartheid.

Indeed, South Africa is in many ways a microcosm of the world and illustrates the world's need for sustainable development. The nation has extreme wealth and extreme poverty, and its wealth is concentrated among a small proportion of its citizens. It possesses urban metropolises with skyscrapers, cell phones, and computer networks but also is home to hunter-gatherer tribes who still live largely by traditional means. It has areas of rich biodiversity and unique endemic species that are increasingly threatened by human impact. South Africa supports its economy with valuable natural resources such as gold and diamonds, although these are nonrenewable and their trade has enriched only a small number of people. The country has become dependent on international trade to supply products it needs, such as fossil fuels. It is struggling with the AIDS crisis that is ravaging Africa and other areas of the world. And it has struggled with a dark history of racial discord and repression, but today is finding ways to leave those old fears and prejudices behind and build a more socially just society.

These issues and more set the stage for the international discussion at the World Summit. The Johannesburg meeting was the third historic international conference on human interactions with the environment. The first took place in Stockholm, Sweden, in 1972, as citizens throughout the developed world pushed their leaders to begin devising solutions to the environmental problems that had been building for years as a result of growth in population and consumption. At the Stockholm conference, nations came together for the first time and broadly agreed that dramatic steps needed to be taken to respond to the increasing degradation of the natural environment.

This meeting was followed two decades later by the U.N. Conference on Environment and Development held in Rio de Janeiro, Brazil. Representatives of the world's nations crystallized the notion that environmental protection can—and should— go hand-in-hand with improving living conditions for the world's people, particularly the world's poor. This linkage of economic and social development with environmental protection set a new course that led eventually to the Johannesburg meeting 10 years later. The program agreed upon at the Rio meeting, called *Agenda 21,* drew a roadmap for nations to improve people's lives and improve environmental conditions.

Few of the goals of Agenda 21 have been fulfilled by the world's nations, and as participants ventured to Johannesburg, they knew that more had to be accomplished. After reaffirming their commitment to Agenda 21, they succeeded in forging new alliances and launching over 300 initiatives worldwide that committed over $200 million toward new cooperative projects between governments and the private sector. Billions more dollars were pledged by many countries toward projects involving water and sanitation, energy, health, agriculture, finance and trade, and biodiversity protection and ecosystem management.

Amid all the pledges and proclamations, many critics said the meeting featured too much talk and not enough action. Others criticized it for focusing more on development than on sustainability. U.N. Secretary-General Kofi Annan stressed that it was too early to judge the success of the conference and that it should be judged by the impacts it generates in coming years. "We have to go out and take action," Annan said. "This is not the end. It's the beginning."

Critics emphasized that action toward sustainability was necessary to accomplish quickly and soon. Experts at the United Nations seem to agree; the new head of the U.N.'s Sustainable Development program in 2003 titled his welcome statement, "No Time to Lose."

Sustainable Development

The United Nations defines sustainable development as "development that meets the needs of the present without compromising the ability of future generations to meet their own needs." The definition is taken from a U.N. commission (called the Brundtland Commission after its chair, Norwegian prime minister Gro Harlem Brundtland) that produced an influential 1987 report entitled *Our Common Future.*

Development may be thought of as the act of making purposeful changes to improve the quality of human life. However, environmental scientists have long pointed out that many forms of development have so degraded the natural environment that they threaten the very improvements for human life that were intended. Human actions on Earth should be sustainable, many have argued; they should not degrade the environment's capacity to allow those actions to continue indefinitely into the future. Thus scientists, policymakers, economists, and advocates arrived at the idea of sustainable development, a notion that is broadly accepted but means many things to many people.

Prior to the Rio conference and the Brundtland Report, most people aware of human impact on the environment might have thought "sustainable development" to be an oxymoron—a phrase that contradicts itself. If human development has negative impacts on the natural environment, they would have asked, and if our natural environment is ultimately limited in its space and its resources, how then can we continue to develop and yet do so sustainably, without depleting our planet's bounty for the future? Indeed, even today most people labor under the impression that protecting the environment is incompatible with serving the needs of people.

The question "Can we develop in a sustainable way?" may well be the single most important question in the world today. It certainly dwarfs in significance the news of politics, terrorism, and celebrity entertainment that dominate the daily headlines. The people who are asking this question have become more hopeful that developing in sustainable ways is possible. At the same time, they recognize that to meet this goal, thoughtful and purposeful action will be needed—and quickly.

What precisely do we want to sustain?

We have said that development aims to improve the quality of human life, but when environmental scientists speak of *sustainable* development, what exactly do they mean to sustain? Certainly they mean to sustain the natural environment, its species, and its systems in a healthy and functional state. They also mean to sustain human civilization in a healthy state. Although the goals of development and environmental sustenance are often cast as being in opposition, environmental scientists recognize that human civilization depends on an intact and functional natural environment; in fact, civilization cannot exist without it. The contributions of biodiversity to human welfare (Chapter 15) and of ecosystem goods and services in general (Chapter 2) to our civilization are tremendous and beyond question. Indeed, they are so fundamental (some would say infinitely valuable, thus literally priceless) that we have long taken them for granted. Our utter dependence on the world around us becomes especially clear once one realizes and accepts that humans do not exist apart from the environment but are part of the environment.

Humans are not separate from the environment

It is common to hear of "humans and the environment" or of "people and nature" being set in contrast, as though they were two completely separate things. Given the limits of language, it is difficult *not* to speak in these terms, as, indeed, we have felt forced to do throughout much of this book. Unfortunately, this constraint in our language, many feel, has led to limitations in our thinking. Some philosophers go as far as to say that the perceived dichotomy between humans and nature is at the root of all our environmental problems.

Part of why we commonly draw this dichotomy is that on a day-to-day basis, it is easy to feel completely disconnected from the natural environment, particularly in developed nations and in large cities. We live in houses; work in shuttered buildings; travel in cars, airplanes, and other enclosed spaces; and generally know little about the plants and animals around us. Millions of urban citizens have never set foot in a wilderness area. Just a few centuries or even decades ago, most of the world's people were able to name and describe the habits of many of the plants and animals that lived around them. They knew exactly where their food, their water, and their clothing came from. Today it seems that water comes from the faucet and food comes from the grocery store, and we rarely think beyond that. It's little wonder we have lost track of the connections that tie us to our environment.

However, this psychological severance, this feeling of isolation, doesn't make our connections to the natural environment any less real. Even though we are conditioned not to think about where our food comes from, it does come from somewhere, and its production does cause real environmental impacts.

Consider one of the most un-"natural" (yet delicious!) inventions of the human species: the banana split. Even in this triumph of human creation, seemingly concocted *de novo* at an ice cream shop, each and every element has distinct ties to the resources of the natural environment. The ice cream, of course, comes partly from the processed milk of cows. These may graze on native or invasive grasses on rangeland or be raised in feedlots on grain grown in monocultures. Sugar that sweetens the ice cream may come from sugarcane plantations such as those that appropriate water from the flow that nourishes the Florida Everglades. The banana has perhaps been shipped (using oil to fuel its transport) thousands of miles from Ecuador or some other tropical country, where it was grown on a plantation where rainforest had formerly stood. The bananas were treated with fertilizers and fungicides (also developed with petroleum) to spur growth and keep pathogens at bay. Fruits and nuts of the banana split may have been grown in the fertile soils of California's Central Valley and irrigated with generous amounts of water piped in from the Sierras. The spoon used to eat the banana split originated in metal ores deep underground that were mined and removed along with thousands of tons of soil, then transported to facilities where they were processed into stainless steel using large inputs of energy from fossil fuels. Or the spoon may be plastic, produced by an industrial process relying on petroleum. One could go on and on for each element of the banana split—and one can go through the same thought experiment for every material item in our lives.

To preserve the pleasure that banana splits can bring to our lives, then, we must preserve the environment that makes their creation possible. Once we learn to think in this way, to consider where the things we use and value every day actually come from, it is easier to see how humans are part of the environment. Once we reestablish this connection, it becomes readily apparent that our own interests are served by preservation or responsible stewardship of the natural systems around us.

Win-win situations are possible

Because what is good for the environment can also be good for people, win-win solutions are very much within reach. We *can* have it both ways, if we learn from what science can teach us, if we think creatively, and if we are willing to act on our ideas. Ultimately, the only way for our species to "win"—to survive and thrive—is to conduct our activities in ways that sustain the processes and resources of our environment. If the environment "loses" through our impacts upon it then so will we.

Furthermore, in a lose-lose scenario, it may be only humans who lose in the very long term. If we change our environment so much that we destroy ourselves, our species will be gone forever, but the environment will bounce back eventually. Biodiversity may recover in its numbers (although with a very different collection of organisms) perhaps 10 million years or so after a mass extinction, paleontologists estimate. Desertified landscapes and eroded and salinized soil will, after millions of years, eventually plunge back into the mantle and be reborn through plate tectonics. A disturbed atmosphere and warmed global climate may eventually cool. In the very long term, then, it is we who will be the losers if we degrade our environment, because once extinct, our species will not return.

If we succeed in sustaining the environmental conditions that have supported our species to this day, however, then we preserve the chance to sustain our species and our phenomenal society. By safeguarding our natural resources and by finding methods of turning our waste products into new resources, we can enable our society to maintain itself and to improve in the future.

Weighing the Issues: Unavoidable Impacts?

Ecology teaches us that when one species is successful, it may exclude other species from resources, and this has clearly been the case with humanity's recent success in spreading across the planet. However, in the long scope of Earth's history, biodiversity has increased, as organisms have evolved and adapted to environmental changes caused by other organisms. Do you think that our continuing existence need necessarily hinder other species? Does our society's development demand some amount of unavoidable harm to other species or to environmental processes? What kinds of impacts might be unavoidable, and what kinds might we be able to avoid?

What accounts for the perceived economy-versus-environment divide?

If win-win possibilities exist, what accounts for the conventional view that we cannot provide for people's needs and protect the environment simultaneously? Perhaps our species' long history has bestowed on us instincts to take as much as we can from the environment. For the thousands of years that we lived as nomadic hunter-gatherers, population sizes were low and consumption and environmental impact were limited by primitive technology. With natural resources in little danger of running out, perhaps humans had little reason to adopt a conservation ethic, and it is difficult for our species to adjust its way of thinking now. Alternatively, perhaps the establishment of sedentary agricultural societies — and later cities — beginning about 10,000 years ago was responsible for causing us to overlook the connections between our economies and our environments.

In our day and age, it could be argued that the recent history of the environmental movement, and the reaction to it, has contributed to the concept of the environment-versus-economy divide. The wave of modern environmentalism that swept much of the developed world in the 1960s and 1970s was a response to what many people viewed as rampant and unthinking human impact on natural systems. Early environmentalists were simply desperate to reduce what they saw as human depredation on the environment, in any way that they could. Citizen outrage inspired politicians to take action, and the easiest course of action was legislation, leading to a regulatory approach. Governments enacted laws and regulations bearing on everything from clean air to clean water to endangered species, and many of these limited what businesses, individuals, and government agencies could do.

Although command-and-control regulation may have been the best way to slow human impacts at the outset of the environmental movement (and has resulted in cleaner air, cleaner water, and many other advances), the regulatory approach has today become so ingrained

Figure 20.1 The northern spotted owl (*Strix occidentalis occidentalis*) has become a symbol of the "jobs-versus-environment" debate. A bird of the Pacific Northwest rainforest, it is considered endangered due to the logging of mature forests. Proponents of logging argue that laws protecting endangered species cause economic harm and job loss. Advocates of endangered species protection argue that unsustainable logging practices pose a larger risk of job loss.

that people often feel it infringes on their liberties. Corporations complain of excessive costs for environmental mitigation, and landowners worry about restrictions placed on use of their land. In developing countries, people have sometimes found it harder to make a living because well-intentioned residents of developed countries, concerned about the global environment, helped set aside land for preservation—land on whose resources local people had depended.

One common perception is that environmental protection measures hurt the economy by costing people jobs. A prime example is the controversy over logging of old-growth forest in the Pacific Northwest. The protection that the U.S. Endangered Species Act has given to vulnerable organisms such as the spotted owl (Figure 20.1) has entailed restrictions on timbering activity in some forested areas. Proponents of logging have rallied opposition to protection of old-growth forests by claiming that the restrictions cost local loggers their jobs. Job loss has occurred to a limited extent in some cases, but the

"jobs-versus-owls" debate overlooks many factors. One is that loggers' jobs are most at risk when timber companies cut timber at a nonrenewable rate and, consequently, are forced to leave a region completely to seek more mature forests elsewhere. As we saw in Chapter 16, this has happened repeatedly in one region after another throughout U.S. history. It is playing out in the Pacific Northwest now, as U.S. timber companies begin moving operations abroad, now that nearly all primary forest is gone.

Environmental protection measures can enhance economic opportunity

The jobs-versus-environment debate in general also overlooks the fact that even as some industries decline, new ones spring up to take their place. As jobs in logging, mining, and manufacturing have decreased in the United States over the past few decades, jobs have greatly increased in computer and other high-technology sectors and in many service-oriented occupations. As we decrease our dependence on fossil fuels in the future, experts predict, jobs and investment opportunities will open up in renewable energy sectors such as wind power and fuel cell technology (Chapter 18).

In addition, people desire to live in areas that have clean air and water, intact forests, and parks and open space. Environmental protection increases the attractiveness of a region, drawing more residents and increasing property values and the tax revenues that help fund social services. As a result, those regions that have acted to protect their environments are those that are generally able to retain and increase their wealth and quality of life.

Thus, it is clear that environmental protection need not lead to economic stagnation and, to the contrary, is often likely to enhance economic opportunities (Table 20.1). A

Table 20.1. Potential Causes of Job Growth Due to Environmental Protection Measures
New environmental technologies create jobs directly and spur new technologies among other products and sectors
Environmental technologies developed by one country can be exported to other countries, generating export income and jobs
Environmental protection enhances the livability of a region, attracting workers and businesses
Government and private-sector environmental investments can boost job growth during recessions

Adapted from *The Trade-Off Myth*, Eban Goodstein, 1999.

Assessing Costs and Benefits of Environmental Regulations

Federal regulations that aim to protect environmental quality often result in clearer air or cleaner water but usually have financial implications for affected parties. Are such regulations worth the costs to industry, businesses, and consumers? In 2003, a federal study determined that the answer to that question is a definitive, "Yes."

The study, from the White House Office of Management and Budget (OMB), weighed the costs of a range of recently created federal regulations against the economic value of the beneficial changes resulting from the regulations. The OMB examined 107 federal regulations enacted in the United States from 1992 to 2002, many of which dealt with environmental issues.

The study found that the reviewed regulations carried a sizable annual cost, estimated at $36 billion to $42 billion. However, their estimated value in terms of public good totaled $146 billion to $230 billion. Thus, the economic benefits of these regulations far exceeded their costs. Moreover, of all the regulations, those dealing with environmental protection were found to be especially cost-effective.

Most regulation benefits were reflected in health and social gains resulting from clean-air. Over the

studied 10-year period, the benefits of clean-air regulations were found to be 6 to 10 times greater than the costs of complying with the regulations. The value of reductions in hospitalization and emergency room visits, premature deaths, and lost workdays resulting from improved air quality were estimated between $118 billion and $177 billion.

The authors of the OMB review relied on cost-benefit analyses previously made by experts in the EPA and other agencies, scrutinizing them for appropriateness and accuracy before using them. A look at one EPA regulation included in the OMB report—an air pollution regulation called the *Heavy Duty Engine/Diesel Fuel Rule*—shows how the EPA performed its own economic analysis and how the OMB double-checked the EPA's findings.

The diesel regulation aimed to address diesel engines that emit fine soot, sulfur compounds, and nitrogen oxides (NO_x). In the 1990s, the EPA regulation was intended to make diesel trucks and buses run more cleanly by lowering the fuel's sulfur content. However, achieving cleaner diesel fuel would mean substantial—and potentially costly—changes to the

way that fuel is made and handled.

To determine potential costs, EPA analysts examined how various parties involved with diesel fuel use—from fuel processors to engine manufacturers to vehicle operators—would need to change their processes or equipment to clean up the fuel. For example, a heavy-duty engine redesigned to run more efficiently on cleaner fuel would cost approximately $4,600 more to operate, the analysts calculated. Total research and development costs for emission control were predicted to exceed $600 million. The EPA asked for input from affected parties, scientists, and the public, finally determining that removing sulfur from diesel fuel would be neither cheap nor quick, with total annualized costs of about $4.2 billion by 2030.

EPA analysts then considered the benefits of reduced diesel pollution. Using air quality data from more than 940 smog monitors around the country, the EPA estimated diesel-related air pollution levels with and without the proposed diesel regulation. The analysts examined public health statistics, especially data related to respiratory illnesses, such as asthma or chronic bronchitis, and estimated how many cases of dis-

recent U.S. government review concluded that the economic benefits of environmental regulations greatly exceeded their economic costs (see The Science behind the Story). This connection is also suggested by the fact that both the U.S. and global economies have expanded rapidly in the past 30 years, the period during which environmental protection measures have proliferated.

Growth versus Progress

Correlating economic expansion with environmental protection measures is one way of addressing whether there is or is not a tradeoff between environmental protection and economic development. However, many environmental scientists who have thought deeply about

ease or premature death might be prevented if people breathed fewer diesel-related pollutants. They determined the economic effects of such health problems by determining the medical, workforce, and social costs. Each premature death was assigned an impact of $6 million, and each case of chronic bronchitis was pegged at $331,000. Many of the estimates were also regionalized to account for different economic conditions in various parts of the country.

In the end, the EPA analysis determined, the benefits of cleaner diesel fuel far exceeded the costs. If sulfur content of diesel was cut from 500 to 15 parts per million, then 2.6 million tons of smog-causing NO_x emissions would be eliminated each year. An estimated 8,300 premature deaths, 5,500 cases of chronic bronchitis, and more than 360,000 asthma attacks would be prevented each year. About 1.5 million lost workdays would be prevented. By 2030, the analysis determined, benefits would reach about $70 billion. Following that analysis, the Heavy Duty Engine/Diesel Fuel Rule took effect in early 2001. Requirements will be phased in gradually through 2010.

Estimates of the Total Annual Benefits and Costs of Major Federal Rules, October 1, 1992 to September 30, 2002 (millions of 2001 dollars)

Agency	Benefits	Costs
Agriculture	3,094 to 6,176	1,643 to 1,672
Education	655 to 813	361 to 610
Energy	4,700 to 4,768	2,472
Health & Human Services	9,129 to 11,710	3,165 to 3,334
Housing & Urban Development	551 to 625	348
Labor	1,804 to 4,185	1,056
Transportation	6,144 to 9,456	4,220 to 6,718
Environmental Protection Agency	120,753 to 193,163	23,359 to 26,604
Total	146,812 to 230,896	36,625 to 42,813

Source: Informing regulatory decisions: 2003 report to Congress on the costs and benefits of federal regulations and unfunded mandates on state, local, and tribal entities, Office of Management and Budget, 2003.

The OMB, in compiling its own review of federal rules, scrutinized the study conducted by the EPA to determine whether the EPA estimates made economic sense. The OMB analysts concluded that most of the estimates did, indeed, make sense. The OMB analysts made some minor changes when calculating their own figures. For example, although the EPA said every 1-ton reduction in NO_x emissions represented a benefit of $10,200, the OMB used its own updated figures and determined NO_x reductions to be worth $5,500 per ton. Overall, however, the OMB confirmed the validity of the EPA estimates and used most of the EPA's economic data in the larger OMB report.

The OMB report, in totaling and highlighting the value of many recent environmental regulations, is now being used to evaluate and support efforts at environmental regulation around the country.

what it will take to achieve sustainability have become troubled by the notion of economic expansion.

It is conventional among economists and the policymakers who heed their advice to speak of economic growth as a goal for a nation, a region, or the world. Many politicians see nurturing an expanding economy as their prime responsibility while in office. In the United States and many other nations, economic growth is largely driven by consumption, the purchase of material goods and services by consumers. The human tendency to believe that bigger is better, faster is better, and more is better, is reinforced by advertisers looking to sell more goods more quickly. Consumption has grown tremendously, with the wealthiest nations leading the way. The United States, with less than 5% of the world's population, now consumes 30% of its energy resources and 40% of its total resources. U.S. houses are larger than ever, sports-utility vehicles are among the most popular passenger vehicles, and many citizens have more material belongings than they know what to do with. We think nothing now of having home computers with high-speed Internet access, let alone the televisions, telephones, refrigerators, and dishwashers that were a marvel just decades ago. Where will such growth in consumption lead?

Can consumption continue growing?

Because most of Earth's natural resources are limited, consumption of nonrenewable resources cannot continue growing forever. In fact, some analysts have estimated that there are not nearly enough resources for the rest of the world's people ever to attain the level of consumption already reached by the average U.S. citizen. If the entire world population consumed at the rate of today's average U.S. citizen, these analysts have calculated, we would need at least another two to three planet Earths. Eventually, if we do not shift to sustainable practices of resource use, per capita consumption will drop for rich and poor alike as resources dwindle.

Cornucopian critics often scoff at the notion that resources are limited, but we must remember that our perspective in time is limited as well. It is difficult for an American who has grown up in the late 20th century amongst relative plenty, for instance, to envision that his or her lifestyle is not representative of all history, past and future. Our short lifetimes allow us to feel viscerally only what we know from experience. Through education, however, we can come to learn the broader historical perspective that encompasses the past and allows us to better predict our possible futures. We can also learn the broader geographical perspective that reveals the deep gulf between the developing and developed worlds. Obtaining a more informed picture is at the very core of environmental science. Such a picture reveals that our consumptive lifestyles represent an extraordinarily brief slice of time in the long course of history (Figure 20.2). Our consumptive lifestyles are a brand-new phenomenon on Earth and may not last much longer. We are living with the greatest material prosperity in all of human history; if we do not find ways to make our wealth sustainable, it could be all downhill from here, fast.

Consuming less can be more satisfying

Material consumption, however, is only one way to measure prosperity, and many people down through the

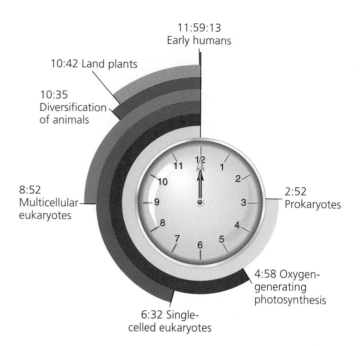

Figure 20.2 Viewing Earth's 4.5-billion-year history as a 12-hour clock helps us obtain a proper sense of relative timescales across the immense span of geological time. *Homo sapiens*, as a species, has come into existence only during the final one or two seconds, around 11:59:59. The agricultural and industrial revolutions that have increased our environmental impacts have taken up only a miniscule fraction of a second.

ages, from ancient philosophers to modern-day environmental scientists, have maintained that it is not necessarily the best way to do so. Consuming goods and services is certainly an important element in making our lives comfortable and happy. However, consumption alone does not reflect a person's quality of life. Sociological studies show that materially wealthier people may on average be more content than poorer people but also show that people with more material wealth do not necessarily live happier lives than those with less. Our modern lives have become filled with goods and services that are bigger, faster, and more numerous, and for many people in the United States and other developed nations, the accumulation of innumerable possessions has not brought contentment. Social critics have even coined a word for the way material goods often fail to bring happiness to people affluent enough to afford them: *affluenza*. Although economic growth is often equated with "progress," many people choose to use a different ethical standard by which to judge progress. For them, true progress consists of an increase in human happiness, not simply growth in material wealth. In the end we are, one would hope, more than simply the sum of what we buy.

Consumption can be greatly reduced even while one's happiness or quality of life increases. Indeed, both outcomes are necessary if we are to develop sustainably. We can squeeze more from less in at least three major ways. One is through improvements in the technology of materials and manufacturing and in the efficiency of manufacturing processes. By improving technology and efficiency, industry can produce goods using fewer natural resources. Another way is by developing a sustainable manufacturing system—one that is circular and based on the principle of recycling, in which the waste from a process becomes raw material for input into that process or into other processes of the same business or other businesses. Such sustainable systems could potentially allow consumption to be sustained at some maximum level indefinitely.

A third way to halt runaway consumption involves the consumer rather than industry; it is that each one of us modify our behavior, attitudes, and lifestyles, and strive to make personal choices that minimize consumption. At the outset such choices may seem like sacrifices, but those who have made them often say the opposite, that freeing oneself of an attachment to excess material possessions can feel tremendously liberating. More and more people in the developed world today are finding that slowing down the pace of their busy lives and lessening their accumulation of material goods is more satisfying than the rampant consumerism that many feel has come to characterize American culture.

By changing the way we measure prosperity and progress, and by adopting sustainable practices in industry, government, and our daily lives, it may be possible to continue increasing our quality of life while reducing or keeping constant our level of consumption and while sustaining the natural environment we depend on. Adopting sustainable agricultural processes that conserve soil could enable farmland to remain productive year in and year out, providing a reliable supply of food crops. Adopting sustainable forestry practices could allow for timber harvesting at a level that prevents deforestation while allowing for continued consumption of valuable forest products. Switching to renewable energy sources might eventually provide us abundant power at relatively low cost, enabling us to maintain consumption at high but sustainable levels.

Weighing the Issues: Renewable Energy and Sustainability

Environmental advocates promoting alternative energy sources such as solar, wind, and geothermal power often argue that, given adequate development, such renewable sources could provide us abundant and cheap energy and help bring about a sustainable future. Do you agree? Or do you think abundant and cheap renewable energy might simply lead to further unsustainable development and cause further environmental impact? How might a shift to renewable energy sources influence the quest for sustainable development?

Human population cannot keep growing

Just as continued growth in consumption is not sustainable, neither is growth in the human population. No population of organisms can continue growing forever. In Chapter 5 we saw how populations may grow exponentially for a time but eventually run up against limiting factors and level off at a carrying capacity. Although we have increased Earth's carrying capacity for humans with the help of our technology, our population growth cannot continue forever; sooner or later, the human population will level off. The question is how: through war, plagues, and famine, or through voluntary means as a result of wealth and education?

The demographic transition (Chapter 7) that many developed nations are undergoing provides reason to expect that population sizes will stabilize and even begin to fall. This transition is already taking place in many developed nations thanks to urbanization, wealth, education, and the empowerment of women. If the demographic transition model holds true for today's developing nations, and if developed nations are able and willing to extend the aid developing nations need to pass through the demographic transition, then there is hope that humanity may reverse its population growth while creating more prosperous and equitable societies. Otherwise, we may be forced to adopt draconian governmental restrictions on personal reproductive choice, as China has done (Chapter 7), to diminish the likelihood that population reduction will come about through plagues, famine, and civil conflict.

Technology can help us toward sustainability

It is largely technology—developed as part of the agricultural revolution, the industrial revolution, and advances in medicine and health—that spurred our population increase. Technology has extended, deepened, and strengthened our impacts on Earth's environmental systems. However, technology also can help us develop ways to reduce environmental impacts and achieve sustainability.

Technology is not an independent actor, but is a tool of human agency. The short-sighted use of technology may have gotten us into this mess, but wiser use of technology can help get us out.

Recall the I=PAT equation from Chapter 7, which summarized human environmental impact by equating overall impact (I) to the interaction of population (P), consumption or affluence (A), and technology (T). Technology can exert either a positive or negative value in this equation. In recent years technology has been increasing human impact on the environment in developing countries. Industrial technologies long-used in the developed world have been exported to poorer nations eager to industrialize.

In developed nations, however, much technology has begun reducing our environmental impact. Catalytic converters on cars have reduced emissions, as have scrubbing devices on power plants' smokestacks. Recycling technology and advances in wastewater treatment are helping reduce our waste output. Solar, wind, and geothermal energy technologies are producing cleaner and increasingly cheaper renewable energy. Satellites are making it easier to track and understand large-scale environmental change. Hydrogen fuel cells and associated technology may soon propel Iceland into a new energy age (Chapter 18), and perhaps the rest of the world not long after. Countless technological advances like these are one reason that people of the United States and western Europe today enjoy cleaner environments, although they consume far more, than people of eastern Europe and rapidly industrializing nations such as China.

Producing responsible and constructive technology that can lead us toward sustainable solutions requires scientific research. Scientific research requires funding. And funding from democratic governments requires public pressure, citizens letting their leaders know what they want. All too often, promising technologies have been suppressed or overlooked because inferior technologies that benefited certain powerful interests were developed first or displaced them. Many of the initial fundamental innovations in renewable energy, for instance, were made not in the past few years but many decades or centuries ago. Partly through the action of misplaced government subsidies and unregulated private corporate interests, technologies that might move us toward sustainable development were ignored while others, such as fossil fuels, were encouraged and went on to flourish. We cannot simply develop technology for technology's sake or assume that all technology will be beneficial. Rather, we must decide what we want from technology and then set out to get it, encouraging our governments to set up incentives that prompt private enterprise to help us reach our technological goals.

Growth does not equal progress

Progress toward sustainability, then, can come from wisely harnessing technology, from halting population growth, and from reducing consumption by reorienting one's perspective as to what comprises true prosperity. Sustainability cannot result from runaway growth. Consider that runaway growth is the mechanism of cancer cells, which, if not halted, eventually kills the patient.

Having too many people on Earth can crowd societies and make life miserable for all. More and more consumption will lead eventually to less and less consumption because of dwindling resources, if we fail to develop sustainable ways of turning waste back into raw materials. Economic growth as conventionally measured cannot continue indefinitely, and degraded environments and reduced resource availability will lead to decreases in the standard economic indicators. None of these types of growth, therefore, is necessary for a happy and healthy society. All of them, if unchecked, can be expected to lead eventually to just the opposite. Sustainable development seeks progress in the human condition through means other than nonstop growth.

Sustainable Solutions

If sustainability involves finding win-win solutions for the human condition and for Earth's environmental systems that are truly lasting, how can these be achieved? Specific sustainable solutions to environmental problems are numerous; there are at least as many solutions as there are problems. The challenges lie in being imaginative enough to think of solutions and then being shrewd and dogged enough to overcome whatever political or economic obstacles may lie in the path of their implementation.

The notion of sustainability has run throughout this book, and if you look back on the many examples of solutions we have offered to the problems we have covered, you will see that most of these fulfill or at least bring us nearer to a sustainable way of living. Rather than present a laundry list here, it is perhaps more helpful to examine some broad approaches that can spawn sustainable solutions to problems in general, whether they be existing problems or future problems that we cannot know or predict today. We will examine several of what many environmental scientists consider the most important principles of sustainable development and sustainability (Table 20.2)

We can redefine our priorities regarding economic growth and quality of life

One pillar of sustainability is the issue we have just examined through the lens of consumption—the idea that economic growth is not the only measure of progress. This is the viewpoint of the school of ecological economics (Chapter 2). Ecological economists hold that viewing human economies as entities that function within natural ecosystems, rather than outside them or encompassing them, leads to a better understanding of economics and of how humans interact with the environment.

Of foremost importance, these economists say, is to incorporate external costs, such as damage to environmental systems and human health, into the market price of goods and services. Currently, goods and services are priced as though the extraction and use of "natural capital," or resources from nature, were free and involved no costs to society. This may have been roughly true decades or centuries ago when human populations were smaller, resources were more plentiful, and more land was uninhabited. In today's world, however, virtually every environmental impact, whether a coal mine or road construction project or industrial plant or housing development, will have some sort of negative consequence for someone.

For instance, a developing country that grants a multinational corporation a forestry concession to clear-cut virgin timber from its forests may be paid an amount of money that, in the short term, may seem reasonable and fair. The price for this export of natural capital, however, has likely not reflected the fact that that swath of land cannot be used for forest products again for many years; may grow back into a different type of forest; may kill or displace thousands of animals, including those that are food resources for local people; may decrease the area's value for tourism or other uses; and may cause erosion that carries off nutrient-rich topsoil, pollutes water for villages downstream, and builds up silt behind hydroelectric dams (Figure 20.3). As far back as 1862, the economist John Ruskin (Chapter 2) wrote, "That which seems to be wealth may in verity be only the gilded index of far-reaching ruin."

If we can make our accounting practices reflect indirect negative consequences and give a clearer view of the full costs and benefits of any particular action or product

Table 20.2.   Some Major Approaches to Sustainability
Redefine our priorities regarding economic growth and quality of life
Mimic natural systems and make human industrial systems circular and recycling-oriented
Base our decisions on long-term thinking
Protect local self-sufficiency and be aware of the effects of globalization
Vote with our wallets
Vote with our ballots
Promote research and education

then the free market itself can become a force for improvement of environmental quality, our economy, and our quality of life. If we can incorporate external costs into our price structures, then market capitalism could potentially become the optimal tool for achieving prosperous and sustainable economies.

The political obstacles to this are great, however, because environmentally and economically destructive taxes and subsidies remain strikingly widespread. These economic incentives for environmental harm total many billions of dollars each year, propping up technologies and practices that might best be replaced with others. Meanwhile, citizens and businesses are generally taxed for things they desire (such as income and property) rather than things they would like to discourage (such as overuse and waste of resources). Implementing green taxes (Chapter 2) and phasing out harmful subsidies could hasten the shift to sustainability by many years or decades, many economists maintain. Such changes will not come about on their own but will require educated citizens to push for them and courageous policymakers to implement them.

We can mimic natural systems and make human industrial systems circular and recycling-oriented

The precepts of ecological economics are based on the notion that economies should be made efficient and sustainable. Indeed, we have an excellent and accessible model: nature itself. Natural systems are sustainable; they've been around far longer than we have. As we saw in Chapter 6 and throughout this book, environmental systems tend to operate in cycles. Nutrient cycles, the rock cycle, the hydrologic cycle—all are made up of feedback loops and the circular flow of materials. In natural systems, output is recycled into input, either for the same system or for another.

Human processes of manufacturing, in contrast, have run on a linear model, in which raw materials are input, go through processing, and create a product, with by-products and waste generated and disposed of (Chapter 19). It is largely a linear process, with a discrete beginning point and ending point.

We could in theory make all our industrial processes sustainable if we could transform linear processes into circular ones, in which waste is recycled and reused as raw materials, as it is in nature. Some such solutions are being implemented by forward-thinking industrialists already. For instance, several companies that produce carpets are now making them with materials that can be retrieved from the consumer when the carpet is worn out, and recycled in order to create new carpeting. The chemical company BASF has designed a type of nylon fiber that appears to be recyclable and reusable in this way. Some automobile manufacturers are in the early stages of designing cars that can be disassembled and recycled into new cars. Proponents of this industrial model see little reason why virtually all appliances and other products cannot be recycled, given the right technology. The ultimate vision is to create industrial processes that involve entirely closed loops, with no waste generated.

The Swiss Zero Emissions Research and Initiatives (ZERI) Foundation sponsors dozens of innovative projects worldwide that attempt to create goods and services without generating waste. Although most are not fully closed-loop systems, they attempt to approach this ideal. In so doing, they cut down on waste while increasing output and income, and often generate new jobs as well. One example involves breweries, currently being pursued in Canada, Sweden, Japan, and Namibia (Figure 20.4). Brewers in these projects take the waste from the beer-brewing process and use it to fuel a series of other processes. Spent grain from brewing is used to make bread and to farm mushrooms. The waste from the mushroom growing, together with the brewery's wastewater, is used to feed pigs, whose waste is digested in specialized containers to form and collect natural gas. Nutrients from the digesting container are used to grow algae to feed fish in fish farms. The brewer as a result can make money from bread, mushrooms, pigs, gas, and fish, as well as beer, all while producing little waste.

Figure 20.3 The exploitation of natural resources for short-term economic gain often ends up having long-term costs that are not included in the price of the original transaction. Here, rescue workers remove a body from the debris of a mudslide following flash floods in 1982 in San Salvador, El Salvador. Deforestation of hillsides for farming and commercial logging in El Salvador and other tropical countries creates conditions that frequently result in such disasters.

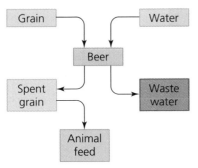

(a) Traditional brewery process

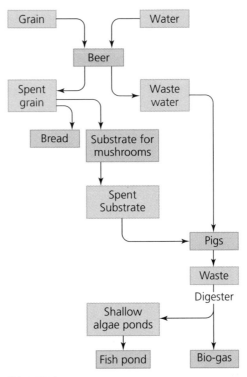

(b) ZERI brewery process

Figure 20.4 Breweries sponsored by Switzerland's Zero Emissions Research and Initiatives (ZERI) Foundation show how byproducts and waste from one industrial process can be used as raw materials for other processes. Traditional breweries (**a**) produce only beer, while generating much waste, some of which goes toward animal feed. ZERI-sponsored breweries (**b**) use their waste grain to make bread and farm mushrooms. Waste from the mushroom farming, along with brewery wastewater, goes to feed pigs. The pigs' waste is digested in containers that capture natural gas and that collect nutrients used to nourish algae for growing fish in fish farms. The brewer derives income from bread, mushrooms, pigs, gas, and fish, as well as beer.

We can base our decisions on long-term thinking

To be sustainable, a solution must work in the long term. Often the best long-term solution is not the best short-term solution, a fact that explains why much of what human societies currently do is not sustainable. Policymakers in democracies very often act for short-term good because they compete to produce immediate, positive results in order to be reelected. This short-term approach has been a major hurdle for addressing environmental dilemmas, because so many of these dilemmas are cumulative, worsen gradually, and can be resolved only in the long term. Often the costs for addressing environmental problems are short-term, while the benefits are long-term, giving a politician little incentive to tackle the problems. For this reason, citizen pressure on policymakers is especially important for issues that involve long-term effects.

Businesses may act according either to long-term or short-term interests. A business that is committed to operating in a certain location or with a certain community for a long period of time has incentive to sustain environmental quality. However, if the business is operating for only a short time, attempting to make a profit and move on, it will have little incentive to invest in environmental protection measures that are at all costly.

In 2002, the U.N. Environment Programme conducted analyses to predict the likely effects of a large-scale shift to sustainable development strategies over the 30 years until 2032. The U.N. scientists compared projections for a variety of environmental and socioeconomic indicators across four different scenarios: (1) a "markets first" scenario that assumed that forces of market capitalism would have the greatest impact on global trends, (2) a "policy first" scenario that assumed that government policy would play the dominant role in attempting to reach social and environmental goals, (3) a "security first" scenario that assumed that responses to conflict within and among nations and classes would guide global trends, and (4) a "sustainability first" scenario that assumed that a new sustainable development paradigm would take hold. For one indicator after another, the results showed that the "sustainability first" approach was predicted to result in the least negative impacts and the greatest positive impacts (Figure 20.5).

We can promote local self-sufficiency and be aware of the effects of globalization

Many proponents of sustainability believe that encouraging local self-sufficiency is an important element of building sustainable societies. When people are tied more closely to the area they live, they tend to value the

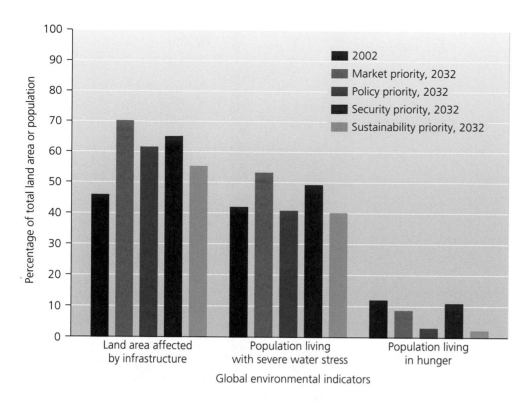

Figure 20.5 U.N. Environment Programme scientists projected the status of a number of environmental and socioeconomic indicators in 2032, across four different scenarios. The "markets first" scenario assumed that forces of market capitalism, reflecting the values and expectations prevailing currently in industrialized nations, would have the greatest impact on global trends. The "policy first" scenario assumed that governments would take strong action to reach social and environmental goals. The "security first" scenario assumed a world of conflict within and among nations and classes and that security would be governments' highest priority. The sustainability first scenario assumed that a new sustainable development paradigm would take hold, encouraging sustainable economic development, environmental protection, and social equity. The sustainability first approach was consistently predicted to result in the fewest negative impacts. Here it is shown to result in the least amount of land affected by infrastructure, the fewest people living under water stress, and the fewest people living with hunger. Data from Petra Döll et al., The global integrated water model 2.1; Polestar Project; and United Nations Environment Programme, Global Methodology for Mapping Human Impacts on the Biosphere.

area more and seek to sustain its environment and its human communities. This line of argument has frequently been made regarding the growing and distribution of food, specifically in encouraging locally based organic or sustainable agriculture. For example, the owner of the Farmers Diner in Barre, Vermont, goes to great lengths to obtain the vast majority of his food from small farmers within 80 km (50 mi) of his restaurant (Figure 20.6). In contrast, most food Americans eat is transported 2,400 km (1,500 mi) or more from the place it was grown to the place it is eaten, the Worldwatch Institute estimates.

Modern enthusiasm for locally grown and organic foods in North America first blossomed on the west side of the continent, when restaurateur Alice Waters established her Berkeley, California, eatery, Chez Panisse. For the past three decades, she has served gourmet menus using only fresh, locally grown food. In 2003 the restaurant brought its outlook to a dining hall at Yale University, training the staff there in preparing a new menu of healthy and organic foods grown locally in Connecticut.

The urge among some people to focus their economic activity locally is complemented by the desire of some others to combat the phenomenon known as **globalization.** Globalization means many things to many people and likely has mixed implications for the pursuit of sustainability.

The prominent ecological economist Herman Daly has untangled two distinct sets of ideas that people have often lumped together under the term *globalization.* Those who view it as a positive phenomenon generally accentuate what Daly instead dubs *internationalization,*

Figure 20.6 The Farmers Diner in Barre, Vermont, serves food prepared from produce mostly grown by small farmers within 80 km (50 mi) of the restaurant. Owner and manager Tod Murphy is a believer in the virtues of local self-sufficiency in food and agriculture.

the process by which people of the world's diverse cultures are increasingly communicating with one another and learning about and partaking in one another's diversity. For instance, technologies from books to airplanes to television have made people of every culture more aware of other cultures and thus more likely to respect and celebrate, rather than fear, differences among cultures. Besides technologies, the activities of the United Nations and many other public and private institutions have helped achieve these ends.

In contrast, people who view the phenomenon globalization in a negative light generally accentuate certain aspects that Daly reserves for his use of the term *globalization*. These comprise the process of homogenization of the world's cultures, with certain few cultures and worldviews displacing many others. For instance, the world's many languages are disappearing at a faster rate than ever; in fact, human languages are going extinct more quickly than are species of plants and animals. Traditional ways of life in many cultures are being abandoned as more people take up the material and cultural trappings, and the worldviews, of a few dominant cultures. In recent years, more people have protested against this homogenization, the spread of American culture, and the growing power of large multinational corporations. In France, farm activist Jose Bove became a popular hero when he used a tractor to wreck a McDonald's restaurant, to protest what many French farmers see as a threat to local French cuisine. In Seattle in 1999, thousands of protesters picketed the meeting of the World Trade Organization, which they viewed as a symbol of the homogenizing effects of Western market capitalism (Figure 20.7).

Figure 20.7 Thousands of protesters picketed the World Trade Organization's meeting in Seattle in 1999, criticizing the homogenizing effects of globalization, as well as relaxations in labor and environmental protections brought about by free trade.

Internationalization will likely foster the pursuit of sustainability, and globalization in itself does not necessarily threaten sustainability. However, Daly and others argue that globalization entails multinational corporations attaining greater and greater power over global trade, while governments retain less and less. Because they consider businesses less likely than governments to encourage environmentally protective measures, opponents of globalization feel that this process will hamper efforts toward sustainability. In addition, the American model of market capitalism, which is most widely promulgated, promotes a high-consumption lifestyle. The promotion of consumption—and the attitude that more, bigger, and faster is always better—does indeed threaten efforts to advance toward sustainable solutions.

On the positive side, the fact that American democracy, as imperfect as it is, still serves as a model for many countries with more repressive governments may help in the global quest for sustainability. Open societies allow for entrepreneurship and for the full flowering of creativity in academic research and in business. Millions of minds thinking and searching independently for solutions to problems will be more likely to come up with sustainable solutions than the minds of a few who hold authoritarian power.

While people may disagree over whether global homogenization of cultures in itself is desirable, it can definitively be said that putting all your eggs in one basket is always a risky strategy. It would seem desirable for all cultures to be able to compare and to be free to take the elements they prefer from others. By this process, cultures could potentially end up homogenizing across the world. If this process occurs by free choice, it may well bring us a better world and hasten the achievement of sustainability. If it is not by choice, however, but involves the imposition or replacement of lifestyles or values by force, constraints, or unfair competition, then diversity would decrease for no good reason and likely cause more harm than good.

Weighing the Issues:
Globalization and Internationalization

From your own experience, what advantages and disadvantages do you see to the processes called globalization and internationalization? Have you personally benefited or been hurt by either in any way? In what ways might promoting local self-sufficiency be helpful for the pursuit of global sustainability, and in what ways might it not? Many Americans are happy to eat at Vietnamese restaurants in the United States, yet many American anti-globalization protesters are quick to criticize the presence of McDonald's restaurants in Vietnam. Do you think this represents a double-standard, or are there reasons the two are not comparable?

We can vote with our wallets

A capitalist free-market system driven by consumerism holds one great asset for sustainability: Consumers can exercise a great deal of power through their choices of what to buy. When products produced through sustainable methods are labeled as such ("green labeling," Chapter 3), consumers can "vote with their wallets" in favor of sustainable products by choosing green-labeled ones over others. The power of consumer choice has helped drive the sales of everything from "dolphin-safe" tuna to recycled paper to organic produce (Figure 20.8)

One recent example is green-labeled wood for furniture. Products produced using sustainable forestry methods are labeled as such so that consumers can make choices and exert power in the marketplace to reward these efforts. Sustainably harvested products are certified by the Forest Stewardship Council or several other organizations. At the end of 2000, 2% of world forests had been accredited as sustainable by these groups, most in Scandinavia, northern Europe, and North America. Several retail outlets, including Home Depot, have committed to certification, purchasing only wood that has been sustainably harvested. With such major businesses signing on, sustainable forestry is growing rapidly. In 2000, 149 countries engaged in nine international initiatives to develop sustainability criteria for managing 85% of the world's forests. Many of the proposed programs involve local residents carrying out day-to-day management and protection in exchange for preferential access to the forest's resources. Thus an American consumer in the course of his or her shopping can have some positive influence on the integrity of a forest and the people who live there half a world away in Indonesia, Thailand, or Malaysia.

We can vote with our ballots

Although economic decisions can make large differences, many of the changes that will be needed to attain sustainable solutions are political, requiring policymakers to usher them through. Policymakers, for the most part, respond to whomever exerts influence. Corporations and interest groups employ lobbyists to push politicians in one direction or another all the time. Citizens in a

Figure 20.8 Products harvested, manufactured, or processed in sustainable ways may be advertised as such. Consumer choice of "green-labeled" products—such as organic vegetables—harnesses the powerful forces of market capitalism to promote sustainability.

democratic republic have the same power, *if* they choose to exercise it. You can exercise this power at the ballot box, and also by attending public hearings, writing letters, and making phone calls to policymakers to give them your opinions on issues that matter to you. You may be surprised how little input policymakers receive from the general public on many issues; sometimes a single letter or phone call can, in fact, make a difference.

Today's major environmental laws came about because of the activity of citizens pressuring their representatives in government to do something about perceived environmental problems. The raft of legislation enacted in the 1960s and 1970s in the United States and other nations might never have come about had ordinary citizens not stepped up and demanded action from policymakers. We owe the same to our children and future generations: to be engaged and to act responsibly now so that they have a better world in which to live. The words of anthropologist Margaret Mead are still worth repeating: "Never doubt that a small group of thoughtful, committed people can change the world. Indeed, it's the only thing that ever has."

We can promote research and education

As helpful as all of these approaches to sustainable solutions may potentially be, none of them will succeed fully if the public is not broadly aware of their importance. An individual's decisions to vote for candidates who support sustainable approaches, or to purchase green-labeled products, or to reduce consumption, will have limited effect unless a great many other people do the same. Individuals can influence large numbers of people, and demonstrate the truth of Margaret Mead's statement, by educating others with information and by serving as role models through their actions. The first step is to educate oneself, and the second step is to educate others. The discipline of environmental science, of course, plays a key role in providing information with which people can make wise decisions about environmental problems. By promoting scientific research and by educating the public about both the process and the results of environmental science, we can all assist in the pursuit of sustainable solutions.

Precious Time

The pace of our lives is getting faster. We do more, consume more, travel more, and often work more. For each minute or hour that is saved by some new technological device, it seems another is taken up by some new responsibility or distraction. The commotion of our lives can often make it hard to give attention to problems we don't need to deal with on a daily basis or that don't seem immediate to us. Indeed, the sheer number of environmental problems we hear of can feel overwhelming; even the best-intentioned among us may feel we have little time to devote to saving the planet.

However, time is getting short, and impacts on the natural systems we depend on are occurring fast. Many human impacts continue to become more severe and widespread, including deforestation, overfishing, land-clearing, wetland-draining, and resource extraction. The window of opportunity for turning some of these trends around is getting short. Even if we can visualize sustainable solutions to our many problems, how can we possibly find the time to implement them all and solve those problems before we have done irreparable damage to our environment and our own future?

We need to reach again for the moon

On May 25, 1961, U.S. President John F. Kennedy told Americans that within the decade the United States would be "landing a man on the moon and returning him safely to the Earth" (Figure 20.9). It was a bold and astonishing statement; the technology to achieve this unprecedented, almost unimaginable, feat did not yet exist. To most people in 1961, the idea of reaching the moon was almost as foreign a concept as it had been centuries earlier.

Kennedy's directive had powerful motivation behind it, however. The United States was caught up with the Soviet Union in the Cold War. In this competition for global hegemony, the two nations, held mutually at bay with nuclear weapons, each tried to prove their mettle through other means. The race to show dominance in space became the centerpiece of the rivalry. Early victories in the "space race" went to the Soviet Union, which sent into orbit humanity's first artificial satellite, *Sputnik*, in 1957, followed it with others, and eventually sent the first human being into space, orbiting cosmonaut Yuri Gagarin around Earth. The United States, meanwhile, was crashing rockets. The prospect of "losing" the space race prompted Kennedy's administration to set a national goal and an ambitious timeline to reach it. Congress followed through with the funding, NASA performed the science and engineering, and in 1969 astronauts walked on the moon. The United States had accomplished this amazing feat, this milestone in human history, by building public support for a goal and by giving its scientists and engineers the wherewithal to focus their efforts on developing technology and strategies for meeting that goal.

The rapid and historic accomplishments of both the United States and the Soviet Union during the space race show what societies can accomplish when they provide focused support for a chosen goal. Similar successes occurred when the United States confronted the Great Depression, when it threw itself into preparing for World War II, and when it conducted the Marshall Plan after the war to help rebuild western Europe.

Today humanity faces a challenge more important than any previous one—the challenge of achieving sustainability. Attaining sustainability is a larger and more complex process than traveling to the moon. However, it is one to which every single person on Earth can contribute; in which government, industry, and citizens can all cooperate; and toward which governments of all nations can work together. Unlike the space race, there is a real time limit in reaching sustainability; we cannot wait much longer before lessening our impact on the planet, or we may do increasing amounts of permanent harm that cannot be reversed. If America was able to reach the moon in a mere eight years, then certainly humanity can begin down the road to sustainability with comparable speed. Human ingenuity is capable of it; it is just a question of rallying public resolve and engaging our governments, institutions, and entrepreneurs in the race.

Figure 20.9 U.S. President John F. Kennedy in 1961 called on Congress to fund a space program to send men to the moon and back again before 1970. Addressing our environmental problems and shifting our political, economic, and social institutions to a paradigm of sustainable development will require at least as much vision, resolve, and commitment. The fact that astronauts reached the moon in 1969 demonstrates the power of human ingenuity in meeting a challenge and provides some hope that we may yet be able to meet the bigger challenge of living sustainably on Earth.

We must pass through the environmental bottleneck

Human ingenuity and human compassion give us ample reason to hope that we may achieve sustainability before doing too much damage to our planet and to our own prospects. However, we must be realistic about the challenges that lie ahead. As we decrease the amount of natural capital on which we can draw, we give ourselves and the rest of the world's creatures less and less room to maneuver; we squeeze ourselves through a progressively tighter space, like being forced through the neck of a bottle. The key question for the future of our species and our planet thus becomes whether we can make it safely through this bottleneck.

Biologist Edward O. Wilson (Chapter 15) has written eloquently and persuasively of this view:

At best, an environmental bottleneck is coming in the twenty-first century. It will cause the unfolding of a new kind of history driven by environmental change. Or perhaps an unfolding on a global scale of more of the old kind of history, which saw the collapse of regional civilizations, going back to the

Environment vs. Economy?

Is environmental protection economically costly? If sustainable development is a goal, how can we best balance the demands of economic development with the demands of environmental protection?

The Tradeoff Myth

Reducing pollution and protecting resources costs money. As a nation we spend over $200 billion each year, or more than 2% of GDP, on environmental protection. However, it is often contested how much environmental protection costs in any particular case. Moreover, forecasting the costs of compliance is difficult because economists have an equally difficult time estimating future technological responses that may lower costs. For example, credible industry estimates for sulfur dioxide reduction from power plants under the 1990 Clean Air Act Amendments were eight times too high; the EPA overestimated costs by a factor of 2 to 4.

One "cost" of environmental protection that is blown out of proportion is job loss. Contrary to popular belief, there is no net "jobs-environment tradeoff" in the economy, only a steady shift of jobs to cleanup work. On one hand, about 2,000 workers in the U.S. lose their jobs each year for environment-related reasons, which is less than $1/10$ of 1% of all layoffs. At the same time, as we spend more on the environment, more jobs are created. Given the industrial nature of much cleanup work, these jobs are also heavily weighted toward manufacturing and construction. Finally, and again contrary to folk wisdom, very few manufacturing plants flee the industrial countries to escape onerous environmental regulation. Plants do leave, but the overwhelming reason is the cost of labor.

What is the best way to balance costs against the benefits of environmental clean-up? Formal benefit-cost analysis is one approach, but it is of limited value when the benefits of environmental protection are highly uncertain, which if often the case, or the costs of resource degradation are borne by a relatively small group. In these situations, the best approach is to define a health or ecological standard for cleanup, and trust democratic processes to ensure that the costs of cleanup do not rise too high.

Eban Goodstein is professor of Economics at Lewis and Clark College in Portland, OR, and a research associate at the Economic Policy Institute in Washington, DC. Professor Goodstein is the author of the textbook *Economics and the Environment* (John Wiley and Sons: 2001), as well as *The Trade-off Myth: Fact and Fiction about Jobs and the Environment* (Island Press: 1999).

Balancing Competing Values

Protecting the environment is not free. That's why economic progress is a prerequisite for environmental quality. Hungry folks don't have the luxury of investing in the preservation of endangered songbirds. The real enemy of the environment is poverty, not affluence.

It's not an accident that nations with the highest per capita incomes have the cleanest environments. Between 1976 and 1997 U.S. GDP increased 158%, energy consumption 45%, and vehicle miles traveled 143%. During that same period, ozone levels-the major contributor to urban smog-decreased 30.9%. Sulfur dioxide — the primary component of acid rain — decreased 66.7%; nitrogen oxides decreased 37.9%; carbon monoxide decreased 66.4%; and lead decreased a dramatic 97.3%. The Department of Interior concluded, "Cleaner air is a direct consequence of . . . the enormous and sustained investments that only a rich nation could afford."

Only when people can provide the basics for themselves and their families (for example, shelter, food, and security) will they increase their demand for environmental quality. "These wild things," Aldo Leopold reminds us in *A Sand County Almanac*, "had little human value until mechanization assured us of a good breakfast."

We protect the environment because of qualities we value. They include human health, natural beauty, and the preservation of other species. But we also value fresh produce in winter, comfortable housing, and fast and convenient transportation. The questions we face all involve balancing competing values. In what combination and in what amounts should we seek the things we want?

The reality is that people must constantly trade off many positive values. Just as people on fixed budgets must choose between buying a new television or a new sofa, societies must choose among more health care, safer roads, or more funding for education. Responsible solutions are more likely when we understand that environmental quality is not exempt from this reality.

Pete Geddes is program director of the Foundation for Research on Economics and the Environment (FREE) and Gallatin Writers. Both are based in Bozeman, MT.

earliest in history, in northern Mesopotamia, and subsequently Egypt, then the Mayan and many others. . . . People died in large numbers, often horribly. Sometimes they were able to emigrate and displace other people, making them die horribly instead. . . . Somehow humanity must find a way to squeeze through the bottleneck without destroying the environments on which the rest of life depends."

We must think of Earth as an island

We began this book with the vision of Earth as an island, and indeed that is what it is—a large one, but an island nonetheless. Islands can be paradise, as Easter Island (Chapter 1) likely was when the Polynesians first reached it. But if people do not live sustainably on islands, they can turn paradises into desolate graveyards. The Europeans who arrived at Easter Island much later witnessed the scene of a civilization that had depleted its resources, degraded its environment, and collapsed as a result. For the few people who remained of the once-mighty culture, life was difficult and unrewarding. They had lost even the knowledge of the history of their ancestors, who had continued cutting trees unsustainably, in effect kicking the base out from beneath their elaborate and prosperous civilization.

As Easter Island's trees disappeared, the people must have known at some level that their resources were vanishing. It is likely there were individuals who spoke out for conservation and for finding ways to live sustainably amid dwindling resources. And likely there were those who ignored those calls and went on extracting more than the land could bear, assuming that

somehow things would turn out all right. Indeed, whoever cut down the very last tree from atop the most remote mountaintop could have looked out across the island and seen that it was the last tree. And yet that person cut it down.

It would be tragic folly to let such a fate occur to our planet as a whole. And yet if human impact on the environment continues growing, it is difficult to see how it can be avoided. By recognizing this, by deciding to shift our individual behavior and our cultural institutions in ways that encourage sustainable practices, and by employing sound science to help us achieve these ends, we may yet be able to live sustainably and happily on our wondrous island, Earth.

Conclusion

In every society facing the dilemma of dwindling resources and environmental degradation, there will be those who raise alarms and those who ignore them. Fortunately, in our global society today we have many thousands of scientists who study Earth's processes and resources closely. For this reason we have access to an accumulated knowledge and an ever-developing understanding of our dynamic planet, what it offers us, and what impacts it can bear. The challenge for our global society today, our one-world island of humanity, is to support that science, and listen to those scientists, so that we can accurately judge false alarms from real problems, and distinguish legitimate concerns from thoughtless denial. This science, this study of Earth and of ourselves, is what offers us hope for our future.

REVIEW QUESTIONS

1. A goal of the World Summit on Sustainable Development in 2002 was to more fully fund and implement initiatives of Agenda 21, the program previously agreed to at the U.N. Conference on Environment and Development in 1992. What actions were taken in this regard? How would you characterize Kofi Annan's remarks about the summit?

2. What do environmental scientists mean by "sustainable development?" What is "development?" What is meant to be sustained?

3. Why does it sometimes appear that development and sustaining the environment are in direct opposition?

4. Explain why the regulatory approach to environmental protection has helped lead to the perception of a dichotomy between economic and environmental interests.

5. Why are many people now living at the very highest level of material prosperity in history? Is this level of consumption sustainable?

6. How can it feel good to consume less? How might we obtain more while consuming less?

7. For what reasons might the human species experience a stabilization of its population size? Why is it impossible for our population to continue to grow indefinitely? How does demographic transition theory offer hope for population stabilization?

8. In what ways can technology help us achieve sustainability?

9. How can the free market help us to create sustainable as well as profitable economies?

10. In what ways do natural processes provide good models of sustainability for manufacturing? Provide two examples.

11. How can short-term thinking by business and political leaders affect average citizens?

12. Why do some people feel that local self-sufficiency is important? What effects of globalization may threaten the environment? How can American-style democracy help to promote sustainability?

13. How can consumers exercise power in the interest of achieving sustainability?

14. How can citizens exercise power in the interest of achieving sustainability?

15. Explain Edward O. Wilson's metaphor of the "environmental bottleneck."

16. How can thinking of Earth as an island help prevent us from repeating the mistakes of other civilizations?

DISCUSSION QUESTIONS

1. Why might we be able to "have it both ways," increasing our quality of life through development and also protecting the integrity of the environment? How might what is good for the environment also be good for humans? Discuss examples from this chapter or other chapters of this book that illustrate that such win-win solutions are possible. Are there cases in which you think win-win solutions will not be possible?

2. Do you feel that your own actions have a significant bearing on the future of the planet? Do you plan to begin or continue recycling? Buying "green-labeled" products? Writing to members of Congress? Taking other actions? Would you be happy voluntarily reducing your consumption of certain goods or services, or would you rather pursue other avenues toward sustainability?

3. Evaluate technology as a strategy that can help guide us toward the goal of sustainability. How can the relationship between technology and globalization be a positive or negative influence on the development of sustainable solutions?

4. Choose one item or product that you enjoy, and consider how it came to be yours. What steps were involved in creating its components, and where did the raw materials come from? How was it manufactured? How was it delivered to you? Try to think of as many details about the item or product as you can, and determine how they were obtained or created.

5. Why do many observers consider the conventional outlook that environmental protection costs jobs to be flawed? Use examples from the chapter to illustrate your response. Are you familiar with any cases in your community or at your college that bear on this issue?

6. Reflect on the experiences of prior human civilizations and how they ultimately came to an end. What is your prognosis for our current human civilization? Do you see a vast world of independent cultures all individually responsible for themselves, or do you consider human civilization to be one great entity? Is Earth an island, or a collection of many islands? If we accept that all humans are dependent on the same environmental systems for sustenance, what resources and strategies do we have to ensure that the actions of a few do not determine the outcome for all and that sustainable solutions are a common global goal?

Media Resources *For further review, go to the website* **www.envscienceplace.com** *or student CD-ROM, where you will find quizzes, flashcards, a glossary, additional interactive exercises, and links to relevant news and research sources. Also, on the website and CD-ROM is* **GRAPH IT**, *a series of interactive graphing tutorials to help you interpret graphs and plot data.*

Appendix A
Some Basics on Graphs

The collection of data to understand trends in various phenomena and to test ideas is a core part of the scientific endeavor. Data are numbers presented within a context, and, as such, help scientists to gain insights and draw conclusions. The basic tool of scientists for expressing data is the graph, and the ability to read graphs and plot data is one that students must cultivate early in their study of the various sciences. This appendix provides basic information on three of the most common types of graphs—line, scatter, and bar—and the rationale for the use of each.

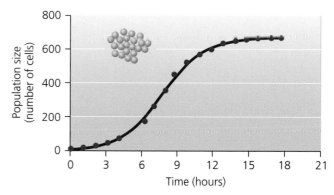

Yeast cells, *Saccharomyces cerevisiae*

Line Plot

A line plot is drawn when the data set involves a sequence of some kind, for example, a sequence through time or across distance (see Figure A.1; see figure 5.12 a and b in the main text). Line plots are appropriate when both the x and the y axes represent continuous, numerical data.

Using a line plot allows you to see increasing or decreasing trends in the data. A curved line indicates that the y variable is increasing or decreasing more quickly for some values of x.

A useful graphing technique is to plot two data sets together on the same graph (Figure A.2; see figure 5.17 in the main text). This allows you to compare the trends in the two data sets to see whether and how they seem to be related.

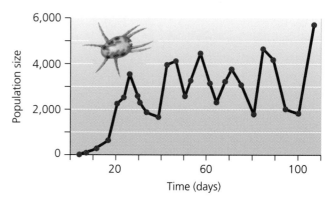

Mite, *Eotetranychus sexmaculatus*

Figure A.1

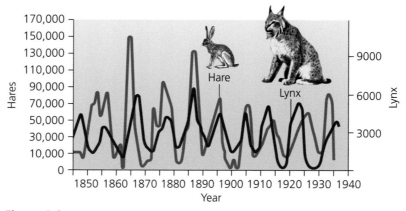

Figure A.2

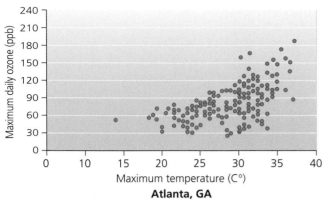

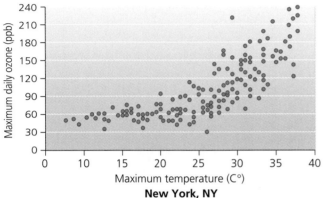

Figure A.3

Scatter Plot

A scatter plot is used most often when there is no sequence in the data set, for example, when both the x and y axes represent continuous data. Each point is separate and has no particular connection to another point (see Figure A.3; see figure 12.14 a and b in the main text).

Using a scatter plot allows you to see broad increasing or decreasing trends in the data.

Bar Chart

A bar chart is used when one of the variables represents a category rather than a numerical value (Figure A.4 below; see page 185, Chapter 6 of the main text). Bar charts allow you to see differences between categories or groups of categories.

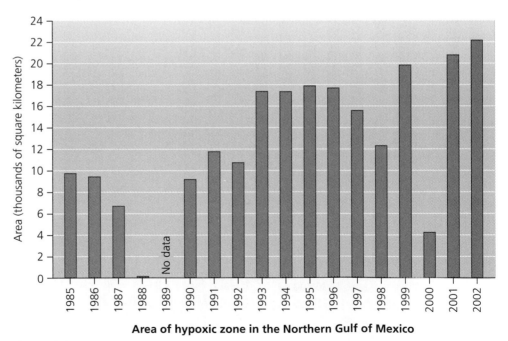

Figure A.4

It is often instructive to graph two variables together to discover patterns and relationships (Figure A.5; see see figure 12.13b in the main text).

Notice that an age pyramid is simply a bar chart turned on its side (Figure A.6; see Figure 5.9 in the main text). The age groups are displayed on the y axis, with the bars for each age group varying in width instead of height as in the previous graphs.

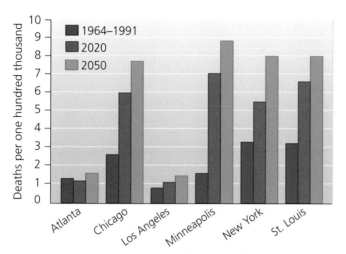

Average summer mortality rates attributed to hot weather episodes

Figure A.5

But don't stop here. Take advantage of the **GRAPH IT** tutorials on the CD-ROM included with this text, or go to the Environmental Science Place web site at www.envscienceplace.com. The **GRAPH IT** tutorials will help you further build and expand your understanding of graphs, and provide you with an opportunity to plot your own data.

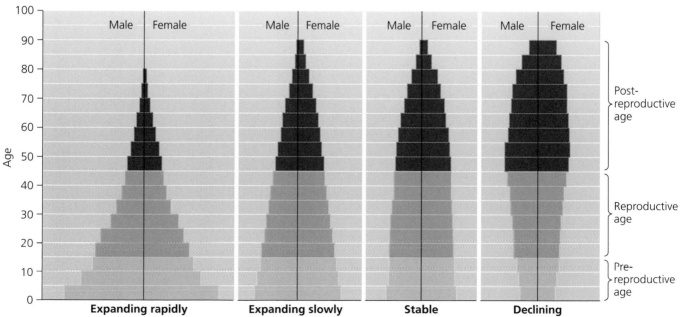

Figure A.6

Appendix B
Periodic Table of the Elements

Representative (main group) elements

Transition metals

Representative (main group) elements

IA																	VIIIA
1 **H** 1.0079	IIA											IIIA	IVA	VA	VIA	VIIA	**2** **He** 4.003
3 **Li** 6.941	**4** **Be** 9.012											**5** **B** 10.811	**6** **C** 12.011	**7** **N** 14.007	**8** **O** 15.999	**9** **F** 18.998	**10** **Ne** 20.180
11 **Na** 22.990	**12** **Mg** 24.305	IIIB	IVB	VB	VIB	VIIB		VIIIB		IB	IIB	**13** **Al** 26.982	**14** **Si** 28.086	**15** **P** 30.974	**16** **S** 32.066	**17** **Cl** 35.453	**18** **Ar** 39.948
19 **K** 39.098	**20** **Ca** 40.078	**21** **Sc** 44.956	**22** **Ti** 47.88	**23** **V** 50.942	**24** **Cr** 51.996	**25** **Mn** 54.938	**26** **Fe** 55.845	**27** **Co** 58.933	**28** **Ni** 58.69	**29** **Cu** 63.546	**30** **Zn** 65.39	**31** **Ga** 69.723	**32** **Ge** 72.61	**33** **As** 74.922	**34** **Se** 78.96	**35** **Br** 79.904	**36** **Kr** 83.8
37 **Rb** 85.468	**38** **Sr** 87.62	**39** **Y** 88.906	**40** **Zr** 91.224	**41** **Nb** 92.906	**42** **Mo** 95.94	**43** **Tc** 98	**44** **Ru** 101.07	**45** **Rh** 102.906	**46** **Pd** 106.42	**47** **Ag** 107.868	**48** **Cd** 112.411	**49** **In** 114.82	**50** **Sn** 118.71	**51** **Sb** 121.76	**52** **Te** 127.60	**53** **I** 126.905	**54** **Xe** 131.29
55 **Cs** 132.905	**56** **Ba** 137.327	**57** **La** 138.906	**72** **Hf** 178.49	**73** **Ta** 180.948	**74** **W** 183.84	**75** **Re** 186.207	**76** **Os** 190.23	**77** **Ir** 192.22	**78** **Pt** 195.08	**79** **Au** 196.967	**80** **Hg** 200.59	**81** **Tl** 204.383	**82** **Pb** 207.2	**83** **Bi** 208.980	**84** **Po** 209	**85** **At** 210	**86** **Rn** 222
87 **Fr** 223	**88** **Ra** 226.025	**89** **Ac** 227.028	**104** **Rf** 261	**105** **Db** 262	**106** **Sg** 263	**107** **Bh** 262	**108** **Hs** 265	**109** **Mt** 266	**110** **Uun** 269	**111** **Uuu** 272	**112** **Uub** 277		**114**		**116**		

Rare earth elements

Lanthanides

58 **Ce** 140.115	59 **Pr** 140.908	60 **Nd** 144.24	61 **Pm** 145	62 **Sm** 150.36	63 **Eu** 151.964	64 **Gd** 157.25	65 **Tb** 158.925	66 **Dy** 162.5	67 **Ho** 164.93	68 **Er** 167.26	69 **Tm** 168.934	70 **Yb** 173.04	71 **Lu** 174.967

Actinides

90 **Th** 232.038	91 **Pa** 231.036	92 **U** 238.029	93 **Np** 237.048	94 **Pu** 244	95 **Am** 243	96 **Cm** 247	97 **Bk** 247	98 **Cf** 251	99 **Es** 252	100 **Fm** 257	101 **Md** 258	102 **No** 259	103 **Lr** 262

The periodic table arranges elements according to atomic number and atomic weight into horizontal rows called periods and 18 vertical columns called groups or families. The elements in the groups are classified as being in either A or B classes.

Elements of each group of the A series have similar chemical and physical properties. This reflects the fact that members of a particular group have the same number of valence shell electrons, which is indicated by the number of the group. For example, group IA elements have one valence shell electron, group IIA elements have two, and group VA elements have five. In contrast, as you progress across a period from left to right, the properties of the elements change in discrete steps, varying gradually from the very metallic properties of groups IA and IIA elements to the nonmetallic properties seen in group VIIA (chlorine and others), and finally to the inert elements (noble gases) in group VIIIA. This change reflects the continual increase in the number of valence shell electrons seen in elements (from left to right) within a period.

Class B elements are referred to as transition elements. All transition elements are metals, and in most cases they have one or two valence shell electrons. (In these elements, some electrons occupy more distant electron shells before the deeper shells are filled.)

In this periodic table, the colors are used to convey information about the phase (solid, liquid, or gas) in which a pure element exists under standard conditions (25 degrees centigrade and 1 atmosphere of pressure). If the element's symbol is solid black, then the element exists as a solid. If its symbol is red, then it exists as a gas. If its symbol is dark blue, then it is a liquid. If the element's symbol is green, the element does not exist in nature and must be created by some type of nuclear reaction.

Appendix C
Environmental Organizations and Agencies

The list of environmental organizations below is in alphabetical order. Both an email address and a web site are listed if available; however, note that some organizations rely solely on the web site for feedback and do not include email addresses as contact information. This list is not meant to be comprehensive; as a consequence, organizations at the local level are not included.

Center for Health, Environment, and Justice (CHEJ)
http://www.chej.org/
P.O. Box 6806
Falls Church, VA 22040
Phone: (703) 237-2249
Email: chej@chej.org

Originally founded in 1981 as the *Citizens Clearinghouse for Hazardous Waste* the CHEJ works to help empower citizens and organizations on the local level to build a healthy and sustainable future.

Earth Island Institute (EII)
http://www.earthisland.org/home_body.cfm
300 Broadway, Suite 28
San Francisco, CA 94133
Phone: (415) 788-3666
Fax: (415) 788-7324

EII was founded in 1982 by well-known gadfly environmentalist David Brower. The goal is to foster provide organizational support for creative conservation, preservation, and restoration projects helping the global environment.

EarthJustice
http://www.earthjustice.org/
426 17th Street, 6th Floor
Oakland, CA 94612-2820
Phone: (510) 550-6700
Fax: (510) 550-6740
Email: eajus@earthjustice.org

A nonprofit pro bono law firm, originally founded in 1971 as Sierra Club Legal Defense Fund (SCLDF). Their mission is to "safeguard public lands, national forests, parks, and wilderness area contamination; and to preserve endangered species and wildlife habitat."

Environmental Defense
http://www.environmentaldefense.org/home.cfm
257 Park Avenue South
New York, NY 10010
Telephone: (212) 505-2100
Fax: (212) 505-2375
Email: members@environmentaldefense.org

A nonprofit organization, founded in 1967, that works to take the best that science, economics, and the law have to offer to solve environmental problems.

Food First/Institute for Food and Development Policy
http://www.foodfirst.org/
398 60th St.
Oakland, CA 94618
Phone: (510) 654-4400
Fax: (510) 654-4551
Email: foodfirst@foodfirst.org

A member-supported, nonprofit, progressive think tank, Food First produces books, reports, articles, and films that highlight root causes and value-based solutions to hunger and poverty around the world. Food First was founded in 1975 by Frances Moore Lappé and Joseph Collins, following the international success of the book, *Diet For a Small Planet*.

Friends of the Earth
http://www.foe.org/
1025 Vermont Ave., NW, Suite 300
Washington, DC 20005
Phone: (877) 843-8687 - toll free
Fax: (202) 783-0444
E-mail: foe@foe.org

FOE is the U.S. voice of an influential international network of grassroots groups in dozens of countries and has been at the forefront of high-profile efforts to create a more healthy, just world. See also: **Friends of the Earth UK:** http://www.foe.co.uk/ and **Friends of the Earth International (FOEI)** at http://www.foei.org/.

Global Exchange
http://www.globalexchange.org/
2017 Mission Street, #303
San Francisco, CA, 94110
Phone: 415.255.7296
Fax: 415.255.7498
Email: membership@globalexchange.org

Global Exchange is an international human rights organization dedicated to promoting environmental, political and social justice. Founded in 1988, it works to increase awareness of global economic/environmental and human rights in the U.S., while building worldwide partnerships.

Greenpeace USA
http://www.greenpeaceusa.org/
P.O. Box 7939
Fredericksburg, VA 22404-9917
Phone: (800) 326-0959
No email, has feedback form on website:
http://www.greenpeaceusa.org/bin/view.fpl/6448.html

Greenpeace, which was founded in 1971, focuses on six issues: saving ancient forests, stopping global warming, eliminating persistent organic pollutants (POPs), protecting ocean life, fighting genetic engineering, and stopping all phases of nuclear production and use.

League of Conservation Voters (LCV)
http://www.lcv.org/
1920 L Street, NW, Suite 800
Washington, DC 20036
Phone: (202) 785-8683
Fax: (202) 835-0491
No email, has feedback form on website: http://www.lcv.org/Feedback/Feedback.cfm

A powerful lobbying group devoted full-time to effecting a pro-environment Congress and White House. LCV runs campaigns to defeat anti-environment candidates, and publishes the popular *National Environmental Scorecard* every Congressional election.

National Audubon Society
http://www.audubon.org/ [Main web page]
700 Broadway
New York, NY 10003
Phone: (212) 979-3000
Fax: (212) 979-3188
Email (Chapter Information): ltennefoss@audubon.org
Email (Education): education@audubon.org
Email (Membership Subscriptions): join@audubon.org

Audubon Society chapters in all 50 U.S. states, D.C., Canada, and internationally, along with community-based nature centers and chapters, scientific and educational programs.

National Parks Conservation Association (NPCA)
http://www.npca.org/about_npca/ [Non-flash]
http://www.npca.org/flash.html [Flash]
1300 19th St., N.W., Suite 300, Washington, DC 20036
phone: (800) 628-7275
fax: (202) 659-0650
Email: npca@npca.org

The NPCA (founded 1919) focuses on protecting and preserving the U.S. National Park System. It has more than 300,000 members and 65,000 activists.

National Wildlife Federation
http://www.nwf.org/
11100 Wildlife Center Dr.
Reston, VA 20190-5362
(800) 822-9919 [for donations]
No email, has feedback form on website: http://www.nwf.org/contact/

Proponents of environmental education, the NWF goes into communities to offer schoolyard, campus, and community programs. They publish *Ranger Rick* and *Your Big Backyard* magazines; and present an online conservation directory.

Natural Resources Defense Council (NRDC)
http://www.nrdc.org/default.asp
40 West 20th Street
New York, NY 10011
Phone: (212) 727-2700
Fax: (212) 727-1773
Email: nrdcinfo@nrdc.org

NRDC focuses on the following issues: clean air and energy; clean water and oceans; wildlife and fish; preserving parks forests and wild lands; healthier cities and green living; and global warming, toxic chemicals, and nuclear weapons and waste. Regional offices are in Washington, San Francisco, and Los Angeles.

Nature Conservancy, The (TNC)
http://nature.org/
4245 North Fairfax Drive, Suite 100
Arlington, VA 22203-1606
Phone: (703) 841-5300
E-mail: comment@tnc.org

TNC's mission is to "preserve the plants, animals and natural communities that represent the diversity of life on Earth by protecting the lands and waters they need to survive."

Occidental Arts and Ecology Center (OAEC)
http://www.oaec.org/
15290 Coleman Valley Rd.
Occidental, CA 95465
Phone: (707) 874-1557 x201
Email: inquiry@oaec.org

OAEC is an organic farm and nonprofit research center in Northern California's Sonoma County. Founded in 1994 by biologists, horticulturists, educators, activists, and artists, OAEC seeks to: "Preserve and restore native biodiversity and healthy ecosystems; support appropriate-scale ecological farming and sustainable, local food systems; and cultivate ecological literacy, among other goals."

Pesticide Action Network North America (PANNA)
http://www.panna.org/
49 Powell St., Suite 500
San Francisco, CA 94102
Phone: (415) 981-1771
Fax: (415) 981-1991
Email: panna@panna.org

A leader in the search for ecologically sound and socially just alternatives to pesticide use, PANNA creates a powerful citizens' action network of local and international consumers, labor, health, environment and agriculture group. It is one of five PAN Regional centers.

Pesticide Action Network (PAN)
http://www.pan-international.org/

PAN is a network of over 600 participating nongovernmental organizations, institutions and individuals in over 60 countries working to replace the use of hazardous pesticides with ecologically sound alternatives.

Rainforest Action Network (RAN)
http://www.ran.org/
221 Pine St., Suite 500
San Francisco, CA 94104
Phone: (415) 398-4404
Fax: (415) 398-2732
Email: rainforest@ran.org

Since 1985, RAN has worked to protect the Earth's rainforests and support the rights of its inhabitants through education, grassroots organizing, and non-violent direct action.

Sea Shepherd Conservation Society
http://www.seashepherd.org/
P.O. Box 2670
Malibu, CA 90265
Tel: (360) 370-5650
Fax: (360) 370-5651
Email: seashepherd@seashepherd.org

This NGO is "involved with the investigation and documentation of violations of international laws, regulations, and treaties protecting marine wildlife species. The Society is also involved with the enforcement of international laws . . . with a particular focus on halting illegal fishing activities."

Sierra Club
http://www.sierraclub.org/
85 Second Street, 2nd Floor
San Francisco, CA 94105
Phone: (415) 977-5500
Fax: (415) 977-5799
Email: information@sierraclub.org

It is America's oldest, largest and most influential grassroots environmental organization started more than a century ago by John Muir. Their mission: "Explore, enjoy, and protect the wild places of the earth."

Surfrider Foundation
http://www.surfrider.org/
P.O. Box 6010
San Clemente, CA 92674-6010
Phone: (949) 492-8170
Fax: (949) 492-8142
Email: enviro@surfrider.org

A non-profit organization with 60 chapters, dedicated to protecting the oceans, waves and beaches around the world.

Tahoe-Baikal Institute (TBI)
http://www.tahoebaikal.org/
P.O. Box 13587
South Lake Tahoe, CA 96151
(also offices in Incline Village, NV, and Irkutsk, Russia)
Phone: (530) 542-5599
Fax : (530) 542-5567
Email: info@tahoebaikal.org

TBI is an international partnership founded in 1991 to help preserve Lake Tahoe in California and Lake Baikal in Siberia, as well as other significant and threatened natural areas around the world. TBI is offers environmental education programs, research, and international exchanges of students and practitioners in science, policy, economics, and other related disciplines.

WISE (World Information Service on Energy) Uranium Project
http://www.antenna.nl/~wise/uranium/index.html
WISE -Amsterdam
P.O. Box 59636
1040 LC Amsterdam
The Netherlands
Email: wiseamster@antenna.nl

WISE is a small anti-nuclear group in the Netherlands offering information about nuclear energy, radioactive waste, radiation, and uranium.

Worldwatch Institute
http://www.worldwatch.org/
1776 Massachusetts Ave., N.W., Washington, D.C. 20036-1904
Phone: 202.452.1999
Fax: 202.296.7365
Email: worldwatch@worldwatch.org

WW is an independent, interdisciplinary research organization working for an environmentally sustainable and socially just society. Its bimonthly *Worldwatch Magazine*; annual *State of the World* report and other publications are essential resources for news on global environmental, social, political, and economic trends. WW was founded by Lester Brown in 1974.

World Wide Fund For Nature (WWF)
http://www.panda.org/
1250 24th Street NW
Washington, DC 20037-1175
Phone: (202) 293 4800
Fax: (202) 293 9211
http://www.panda.org/about_wwf/who_we_are/offices/offices_n_america.cfm

The WWF is a global network acting through local offices, where they "do all they can to halt the accelerating destruction of our natural world." In the U.S., it's still called "World Wildlife Fund."

Online Resources

Amazing Environmental Organization Webdirectory!
http://www.webdirectory.com/
Email: sales@webdirectory.com

Billed as the "Earth's Biggest Environment Search Engine," this for-profit directory lists thousands of sites from more than 100 countries, with such subdirectories as: Agriculture, animals, disasters, energy, forestry, general environmental interest, governmental agencies, health, land conservation, parks, pollution, recycling, sciences, sustainable development, transportation, water, and wildlife/endangered species.

EnviroLink: The Online Environmental Community
http://www.envirolink.org/
EnviroLink Network
P.O. Box 8102
Pittsburgh, PA 15217
Email: websupport@envirolink.org

A non-profit on-line environmental information service providing access to thousands of online environmental resources since 1991. Offers news, environmental resources by topic, online forums, and the EnviroLink U.S. Atlas™: a group of interactive maps of U.S.-based resources

National Environmental Directory, The
http://www.environmentaldirectory.net/default.htm
Harbinger Communications
616 Sumner St.
Santa Cruz, CA 95062
Phone: (831) 457-0130
Email: info@environmentaldirectory.net

Lists more than 13,000 organizations in the United States concerned with environmental issues and environmental education. Directory also available as software.

National Wildlife Federation Online Conservation Directory
http://www.nwf.org/conservationdirectory/

Lists more than 4,000 organizations and governmental agencies

Forest Conservation Portal
Vast Rainforest, Forest and Biodiversity Conservation News & Information
http://forests.org/

The mission of Forests.org is to contribute to ending deforestation, preserving old-growth forests, conserving all forests, maintaining climatic systems and commencing the age of ecological restoration

Climate Ark—The Premier Climate Change & Renewable Energy Portal
http://www.climateark.org/

Eco-Portal—The EnvironmentalSustainability.Info Source
An Information Gateway Empowering the Movement for Environmental Sustainability
http://www.environmentalsustainability.info/

WaterConserve—A Water Conservation Portal
http://www.waterconserve.info/

Environmental News Network
http://www.enn.com/index.asp

All environmental news from all over the world.

Appendix D
The Metric System

Measurement	Unit and Abbreviation	Metric Equivalent	Metric to English Conversion Factor	English to Metric Conversion Factor
Length	1 kilometer (km)	= 1000 (10^3) meters	1 km = 0.62 mile	1 mile = 1.61 km
	1 meter (m)	= 100 (10^2) centimeters = 1000 millimeters	1 m = 1.09 yards 1 m = 3.28 feet 1 m = 39.37 inches	1 yard = 0.914 m 1 foot = 0.305 m
	1 centimeter (cm)	= 0.01 (10^{-2}) meter	1 cm = 0.394 inch	1 foot = 30.5 cm 1 inch = 2.54 cm
	1 millimeter (mm)	= 0.001 (10^{-3}) meter	1 mm = 0.039 inch	
	1 micrometer (μm) [formerly micron (μ)]	= 0.000001 (10^{-6}) meter		
	1 nanometer (nm) [formerly millimicron (mμ)]	= 0.000000001 (10^{-9}) meter		
	1 angstrom (Å)	= 0.0000000001 (10^{-10}) meter		
Area	1 square meter (m^2)	= 10,000 square centimeters	1 m^2 = 1.1960 square yards 1 m^2 = 10.764 square feet	1 square yard = 0.8361 m^2 1 square foot = 0.0929 m^2
	1 square centimeter (cm^2)	= 100 square millimeters	1 cm^2 = 0.155 square inch	1 square inch = 6.4516 cm^2
Mass	1 metric ton (t)	= 1000 kilograms	1 t = 1.103 ton	1 ton = 0.907 t
	1 kilogram (kg)	= 000 grams	1 kg = 2.205 pounds	1 pound = 0.4536 kg
	1 gram (g)	= 1000 milligrams	1 g = 0.0353 ounce 1 g = 15.432 grains	1 ounce = 28.35 g
	1 milligram (mg)	= 0.001 gram	1 mg = approx. 0.015 grain	
	1 microgram (μg)	= 0.000001 gram		
Volume (solids)	1 cubic meter (m^3)	= 1,000,000 cubic centimeters	1 m^3 = 1.3080 cubic yards 1 m^3 = 35.315 cubic feet	1 cubic yard = 0.7646 m^3 1 cubic foot = 0.0283 m^3
	1 cubic centimeter (cm^3 or cc)	= 0.000001 cubic meter = 1 milliliter	1 cm^3 = 0.0610 cubic inch	1 cubic inch = 16.387 cm^3
	1 cubic millimeter (mm^3)	= 0.000000001 cubic meter		
Volume (liquids) and gases	1 kiloliter (kl or kL)	= 1000 liters	1 kL = 264.17 gallons	1 gallon = 3.785 L
	1 liter (l or L)	= 1000 milliliters	1 L = 0.264 gallons 1 L = 1.057 quarts	1 quart = 0.946 L
	1 milliliter (ml or mL)	= 0.001 liter = 1 cubic centimeter	1 ml = 0.034 fluid ounce 1 ml = approx. $\frac{1}{4}$ teaspoon 1 ml = approx. 15–16 drops (gtt.)	1 quart = 946 ml 1 pint = 473 ml 1 fluid ounce = 29.57 ml 1 teaspoon = approx. 5 ml
	1 microliter (μl or μL)	= 0.000001 liter		
Time	1 second (s)	= $\frac{1}{60}$ minute		
	1 millisecond (ms)	= 0.001 second	1 second (s)	= $\frac{1}{60}$ minute
Temperature	Degrees Celsius (°C)	°F = $\frac{9}{5}$°C + 32	°C = $\frac{5}{9}$(°F − 32)	
Energy and Power	1 kilowatt hour	= 34113 BTUs = 860,421 calories		
	1 watt	= 3.413 BTU/hr = 14.34 calorie/min		
	1 calorie	= the amount of heat necessary to raise the temperature of 1 gram (1cm^3) of water 1 degree Celsius		
	1 horsepower	= 7.457 × 102 watts		
	1 joule	= 9.481 × 10^{-4} BTU = 0.239 cal = 2.778 × 10^{-7} kilowatt-hour		
Pressure	1 Pound per Square inch (psi)	= 6894.757 Pascal (Pa) = 0.068045961 Atmosphere (atm) = 51.71493 Millimeters of Mercury (mm hg = Torr) = 68.94757 millibars (mbar) = 68.94757 (hectopascal hPa) = 6.894757 Kilopascal (kPa) = 0.06894757 Bar (bar)		
	1 Atmosphere (atm)	= 101.325 kilopascal (kPa)		

Source: http://hemsidor.torget.se/users/b/bohjohan/convert/conv2_e.htm#?

AREER PROFILE Environmental Education

If you've ever been interested in becoming a teacher, you may want to consider a growing segment in the field of education that can offer more flexibility and diversity than teaching in a traditional classroom. Environmental educators are partnering with traditional teachers to boost science programs and make classes more fun and interesting for students.

Always a unique experience, environmental education runs the gamut from nature study to conservation education. More than just a theoretical learning experience, environmental science focuses on getting students personally involved in scientific applications that spark their interest and enthusiasm. For many students, environmental science programs provide their first exposure to outdoor learning and foster an appreciation of the environment.

Marilou Seiff is the program manager of the Marine Science Institute in Redwood City, California. As an environmental educator, she, along with her colleagues, offers students from neighboring schools authentic science experiences that go beyond books or classrooms. "We reach out to students of all ages," says Seiff. "We have a hands-on way of teaching them science and about what's in their own backyards. We teach them about responsible stewardship. Until you've actually touched it [nature] or felt it, you won't be emotionally involved with it."

As informal educators working to supplement schoolteachers' formal lesson plans, environmental educators teach all aspects of science. "We teach about water quality, natural cycles, interactions in the environment and interactions between species," says Seiff. "We teach about global species adaptation. We look at everything living and in nature."

In addition to teaching on shore, the educators at the Marine

> **"We reach out to students of all ages. . . teaching them science and about what's in their own backyards."**

Science Institute routinely take students out on a 90-foot research vessel to collect samples. Students see the Pacific Coast's intratidal areas, San Francisco Bay, mudflats, and sandy beach areas. In these settings, students collect fish and invertebrates, as well as water, soil, and sediment samples. This experience also provides an opportunity for the students to learn about human impact, food chains, and biodiversity.

"We're able to hook students on science here and have them excel because rather than lecturing to them, we have them learn by doing," says Seiff. "They become involved rather than just visiting a center."

Becoming an environmental educator requires a minimum of a bachelor's degree in any of a wide range of sciences, including biology, physics, and chemistry. A strong working knowledge of ecology, natural history, and even recreational resource management skills is helpful as well.

"Most environmental educators have a science degree in biology, geology, or other related fields," says Seiff. "There is also an increase in environmental programs that offer a combination of teaching credentials and a master's degree."

Most environmental educators are employed by aquariums, natural history museums, and science centers around the nation. Each specializes in a different field tailored to its environment. The Marine Science Institute has 20 environmental educators on staff and carries out much of its own training for incoming educators. They are taught about the specific environment that the Marine Science Institute highlights, and they are taught how to identify diverse populations.

"I really enjoy teaching kids and seeing a new world open up for them," says Seiff. "When you take kids who've never been out there and you see things through their eyes, it's just a lot of fun."

For more information, contact:
The North American Association for Environmental Educators
410 Tarvin Road
Rock Spring, GA 30739
Telephone: (706)764-2708
Fax: (706)764-2094
http://www.naaee.org

Also:
Environmental Education and Training Partnership
College of Natural Resources
University of Wisconsin-Stevens Point
Stevens Point, WI 54481
Telephone: (715)346-4958
Fax: (715)346-4385
http://www.eetap.org

CAREER PROFILE

Environmental Chemist

Every day, countless organic and inorganic compounds find their way into our air, water, and soil. These compounds can have a direct impact on our health and the health of our environment. An environmental chemist is often our first line of defense in diagnosing potential problems.

Ted Lyter is such a chemist. Lyter is bureau chief and specializes in inorganic chemistry for the Bureau of Laboratories within the Department of Environmental Protection for the State of Pennsylvania. Lyter and his staff collect and analyze air, water, and soil samples and raise the alarm when something is amiss. "We monitor streams, lakes, and drinking water. We'll handle abandoned mine drainage, active mine drainage, sewage treatment, and effluence," explains Lyter. "Mostly, we test drinking water supplies and investigate water that's been brought in by private citizens. We analyze soil water, and solid waste samples, as well as air samples that may contain trace metals. Although sampling water is a large part of what we do, we're really a general environmental testing lab."

Environmental chemistry is a broad field. Although much of the work is focused on the qualities of water and sediments, it might also involve studying the transformation of chemicals by living organisms, comparing one aquatic system to another, or collaborating with other scientists to see how global warming is affecting the carbon cycle. Lyter's lab, conducts testing for various state and local agencies, including two welfare facilities, and acts as a private lab for a municipal wastewater treatment plant. "Most of our projects are ongoing," says Lyter. "We're currently working with an engineering firm that is testing water treatment plants to remove arsenic. Our staff looks at the testing plan and puts together the samples, analysis, and documentation."

> **"Communications skills are a big plus. . . . It's our job to come up with a game plan."**

Environmental chemists are employed by a variety of organizations, including wastewater treatments plants; private labs; conservation organizations; state, federal and local agencies; private industries that are concerned about pollution; and even food and beverage manufacturers. Some environmental chemists become teachers and researchers at the university level.

For a career as an environmental chemist, a good foundation in one or more of the basic science disciplines is necessary. In addition to chemistry, knowledge of biology, geology, physics, mathematics, or computers is helpful. Communication skills and the ability to interpret documentation are also important. Proficiency in all these areas makes an environmental chemist an exceptionally well-rounded professional. This quality is crucial because environmental chemists often must combine their efforts with those of many other specialists.

"Communications skills are a big plus. We have to coordinate with field and lab staff on projects, and we deal with program staff and the people we service. We also work closely people whose lab expertise is in hydrology, aquatic pollution, and field sampling. It's our job to come up with a game plan."

Although it may be necessary to attain a master's degree or even a doctorate for an upper-level position, Lyter started off with a bachelor of arts degree in an entry-level chemistry position (Chemist I) and worked his way up. Thirty years later, he runs his lab section. "You can start off and work your way through a Chemist I, II, or III position with an undergraduate degree, but anything above that usually requires more education," says Lyter. "Nowadays, it's a little harder to advance to those levels the way that I did, but it happens." Additionally, some states require accreditation as an environmental chemist.

Lyter's favorite aspect of the is the diversity and camaraderie that exists among his staff. "Since 9/11, there have been a couple of occasions when we suspected a potential contamination of the water supply," explains Lyter. "We had the samples submitted and turned around within a couple of hours. Thankfully, they were false alarms, but it was a good feeling to pull the staff together after working hours."

For more information on a career in environmental chemistry, contact:
The American Chemical Association
1155 16th Street NW
Washington, D.C. 20036
Telephone: (202) 872-4615 or (800) 227-5558
Fax: (202) 872-4616
http://www.chemistry.org

AREER PROFILE | Ecotourism Specialist

Tourism with a conscience? Ecotourism is just that: responsible travel to natural areas—travel that conserves the environment and preserves local cultures. Ecotourism is travel with a social motive.

One of the big advantages of ecotourism is that it benefits local economies directly. The monies generated go, not to big tour companies and outfitters, but instead to locally owned lodges and homegrown tour guides and artisans. Tourist dollars also help fund the management of the natural areas that are visited. Equally important to supporting local economies is the way ecotourists see the areas they are visiting. Small groups stay at "eco-lodges" and "green hotels" (environmentally friendly lodging), travel with educational guides, and most important, stay on the trails, where, ecotourism manager Eileen Gutierrez says, "they take only pictures and leave no footprints."

Although ecotourism is still a relatively new industry, the term was coined in the early 1980s, when a handful of conservation biologists asserted that local economies could benefit from tourist dollars while maintaining local cultures and environments. Bringing tourists to visit an area responsibly has become a popular way to conserve natural areas and to allow travelers a chance to see destinations that they might otherwise never have a chance to see. In 2002, a United Nations summit in Quebec named that year "The International Year of Ecotourism," signaling the official worldwide acceptance of the trade and adding significantly to ecotourism's legitimacy as a beneficial and responsible way to see the world.

Eileen Gutierrez works for Conservation International, based in Washington, D.C. As an ecotourism manager whose regional specialty is mainland Asia, her job is to promote economic development and conservation in those areas by creating ecotourism initiatives. "We work in conservation. Our foundation deals with

> 66 [Ecotourists] take only pictures and leave no footprints. 99

diverse areas worldwide that we call 'hot spots'—highly biodiverse but threatened," explains Gutierrez. "We work with communities in and around protected areas whose threats are socioeconomic in nature. We help provide alternatives to their livelihood that can benefit conservation."

As an ecotourism expert, Gutierrez works on a policymaking level, designing and developing fundraising for ecotourism efforts. These tourism and development guidelines are carefully initiated because although they are designed to benefit communities, they could also easily threaten them. "Ecotourism plans for parks and protected areas and provides guidelines on development and how to zone for ecotourism. We design facilities (eco-lodges and hotels) to be more harmonious with the environment. We also analyze how community involvement and participation might take place and be encouraged," says Gutierrez.

Training for ecotourism is diverse. More schools are offering specialized degrees in environmental sustainability and even in ecotourism itself. According to Gutierrez, the variety of back-grounds among people working to further ecotourism's cause is almost endless. "Ecotourism employs biologists, social anthropologists, and people with business administration degrees working within the field," says Gutierrez. "The four key pillars for success in ecotourism are knowledge of international development issues, a solid foundation in business and economics, an understanding of community development issues, and an understanding of biodiversity and conservation."

These broad requirements mean that anyone interested in ecotourism must be prepared to get a broad education. Working in ecotourism might mean managing endangered areas, as Gutierrez does, or it might involve such varied jobs as managing an eco-lodge or conducting tours and educational programs. For Gutierrez, this kind of variety in her day-to-day job is exactly what motivates her. "Although ecotourism is not an extremely well defined field right now, there's a lot of groundwork that's being done, and the creativity involved in that is great. I've had the chance to develop methodologies for ecotourism assessments, and it has been exciting to be a part of that."

For more information on a career in ecotourism, contact:
The International Ecotourism Society
733 15th Street NW, Suite 100
Washington, D.C. 20005
Telephone: (202) 347-9203
Fax: 202-387-7915
http://www.ecotourism.org

Also
United Nations Environmental Programme (for further information on conserving natural environments and preserving biodiversity)
http://www.unep.org

AREER PROFILE Organic Farming

When you take a bite out of a crisp, juicy apple or enjoy a leafy salad, do you ever wonder how it was grown? Have you ever given any thought whether or not the produce you eat is organic? If so, you're not alone. Organic food and agriculture is on the rise in the United States, and the U.S. Department of Agriculture (USDA) estimates that the numbers of organic farmers are increasing at a brisk 12 percent a year. It was one of the fastest-growing segments of U.S. agriculture during the 1990s.

Most organic farmers like Tim Vos are small-scale producers. Since 1980, Vos has farmed organically and has owned and operated Blue Heron Farms in Corralitos, California for nearly 20 years. For Vos, growing his vegetables organically was the only option in response to an agricultural industry he found highly toxic and growing more so by the day. "For me, I think that taking up organic farming has to do with environmental, political and philosophical concepts," says Vos. "I've always wanted to do something different and show that it was possible to do it well and make a living at it."

Organic refers to the way agricultural products are grown, handled and processed. Before finding its way onto your dinner table, organic produce must pass stringent requirements that are set and imposed by the USDA. The basis of organic food production rests on a system of farming that maintains the integrity of the soil. Its fertility is safeguarded without the use of toxic fertilizers and pesticides.

Prevention is the first and most important line of defense for an organic farmer faced with weeds, disease and insect control. Because they do not enjoy the same control over these circumstances that traditional farmers do, building a healthy soil is of paramount importance for organic farmers. Vos and his peers rely on natural fertilizers and soil replenishers such as turkey and cow manure from organic farms, weeding, crop rotation, cover cropping and

> **❝**Organic food and agriculture is on the rise. . . .**❞**

composting. Organic farmers achieve these goals with a minimum of off-farm input.

It's certainly not the easiest route for farmers, but it's one that has always made sense to Vos. "You don't have the same technology packages in organic farming that traditional farmers do," says Vos. "You have to be more innovative and creative."

Since the field of organic farming is still relatively in its infancy, there are only a handful of schools around the nation that offer degrees or certificate programs in alternative agriculture. The Center for Agroecology and Sustainable Food Systems (CASFS) at the University of California Santa Cruz is one such school. The oldest and most established in the country, it pioneered a certificate program in organic farming. The six-month program's curriculum is a combination of 300 hours of academic classroom instruction, and 700 hours of in-field training and hands-on experience at an on-site student farm and an internship at an organic farm. Although students are not required to have an undergraduate degree in science to participate in the program, most have a natural science background with a BA in Environmental Studies that emphasizes on agroecology or ecology.

"The students learn about soil science, integrative pest management, botany, horticulture, propagation, entomology, bed preparation and natural fertilizers," explains Vos who also teaches in UCSC's Department of Environmental Studies. "It's really a lot of biology and horticulture you need. You need to fully understand ecology, agroecology and hydrology because all the different levels of living systems come together in a farming system."

Vos says that science will come to play a very important part in this agrarian occupation as the field of organic farming matures and takes hold. "You have to really understand plant biology and ecology," says Vos. And although organic farming has its challenges, it also has its benefits. "The most rewarding aspect is to harvest a crop, prepare it, present something special at the farmer's market and have people say how beautiful it is," says Vos. "It's about the fruit of the whole process. You're feeding people. You're putting something out in the world that's really beneficial."

For a series of links to schools that offer programs and access to alternative farming education, visit http://www.agsci.ubc.ca/ubcfarm/links.php.

For more information, contact:
California Certified Organic Farmers
1115 Mission Street
Santa Cruz, CA 95060
(831) 423-2263
(888) 423-2263
(831) 423-4528
http://www.ccof.org

CAREER PROFILE Soil Scientist

It is literally the ground that you're standing on. Soil is a key element at the foundation of any biological system, so what impacts soil directly impacts our lives. Healthy soil is vital to agriculture, which enables us to feed, clothe, and house ourselves. But who monitors the health of soil?

Scott Stevens is a soil scientist with Soil Science and Environmental Services Incorporated in Connecticut. Because much of his state is made up of wetlands, a large part of his job involves identifying this soil type for his clients. "I identify wetlands, classify soil types, and assess lands for environmental contamination," says Stevens. "We work mostly for private industries, such as developers, engineers, and surveyors. Sometimes, regular people who want to build on a piece of land will call us in to analyze its soil."

When a property is slated for development—whether for public, private, or recreational use—it makes sense to consult a soil scientist to determine the condition of the land and the presence of any hazards. For example, building too close to a wetlands area, where the soil may be unstable, can be disastrous. A soil scientist assesses a piece of land for moisture retention and drainage, sustainability, and even for the environmental impact of construction.

"We look at soil textures, depth of modeling, and depth of bedrock," explains Stevens. "We carry out our examinations mostly through visual observation by tak-ing a shovel and unearthing a deep layer of soil. Occasionally we also collect groundwater and send it along with a soil sample for laboratory testing. We work closely with biologists, hydrologists and hydrogeologists."

A soil scientist might work as a consultant, researcher, or technical expert in a variety of private in-

> **"You're like an environmental policeman."**

dustries, public agencies, academic institutions, or nonprofit environmental organizations. A growing awareness of environmental issues has increased demand for the services of such scientists. "Although we do work for a lot of private industries, we also work for institutions and the state, and we do subcontract work for engineering firms," says Stevens. "But most soil scientists are either flagging wetlands or working with septic systems. That's what drives our business."

Understanding soil requires understanding many complex systems. A career as a soil scientist requires a bachelor's degree in environmental science or a closely related field, such as natural resources or agronomy, along with considerable soils-related coursework to provide a solid knowledge of soil environment, soil chemistry, physics, biology, and soil morphology. "I took about 24 credits in soil courses and an additional 30 credits in biology, hydrology, and other science courses. I also have a master's degree in environmental science," says Stevens. Additionally, most soil scientists choose to become registered as such; other titles that offer some recognition of expertise in the field include soil specialist and soil classifier. Often additional testing is required for these certifications, but the extra effort can result in higher salaries.

"It's a tough job, actually," says Stevens, but that seems to be part of the appeal. "You're out there dealing with a lot of mosquitoes, ticks, and snakes . . . really communing with nature." For Stevens, his career is enormously rewarding. "It's always something new. When you know you're out there protecting different habitats for plants and animals, it feels great. You're like an environmental policeman."

For more information, contact:
The National Society of Consulting Soil Scientists
PMB 700, 325 Pennsylvania Avenue SE
Washington DC, 20003
Telephone: (800) 535-7148
http://www.nscss.org

Also
Agronomy Crop and Soil Science Societies / Agronomic Science Foundation
677 South Segoe Road
Madison, WI 53711
Telephone: (608) 273-8095 or (608) 273-8080
Fax: (608) 273-2021
http://www.soils.org
http://www.agronomy.org

Another helpful link:
www.soils.usda/gov

CAREER PROFILE

Forester

From the framework of your house, to the method you might use to heat that house, to the very paper this text is printed on, wood is one of our most useful and valuable resources. Protecting and managing our nation's forests is an important and dynamic job. And that's what a forester does. Considered keepers of our forests, foresters are involved in nearly every aspect of forest protection, maintenance, and planning.

Meet Jim Dalton. As a forester for the state of Wisconsin, he is responsible for the administration of the forest laws of his state and county. Dalton also provides services for landowners as part of a private assistance program. Not all forest land is public land: Private individuals own nearly 54% of America's forests. This means that "private assistance" is a major part of any forester's job. "If a landowner owns 10 or more acres of woodlands, they can request forest inventory and management from the county," explains Dalton. "I walk through their forest lands and take science-related inventory and talk to the landowners about what they can do in relation to forest land capabilities."

The concept of preserving and managing America's woodlands through the help of foresters was developed at the turn of century. As cities and urban communities grew, so did the need for wood products to support a boom in construction. According to Dalton, Wisconsin was an early leader in providing resource-based forestry information. "We wanted to sustain forest resources in the state and keep them healthy," says Dalton. "Much of northern Wisconsin, Minneapolis, Chicago, and Milwaukee were built from Wisconsin's wood."

Through Wisconsin's Stewardship Forestry Plan, Jim Dalton and other foresters help landowners realize their goals and objectives for their land. The consulting forester assesses the forests' composition and outlines practices that will en-

> **"Private individuals own nearly 54% of America's forests."**

courage growth, productivity, health, abundance of wildlife population, and erosion control. "It's up to the landowners to take it to the next level," says Dalton. "Many owners know that they should harvest their timber only when it's mature, but we help them harvest it thoughtfully with the next generation in mind."

Becoming a forester requires a bachelor's degree in one of the sciences in a program accredited by the Society of American Foresters. "Lots of pure science, physics, chemistry, botany and zoology are required," says Dalton. "Mathematics is also important because you will need to gather a lot of data and understand it. There's also a strong social science aspect, because you can't have forests without people. Skills in forest economics, writing, and communications are all important because you are likely to have to give public presentations. Writing is a major emphasis now in the field and has really mushroomed over the years."

Dalton notes that landing a position as a forester isn't always easy. Experience counts; summer jobs and internships will help set an aspiring forester apart from the competition. "Lots of graduates have a lot of degrees, but nothing to back them up," says Dalton. "Even volunteering in a forestry-type job is helpful. Try to find a job that involves not just manual labor, but making management decisions. The competition for natural resources jobs is fierce. Although a master's degree is not necessary, it will give you a leg up on the competition."

Foresters are employed in a number of industries, including sawmills, government paper mills, the military, and even environmental groups, such as the Sierra Club. Each state requires that these professionals undergo an accreditation process. As for Dalton, the most fulfilling aspect of his job is educating people on the importance of conservation. "My favorite thing is hiking around all tracks and forest lands people have called me about and being able to give them good news," says Dalton. "When a landowner trusts me and wants to make use of my programs, that's the best thing. When we're able to supply them with programs, landowners reach a level of understanding and interest that will ultimately serve us all."

For more information, contact:
The Society of American Foresters
5400 Grosvenor Lane
Bethesda, MD 20814
Telephone: (301) 897-8720
Fax: (301) 897-3690
http://www.safnet.org

CAREER PROFILE | Air Quality Specialist

Clean air is essential. Cleaning up our air and working to keep air pollution in check requires the skills of a specialist—an air quality specialist—who works to counteract the damage that our industries, our cars, and other aspects of daily life inflict on our air.

Travis Epes of Environmental Consulting & Technology (ECT) in Gainesville, Florida, is such a specialist. His firm monitors and consults with industries to make sure they comply with air quality standards. "If a company would like to build a facility and the facility has the potential to produce pollution, we evaluate the process, the emission sources, and how much and what kind of pollutants will be emitted. Then we compare the results to industry standards," says Epes. "We apply for permits from state and local agencies and fill out the paperwork for our clients."

Air quality specialists such as Epes and his colleagues conduct a battery of field tests, interpret the results, and determine whether a problem exists. When a problem does come up, Epes uses sophisticated measuring instruments and computer programs to predict the air quality around the industrial operation. This is a complicated process, even more so when it is carried out in an urban area. "You have to consider all sorts of factors, such as traffic patterns, climate, and even the pollution from housing developments."

First, the team must identify emission sources and determine how much pollution is actually being emitted. Epes explains, "We use best achievable control technology (BACT) to evaluate each pollution source and figure out which practices will give us the most efficient control at the most reasonable cost. The last step is an

> **"My favorite part of the job is being able to look at a new area or project and visualize how it creates an environmental impact."**

air dispersion/impact analysis. We take emissions and put them in an atmospheric dispersion model. This process enables us to determine how the facility is impacting the air and whether it is meeting standards."

Another part of the job is researching new environmental technologies that will improve quality. Some air quality specialists are designing equipment and new systems to combat pollution, for example, in pollution control devices for cars and scrubbers on industrial smokestacks.

Epes and ECT work with private companies in matters regarding compliance and permits. Although ECT has managed some smaller county projects, its focus is on large industries that pose a risk of polluting the air. "We work with Duke Energy, Calpine, Anheuser-Busch, Connective, CSX Transportation, and many other firms."

Becoming an air quality specialist requires a bachelor's degree in chemistry, environmental science, or physics. This is usually the minimum requirement for an entry-level job. Individuals with engineering degrees might focus on positions that require a background in civil engineering and environmental engineering. "I am a Staff Engineer II," explains Epes. "I have a bachelor of science in environmental engineering and a master's degree in air specialty. I did take lots of biology, organic chemistry, and core engineering classes." Accreditation is also required as a professional engineer. A test administered on a state-by-state basis and a few years of work experience are the typical requirements.

For Epes, the most exciting part of his job is the variety. "My favorite part of the job is being able to look at a new area or project and visualize how it creates an environmental impact and then to see whether it's detrimental or not."

For more information on a career as an air quality specialist, contact The Air & Waste Management Association
One Gateway Center, 3rd Floor
420 Fort Duquesne Blvd.
Pittsburgh, PA 15222
Telephone: (412) 232-3444
Fax: (412) 232-3450
http://www.awma.org

CAREER PROFILE

Hydrogeologist

It's an incredible journey. Consider the character, source, occurrence, movement, availability, and use of . . . water. Nearly 97% of the planet's fresh drinking water is in the form of groundwater. Where does it come from? How does it get there, and what happens to it along the way? Whereas geologists study rocks and structures that are formed over large periods of time, hydrogeologists study how water interacts with other geological systems.

James F. Strandberg is a hydrogeologist with Malcolm Pirnie, a large national environmental consulting firm. His job is to deal with issues surrounding the supply and management of groundwater. Increasingly, he notes, much of his work is related to assessing and cleaning up groundwater contamination.

"I test groundwater near airports to see whether there's jet fuel in it. I test industrial sites. I look for chlorinated sulfites and NTBE from gasoline pumps, " says Strandberg. "I also look for nitrites (pesticides) from agricultural applications. I test for contaminants at landfill. These can get into the groundwater and travel some distance. I work to determine where they've gone and compare the concentrations to acceptable standards."

As a consulting hydrogeologist, Strandberg offers guidelines to Pirnie's clients on how to avoid pollution, and when necessary, how to clean it up. Sometimes, the release of by-products is unavoidable, so hydrogeologists determine which materials yield the least toxicity.

"I determine the potential volatilization of chemicals, " says Strandberg. "Sometimes, these migrate up the soil and get into buildings and present a risk of indoor air contamination. I collect data and judge what's happening."

When the concentrations of pollutants have gone too far for the groundwater to be salvaged,

> **❝Hydrogeology is multidisciplinary. You get to work with a wide variety of people.❞**

hydrogeologists must use drastic methods to solve the problem. "With a method called pump and treat, we extract the water, treat it, and then reinject it in a stream or river," says Strandberg.

The requirements for a career as a hydrogeologist are similar to those for a career as a geologist. Students must attain a degree in geology, hydrogeology, physical or natural science, engineering, or environmental sciences with at least 37 hours in courses related to hydrogeology. Although the course load involves a math, physics, and chemistry, communication skills are important, too.

"Hydrogeology is multidisciplinary. You get to work with a wide variety of people. It's pretty rare that you work alone," says Strandberg. "You need good communication skills—both written and oral—because you're dealing with lots of different people."

Hydrogeologists need at least five years of postgraduate professional work experience and must also fulfill state requirements for licensure as a geologist. Additionally, hydrogeologist licensing is required nationwide.

The best part of Strandberg's job? Deciding how to solve problems that open doors to other markets and clients. "I get to be outdoors," says Strandberg. "I like working on different projects. Each client has different objectives, attitudes, and behaviors. Each and every project is different."

For more information on a career in hydrogeology, contact:
The National Ground Water Association
601 Dempsey Road
Westerville, OH 43081
Telephone: (800) 551-7379 or (614) 898-7791
Fax: (614) 898-7786
http://www.ngwa.org
Or
The American Institute of Hydrology
2499 Rice Street, Suite #135
St. Paul, MN 55113
Telephone: (651) 484-8169
Fax: (651) 84-8357
http://www.aihydro.org

CAREER PROFILE

Hazardous Waste Specialist

Hazardous wastes come in any of several forms: liquid, solid, or sludge. Waste is labeled as hazardous if it possesses particular qualities, such as ignitability, reactivity, corrosivity, or toxicity. These qualities require that hazardous waste be treated differently and disposed of properly. Treatment can be any method, technique, or process that alters the physical, biological, or chemical composition of the waste. In some cases, reducing or removing its toxic properties is also an option.

Patrick Callahan is a hazardous waste specialist for Clean Venture in Baltimore, Maryland. Identifying, removing, and disposing of hazardous wastes in a prompt and safe manner is what Callahan does day in and day out. "A lot of hazardous waste specialists are chemists, " says Callahan. "They characterize the waste from the source by doing field tests and analyses. Then they pack the waste appropriately and label it for disposal."

Wastes that are considered hazardous and need special care in their disposal include flammable solvents from painting operations, acids generated by metal plating operations, residues from pharmaceutical industries, oil sludge from manufacturing processes that use lubricants, and filter cakes from manufacturing facilities that remove solids from their own wastewater treatment systems. Once a hazardous waste specialist is called on the scene, they determine the best way to deal with possible pollution, contamination, and allaround volatility of the substance. "I work with customers and handle special projects for cleaning up waste that has been spilled, sometimes involving equipment that needs decontamination or soils that need to be excavated," reveals Callahan.

"How we treat the waste depends on its chemical composition. If we're dealing with an acid, we might neutralize it with sodium bicarbonate, for example. With

> **"It's an exciting job that always offers a new challenge."**

solvents, we might have to remove them and wash them, or they might go to cement kilns, where they are burned for fuel as an alternative to natural gas. Oil that has made its way into the soil might need to be excavated and, if possible, recycled, " says Callahan.

Once a hazardous waste has been crated and labeled, Callahan takes it to a certified disposal facility or to Clean Venture's own facility, if the waste is one that Clean Venture's facility is equipped to handle. Most of Clean Venture's customers are government facilities, small and large businesses, including automotive, chemical, and manufacturing facilities, are also among the company's customers. "Our most common spills are petroleum spills, leaking tanks, and lead or chrome contamination," says Callahan. "These are hazardous and have to be cleaned up. Contaminants are either excavated, recycled, or sent to a landfill."

A career in hazardous waste management can include donning many different hats. A hazardous waste specialist might assess and clean up the waste, dispose of it, or even work as an inspector. Working as an inspector requires knowledge of the regulatory arena to make sure that polluters and generators of hazardous wastes are in compliance with federal laws and regulations.

A bachelor's degree in science is crucial for work in this field. Chemistry and environmental science are obvious choices, but a degree in geology, biology or even chemical engineering can also serve as an entry to the field. According to Callahan, the best knowledge is knowledge that comes from experience. "You have to be 18 to work in the field, so becoming a laborer or intern for the summer is a good idea, " says Callahan. "Any experience is a huge plus. There are all sorts of people in this field, both with degrees and without, but having a degree will get you a job right away with good pay."

The job can require some physical labor. And, according to Callahan, it can also involve quite a bit of hauling and getting dirty—no matter how long you've been on the job. "You will do physical labor at least part of the time," says Callahan. "You will be asked to move drums and possibly even drive the box truck that hauls the drums away. It's an exciting job that always offers a new challenge."

For more information on hazardous waste specialists, contact:
The Environmental Protection Agency
Ariel Rios Building
1200 Pennsylvania Avenue, Northwest
Washington, D.C. 20460
Phone: (202) 272-0167
http://www.epa.gov

Also
Solid Waste Association of North America
http://www.swana.org

Glossary

abiotic factor Any nonliving component of the *environment*. Compare *biotic factor.*

acid precipitation *Precipitation* (such as rain, snow, or fog) that has increased acidity caused by the mixing of pollutants, mostly sulfur dioxide and nitrogen oxides from the combustion of *fossil fuels,* with water, oxygen, and oxidants.

acidic The property of a solution in which the concentration of hydrogen (H^+) *ions* is greater than the concentration of hydroxide (OH^-) ions. Compare *basic.*

active solar energy collection An approach to solar *energy* collection in which technological devices are used to focus, move, or store solar energy. Compare *passive solar energy collection.*

acute exposure Exposure to a *toxicant* occurring in high amounts for short periods of time. Compare *chronic exposure.*

adaptive radiation An explosion of diversity that occurs when an ancestral *species* gives rise to many species that fill empty *niches;* each species adapts to its niche by *natural selection.*

adaptive trait (adaptation) A trait that confers greater *fitness.*

administrative agency An agency within the *executive branch* assigned to elaborate and enforce statutory laws. It may be established by Congress or by presidential order.

aerobic Occurring in an *environment* where oxygen is present. For example, the decay of a rotting log proceeds by aerobic decomposition. Compare *anaerobic.*

aerosol A suspension of microscopic particles or droplets in a gas.

age distribution The relative numbers of organisms of each age within a population. Age distributions can have a strong effect on rates of population growth or decline and are often expressed as a ratio of age classes, consisting of organisms (1) not yet mature enough to reproduce, (2) capable of reproduction, and (3) beyond their reproductive years.

age structure diagram A visual tool that uses horizontal bars to show the *age distribution* of a population. The width of each bar represents the relative size of each age class.

age structure See *age distribution.*

agriculture The practice of cultivating *soil,* producing crops, and raising livestock for human use and consumption.

agroforestry A practice that combines *strip cropping* with *shelterbelts* in which fields planted in rows of mixed crops are surrounded by or interspersed with rows of trees that may provide fruit, wood, or protection from wind.

agronomist A scientist who combines *soil* and plant *science* to manage agricultural *systems.*

A horizon A layer of *soil* found in a typical *soil profile.* It forms the top layer or lies below the *O horizon* (if one exists). It consists of mostly inorganic mineral components such as weathered substrate, with some organic matter and *humus* from above mixed in. The A horizon is often referred to as *topsoil.* Compare *B horizon; C horizon; E horizon; R horizon.*

air pollution Material added to the *atmosphere* that can affect *climate* and harm organisms.

allergen A *toxicant* that overactivates the immune system, causing an immune response when one is not necessary.

allopatric speciation The formation of *species* in separate locations when populations are physically separated over a geographic distance. Compare *sympatric speciation.*

alternative hypothesis A different potential answer to the question of interest.

amensalism A relationship between members of different *species* in which one organism is harmed and the other is unaffected. An example is large trees shading smaller trees. Compare *commensalism.*

anaerobic Occurring in an *environment* that has little or no oxygen. The conversion of organic matter to *fossil fuels* (crude oil, coal, natural gas) at the bottom of a deep lake, swamp, or shallow sea is an example of anaerobic decomposition. Compare *aerobic.*

anthropic principle The idea that only those rare planets that manage to support life for billions of years evolve observers able to reflect on their own planet.

anthropocentrism A human-centered view of our relationship with the *environment.*

anthropogenic Human-caused.

aquaculture The raising of aquatic organisms for food in controlled *environments.*

aquifer An underground water reservoir.

aquifer recharge zone A geographic area where water infiltrates Earth's surface and reaches an *aquifer* below.

area effect The influence of island size on the number of *species* found on an island. Very roughly, all else being equal, the number of species on an island is expected to double as island size increases 10 times. See *distance effect; equilibrium theory of island biogeography.*

artificial selection *Natural selection* conducted under human direction.

atmosphere The thin layer of gases surrounding planet Earth. Compare *biosphere; hydrosphere; lithosphere.*

atmospheric pressure The weight per unit area produced by a column of air.

atom The smallest component of an *element* that maintains the chemical properties of that element.

autotroph (primary producer) An organism that can use the energy from sunlight to produce its own food. Includes green plants and cyanobacteria.

Bacillus thuringiensis (Bt) A naturally occurring *soil* bacterium that produces a protein that kills many pests, including caterpillars and the larvae of some flies and beetles.

background rate of extinction The rate of *extinctions* that occurred before the appearance of humans. For example, the *fossil* record indicates that for both birds and mammals, one *species* in the world typically became extinct every 500–1,000 years.

basic The property of a solution in which the concentration of hydroxide (OH⁻) *ions* is greater than the concentration of hydrogen (H⁺) ions. Compare *acidic*.

bedrock The continuous mass of solid rock that makes up Earth's *crust*.

benthic Of, relating to, or living on the ocean floor. Compare *pelagic*.

B horizon The layer of *soil* that lies below the *E horizon* and above the *C horizon*. The minerals that leach out of the E horizon are carried down into the B horizon (or subsoil) and accumulate there. The B horizon is sometimes called the "zone of accumulation" or "zone of deposition." Compare *A horizon; O horizon; R horizon*.

bill A draft law.

bioaccumulation The buildup of *toxicants* in the tissues of an animal.

biocentrism A philosophy that ascribes relative values to actions, entities, or properties on the basis of their effects on all living things or on the integrity of the biotic realm in general. The biocentrist would evaluate an action in terms of its overall impact on living things, including—but not exclusively focusing on—human beings.

biodiversity See *biological diversity*.

biodiversity hotspot An area that supports an especially great diversity of *species,* particularly species that are *endemic* to the area

biogenic gas *Natural gas* created at shallow depths by the *anaerobic* decomposition of organic matter by bacteria. Compare *thermogenic gas*.

biogeochemical cycle See *nutrient cycle*.

biological control (biocontrol) The attempt to battle pests and weeds with other organisms that eat or infect them, rather than by using *pesticides*.

biological diversity (biodiversity) The sum total of all organisms in an area, taking into account the diversity of *species,* their *genes,* their populations, and their *communities*.

biological weathering The breakdown of *parent material* into smaller particles through the activities of living things.

biomagnification The magnification of the concentration of *toxicants* in an organism caused by its consumption of other organisms in which toxicants have *bioaccumulated*.

biomass Matter contained in living organisms. Organic substances produced by recent *photosynthesis,* unlike *fossil fuels,* which received their photosynthetic sustenance millions of years ago.

biome A major regional complex of similar plant *communities;* a large ecological unit defined by its dominant plant type and vegetation structure.

biophilia A hypothetical phenomenon that E. O. Wilson defined as "the connections that human beings subconsciously seek with the rest of life."

biosphere The sum total of all the planet's living organisms and the *abiotic* portions of the *environment* with which they interact. Compare *atmosphere; hydrosphere; lithosphere*.

biotechnology The material application of biological *science* to create products derived from organisms. The creation of *transgenic organisms* is one type of biotechnology.

biotic factor Any living component of the *environment*. Compare *abiotic factor*.

biotic potential The innate reproductive capacity of a *species*.

by-catch That portion of a commercial fishing catch consisting of animals caught unintentionally. By-catch kills many thousands of fish, sharks, marine mammals, and birds each year.

calcium carbonate (CaCO₃) An essential ingredient in the skeletons and shells of microscopic marine organisms.

capitalist market economy An *economy* in which buyers and sellers interact to determine which *goods* and *services* to produce, how much to produce, and how goods and services should be produced and distributed. Compare *centrally planned economy*.

carbohydrate An *organic compound* consisting of *atoms* of carbon, hydrogen, and oxygen.

carbon cycle A major *nutrient cycle* consisting of the routes that carbon *atoms* take through the nested networks of environmental *systems*.

carcinogen A chemical or type of radiation that causes cancer.

carnivore An animal that eats other animals. Compare *herbivore; omnivore*.

carrying capacity The maximum *population size* that a given *environment* can sustain.

case law A body of law consisting of decisions rendered by the courts. See *judicial branch*.

cell The most basic organizational unit of organisms.

cellular respiration The process by which a *cell* uses the chemical reactivity of oxygen to split glucose into its constituent parts, water and carbon dioxide, and thereby release *chemical energy* that can be used to form chemical bonds or to perform other tasks within the cell. This extraction of energy from glucose occurs in the *autotrophs* that created the glucose and also in the animals that gain glucose by consuming autotrophs. Compare *photosynthesis*.

centrally planned economy An *economy* in which a nation's government determines how to allocate resources in a top-down manner. Also called a state socialist economy. Compare *capitalist market economy*.

chaparral A *biome* consisting mostly of densely thicketed evergreen shrubs occurring in limited small patches. The "Mediterranean" *climate* of mild, wet winters and warm, dry summers is induced by oceanic influences. In addition to ringing the Mediterranean Sea, chaparral occurs along the coasts of California, Chile, and southern Australia.

character displacement The *evolution* of the physical characteristics of *species* that reflect their reliance on the portion of the resource they use. Character displacement can result from *resource partitioning*.

chemical energy *Potential energy* held in the bonds between atoms.

chemical weathering   *Weathering* that results from the chemical interaction of water, atmospheric gases, and other substances with *parent material*. Compare *physical weathering*.

chemosynthesis   The process by which bacteria in *hydrothermal vents* use the *chemical energy* of hydrogen sulfide (H_2S) to transform inorganic carbon into *organic compounds*. Compare *photosynthesis*.

C horizon   The layer of *soil* that lies below the *B horizon* and above the *R horizon*. It contains rock particles that are larger and less weathered than the layers above. It consists of *parent material* that has been altered only slightly or not at all by the processes of soil formation. Compare *A horizon; E horizon; O horizon*.

chlorofluorocarbon (CFC)   One of a group of human-made compounds derived from simple *hydrocarbons,* such as ethane and methane, in which hydrogen *atoms* are replaced by chlorine, bromine, or fluorine. CFCs deplete the protective *ozone layer* in the *stratosphere*.

chronic exposure   Exposure to a *toxicant* occurring in low amounts for long periods of time. Compare *acute exposure*.

city planning   The professional pursuit that attempts to design cities in such a way as to maximize their efficiency, functionality, and beauty

clay   *Sediment* consisting of particles less than 0.002 mm in diameter. Compare *sand; silt*.

clear-cutting   The harvesting of timber by cutting all the trees in an area, leaving only stumps. Although it is the most cost-efficient method, clear-cutting is also the most environmentally damaging.

climate   The pattern of atmospheric conditions found across large geographic regions over long periods of time. Compare *weather*.

climate diagram (climatograph)   A visual representation of a region's average monthly temperature and precipitation.

climax community   A *community* that remains in place, with minimal modification, until some environmental change alters or displaces it.

closed system   A *system* that is isolated and self-contained. Scientists may treat a system as closed to simplify some question they are investigating, but no natural system is truly closed. Compare *open system*.

cloud forest   A *lower montane forest*. So called because much of the moisture it receives arrives in the form of low-moving clouds that blow inland from the sea.

clumped distribution   A *population distribution* in which organisms arrange themselves according to the availability of the resources they need to survive. This is the pattern most common in nature. Compare *random distribution; uniform distribution*.

coal   A *fossil fuel* composed of organic matter that was compressed under very high pressure to form a dense, solid carbon structure.

cogeneration   A practice in which the extra heat generated in the production of electricity is captured and put to use heating workplaces and homes, as well as producing other kinds of power.

cold front   The boundary where a mass of cold air displaces a mass of warmer air. Compare *warm front*.

command and control   An approach to environmental protection that sets strict legal limits and threatens punishment for violations of those limits.

commensalism   A relationship between members of different *species* in which one organism benefits and the other is unaffected. An example is an epiphytic plant (one that lives on the surface of another plant) and its *host*. Compare *amensalism*.

communicable disease   See *infectious disease*.

community   A group of populations of organisms that live in the same place at the same time.

community diversity   The number and variety of *community* types in some specified area.

competition   A relationship in which multiple organisms seek the same limited resource.

competitive exclusion   The exclusion from a *community* of a *species* by another species as a result of *interspecific competition*. Compare *species coexistence*.

compost   A mixture produced when *decomposers* break down organic matter, including food and crop waste, in a controlled *environment*.

composting   The conversion of organic *waste* into mulch or *humus* by encouraging, in a controlled manner, the natural biological processes of decomposition.

compound   A *molecule* whose *atoms* are composed of two or more *elements*.

confined (artesian) aquifer   A water-bearing, porous layer of rock, sand, or gravel that is trapped between an upper and lower layer of less permeable substrate, such as clay. The water in a confined aquifer is under pressure because it is trapped between two impermeable layers. Compare *unconfined aquifer*.

conservation biology   A scientific discipline devoted to understanding the factors, forces, and processes that influence the loss, protection, and restoration of *biological diversity* within and among ecosystems.

conservation concession   A conservation strategy pioneered by Conservation International, in which the organization pays nations for concessions for conservation rather than for resource extraction. The nation gets the money *and* keeps its natural resources intact.

conservation district   One of many county-based entities created by the Soil Conservation Service (now the Natural Resources Conservation Service) to promote *soil conservation* practices.

conservation ethic   An ethic holding that humans should put natural resources to use but that we have a responsibility to manage them wisely. Compare *preservation ethic*.

consumptive use   *Freshwater* use in which water is removed from a particular *aquifer* or surface water body and is not returned to it. *Irrigation* for *agriculture* is an example of consumptive use. Compare *nonconsumptive use*.

continental shelf   The gently sloping, underwater edge of a continent, varying in width from 100 m (330 ft) to 1,300 km (800 mi), with an average slope of 1.9 m/km (10 ft/mi).

contingent valuation   A technique that uses surveys to determine how much people would be willing to pay to protect a resource or to restore it after damage has been done.

contour farming   The practice of plowing furrows sideways across a hillside, perpendicular to its slope, to help prevent the formation of rills and gullies. The technique is so named because the furrows follow the natural contours of the land.

controlled experiment   An *experiment* in which the effects of all *variables* are controlled, except the one whose effect is being tested.

convection current   A circular *current* in which a warm fluid rises and is replaced by a colder fluid descending.

conventional law   International law that arises from conventions, or treaties, that nations agree to enter into. Compare *customary law.*

Convention on Biological Diversity   An international treaty that aims to conserve *biodiversity,* use biodiversity in a sustainable manner, and ensure the fair distribution of biodiversity's benefits. Although many nations have agreed to the treaty (as of March 2002, 183 nations had become parties to it), several others, including the United States, have not.

Convention on International Trade in Endangered Species of Wild Fauna and Flora (CITES)   A 1973 treaty facilitated by the United Nations that protects endangered *species* by banning the international transport of their body parts.

coral reef   A mass of *calcium carbonate* composed of the skeletons of tiny colonial marine organisms called corals.

core   The innermost part of the Earth, made up mostly of iron, that lies beneath the *crust* and *mantle.*

Coriolis effect   The apparent deflection of north-south air *currents* to a partly east-west direction, caused by the faster spin of regions at the equator than of regions at the poles as a result of Earth's rotation.

correlation   A relationship among *variables.*

corridor   A passageway of protected land established to allow animals to travel between islands of protected *habitat.*

covalent bond   A chemical bond in which the uncharged *atoms* in a *molecule* share *electrons.* For example, the uncharged atoms of carbon and oxygen in carbon dioxide form a covalent bond. Compare *ionic bond.*

cropland   Land used to raise plants for human food.

crop rotation   The practice of alternating the kind of crop grown in a particular field from one season or year to the next

crude oil (petroleum)   A *fossil fuel* produced by the conversion of organic compounds by heat and pressure. Crude oil is a mixture of hundreds of different types of *hydrocarbon* molecules characterized by carbon chains of different length.

crust   The lightweight outer layer of the Earth, consisting of rock that floats atop the malleable *mantle,* which in turn surrounds a mostly iron *core.*

culture   The overall ensemble of knowledge, beliefs, values, and learned ways of life shared by a group of people.

current   The flow of a fluid (liquid or gas) in a certain direction.

customary law   International law that arises from long-standing practices, or customs, held in common by most *cultures.* Compare *conventional law.*

Cuyahoga River   A river in northeastern Ohio that was so polluted with oil and industrial *waste* that the river itself caught fire near Cleveland more than half a dozen times during the 1950s and 1960s. This spectacle, coupled with an enormous oil spill off the Pacific coast near Santa Barbara, California, in 1969, moved the public to prompt Congress and the president to do more to protect the *environment.*

debt-for-nature swap   A conservation strategy in which a nongovernmental organization raises money and then offers to pay off a portion of a developing country's international debt, in exchange for a promise by the recipient country to set aside reserves, fund environmental education, and better manage protected areas.

decomposer   An organism such as a fungus or bacterium that breaks down nonliving matter into simpler constituents that can then be taken up and used by plants. Compare *detritivore.*

deep ecology   A philosophy established in the 1970s based on principles of self-realization (the awareness that humans are inseparable from nature) and *biocentric* equality (the precept that all living beings have equal value). Because we are truly inseparable from our *environment,* we must protect all other living things as we would protect ourselves.

deep-well injection   A *hazardous waste* disposal method in which a well is drilled deep beneath an area's *water table* into porous rock below an impervious *soil* layer. Wastes are then injected into the well, so that they will be absorbed into the porous rock and remain deep underground, isolated from *groundwater* and human contact. Compare *surface impoundment.*

deforestation   The clearing and loss of forests.

demand   The amount of a *good* or *service* that people will buy at a given price if they are free to do so. Compare *supply.*

demographic transition model   A theoretical *model* of economic and cultural change that explains the declining death rates and birth rates that occurred in Western nations as they became industrialized. The model holds that industrialization caused these rates to fall naturally by decreasing mortality and by lessening the need for large families. Parents would thereafter choose to invest in quality of life rather than quantity of children.

demography   A *social science* that applies the principles of population *ecology* to the study of statistical change in human populations.

denitrifying bacteria   Bacteria that convert the nitrates in *soil* or water to gaseous nitrogen and release it back into the *atmosphere.*

density-dependent factor   A *limiting factor* whose effects on a population increase or decrease depending on the *population density.* Compare *density-independent factor.*

density-independent factor   A *limiting factor* whose effects on a population are constant regardless of *population density.* Compare *density-dependent factor.*

deoxyribonucleic acid   See *DNA.*

dependent variable   The *variable* that is affected by manipulation of the *independent variable.*

deposition   The arrival and accumulation of eroded material from another location. Contrast *erosion.*

desalination   The removal of salt from seawater.

desert   The dries *biome* on Earth. Annual *precipitation* is less than 25 cm, much of it falling during isolated storms that may occur months or years apart. Because deserts have relatively little vegetation to insulate them from temperature extremes, sunlight readily heats them in the daytime, but daytime heat is quickly lost at night, so temperatures vary widely from day to night and in different seasons.

desertification   A loss of more than 10% of a land's productivity due to *erosion, soil* compaction, forest removal, *overgrazing,* drought, *salinization, climate* change, depletion of water sources, and an array of other factors. Severe desertification can result in the actual expansion of desert areas or creation of new ones in areas that once supported fertile land.

detritivore   An organism such as a millipede or soil insect that eats the waste products or the dead bodies of other *community* members. Compare *decomposer.*

directional selection   Selection that acts to drive a feature in one direction rather than another; for example, toward larger or smaller, faster or slower. Compare *disruptive selection; stabilizing selection.*

disruptive selection   Selection that acts to produce extreme traits. Compare *directional selection; stabilizing selection.*

dissolved oxygen content   A chemical property of water denoting its concentration of dissolved oxygen. Dissolved oxygen is an indication of aquatic ecosystem health because surface waters low in dissolved oxygen (see *hypoxia*) are less capable of supporting aquatic life.

distance effect   The influence of the distance of an island from a continent, or from a source of immigrants, on the number of *species* found on the island. The farther the island is from a continent or other source of immigrants, the fewer species live on the island. See *area effect; equilibrium theory of island biogeography.*

DNA (deoxyribonucleic acid)   A double-stranded *nucleic acid* composed of four nucleotides, each of which contains a sugar (deoxyribose), a phosphate group, and a nitrogenous base. DNA carries the hereditary information for living organisms and is responsible for passing traits from parents to offspring. Compare *RNA.*

dose-response curve   A curve produced by plotting the *toxicant* dose versus the response of test animals, generally quantified by measuring the proportion of animals exhibiting negative effects.

downwelling   In the ocean, the flow of warm surface water toward the ocean floor. Downwelling occurs where surface *currents* converge. Compare *upwelling.*

Dust Bowl   (1) An arid area that loses huge amounts of *topsoil* to wind *erosion* as a result of drought and/or human impact. (2) The region in the Great Plains severely affected by drought and topsoil loss in the 1930s.

dynamic equilibrium   The state reached when processes within a *system* are moving in opposing directions at equivalent rates so that their effects balance out.

Earth Day   A worldwide celebration of support for environmental protection that takes place annually in April. It consists of thousands of locally based events featuring such activities as speeches, lectures, demonstrations, hikes, and birdwalks.

ecocentrism   A philosophy that considers actions in terms of their damage or benefit to the integrity of whole ecological *systems,* including both *biotic* and *abiotic* elements. For an ecocentrist, the well-being of an individual organism—human or otherwise—is less important than the long-term well-being of a larger integrated ecological system.

ecofeminism   A philosophy holding that the patriarchal (male-dominated) structure of society is a root cause of both social and *environmental problems.* Ecofeminists hold that a *worldview* traditionally associated with women, which interprets the world in terms of interrelationships and cooperation, is more in tune with nature than a worldview traditionally associated with men, which interprets the world in terms of hierarchies and competition.

ecolabeling   The practice of designating on a product's label how the product was grown, harvested, or manufactured, so that consumers buying it are aware of the processes involved and can differentiate between brands that use processes believed to be environmentally beneficial (or less harmful than others) and those that do not.

ecological economics   A recent *theory* of *economics* that applies the principles of *ecology* and *systems* thinking to the description and analysis of *economies.* Compare *environmental economics; neoclassical economics.*

ecological footprint   The cumulative amount of land and water required to provide the raw materials a person or population consumes and to dispose of or recycle the *waste* that is produced.

ecological restoration   Efforts to reverse the effects of human disruption and restore *communities* to their natural state.

ecology   The *science* that deals with the distribution and abundance of organisms, the interactions among them, and the interactions between organisms and their *abiotic environments.*

economics   The study of how we decide to use scarce resources to satisfy the *demand* for *goods* and *services.*

economy   A social *system* that converts resources into *goods* and *services.*

ecosystem diversity   The number and variety of ecosystems in some specified area.

ecosystem management   The attempt to protect, recover, or restore whole plant and animal *communities* that have been lost or degraded.

ecosystem service   An essential service the ecosystem provides that supports the life that makes economic activity possible. For example, the ecosystem naturally purifies air and water, cycles *nutrients,* provides for plants to be pollinated by animals, and serves as a receptacle and *recycling* system for the *waste* generated by our economic activity.

ecotone   A transitional zone where ecosystems meet, in which *biotic* and *abiotic* elements of the ecosystems mix.

ecotourism The touring of natural *habitats* in a manner meant to minimize ecological impact.

ectoparasite A *parasite* that lives on the exterior of its *host*. Compare *endoparasite*.

effective dose–50% (ED$_{50}$) The amount of a *toxicant* it takes to affect 50% of a population of test animals. Compare *threshold dose; lethal dose–50%.*

E horizon (eluviation horizon) The layer of *soil* that lies below the *A horizon* and above the *B horizon. Eluviation* means "loss," and the E horizon is characterized by the loss of certain minerals through *leaching.* It is sometimes called the "zone of leaching." Compare *C horizon; O horizon; R horizon.*

El Niño The exceptionally strong warming of the eastern Pacific Ocean that occurs every 2 to 7 years and hurts local fish and bird populations by altering the marine *food web* in the area. Originally, the name that Spanish-speaking fishermen gave to an unusually warm surface current that sometimes arrived near the Pacific coast of South America around Christmas time. Compare *La Niña.*

electron A negatively charged particle that surrounds the nucleus of an *atom.*

element A fundamental type of matter; a chemical substance with a given set of properties, which cannot be broken down into substances with other properties. Chemists currently recognize 92 elements that occur in nature, as well as more than 20 others that have been artificially created.

emergent property A characteristic that is not evident in a *system*'s components.

Emerson, Ralph Waldo (1803–1882) American author, poet, and philosopher who espoused *transcendentalism* and promoted a holistic view of nature among the public.

emigration The departure of individuals from a population.

Endangered Species Act (ESA) The primary legislation, enacted in 1973, for protecting *biodiversity* in the United States. It forbids the government and private citizens from taking actions (such as developing land) that would destroy endangered *species* or their *habitats* and prohibits trade in products made from endangered species.

endemic Native or restricted to a particular geographic region. An endemic species occurs in one place and nowhere else on Earth.

endocrine disruptor A *toxicant* that interferes with the endocrine system.

endoparasite A *parasite* that lives inside its *host*. Compare *ectoparasite.*

energy conservation The practice of reducing *energy* use as a way of extending the lifetime of our *fossil fuel* supplies, of being less wasteful, and of reducing our environmental impact.

energy An intangible phenomenon that can change the position, physical composition, or temperature of matter.

entropy The degree of disorder in a substance, *system,* or process. See *second law of thermodynamics.*

environment The sum total of our surroundings, including all of the living things and nonliving things with which we interact.

environmental economics A recent *theory* of *economics* based on modifying the principles of *neoclassical economics* to address environmental challenges. An environmental economist believes that we can attain sustainability within our current economic *systems.* Whereas ecological economists call for revolution, environmental economists call for reform. Compare *ecological economics; neoclassical economics.*

environmental ethics The application of ethical standards to environmental questions.

environmental health Environmental factors that influence human health and quality of life and the health of ecological systems essential to environmental quality and long-term human well-being.

environmental impact statement (EIS) A report of results from detailed studies that assess the potential effects on the *environment* that would likely result from development projects or other actions undertaken by the government.

environmental justice A movement based on a moral sense of fairness and equality that seeks to expand society's domain of ethical concern from men to women, from humans to nonhumans, from rich to poor, and from majority races and ethnic groups to minority ones.

environmental problem Any undesirable change in the *environment.*

Environmental Protection Agency (EPA) An *administrative agency* created by executive order in 1970. The EPA is charged with conducting and evaluating research, monitoring environmental quality, setting standards, enforcing those standards, assisting the states in meeting standards and goals, and educating the public.

environmental resistance The sum total of the *limiting factors* in an *environment.* The environmental resistance eventually stabilizes *population size* at its *carrying capacity.*

environmental toxicology The study of *toxicants* that come from or are discharged into the *environment,* including the study of health effects on humans, other animals, and ecosystems.

environmentalism A social movement dedicated to protecting the natural world.

enzyme A *molecule* that catalyzes, or promotes, certain chemical reactions.

epidemiological study A study that involves large-scale comparisons among groups of people, usually contrasting a group known to have been exposed to some *toxicant* and a group that has not.

equilibrium A stable point at which the *population size* of each *species* in a relationship of *species coexistence* remains fairly constant over time.

equilibrium theory of island biogeography A *theory* that was initially applied to oceanic islands to explain how *species* come to be distributed among them. Since its development, researchers have increasingly applied the theory to other types of islands, including islands of *habitat* (patches of one type of habitat isolated within vast "seas" of others). Important tenets of the theory include *immigration* and *extinction* rates, the effect of island size (see *area effect*), and the effect of distance from the mainland (see *distance effect*).

erosion The removal of material from one place and its transport to another by the action of wind or water. See *gully erosion; rill erosion; sheet erosion; splash erosion*. Contrast *deposition*.

estuary An area where a river flows into the ocean, mixing *fresh water* with salt water.

ethics The study of good and bad, right and wrong. The term can also refer to a person's or group's set of moral principles or values.

eukaryote A multicellular organism. The cells of eukaryotic organisms consist of a membrane-enclosed *nucleus* that houses DNA, an outer membrane of lipids, and an inner fluid-filled chamber containing *organelles*. Compare *prokaryote*.

euphotic zone The well-lit top layer of the oceans.

eutrophication The process of *nutrient* enrichment, increased production of organic matter, and subsequent ecosystem degradation.

evaporation The conversion of a substance from a liquid to a gaseous form.

evolution Genetically based change in the appearance, functioning, and/or behavior of organisms across generations, often by the process of *natural selection*.

evolutionary arms race A duel of escalating adaptations in which *hosts* and *parasites* repeatedly evolve new responses to the other's latest advance.

executive branch The branch of the U.S. government, headed by the president, that enforces legislation. Among other powers, the president may approve (enact) or reject (veto) legislation and issue *executive orders*. Compare *judicial branch; legislative branch*.

executive order A specific legal instruction for government agencies issued by the president.

experiment An activity designed to test the validity of a *hypothesis* by manipulating *variables*.

exponential growth The increase of a population (or of anything) by a fixed percentage each year.

external cost A negative *externality*; a cost borne by someone not involved in a transaction.

externality A cost or benefit of a transaction that involves people other than the buyer or seller.

extinction The disappearance of an entire *species* from the face of the Earth. Compare *extirpation*.

extirpation The disappearance of a particular population from a given area, but not the entire *species* globally. Compare *extinction*.

extrusive igneous rock An *igneous rock* formed by rapid cooling of *lava* at Earth's surface. Basalt is the most common extrusive igneous rock. Compare *intrusive igneous rock*.

feedback loop A circular process in which a system's output serves as input to that same *system*. See *negative feedback loop; positive feedback loop*.

feedlot A huge barn or outdoor pen designed to deliver energy-rich food to animals living in extremely high densities. Also called a factory farm or concentrated animal feeding operation (CAFO).

Ferrel cell One of a pair of *convection currents* near 60° north and south latitude, in which air is forced upward, creating *precipitation*. Compare *Hadley cell; polar cell*.

fertilizer A substance that contains essential *nutrients*.

first law of thermodynamics *Energy* can change from one form to another, but it cannot be created or lost. The total energy in the universe remains constant and is said to be conserved.

fission The conversion of the *energy* within an *atom*'s nucleus to usable thermal energy by splitting apart atomic nuclei. Compare *fusion*.

fitness (1) The likelihood that an individual will reproduce. (2) The number of offspring an individual produces over its lifetime.

floodplain The region of land over which a river has historically wandered and periodically floods.

food chain A simplified linear visual representation of *community* feeding interactions. Compare *food web*.

food security An adequate, reliable, and available food supply to all people at all times.

food web A relatively complicated visual representation of *community* feeding interactions that shows an array of relationships between organisms at different *trophic levels*. Compare *food chain*.

fossil The remains, impression, or trace of an animal or plant of past geological ages that has been preserved in the Earth's crust.

fossil fuel A *nonrenewable energy resource*, such as *crude oil, natural gas*, or *coal*, produced by the decomposition and fossilization of ancient life.

fresh water Water that is relatively pure, holding very few dissolved salts.

front The boundary between two air masses that differ in temperature and density. See *cold front; warm front*.

fundamental niche The full *niche* of a *species*. Compare *realized niche*.

fusion The conversion of the *energy* within an *atom*'s nucleus to usable thermal energy by forcing together the small nuclei of lightweight *elements*. Compare *fission*.

Gaia hypothesis James Lovelock's proposition that Earth behaves like a single self-regulating *system* or superorganism and that living things affect the *environment* in ways that stabilize the *climate* and make it possible for life to persist and flourish.

gene A unit of hereditary information.

gene bank See *seed bank*.

genetically modified (GM) organism An organism that has been *genetically engineered* using a technique called *recombinant DNA* technology.

genetic diversity The differences in *DNA* composition among individuals within a given *species*.

genetic engineering   Any process scientists use to manipulate an organism's genetic material in the lab, by adding, deleting, or changing segments of its *DNA*.

Genuine Progress Indicator (GPI)   An economic indicator introduced in 1995 that attempts to differentiate between desirable and undesirable economic activity. The GPI accounts for benefits such as volunteerism and for costs such as environmental degradation and social upheaval. Compare *Gross Domestic Product (GDP)*.

geothermal energy   Renewable *energy* that is generated deep within the Earth. The radioactive decay of elements amid the extremely high pressures and temperatures at depth generate heat that rises to the surface through magma and through fissures and cracks. Where this energy heats *groundwater,* natural eruptions of heated water and steam are sent up from below. See *hydrothermal vent.*

global climate change   Changes in Earth's *climate,* such as temperature, *precipitation,* and storm intensity.

globalization   The development of an increasingly integrated global *economy* through free trade of *goods* and *services* and the free flow of capital, technology, and labor throughout the world. Those who favor globalization define it as the process by which people of the world's diverse *cultures* are increasingly communicating with one another and learning about and partaking in one another's diversity. Those who oppose globalization define it as the process of homogenization of the world's cultures, with certain few cultures and *worldviews* displacing many others.

global warming   An increase in Earth's average temperature, by whatever cause.

good   A material commodity manufactured for and bought by individuals and businesses.

greenhouse effect   The warming of Earth's surface and *atmosphere* (especially the *troposphere*) caused by the *energy* emitted by *greenhouse gases.*

greenhouse gas   A gas that effectively absorbs infrared radiation released by Earth's surface and later warms the surface by emitting energy, giving rise to the *greenhouse effect.* Greenhouse gases include carbon dioxide (CO_2), water vapor, ozone (O_3), nitrous oxide (N_2O), *halocarbon gases,* and methane (CH_4).

green manure   Fresh vegetation.

green revolution   An intensification of the industrialization of *agriculture* that dramatically increases the crops produced per acre of farmland. New practices include devoting large areas to identical crops specially bred for high yields and rapid growth; heavy use of *fertilizers, pesticides,* and *irrigation* water; and sowing and harvesting on the same piece of land more than once per year or per season.

green tax   A charge on environmentally harmful activities and products aimed at providing a market-based incentive to correct for *market failure.* Compare *subsidy.*

Gross Domestic Product (GDP)   The total monetary value of final *goods* and *services* produced in a country each year. The GDP sums all economic activity, whether good or bad, and does not account for benefits such as volunteerism or for *external costs* such as environmental degradation and social upheaval. Compare *Genuine Progress Indicator (GPI).*

groundwater   The *precipitation* that infiltrates Earth's land surface and percolates into the *soil* layers.

growth rate   The net change in a population's size, per 1,000 individuals. Calculated by adding the crude birth rate to the *immigration* rate and then subtracting the crude death rate and the *emigration* rate, each expressed as the number per 1,000 individuals per year.

gully erosion   A type of water *erosion* in which rills merge to form larger and larger channels and eventually gullies. Such erosion causes the most dramatic and visible changes in the landscape. Compare *rill erosion; sheet erosion; splash erosion.*

habitat   The specific *environment* in which an organism lives, including both *biotic* and *abiotic factors.*

habitat diversity   The number and variety of *habitats* in some specified area.

habitat selection   The process by which organisms select *habitats* in which to live from among the range of options they encounter.

Hadley cell   One of a pair of *convection currents* near the equator, in which surface air is warmed by the intense sunlight, rising and expanding in two giant columns, one heading toward each pole. Compare *Ferrel cell; polar cell.*

halocarbon gas   Any of a diverse group of *greenhouse gases* that include *chlorofluorocarbons (CFCs)* and hydrochlorofluorocarbons (HFCs).

hardness   A chemical property of water denoting its concentration of calcium and magnesium ions. Hard water prevents soap from lathering and leaves hard, chalky deposits behind when heated or boiled.

harmful algal bloom   A population explosion of toxic algae caused by excessive *nutrient* concentrations.

hazardous waste   *Waste* that is toxic, chemically reactive, flammable, or corrosive. Compare *industrial solid waste; municipal solid waste.*

herbivore   A *primary consumer;* an organism that consumes *autotrophs.* Compare *carnivore; omnivore.*

herbivory   The eating of plants by animals.

heritable   Able to be passed on from generation to generation.

heterotroph   An organism that consumes *autotrophs.* Includes most animals, as well as fungi and microbes that decompose organic matter.

high-pressure system   An air mass with elevated *atmospheric pressure,* containing air that descends, typically bringing fair *weather.* Compare *low-pressure system.*

HIPPO   An acronym denoting the five primary causes of *species* decline and *extinction:* Habitat alteration, Invasive species, Pollution, Population, and Overexploitation.

historical science   A *science,* such as cosmology or paleontology, that studies the causes and consequences of particular historical events rather than the behavior of general constants. Compare *natural science; social science.*

homeostasis The tendency of a *system* to maintain constant or stable internal conditions.

horizon A distinct layer of *soil*. See *A horizon; B horizon; C horizon; E horizon; O horizon; R horizon.*

host An organism that affords nourishment or some other benefit to another organism (the *parasite*) and is harmed by the parasite.

humus A dark, spongy, crumbly mass of undifferentiated material made up of complex *organic compounds*. Humus results from the partial decomposition of organic matter.

hydrocarbon An *organic compound* consisting solely of hydrogen and carbon *atoms*.

hydrologic cycle The flow of water—in liquid, gaseous, and solid forms—through our *biotic* and *abiotic environment*.

hydrosphere All water—salt or fresh, liquid, ice, or vapor—in surface bodies, underground, and in the *atmosphere*. Compare *biosphere; lithosphere.*

hydrothermal vent A jet of geothermally heated water that emerges in the cold depths of the ocean. See *geothermal energy.*

hypothesis An educated guess that explains a phenomenon or answers a scientific question. Compare *theory.*

hypoxia The condition of extremely low dissolved oxygen concentrations in a body of water.

ice core A long column of ice extracted by drilling into Earth's ice caps or glaciers. By examining the tiny bubbles of ancient *atmosphere* trapped within the cores, scientists can determine the atmospheric composition, *greenhouse gas* concentrations, and temperature trends of earlier geologic ages.

igneous rock One of the three great groups of rock. A rock formed from cooling *magma*. Compare *metamorphic rock; sedimentary rock.*

immigration The arrival of individuals from outside a population.

incineration A controlled process of burning in which mixed garbage is combusted at very high temperatures.

independent variable The *variable* that the scientist manipulates in a *manipulative experiment*.

industrial ecology A holistic approach to industry that integrates principles from engineering, chemistry, *ecology, economics*, and other disciplines and seeks to redesign industrial systems in order to reduce resource inputs and minimize inefficiency.

industrialized agriculture A form of *agriculture* that uses large-scale *fossil fuel* combustion and mechanization, enabling farmers to replace horses and oxen with faster and more powerful means of cultivating, harvesting, transporting, and processing crops. Other advances facilitate *irrigation* and the use of *fertilizers*; the use of chemical herbicides and *pesticides* reduces competition from weeds and *herbivory* by insects. Compare *traditional agriculture.*

industrial revolution The shift in the mid-1700s from rural life, animal-powered agriculture, and manufacturing by craftsmen to an urban society powered by *fossil fuels* such as *coal* and oil.

industrial smog Gray-air *smog* caused by the incomplete combustion of *coal* or oil when burned. Compare *photochemical smog.*

industrial solid waste Nonliquid *waste* that is not especially hazardous and that comes from production of consumer goods, mining, petroleum extraction and refining, and *agriculture*. Compare *hazardous waste; municipal solid waste.*

industrial stage The third stage of the *demographic transition model*, characterized by falling birth rates that close the gap with falling death rates and reduce the rate of population growth. Compare *pre-industrial stage; postindustrial stage; transitional stage.*

infectious disease A disease in which a pathogen attacks a host.

inorganic fertilizer A *fertilizer* that consists of mined or synthetically manufactured mineral supplements. Inorganic fertilizers are generally more susceptible than *organic fertilizers* to *leaching* and *runoff* and may be more likely to cause unintended off-site impacts.

integrated pest management (IPM) The use of many different techniques to achieve long-term suppression of pests, including *biocontrol*, use of *pesticides*, close monitoring of populations, *habitat* alteration, *crop rotation*, transgenic crops, alternative tillage methods, and mechanical pest removal.

intensive traditional agriculture A form of *traditional agriculture* in which a farming family sometimes uses draft animals and employs significant quantities of *irrigation* water and *fertilizer*, but stops short of using *fossil fuels*. This type of agriculture aims to produce excess food to sell in the market, in addition to the food required for the farming family itself. Compare *subsistence agriculture.*

interdisciplinary field A field that borrows techniques from several more traditional fields of study and brings together research results from these fields into a broad synthesis.

interspecific competition *Competition* that takes place between members of two or more different *species*. Compare *intraspecific competition.*

intertidal Of, relating to, or living along shorelines between the farthest reach of the highest *tide* and the lowest reach of the lowest tide.

intraspecific competition *Competition* that takes place between members of the same *species*. Compare *interspecific competition.*

intrusive igneous rock An *igneous rock* formed by slow cooling of *magma* while it is still below Earth's surface. Granite is the best-known intrusive igneous rock. Compare *extrusive igneous rock.*

invasive species A *species* that spreads widely and rapidly becomes dominant in a *community*, interfering with the community's normal functioning..

ion An electrically charged *atom* or combination of atoms.

ionic bond A chemical bond in which oppositely charged *ions* are held together by electrical attraction. Compare *covalent bond.*

ionic compound (salt) An association of *ions* that are bonded electrically in an *ionic bond*.

IPAT model   A formula that represents how humans' total impact (I) on the *environment* results from the interaction among three factors: population (P), affluence (A), and technology (T).

irrigation   The artificial provision of water to support *agriculture.*

isotope   One of several forms of an *element* having differing numbers of *neutrons* in the nucleus of its *atoms.* Chemically, isotopes of an element behave almost identically, but they have different physical properties because they differ in mass.

judicial branch   The branch of the U.S. government, consisting of the Supreme Court and various lower courts, that is charged with interpreting the law. Compare *executive branch; legislative branch.*

kelp   Large brown algae or seaweed.

keystone species   A *species* that has an especially far-reaching effect on a *community*

kinetic energy   *Energy* of motion. Compare *potential energy.*

K–strategist   A *species* with low *biotic potential* whose members produce a small number of offspring and take a long time to gestate and raise each of their young. Populations of K–strategists are generally regulated by *density-dependent factors.* Compare *r–strategist.*

Kyoto Protocol   An agreement drafted in 1997 that would, by 2012, reduce emissions of six *greenhouse gases* to levels lower than their levels in 1990. Although the United States has refused to ratify the protocol, it had been ratified by 111 nations as of late 2003 and was awaiting only the ratification of Russia to take effect.

La Niña   The exceptionally strong cooling of surface water in the equatorial Pacific Ocean. Compare *El Niño.*

land use   The ways that areas of land are apportioned for different functions.

landscape diversity   The variety and geographic arrangement of *habitats, communities,* or ecosystems over a wide area, including the sizes, shapes, and interconnectedness of patches of these entities.

landscape ecology   An approach to the study of organisms and their *environments* at landscape scale, focusing on geographical areas that include multiple ecosystems.

latitudinal gradient   The variation in diversity with latitude.

lava   *Magma* that is released from the *lithosphere* and flows or spatters across Earth's surface.

lawsuit   A request for judicial action by the courts.

leachate   Liquids that seep through liners and leach through the *soil* underneath.

leaching   The process by which solid materials such as minerals are dissolved in a liquid (usually water) and transported to another location.

legislative branch   The branch of the U.S. government that passes laws; it consists of Congress, which includes the House of Representatives and the Senate. Compare *executive branch; judicial branch.*

Leopold, Aldo (1887–1949)   American scientist, scholar, philosopher, and author. His book *The Land Ethic* argued that humans should view themselves and the land itself as members of the same community and that humans are obliged to treat the land ethically.

lethal dose–50% (LD_{50})   The amount of a *toxicant* it takes to kill 50% of a population of test animals. Compare *effective dose–50%; threshold dose.*

lichen   A *mutualistic* aggregation of fungi and algae.

life expectancy   The average number of years that individuals in particular age groups are likely to continue to live.

limiting factor   A physical, chemical, or biological characteristic of the *environment* that restrains population growth.

lipid   One of a chemically diverse group of compounds that are classified together because they do not dissolve in water. Lipids include fats, phospholipids, waxes, pigments, and steroids.

lithosphere   The solid part of the Earth, including the rocks, *sediment,* and *soil* at the surface and extending down many miles underground. Compare *atmosphere; biosphere; hydrosphere.*

littoral   See *intertidal.*

loam   *Soil* with a relatively even mixture of *clay-, silt-,* and *sand-*sized particles.

lobbying   The expenditure of time or money in an attempt to change an elected official's mind.

logistic growth curve   A plot that shows how the initial *exponential growth* of a population is slowed and finally brought to a standstill by *limiting factors.*

lower montane rainforest   A type of rainforest that occurs on high mountains in the tropics. Examples include some of the lush forests above Monteverde, Costa Rica, which begin at elevations of around 1,600 m.

low-input agriculture   *Agriculture* that uses smaller amounts of *pesticides, fertilizers,* growth hormones, water, and *fossil fuels* than are currently used in *industrialized agriculture.* Compare *organic agriculture.*

low-pressure system   An air mass in which the air moves toward the low *atmospheric pressure* at the center of the system and spirals upward, typically bringing clouds and *precipitation.* Compare *high-pressure system.*

macromolecule   A very large molecule, such as a *protein, nucleic acid, carbohydrate,* or *lipid.*

macronutrient   A *nutrient* that organisms require in relatively large amounts. Compare *micronutrient.*

magma   Molten, liquid rock.

malnutrition   The lack of nutritional *elements* the body needs, including a complete complement of vitamins and minerals. Malnutrition can occur in both undernourished and overnourished individuals.

mangrove   A tree with a unique type of roots that curve upward to obtain oxygen, which is lacking in the mud in which they grow, and that serve as stilts to support the tree in changing water levels. Mangrove forests grow on the coastlines of the tropics and subtropics.

manipulative experiment   An *experiment* in which the researcher actively chooses and manipulates the *independent variable*. Compare *natural experiment.*

mantle   The malleable layer of rock that lies beneath Earth's *crust* and surrounds a mostly iron *core.*

marine protected area (MPA)   An area of the ocean set aside to protect the marine life from fishing pressures. An MPA may be protected from some human activities but open to others. Compare *marine reserve.*

marine reserve   An area of the ocean designated as a "no-fishing" zone. Compare *marine protected area.*

marketable emissions permit   A permit issued to polluters that allows them to emit a certain fraction of the total amount of pollution the government will allow an entire industry to produce. Polluters are then allowed to buy, sell, and trade these permits with other polluters.

market failure   The breakdown of a market that can occur when it does not take into account the environment's good effects on *economies* (for example, *ecosystem services*) or does not reflect the bad effects of economic activity on the *environment* and thereby on people (*external costs*).

mass extinction   The dying off of many *species* at once in a catastrophic event.

maximum sustainable yield   The maximal harvest of a particular fish population that can be accomplished while still keeping fish available for the future.

mesosphere   The layer of the *atmosphere* above the *stratosphere* and below the *thermosphere*; it extends from 50 km (31 mi) to 90 km (56 mi) above sea level. See also *troposphere.*

metamorphic rock   One of the three great groups of rock. A rock formed by great heat and/or pressure. The forces that transform rock into metamorphic rock occur at temperatures lower than the melting point of the rock but high enough to reshape the crystals within it and change its appearance and physical properties. Common metamorphic rocks include marble and slate. Compare *igneous rock; sedimentary rock.*

micronutrient   A *nutrient* that organisms require in relatively small amounts. Compare *macronutrient.*

Milankovitch cycle   One of three types of variations in Earth's rotation and orbit around the Sun that result in slight changes in the relative amount of solar radiation reaching Earth's surface at different latitudes. As the cycles proceed, they change the way solar radiation is distributed over Earth's surface and contribute to changes in atmospheric heating and circulation that have triggered the onset and end of the ice ages and other more subtle *climate* changes.

Mill, John Stuart (1806–1873)   British philosopher who believed that as resources become harder to find and extract, economic growth will slow and eventually stabilize into a *steady-state economy.*

mitochondrion   An *organelle* that extracts energy from sugars and fats.

model   A deliberately simplified representation of a complex natural (or social) *system.* Models synthesize known information and attempt to use it to represent how complex processes operate. Compare *hypothesis; theory.*

molecule   A combination of two or more *atoms.*

monoculture   The uniform planting of a single crop. Characterizes *industrialized agriculture.* Compare *polyculture.* monthly temperature and precipitation.

Muir, John (1838–1914)   Scottish immigrant to the United States who eventually settled in California and made the Yosemite Valley his wilderness home. Today, he is most strongly associated with the *preservation ethic.* He argued that nature deserved protection for its own inherent values (an *ecocentrist* argument) but also claimed that nature played a large role in human happiness and fulfillment (an *anthropocentrist* argument).

municipal solid waste   Nonliquid *waste* that is not especially hazardous and that comes from homes, institutions, and small businesses. Compare *hazardous waste; industrial solid waste.*

mutagen   A *toxicant* that causes *mutations* in the *DNA* of organisms

mutation   An accidental change in *DNA* that may range in magnitude from the deletion, substitution, or addition of a single nucleotide to mistakes affecting entire sets of chromosomes.

mutualism   A relationship in which all participating organisms benefit from their interaction. Compare *parasitism.*

National Environmental Policy Act (NEPA)   A law enacted on January 1, 1970, that created an agency called the Council on Environmental Quality and required that an *environmental impact statement* be prepared for any major federal action.

national forest   Public lands consisting of 191 million acres (more than 8% of the nation's land area) in many tracts spread across all but a few states.

national park   A scenic area set aside for recreation and enjoyment by the public. The national park system today numbers 388 sites totaling 78.8 million acres and includes national historic sites, national recreation areas, national wild and scenic rivers, and other types of areas.

national wildlife refuge   An area set aside to serve as a haven for wildlife and also sometimes to encourage hunting, fishing, wildlife observation, photography, environmental education, and other public uses.

natural experiment   An *experiment* in which the researcher cannot control the *variables* and therefore must observe nature and interpret the results. Compare *manipulative experiment.*

natural gas   A *fossil fuel* composed primarily of methane (CH_4), produced as a by-product when bacteria decompose organic material under *anaerobic* conditions.

natural rate of population change   The population change resulting from birth and death rates alone, excluding migration.

natural resource   Any of the various substances and *energy* sources we need to survive.

natural science   A discipline that studies the natural world. Compare *historical science; social science.*

natural selection   A natural process resulting in the *evolution* of organisms that are best adapted to their *environment.* Unscientifically, "survival of the fittest." See *fitness.*

negative feedback loop A *feedback loop* in which output of one type acts as input that moves the *system* in the opposite direction. The input and output essentially neutralize each other's effects, stabilizing the system. Compare *positive feedback loop.*

neoclassical economics A *theory* of *economics* that focuses on consumer choices and the psychological factors underlying those choices and explains market price in terms of consumer preferences for units of particular commodities. Buyers desire the lowest possible price whereas sellers desire the highest possible price. This conflict between buyers and sellers results in a compromise price being reached and the "right" quantity of commodities being bought and sold. Compare *ecological economics; environmental economics.*

neolithic (agricultural) revolution The transition from a hunter-gatherer lifestyle to an agricultural way of life that occurred around 10,000 years ago.

net primary productivity The rate at which *biomass* available to consumers is produced. *Wetlands* and *tropical rainforests* tend to have the highest net primary productivities. See *primary productivity.*

neurotoxin A *toxicant* that assaults the nervous system. Neurotoxins include heavy metals, *pesticides,* and some chemical weapons developed for use in war.

neutron An electrically neutral (uncharged) particle in the nucleus of an *atom.*

niche The functional role of a *species* in a *community.* See *fundamental niche; realized niche.*

nitrification The conversion by bacteria of ammonium (NH_4^+) ions first into nitrite (NO_2^-) ions and then into nitrate (NO_3^-) ions.

nitrogen cycle A major *nutrient cycle* consisting of the routes that nitrogen *atoms* take through the nested networks of environmental *systems.*

nitrogen fixation The process by which inert nitrogen gas combines with hydrogen to form ammonium (NH_4^+) ions, which are chemically and biologically active and can be taken up by plants.

nitrogen-fixing bacteria Bacteria that live in a *mutualistic* relationship with many types of plants and provide *nutrients* to the plants by converting nitrogen to a usable form.

nonconsumptive use *Freshwater* use in which the water from a particular *aquifer* or surface water body either is not removed or is removed only temporarily and then returned. The use of water to generate electricity in hydroelectric dams is an example of nonconsumptive use. Compare *consumptive use.*

nonmarket value A value that is not usually included in the price of a *good* or *service.*

non-point source A diffuse source of pollutants, often consisting of many small sources. Compare *point source.*

nonrenewable natural resource A resource that is in limited supply and is formed much more slowly than we use it. Compare *renewable natural resource.*

nucleic acid A *macromolecule* that directs the production of *proteins.*

nutrient An *element* or *compound* that organisms consume and require for survival.

nutrient cycle The cycle by which *nutrients* move through the *environment.*

oceanography The study of the physics, chemistry, and geology of the oceans.

O horizon The top layer of *soil* in some *soil profiles,* made up of organic matter, such as decomposing branches, leaves, crop residue, and animal waste. Compare *A horizon; B horizon; C horizon; E horizon; R horizon.*

omnivore An animal that eats both plants and other animals. Compare *carnivore; herbivore.*

open system A *system* that exchanges *energy,* matter, and information with other systems. If we look at any system closely enough or wait long enough, we will see such exchanges, so all systems are actually open. Compare *closed system.*

organelle A structure such as a *ribosome* or *mitochondrion* inside the *cell* that performs specific functions.

organic agriculture *Agriculture* that uses no synthetic *fertilizers* or *pesticides* but instead relies on biological approaches such as composting and *biocontrol.* Compare *low-input agriculture.*

organic compound A *compound* made up of carbon *atoms* (and, generally, hydrogen atoms) joined by *covalent bonds* and sometimes including other *elements,* such as nitrogen, oxygen, sulfur, or phosphorus. The unusual ability of carbon to build elaborate molecules has resulted in millions of different organic compounds showing various degrees of complexity.

organic fertilizer A *fertilizer* made up of natural materials (largely the remains or wastes of organisms), including animal manure; crop residues; fresh vegetation, and *compost.* Compare *inorganic fertilizer.*

organic molecule A *molecule* that contains carbon.

overgrazing The consumption by too many animals of plant cover, impeding plant regrowth and preventing the replacement of *biomass.* Overgrazing has been shown to cause a number of impacts, some of which give rise to *positive feedback cycles* that exacerbate damage to *soils,* natural *communities,* and the land's productivity for grazing.

overland flow See *sheet erosion.*

overnutrition A condition of excessive food intake in which people receive more than their daily caloric needs.

ozone hole A thinned ozone concentration in the *ozone layer,* caused by the release of *CFCs* into the *atmosphere.*

ozone layer A portion of the *stratosphere,* roughly 17–30 km (10–19 mi) above sea level, that contains most of the ozone in the *atmosphere.*

paradigm A dominant philosophical and theoretical framework of a *science* or discipline.

parasite An organism that depends on another organism (the *host*) for nourishment or some other benefit while simultaneously harming the host.

parasitism   A relationship in which one organism, the *parasite*, depends on another, the *host*, for nourishment or some other benefit while simultaneously doing the host harm. Compare *mutualism*.

parent material   The base geological material in a particular location.

particulate matter   A solid or liquid particle small enough to be carried aloft into the *atmosphere*.

passive solar energy collection   An approach to solar *energy* collection in which buildings are designed and building materials are chosen to maximize their direct absorption of sunlight in winter, even as they keep the interior cool in the summer. Compare *active solar energy collection*.

peat   A kind of precursor stage to *coal*, produced when organic material that is broken down by *anaerobic* decomposition remains wet, near the surface, and not well compressed

pelagic   Of, relating to, or living between the surface and floor of the ocean. Compare *benthic*.

pesticide   An artificial chemical used to kill insects (insecticide), plants (herbicide), or fungi (fungicide).

photochemical smog   Brown-air *smog* caused by light-driven reactions of *primary pollutants* with normal atmospheric compounds that produce a mix of over 100 different chemicals, ground-level ozone often being the most abundant among them. Compare *industrial smog*.

photoelectric effect   The effect produced when light strikes one of a pair of negatively charged metal plates in a *photovoltaic cell*, causing it to release *electrons*. The electrons are attracted by electrostatic forces to the other plate. The repeated flow of electrons from one plate to the other forms an electrical current, which can be converted into alternating current (AC) and used for residential and commercial electrical power.

photosynthesis   The process by which *autotrophs* produce their own food. Sunlight powers a series of chemical reactions that convert carbon dioxide and water into sugar (glucose), thus transforming low-quality *energy* from the sun into high-quality energy the organism can use. Compare *cellular respiration*.

photovoltaic (PV) cell   A device designed to collect sunlight and convert it to electrical *energy* directly by making use of the *photoelectric effect*.

phylogenetic tree   A treelike diagram that represents the history of divergence of *species*.

physical (mechanical) weathering   *Weathering* that breaks rocks down into smaller particles without triggering a chemical change in the *parent material*. Compare *chemical weathering*.

Pinchot, Gifford (1865–1946)   The first American professionally trained forester. Today, he is the person most closely associated with the *conservation ethic*.

pioneer species   The first *species* to arrive in a terrestrial or aquatic *system* and begin a new *community*.

plate tectonics   The theory that Earth's surface consists of about 15 major tectonic plates, each including some combination of oceanic and continental *crust*. The lightweight, thin crust of rock floats atop a malleable *mantle*, which in turn surrounds a molten heavy *core* made mostly of iron. Earth's internal heat drives *convection currents* that flow in loops in the mantle, pushing the mantle's soft rock cyclically upward (as it warms) and downward (as it cools), like a gigantic conveyor belt. As the mantle material moves, it drags the tectonic plates along its surface edge.

plowpan   A hard layer of *soil* that resists both the infiltration of water and the penetration of roots and that can make cultivation more difficult. It is caused by continual plowing of the same field at the same depth.

point source   A specific spot—such as a factory's smokestacks—where large quantities of pollutants are discharged. Compare *non-point source*.

polar cell   One of a pair of *convection currents* near 30° north and south latitude and in the polar regions, in which air is forced downward, creating arid *climates*. Compare *Ferrel cell; Hadley cell*.

policy   A rule or guideline that directs individual, organizational, or societal behavior.

policy cycle   The full process of creating a *policy*, both before and after enactment, including (1) agenda setting, (2) policy formulation, (3) policy legitimation, and (4) policy implementation.

pollination   An interaction in which one organism (for example, bees) transfers pollen (male sex cells) from one flower to the ova (female cells) of another, fertilizing the female flower, which subsequently grows into a fruit.

polyculture   The planting of many different crops together. Characterizes *traditional agriculture*. Compare *monoculture*.

polymer   A *chemical compound* or mixture of compounds consisting of long chains of repeated *molecules*. Some polymers play key roles in the building blocks of life.

population density   The number of individuals within a population per unit area. Compare *population size*.

population dispersion   See *population distribution*.

population distribution   The spatial arrangement of organisms within a particular area.

population growth curve   A plot of population increase or decrease over time.

population size   The number of individual organisms present at a given time.

positive feedback loop   A *feedback loop* in which output of one type acts as input that moves the *system* in the same direction. The input and output drive the system further toward one extreme or another. Compare *negative feedback loop*.

postindustrial stage   The fourth and final stage of the *demographic transition model*, in which both birth and death rates have fallen to a low level and remain stable there, and populations may even decline slightly. Compare *industrial stage; pre-industrial stage; transition stage*.

potential energy   *Energy* of position. Compare *kinetic energy*.

precautionary principle   The idea that one should not undertake a new action until the ramifications of that action are well understood.

precedent A previous court ruling that serves as a legal guide for later cases.

precipitation The water that condenses out of the *atmosphere* and falls to Earth in droplets or crystals.

predation The process in which one *species* (the *predator*) hunts, tracks, captures, and ultimately kills its *prey*.

predator A member of a *species* that hunts, tracks, captures, and ultimately kills its *prey*.

prediction A specific statement that can be tested directly and unequivocally.

pre-industrial stage The first stage of the *demographic transition model,* characterized by conditions that defined most of human history. In pre-industrial societies, both death rates and birth rates are high. Compare *industrial stage; postindustrial stage; transitional stage.*

preservation ethic An ethic holding that we should protect the natural *environment* in a pristine, unaltered state. Compare *conservation ethic.*

prey A member of a species that is hunted, tracked, captured, and ultimately eaten by a *predator.*

primary consumer A plant-eating animal; an organism that consumes *autotrophs* (primary producers). Primary consumers constitute the second *trophic level* in the *food chain* (below autotrophs). Compare *secondary consumer; tertiary consumer.*

primary extraction The initial drilling and pumping of available *crude oil.* Compare *secondary extraction.*

primary pollutant A hazardous substance, such as soot or carbon monoxide, that is emitted into the *troposphere* in a form that is directly harmful. Compare *secondary pollutant.*

primary producer See *autotroph.*

primary productivity The ability to convert solar energy (sunlight) to *biomass* rapidly. *Coral reefs* are examples of organisms with high primary productivity.

primary succession The development of an ecosystem in an area that has never had a *community.* In terrestrial *systems,* for example, primary succession begins when a bare expanse of rock, sand, or sediment becomes newly exposed to the atmosphere and *pioneer species* arrive. Compare *secondary succession.*

primary treatment A *wastewater* treatment in which contaminants are physically removed. The wastewater flows into a series of tanks in which matter such as sewage solids, grit, and particulate matter are allowed to settle to the bottom. Greases and oils float to the surface and can be skimmed off. Compare *secondary treatment.*

probability A numerical expression of the likelihood that a particular conclusion is true.

prokaryote A typically unicellular organism. The *cells* of prokaryotic organisms lack *organelles* and a *nucleus.* All bacteria are prokaryotes. Compare *eukaryote.*

protein A *macromolecule* made up of long chains of amino acids.

proton A positively charged particle in the nucleus of an *atom.*

proven recoverable reserve The amount of a given *fossil fuel* in a deposit that is technologically and economically feasible to remove under current conditions

qualitative data Information that cannot be expressed using numbers. Compare *quantitative data.*

quantitative data Information that can be expressed using numbers. Compare *qualitative data.*

rainshadow The dry leeward side of a mountain range. Air currents forced upward by mountain ranges cause water vapor to condense and fall as rain on the windward side, leaving the leeward side to receive dry air depleted of its moisture.

random distribution A *population distribution* in which individuals are located haphazardly in space in no particular pattern. This type of distribution can occur when the resources an organism depends on are found throughout an area and other organisms do not exert a strong influence on where members of a population settle. Compare *clumped distribution; uniform distribution.*

rangeland Land used for grazing livestock.

realized niche The portion of the *fundamental niche* that is fully realized (used) by a *species.*

recombinant DNA DNA that has been patched together from the DNA of multiple organisms in an attempt to produce desirable traits (such as rapid growth, disease and pest resistance, or higher nutritional content) in organisms lacking those traits.

recycling The collection of materials that can be broken down and reprocessed to manufacture new items.

Red List An updated list of *species* facing unusually high risks of *extinction.* The list is maintained by the World Conservation Union.

red tide A *harmful algal bloom* consisting of algae that produce reddish pigments that discolor surface waters.

regolith Unconsolidated material derived from rock. Compare *soil.*

regulation A specific rule issued by an *administrative agency,* based on the more broadly written statutory law passed by Congress and enacted by the president.

regulatory taking The deprivation of a property's owner, by means of a law or regulation, of most or all economic uses of that property.

relative humidity The ratio of the water vapor a given volume of air contains to the maximum amount it could contain, for a given temperature.

renewable natural resource A natural resource that is virtually unlimited or that is replenished by the *environment* over relatively short periods of hours to weeks to years. Compare *nonrenewable natural resource.*

replacement fertility The *total fertility rate (TFR)* that maintains a stable population size.

resource management Strategic decision making about who should extract resources and in what ways, so that resources are used wisely and not wasted.

resource partitioning   The process in which *species* adapt to *competition* by evolving to use slightly different resources, or to use their shared resources in different ways, to avoid interference with one another.

restoration ecology   The study of the historical conditions of ecological *communities* as they existed before humans altered them.

revolving door   The movement of powerful officials between the private sector and government agencies.

R horizon   The bottommost layer of *soil* in a typical *soil profile*. Also called *bedrock*. Compare *A horizon; B horizon; C horizon; E horizon; O horizon.*

ribonucleic acid   See *RNA.*

ribosome   An *organelle* that synthesizes proteins.

rill erosion   A type of water *erosion* that takes place when surface water runs along small contours on the surface of the *topsoil*, gradually deepening and widening the contours into rills, or small channels. Compare *gully erosion; sheet erosion; splash erosion.*

risk   The mathematical probability that some harmful outcome (for instance, injury, death, environmental damage, or economic loss) will result from a given action, event, or substance.

risk assessment   The quantitative measurement of *risk* and the comparison of risks involved in different activities or substances.

risk management   Decisions made and strategies created to minimize *risk.*

RNA (ribonucleic acid)   A usually single-stranded *nucleic acid* composed of four nucleotides, each of which contains a sugar (ribose), a phosphate group, and a nitrogenous base. RNA carries the hereditary information for living organisms and is responsible for passing traits from parents to offspring. Compare *DNA.*

rock cycle   The very slow process in which rocks and the minerals that make them up are heated, melted, cooled, broken, and reassembled

r–strategist   A *species* with high *biotic potential* whose members produce a large number of offspring in a relatively short time and do not care for their young after birth. Populations of r–strategists are generally regulated by *density-independent factors.* Compare *K–strategist.*

runoff   The water from *precipitation* that flows into streams, rivers, lakes, and ponds, and (in many cases) eventually to the ocean.

Ruskin, John (1819–1900)   British art critic, poet, and writer who criticized industrialized cities and their pollution of the *environment* and worked to mitigate the effects of industrialization. Ruskin also believed that people had come to value the material benefits the environment could provide but no longer appreciated its spiritual or aesthetic benefits.

salinization   The buildup of salts in surface *soil* layers.

salt   See *ionic compound.*

salt marsh   Flat land that is intermittently flooded by the ocean where the *tide* reaches inland. Salt marshes cover large areas of coastline in temperate regions and are thickly vegetated with grasses, rushes, shrubs, and other herbaceous plants.

sand   *Sediment* consisting of particles 0.005–2.0 mm in diameter. Compare *clay; silt.*

sanitary landfill   A site at which *waste* is buried in the ground or piled up in large mounds, every effort being made to prevent the waste from contaminating the *environment.*

savanna   A *biome* characterized by regions of grasslands interspersed with clusters of acacias and other trees. The savanna biome is found across large stretches of Africa (where it was the ancestral home of our *species*), South America, Australia, India, and other dry tropical regions.

science   A systematic process for learning about the world and testing our understanding of it.

scientific method   A technique for testing ideas with observations that involves several assumptions and a more or less consistent series of interrelated steps.

secondary consumer   An organism that preys on *primary consumers.* Secondary consumers constitute the third *trophic level* in the *food chain* (below *autotrophs* and primary consumers). Compare *tertiary consumer.*

secondary extraction   The extraction of additional *crude oil* by using solvents or by flushing underground rocks with water or steam. Compare *primary extraction.*

secondary pollutant   A hazardous substance produced through the reaction of substances added to the *atmosphere* with chemicals normally found in the atmosphere. Compare *primary pollutant.*

secondary succession   The development of an ecosystem that begins when some event disrupts or dramatically alters an existing *community.* Compare *primary succession.*

secondary treatment   A *wastewater* treatment in which biological means are used to remove further contaminants. The wastewater is stirred up in the presence of *aerobic* bacteria, which degrade organic pollutants in the water. The wastewater then passes to another settling tank, where remaining solids drift to the bottom. Compare *primary treatment.*

second law of thermodynamics   The nature of *energy* will tend to change from a more-ordered state to a less-ordered state; that is, *entropy* increases.

sediment   The eroded remains of rocks.

sedimentary rock   One of the three great groups of rock. A rock formed when dissolved minerals seep through *sediment* layers and act as a kind of glue, crystallizing and binding sediment particles together. Sandstone and shale are examples of common sedimentary rocks. Compare *igneous rock; metamorphic rock.*

seed bank   A storehouse for samples of the world's crop diversity.

service   Work done for others as a form of business.

sex ratio   The proportion of males to females in a population.

sheet erosion   A type of water *erosion* that occurs when surface water—whether from *precipitation* or *irrigation*—flows downhill, washing *topsoil* away in relatively uniform layers. Compare *gully erosion; rill erosion; splash erosion.*

shelterbelt   A row of trees or other tall perennial plants that are planted along the edges of farm fields to break the wind and thereby minimize wind *erosion.*

Silent Spring   A 1962 book by the American scientist and writer Rachel Carson that awakened the American public to the negative ecological and health effects of *pesticides* and industrial chemicals. The title refers to Carson's warning that pesticides might kill so many birds that few would be left to sing in future springs.

silt   *Sediment* consisting of particles 0.002–0.005 mm in diameter. Compare *clay; sand.*

sinkhole   An area where the ground gives way with little warning as a result of subsidence caused by depletion of water from an *aquifer.*

SLOSS (Single Large or Several Small) dilemma   The debate over whether it is better to make reserves large in size and few in number or many in number but small in size. Large *species* that roam great distances benefit more from the "single large" approach to reserve design, whereas insects that live as larvae in small areas may do well in a number of small isolated reserves.

Smith, Adam (1723–1790)   Scottish philosopher known today as the father of classical *economics.* He believed that when people are free to pursue their own economic self-interest in a competitive marketplace, the marketplace will behave as if guided by "an invisible hand" that ensures their actions will benefit society as a whole.

smog   A fog polluted with smoke and chemical fumes.

social science   A discipline that studies human interactions and institutions. Compare *historical science; natural science.*

soil porosity   A measure of the size of spaces between *soil* particles.

soil profile   The cross-section of a *soil* as a whole, from the surface to the *bedrock.*

soil   A complex plant-supporting *system* consisting of disintegrated rock, organic matter, air, water, *nutrients,* and microorganisms.

source reduction   The reduction of the amount of material that enters the *waste* stream to avoid the costs of disposal and *recycling,* help conserve resources, minimize pollution, and save consumers and businesses money.

speciation   The process by which new *species* are generated.

species   A population or group of populations of a particular type of organism, whose members share certain characteristics and can breed freely with one another and produce fertile offspring.

Species at Risk Act (SARA)   Canada's national endangered *species* legislation, enacted in 2002. To avoid the kind of controversy that has occurred in the United States over the *Endangered Species Act,* Canada's law stresses cooperation with landowners and provincial governments.

species coexistence   A relationship between two *species* in which no outside influence disturbs the relationship and the species live side by side in a certain ratio of *population sizes.* Compare *competitive exclusion.*

species diversity   The number of *species* in the world or in a particular region.

splash erosion   A type of water *erosion* that occurs when rain striking the *soil* surface breaks aggregates into smaller sizes. Soil particles are released in the process and fill in gaps between the remaining clumps, decreasing the soil's ability to absorb water. Compare *sheet erosion; rill erosion; gully erosion.*

sprawl   The phenomenon of urban and suburban spread across the landscape.

stabilizing selection   Selection that acts to produce intermediate traits, in essence preserving the status quo. Compare *directional selection; disruptive selection.*

steady-state economy   An *economy* that does not grow or shrink but remains stable.

stratosphere   The layer of the *atmosphere* above the *troposphere* and below the *mesosphere;* it extends from 11 km (7 mi) to 50 km (31 mi) above sea level. See also *thermosphere.*

strip cropping   The practice of planting alternating bands of different types of crops across a slope.

strip-mining   The use of heavy machinery to remove huge amounts of earth to expose *coal* or minerals, which are mined out directly.

subduction   The process by which denser ocean *crust* slides beneath lighter continental crust.

subsidy   A government incentive (a giveaway of cash or publicly owned resources, or a tax break) intended to encourage a particular activity. Compare *green tax.*

subsistence agriculture   A form of *traditional agriculture* in which the members of a farming family produce only enough food for themselves and do not make use of large-scale *irrigation, fertilizers,* or large teams of laboring animals. Compare *intensive traditional agriculture.*

subsistence economy   A survival *economy,* one in which people meet most or all of their daily needs directly from nature and do not purchase or trade for most of life's necessities.

succession   Change in the composition and structure of a *community* as the available competing organisms respond to and modify the *environment.* See *primary succession; secondary succession.*

Superfund   A program administered by the *Environmental Protection Agency* in which experts identify sites polluted with hazardous chemicals, protect *groundwater* near these sites, and clean up the pollution.

supply   The amount of a product offered for sale at a given price. Compare *demand.*

surface impoundment   A *hazardous waste* disposal method in which a shallow depression is dug and lined with impervious material, such as clay. Water containing small amounts of hazardous waste is placed in the pond and allowed to evaporate, leaving a residue of solid hazardous waste on the bottom. Compare *deep-well injection.*

sustainable agriculture   *Agriculture* that does not deplete *soils* faster than they form.

sustainable development   The use of *renewable natural resources* and *nonrenewable natural resources* in a manner that satisfies our current needs without compromising the future availability of resources.

symbiosis A *parasitic* or *mutualistic* relationship between different *species* of organisms that live closely together.

sympatric speciation The formation of *species* together in a single location without geographical separation. Compare *allopatric speciation*.

synergistic effect An interactive effect of *toxicants* that is more than or different from the simple sum of their constituent effects.

system A network of relationships among a group of parts, elements, or components that interact with and influence one another through the exchange of *energy*, matter, and/or information.

taiga A *biome* of northern coniferous forest that stretches in a broad band across much of Canada, Alaska, Russia, and Scandinavia. It consists of a limited number of *species* of evergreen trees, such as black spruce, that dominate large regions of forests interspersed with occasional bogs and lakes.

takings clause The Fifth Amendment to the U.S. Constitution; it ensures, in part, that private property shall not "be taken for public use without just compensation."

temperate deciduous forest A *biome* consisting of mid-latitude forests characterized by broad-leafed trees that lose their leaves each fall and remain dormant during winter. These forests occur in areas where *precipitation* is spread relatively evenly throughout the year: much of Europe, eastern China, and eastern North America. The biome generally consists of far fewer tree species than are found in *tropical rainforests*.

temperate grassland A *biome* where temperature differences between winter and summer are more extreme and there is less rainfall than in *temperate deciduous forests*. The amount of *precipitation* can support grasses more easily than trees.

temperate rainforest A *biome* in the Pacific Northwest of North America consisting of tall coniferous trees such as cedars, spruces, and Douglas fir; the forest interior is darkly shaded and damp.

temperature inversion A departure from the normal temperature distribution in the *atmosphere*, in which a pocket of relatively cold air occurs near the ground, with warmer air above it. The cold air, denser than the air above it, traps pollutants near the ground and causes a buildup of *smog*.

teratogen A *toxicant* that causes harm to the unborn.

terracing The cutting of level platforms, sometimes with raised edges, into steep hillsides to contain water from *irrigation* and *precipitation*. Terracing transforms slopes into series of steps like a staircase, enabling farmers to cultivate hilly land without losing as much *soil* to water *erosion* as they would otherwise.

tertiary consumer A *predator* that eats at higher *trophic levels* than *carnivores* and *omnivores*. An example is a hawk that eats rodents that have eaten herbivorous grasshoppers. Compare *primary consumer*; *secondary consumer*.

theory A widely accepted, well-tested explanation of one or more cause-and-effect relationships that has been extensively validated by a great amount of research. Compare *hypothesis*; *model*.

thermal inversion See *temperature inversion*.

thermogenic gas *Natural gas* formed deep in the Earth as geothermal heating separates *hydrocarbons* from organic material. Compare *biogenic gas*.

thermosphere The topmost layer of the *atmosphere* above the *mesosphere*; it extends from 90 km (56 mi) to 500 km (300 mi) above sea level. See also *stratosphere*; *troposphere*.

Thoreau, Henry David (1817–1862) American author, poet, and philosopher who viewed all of nature as divine. His book *Walden*, recording his observations and thoughts while he lived at Walden Pond away from the bustle of urban Massachusetts, remains a classic of American literature.

threshold dose The amount of a *toxicant* at which it begins to affect a population of test animals. Compare *effective dose–50%*; *lethal dose–50%*.

tide The periodic rise and fall of the ocean's height at a given location, caused by the gravitational pull of the moon and sun.

topsoil That portion of the *soil* that is most nutritive for plants and is thus of the most direct importance to ecosystems and to *agriculture*.

total fertility rate (TFR) The average number of children born per female member of a population during her lifetime.

toxicant A substance that acts as a poison on humans or wildlife.

toxicity The degree of harm a substance can inflict.

toxicology The *science* that examines the effects of poisonous chemicals and other agents on humans and other organisms.

traditional agriculture Biologically powered *agriculture*, in which human and animal muscle power, along with hand tools and simple machines, perform the work of cultivating, harvesting, storing, and distributing crops. See *intensive traditional agriculture*; *subsistence agriculture*. Compare *industrialized agriculture*.

transboundary International.

transcendentalism A philosophy that views nature as a direct manifestation of the divine, emphasizing the soul's oneness with nature and with God. It flourished in the first half of the nineteenth century; some proponents include *Ralph Waldo Emerson* and *Henry David Thoreau*.

transgenic organism An organism that contains *DNA* from another *species*.

transitional stage The second stage of the *demographic transition model*, which occurs during the transition from the *pre-industrial stage* to the *industrial stage*. It is characterized by declining death rates but continued high birth rates. See also *postindustrial stage*.

transmissible disease See *infectious disease*.

transpiration The release of water vapor by plants through their leaves.

trophic level Rank in the feeding hierarchy.

tropical dry forest   A *biome* that consists of deciduous trees and occurs where wet and dry seasons each span about half the year. Tropical dry forests are widespread in India, Africa, South America, and northern Australia.

tropical rainforest   A *biome* characterized by year-round rain and uniformly warm temperatures. Tropical rainforests are found in Central America, South America, southeast Asia, west Africa, and other tropical regions. They have dark, damp interiors; lush vegetation; and highly diverse *biotic communities*, with greater numbers of *species* of insects, birds, amphibians, and various other animals than any other biome.

tropopause   A boundary at the top of the *troposphere* where temperatures suddenly cease to decline with altitude.

troposphere   The bottommost layer of the *atmosphere*; it extends to 11 km (7 mi) above sea level. See also *mesosphere; stratosphere; thermosphere.*

tundra   A *biome* that is nearly as dry as *desert* but is located at very high latitudes along the northern edges of Russia, Canada, and Scandinavia. Extremely cold winters with little daylight and moderately cool summers with lengthy days characterize this landscape of *lichens* and low scrubby vegetation.

turbidity   A physical characteristic of water denoting the density of suspended particles in a water sample.

unconfined aquifer   A water-bearing, porous layer of rock, sand, or gravel that lies atop a less permeable substrate. The water in an unconfined aquifer is not under pressure because there is no impermeable upper layer to confine it. Compare *confined aquifer.*

undernutrition   A condition of insufficient *nutrition* in which people receive less than 90% of their daily caloric needs.

uniform distribution   A *population distribution* in which individuals are evenly spaced out. This type of distribution can occur when individuals hold territories or otherwise compete for space. Compare *clumped distribution; random distribution.*

upwelling   In the ocean, the flow of cold, deep water toward the surface. Upwelling occurs in areas where surface *currents* diverge. Compare *downwelling.*

variable   In an *experiment,* a condition that can change. See *dependent variable* and *independent variable.*

vector   An organism that transfers the pathogen to the host in an *infectious disease.*

volatile organic compound (VOC)   One of a large group of potentially harmful organic chemicals used in industrial processes.

warm front   The boundary where a mass of warm air displaces a mass of colder air. Compare *cold front.*

waste   Any unwanted product that results from a human activity or process.

waste-to-energy (WTE) facility   An incinerator that uses heat from the furnace to boil water to create steam that drives electricity generation or that fuels heating systems.

wastewater   Any water that is used in households, businesses, industries, or public facilities and is drained or flushed down pipes, as well as the polluted *runoff* from streets and storm drains.

waterlogging   The saturation of *soil* by water, in which the *water table* is raised to the point that water bathes plant roots. Waterlogging deprives roots of access to gases, essentially suffocating them and eventually damaging or killing the plants.

water mining   The withdrawal of water from an *aquifer* faster than the aquifer can be recharged.

watershed   All the land from which water drains into a river.

water table   The upper limit of *groundwater* held in an *aquifer.*

weather   The local physical properties of the *troposphere,* such as temperature, pressure, humidity, cloudiness, and wind, over relatively short time periods. Compare *climate.*

weathering   The physical, chemical, and biological processes that break down rocks and minerals, turning large particles into smaller particles.

wetland   Land whose *soil* contains much moisture, such as a swamp, marsh, bog, or river floodplain.

wind farm   A collection of *wind turbines.*

wind turbine   A mechanical assembly that converts the wind's *kinetic energy,* or energy of motion, into electrical energy.

wise use movement   A loose confederation of individuals and groups that coalesced in the 1980s and 1990s as a response to the increasing success of environmental advocacy. The movement favors extracting more resources from public lands, obtaining greater local control of lands, and obtaining greater motorized recreational access to public lands.

worldview   A way of looking at the world that reflects a person's (or a group's) beliefs about the meaning, purpose, operation, and essence of the world.

zooxanthellae   *Symbiotic* algae that inhabit the bodies of corals and provide them with nourishment.

Photo Credits

Unit Opening Photos: Unit I Joel W. Rogers/CORBIS **Unit II** Natalie Fobes/CORBIS

Chapter 1 opening photo NASA/Johnson Space Center **1.2b** Charles O'Rear/CORBIS **1.3a** Art Resource **1.3b** Bettman/CORBIS, **1.4** John N. Smith **1.5** George Konig/Hulton Archive Photos **1.7** Joel W. Rogers/CORBIS **1.8a** George Bukenhofer **1.8b** Reuters NewMedia Inc./CORBIS **1.12** CORBIS **1.13** Steven Zazlowski/Peter Arnold, Inc. **1.14** Reuters **1.15** AFP/CORBIS **1.16** Justin Sullivan/Getty Images **1.17** Jon Hrusa/AP Photo **The Science behind the Story: Easter Island** Richard T. Nowitz/CORBIS

Chapter 2 opening photo Reuters NewMedia Inc./CORBIS **Central Case** The Gundjehmi Aboriginal Corporation **2.2a, b** The Gundjehmi Aboriginal Corporation **2.5** Library of Congress **2.6** CORBIS **2.7** CORBIS **2.8** Bettmann/CORBIS **2.10** AP/Wide World Photos **2.13** AFP/CORBIS **2.14** Larry Lee Photography/CORBIS **2.15a** Konrad Wothe/Minden Pictures **2.15b** Bill Hatcher/National Geographic Image Collection **2.15c** Bruce Forster/The Image Bank **2.15d** CORBIS **2.15e** Charles O'Rear/CORBIS **2.15f, g** Frans Lanting/Minden Pictures

Chapter 3 opening photo CORBIS **Central Case** Denis Poroy/Associated Press **3.2** Lori Saldana **3.3** Annie Griffiths Belt/CORBIS **3.6** Kevin Fleming/CORBIS **3.7a** Bettmann/CORBIS **3.7b** Museum of History & Industry/CORBIS **3.7c** University of Washington Libraries **3.8** Erich Hartmann/Magnum Photos, Inc. **3.9** Bettmann/CORBIS **3.10a** CWH, Associated Press **3.10b** Binod Joshi/Associated Press **3.11** Bettmann/CORBIS **3.13** Museum of the City of New York/CORBIS **3.14** Charles Mauzy/CORBIS **3.18** Lori Saldana **The Science behind the Story: Spotting Sewage via Satellite** NASA

Chapter 4 opening photo AFP/CORBIS **Central Case** Exxon Corporation **4.5a** Digital Vision/Picture Quest **4.11** Dorling Kindersley **4.13 left** Daniel Zheng/CORBIS **4.13 right** Anne-Marie Weber/Taxi **4.17** Jack Dykinga/Stone **4.18a** Ken MacDonald/SPL/Photo Researchers, Inc. **4.18b** Woods Hole Oceanographic Institute **4.19** Chip Clark **4.20** Dorling Kindersley **4.22** European Space Agency **4.23a** Bishop Museum, Honolulu **4.23b** PhotoDisc **The Science behind the Story: Student Chemist** Mark Burrell **The Science behind the Story: Isotopes** PhotoDisc

Chapter 5 opening photo Michael & Patricia Fogden/CORBIS **Central Case** Michael & Patricia Fogden/CORBIS **5.1a** Steve Winter/National Geographic Image Collection **5.1b** Michael Fogden/DRK Photo **5.1c** Michael & Patricia Fogden/CORBIS **5.1d** Buddy Mays/CORBIS **5.5** The National History Museum, London, painting by J. Sibbick **5.6** Chip Clark **5.8a** Kennan Ward/CORBIS **5.8b** Art Wolfe/The Image Bank **5.8c** PhotoDisc Green **5.16** Michael & Patricia Fogden/CORBIS **5.18a** Peter Johnson/CORBIS **5.18b** Tom Brakefield/CORBIS **5.18c** Michael & Patricia Fogden/CORBIS **5.21** Anthony Bannister, Gallo Images/CORBIS **5.22** Michael & Patricia Fogden/CORBIS **5.25** F. Nibling/USGS **5.26** Matthias Clamer/Stone **5.27** Gary Braasch/CORBIS **The Science behind the Story: Mass Extinction** Benjamin Cummings

Chapter 6 opening photo NASA **Central Case** Owen Franken/CORBIS **6.9a** Pat O'Hara/CORBIS **6.10a** Philip Gould/CORBIS **6.11a** Charles Mauzy/CORBIS **6.12a** David Samuel Robbins/CORBIS **6.13a** O. Alamany & E. Vicens/CORBIS **6.14a** Wolfgang Kaehler/CORBIS **6.15a** Joe McDonald/CORBIS **6.16a** Darrell Gulin/CORBIS **6.17a** Liz Hymans/CORBIS **6.18a** Charles Mauzy/CORBIS **6.20 all** Dorling Kindersley **6.24** Reuters NewMedia Inc./CORBIS **The Science behind the Story: Biosphere 2** Roger Ressmeyer/CORBIS

Chapter 7 opening photo Robert Semeniuk/CORBIS **Central Case** Louise Gubb/The Image Works **7.1** Reuters NewMedia Inc./CORBIS **7.8 top left** Galen Rowell/CORBIS **7.8 bottom right** Macduff Everton/CORBIS **7.11c** SETBOUN/CORBIS **7.14** David Turnley/CORBIS **7.17b** Liba Taylor/CORBIS **7.21** Tiziana and Gianni Baldizzone/CORBIS **7.25a** Elie Bernager/Stone **7.25b** Ed Kashi/IPN/AURORA **7.27** Getty Images

Chapter 8 opening photo Stephanie Maze/CORBIS **Central Case** Joanna

B. Pinneo/AURORA **8.9** Dorling Kindersley **8.11a** Barry Runk/Stan/Grant Heilman Photography **8.11b** Grant Heilman/Grant Heilman Photography **8.11c, d** U.S. Department of Agriculture **8.12** Yevgeny Khaldei/CORBIS **8.14** Library of Congress **8.15** Ted Spiegel/CORBIS **8.16a** Sylvan Wittwer/Visuals Unlimited **8.16b** Kevin Horan/Stone **8.16c** Ron Giling/Peter Arnold **8.16d** Keren Su/Stone **8.16e** Yann Arthus-Bertrand/CORBIS **8.16f** U.S. Department of Agriculture **8.17a** PhotoDisc Blue **8.17b** Carol Cohen/CORBIS **8.18** Richard Hamilton Smith/CORBIS **8.22** W. Perry Conway/CORBIS **8.23** Natalie Fobes/CORBIS **The Science behind the Story: Overgrazing and Fire Suppression** Malpai Borderlands Group

Chapter 9 opening photo Steve Satushek/Image Bank **Central Case** Macduff Everton/CORBIS **9.2** Alexandra Avakian/CORBIS **9.4a** Jack Dykinga/Image Bank **9.4b** Barry Runk/Stan/Grant Heilman Photography **9.7** Don Herbison-Evans **9.9** PhotoDisc **9.10** Bob Rowan, Progressive Image/CORBIS **9.14a** Native Seeds/SEARCH **9.14b** Hal Fritts, Native Seeds/SEARCH **9.16** James A. Sugar/CORBIS **9.20** Aqua Bounty Farms/AP **9.21** Martin Bourque **The Science behind the Story: Native Maize** Peg Skorpinski

Chapter 10 opening photo Howard K. Suzuki **Central Case** Howard K. Suzuki **10.1** Joel W. Rogers/CORBIS **10.2** Peg Skorpinski **10.3** Bettmann/CORBIS **10.4** Bettmann/CORBIS **10.12** Steve Helber/AP/Wide World Photos **The Science behind the Story: Bisphenol A** both From Hunt, PA, KE Koehler, M Susiarjo, CA Hodges, A Ilagan, RC Voight, S Thomas, BF Thomas and TJ Hassold. 2003. Bisphenol A exposure causes meiotic aneuploidy in the female mouse. *Current Biology 13: 546–553* **The Science behind the Story: Pesticides and Child Development** Elizabeth A. Guillette

Chapter 11 opening photo Photodisc **Central Case** Hulton Archive Photos **11.4** Norbert Rosing/National Geographic **11.11a** NASA Earth Observatory **11.11b** Bettmann/CORBIS **11.11c** Mark E. Gibson/CORBIS **11.14b** Pittsburgh Post-Gazette **11.15a** Allen Russell/Index Stock Imagery **11.17b** NASA/Goddard Space Flight Center **11.19** Erik Schaffer/Ecoscene **11.21a** David Turnley/CORBIS **The Science behind the Story: Acid Rain** Will & Deni McIntyre/CORBIS

Chapter 12 opening photo Rossi, Guido Alberto/Image Bank **Central Case** Adam Woolfitt/CORBIS **12.5a** Ted Spiegel/CORBIS **12.5b** David Whillas, CSIRO Atmospheric Research **12.10a** Macduff Everton/CORBIS **12.10b** Michael Maslan Historic Photographs/CORBIS **12.15** Bettmann/CORBIS **12.16** Simon Fraser/Photo Researchers, Inc. **12.17** CORBIS **12.19** Robert Brenner/PhotoEdit **12.21a** Yann Arthus-Bertrand/CORBIS **12.21b** Kevin Fleming/CORBIS **The Science behind the Story: Scientists Use Pollen** Lowell Georgia/CORBIS

Chapter 13 opening photo Harvey Lloyd/Taxi **Central Case** Jeff Hunter/The Image Bank **13.07** Alcoa Technical Center **13.10** Laguna Design/Science Photo Library/Photo Researchers, Inc. **13.11** Bruce Robison/Minden Pictures **13.12** Ralph A. Clevenger/CORBIS **13.13a** Jeffrey L. Rotman/CORBIS **13.14a** Stephen Frink/CORBIS **13.14b** NOAA **13.14c** CORBIS **13.15b** Carol Cohen/CORBIS **13.16** Kevin Fleming/CORBIS **13.17** Mark A. Johnson/CORBIS **13.18** Nik Wheeler/CORBIS **13.20 left** David M. Phillips/Visuals Unlimited **13.20 right** Bill Bachman/Photo Researchers, Inc. **13.26** Keith Meyers/New York Times

Chapter 14 opening photo Lester Lefkowitz/CORBIS **Central Case** Galen Rowell/CORBIS **14.6a** William Manning/CORBIS **14.6b** Onne van der Wal/CORBIS **14.6c** Hanan Isachar/CORBIS **14.6d** Frans Lanting/Minden Pictures **14.9** Stapleton Collection, UK/Bridgeman Art Library **14.10b** High Country News **14.12a** Ian Berry/Magnum Photos **14.12b** Reuters NewMedia Inc./CORBIS **14.13a** Gilles Saussier/Liaison **14.13b** NASA Visible Earth **14.14** Jim Tuten/Black Star **14.18b** David Muench/CORBIS **14.18c** Andrew Brown, Ecoscene/CORBIS

Chapter 15 opening photo Maurice Hornocker **Central Case** Maurice Hornocker **15.1** AP/Wide World Photos **15.13** Natalie Fobes/CORBIS **15.17** Corbis Sygma **15.18** Tom & Pat Leeson/Photo Researchers, Inc. **The Science behind the Story: Amphibian Diversity** S. D. Biju **The Science behind the Story: Island Biogeography** Stock Connection, Inc./Alamy

Selected Sources and References for Further Reading

Chapter 1

Bahn, Paul, and John Flenley. 1992. *Easter Island, Earth island*. Thames and Hudson, London.

Bowler, Peter J. 1992. *The Norton history of the environmental sciences*. W.W. Norton, New York.

Ehrlich, Paul. 1968. *The population bomb*. 1997 reprint, Buccaneer Books, Cutchogue, New York.

Gardner, Gary. 2001. "Accelerating the shift to sustainability." Pp. 189–206 in *State of the World 2001*, Worldwatch Institute, Washington, D.C., and W.W. Norton, New York.

Goudie, Andrew. 2000. *The human impact on the natural environment*, 5th ed. MIT Press, Cambridge, Massachusetts.

Hardin, Garrett. 1968. The tragedy of the commons. *Science* 162: 1243–1248.

Katzner, Donald W. 2001. *Unmeasured information and the methodology of social scientific inquiry*. Kluwer, Boston.

Kuhn, Thomas S. 1962. *The structure of scientific revolutions*. 2nd ed., 1970. University of Chicago Press, Chicago.

Lomborg, Bjorn. 2001. *The skeptical environmentalist: Measuring the real state of the world*. Cambridge University Press, Cambridge.

Malthus, Thomas R. *An essay on the principle of population*. 1983 ed., Penguin USA, New York.

Ponting, Clive. 1991. *A green history of the world: the environment and the collapse of great civilizations*. Penguin Books, New York.

Popper, Karl R. 1959. *The logic of scientific discovery*. Hutchinson, London.

Porteous, Andrew. 2000. *Dictionary of environmental science and technology*, 3rd ed. John Wiley & Sons, Hoboken, New Jersey.

Sagan, Carl. 1997. *The demon-haunted world: Science as a candle in the dark*. Ballantine Books, New York.

Schneiderman, Jill S., ed. 2000. *The Earth around us: Maintaining a livable planet*. W.H. Freeman, New York.

Siever, Raymond. 1968. Science: observational, experimental, historical. *American Scientist* 56: 70–77.

Valiela, Ivan. 2001. *Doing science: Design, analysis, and communication of scientific research*. Oxford University Press, Oxford.

Worldwatch Institute. 2002. *Vital Signs 2002*. W.W. Norton, New York.

Chapter 2

Balmford, Andrew, et al. 2002. Economic reasons for conserving wild nature. *Science* 297: 950–953.

Barbour, Ian G. 1992. *Ethics in an age of technology*. Harper Collins, San Francisco.

Brown, Lester. 2001. *Eco-economy: Building an economy for the Earth*. Earth Policy Institute and W.W. Norton, New York.

Carson, Richard T., Leanne Wilks, and David Imber. 1994. Valuing the preservation of Australia's Kakadu Conservation Zone. *Oxford Economic Papers* 46: 727–749.

Cole, Luke W., and Sheila R. Foster. 2001. *From the ground up: Environmental racism and the rise of the environmental justice movement*. New York University Press, New York.

Costanza, Robert, et al. 1997. The value of the world's ecosystem services and natural capital. *Nature* 387: 253–260.

Costanza, Robert, et al. 1997. *An introduction to ecological economics*. St. Lucie Press, Boca Raton, Florida.

Daily, Gretchen. 1997. *Nature's services: Societal dependence on natural ecosystems*. Island Press, Washington, D.C.

Daly, Herman E. 1996. *Beyond growth*. Beacon Press, Boston.

Elliot, Robert, and Arran Gare, eds. 1983. *Environmental philosophy: A collection of readings*. Pennsylvania State University Press, University Park.

Fox, Stephen. 1985. *The American conservation movement: John Muir and his legacy*. University of Wisconsin Press, Madison.

Goodstein, Eban. 1999. *The tradeoff myth: Fact and fiction about jobs and the environment*. Island Press, Washington, D.C.

Gundjehmi Aboriginal Corporation. "Welcome to the Mirrar site." http://www.mirrar.net/.

Hawken, Paul, Amory Lovins, and L. Hunter Lovins. 1999. *Natural capitalism*. Little, Brown, and Co., Boston.

Kolstad, Charles D. 2000. *Environmental economics*. Oxford University Press, Oxford.

Leopold, Aldo. 1949. *A Sand County almanac, and sketches here and there*. Oxford University Press, New York.

Nash, Roderick F. 1989. *The rights of nature*, University of Wisconsin Press, Madison.

Nash, Roderick F. 1990. *American environmentalism: Readings in conservation history*. 3rd ed. McGraw-Hill, New York.

O'Neill, John O., R. Kerry Turner, and Ian J. Bateman, eds. 2001. *Environmental ethics and philosophy*. Elgar, Cheltenham, U.K.

Pearson, Charles S. 2000. *Economics and the global environment*. Cambridge University Press, Cambridge.

Singer, Peter, ed. 1993. *A companion to ethics*. Blackwell Publishers, Oxford.

Smith, Adam. 1776. *An inquiry into the nature and causes of the wealth of nations*. 1993 ed., Oxford University Press, Oxford.

Sterba, James P., ed. 1995. *Earth ethics: Environmental ethics, animal rights, and practical applications.*" Prentice Hall, Upper Saddle River, New Jersey.

Turner, R. Kerry, David Pearce, and Ian Bateman. 1993. *Environmental economics: an elementary introduction*. Johns Hopkins Univ. Press, Baltimore.

Wenz, Peter S. 2001. *Environmental ethics today*. Oxford University Press, Oxford.

Chapter 3

Clark, Ray, and Larry Canter. 1997. *Environmental policy and NEPA: Past, present, and future*. St. Lucie Press, Boca Raton, Florida.

Fogleman, Valerie M. 1990. *Guide to the National Environmental Policy Act*. Quorum Books, New York.

Fox, Stephen. 1985. *The American conservation movement: John Muir and his legacy*. University of Wisconsin Press, Madison.

French, Hilary. 2000. "Environmental treaties gain ground." Pp. 134–135 in *Vital Signs 2000*. Worldwatch Institute, Washington, D.C., and W.W. Norton, New York.

Green Scissors. *Green Scissors 2002: Cutting wasteful and environmentally harmful spending*. Friends of the Earth.

Herzog, Lawrence A. 1990. *Where North meets South: Cities, space, and politics on the U.S.-Mexico border*. Center for Mexican-American Studies, University of Texas at Austin.

Kraft, Michael E. 2001. *Environmental policy and politics*. 2nd ed. Addison-Wesley, Longman, New York.

Kubasek, Nancy, and Gary Silverman. 2000. *Environmental law*, 3rd ed. Prentice Hall, Upper Saddle River, New Jersey.

Myers, Norman, and Jennifer Kent. 2001. *Perverse subsidies: How misused tax dollars harm the environment and the economy*. Island Press, Washington, D.C.

The National Environmental Policy Act of 1969, as amended (Pub. L. 91–190, 42 U.S.C. 4321–4347, January 1, 1970, as amended by Pub. L. 94-52, July 3, 1975, Pub. L. 94–83, August 9, 1975, and Pub. L. 97–258, § 4(b), Sept. 13, 1982). http://ceq.eh.doe.gov/nepa/regs/nepa/nepaeqia.htm.

Shafritz, Jay M. 1993. *Dictionary of American government and politics*. Concise edition. Harper Collins, New York.

Southwest Center for Environmental Research and Policy, and San Diego State University. 2000. "Tijuana River Watershed Atlas Project, 1998–2000." Accessed, 15 July, 2002, at: http://typhoon.sdsu.edu/twrp/tjatlas.html.

Steel, Brent S., Richard L. Clinton, and Nicholas P. Lovrich. 2002. *Environmental politics and policy*. McGraw-Hill, New York.

Turner, R. Kerry, David Pearce, and Ian Bateman. 1993. *Environmental economics: an elementary introduction*. Johns Hopkins University Press, Baltimore.

United States Congress. House. H.R. 3378. 2000. "The Tijuana River Valley Estuary and Beach Sewage Cleanup Act of 2000."

Vig, Norman J., and Michael E. Kraft, eds. 2000. *Environmental policy: New directions for the twenty-first century*. 4th ed. CQ Press, Congressional Quarterly, Inc., Washington, D.C.

Wilkinson, Charles F. 1992. *Crossing the next meridian: Land, water, and the future of the West*. Island Press, Washington, D.C.

Chapter 4

Alaska Department of Environmental Conservation. 1993. *The Exxon Valdez oil spill: Final report, State of Alaska response*. June, 1993.

Allen, K.C., and D.E.G. Briggs, eds. 1989. *Evolution and the fossil record*. John Wiley & Sons, Hoboken, New Jersey.

Atlas, Ronald M. 1995. Petroleum biodegradation and oil spill bioremediation. *Marine Pollution Bulletin* 31: 178–182.

Atlas, Ronald M., and Carl E. Cerniglia. 1995. Bioremediation of petroleum pollutants. *Bioscience* 45: 332–338.

Berry, R. Stephen. 1991. *Understanding energy: Energy, entropy and thermodynamics for every man*. World Scientific Publishing Co.

Bragg, James R., et al. 1994. Effectiveness of bioremediation for the *Exxon Valdez* oil spill. *Nature* 368: 413–418.

Campbell, Neil A., and Jane B. Reece. 2002. *Biology*. 6th Edition. Benjamin Cummings, San Francisco.

Darwin, Charles. 1859. *The origin of species by means of natural selection*. John Murray, London.

Endler, John A. 1986. *Natural selection in the wild*. Monographs in Population Biology 21, Princeton University Press, Princeton.

Fenchel, Tom. 2003. *Origin and early evolution of life*. Oxford University Press, Oxford.

Fortey, Richard. 1998. *Life: A natural history of the first four billion years of life on Earth*. Alfred Knopf, New York.

Freeman, Scott, and Jon C. Herron. 1998. *Evolutionary analysis*. Prentice Hall, Upper Saddle River, New Jersey.

Futuyma, Douglas J. 1998. *Evolutionary biology*, 3rd ed. Sinauer Associates, Sunderland, Mass.

Manahan, Stanley E. 2000. *Environmental chemistry*. 7th ed. Lewis Publishers, CRC Press, Boca Raton, Florida.

McMurry, John E. *Organic chemistry*, 6th ed. Brooks/Cole,

National Response Team. *NRT fact sheet: Bioremediation in oil spill response*. U.S. EPA, www.epa.gov/oilspill/pdfs/biofact.pdf.

Nealson, Kenneth H. 2003. Harnessing microbial appetites for remediation. *Nature Biotechnology* 21: 243–244.

Ridley, Mark. 1996. *Evolution*, 2nd ed. Blackwell Science, Cambridge, Mass.

United States Environmental Protection Agency. 2003. "Oil program." http://www.epa.gov/oilspill/

Ward, Peter D., and Donald Brownlee. 2000. *Rare Earth: Why complex life is uncommon in the universe*. Copernicus, New York.

Wassenaar, Leonard I., and Keith A. Hobson. 1998. Natal origins of migratory monarch butterflies at wintering colonies in Mexico: New isotopic evidence. *Proceedings of the National Academy of the United States of America* 95: 15436–15439.

Williams, George C. 1966. *Adaptation and natural selection*. Princeton University Press, Princeton.

Withgott, Jay. 1999. Stalking the wild isotope. *HMS Beagle* 64, 15 October 1999. http://news.bmn.comhmsbeagle/64/notes/profile

Withgott, Jay. 2003. Student chemist gets plants to do the dirty work. *ChemMatters* Apr 2003: 4–6.

Chapter 5

Alvarez, Luis W., et al. 1980. Extraterrestrial cause for the Cretaceous-Tertiary extinction. *Science* 208: 1095–1108.

Barbour, Michael G., et al. 1998. *Terrestrial plant ecology*, 3rd ed. Benjamin/Cummings, Menlo Park, Calif.

Begon, Michael, and Martin Mortimer. 1986. *Population ecology: A unified study of animals and plants*, 2nd ed. Blackwell Scientific, Oxford.

Campbell, Neil A., and Jane B. Reece. 2002. *Biology*. 6th Edition. Benjamin Cummings, San Francisco.

Clark K.L., et al. 1998. Cloud water and precipitation chemistry in a tropical montane forest, Monteverde, Costa Rica. *Atmospheric Environment* 32: 1595–1603.

Crump, L. Martha, et al. 1992. Apparent decline of the golden toad: Underground or extinct? *Copeia* 1992: 413–420.

Darwin, Charles. 1859. *The origin of species by means of natural selection*. John Murray, London.

Freeman, Scott, and Jon C. Herron. 1998. *Evolutionary analysis*. Prentice Hall, Upper Saddle River, New Jersey.

Futuyma, Douglas J. 1998. *Evolutionary biology*, 3rd ed. Sinauer Associates, Sunderland, Mass.

Krebs, Charles J. 2001. *Ecology: The experimental analysis of distribution and abundance*. 5th ed. Benjamin Cummings, San Francisco.

Lawton, Robert O., et al. 2001. Climatic impact of tropical lowland deforestation on nearby montane cloud forests. *Science* 294: 584–587.

Molles, Manuel C. Jr. 2002. *Ecology: Concepts and applications*. 2nd ed. McGraw-Hill, Boston.

Nadkarni, Nalini M., and Nathaniel T. Wheelwright, eds. 2000. *Monteverde: Ecology and conservation of a tropical cloud forest*. Oxford University Press, New York.

Pounds, J. Alan, et al. 1997. Tests of null models for amphibian declines on a tropical mountain. *Conservation Biology* 11: 1307–1322.

Pounds, J. Alan. 2001. Climate and amphibian declines. *Nature* 410: 639.

Pounds, J. Alan, and Martha L. Crump. 1994. Amphibian declines and climate disturbance: The case of the golden toad and the harlequin frog. *Conservation Biology* 8: 72–85.

Pounds, J. Alan, Michael P.L. Fogden, and John H. Campbell. 1999. Biological response to climate change on a tropical mountain. *Nature* 398: 611–615.

Powell, James L. 1998. *Night comes to the Cretaceous: Dinosaur extinction and the transformation of modern geology*. W.H. Freeman, New York.

Raup, David M. 1991. *Extinction: Bad genes or bad luck?* W.W. Norton, New York.

Ricklefs, Robert E., and Dolph Schluter, eds. 1993. *Species diversity in ecological communities*. University of Chicago Press, Chicago.

Ridley, Mark. 1996. *Evolution*, 2nd ed. Blackwell Science, Cambridge, Mass.

Savage, Jay M. 1966. An extraordinary new toad (*Bufo*) from Costa Rica. *Revista de Biologia Tropical* 14: 153–167.

Savage, Jay M. 1998. The "brilliant toad" was telling us something. *Christian Science Monitor*, 14 September 1998: 19.

Smith, Robert L., and Thomas M. Smith. 2003. *Elements of ecology*. 5th ed. Benjamin Cummings, San Francisco.

Ward, Peter. 1994. *The end of evolution*. Bantam Books, New York.

Wilson, Edward O. 1992. *The diversity of life*. W.W. Norton, New York.

Chapter 6

Alling, Abigail, Mark Nelson, and Sally Silverstone. 1993. *Life under glass: The inside story of Biosphere 2*. Biosphere Press.

Barlow, Connie, ed. 1994. Evolution extended. MIT Press, Cambridge, Mass.

Bennett, Jeffrey, et al. 1999. The cosmic perspective. Addison-Wesley, Reading, Massachusetts.

Breckle, Siegmar-Walter. 1999. *Walter's vegetation of the Earth: The ecological systems of the geo-biosphere*, 4th ed. Springer-Verlag, Berlin, 1999.

Campbell, Neil A., and Jane B. Reece. 2002. *Biology*. 6th Edition. Benjamin Cummings, San Francisco.

Capra, Fritjof. 1996. *The web of life: A new scientific understanding of living systems*. Anchor Books Doubleday, New York.

Carpenter, Edward J., and Douglas G. Capone, eds. 1983. *Nitrogen in the marine environment*. Academic Press, New York.

Krebs, Charles J. 2001. *Ecology: The experimental analysis of distribution and abundance*. 5th ed. Benjamin Cummings, San Francisco.

Mitsch, William J., et al. 2001. Reducing nitrogen loading to the Gulf of Mexico from the Mississippi River Basin: Strategies to counter a persistent ecological problem. *BioScience* 51: 373–388.

Molles, Manuel C. Jr. 2002. *Ecology: Concepts and applications*. 2nd ed. McGraw-Hill, Boston.

Montgomery, Carla. 1997. *Environmental Geology*, 5th ed. McGraw-Hill, New York.

National Oceanic and Atmospheric Administration: National Ocean Service. "Hypoxia in the Gulf of Mexico: Progress Toward the Completion of an Integrated Assessment." http://www.nos.noaa.gov/products/pubs_hypox.html.

Rabalais, Nancy N., R.E. Turner, and D. Scavia. 2002. Beyond science into policy: Gulf of Mexico hypoxia and the Mississippi River. *Bioscience* 52: 129–142.

Rabalais, Nancy N., R.E. Turner, and W.J. Wiseman Jr. 2002. Hypoxia in the Gulf of Mexico, a.k.a. "The dead zone." *Annual Review of Ecology and Systematics* 33: 235–263.

Skinner, Brian J., and Stephen C. Porter. 2003. *The dynamic earth: An introduction to physical geology*, 5th ed. John Wiley and Sons, Hoboken, New Jersey.

Smith, Robert L., and Thomas M. Smith. 2003. *Elements of ecology.* 5th ed. Benjamin Cummings, San Francisco.

Vitousek, Peter M., et al. 1997. Human alteration of the global nitrogen cycle: Sources and consequences. *Ecological Applications* 7: 737–750.

Whittaker, Robert H., and William A. Niering. 1965. Vegetation of the Santa Catalina Mountains, Arizona: A gradient analysis of the south slope. *Ecology* 46: 429–452.

Chapter 7

Cohen, Joel E. 1995. *How many people can the Earth support?* W.W. Norton, New York.

Eberstadt, Nicholas. 2000. China's population prospects: Problems ahead. *Problems of Post-Communism* 47: 28.

Ehrlich, Paul R., and John P. Holdren. Impact of population growth: Complacency concerning this component of man's predicament is unjustified and counterproductive. *Science* 171: 1212–1217.

Ehrlich, Paul R., and Anne H. Ehrlich. 1990. The population explosion. Touchstone, New York.

Engelman, Robert, Brian Halweil, and Danielle Nierenberg. 2002. "Rethinking population, improving lives." Pp. 127–148 in *State of the World 2002*, Worldwatch Institute, Washington, D.C., and W.W. Norton, New York.

Greenhalgh, Susan. 2001. Fresh winds in Beijing: Chinese feminists speak out on the one-child policy and women's lives. *Signs: Journal of Women in Culture & Society* 26: 847–887.

Harrison, Paul, and Fred Pearce, eds. 2000. *AAAS Atlas of Population & Environment.* University of California Press, Berkeley.

Holdren, John P. and Ehrlich, Paul R. 1974. Human population and the global environment. *American Scientist* 62: 282–292.

Kane, Penny. 1987. *The second billion: Population and family planning in China.* Penguin Books, Australia, Ringwood, Victoria.

Kane, Penny, and Ching Y. Choi. 1999. China's one child family policy. *British Medical Journal* 319: 992.

McDonald, Mia, with Danielle Nierenberg. 2003. "Linking population, women, and biodiversity." Pp. 38–61 in *State of the World 2003*, Worldwatch Institute, Washington, D.C., and W.W. Norton, New York.

Meadows, Donella H., Dennis L. Meadows, and Jørgen Randers. 1992. *Beyond the Limits: Confronting global collapse, envisioning a sustainable future.* Earthscan Publications, London.

Notestein, Frank. 1953. Economic problems of population change. Pp. 13–31 in *Proceedings of the Eighth International Conference of Agricultural Economists.* Oxford University Press, London.

O'Brien, Stephen J., and Michael Dean. 1997. In search of AIDS-resistance genes. *Scientific American* 277: 44–51.

Population Reference Bureau. 2002. 2002 *World Population Data Sheet.* Population Reference Bureau, Washington, D.C., and John Wiley & Sons, Hoboken, New Jersey.

Redefining Progress. "Sustainability indicators program: Ecological footprint accounts." http://www.redefiningprogress.org/programs/sustainabilityindicators/ef/.

Sheehan, Molly O. 2000. "Urban Population Continues to Rise." Pp. 104–105 in *Vital Signs 2000.* Worldwatch Institute, Washington, D.C., and W.W. Norton, New York.

Sheehan, Molly O. 2000. "Population growth slows." Pp. 66–67 in *Vital Signs 2000.* Worldwatch Institute, Washington, D.C., and W.W. Norton, New York.

UNAIDS and World Health Organization. 2002. *AIDS epidemic update: December 2002.* UNAIDS and WHO, New York.

United Nations Economic and Social Commission for Asia and the Pacific. 2002. *Population and development indicators for Asia and the Pacific, 2002.* UNESCAP, New York.

United Nations Environment Programme. *Africa environment outlook: Past, present, and future perspectives.* UNEP, New York.

United Nations Population Division. 2002. *World population prospects: The 2002 revision.* UNPD, New York.

United Nations Population Fund. "ICPD—ICPD+5—10th anniversary and millennium development goals." UNFPA. http://www.unfpa.org/icpd/.

United Nations Population Fund. *State of World Population 2002.* UNFPA, New York.

United States Census Bureau. http://www.census.gov/.

Chapter 8

Absalom, J.P., N.M.J. Crout, and S.D. Young. 1996. Modeling radiocesium fixation in upland organic soils of northwest England. *Environmental Science and Technology* 30: 2735–2741.

Ashman, Mark R., and Geeta Puri. 2002. *Essential soil science: A clear and concise introduction to soil science.* Blackwell Publishing, Malden, Massachusetts.

Brown, Lester R. 2002. World's rangelands deteriorating under mounting pressure. *Eco-Economy Update* #6, Earth Policy Institute. 5 February 2002.

Charman, P.E.V., and Brian W. Murphy. 2000. *Soils: Their properties and management.* Oxford University Press, South Melbourne, Australia.

Cook, Ray L., and Boyd G. Ellis. 1987. *Soil management: A world view of conservation and production.* John Wiley & Sons, New York.

Curtin, Charles G. 2002. Integration of science and community-based conservation in the Mexico/U.S. borderlands. *Conservation Biology* 16: 880–886.

Diamond, Jared. 1999. *Guns, germs, and steel: The fates of human societies.* W.W. Norton, New York.

Diamond, Jared, and Peter Bellwood. 2003. Farmers and their languages: the first expansions. *Science* 300: 597–603.

Food and Agriculture Organization of the United Nations. 2001. Conservation agriculture: Case studies in Latin America and Africa. *FAO Soils Bulletin No. 78.* FAO, Rome.

Glanz, James. 1995. *Saving our soil: Solutions for sustaining Earth's vital resource.* Johnson Books, Boulder, Colorado.

Goudie, Andrew. 2000. *The human impact on the natural environment*, 5th ed. MIT Press, Cambridge, Massachusetts.

Fox, Stephen. 1985. *The American conservation movement: John Muir and his legacy*. University of Wisconsin Press, Madison.

Halweil, Brian. 2002. "Farmland quality deteriorating." Pp. 102–103 in *Vital Signs 2002*. Worldwatch Institute, Washington, D.C., and W.W. Norton, New York.

Halweil, Brian. 2003. "Grain production drops." Pp. 28–29 in *Vital Signs 2003*. Worldwatch Institute, Washington, D.C., and W.W. Norton, New York.

Harrison, Paul, and Fred Pearce, eds. 2000. *AAAS Atlas of Population & Environment*. University of California Press, Berkeley.

Helms, Douglas. 1998. "Natural Resources Conservation Service," Pp. 434–439 in *A historical guide to the U.S. government*. Oxford University Press, New York.

Hudson, Norman W., et al. 1993. Field measurement of soil erosion and runoff. *FAO Soils Bulletin 68*. FAO and Silsoe Associates, Ampthill, Bedford, U.K.

Jenny, Hans. 1941. *Factors of soil formation: A system of quantitative pedology*. McGraw-Hill, New York.

Larsen, Janet. 2003. Deserts advancing, civilization retreating. *Eco-Economy Update #23*, Earth Policy Institute. 27 March 2003.

Morgan, R.P.C. 1995. *Soil erosion and conservation*, 2nd ed. Longman, Harlow, Essex, England.

Natural Resources Conservation Service. "NRCS soils web site." USDA. http://soils.usda.gov.

Pieri, Christian, et al. 2002. *No-till farming for sustainable rural development*. Agriculture & Rural Development Working Paper. International Bank for Reconstruction and Development, Washington, D.C.

Pierzynski, Gary M., George F. Vance, and James T. Sims. 2000. *Soils and environmental quality*. CRC Press, Boca Raton, Florida.

Pretty, Jules, and Rachel Hine. 2001. *Reducing food poverty with sustainable agriculture: A summary of new evidence*. Center for Environment and Society, University of Essex. *Occasional Paper 2001–2*.

Richter, Daniel D. Jr., and Daniel Markewitz. 2001. *Understanding soil change: Soil sustainability over millennia, centuries, and decades*. Cambridge University Press, Cambridge.

Ritchie, Jerry C. 2000. Combining Cesium-137 and topographic surveys for measuring soil erosion/deposition patterns in a rapidly accreting area. TEKTRAN, United States Department of Agriculture, Division of Agricultural Research, January 14, 2000.

Shaxson, T.F. 1999. The roots of sustainability, concepts and practice: Zero tillage in Brazil. *ABLH Newsletter ENABLE; World Association for Soil and Water Conservation (WASWC) Newsletter*.

Soil Science Society of America. 2001. "Internet glossary of soil science terms." http://www.soils.org/sssagloss/.

Troeh, Frederick R., J. Arthur Hobbs, and Roy L. Donahue. 1998. *Soil and water conservation: Productivity and environmental protection*. Prentice Hall, Upper Saddle River, New Jersey.

United Nations Environment Programme. 2002. "Land." Pp. 62–89 in *Global environment outlook 3 (GEO-3)*. UNEP and Earthscan Publications, Nairobi, Kenya, and London.

Uri, Noel D. 2001. The environmental implications of soil erosion in the United States. *Environmental Monitoring and Assessment* 66: 293–312.

Chapter 9

Bazzaz, Fakhri A. 2001. Plant biology in the future. *Proceedings of the National Academy of the United States of America* 98: 5441–5445.

Buchmann, Stephen L., and Gary Paul Nabhan. 1996. *The forgotten pollinators*. Island Press/Shearwater Books, Washington, D.C./Covelo, California.

[Correspondence to *Nature*, various authors]. 2002. *Nature* 416: 600–602, and 417: 897–898.

Economic Research Service, United States Department of Agriculture. "Briefing room: Food security in the United States." http://www.ers.usda.gov/briefing/foodsecurity/.

Food and Agriculture Organization of the United Nations. 2002. *The state of world fisheries and aquaculture, 2002*. FAO, Rome.

Halweil, Brian. 2000. "Organic farming thrives worldwide." Pp. 120–121 in *Vital Signs 2000*. Worldwatch Institute, Washington, D.C., and W.W. Norton, New York.

Halweil, Brian. 2001. "Farm animal populations soar." Pp. 100–101 in *Vital Signs 2001*. Worldwatch Institute, Washington, D.C., and W.W. Norton, New York.

Halweil, Brian. 2001. "Growth in transgenic area slows." Pp. 102–103 in *Vital Signs 2001*. Worldwatch Institute, Washington, D.C., and W.W. Norton, New York.

Halweil, Brian. 2002. "Farming in the public interest." Pp. 51–74 in *State of the World 2002*, Worldwatch Institute, Washington, D.C., and W.W. Norton, New York.

Harrison, Paul, and Fred Pearce, eds. 2000. *AAAS Atlas of Population & Environment*. University of California Press, Berkeley.

International Food Information Council. 2003. "Food biotechnology." IFIC, Washington, D.C. http://www.ific.org/food/biotechnology/index.cfm

James, Clive. 2002. *Global status of commercialized transgenic crops: 2001*. Brief #27–2002, International Service for the Acquisition of Agri-biotech Applications.

Kuiper, Harry A. 2000. Risks of the release of transgenic herbicide-resistant plants with respect to humans, animals, and the environment. *Crop Protection* 19: 773.

Liebig, Mark A., and John W. Doran. 1999. Impact of organic production practices on soil quality indicators. *Journal of Environmental Quality* 28: 1601–1609.

Losey, John E., Raynor, Linda S., and Carter, Maureen E. 1999. Transgenic pollen harms monarch larvae. *Nature* 399: 214.

Maeder, Paul, et al. 2002. Soil fertility and biodiversity in organic farming. *Science* 296: 1694–1697.

Mann, Charles C . 2002. Transgene data deemed unconvincing. *Science* 296: 236–237.

Manning, Richard. 2000. *Food's frontier: The next green revolution*. North Point Press, New York.

Nestle, Marion. 2002. *Food politics: How the food industry influences nutrition and health*. University of California-Berkeley Press.

Norris, Robert F., Edward P. Caswell-Chen, and Marcos Kogan. 2002. *Concepts in integrated pest management*. Prentice Hall, Upper Saddle River, New Jersey.

Paoletti, Maurizio G. and David Pimentel. 1996. Genetic engineering in agriculture and the environment: Assessing risks and benefits. *Bioscience* 46: 665–673.

Pearce, Fred. 2002. The great Mexican maize scandal. *New Scientist* 174(2347): 14 (15 June 2002).

Pedigo, Larry P. 2001. *Entomology and pest management*, 4th ed. Prentice Hall, Upper Saddle River, New Jersey.

Pimentel, David. 1999. Population growth, environmental resources, and the global availability of food. *Social Research*, Spring 1999.

Quist, David, and Ignacio H. Chapela. 2001. Transgenic DNA introgressed into traditional maize landraces in Oaxaca, Mexico. *Nature* 414: 541–543.

Ruse, Michael, and David Castle, eds. 2002. *Genetically modified foods: Debating technology*. Prometheus Books, Amherst, New York.

Schmeiser, Percy. "Monsanto vs. Schmeiser." http://www.percyschmeiser.com/

Shiva, Vandana. 2000. *Stolen harvest: The hijacking of the global food supply*. South End Press, Cambridge, Massachusetts.

Smil, Vaclav. 2001. *Feeding the world: A challenge for the twenty-first century*. MIT Press, Cambridge, Massachusetts.

Teitel, Martin, and Kimberly Wilson. 2001. *Genetically engineered food: Changing the nature of nature*. Park Street Press.

The Farm Scale Evaluations of spring-sown genetically modified crops. 2003. A themed issue from *Philosophical Transactions of the Royal Society of London B: Biological Sciences* 358(1439), 29 November 2003.

Tuxill, John. 1999. "Appreciating the benefits of plant biodiversity." Pp. 96–114 in *State of the World 1999*, Worldwatch Institute, Washington, D.C., and W.W. Norton, New York.

Westra, Lauren. 1998. Biotechnology and transgenics in agriculture and aquaculture: The perspective from ecosystem integrity. Environmental Values 7: 79.

Wolfenbarger, L. LaReesa 2000. The ecological risks and benefits of genetically engineered plants. *Science* 290: 2088.

Chapter 10

Ames, Bruce N., Margie Profet, and Lois Swirsky Gold. 1990. Nature's chemicals and synthetic chemicals: Comparative toxicology. *Proceedings of the National Academy of the United States of America* 87: 7782–7786.

Cagen, S.Z., et al. 1999. Normal reproductive organ development in wistar rats exposed to bisphenol A in the drinking water. *Regulatory Toxicology and Pharmacology* 30: 130–139.

Carlsen, Elisabeth, et al. 1992. Evidence for decreasing quality of semen during past 50 years. *British Medical Journal* 305: 609–613.

Carson, Rachel. 1962. *Silent spring*. Houghton Mifflin, Boston.

Colburn, Theo, Dianne Dumanoski, and John P. Myers. 1996. *Our stolen future*. Penguin USA, New York.

Crain, D. Andrew, and Louis J. Guillette Jr. 1998. Reptiles as models of contaminant-induced endocrine disruption. *Animal Reproduction Science* 53: 77–86.

Guillette, Elizabeth A., et al. 1998. An anthropological approach to the evaluation of preschool children exposed to pesticides in Mexico. *Environmental Health Perspectives* 106: 347–353.

Guillette, Louis J. Jr., et al. 1999. Plasma steroid concentrations and male phallus size in juvenile alligators from seven Florida lakes. *General and Comparative Endocrinology* 116: 356–372.

Guillette, Louis J. Jr., et al. 2000. Alligators and endocrine disrupting contaminants: A current perspective. *American Zoologist* 40: 438–452.

Halweil, Brian. 1999. "Sperm counts dropping." Pp. 148–149 in *Vital Signs 1999*. Worldwatch Institute, Washington, D.C., and W.W. Norton, New York.

Hayes, Tyrone, et al. 2003. Atrazine-induced hermaphroditism at 0.1 PPB in American leopard frogs (*Rana pipiens*): Laboratory and field evidence. *Environmental Health Perspectives* 111: 568–575.

Hunt, Patricia A., et al. 2003. Bisphenol A exposure causes meiotic aneuploidy in the female mouse. Current Biology 13: 546–553, 1 April 2003.

Kolpin, Dana W., et al. 2002. Pharmaceuticals, hormones, and other organic wastewater contaminants in U.S. streams, 1999-2000: A national reconnaissance. *Environmental Science and Technology* 36: 1202–1211.

Landis, Wayne G., and Ming-Ho Yu. 1998. *An introduction to environmental toxicology: Impacts of chemicals on ecological systems*, 2nd ed. Lewis Publishing, Boca Raton, Florida.

Loewenberg, Samuel. 2003. E.U. starts a chemical reaction. *Science* 300: 405.

Manahan, Stanley E. 2000. *Environmental chemistry*, 7th ed. Lewis Publishers, CRC Press, Boca Raton, Florida.

McGinn, Anne Platt. 2000. "Endocrine disrupters raise concern." Pp. 130–131 in *Vital Signs 2000*. Worldwatch Institute, Washington, D.C., and W.W. Norton, New York.

McGinn, Anne Platt. 2002. "Reducing our toxic burden." Pp. 75–100 in *State of the World 2002*, Worldwatch Institute, Washington, D.C , and W.W. Norton, New York.

McGinn, Anne Platt. 2003. "Combating malaria." Pp. 62–84 in *State of the World 2003*, Worldwatch Institute, Washington, D.C., and W.W. Norton, New York.

Nagel, S.C., et al. 1997. Relative binding affinity-serum modified access (RBA-SMA) assay predicts in vivo bioactivity of the xenoestrogens Bisphenol A and Octylphenol. *Environmental Health Perspectives* 105: 70–76.

National Center for Environmental Health, U.S. Centers for Disease Control and Prevention. 2003. *Second national report on human exposure to environmental chemicals*. January 2003, revised March 2003. NCEH Pub. No. 02-0716, Atlanta.

Renner, Rebecca. 2002. Conflict brewing over herbicide's link to frog deformities. *Science* 298: 938–939.

Rodricks, Joseph V. 1994. *Calculated risks: Understanding the toxicity of chemicals in our environment*. Cambridge University Press, Cambridge.

Spiteri, I. Daniel, Louis J. Guillette Jr., and D. Andrew Crain. 1999. The functional and structural observations of the neonatal reproductive system of alligators exposed *in ozo* to atrazine, 2,4-D, or estradiol. *Toxicology and Industrial Health* 15: 181–186.

Stancel, George, et al. 2001. "Report of the bisphenol A subpanel." Chapter 1 in *National Toxicology Program's report of the endocrine disruptors low-dose peer review*. U.S. EPA and NIEHS, NIH.

Stockholm Convention on Persistant Organic Pollutants. http://www.pops.int/.

United States Environmental Protection Agency. 2003. "Pesticide registration program." http://www.epa.gov/ pesticides/factsheets/registration.htm.

United States Environmental Protection Agency. 2003. "Toxic Substances Control Act." http://www.epa.gov/region5/defs/html/tsca.htm.

United States Environmental Protection Agency. 2003. *EPA's draft report on the environment*. EPA 600-R-03-050. Washington, D.C., June 2003.

Williams, Phillip L., Robert C. James, and Stephen M. Roberts, eds. 2000. *The principles of toxicology: Environmental and industrial applications*, 2nd ed. Wiley-Interscience, New York.

Chapter 11

Ahrens, C. Donald. 1998. *Essentials of meteorology*, 2nd ed., Wadsworth, Belmont, California.

Bell, Michelle L., and Devra L. Davis. 2001. Reassessment of the lethal London fog of 1952: Novel indicators of acute and chronic consequences of acute exposure to air pollution. *Environmental Health Perspectives* 109(Suppl 3): 389–394.

Bernard, Susan M., et al. 2001. The potential impacts of climate variability and change on air pollution-related health effects in the United States. *Environmental Health Perspectives* 109(Suppl 2): 199–209.

Biscaye, Pierre E., et al. 2000. Eurasian air pollution reaches eastern North America. *Science* 290: 2258–2259.

Boubel, Richard W., ed. 1994. *Fundamentals of air pollution*, 3rd ed. Academic Press, San Diego, California.

Bruce, Nigel, Rogelio Perez-Padilla, and Rachel Albalak. 2000. Indoor air pollution in developing countries: A major environmental and public health challenge. *Bulletin of the World Health Organization* 78: 1078–1092.

Cooper, C. David, and F.C. Alley. 2002. *Air pollution control*, 3rd ed. Waveland Press.

Davis, Devra. 2002. *When smoke ran like water: Tales of environmental deception and the battle against pollution*. Basic Books, New York.

Davis, Devra L., Michelle L. Bell, and Tony Fletcher. 2002. A look back at the London smog of 1952 and the half century since. *Environmental Health Perspectives* 110: A734

Driscoll, Charles T., et al. 2001. *Acid rain revisited: Advances in scientific understanding since the passage of the 1970 and 1990 Clean Air Act Amendments*. Hubbard Brook Research Foundation. Science Links™ Publication. Vol. 1, no.1.

Ezzati Majid, and Daniel M. Kammen. 2001. Quantifying the effects of exposure to indoor air pollution from biomass combustion on acute respiratory infections in developing countries. *Environmental Health Perspectives* 109: 481–488.

Freedman, Bill. 1995. *Environmental ecology: The ecological effects of pollution, disturbance, and other stresses*. Academic Press, San Diego, California.

Hunt, Andrew, et al. 2003. Toxicologic and epidemiologic clues from the characterization of the 1952 London smog fine particulate matter in archival autopsy lung tissues. *Environmental Health Perspectives* 111: 1209–1214.

Jacobson, Mark Z. 2002. *Atmospheric pollution: History, science, and regulation*. Cambridge University Press, New York.

Kunzli, Nino, et al. 2000. Public-health impact of outdoor and traffic-related air pollution: A European assessment. *Lancet* 356: 795–801.

Lelieveld, Jos, et al. 2001. The Indian Ocean experiment: Widespread air pollution from South and Southeast Asia. *Science* 291: 1031–1036.

London, Stephanie J., and Isabelle Romieu. 2000. Health costs due to outdoor air pollution by traffic. *Lancet* 356: 782–783.

Molina, Mario J., and F. Sherwood Rowland. 1974. Stratospheric sink for chlorofluoromethanes: chlorine atom catalyzed destruction of ozone. *Nature* 249: 810–812.

O'Meara, Molly. 1999. "Urban air taking lives." Pp. 128–129 in *Vital Signs 1999*. Worldwatch Institute, Washington, D.C., and W.W. Norton, New York.

Office of Global Programs, National Oceanic and Atmospheric Administration. "Reports to the nation on our changing planet: Our ozone shield." http://www.ogp.noaa.gov/library/rtnf92.htm

Parson, Edward A. 2003. *Protecting the ozone layer: Science and strategy*. Oxford University Press, Oxford.

United Nations Environment Programme. "The Montreal Protocol on Substances that deplete the Ozone Layer." http://www.unep.org/ozone/montreal.shtml.

United Nations Environment Programme. 2002. "Atmosphere." Pp. 210–239 in *Global environment outlook 3 (GEO-3)*. UNEP and Earthscan Publications, Nairobi, Kenya, and London.

United States Environmental Protection Agency. 2003. *EPA's draft report on the environment*. EPA 600-R-03-050. Washington, D.C., June 2003.

United States Environmental Protection Agency. 2003. *Latest findings on national air quality: 2002 status and trends*. EPA 454/K-03-001, Aug. 2003.

World Health Organization. 2000. "Air pollution: Including WHO's 1999 guidelines for air pollution control." http://www.who.int/inf-fs/en/fact187.html

Chapter 12

Alley, Richard B. 2000. The two-mile time machine: Ice cores, abrupt climate change, and our future." Princeton University Press, Princeton, New Jersey.

Drake, Frances. 2000. *Global warming: the science of climate change*. Oxford University Press, Oxford.

Dunn, Seth. 2001. "Decarbonizing the energy economy." Pp. 83–102 in *State of the World 2001*, Worldwatch Institute, Washington, D.C., and W.W. Norton, New York.

Dunn, Seth, and Christopher Flavin. 2002. "Moving the climate change agenda forward." Pp. 24–50 in *State of the World 2002*, Worldwatch Institute, Washington, D.C., and W.W. Norton, New York.

Intergovernmental Panel on Climate Change. 2001. *IPCC third assessment report—Climate change 2001: The scientific basis*. World Meteorological Organization and United Nations Environment Programme.

Intergovernmental Panel on Climate Change. 2001. *IPCC third assessment report—Climate change 2001: Impacts, adaptations, and vulnerability*. World Meteorological Organization and United Nations Environment Programme.

Intergovernmental Panel on Climate Change. 2001. *IPCC third assessment report—Climate change 2001: Mitigation*. World Meteorological Organization and United Nations Environment Programme.

Intergovernmental Panel on Climate Change. 2001. *IPCC third assessment report—Climate change 2001: Synthesis report*. World Meteorological Organization and United Nations Environment Programme.

Intergovernmental Panel on Climate Change. 2001. *Technical Summary of the Working Group 1 report*. http://www.ipcc.ch/pub/wg1TARtechsum.pdf.

Intergovernmental Panel on Climate Change. 2003. http://www.ipcc.ch/.

Kareiva, Peter M., Joel G. Kingsolver, and Raymond B. Huey, eds. 1993. *Biotic interactions and global change*. Sinauer Associates, Sunderland, Mass.

Lomborg, Bjorn. 2001. *The skeptical environmentalist: Measuring the real state of the world*. Cambridge University Press, Cambridge.

Mastny, Lisa. 2000. "Ice cover melting worldwide." Pp. 126–127 in *Vital Signs 2000*. Worldwatch Institute, Washington, D.C., and W.W. Norton, New York.

Mayewski, Paul A., and Frank White. 2002. *The ice chronicles: The quest to understand global climate change*. University Press of New England, Hanover, New Hampshire.

McLachlan, Jason S., and Linda B. Brubaker. 1995. Local and regional vegetation change on the northeastern Olympic Peninsula during the Holocene. *Canadian Journal of Botany* 73: 1618–1627.

National Assessment Synthesis Team. 2000. *Climate change impacts on the United States: The potential consequences of climate variability and change*. U.S. Global Change Research Program. Cambridge University Press, 2000.

National Research Council, Board on Atmospheric Sciences and Climate, Commission on Geosciences, Environment, and Resources. 1998. *The atmospheric sciences: Entering the twenty-first century*. National Academy Press, Washington, D.C.

National Research Council, Committee on the Science of Climate Change, Division of Earth and Life Studies. 2001. *Climate change science: An analysis of some key questions*. National Academy Press, Washington, D.C.

National Science Foundation. 2001. *Climate change impacts on the United States: The potential consequences of climate variability and change: A report of the National Assessment Synthesis Team, U.S. Global Change Research Program*. Cambridge University Press, Washington/Cambridge/New York.

Nordhaus, William D. 1998. "Assessing the economics of climate change: An introduction," in *Economic and policy issues in climate change* (William D. Nordhaus, ed.). Resources for the Future Press, Washington D.C.

Parmesan, Camille, and Gary Yohe. 2003. A globally coherent fingerprint of climate change impacts across natural systems. *Nature* 421: 37–42.

Root, Terry L., et al. 2003. Fingerprints of global warming on wild animals and plants. *Nature* 421: 57–60.

Schneider, Stephen H. and Terry L. Root, eds. 2002. *Wildlife responses to climate change: North American case studies*. Island Press, Washington, D.C.

Seinfeld, John H., and Spyros N. Pandis. 1998. *Atmospheric chemistry and physics*. Wiley-Interscience, New York.

Shapiro, Robert J., Kevin A. Hassett, and Frank S. Arnold. 2002. *Conserving energy and preserving the environment: The role of public transportation*. American Public Transportation Association, July 2002.

Sheehan, Molly O. 2003. "Carbon emissions and temperature climb." Pp. 40–41 in *Vital Signs 2003*. Worldwatch Institute, Washington, D.C., and W.W. Norton, New York.

Stevens, William K. 1999. *The change in the weather: People, weather and the science of climate*. Delta Trade Paperbacks, New York.

Taylor, David. 2003. "Small islands threatened by sea level rise." Pp. 84–85 in *Vital Signs 2003*. Worldwatch Institute, Washington, D.C., and W.W. Norton, New York.

United Nations. "The Convention and Kyoto Protocol." http://unfccc.int/resource/convkp.html.

United States Congress. Senate. Committee on Foreign Relations. 2000. "Climate change: Status of the Kyoto Protocol after three years." Joint hearing before the Committee on Foreign Relations and the Committee on Energy and Natural Resources, United States Senate, One Hundred Sixth Congress, second session, 28 September 2000.

United States Congress. House. Committee on Science. 2001. "Climate change: The state of the science." Hearing before the Committee on Science, House of Representatives, One Hundred Seventh Congress, first session, 14 March 2001.

Chapter 13

Ault, J.S., Bohnsack, J.A. and G.A. Meester. 1998. A retrospective (1979–1996) multi-species assessment of coral reef fish stocks in the Florida Keys. *Fishery Bulletin* 96: 395–414.

Baker, Beth. 1999. First aid for an ailing reef: Research in the Florida Keys National Marine Sanctuary. *Bioscience* 49: 173–178.

Castro, Peter, and Michael E. Huber. 1992. Marine Biology. Mosby-Year Book.

Causey, Billy D., Joanne Delaney, and Brian D. Keller. 2001. *The status of coral reefs of the Florida Keys.* Florida Keys National Marine Sanctuary.

[Correspondence to *Science*, various authors]. 2001. *Science* 295: 1233–1235.

Embassy of the People's Republic of China. 2001. "Fishing statistics "basically correct," ministry says." Press release, 18 December 2001. http://saup.fisheries.ubc.ca/Media/Chinese_Embassy_18_Dec_2001.pdf

Food and Agriculture Organization of the United Nations. 2002. *The state of world fisheries and aquaculture, 2002.* FAO, Rome.

Food and Agriculture Organization of the United Nations. 2002. "Fishery statistics: Reliability and policy implications." FAO Fisheries Department, Rome. http://www.fao.org/fi/statist/nature_china/30jan02.asp

Garrison, Tom. 1999. *Oceanography: An invitation to marine science*, 3rd ed. Wadsworth, Belmont, California.

Gell, Fiona R., and Callum M. 2003. Benefits beyond boundaries: The fishery effects of marine reserves. *Trends in Ecology and Evolution* 18: 448–455.

Halpern, Benjamin S., and Robert R. Warner. 2002. Marine reserves have rapid and lasting effects. *Ecology Letters* 5: 361–366.

Halpern, Benjamin S., and Robert R. Warner. 2003. Matching marine reserve design to reserve objectives. *Proceedings of the Royal Society of London B*: DOI 10.1098/ rspb.2003.2405.

Jackson, Jeremy B.C., et al. 2001. Historical overfishing and the recent collapse of coastal ecosystems. *Science* 293: 629–638.

Mastny, Lisa. 2001. "World's coral reefs dying off." Pp. 92–93 in *Vital Signs 2001.* Worldwatch Institute, Washington, D.C., and W.W. Norton, New York.

Myers, Ransom A., and Boris Worm. 2003. Rapid worldwide depletion of predatory fish communities. *Nature* 423: 280–283.

National Academy of Public Administration. 1999. *Protecting Our National Marine Sanctuaries.* Center for the Economy and the Environment, NAPA, Washington, D.C.

National Center for Ecological Analysis and Synthesis (NCEAS) and Communication Partnership for Science and the Sea (COMPASS), sponsors. 2001. *Scientific consensus statement on marine reserves and marine protected areas.* Available online at: http://www.nceas.ucsb.edu/consensus.

National Oceanic and Atmospheric Administration (NOAA). "Florida Keys National Marine Sanctuary." http://www.fknms.nos.noaa.gov/.

Palumbi, Stephen. 2003. *Marine reserves: A tool for ecosystem management and conservation.* Pew Oceans Commission.

Pauly, Daniel, et al. 2002. Towards sustainability in world fisheries. *Nature* 418: 689–695.

Pinet, Paul R. 1996. *Invitation to oceanography.* West Publishing Company.

Richardson, L.L. 1998. Coral diseases: What is really known? *Trends in Ecology and Evolution* 13: 438–443.

Roberts, Callum M., et al. 2001. Effects of marine reserves on adjacent fisheries. *Science* 294: 1920–1923.

Sumich, James L. 1992. *An introduction to the biology of marine life.* William C. Brown, Dubuque, Iowa.

Thurman, Harold V. 1990. *Essentials of oceanography*, 4th ed. MacMillan Publishing, New York.

United Nations Environment Programme. 2002. "Coastal and marine areas." Pp. 180–209 in *Global environment outlook 3 (GEO-3).* UNEP and Earthscan Publications, Nairobi, Kenya, and London.

United States Department of Commerce. 1996. *Strategy for stewardship: Florida Keys National Marine Sanctuary final management plan/environmental impact statement.* 3 vol. Dept. of Commerce, Washington, D.C.

United States Department of Commerce and United States Department of the Interior. "Marine protected areas of the United States. http://www.mpa.gov.

Watson, Reginald, Lillian Pang, and Daniel Pauly. 2001. *The marine fisheries of China: Development and reported catches.* Fisheries Centre Research Reports 9(2). Fisheries Centre, University of British Columbia, Canada.

Watson, Reginald, and Daniel Pauly. 2001. Systematic distortions in world fisheries catch trends. *Nature* 414: 534–536.

Withgott, Jay. 2001. Send in the marine reserves. *BioMedNet News, Special feature.* http://news.bmn.com.

Chapter 14

American Rivers. 2002. *The ecology of dam removal: A summary of benefits and impacts.* American Rivers, Washington D.C., February 2002.

Bolduc, Daniel and James Fitch. 2002. Practical engineering combined with sound operations optimizes phosphorus removal. *Water Engineering & Management*, Des Plaines; 2002.

British Geographical Society and Bangladesh Department of Public Health Engineering. 2001. *Arsenic contamination of groundwater in Bangladesh*. Technical Report WC/00/19, Volume 1: Summary.

De Villiers, Marq, 2000. *Water: The fate of our most precious resource*. Mariner Books.

Fitch, James, and Bolduc, Daniel. (2002). Practical engineering combined with sound operations optimizes phosphorus removal. *Water Engineering & Management*, Apr. 2002.

Gleick, Peter. 2002. *The world's water 2002–2003: The biennial report on freshwater resources*. Island Press, Washington D.C.

Harrison, Paul, and Fred Pearce, eds. 2000. *AAAS Atlas of Population & Environment*. University of California Press, Berkeley.

Harvey, Charles F., et al. 2002. Arsenic mobility and groundwater extraction in Bangladesh. *Science* 298: 1602–1606.

Institute of Governmental Studies, University of California, Berkeley. 2003. "Imperial Valley-San Diego water transfer controversy." www.igs.berkeley.edu/library/htImperial Water Transfer2003.html

Jenkins, Matt. 2002. The royal squeeze. *High Country News* 35(1), January 20, 2003.

Jenkins, Matt. 2003. California's water binge skids to a halt. *High Country News* 34(17), September 16, 2002.

Larsen, Janet. 2001. "Hydrological poverty worsening." Pp. 94–95 in *Vital Signs 2001*. Worldwatch Institute, Washington, D.C., and W.W. Norton, New York.

Lloyd, J.W. 2001. Groundwater vulnerability and the reuse of water, with reference to Saudi Arabia. *International Journal of Water Resources Development* 17: 179–184.

Marston, Ed. 2001. Quenching the big thirst. *High Country News* 33(10), May 21, 2001.

Nickson, Ross, et al. 1998. Arsenic poisoning of Bangladesh groundwater. *Nature* 395: 338.

Postel, Sandra. 1999. *Pillar of sand: Can the irrigation miracle last?* W.W. Norton, New York.

Postel, Sandra. 2000. "Groundwater depletion widespread." Pp. 122–123 in *Vital Signs 2000*. Worldwatch Institute, Washington, D.C., and W.W. Norton, New York.

Postel, Sandra. 2002. "Freshwater species at increasing risk." Pp. 106–107 in *Vital Signs 2002*. Worldwatch Institute, Washington, D.C., and W.W. Norton, New York.

Reisner, Marc. 1986. *Cadillac desert: The American West and its disappearing water*. Viking Penguin, New York.

Sampat, Payal. 2000. "Groundwater quality deteriorating." Pp. 124–125 in *Vital Signs 2000*. Worldwatch Institute, Washington, D.C., and W.W. Norton, New York.

Sampat, Payal. 2001. "Uncovering Groundwater Pollution." Pp. 21–42 in *State of the World 2001*, Worldwatch Institute, Washington, D.C., and W.W. Norton, New York.

Sampat, Payal. 2000. "Hydrological poverty worsening." Pp. 94–95 in *Vital Signs 2001*. Worldwatch Institute, Washington, D.C., and W.W. Norton, New York.

Sibley, George. 1997. A tale of two rivers: The desert empire and the mountain. *High Country News* 29(21), November 10, 1997.

Smith, Lingas, Rahman. 2000. Contamination of drinking water by arsenic in Bangladesh: A public health emergency. *Bulletin of the World Health Organization*, 78(9).

Stone, Richard. 1999. Coming to grips with the Aral Sea's grim legacy. *Science* 284: 30–33.

United Nations Commission on Sustainable Development. 2002. "Comprehensive assessment of the freshwater resources of the world: Report of the Secretary-General." 4 February 1997. http://www.un.org/esa/sustdev/freshwat.htm.

United Nations Environment Programme. 2002. "Freshwater." Pp. 150–179 in *Global environment outlook 3 (GEO-3)*. UNEP and Earthscan Publications, Nairobi, Kenya, and London.

United States Bureau of Reclamation, Lower Colorado Regional Office. http://www.usbr.gov/lc/region.

United States Environmental Protection Agency. 2003. *EPA's draft report on the environment*. EPA 600-R-03-050. Washington, D.C., June 2003.

World Health Organization. 2000. *Global water supply and sanitation assessment 2000 report*. WHO.

Chapter 15

Balmford, Andrew, et al. 2002. Economic reasons for conserving wild nature. *Science* 297: 950–953.

Baskin, Yvonne. 1997. *The work of nature: How the diversity of life sustains us*. Island Press, Washington, D.C.

Bright, Chris. 1998. *Life out of bounds: Bioinvasion in a borderless world*. Worldwatch Institute and W.W. Norton, Washington D.C. and New York.

CITES Secratariat. "Convention on International Trade in Endangered Species of Wild Fauna and Flora. http://www.cites.org/.

Convention on Biological Diversity. http://www.biodiv.org/.

Daily, Gretchen C., ed. 1997. *Nature's services: Societal dependence on natural ecosystems*. Island Press, Washington, D.C.

Ehrenfeld, David W. 1970. *Biological conservation*. International Thomson Publishing, London.

Groombridge, Brian, and Martin D. Jenkins. 2002. *Global biodiversity: Earth's living resources in the 21st century*. UNEP-World Conservation Monitoring Centre, Cambrdige, U.K. Aventis Foundation.

Groombridge, Brian, and Martin D. Jenkins. 2002. *World atlas of biodiversity: Earth's living resources in the 21st century*. University of California Press, Berkeley.

Grumbine, R. Edward. 1992. *Ghost bears: Exploring the biodiversity crisis*. Island Press, Washington, D.C.

Hanken, James. 1999. Why are there so many new amphibian species when amphibians are declining? *Trends in Ecology and Evolution* 14: 7–8.

Harris, Larry D. 1984. *The fragmented forest: Island biogeography theory and the preservation of biotic diversity*. University of Chicago Press, Chicago.

Harrison, Paul, and Fred Pearce, eds. 2000. *AAAS Atlas of Population & Environment.* University of California Press, Berkeley.

MacArthur, Robert H., and Edward O. Wilson. 1967. *The theory of island biogeography.* Princeton University Press, Princeton.

Mackay, Richard. 2002. *The Penguin atlas of endangered species: A worldwide guide to plants and animals.* Penguin, New York.

Maehr, David S., Reed F. Noss, and Jeffrey Larkin, eds. 2001. *Large mammal restoration: Ecological and sociological challenges in the 21ˢᵗ century.* Island Press, Washington, D.C.

Matthiessen, Peter. 2000. *Tigers in the snow.* North Point Press, New York.

Meegaskumbura, Madhava, et al. 2002. Sri Lanka: An amphibian hot spot. *Science* 298: 379.

Meffe, Gary K., et al. 1997. *Principles of conservation biology,* 2ⁿᵈ ed. Sinauer Associates, Sunderland, Mass.

Miquelle, Dale, Howard Quigley, and Maurice Hornocker. 1999. A habitat protection plan for Amur Tiger conservation: A proposal outlining habitat protection measures for the Amur Tiger. Hornocker Wildlife Institute.

Mooney, Harold A. and Richard J. Hobbs, eds. 2000. *Invasive species in a changing world.* Island Press, Washington, D.C.

Newmark, William D. 1987. A land-bridge perspective on mammal extinctions in western North American parks. *Nature* 325: 430.

Quammen, David. 1996. *The song of the dodo: island biogeography in an age of extinction.* Touchstone, New York.

Relyea, Rick, and Nathan Mills. 2001. Predator-induced stress makes the pesticide carbaryl more deadly to gray treefrog tadpoles. *Proceedings of the National Academy of Sciences, USA* 98: 2491–2496.

Rosenzweig, Michael L. 1995. *Species diversity in space and time.* Cambridge University Press, Cambridge.

Simberloff, Daniel S. 1969. Experimental zoogeography of islands: A model for insular colonization. *Ecology* 50: 296–314.

Simberloff, Daniel S., and Edward O. Wilson. 1969. Experimental zoogeography of islands: The colonization of empty islands. *Ecology* 50: 278–296.

Simberloff, Daniel S., and Edward O. Wilson. 1970. Experimental zoogeography of islands: A two-year record of colonization. *Ecology* 51: 934–937.

Soulé, Michael E. 1986. *Conservation biology: The science of scarcity and diversity.* Sinauer Associates, Sunderland, Mass.

Sparling, Donald W., Gary M. Fellers, and Laura McConnell. 2001. Pesticides and amphibian declines in California, USA. *Environmental Toxicology and Chemistry* 20: 1591–1595.

Takacs, David. 1996. *The idea of biodiversity: Philosophies of paradise.* Johns Hopkins University Press, Baltimore.

United Nations Environment Programme. 2002. "Biodiversity." Pp. 120–149 in *Global environment outlook 3 (GEO-3).* UNEP and Earthscan Publications, Nairobi, Kenya, and London.

United Nations Environment Programme. 2003. "Sustaining life on Earth: How the Convention on Biological Diversity promotes nature and human well-being." http://www.biodiv.org/doc/publications/guide.asp.

United States Fish and Wildlife Service. 2003. *The endangered species act of 1973.* Accessible online at: http://endangered.fws.gov/csa.html.

Wilson, Edward O. 1984. *Biophilia.* Harvard University Press, Cambridge, Mass.

Wilson, Edward O. 1992. *The diversity of life.* W.W. Norton, New York.

Wilson, Edward O. 1994. *Naturalist.* Island Press, Shearwater Books, Washington D.C.

Wilson, Edward O. 2002. *The future of life.* Alfred A. Knopf, New York.

Wilson, Edward O., and Daniel S. Simberloff. 1969. Experimental zoogeography of islands: Defaunation and monitoring techniques,. *Ecology* 50: 267–278.

World Conservation Union. 2003. IUCN Red List. http://www.iucnredlist.org/.

Chapter 16

Abramovitz, Janet. 2002. "Forest loss unchecked." Pp. 104–105 in *Vital Signs 2002.* Worldwatch Institute, Washington, D.C., and W.W. Norton, New York.

Bachrach, Julia Sniderman. 2001. Jens Jensen: Friend of the native landscape. *Chicago Wilderness,* Spring 2001 (http://www.chicagowildernessmag.org/issues/spring2001/jensjensen.html).

Christy, Stephen F. Jr. 1999. To preserve and protect . . . *Chicago Wilderness,* Winter 1999 (http://www.chicagowildernessmag.org/issues/winter1999/preserve.html).

Clary, David. 1986. *Timber and the Forest Service.* University Press of Kansas, Lawrence.

Cronon, William. 1991. *Nature's metropolis: Chicago and the great West.* W.W. Norton, New York.

Harrison, Paul, and Fred Pearce, eds. 2000. *AAAS Atlas of Population & Environment.* University of California Press, Berkeley.

Jacobs, Lynn. 1991. *Waste of the West: Public lands ranching.* Lynn, Jacobs, Tucson, Ariz.

Landres, Peter, David R. Spildie, and Lloyd P. Queen. 2001. GIS applications to wilderness management: Potential uses and limitations. Gen. Tech. Rep. RMRS-GTR-80. U.S. Department of Agriculture, Forest Service, Rocky Mountain Research Station, Fort Collins, Colo.

Logan, Michael F. 1995. *Fighting sprawl and city hall.* University of Arizona Press, Tucson.

Myers, Norman, and Jennifer Kent. 2001. *Perverse subsidies: How misused tax dollars harm the environment and the economy.* Island Press, Washington, D.C.

Rosenzweig, Michael L. 2003. *Win-win ecology: How Earth's species can survive in the midst of human enterprise.* Oxford University Press, Oxford.

Runte, Alfred. 1979. *National parks and the American experience*. University of Nebraska Press, Lincoln.

Sedjo, Robert A. 2000. *A vision for the US Forest Service*. Resources for the Future, Washington, D.C.

Sheehan, Molly O'Meara. 2001. "Making better transportation choices." Pp. 103–122 in *State of the World 2001*, Worldwatch Institute, Washington, D.C., and W.W. Norton, New York.

Sheehan, Molly O. 2002. "Sprawling cities have global effects." Pp. 152–153 in *Vital Signs 2002*. Worldwatch Institute, Washington, D.C., and W.W. Norton, New York.

Singh, Ashbindu, et al. 2001. An assessment of the status of the world's remaining closed forests. United Nations Environmental Program, UNEP/DEWA/TR 01-2l, August 2001.

Smith, W. Brad, et al. 2001. Forest resources of the United States, 1997. General Technical Report NC-219, North Central Research Station, USDA Forest Service, St. Paul, Minnesota, August 2001.

Soulé, Michael E. and John Terborgh, eds. 1999. *Continental conservation*. Island Press, Washington, D.C.

Stegner, Wallace. 1954. *Beyond the hundredth meridian: John Wesley Powell and the second opening of the West*. Houghton Mifflin, Boston.

Thornton, Ralph C. The early history of the Forest Preserve District of Cook County, 1869–1922. http://users.rcn.com/clonk/CCFPD/Origins.html, and at http://www.fpdcc.com/tier3.php?content_id=4

United Nations Environment Programme. 2002. "Forests." Pp. 90–119 in *Global environment outlook 3 (GEO-3)*. UNEP and Earthscan Publications, Nairobi, Kenya, and London.

United Nations Environment Programme. 2002. "Urban areas." Pp. 240–269 in *Global environment outlook 3 (GEO-3)*. UNEP and Earthscan Publications, Nairobi, Kenya, and London.

USDA Forest Service. 2001. *U.S. forest facts and historical trends*. FS-696, March 2001.

U.S. National Park Service. 2002. *National Park Service statistical abstract 2002*. NPS Public Use Statistics Office, U.S. Department of the Interior, Denver, Colorado.

Chapter 17

ASEA Brown Boveri Environmental Systems, for the U.S. Department of Energy. 1996. SNOX demonstration project, final report, volume II. Published online at http://www.lanl.gov/cgi-bin/byteserver.pl/projects/cctc/resources/pdfs/snox/M96050314.pdf. July 1996.

British Petroleum. 2003. *BP statistical review of world energy*. London, June 2003.

Campbell, Colin J. 1997. *The coming oil crisis*. Multi-Science Publishing Co., Essex, U.K.

Chandler, David. 2003. America steels itself to take the nuclear plunge. *New Scientist*, August 9, 2003 p. 10–13.

Deffeyes, Kenneth S. 2001. *Hubbert's peak : The impending world oil shortage*. Princeton University Press, Princeton, New Jersey.

Dunn, Seth. 2001. "Decarbonizing the energy economy." Pp. 83–102 in *State of the World 2001*, Worldwatch Institute, Washington, D.C., and W.W. Norton, New York.

Energy Information Administration, U.S. Department of Energy. http://www.eia.doe.gov.

Hellman, Karl H., and Robert M. Heavenrich. 2003. Light-duty automotive technology and fuel economy trends: 1975 through 2003. U.S. Environmental Protection Agency. EPA420-S-03-004, April 2003.

Holmes, Bob, and Nicola Jones. 2003. Brace yourself for the end of cheap oil. *New Scientist*, August 2, 2003 p. 9–11.

International Energy Agency. 2002. *Key world energy statistics from the IEA*. IEA Publications, Paris.

International Energy Agency. 2002. *World energy outlook 2002*. IEA Publications, Paris.

Kraushaar, Jack J. and Robert A. Ristinen. 1984. *Energy and problems of a technical society*. John Wiley and Sons, New York.

Lenssen, Nicholas. 2003. "Nuclear power rises." Pp. 36–37 in *Vital Signs 2003*. Worldwatch Institute, Washington, D.C., and W.W. Norton, New York.

Nellemann, Christian, and Raymond D. Cameron. 1998. Cumulative impacts of an evolving oil-field complex on the distribution of calving caribou. *Canadian Journal of Zoology* 76: 1425–1430.

Pelley, Janet. 2001. Will drilling for oil disrupt the Arctic National Wildlife Refuge? *Environmental Science and Technology* 35: 240–247.

Powell, Stephen G. 1990. Arctic National Wildlife Refuge: How much oil can we expect? *Resources Policy* Sept. 1990: 225–240.

Skinner, Brian J., and Stephen C. Porter. 2003. *The dynamic earth: An introduction to physical geology*, 5th ed. John Wiley and Sons, Hoboken, New Jersey.

United States Department of Energy and National Energy Technology Laboratory. 2000. SNOX flue gas cleaning demonstration project: A DOE assessment. Published online at http://www.lanl.gov/projects/cctc/resources/pdfs/snox/SNOX2.pdf. June 2000.

United States Fish and Wildlife Service. 2001. "Potential impacts of proposed oil and gas development on the Arctic Refuge's coastal plain: Historical overview and issues of concern." Web page of the Arctic National Wildlife Refuge, Fairbanks, Alaska. http://arctic.fws.gov/issues1.html

United States Geological Survey. 2001. *The National Petroleum Reserve-Alaska (NPRA) data archive*. USGS Fact Sheet FS-024-01, March 2001.

United States Geological Survey. 2001. *Arctic National Wildlife Refuge, 1002 Area, petroleum assessment, 1998, including economic analysis*. USGS Fact Sheet FS-028-01, April 2001.

Walker, Donald A. 1996. Disturbance and recovery of arctic Alaskan vegetation. Pp. 35–71 in *Landscape function and disturbance in arctic tundra*, J.F. Reynolds and J.D. Tenhunen, eds. *Ecological Studies*, Vol. 120. Springer-Verlag, Berlin/Heidelberg.

Walker, Donald A. 1997. Arctic Alaskan vegetation disturbance and recovery. Pp. 457–479 in *Disturbance and recovery in Arctic lands*, R.M.M. Crawford, ed. Kluwer Academic Publishers, Dordrecht, Netherlands.

Chapter 18

Alternative Energy Institute webpage. 2003. AEI, http://www.altenergy.org/.

Ananthaswamy, Anil. 2003. Reality bites for the dream of a hydrogen economy. *New Scientist*, November 15, 2003 p. 6–7.

Arnason, Bragi, and and Thorsteinn I. Sigfusson. 2000. Iceland—a future hydrogen economy. *International Journal of Hydrogen Energy* 25: 389–394.

Ásmundsson, Jón Knútur. 2002. Will fuel cells make Iceland the 'Kuwait of the North?' *World Press Review* 15 February 2002.

Burkett, Elinor. 2003. A mighty wind. *New York Times magazine*. June 15, 2003.

DaimlerChrysler. 2003. *360 DEGREES / DaimlerChrysler Environmental Report 2003*. DaimlerChrysler AG, Stuttgart, Germany.

Dunn, Seth. 2000. The hydrogen experiment. *World Watch* 13: 14–25.

Energy Information Administration, U.S. Department of Energy. http://www.eia.doe.gov.

Flavin, Christopher, and Seth Dunn. 1999. A new energy paradigm for the 21st century. *Journal of International Affairs* 53: 167–190.

Hirsch, Tim. 2001. Iceland launches energy revolution. *British Broadcasting Corporation News*, 24 December 2001.

Hydrogen & Fuel Cell Letter. 2003. World's first commercial hydrogen station opens in Iceland. *Hydrogen & Fuel Cell Letter* May 2003. http://www.hfcletter.com/letter/May03/

Idaho Wind Power Working Group for the Idaho Department of Water Resources Energy Division. 2002. *Idaho wind power development strategic plan*. Boise, Idaho. Published online at http://www.idwr.state.id.us/energy/wind/stragic_plan.pdf, January 2002.

Idaho Wind Power Working Group for the Idaho Department of Water Resources Energy Division. 2002. *Wind power potential in Idaho by county*. Boise, Idaho. Published online at http://www.idwr.state.id.us/energy/wind/id_wind_breakdowns.pdf, 2002.

International Energy Agency. 2001. *Toward a sustainable energy future*. IEA Publications, Paris.

International Energy Agency. 2002. *Renewables in global energy supply: An IEA fact sheet*. IEA Publications, Paris, November 2002.

International Energy Agency Renewable Energy Working Party. 2002. *Renewable energy . . . into the mainstream*. SITTARD, The Netherlands, October 2002.

Kraushaar, Jack J. and Robert A. Ristinen. 1984. *Energy and problems of a technical society*. John Wiley and Sons, New York.

Melis, Anastasios, et al. 2000. Sustained photobiological hydrogen gas production upon reversible inactivation of oxygen evolution in the green alga *Chlamydomonas reinhardtii*. *Plant Physiology* 122: 127–135.

National Renewable Energy Lab webpage. 2003. U.S. Department of Energy. http://www.nrel.gov/.

Office of Energy Efficiency and Renewable Energy, U.S. Department of Energy http://www.eere.energy.gov/windpoweringamerica/.

Randerson, James. 2003. The clean green energy dream. *New Scientist*, August 16, 2003 p. 8–11.

Reeves, Ari, with Fredric Beck. 2003. *Wind energy for electric power: A REPP issue brief*. Renewable Energy Policy Project, Washington, D.C., July 2003.

Renewable Energy Policy Project webpage. 2003. REPP, Washington, D.C. http://solstice.crest.org/index.html.

Rocky Mountain Institute webpage. 2003. "Energy." RMI, Snowmass, Colo. http://www.rmi.org/sitepages/pid17.php.

Sawin, Janet. 2003. "Charting a new energy future." Pp. 85–109 in *State of the World 2003*, Worldwatch Institute, Washington, D.C., and W.W. Norton, New York.

Sawin, Janet L. 2003. "Wind power's rapid growth continues." Pp. 38–39 in *Vital Signs 2003*. Worldwatch Institute, Washington, D.C., and W.W. Norton, New York.

Sheehan, Molly O. 2002. "Solar cell use rises quickly." Pp. 44–45 in *Vital Signs 2002*. Worldwatch Institute, Washington, D.C., and W.W. Norton, New York.

U.S. Department of Energy National Laboratory directors. 1997. *Technology opportunities to reduce U.S. greenhouse gas emissions*. DOE, Washington, D.C., October 1997.

Weisman, Alan. 1998. *Gaviotas: A village to reinvent the world*. Chelsea Green Publishing Co., White River Junction, Vermont.

Chapter 19

Abramovitz, Janet N. 2000. "Paper recycling remains strong." Pp. 132–133 in *Vital Signs 2000*. Worldwatch Institute, Washington, D.C., and W.W. Norton, New York.

Allen, G.H. and R.A. Gearheart, eds. 1988. *Proceedings of a conference on wetlands for wastewater treatment and resource enhancement*. Humboldt State University, Arcata, Calif.

Ayres, Robert U., and Leslie W. Ayres. 1996. *Industrial ecology: Towards closing the materials cycle*. Edward Elgar Press, Cheltenham.

Diesendorf, Mark, and Clive Hamilton. 1997. *Human ecology, human economy*. Allen and Unwin, St. Leonards.

Edmonton, Alberta, City of. 2003. Co-Composting Facility. http://www.gov.edmonton.ab.ca/am_pw/waste_management/compost_facility.html.

Energy Information Administration, U.S. Department of Energy. 2003. "Municipal solid waste." EIA, DOE, Washington, D.C. http://www.eia.doe.gov/cneaf/solar.renewables/page/mswaste/msw.html

Institute for Local Self-Reliance. 2000. *Wasting and recycling in metropolitan Manila, Philippines.* Washington, D.C., for Greenpeace Southeast Asia, Quezon City, Philippines, October 2000.

Integrated Waste Services Association. 2003. "WTE: About waste-to-energy." http://www.wte.org/waste.html. IWSA, Washington, D.C.

Manahan, Stanley E. 1999. *Industrial ecology: Environmental chemistry and hazardous waste.* Lewis Publishers, CRC Press, Boca Raton, Florida.

McDonough, William, and Michael Braungart. 2002. *Cradle to cradle: Remaking the way we make things.* North Point Press, New York.

McGinn, Anne Platt. 2002. "Toxic waste largely unseen." Pp. 112–113 in *Vital Signs 2002.* Worldwatch Institute, Washington, D.C., and W.W. Norton, New York.

Ng, Roli. 2000. Threat of infection grows at Manila garbage site. *Reuters,* July 12, 2000.

Office Four Production. 2001–2002. "God's children." http://www011.upp.so-net.ne.jp/office4pro/kaminoko/ en/e_index.htm. Tokyo, Japan.

Oregon Department of Environmental Quality. 2001. "Bottle bill turns 30!" http://www.deq.state.or.us/wmc/solwaste/BottleBill30.html.

Rathje, William. 1997. A Bum Wrap? *The ULS Report* 4(2), March-April 1997, published online at http://www.cygnus-group.com/use-less-stuff/Archive/No.19_MA97.html.

Rathje, William, and Colleen Murphy. 2001. *Rubbish! The archeology of garbage.* University of Arizona Press, March 2001.

Smith, Ronald S. 1998. *Profit centers in industrial ecology.* Quorum Books, Westport.

Socolow, Robert H., et al., eds. 1994. *Industrial ecology and global change.* Cambridge University Press, Cambridge.

United Nations Environment Programme. 2000. *International source book on environmentally sound technologies (ESTs) for municipal solid waste management (MSWM).* UNEP IETC, Osaka, Japan.

United States Environmental Protection Agency, Office of Wetlands, Oceans, and Watersheds. 1993. Constructed wetlands for wastewater treatment and wildlife habitat. EPA, Washington D.C. http://www.epa.gov/owow/wetlands/construc/content.html.

United States Environmental Protection Agency. 1998. *Wastewater primer.* EPA 833-K-98-001, Office of Wastewater Management, May 1998.

United States Environmental Protection Agency. 2002. *Municipal solid waste in the United States: 2000 facts and figures.* EPA530-R-02-001, Office of Solid Waste and Emergncy Response, June 2002.

United States Environmental Protection Agency. 2003. "Municipal solid waste." http://www.epa.gov/epaoswer/non-hw/muncpl/index.htm.

World Bank Group. 2001. *Philippines environment monitor 2001.* World Bank, Manila and Washington, D.C.

Chapter 20

Bahn, Paul, and John Flenley. 1992. *Easter Island, Earth island.* Thames and Hudson, London.

Brower, Michael and Warren Leon. 1999. *The consumer's guide to effective environmental choices: practical advice from the Union of Concerned Scientists.* Three Rivers Press, New York.

Brown, Lester. 2001. *Eco-economy: Building an economy for the Earth.* Earth Policy Institute and W.W. Norton, New York.

Daly, Herman E. 1996. *Beyond growth.* Beacon Press, Boston.

Dasgupta, Partha, Simon Levin, and Jane Lubchenco. 2000. Economic pathways to ecological sustainability. *BioScience* 50: 339–345.

De Graaf, John, David Wann, and Thomas Naylor. 2002. *Affluenza: The all-consuming epidemic.* Berrett-Koehler Publishers, San Francisco.

Dodson, Jessica. 2002. "Rio treaties post some success." Pp. 114–115 in *Vital Signs 2002.* Worldwatch Institute, Washington, D.C., and W.W. Norton, New York.

Durning, Alan. 1992. *How much is enough?: The consumer society and the future of the Earth.* Worldwatch Institute, Washington, D.C.

Erickson, Jon D., and John M. Gowdy. 2002. The strange economics of sustainability. *BioScience* 52: 212.

Gardner, Gary. 2001. "Accelerating the shift to sustainability." Pp. 189-206 in *State of the World 2001,* Worldwatch Institute, Washington, D.C., and W.W. Norton, New York.

Goodstein, Eban. 1999. *The tradeoff myth: Fact and fiction about jobs and the environment.* Island Press, Washington, D.C.

Hawken, Paul. 1994. *The ecology of commerce: A declaration of sustainability.* HarperBusiness, New York.

Heinkel, Robert, Alan Kraus, and Josef Zechner. 2001. The effect of green investment on corporate behavior. *Journal of Financial and Quantitative Analysis* 36: 431–449.

Mastny, Lisa. 2002. "Ecolabeling gains ground." Pp. 124–125 in *Vital Signs 2002.* Worldwatch Institute, Washington, D.C., and W.W. Norton, New York.

Office of Management and Budget, Executive Office of the President of the United States, Washington, D.C. 2003. *Informing regulatory decisions: 2003 report to Congress on the costs and benefits of federal and unfunded mandates on state, local, and tribal entities.* Washington, D.C., September 2003.

Rees, Williams E., Phil Testemale, and Mathis Wackernagel. 1995. *Our ecological footprint: Reducing human impact on the Earth.* New Society Publishers, Gabriola Island, British Columbia.

Sanderson, Eric W., et al. 2002. The human footprint and the last of the wild. *Bioscience* 52: 891–904.

Schor, Juliet B., and Betsy Taylor, eds. 2002. *Sustainable planet: Solutions for the twenty-first century.* The Center for a New American Dream. Beacon Press, Boston.

United Nations. 2002. *Report of the World Summit on Sustainable Development, Johannesburg, South Africa, 26 August–4 September 2002*. United Nations, New York.

United Nations. 2002. *The road from Johannesburg: What was achieved and the way forward*. United Nations, New York.

United Nations Development Programme. 2002. *Human development report 2002*. Oxford University Press, Oxford.

United Nations Division for Sustainable Development. 1990. *Agenda 21*. Accessible online at: http://www.un.org/esa/sustdev/documents/agenda21/index.htm.

United Nations Environment Programme. "Declaration Of The United Nations Conference On The Human Environment." 16 June 1972, Stockholm. http://www.unep.org/Documents/Default.asp?DocumentID=97&ArticleID=1503.

United Nations Environment Programme. 2002. "Outlook: 2002-2032." Pp. 319–400 in *Global environment outlook 3 (GEO-3)*. UNEP and Earthscan Publications, Nairobi, Kenya, and London.

United States Environmental Protection Agency. 2000. *Regulatory impact analysis: Heavy-duty engine and vehicle standards and highway diesel fuel sulfur control requirements*. EPA420-R-00-026. Washington, D.C., December 2000.

Wackernagel, Mathis, Lillemor Lewan, and Carina Borgström-Hansson. 1999. Evaluating the use of natural capital with the ecological footprint. *Ambio* 28: 604.

Wilson, Edward O. 1998. *Consilience: The unity of knowledge*. Alfred A. Knopf, New York.

Wilson, Edward O. 2002. *The future of life*. Alfred A. Knopf, New York.